Free Online Access

Plunkett's Real Estate & _____ Almanac 2016

Your purchase includes access to Book Data and Exports online

As a book purchaser, you can register for free, 1-year, 1-seat online access to the latest data for your book's industry trends, statistics and company profiles. This includes tools to export company data. Simply send us this registration form, and we will send you a user name and password. In this manner, you will have access to our continual updates during the year. Certain restrictions apply.

_____ YES, please register me for free online access. I am the actual, original purchaser. (Proof of purchase may be required.)

Customer Name _____

Title_____

Organization _____

Address _____

City_____State_____Zip_____

Country (if other than USA) _____

Phone_____Fax _____

E-mail _____

Return to: # Plunkett Research®, Ltd.

Attn: Registration
P.O. Drawer 541737, Houston, TX 77254-1737 USA
713.932.0000 · Fax 713.932.7080 · www.plunkettresearch.com
customersupport@plunkettresearch.com

* Purchasers of used books are not eligible to register. Use of online access is subject to the terms of the end user license agreement.

PLUNKETT'S REAL ESTATE & CONSTRUCTION INDUSTRY ALMANAC 2016

The only comprehensive guide to the real estate & construction industry

Jack W. Plunkett

Published by:
Plunkett Research®, Ltd., Houston, Texas
www.plunkettresearch.com

PLUNKETT'S REAL ESTATE & CONSTRUCTION INDUSTRY ALMANAC 2016

Editor and Publisher:
Jack W. Plunkett

Executive Editor and Database Manager:
Martha Burgher Plunkett

Senior Editors and Researcher:
Isaac Snider

Editors, Researchers and Assistants:
Ashley Bass
Allan Butler
Anushree Gogate
Gina Sprenkel
Suzanne Zarosky
Shuang Zhou

E-Commerce & Enterprise Accounts Manager:
Jillian Claire Lim

Information Technology Manager:
Seifelnaser Hamed

Video Production:
Uriel Rios

Special Thanks to:
National Association of Realtors
National Association of Home Builders
National Bureau of Statistics of China
Real Capital Analytics
U.S. Bureau of Economic Analysis
U.S. Bureau of Labor Statistics
U.S. Census Bureau
U.S. Federal Reserve

Plunkett Research®, Ltd.
P. O. Drawer 541737, Houston, Texas 77254 USA
Phone: 713.932.0000   Fax: 713.932.7080
www.plunkettresearch.com

Plunkett Research®, Ltd.
P. O. Drawer 541737
Houston, Texas 77254-1737
Phone: 713.932.0000, Fax: 713.932.7080 www.plunkettresearch.com

ISBN13 # 978-1-62831-395-6 (eBook Edition # 978-1-62831-728-2)

Limited Warranty and Terms of Use:

PLUNKETT'S REAL ESTATE & CONSTRUCTION INDUSTRY ALMANAC 2016

CONTENTS

Continued on next page

INTRODUCTION

PLUNKETT'S REAL ESTATE & CONSTRUCTION INDUSTRY ALMANAC, the thirteenth edition of our guide to the real estate and construction field, is designed as a general source for researchers of all types.

The data and areas of interest covered are intentionally broad, ranging from the various types of businesses involved in real estate and construction, to the online services and technologies that are changing the real estate and mortgage sectors, to an in-depth look at the major firms (which we call "THE REAL ESTATE 450") within the many segments that make up the real estate and construction industry. Our definition of the types of businesses involved in real estate and construction is applied in a liberal sense. Accordingly, this book includes real estate brokerage, investment, development and management firms (both commercial and residential). It also covers companies involved in construction of all types, mortgages, hotels, shopping centers and apartments. In addition, this book covers companies that provide important services to the real estate industry, such as firms that write title insurance or operate industry databases and web sites.

This reference book is designed to be a general source for researchers. It is especially intended to assist with market research, strategic planning, employment searches, contact or prospect list

creation and financial research, and as a data resource for executives and students of all types.

PLUNKETT'S REAL ESTATE & CONSTRUCTION INDUSTRY ALMANAC takes a rounded approach for the general reader and presents a complete overview of the real estate and construction field (see "How To Use This Book").

THE REAL ESTATE 450 is our unique grouping of the biggest, most successful corporations in all segments of the real estate and construction industry. Tens of thousands of pieces of information, gathered from a wide variety of sources, have been researched and are presented in a unique form that can be easily understood. This section includes thorough indexes to THE REAL ESTATE 450, by geography, industry, sales, brand names, subsidiary names and many other topics. (See Chapter 4.)

Especially helpful is the way in which PLUNKETT'S REAL ESTATE & CONSTRUCTION INDUSTRY ALMANAC enables readers who have no business background to readily compare the financial records and growth plans of real estate and construction companies and major industry groups. You'll see the mid-term financial record of each firm, along with the impact of earnings, sales and strategic plans on each company's potential to fuel growth, to serve new markets and to provide investment and employment opportunities.

No other source provides this book's easy-to-understand comparisons of growth, expenditures, technologies, corporations and many other items of great importance to people of all types who may be studying this, one of the largest industry sectors in the world today.

By scanning the data groups and the unique indexes, you can find the best information to fit your personal research needs. The major companies in real estate and construction are profiled and then ranked using several different groups of specific criteria. Which firms are the biggest employers? Which companies earn the most profits? These things and much more are easy to find.

In addition to individual company profiles, a thorough analysis of trends in real estate and construction sectors is provided. These trends include the growth of discount residential brokerage firms, online real estate services and the changing nature of the mortgage industry. This book's job is to help you sort through easy-to-understand summaries of today's trends in a quick and effective manner.

Whatever your purpose for researching the real estate and construction field, you'll find this book to be a valuable guide. Nonetheless, as is true with all resources, this volume has limitations that the reader should be aware of:

- Financial data and other corporate information can change quickly. A book of this type can be no more current than the data that was available as of the time of editing. Consequently, the financial picture, management and ownership of the firm(s) you are studying may have changed since the date of this book. For example, this almanac includes the most up-to-date sales figures and profits available to the editors as of early 2016. That means that we have typically used corporate financial data as of late-2015.

- Corporate mergers, acquisitions and downsizing are occurring at a very rapid rate. Such events may have created significant change, subsequent to the publishing of this book, within a company you are studying.

- Some of the companies in THE REAL ESTATE 450 are so large in scope, and in variety of business endeavors conducted within a parent organization, that we have been unable to completely list all subsidiaries, affiliations, divisions and activities within a firm's corporate structure.

- This volume is intended to be a general guide to a vast industry. That means that researchers should look to this book for an overview and, when conducting in-depth research, should contact the specific corporations or industry associations in question for the very latest changes and data. Where possible, we have listed contact names, toll-free telephone numbers and Internet site addresses for the companies, government agencies and industry associations involved so that the reader may get further details without unnecessary delay.

- Tables of industry data and statistics used in this book include the latest numbers available at the time of printing, generally through 2015. In a few cases, the only complete data available was for earlier years.

- We have used exhaustive efforts to locate and fairly present accurate and complete data. However, when using this book or any other source for business and industry information, the reader should use caution and diligence by conducting further research where it seems appropriate. We wish you success in your endeavors, and we trust that your experience with this book will be both satisfactory and productive.

Jack W. Plunkett
Houston, Texas
May 2016

HOW TO USE THIS BOOK

The two primary sections of this book are devoted first to the real estate and construction industry as a whole and then to the "Individual Data Listings" for THE REAL ESTATE 450. If time permits, you should begin your research in the front chapters of this book. Also, you will find lengthy indexes in Chapter 4 and in the back of the book.

> **🎥 Video Tip**
> For our brief video introduction to the real estate and construction industry, see www.plunkettresearch.com/video/realestate.

THE REAL ESTATE & CONSTRUCTION INDUSTRY

Glossary: A short list of real estate and construction industry terms.

Chapter 1: Major Trends Affecting the Real Estate & Construction Industry. This chapter presents an encapsulated view of the major trends that are creating rapid changes in the real estate and construction industry today.

Chapter 2: Real Estate & Construction Industry Statistics. This chapter contains an extensive set of industry statistics.

Chapter 3: Important Real Estate & Construction Industry Contacts – Addresses, Telephone Numbers and Internet Sites. This chapter covers contacts for important government agencies and trade groups. Included are numerous important Internet sites.

THE REAL ESTATE 450

Chapter 4: THE REAL ESTATE 450: Who They Are and How They Were Chosen. The companies compared in this book (the actual count is 445) were carefully selected from the real estate and construction industry, largely in the United States. 118 of the firms are based outside the U.S. For a complete description, see THE REAL ESTATE 450 indexes in this chapter.

    Individual Data Listings:
    Look at one of the companies in THE REAL ESTATE 450's Individual Data Listings. You'll find the following information fields:
    Company Name:
    The company profiles are in alphabetical order by company name. If you don't find the company you are seeking, it may be a subsidiary or division of one of the firms covered in this book. Try looking it up in the Index by Subsidiaries, Brand Names and Selected Affiliations in the back of the book.

Industry Code:

Industry Group Code: An NAIC code used to group companies within like segments.

Types of Business:

A listing of the primary types of business specialties conducted by the firm.

Brands/Divisions/Affiliations:

Major brand names, operating divisions or subsidiaries of the firm, as well as major corporate affiliations—such as another firm that owns a significant portion of the company's stock. A complete Index by Subsidiaries, Brand Names and Selected Affiliations is in the back of the book.

Contacts:

The names and titles up to 27 top officers of the company are listed, including human resources contacts.

Growth Plans/ Special Features:

Listed here are observations regarding the firm's strategy, hiring plans, plans for growth and product development, along with general information regarding a company's business and prospects.

Financial Data:

Revenue (2015 or the latest fiscal year available to the editors, plus up to five previous years): This figure represents consolidated worldwide sales from all operations. These numbers may be estimates.

R&D Expense (2015 or the latest fiscal year available to the editors, plus up to five previous years): This figure represents expenses associated with the research and development of a company's goods or services. These numbers may be estimates.

Operating Income (2015 or the latest fiscal year available to the editors, plus up to five previous years): This figure represents the amount of profit realized from annual operations after deducting operating expenses including costs of goods sold, wages and depreciation. These numbers may be estimates.

Operating Margin % (2015 or the latest fiscal year available to the editors, plus up to five previous years): This figure is a ratio derived by dividing operating income by net revenues. It is a measurement of a firm's pricing strategy and operating efficiency. These numbers may be estimates.

SGA Expense (2015 or the latest fiscal year available to the editors, plus up to five previous years): This figure represents the sum of selling, general and administrative expenses of a company, including costs such as warranty, advertising,

interest, personnel, utilities, office space rent, etc. These numbers may be estimates.

Net Income (2015 or the latest fiscal year available to the editors, plus up to five previous years): This figure represents consolidated, after-tax net profit from all operations. These numbers may be estimates.

Operating Cash Flow (2015 or the latest fiscal year available to the editors, plus up to five previous years): This figure is a measure of the amount of cash generated by a firm's normal business operations. It is calculated as net income before depreciation and after income taxes, adjusted for working capital. It is a prime indicator of a company's ability to generate enough cash to pay its bills. These numbers may be estimates.

Capital Expenditure (2015 or the latest fiscal year available to the editors, plus up to five previous years): This figure represents funds used for investment in or improvement of physical assets such as offices, equipment or factories and the purchase or creation of new facilities and/or equipment. These numbers may be estimates.

EBITDA (2015 or the latest fiscal year available to the editors, plus up to five previous years): This figure is an acronym for earnings before interest, taxes, depreciation and amortization. It represents a company's financial performance calculated as revenue minus expenses (excluding taxes, depreciation and interest), and is a prime indicator of profitability. These numbers may be estimates.

Return on Assets % (2015 or the latest fiscal year available to the editors, plus up to five previous years): This figure is an indicator of the profitability of a company relative to its total assets. It is calculated by dividing annual net earnings by total assets. These numbers may be estimates.

Return on Equity % (2015 or the latest fiscal year available to the editors, plus up to five previous years): This figure is a measurement of net income as a percentage of shareholders' equity. It is also called the rate of return on the ownership interest. It is a vital indicator of the quality of a company's operations. These numbers may be estimates.

Debt to Equity (2015 or the latest fiscal year available to the editors, plus up to five previous years): A ratio of the company's long-term debt to its shareholders' equity. This is an indicator of the overall financial leverage of the firm. These numbers may be estimates.

Address:

The firm's full headquarters address, the headquarters telephone, plus toll-free and fax numbers where available. Also provided is the World Wide Web site address.

Stock Ticker, Exchange: When available, the unique stock market symbol used to identify this firm's common stock for trading and tracking purposes is indicated. Where appropriate, this field may contain "private" or "subsidiary" rather than a ticker symbol. If the firm is a publicly-held company headquartered outside of the U.S., its international ticker and exchange are given.

Total Number of Employees: The approximate total number of employees, worldwide, as of 2015 (or the latest data available to the editors).

Parent Company: If the firm is a subsidiary, its parent company is listed.

Salaries/Bonuses:

(The following descriptions generally apply to U.S. employers only.)

Highest Executive Salary: The highest executive salary paid, typically a 2015 amount (or the latest year available to the editors) and typically paid to the Chief Executive Officer.

Highest Executive Bonus: The apparent bonus, if any, paid to the above person.

Second Highest Executive Salary: The next-highest executive salary paid, typically a 2015 amount (or the latest year available to the editors) and typically paid to the President or Chief Operating Officer.

Second Highest Executive Bonus: The apparent bonus, if any, paid to the above person.

Other Thoughts:

Estimated Female Officers or Directors: It is difficult to obtain this information on an exact basis, and employers generally do not disclose the data in a public way. However, we have indicated what our best efforts reveal to be the apparent number of women who either are in the posts of corporate officers or sit on the board of directors. There is a wide variance from company to company.

Hot Spot for Advancement for Women/Minorities: A "Y" in appropriate fields indicates "Yes." These are firms that appear either to have posted a substantial number of women and/or minorities to high posts or that appear to have a good record of going out of their way to recruit, train, promote and retain women or minorities. (See the Index of Hot Spots For Women and Minorities in the back of the book.) This information may change frequently and can be difficult to obtain and verify. Consequently, the reader should use caution and conduct further investigation where appropriate.

Glossary: A short list of real estate and construction industry terms.

Chapter 1

MAJOR TRENDS AFFECTING THE REAL ESTATE & CONSTRUCTION INDUSTRY

Major Trends Affecting the Real Estate and Construction Industry:

1) Introduction to the Real Estate and Construction Industry
2) Smaller Down Payments/Easier Loan Qualifications Change Mortgage Market
3) Online Competition Changes the Mortgage Industry
4) Home Sales May Stall If Prices and Interest Rates Rise
5) Real Estate Online Services Continue to Grow
6) Internet-Based Home Sales and Cheap Commissions Rock Residential Brokers
7) Homes and Commercial Buildings Seek Green Certification
8) Prefabricated Houses and Buildings Evolve
9) Baby Boomers Become a Strong Influence in the Housing Market/Universal Design Catches On
10) Multigenerational Families Are Increasingly Living Under One Roof
11) Commercial and Residential Construction Starts Expected to Rise
12) Real Estate Markets in China and India Face Challenges

13) Mixed-Use Developments Go Vertical
14) Retail Center Occupancy Is High, with Sales Rising in Upscale Malls/Online Sales Hurt Stores
15) Malls Remodel to Boost Sales and Attract Shoppers/Store Visitor Counts Are Disappointing
16) Apartment House Occupancy Rates Fall as New Construction Soars
17) Tiny Apartments and Condos Proliferate, Solve Affordability Problem
18) Condo Market Rebounds
19) Hotel Occupancy, Profits and New Construction Grow
20) Hotel Mergers Enable Chains to Claim Market Share, Add Unique Properties
21) New Urbanism and Traditional Neighborhood Developments Are Retro Trends
22) Remodeling Market Is Strong
23) The Future of Housing/The Future of Cities

1)   Introduction to the Real Estate and Construction Industry

📹 Video Tip

For our brief video introduction to the Real Estate and Construction industry, see www.plunkettresearch.com/video/realestate.

The real estate and construction sectors, including the many professions and fields associated with them, are among the larger components of the global economy. As of early 2016, the U.S. Bureau of Labor estimated that 6.22 million Americans were employed in the construction industry (down from 6.45 million at the beginning of 2015, and down dramatically from a peak of 7.6 million in 2007). The agency also estimated that 1.5 million Americans were employed in the real estate industry as of early 2016 (this figure is the same as of the beginning of 2015, and equal to the 1.5 million of 2007).

There was $13.8 trillion in all outstanding mortgage debt in America at year-end 2015, up slightly from $13.4 trillion one year earlier including homes and commercial projects. (Many houses are being purchased with cash, particularly when the purchaser is an investor. This mortgage total included $10.0 trillion in home mortgages, up slightly from $9.9 trillion in 2014).

Home prices have rebounded dramatically from the depths of the recent recession, but in some markets they remain below their 2006-07 peaks. However, prices have risen to the point that bargains are hard to come by. More than a few observers are concerned that another housing bubble is developing in the most popular markets.

Meanwhile, real estate had enjoyed a significant boom over the recent past in China, Canada, the UK, Australia and a few other select spots. By mid- to late 2015, however, many international real estate markets were off their recent highs, including significant softness in China.

In the U.S., home sales volume has picked up substantially from the dismal years of the 2007-09 housing bust and recession, resulting in rapidly rising prices in many markets and a shortage of available homes in the most desirable neighborhoods. This big turnaround is being fueled by many factors. The most significant may be the continued efforts by the U.S. Federal Reserve to hold interest rates down to incredibly low levels, which significantly reduces the monthly payments of home owners who take out mortgages. Another boost to the market is new

household formation by members of Generation Y. The oldest members of this segment were turning 34 years of age by 2016. Many had been living with their parents for much longer than normal, due to the recession. While it is sometimes difficult for first time homebuyers to find an affordable home and qualify for a mortgage, Generation Y is boosting the market, both as buyers and as new renters. (Relatively new mortgage programs offering down payments as low as 3% will accelerate sales to first-time buyers.)

An additional factor is that the market has seen a steady decline in the number of foreclosed, bank-owned homes. Massive numbers of them were acquired by investors. A lower supply of bargain-priced foreclosures means higher prices and a stronger market for houses overall. Potential homebuyers are also feeling more comfortable financially, as they have paid down credit card debts over the past few years, and the values of their stock market holdings have risen substantially in the recent past.

Construction of all types, including commercial, has enjoyed a rebound. About $1.14 trillion in new American construction was expected to be put into place for 2016, a seasonally adjusted annual rate as of February, according to the U.S. Bureau of the Census. This is very near the 2006 peak of $1.16 trillion.

Home builders in the U.S. are once again enjoying robust sales and profits. Many are challenged in finding enough new lots for future building. In many markets, they are also challenged in creating new homes at price points that are attractive to first-time buyers, as lot prices are high and cities are often charging high infrastructure fees for each new lot.

Retail centers in the U.S. were seriously overbuilt during the last boom, through 2007, and there were too many stores trying to sell to too few shoppers. The end result of the glut was that large numbers of retail store chains took bankruptcy during the recession, and many others have either curtailed expansion plans or are opening much smaller stores than in the past. However, by late 2012 and into 2016, shopping center occupancy rates were improving in general. Nonetheless, retail stores are facing dramatic competition from sales made over the Internet. Online commerce is growing at stellar rates while mall traffic and store revenues are generally disappointing. (Malls and shopping centers that cater to wealthier customers have seen better results.)

Apartment house operators are enjoying brisk business in general. For the fourth quarter of 2015, the vacancy rate at U.S. apartments rose to 4.4%, compared to 4.3% in the fourth quarter of 2014. These extremely low vacancy rates are encouraging the construction of new apartment buildings at a rapid rate.

Over the long term, on a global basis, there will be continuing demand from the health care sector for new or remodeled properties as the percentage of the population over age 65 continues to grow in many nations from North America to Asia to Europe, boosting the need for medical care and assisted living centers. This trend has been accelerated in America by the implementation of the Affordable Care Act.

Another growing trend in construction in major economies is to incorporate a higher number of energy conservation technologies in new buildings. This is true in both residential and commercial construction. Several "green" building certification plans are now in place, so that architects and builders may seek to attain certain energy conservation and eco-friendly standards.

2) Smaller Down Payments/Easier Loan Qualifications Change Mortgage Market

In response to the recession of 2008-09 as well as stricter banking regulations, mortgage lenders tightened their requirements for down payments and generally made loans much harder to get in most of the world. In the U.S., borrowers were required to put 20% down, or the lending banks would have to hold 5% of the loan's risk once it was packaged and sold to investors. However, by early 2014, home sales were once again slowing, and lender regulating agencies including Fannie Mae and Freddie Mac reduced down payment requirements with the goal of stimulating borrowing. Meanwhile, private mortgage lenders have become more anxious than in 2008-12 to make jumbo mortgages, low down-payment mortgages and ARMs (adjustable rate mortgages). Some loans are even being made once again with interest-only terms for the first several years. These are practices not generally seen since the recession and the housing market crash. Nonetheless, borrowers with low credit scores may have little to no luck if they seek a mortgage—a major change from the pre-crash market.

By late 2012 and in to 2016, the housing market and related prices had clearly rebounded in nearly all markets around the U.S. Low mortgage interest rates combined with the fact that investors purchased large numbers of homes at what they considered to be compelling bargains, helped to push the previously built home market ahead. At the same time, new household formation surged, as many young people moved out of their parents' homes and purchased or rented houses. The inventory of existing homes for sale inventory plummeted. While the end of 2011 saw an 8.3 months' supply of existing homes on the market in the U.S. (that is, the length of time it would take to sell all homes at the current monthly sales rate), inventory had shrunk to 5.8 months as of March 2016, according to the Federal Reserve Bank of St. Louis.

Exceptionally low mortgage interest rates have been fueling the rebound in home sales, as monthly payments can be very low for new buyers. The National Association of Realtors reported more than 4 million existing homes in the inventory (plus about 500,000 new homes) at the peak in mid-2007. By March 2016, the existing homes inventory had dropped to 1.98 million. All of these factors have very positive implications for America's ability to clear out what remains of potential foreclosures and return the housing market to reasonable stability.

By 2014, the Federal Reserve, Securities and Exchange Commission (SEC) and the Department of Housing and Urban Development had approved more lenient rules. Specifically, banks are able to avoid the 5% risk-retention requirement by verifying a borrower's ability to repay the loan and other requirements (such as debt payments not exceeding 43% of income). This rule change is significant, but it nonetheless affects only loans that are made without federal backing through such agencies as FannieMae, estimated to be less than 2% of residential mortgages. Loans sold to Fannie Mae and Freddie Mac were already exempt from the risk-retention requirement.

The Federal Reserve reported a total outstanding mortgage debt in the U.S. of $13.80 trillion in 2015, up from $13.43 in 2014. However, home ownership in the U.S. fell from 64.5% in 2014 to 63.8% in 2015, the lowest since 1994 according to the U.S. Census Bureau.

Another problem is a significant number of U.S. homeowners that are still suffering from home values that are lower than their mortgage balances in the aftermath of the global economic recession. RealtyTrac reported 6.44 million U.S. residential properties were "underwater" at the end of 2015 (11.5% of all properties with a mortgage, down from 2012's peak of 31.4%).

The bottom line is that as of early 2016 the housing market was very definitely improved and much healthier than that of 2008-12, but it seemed to be reaching a plateau, at least temporarily, as buyers appeared to be nearing the limit of how much they could (or were willing to) pay. The results will include continued high occupancy rates in apartments and rental homes, and a possible switch to construction of smaller, more affordable homes by builders willing to change strategy. Another result may be more aggressive and creative tactics by lenders, including adjustable mortgages with low initial payments.

In 2015, both FannieMae and Freddie Mac began offering mortgages with down payments as little as 3% of the home's purchase price. FannieMae's My Community Mortgage program is open to first time buyers with a minimum credit score of 620. Freddie Mac's Home Possible Advantage mortgages are available to first time and other qualified borrowers. The initiative comes after the National Association of Realtors (NAR) reported that as of late 2014, first-time home buyers made up just 29% of home sales, compared to a historical rate of 40%.

In 2015, Bank of America also began offering a mortgage product with down payments a little as 3%. The loans, which are backed in a partnership with FreddieMac and the SelfHelp Ventures Fund, are available to borrowers with a credit score of at least 660, higher than the Federal Housing Administration's (FHA's) requirement.

Another trend offering relief to consumers with less than perfect credit is a rent-to-own program. Home Partners of America (www.homepartners.com) offers a lease with a right to purchase plan. The company received a $500 million equity investment from a unit of BlackRock, Inc. in 2015. By early 2016, Home Partners had invested more than $1 billion in homes that it leased and later sold to 9,000 customers.

3)   Online Competition Changes the Mortgage Industry

The Internet has created a much more competitive landscape for home mortgages. Sites such as LendingTree and Quicken Loans offer low origination fees and make it simple to complete applications online, while the Internet itself makes it much simpler for consumers to compare interest rates and mortgage options. At Bank of America's user-friendly web site, prospective borrowers can apply online, view a breakdown of proposed fees, compare mortgage options and check the status of a loan in progress.

Mortgage competition online is really part of a much larger, global trend: fierce competition among financial services providers of all types. Firms throughout the consumer finance sector, from banks to insurance companies to stock brokerages, are fiercely battling to attract, retain and cross-market new services to consumers

Experts at the Mortgage Bankers Association (MBA) estimate that there were $1.63 trillion in residential mortgage originations in 2015, up significantly from 2014's $1.26 trillion. For 2016, the MBA expects mortgage originations to fall to $1.48 trillion.

Internet Research Tip: Online Mortgage Tools
For mortgage loan and loan provider advice, see
1) www.mtgprofessor.com, a site run by Jack M. Guttentag, Professor of Finance Emeritus at the Wharton School of the University of Pennsylvania.
2) www.myfico.com, a site operated by the credit analysis firm Fair Isaac Corp.
3) www.bankrate.com, a site that provides financial rate information.
4) www.hsh.com, a site run by financial publisher HSH Associates.

---

4)   Home Sales May Stall If Prices and Interest Rates Rise

According to the National Association of Realtors (NAR), sales of existing homes fell from 7.08 million units in 2005 at the peak, to 4.12 million in 2008 at the bottom for recent years. For 2015, 5.25 million units were sold, and 5.38 million are projected for 2016.

In response to crashing demand during and immediately after the 2008-09 recession, home prices fell significantly. NAR reported a drop in median existing U.S. home sales prices to $218,900 for 2007 (from the 2006 level of $221,900), the first annual fall since the Great Depression. The median price dropped precipitously thereafter, to less than $160,000. By 2012, however, median existing home prices were climbing, and hit $177.200 by year end, rising further to $197,100 in 2013 and $208,300 in 2014. In 2015, a median price of $222,400 was reported, and for 2016, $231,700 was forecasted. This means that affordability is becoming an issue, as higher prices mean higher monthly mortgage payments.

Meanwhile, there is concern that interest rates may be rising sometime in 2016, as the U.S. Federal Reserve has indicated that it may eventually let rates rise in several steps. Here again, higher interest rates would create an affordability issue, since higher rates mean higher monthly mortgage payments.

The U.S. Census reported homeownership rates (that is, the percentage of households living in homes that they own) falling from 65.2% at the end of 2013 to 64.5% during 2014 and 63.8% in 2015. 2006 saw homeownership at its highest in history at 68.9% or 76.5 million households.

The Standard & Poor's/Case-Shiller home-price index (which covers 20 major U.S. cities) rose by 13.4% for all of 2013. Price increases in 2014 slowed to 4.43%. For 2015, a slightly higher increase of 5.2% was reported.

It's interesting to note that the average home size is also on the rise. According to the National Association of Home Builders, the average new home size rose to 2,720 square feet in 2015, up from 2,660 square feet in 2014.

Positive factors supporting a rebound in home sales and prices in America include:
- Generation Y (Millennials, born from 1982 through 2002) is maturing and graduating from colleges and universities in large numbers, leading to significant new household formation, including apartment renters and some first-time home buyers. However, high levels of student debt are a deterrent to home purchases.
- Low gasoline prices free purchasing power for other areas such as real estate.

The most significant negative factors for the housing market, as of early 2016, include:
- Many underemployed, would-be homebuyers continue to have difficulty finding jobs with substantial pay or job security.
- Economic growth remains relatively slow.
- Affordability has become a significant issue, with home prices up considerably in recent years.

The U.S. government is attempting to boost sales on the low- to mid-priced end of the market by encouraging FHA to make large quantities of mortgages with very small down payments. These loans come with strict income requirements. Buyers must submit proof of income and document employment history with two years of pay stubs and W-2 forms. Borrowers are typically limited to amounts that are about 31% of their income, or 43%

when other debt is taken into account. Extremely low down payments of at least 3.0% are required. The FHA's market share of home purchase financing has increased substantially. (While the FHA programs may boost demand and borrowing, they also can lead to defaults by those buyers who fall behind financially and feel little regret at walking away from their very small down payments.)

5) Real Estate Online Services Continue to Grow

According to a 2014 National Association of Realtors (NAR) survey, 92% of homebuyers use the Internet as a primary house research tool, and about 50% of real estate-related searches occurred on mobile devices. Among the busiest sites are Realtor.com, Move.com, and Yahoo! Real Estate, along with the sites operated by RE/MAX and Coldwell Banker.

Sites such as HomeGain, ZipRealty and Move.com provide home listings as well as tools for calculating mortgages and researching schools in different neighborhoods. The sites make money by selling advertising, selling customer leads to real estate agents and placing home listings on their sites.

Home finance is also a hot Internet area. A large percent of homebuyers apply for mortgage pre-approval before deciding upon a house. Many of those buyers make their applications online. Sites such as LendingTree and E-Loan, Inc. make applying for a mortgage online simple and competitive, because they maintain relationships with a wide variety of lending institutions and can offer several options.

LendingTree, for example, sends each mortgage application to up to five different lenders. In return, the lenders offer competing interest rates, closing costs and terms. LendingTree earns a fee from each loan's closing. The site also generates revenue through matching potential homebuyers with Realtors.

Even eBay is having success selling real estate online. It's doing a significant volume of business auctioning houses, land, commercial property and time-share interests.

Zillow, www.zillow.com , is an extremely popular site launched in 2005. It provides online access to aerial home photos, and enables users to estimate home values based on a variety of factors, using a proprietary algorithm. Zillow is now a publicly-held firm, with revenues soaring to $644.7 million for 2015, up from $325.9 million the previous year.

Internet Research Tip: Public Real Estate Data
For home value estimates, recent sales activity, tax information, title history and more, see:

Trulia	www.Trulia.com
Property Shark	www.propertyshark.com
RealEstateABC	www.realestateabc.com
Zillow	www.zillow.com

Online listings are typically generated by Multiple Listing Services (MLS), which are run by local groups of Realtors. The services feed information about homes for sale to brokers' web sites and national real estate sites, as well as sites representing a geographic area (most cities have real estate sites such as www.har.com which is operated by the Houston Association of Realtors in Houston, Texas). Most prospective buyers hope to see all available listings on these sites, as opposed to listings for a particular Realtor.

Savvy real estate agents are utilizing web sites such as YouTube and Facebook, blogs, tweets and text messages to reach younger, first time buyers. Some realty firms are hiring social media workers.

Mergers and acquisitions are big news among real estate sites. Zillow purchased competitor Trulia in July 2014 for $3.5 billion. In September 2014, News Corp. acquired the parent company of Realtor.com for $950 million (News Corp. already owned real estate sites in Australia and Southeast Asia).

SPOTLIGHT: Opendoor

Opendoor is a San Francisco, California-based online real estate portal that purchases homes from users who answer basic questions about the property they wish to sell. The company quickly makes offers via email based on comparative market analysis and then schedules free inspections to confirm a property's sellable condition. Closings are scheduled at the user's convenience, usually between three and 60 days after an offer is agreed upon. Opendoor then makes improvements or renovates the purchased properties and lists them for sale. As of early 2016, Opendoor offered its services in the Phoenix, Arizona area and the Dallas/Fort Worth, Texas area, and planned to launch in Portland, Oregon and Denver, Colorado.

6) Internet-Based Home Sales and Cheap Commissions Rock Residential Brokers

From 2004 through 2007, residential real estate agents enjoyed a golden age. (In fact, making money by brokering real estate looked so easy that membership in the National Association of Realtors grew from about 750,000 in 2000 to more than 1,300,000 in 2007.) At the same time, the amount of time or effort necessary to sell a house declined significantly in many markets before the real estate bubble burst, as eager buyers snapped up houses as soon as they came on the market. Many residential agents had been earning huge sums—sellers were paying ever-larger fees as home prices escalated.

Today, however, the real estate broker as an intermediary is facing intense competition from Internet sites that connect buyers and sellers directly. A good analogy can be found in the travel agent industry, in which travel consumers have turned to the Internet for a vast portion of their purchases and information needs, and travel agents have been forced to downsize or become providers of enhanced services, unique expertise and value-added packages.

The days of 6% commissions for real estate brokers are under pressure thanks to maverick real estate brokers such as Foxtons (www.foxtons.co.uk) and ZipRealty (www.ziprealty.com). Instead of the usual 6% commission, these upstart brokers' fees are as little as 3% and up to 5% of the selling price. They are able to cut costs by capitalizing on the power of the Internet to advertise homes for sale, listing homes with accompanying photographs and virtual tours, but not always including services such as conducting in-person showings and open houses or negotiating prices.

ZipRealty offers sellers a reduced commission structure that is about 5% of the sales price, compared to standard commissions of 6% elsewhere. At the same time, it attracts buyers with rebates of up to 20% of its commissions. For example, the seller of a $200,000 home might save $2,000 in commissions. When ZipRealty represents a buyer of a $200,000 home listed by another real estate firm, the buyer might receive a rebate of $1,200, assuming ZipRealty earns a co-brokerage commission of 3% from the listing broker or Realtor. (ZipRealty's 3% commission would be $6,000 in this case, and the buyer would receive 20% of that amount, or $1,200.) Certain restrictions apply. ZipRealty was acquired by Realogy Holdings Corp. in 2014.

Competing against the extremely well-entrenched Realtors group is difficult at best. Discount brokerage Foxtons, which originated in the UK, at one time charged sellers as little as 2% for listing their homes, provided the sellers did the showings and other tasks themselves. In 2000, Foxtons opened operations in the U.S., focused on the New York City area. In May 2007, before the current housing slump had gained much momentum in the UK, Foxtons was acquired by a private equity firm for about $770 million. The acquisition excluded the U.S. operations. By September 2007, the U.S. unit filed for bankruptcy, despite the fact that it had listed about 4,400 homes for sale at the time. As of 2016, Foxtons continues to operate in London, Middlesex and Surrey in the UK.

SPOTLIGHT: Redfin

Redfin is a technology-powered real estate brokerage firm which provides information on real estate listings, home appraisal and past record sales while allowing buyers, sellers and renters to connect with real estate and mortgage professionals. In contrast to traditional firms, Redfin agents do not receive a full commission from transactions completed. Instead, a portion of the commission is refunded to the customer and usually applied to closing costs. On the firm's web page, consumers can also browse, share and create albums of houses for sale through the free service Redfin Collections. The Redfin Home Value Tool gives home owners real-time data on the estimated value of their home. The company offers services in 80 markets across 42 states and Washington, D.C. and partners with many other real estate agents across the country to service areas not covered. Plunkett Research estimates Redfin's revenues at $125 million for 2015.

Another discount alternative for a seller is to pay a flat fee (usually around $500) to a discount broker to list their home in MLS. Sellers then do the showing and negotiating themselves. A different tweak to this option is ForSaleByOwner.com, where owners choose from basic packages priced at about $80 for a one-month listing on its own site to about $689 six-month listing on MLS (owners using this service often offer a 2% to 3% commission to Realtors who are representing buyers).

Some online home-selling sites regularly mine MLS listings and then post those listings on their own sites. In order to have access to MLS data, a broker employed by the site must be a member of the local Realtors organization. Some online sites have been known to hire a member Realtor to feed MLS data to them. Disputes frequently break out between local boards of Realtors and such sites as to whether MLS data is proprietary.

The National Association of Realtors (NAR) has spent millions on television and radio campaigns that highlight the benefits of using a Realtor and the difficulties homeowners face when selling on their own.

7) Homes and Commercial Buildings Seek Green Certification

In a growing trend, many homebuilders across the U.S. are constructing homes in accordance with the National Association of Home Builders' (NAHB) "green" specifications.

The NAHB's green specifications require resource-efficient design, construction and operation, focusing on environmentally friendly materials. An effort to save energy and reduce waste is spurring this trend. In addition, local building codes in many cities, such as Houston, Texas, are requiring that greater energy efficiency be incorporated in designs before a building permit will be issued.

There are several advantages to building along eco-sensitive lines. Lower operating costs are incurred because buildings built with highly energy-efficient components have superior insulation and require less heating and/or cooling. These practices include the use of oriented strand board instead of plywood; vinyl and fiber-cement sidings instead of wood products; and well insulated foundations, windows and doors. Heating and cooling equipment with greater efficiency is being installed, as well as dishwashers, refrigerators and washing machines that use between 40% and 70% less energy than their 1970s counterparts. Some builders are opting for high efficiency geothermal heating and cooling, while some home and building owners want solar electricity generation.

In addition to energy concerns, plumbing and water efficiency are vital goals in green buildings. This trend will accelerate due to deep concerns about the availability of water in populous regions ranging from California to China. Wastewater heat recovery systems use wastewater to heat incoming water. Toilets are more efficient. Current models use a mere 1.28 gallons of water per flush, as opposed to four gallons in the 70's. Landscaping is likewise being designed for much lower water usage.

The U.S. Environmental Protection Agency (EPA) established the WaterSense certification, a voluntary program to promote water-efficient products and services. For example, WaterSense certifies low-flow toilets that use a mere 1.28 gallons per flush, creates standards for bathroom-sink faucets that flow at no more than 1.5 gallons per minute and offers a certification program for irrigation companies that use water-efficient practices.

The main disadvantage is that green building is more expensive than traditional construction methods. Added building costs often reach 10% to 20% and more per home; however, some homebuyers are willing to pay the increased price for future savings on utilities and maintenance. As energy prices increased over recent decades, builders became more amenable to constructing homes with energy-savings measures. In addition, some consumers are inclined to spend more when they feel they are buying environmentally friendly products, including homes. (Marketing analysts refer to this segment as "LOHAS," a term that stands for "Lifestyles of Health and Sustainability." It refers to consumers who choose to purchase items that are natural, organic, less polluting and so forth. Such consumers may also prefer products powered by alternative energy, such as hybrid cars.)

The U.S. government and all 50 states offer tax incentives in varying amounts to builders using solar technology. Zero Energy Design ("ZED") is slowly catching on. A handful of "zero energy homes ("ZEH")" that produce approximately as much electricity as they use are being built. In 2014, the U.S. Department of Energy certified 370 homes as ZEH.

Internet Research Tip, Zero Energy Homes:
Zero Energy Building Designs
http://zeb.buildinggreen.com
Passive House Institute www.phius.org/home-page
HomeInnovation Research Labs
www.homeinnovation.com

By installing photovoltaic panels or other renewable sources to generate electricity, and using improved insulation and energy-efficient appliances and lighting, the zero-energy goal may be achieved, at least in sunny climates such as those in the American West and Southwest.

The state of California revised its energy efficiency requirements recently, which took effect in 2014. Requirements include solar-ready roofs that include space for optional solar panels, hot water pipe insulation and the verification by an independent inspector that all air conditioning units are properly installed for maximum efficiency. The state code goes a step farther, recommending whole-house fans to displace warm air with cooler night air in the

summer seasons, improved windows and better insulation. State regulators estimate that these changes will make residential and commercial buildings between 25% and 30% more energy efficient.

In the commercial sector, businesses may have several reasons to build greener, more energy-efficient buildings. To begin with, long-term operating costs will be lower, which will likely more than offset higher construction costs. Next, many companies see great public relations benefit in the ability to state that their new factory or headquarters building is environmentally friendly. Many office buildings, both public and private, are featuring alternative energy systems, ultra-high-efficiency heating and cooling, or high-efficiency lighting. In California, many public structures are incorporating solar power generation.

Even building maintenance is getting involved—building owners are finding that they can save huge amounts of money by scheduling janitorial service during the day, instead of the usual after-hours, after-dark schedule. In this manner, there is no need to leave lighting, heating or cooling running late at night for the cleaning crews.

An exemplary green office building is Bank of America Tower (formerly One Bryant Park), a 54-story skyscraper on the Avenue of the Americas in New York City. Completed in 2009, the $1.2-billion project is constructed largely of recycled and recyclable materials. Rainwater and wastewater is collected and reused, and a lighting and dimming system reduces electrical light levels when daylight is available. The building supplies about 70% of its own energy needs with an on-site natural gas burning power plant. It was the first skyscraper to rate platinum certification by adhering to the Leadership in Energy and Environmental Design (LEED) standards, set by the U.S. Green Building Council in 2000 (see www.usgbc.org).

The Pearl River Tower, a 71-story skyscraper which opened in 2011 in Guangzhou, China, was designed to be one of the first major zero-energy buildings. Designed by Chicago architecture firm SOM, the tower was planned to be 58% more energy efficient than traditional skyscrapers by using solar roof panels, novel wind turbines embedded in four openings spaced throughout the tower and walls with eight-inch air gaps that trap heat which then rises to power heat exchangers for use in cooling systems. The building encompasses about 2.3 million square feet of floor space.

A growing number of buildings are being retrofitted to use energy more efficiently. One example is the initiative underway at Citigroup, Inc. The banking firm is turning off lobby escalators, incorporating more natural light and using recycled materials in dozens of its properties around the world. Citigroup says it can save as much as $1 per square foot of building per year by making its offices more efficient. Elsewhere, Google, Inc. installed a solar rooftop at its California headquarters as early as 2007, and retail chains such as Wal-Mart and Kohl's are installing solar panels on their California stores. In Wal-Mart's case, it had more than 150 solar installations in the U.S. by mid-2012, and planned to have 1,000 solar-powered locations by 2020.

Sports stadiums are also going green in a big way. Lincoln National Field, the home of the Philadelphia Eagles, installed 2,500 solar panels, 80 wind turbines (each measuring 20 feet high) and a natural gas and biodiesel-burning generator. The field contracted with Florida-based Solar Blue, which spent $30 million to install the equipment. In return, the Eagles are paying Solar Blue fixed amounts for energy with increases of 3% per year for a period of 20 years. Solar Blue is free to sell excess energy created in the stadium to the local utility. Staples Center in Los Angeles and New Meadowlands Stadium in New York also have significant green initiatives underway.

LEED standards have been adopted by companies such as Ford, Pfizer, Nestlé and Toyota, which have all built LEED-certified structures in the U.S. In addition, the standards have been adopted by 25 states and 48 cities for government-funded projects, including New York, Los Angeles and Chicago. Industry analysts estimate the value of government-financed construction projects at $200 billion per year. One of the world's largest green complexes is the campus of King Abdullah University of Science and Technology (KAUST) in Saudi Arabia. The campus spans more than 118 million square feet of classrooms, laboratories and a coral reef ecosystem, and features more than 13,500 square feet of solar thermal panels and upwards of 54,300 square feet of photovoltaic arrays.

LEED is not without competition. Another green verification program called Green Globes is backed by the Green Building Initiative in the U.S. Green Building Initiative is a group led by a former timber company executive and funded by several timber and wood products firms. Several U.S. states have adopted Green Globes guidelines instead of those supported by LEED for government-subsidized building projects. In Canada, a version of Green Globes for existing buildings is overseen by the Building Owners and Managers Association of Canada (BOMA Canada) under the brand "BOMA Best." Green Globes is more wood friendly than LEED, which is not surprising considering the involvement of the timber industry. It promotes the use of wood and wood products in construction with fewer restrictions than LEED, which approves of wood if it comes from timber grown under sustainable forestry practices approved by the Forest Stewardship Council, an international accrediting group.

In one ambitious project, a $30 million office building in Seattle, Washington was completed in April 2013, spent its first year in a kind of sustainability test. The Living Building Challenge (living-future.org) established and overseen by the International Living Future Institute, will measure the building's sustainability in seven areas: site, water, energy, health, materials, equity and beauty. Those areas were tested by a list of 20 requirements such as net zero use of water and energy, operable windows and car free living. The new six-story building in Seattle is called Bullitt Center. At the time of its construction, only three buildings in the U.S. that had been certified by the Living Building Challenge.

Retail giant Wal-Mart has attained remarkable achievements in reducing energy use in stores, cutting waste in packaging and increasing the efficiency of its massive trucking and distribution system. It recently announced a goal to double the generation of solar energy on the roof tops of its buildings from 2013 through 2020, when it hopes to be generating 7 billion kWh of renewable energy. Its 2020 commitments will save approximately $1 billion yearly in energy costs. As of 2015, it already got 26% of its global power from renewable sources, compared to the 13% share of U.S. generation.

Walgreen Company built a net zero energy drug store near Chicago with 800 solar panels on its roof during 2013. This building is also powered by two 35-foot wind turbines and an underground geothermal system. The store's engineers hope that it will generate more than 450,000 kilowatt-hours of power each year, using only about 200,000 to run the facility. Throughout its chain of stores, Walgreen hopes to reduce energy use by 20% by 2020.

SPOTLIGHT: Solar Power Direct from Roofing Shingles

Dow Chemical has invested $100 million (plus a $10 million grant from the Department of Energy) in researching new plastic photovoltaic roof panels using thin-film solar cells. The product, called Powerhouse, was available in 17 U.S. states as of early 2015. Powerhouse (and products like it) cost a homeowner about $31,000, after government subsidies and tax rebates, for approximately 3,000 square feet of roofing material. This compares to about $12,000 for traditional asphalt shingles, but Dow claims that homeowners will save $76,200 in energy costs over 25 years and increase a home's value by $22,000. In addition, the installation of these shingles may qualify for tax credits.

In Europe, the EU has mandated that member states revisit building codes every five years and create standards of energy efficiency. Buildings are also required to submit an energy certificate that can be shown to prospective buyers and renters. Elsewhere, nations such as Japan that are focused on becoming much more energy-efficient are emphasizing the use of green methods in new construction.

8) Prefabricated Houses and Buildings Evolve

In this increasingly technology-driven era, it is interesting to note that homebuilding is one of the last industries to remain focused on manufacturing by hand using traditional methods. For several decades, entrepreneurs have attempted to commercialize various kinds of prefabricated housing with very little success. For example, HUD (the U.S. Department of Housing and Urban Development) pushed a well-funded program in the late 1960s that sought to encourage modestly-priced, high-quality housing through factory fabrication in materials that included steel as well as wood. The Sears Roebuck catalog sold home kits as early as 1908 (these now historic homes are considered highly desirable today). Post-World War II, William J. Levitt, a homebuilding leader of great fame, introduced modular housing to lukewarm response, which petered out once the housing boom subsided.

Nonetheless, until recently, the market for prefabricated housing was tiny, and most homes are still constructed on-site by hand. This is also true in commercial buildings.

Enter "panelization," the practice of building and assembling elements such as foundations, walls and staircases off-site at a factory and then trucking them to a home site for assembly. PulteGroup, one of the largest homebuilders in the U.S., is embracing the practice, as are many major builders across the nation.

Today's panelization is more far-reaching than Levitt's historic practice, which merely assembled basic framing off-site. The quality of panelized flooring, walls and staircases has risen remarkably thanks to new techniques, and building these elements at a factory offers many advantages to working on-site. A foundation can be poured and cured indoors, thereby eliminating the need to wait for good weather at the home site. Substantial savings in construction time are a plus, and panelization also addresses the problem of the scarcity of skilled laborers in many areas. Homebuilders who have embraced at least some aspect of the prefab component trend include PulteGroup, Toll Brothers, Inc. and Beazer Homes USA, Inc.

A startup in San Francisco called Project Frog (www.projectfrog.com) specializes in prefab commercial buildings. The firm had constructed a number of buildings (including a convenience store for 7-Eleven, a small conference center and small school facilities). Project Frog's buildings, such as the Parkway Heights Middle School planned for South San Francisco, California in 2015, average about $210 per square foot in construction costs. The company hopes to build for the lucrative health care market over the near term,

China's Broad Sustainable Building (BSB, formerly the Broad Group) is taking prefabrication to a whole new level. It builds modules that it calls "main boards" used in multistory buildings such as the T30, a 30-story hotel which was completed on its final site in Hunan Province in only 15 days. The main boards are 12.8 feet by 51.2 feet and include water pipes, ventilation shafts and even finishes such as floor tiles, lights and plumbing fixtures. In addition to the amazing building speeds, BSB also touts the sustainability of its structures, claiming that they create 1% of the waste of traditional buildings and are five times more energy efficient due to pre-installed thermal installation and four-paned windows. BSB erected a 30-story hotel in Changsha in 2012 in only 15 days. Its most ambitious project, the 202-story Sky City proposed for Changsha, the capital of the Hunan province, broke ground in July 2013, and was promised to be completed in only four months. However, the building is mired in problems and delays, and construction stopped in September of

2013. As of early 2016, it was still on hold pending safety examinations and building permits. BSB did complete the 57-story Mini Sky City in Changsha in south-central China in mid-2015, finishing the final 37 stories in 18 days.

SPOTLIGHT: Resolution: 4 Architecture

Bucking a long-term, widely held distaste for prefabricated housing, a maverick firm called Resolution: 4 Architecture (www.re4a.com) is promoting manufactured, modular housing. Starting at around $200,000, modules can be manufactured in a factory and assembled into a completed house in one of several dozen designs.

The company won a competition sponsored by *Dwell*, a home design magazine, to design a house near Chapel Hill, North Carolina. In addition, the firm has designed a beach house in Ventura, California, a vacation home in East Hampton, New York and a 1,725-square-foot suburban home in Long Island, New York. Contemporary and hip in design, the homes generally appeal to young, cost-conscious homebuyers with cutting-edge tastes.

Another option in prefab building is refurbishing shipping containers into living spaces. Montainer (www.montainer.org) is a Missoula, Montana-based firm that offers a variety of rehabbed container options from a $65,000 Nomad base module to multi-container expansion modules that run between $20,000 and $30,000 more. Prices include design and permit review, manufacturing and delivery and installation. Containers are outfitted with windows, interior walls, kitchenettes and bathrooms.

9) Baby Boomers Become a Strong Influence in the Housing Market/Universal Design Catches On

Senior citizens make up a very fast-growing segment of the population, in the U.S. and in most of the world including China. The year 2011 saw the first Baby Boomers turning age 65. The term "Baby Boomer," generally referring to people born from 1946 to 1964, has evolved to include the children of soldiers and war industry workers who were involved in World War II. When those veterans and war industry workers returned to civilian life, they started or added to families in large numbers. As a result, the Baby Boom generation, originally 78 million people in America, is one of the largest demographic segments in the U.S. As a result, an estimated 10,000 Americans turn age 65 every day. The Baby Boom was not limited to the United States by any means.

High rates of new household formation and high birth rates were common throughout much of the world in the post-war period, with a very significant effect on today's growth in the senior segment of the population.

The real estate industry is quickly evolving to meet the needs and tastes of seniors, including the Baby Boomers who are hitting the traditional retirement age of 65 and members of older generations who require increasingly specialized homes and services as they hit 80 years and older. Housing options that cater to aging homeowners include retirement communities (also referred to as "independent living"), continuing care (which provides a broad spectrum of living options from independent living to nursing homes), assisted living (which offers assistance in daily tasks such as shopping, cleaning, etc. but does not provide onsite medical care) and communities that provide medical care for specific conditions such as Alzheimer's disease.

Another modest trend in senior housing is cooperative or communal living, in which private dwellings are grouped around a communal area or "common house" which usually contains a living room, kitchen and dining area for group dinners. Some communities of this type also include housing for on-site medical caregivers.

Today's seniors are significantly different from those of the past. Baby Boomers in particular will enjoy much longer lifespans, will be much more affluent, will be more physically active and will be much more discerning of the specific qualities, features and services that they want in their living environments than previous generations. Watch for substantial growth in the senior living industry for many years to come.

SPOTLIGHT: Universal Design

Universal design (UD) creates living spaces that are comfortably livable for the largest variety of people, including those who are older or dealing with physical challenges. As many homeowners age, they look for a house with elements such as wide, step-less doorways, sturdy hand rails, automatic faucets and stacked closets that can be converted to elevators. The Center for Universal Design promotes UD throughout the U.S. Located at North Carolina State University in Raleigh, North Carolina, the center conducts research and collaborates with builders and manufacturers. For more information, see www.ncsu.edu/ncsu/design/cud.

10) Multigenerational Families Are Increasingly Living Under One Roof

American families are shifting their living arrangements in response to economic conditions, longer life spans and the influence of different cultures. A Gallup survey conducted in late 2013 found that 29% of American adults under age 35 are living (or have recently lived) with their parents, while 51% of adults aged 18 through 23 are living at home. Homebuilder PulteGroup conducted a survey in 2012 of 511 homeowners across the U.S. which found that 15% of homeowners with aging parents already had those parents living with them, and 32% expect to eventually share their homes with their parents. In addition to a multigenerational concept within one family, some homeowners may want space in which to house room-mates or renters to help pay the mortgage.

This may have a significant effect on the needs of future home buyers. Builders are coming up with alternatives in their residential floorplans to traditional layouts. For example, U.S. homebuilder Lennar offers the Next Gen plan, which is a three bedroom house with an extra suite featuring an additional bedroom, eat-in kitchenette and a living room. The suite can have its own entrance for extra privacy, and even a separate additional garage. Such an arrangement can be ideal for an aging member of the family, particularly if combined with features suited to older residents, such as wide doorways, few or no stairs and easy to use handles on doors, cabinets and bath fixtures. Toll Brothers offers the multigenerational option in all of its 300 U.S. locations in 50 markets in 19 states. The option accounted for approximately 5% of its 2015 revenues, compared to 1% in 2012.

Many communities have zoning laws that have been in place for years that limit construction to traditional single family homes. A notable exception is the state of California which has a state-wide law stipulating that homeowners have the right to build extra units in single family homes. Newer plans such as Lennar's Next Gen avoid zoning hassles because the home runs on a single electric meter, has only a microwave oven in the kitchenette and looks from the outside as if it is a single-family home.

11) Commercial and Residential Construction Starts Expected to Rise

The years 2011 through early 2015 saw commercial property markets that were very strong in a handful of markets, such as Houston, Texas and the Silicon Valley, while rent prices and construction has picked up globally from the depths of the recession. For the fourth quarter of 2015, analysts at CBRE reported that commercial real estate across America had vacancy rates of 13.2%, compared to 13.9% for the same quarter in 2014 and down from 16.9% in the fourth quarter of 2013. Average asking rents rose in the 2015 third quarter to $29.36 per square foot, up 4.8% from the same quarter in 2014.

Commercial markets continue to show signs of a good rebound. CBRE reported that construction completions reached 35.7 million sq. ft. in 2015, the highest total since 2009 but still less than half of the previous 2008 peak of 75.9 million sq. ft. Of the 57 markets tracked by CBRE Research, 11 had no construction underway at year-end, down from 16 markets one year earlier.

On the residential side, the construction market has improved as well. 2011 saw 302,000 new single family home sales, according to the Bureau of the Census, which rose to 501,000 in 2015. The value of private residential U.S. construction put into place was $280.3 billion in 2012, rising to $349.9 billion in 2015. For 2016, $447.9 billion is expected.

Residential construction in the U.S. is expected to improve significantly in 2016 thanks to continued job growth, better consumer confidence and rising incomes. The U.S. Department of Commerce and the Department of Housing and Urban Development (HUD) reported housing starts up 10.8% in 2015 to reach 1.11 million units. According to the National Association of Home Builders, the average new home size rose to 2,720 square feet in 2015, up from 2,660 square feet in 2014.

However, affordability has become a real issue for first-time homebuyers. Texas-based home builder D.R. Horton, Inc. launched a new division in early 2014 called Express Homes to build smaller, cheaper houses to attract first time buyers and other budget conscious buyers who have previously been priced out of the new home market. The Express Homes division of D.R. Horton offers small, no-frills units in four U.S. states with prices ranging from $120,000 to $150,000, considerably less than the company's overall average price of $278,900.

12) Real Estate Markets in China and India Face Challenges

A recent boom in Chinese new home and condominium construction had many analysts concerned about inflation in home prices, an overextended market and the potential for the bubble to burst. However, in September 2015, new home prices were up 0.1% compared to September in 2014, and October saw a rise of 0.2% over the previous year. The Chinese government cut interest rates in November 2014 for the first time since 2012, and continued to cut rates five more times by October 2015. Government regulations on home buying are also softening, including China's central bank dropping the minimum mortgage down payment for first-time buyers from 25% to 20% in early 2016.

There is a critical difference between the housing markets in major Chinese cities such as Beijing, Shanghai and Shenzhen and smaller, tier-two cities. Large cities are seeing significant growth in property prices. Beijing reported new home prices up 6.5% in October 2015 over the same month in 2014, while Shanghai had 10.9% growth and Shenzhen has 40% according to data from Reuters. Conversely, small cities elsewhere in China have seen tremendous numbers of newly constructed homes that sit empty. Investors are often happy to own them in this condition, as real estate is seen as an investment with greater growth potential than bank deposits and less volatility than the stock market. Speculators and individual investors have been prime drivers of the Chinese real estate market.

Meanwhile, official support of urbanization will remain a powerful force in China for decades to come. Some 300 million people will move from rural areas to China's cities from 2013 through 2030, creating intense demand for affordable housing.

Some builders are focusing on first time buyers and those of modest means by promoting tiny, affordable living spaces as small as 160 square feet. With a bed that folds to provide seating and a miniscule tube of a shower next to the front door, costs are kept to modest levels more likely to be within the means of millions of young urban professionals.

In India, Prime Minister Narendra Modi promised to build 100 smart cities in the country by 2020. The first effort is Gujarat International Finance Tec-City (GIFT), which began construction in 2012. By mid-2015, two 29-story steel and glass office buildings had been completed totaling 1.6-million-square-feet (with only part of one building occupied). By 2024, the city is planned to have 110 office and residential towers, a metro, 25,000 apartments, hotels and hospitals, which is hoped will create 1 million jobs. Funding for these grand projects is to come in part from the government ($945 million was budgeted for 2015).

SPOTLIGHT: Another Boom in Dubai?

After a devastating real estate market bust in 2009 in Dubai, with neighboring Abu Dhabi bailing out a number of defaulting construction projects, this ultra-modern city in the United Arab Emirates is once again riding a dizzying wave of new multi-billion-dollar development projects. The government is backing the $32.7 billion Dubai World Central airport which will house leading global carrier Emirates Airline (the project will be larger than London's Heathrow and Chicago's O'Hare combined).

The government is also funding the Mall of the World, a $6.8 billion enclosed retail shopping district. Spurring the growth is the upcoming 2020 World Expo, which is expected to attract 20 million visitors. Developers from around the world are investing in Dubai projects, including 40,000 new hotel rooms in addition to the 80,000 existing rooms as of 2015.

However, the freefall in global oil prices have hit Dubai's property values hard. In addition, regulators introduced caps on mortgage size and doubled transactions fees to avoid speculation. After real estate firm Knight Frank tracked 27.7% growth between March 2013 and March 2014, 2015 was a different story. For the first six months of the year, Dubai property prices fell by 12.2%.

---

13) Mixed Use Developments Go Vertical

Mixing office, retail, residential and entertainment space in one location is an idea on the rise, literally. For example, take the Time Warner Center in New York City. The 80-story, $1.7-billion building houses a 2.8-million-square-foot combination of condominiums, office space, a Mandarin Oriental hotel, CNN television studios, performance space, restaurants and retail space that includes a Whole Foods Market. It's a stupendous urban version of mixed-use developments that have cropped up in suburbs across the U.S. since the early 90s.

These spaces often outperform standard suburban real estate in office and retail lease rates, residential rents, retail sales, hotel room occupancy rates and property values, both on-site and in surrounding areas (according to a study by Charles Lockwood, a real estate historian and author). Many retailers in the Time Warner Center reported sales as high as 70% to 80% above their initial projections, before the recent economic crisis. All this multi-use bounty comes at a price, however, since building these 24/7 communities costs more. Design is more complex because retail needs are often at odds with residential needs. Restaurants have different logistical requirements than office space. It's also difficult to find the right mix of high- and low-end retail tenants. Residents need groceries and dry cleaning far more often than luxury jewelry or clothing, making the right mix of tenants a top priority for developers. Nonetheless, in close-in urban areas or highly desirable waterfront or scenic locations, the extremely high cost of land often dictates dense, vertical mixed use development.

The vertical trend is expected to dominate built-out suburban cities with set boundaries over the mid-term and beyond. Michael Beyard at the Urban Land Institute projects that vertical, multi-use expansion of aging strip malls that are on prime land will promote significant reinvestment and tax base expansion.

Another twist to the vertical trend is to build "live, work, play" communities around sports arenas. Take the 75-acre Victory Park project in Dallas, Texas, for example (www.victorypark.com). Built around the American Airlines Center, home of the NBA Mavericks, Victory Park currently boasts the W Dallas Victory Hotel and Residences, The House (a 28-story condominium high-rise) and two apartment complexes, Cirque and The Vista. Developed by Hillwood and partly owned by Hicks Holding LLC, the multi-billion dollar project includes large amounts of office, hotel, residential and retail space.

A number of vertical developments are cropping up around light rail line terminals such as a project in Carrollton, Texas near Dallas. The city approved a $38 million mixed-use development next to a commuter rail station that links Carrollton with Dallas. Commuters using rail are likely to look for housing, shopping and entertainment venues close to stations. Many communities are using funds from bonds and government subsidies to build new developments. Examples of recent developments include the $1 billion revitalization of Union Station in Denver, Colorado and a $2.3 billion ($1.5 billion from private investors and $800 million in municipal, county, state and federal funds) redevelopment of central Columbus, Ohio.

14) Retail Center Occupancy Is High, with Sales Rising in Upscale Malls/Online Sales Hurt Stores

The largest enclosed shopping malls are sometimes referred to as "super-regional centers" (malls which include at least three department stores and at least 800,000 square feet of retail space.) Such malls often contain more than 1.2 million square feet.

However, projects have become somewhat smaller since the recession of 2007-09, and it is much more difficult today for landlords to find multiple department store anchors.

Other types of shopping center properties, such as power centers and lifestyle centers, have been developing rapidly in recent years, in many cases robbing traffic from traditional malls. A "power center" is typically an open-air complex of category-dominant anchors such as category-killers, home improvement stores, discount stores and warehouse clubs. A "lifestyle center" is an open-air, highly landscaped configuration of approximately 50+ stores. Generally located near upscale neighborhoods, lifestyle centers offer leasable retail areas of 150,000 to 500,000 square feet, with at least 50,000 square feet of space typically dedicated to upscale national specialty stores such as Williams-Sonoma.

There are also hybrid centers, and value-oriented centers. Hybrids have some of the features of enclosed malls and lifestyle centers. That is, they have both open-air sections and enclosed sections. Value-oriented centers are built on formats that emphasize discounted prices. These include outlet malls. Many new value-oriented centers feature significant entertainment segments.

Cushman & Wakefield, a leading brokerage firm, reported shopping center completions in the U.S. in 2013 totaling 400 new centers and more than 129 million square feet, the first time shopping center completions rose since 2007. Up until then, so little retail space had been built that occupancy rates and rentals were improving in many markets. By the first quarter of 2016, analysts at Reis reported that the vacancy rate for regional malls was holding steady at 7.8%, down from a peak of 9.4% in the third quarter of 2011. For Neighborhood and Community malls (strip malls), the vacancy rate remained at 10.0%, down from the 11.1% peak in the third quarter of 2011.

Chain store executives are generally more cautious about opening new stores today. Many chains have significantly downsized their latest stores. While online sales have been soaring, retail store traffic and sales continued to be disappointing for some chains in 2014, including Wal-Mart's U.S. stores. National chain Coldwater Creek took bankruptcy in 2014, closing hundreds of stores. In 2015, American Apparel filed for Chapter 11 bankruptcy, as did Wet Seal and Quiksilver.

Analysts at eMarketer estimated American e-commerce sales in 2015 at $349.06 billion (up 14%

from $305 billion in 2014). This figure includes online retail sales, travel sales and digital downloads, but not sales of tickets to events or online gambling.

Shopping mall performance in recent years was sharply delineated by economic scale. Low-end malls showed weak sales per square foot while high-end malls were the most successful. Many malls that fit into the low-end category are old and outdated. CoStar Group, a real estate information firm, reported in 2015 that the national retail vacancy rate was 6.2% during the first quarter and continues to fall, although shopping centers that were 60% or more occupied had a vacancy rate of only 2.7% on average.

Malls catering to wealthier customers cost more to build and operate, and are able to charge higher rental rates. Such malls have seen the best retail sales in recent years. For example, Taubman Centers, Inc., a luxury shopping center developer that owns malls in choice spots throughout the U.S., reported average sales per square foot of mall space of $800 in 2015, a figure much higher than typical malls. Average rent per square foot for 2015 was $60.38, up from $59.14 in 2014.

Simon Property Group, which owns, develops and manages major shopping centers throughout the U.S. and Puerto Rico, reported that regional mall sales per square foot improved slightly in 2015, reaching $607, up from $603 in 2014. Base minimum rent per square foot reached $48.96 in 2015, up from $47.01 in 2014. These are rolling 12 month comparable sales per square foot for mall stores, less than 10,000 square feet in regional malls and all owned gross leasable area in premium outlets.

A number of retail chains, including Bloomingdale's and Nike, are experimenting with downsizing. Bloomingdale's store in Santa Monica, California, for example, is about one-eighth the size of its New York flagship store at only 105,000 square feet. The chain is gambling that shoppers are overwhelmed in larger stores that have greater volumes of merchandise from which to choose. Smaller spaces with few items could prove a cozier alternative, making shoppers more comfortable and therefore (hopefully) spending more. The Bloomingdale's store creatively utilizes its pared-down spaces by dropping dressing rooms when needed from recessed areas in the ceiling. The pod-like "rooms" look like Japanese lanterns. The 22,000-square-foot Nike store (compared to typical Nike-towns which are 50,000 square feet) in Santa Monica also uses flexible layout designs with cash registers mounted on moveable counters and interchangeable hardware fixtures that can be moved about at will.

SPOTLIGHT: Mall of America

The largest mall in the U.S. is the Mall of America (www.mallofamerica.com), located in Bloomington, Minnesota. First opened in 1992, it contains 4.2 million square feet of gross building area (2.5 million leasable) and is visited by an average of 110,000 people daily (or 40 million people per year). Entertainment is everywhere in this mall: 14 movie screens, a comedy club, night clubs, a 1.2-million-gallon walk-through aquarium, a seven-acre amusement park, the Nickelodeon Universe theme park, LEGO building center, A.C.E.S. Flight Simulation and over 520 specialty shops.

Total employment in the center ranges from 11,000 in normal times and up to 13,000 during holiday and summer seasons. (Not everyone comes here just to shop: the Mall of America reports that more than 5,500 couples have exchanged wedding vows in the Mall's "Chapel of Love" wedding chapel, and the mall's mall-walking list, the "Walksport Mall Stars!" has registered more than 4,000 people.)

With the addition of a 306,000-square-foot Ikea home furnishings store, the mall enters its newest phase of construction. Shoppers from far and wide visit Bloomington and stay several days specifically to shop this mammoth mall. International visitors are common. Overall, visitors from outside the mall's 150-mile radius hover at around 35%.

The Ikea store, while not directly connected to the original mall, is part of a 42-acre new Phase II development just north of the original site. Tentative plans for new tenants include several hotels, an office building, a spa and entertainment attractions, which may include a performing arts theater. The Phase II mixed-use complex is zoned for up to 5.6 million square feet of new development.

---

15) Malls Remodel to Boost Sales and Attract Shoppers/Store Visitor Counts Are Disappointing

The face of retailing continues to evolve quickly. Malls and their stores face daunting competition from web sites, giant discount stores and wholesale clubs such as Sam's. At the same time, many consumers have lost enthusiasm for frivolous shopping and needless credit card debt. Analysts at ShopperTrak report that retail store traffic in general is down. Malls are forced to adjust in order to stay in business.

According to Green Street Advisors, department store chains (which serve as anchor tenants in most malls) are suffering from slowing foot traffic. Such stores would have to close approximately 500 locations (about 15% of all anchor space in U.S. malls) to generate the same sales per square foot as they did in 2006. Closings are already happening, such as the departure of the Macy's store in the Ridgmar Mall in Fort Worth, Texas in 2016 and another Macy's store closing at Quail Springs Mall in Oklahoma City, Oklahoma.

Mall owners and developers are finding that investing in significant upgrades to existing malls can yield exceptional financial rewards. General Growth Properties invested $1.5 billion in redevelopment, commencing work on 17 properties in 2012 and acquired five interests in retail properties (totaling 1.3 million square feet) in 2014. Westfield Group spent $3 billion between 2012 and 2015 (compared to the $800 million it spent on redevelopment from 2008 through 2011). More and more investment is being spent on opening restaurants, movie theaters and grocery stores that bring in new customer traffic. Entertainment is an important part of successful malls. Most redevelopments include enhanced movie theaters with full menus and lounge seating, children's play areas and open pavilions for live entertainment.

Top malls typically have more than 1 million square feet. Luxury stores such as Tiffany, Prada and Ralph Lauren help to draw customers. Shoppers also want cutting-edge technology stores such as those operated by Apple, Microsoft or Sony.

One way to attract customers is to raise the roof—literally tear off the connecting roofs between stores to create an open-air environment. Parking garages are coming down as well so shoppers can park directly by the entrance of their favorite stores. Typically built to have a single floor, the stand-alone locations have plenty of close-in parking, and offer shopping carts and centralized checkout stands for convenience.

For example, Bella Terra in Huntington Beach, California was an enclosed shopping mall built in 1966. Builder J.H. Snyder Company spent $170 million to tear off the roof and knock down walls, transforming the mall into an open-air lifestyle center with 74 shops and restaurants and a 4,000-seat, 20-screen cinema megaplex. The rebuilt center features Italian architecture and design accents. The Santa Monica Place mall underwent a two-year, $265 million renovation, reopening in late 2010. The mall, formerly an enclosed, three-story structure built in the 1980s, now has roof-free spaces between stores which converge on a central, open-air plaza. In 2012, the Springfield Mall in Springfield, Virginia began a multi-year, $200 million renovation that includes building one centralized entrance, a state-of-the-art movie theater complex and a food court in the initial phase, as well as a 225-room hotel, pedestrian plazas, recreational facilities and 2,000 residential units to be part of later phases. In 2016, Houston's Galleria completed work on updating its Galleria III section, moving anchor Saks Fifth Avenue to a larger 198,000 square foot space and closing one of the mall's two Macy's stores. The project includes a standalone structure in the current parking lot and a residential tower.

Some operators of enclosed malls are scrambling to find alternative tenants to fill empty spaces. Wal-Mart, for example, opened a two-story store in a mall in a suburb of New York City and a three-story location in a Los Angeles mall formerly occupied by a Macy's department store. The pluses for Wal-Mart in these locations are access to mall shoppers without having to acquire real estate that is scarce in congested areas, as well as favorable leases with mall owners who are anxious to fill large amounts of space. On the other hand, Wal-Mart faces increased security and logistics costs and the difficulty of having to switch from a single-floor sprawl to multi-story confusion. Costco is also experimenting with former department-store spaces.

Tenants in marginal malls are morphing from strictly retail to include just about any entity that can pay the rent. Churches, government offices, schools, medical clinics and law firms are now often found in malls. Also, a number of cities are looking for cost-effective ways to use prime real estate that holds failing malls. Columbus, Ohio, for example, knocked down the City Center mall and built a park.

16) Apartment House Occupancy Rates Fall as New Construction Soars

Occupancy rates in U.S. apartment houses are slowly rising while rents remain high. According to analysts at Reis, for the first quarter of 2016, the vacancy rate at U.S. apartments rose to 4.5%, up from 4.4% in the fourth quarter of 2015 and 4.3% in the first quarter of 2015. Historically, the vacancy rate peaked in late 2009 at 8.0% and bottomed out at 4.2% in 2014 and early 2015. Rents reached record high levels in 2015. For the 12 months ending March 31, 2015, rates rose an average of 3.5% nationwide, to an average monthly rate of $1,131.72. By the first quarter of 2016, rents were still rising but at a softened pace, up 0.5% in the first quarter as new apartments opened. The year 2014 saw the completion of 161,518 new apartments nationwide and another 188,306 units were completed in 2015 (the highest since 2009's 188,870) according to Reis. Construction continues, with 2016 completions expected to exceed the number in 2015. Watch for rents to fall as these new units hit the market.

A new niche in the apartment industry is opening, as tech startups enter the market to offer tools to prospective tenants to find apartments and connect them directly with landlords, bypassing brokers and agents. Padspin (www.padspin.com), RadPad (www.onradpad.com) and Zumper (www.zumper.com) are examples. RadPad, based in Los Angeles, California, had approximately 1.2 million listings throughout the U.S. in early 2016, which are presorted to allow renters to find the best listings easily. Listings are given scores based on the quality of listing photos, landlord responsiveness and listing address accuracy.

Internet Research Tip: Apartments
Find out everything you ever wanted to know about apartments at the web site of the highly regarded National Apartment Association, www.naahq.org. Particularly useful are its archive of historical statistics and its commentary on the rental market.

17) Tiny Apartments and Condos Proliferate, Solve Affordability Problem

Apartment sizes are shrinking, with new apartments in San Francisco averaging 18% smaller than a decade ago, while units are 22% smaller in Boston, according to Axiometrics, Inc. A number of urban areas in the U.S. are seeing the construction of micro-apartments, typically 300 square feet in size (about the size of a hotel room). CoStar Group reported 26 micro-apartment projects totaling 2,000 units built or under construction in San Francisco, Washington, Los Angeles, Boston, New York and Seattle as of early 2014. In 2015, CoStar reported more than 2,200 additional units under construction or proposed in the U.S. Rents for the tiny units are considerably less than for standard one-bedroom units, making them attractive to Generation Y renters who are moving out of their parents' homes for the first time and want to live and work in the heart of urban areas. Asking rents for units of 400 square feet or less rose from $600 per month in 2011 to $1,143 in 2015, while the vacancy rate fell from

approximately 16% in 2011 to 6.6% in 2015, according to CoStar.

U.S. living spaces are typically much larger than those in other countries. However, unit size in congested urban areas such as Hong Kong are shrinking to minute proportions. In 2015, a 180-square foot apartment went under contract in a luxury development called High Place for $516,000. Furniture for the unit had to be made to order. Units this tiny are often referred to as "mosquito" units.

SPOTLIGHT: Co-Living
In major U.S. cities such as New York and San Francisco, apartment rents are so high that developers are testing "co-living" spaces where tiny apartments share kitchens, lounges and communal atmospheres. It is basically a dorm room for adults. WeWork Cos., which specializes in shared office space, launched its WeLive co-living project (www.welive.com) in lower Manhattan and Washington D.C. in 2016. WeLive's furnished spaces are available from month to month with no long-term lease required. At its New York location, shared beds start at $1,375 per month and private units start at $2,550. Linens, towels, kitchen utensils, Wi-Fi, housekeeping services and general utilities are included, in addition to perks such as We Community events and free coffee, tea and beer. At one point, the company announced plans to build 70 WeLive locations and have more than 30,000 residents by 2018. Other co-living startups include Common (which has a 19-bedroom building in Brooklyn, New York) and Stage 3 (which is working to open its Ollie branded 425-bed project in Long Island City near Manhattan).

18) Condo Market Rebounds

New condo projects in major cities such as Houston, Dallas and Denver are booming. In New York City, massive buildings are being built that are reaching record heights and unit prices. At 432 Park Avenue in Manhattan, one of the world's tallest residential towers was completed in 2016, at 84 stories. Units sold briskly, including two penthouses in the $90 million price range. One was reportedly being purchased as an investment, not as a personal residence. Buyers of other units, mostly in the $9 million to $50 million range, included wealthy people from outside the U.S. who desire a fashionable New York address—one that may turn out to be a good investment as well. In this record-setting building, small apartments on the 28th and 29th floors were designed available as optional extras at $1.5 million to $3.9 million, intended to be used as space for maids and other personal staff. Individual wine cellars are also for sale, ranging from $158,000 to $378,000. When fully sold, the building will have produced about $3 billion in total sales.

Another recent luxury development in New York City is 56 Leonard, in the TriBeCa neighborhood. This property reportedly sold 70% of its 140 units in the initial 10 weeks. Penthouses were priced at more than $20 million. Numerous other very high-end condo towers are underway or recently completed in Manhattan.

19) Hotel Occupancy, Profits and New Construction Grow

The hotel industry has been enjoying significant growth in recent years. For 2015, PricewaterhouseCoopers reported that average U.S. hotel occupancy reached 65.5% (compared to only 54.6% in 2009), and projected the rate to grow to 65.7% for 2016. Equally important, REVPAR (revenue per available room) had soared to nearly $75 by the beginning of 2015, up from only about $55 in 2009.

Marriott reported a 5.2% worldwide REVPAR increase during 2015, to reach $112.25. Average daily rates were up 4.1% on a constant dollar basis to $152.30, while occupancy increased 0.8%, over the previous year, to 73.7%. The company added 51,547 rooms worldwide during the year (including 9,590 rooms from the acquisition of Delta Hotels and Resorts). By the beginning of 2016, the firm had 270,000 rooms under development worldwide.

Companies like global hotel giants Marriott and Hilton are reporting increasing numbers of business and leisure travelers. Marriott saw strengthening in properties in Asia, Europe, the Caribbean, Latin America, and in its luxury properties around the world. Hotels in Asia particularly benefitted from strong economic growth in that region.

Improving business, rising occupancy levels and rising room rates have recently led to both the remodeling of existing properties and the construction of new hotels and motels. Lodging Econometrics estimated 742 new hotels opened in 2015 in the U.S., with another 845 hotels in the pipeline for 2016 openings and 998 for 2017.

20) Hotel Mergers Enable Chains to Claim Market Share, Add Unique Properties

A combination of low interest rates (making it attractive to issue corporate debt) and a soaring travel market (climbing steadily from the end of the recent recession through early 2016) have led to very significant levels of hotel chain mergers and acquisitions. Large chains want to acquire smaller chains for any of several potential reasons. These may include building market share, moving into underserved parts of the world, purchasing unique and innovative companies that are well positioned for future growth, or simply growing the total size of the parent company.

For example, hotel giant IHG agreed to acquire Kimpton Hotels & Restaurants at the end of 2014. This gave IHG ownership of one of the most innovative and talked-about boutique, upscale chains in America. The purchase price was $430 million in cash. At the time, Kimpton operated 62 hotels and had another 17 in the pipeline. Kimpton's expertise in converting existing properties into fun, hip hotels is extraordinary. While its hotel rooms have relatively expensive rates, it offers a design that appeals perfectly to younger travelers seeking a place to stay while on business, or an exciting place for a wedding. Kimpton also is experienced in building significant new properties, as evidenced by its 2015 opening of a high rise, new hotel in downtown Austin, Texas in 2015. Kimpton operates a unique loyalty program that offers a lot to frequent travelers, and it features amenities such as cocktail or wine events in its lobbies each day. It is also 100% dog friendly. Its restaurants and bars are careful to feature locally-sourced food and drink. For example, its downtown Portland, Oregon property features wines made nearby in the Willamette Valley.

The biggest battle ever seen in the hotel industry was settled in early 2016 when Marriott beat back other suitors to agree to acquire hotel giant Starwood Hotels & Resorts. This is a major gain for Marriott, as Starwood will bring 5,500 hotels and resorts in more than 100 nations to the Marriott portfolio of properties. When completed, the merger will create the world's largest hotel chain by far. Starwood properties include Westin, W and Sheraton, as well as luxury chains St. Regis and Le Meridien.

21) New Urbanism and Traditional Neighborhood Developments Are Retro Trends

New Urbanism, or traditional neighborhood development (TND) represent a strong trend among developers in many parts of America. TND is an attempt to create an old-fashioned aura in a newly built neighborhood. Typical elements include small lots, front porches and other features that encourage being outdoors, meeting your neighbors and walking or cycling (rather than driving) through the neighborhood. The architecture is frequently pre-1940s in style, and easy-to-walk-to retail spaces are often incorporated, sometimes in a town-center style. Also, live-work buildings are frequently placed in special parts of the neighborhood, encouraging entrepreneurs or professionals to work downstairs and live upstairs.

The most noted TND is a community in Florida called Seaside, www.seasidefl.com, which has been in place for more than a decade. Many other TNDs are attempting to emulate Seaside's success, which is based partly on its beachfront location, a tried-and-true draw for the retiree crowd and vacationers looking for something different from life back home.

TNDs are not necessarily instant successes, and they may not be suitable for development in many areas. To begin with, while many developers like the concept because municipal governments are likely to approve smaller lots in higher density arrangements, some elements of TNDs are more expensive to build, such as period streetlights, cobblestone streets or dense landscape. Also, the retail space offered in town-center sections of TNDs is often difficult to rent, resulting in costly high vacancy rates and unsightly empty storefronts. These high costs were especially problematic in the recent economic crisis, and caused new projects to be delayed or shelved altogether.

22) Remodeling Market Is Strong

Home remodeling in the U.S. reached a peak in 2007 at $575.3 billion, according to the Joint Center for Housing Studies of Harvard University. By 2009, it had fallen to $458.3 billion. Fortunately, the remodeling market was on the rise again in 2012, reaching $476.4 billion, and $534.4 billion in 2013. The center forecasted that $598 billion would be spent on remodeling in 2014, and expected 2015 to be a record year.

Many homeowners are remodeling in order to increase energy efficiency, while others capitalized on tax credits for new appliances and other energy saving improvements. Yet another reason Americans are remodeling is to add space for live-in relatives such as grown children or aging parents.

The improving market for both new construction and remodeling of existing homes can be clearly seen in the financial results of home improvement chains Home Depot and Lowe's. Home Depot reported revenues of $83.1 billion for the fiscal year ending in February 2015, up from $78.8 billion in fiscal 2014. Competitor Lowe's saw revenues increase to $56.2

billion for the fiscal year ending in February 2015, up from $53.4 billion in fiscal 2014.

Important trends in remodeling:

- Many cost-conscious consumers are in do-it-yourself (DIY) mode to reduce costs. In particular, younger homeowners are keen on DIY. One interesting result is that the short-term rental of construction tools and equipment has grown.
- The largest demographic segment for remodeling, in terms of dollars spent, is homeowners 50+ years in age. Much of their remodeling is done for cash. Financing for remodeling via home equity loans is harder to obtain than in the past, and consumers are reluctant to incur credit card balances.
- There is a growing focus on green, energy-efficient remodeling projects.
- Federal and state government incentives, such as tax credits, fuel investments in home solar projects, weather stripping, insulation and energy-efficient appliances and windows.
- As America's Baby Boomers age, investments in remodeling for easier access for seniors will boom. Features such as wider doors for wheelchair access, firm handholds in bathrooms and better kitchen accessibility will be a focus. "Universal Design," a concept that focuses on ease of use by residents of limited physical ability, will grow.

Source: Plunkett Research, Ltd.

---

23) The Future of Housing/The Future of Cities

One thing is clear for the future: the world's biggest metropolitan areas will be under intense pressure to provide services and housing that will meet ever-changing needs of residents and businesses. Infrastructure investment will be needed in massive amounts for improvement and replacement of water, sewer, transportation and other public service systems. Many city planners are assuming that there is a practical limit to the growth of giant cities. In China and India, for example, vast migration from low-income rural areas to cities with better economic prospects has created stellar growth in the largest cities. The result, in many cases, has been traffic congestion and unbearable air pollution.

China is attempting to control this growth by creating entirely new population centers. In China's Five-Year Plan for 2016-20, the current 21.7 million population of Beijing is slated to be capped at 23 million. Part of the plan is the creation of one massive population center out of a nine city region in the Pearl River Delta. Infrastructure investment, such as high speed railways, will be continued in a way that will encourage growth in specific areas. Beijing itself will be the economic center of an 82,000-square-mile megaregion called Jing-Jin-Ji that will eventually hold 130 million people.

In the U.S., the Metropolitan Institute (MI) at Virginia Tech University has done extensive research into what it calls "megapolitans;" that is, regional concentrations of population totaling 10 million people (now or by 2040) that comprise two or more contiguous metropolitan areas and feature other selected characteristics.

MI finds that 10 megapolitan areas exist within the U.S. today, comprising less than 20% of all land area in the lower 48 states, but holding almost 200 million people, about 66% of total U.S. population. By 2040, MI projects that these 10 areas will add 83 million additional residents and will attract more than 75% of all private real estate development investments from 2003 to 2040, or about $33 trillion.

MI lists the 10 megapolitans as 1) Cascadia in the Seattle/Portland area of the Pacific Northwest; 2) NorCal in Northern California; 3) Southland in Southern California; 4) Valley of the Sun in the Phoenix/Tucson area; 5) I-35 Corridor stretching from Dallas-Ft. Worth through central Oklahoma and parts of Kansas; 6) Gulf Coast comprising the coastal areas of Texas, Louisiana, Mississippi and Alabama; 7) Peninsula comprising all of Southern and Central Florida; 8) Piedmont in Central Georgia, the Carolinas, and nearby metro areas; 9) Northeast in the Northeast Corridor of the U.S.; and 10) Midwest in Chicago and the surrounding Midwestern metro areas.

Urban vs. Suburban: Dense population areas where people live and work close together foster the use of bicycles, public transportation and walking to get from place to place. They offer compact opportunities for shopping, dining, education, sports events and cultural events.

Some analysts have projected that aging Baby Boomers will desire smaller, urban homes as they age—homes that are easier to care for and less expensive to operate than their suburban houses on large lots. The concern is that such a trend could lead to mass migration from the suburbs into the city, leaving millions of empty homes on large lots. Likewise, some analysts have projected that younger people will forever shun the old model of large,

single-family, suburban homes, opting instead to live in cities.

On the other hand, there is the possibility that urban living may not appeal to such a vast swath of the population. Advanced communication technologies centered on the Internet will continue to evolve quickly, making it ever more effective to work remotely while collaborating with associates, customers and suppliers who are far away. People are also able to use advances in communications to stay in touch with far-flung family and friends. This will encourage many people to live in the suburbs, or in communities with smaller populations that may offer better recreation, cleaner air, lower crime rates, better weather and/or better public education than they would find in major cities.

This could cause large numbers of knowledge workers to relocate to cities such as Boise, Idaho or the Spokane, Washington area. Communities that are even more rural, such as Taos, New Mexico; Grand Junction, Colorado or College Station, Texas could find an influx of knowledge workers who realize that they can live nearly anywhere that suits their unique individual tastes and budgets, as long as they have very fast Internet access and reasonable proximity to a major airport.

While a large segment of younger Americans does seek out the employment and social activities of dense urban living while they are single, they also have the potential to fuel immense demand for suburban homes, with larger lots and access to public schools and public safety that are often much better than those found in major cities. In recent surveys, many of these Americans state that they long to move to the suburbs.

Several major cities that have long been automobile transportation-intense are now experimenting with advanced urban light rail, with good success. Young people, aged 18 to 34, have a strong interest in public transportation, according to studies by the National Association of Realtors and Portland State University. Cities like Los Angeles and Denver are developing highly popular urban train systems. Part of the success in this trend lies in municipal encouragement of mixed-use real estate development near train stations, giving residents the ability to walk to shopping, dining and other amenities, from dense housing units clustered around train stops. Los Angeles alone plans to invest $9 billion in expanded trains by 2026.

Mega-Regions: On a global scale, the Martin Prosperity Institute at the University of Toronto defines 40 mega-regions around the world that collectively power the global economy. While home to one-fifth of the world's population, these mega-regions produce two-thirds of global economic output and more than 85% of global innovation. Major mega-regions include Greater Tokyo (55 million people and $2.5 trillion in economic activity) and the 500-mile Boston-Washington corridor (54 million people and $2.2 trillion in output). Other regions of note include Chicago to Pittsburgh, Los Angeles to San Diego and the areas around global centers such as Amsterdam, London and Bangalore and Mumbai.

Green Construction and Factory-Built House Components: Sustainability and energy efficiency will be of ever-growing importance in the residential and commercial buildings of the future. This will be driven by several imperatives, including consumer interest in green initiatives; local, regional and national legislation and tax incentives; along with what is often seen to be an excellent return on investment for properties that use less energy and water, create less waste and have a more positive impact on the environment. Both building design and the growing availability of ever more sustainable building materials, appliances, heating and air conditioning systems will boost this trend.

Prefabricated building components, created in factories and assembled on site, are likely to become more common, as are large portions of houses that are factory-built and then bolted together on the site to quickly build a finalized home. "Panelization" is the practice of building and assembling elements such as walls and staircases off-site at a factory and then trucking them to a home site for assembly. PulteGroup, one of the largest homebuilders in the U.S., is embracing the practice, as are many major builders across the nation.

Meanwhile, the average housing unit is likely to become more compact. In many nations with mature economies, including the U.S., household sizes are slowly shrinking, and the ratio of single-occupant homes is rising.

A distinct contrast to this demographic trend, however, is increasing demand in the U.S. for homes that can readily accommodate several generations. With Baby Boomers reaching retirement age while enjoying lengthy expected life spans, many families are seeing the economic and support advantages of having the oldest generation living on the same property with a younger generation (possibly with children/grandchildren at home, as well). Builders and zoning authorities in many cities are now creating designs specifically for these situations. An example would be a main house that has a smaller, fully equipped unit attached, where the oldest generation could enjoy its own living space and some privacy, while they could also walk easily through the door to join the children and grandchildren on a regular basis.

At the same time, very small housing units will grow in number in dense urban settings, often less than 300 square feet in size. These tiny dwellings can solve several housing challenges at once, including affordability, energy efficiency and tight supply of land available for new construction.

Chapter 2

REAL ESTATE & CONSTRUCTION INDUSTRY STATISTICS

CONTENTS:

Real Estate & Construction Industry Statistics and Market Size Overview

	Quantity	Unit	Date	Source
Existing Home Sales, 2015, U.S.[1]	5.25	Million	2015	NAR
Existing Home Sales, 2016, U.S.[1, 2]	5.38	Million	2016	NAR
New Single Family Home Sales, U.S.	501	Thousand	2015	Census
New Single Family Homes For Sale, U.S., End of Period	234	Thousand	2015	Census
Single & Multi-family Home Housing Starts (Private), U.S.	1,111.8	Thousand	2015	Census
Median Sales Price of New Single-Family Houses Sold, U.S.	293,600	US$	2015	Census
Average Sales Price of New Single-Family Houses Sold, U.S.	355,500	US$	2015	Census
Median Existing Home Price, 2015, U.S.	222,400	US$	2015	NAR
Median Existing Home Price, 2016, U.S.[2]	231,700	US$	2016	NAR
Size of Median Newly Built Single Family Home, U.S.	2,445	Sq. Ft.	2015 Q3	NAHB
Annual Spending for Single-family Construction (Seasonally Adjusted)	226	Bil. US$	2015	NAHB
Annual Spending for Multifamily Construction (Seasonally Adjusted)	58	Bil. US$	2015	NAHB
Value of U.S. Construction Put into Place	1144.0	Bil. US$	2016[3]	Census
Private	846.2	Bil. US$	2016[3]	Census
Residential	447.9	Bil. US$	2016[3]	Census
Non-Residential	398.3	Bil. US$	2016[3]	Census
Public	297.8	Bil. US$	2016[3]	Census
State & Local	274.5	Bil. US$	2016[3]	Census
Federal	23.2	Bil. US$	2016[3]	Census
Average 5-year Treasury-Indexed Hybrid Adjustable-Rate	2.90	%	Mar. 2016	Freddie Mac
Average 15-Year Fixed-Rate Mortgage Interest Rate	2.98	%	Mar. 2016	Freddie Mac
Average 30-Year Fixed-Rate Mortgage Interest Rate	3.71	%	Mar. 2016	Freddie Mac
Total Outstanding Mortgage Debt, U.S.	13,795.8	Bil. US$	2015	Federal Reserve
One- to Four- Family Residences Mortgage Debt, U.S.	9,986.0	Bil. US$	2015	Federal Reserve
Multifamily Residences Mortgage Debt, U.S.	1,098.8	Bil. US$	2015	Federal Reserve
Commercial Mortgage Debt, U.S.	2,505.9	Bil. US$	2015	Federal Reserve
Farm Mortgage Debt, U.S.	205.1	Bil. US$	2015	Federal Reserve
Homeownership Rate (% of Households), U.S.	63.8	%	2015	Census
Gross Vacancy Rate of Total Housing Inventory, U.S.	12.9	%	2015	Census
Estimated Construction Industry Employment (NAICS 23), U.S.	6.22	Million	Feb. 2016	BLS
Estimated Real Estate Industry Employment (NAICS 531), U.S.	1.52	Million	Feb. 2016	BLS

[1] Existing home sales of single-family homes and condo/coops.

[2] Forecast.

[3] Seasonally-adjusted annual rate as of Feb. 2016.

NAR = National Association of Realtors   Census = U.S. Census Bureau

BLS = U.S. Bureau of Labor Statistics   NAHB = National Association of Home Builders

Plunkett's Real Estate & Construction Industry Almanac 2016

Value of U.S. Private Construction Put in Place: 2009-2015

(In Millions of US$)

Type of Construction	2009	2010	2011	2012	2013	2014	2015
Total Private Construction[1]	591,648	505,290	501,925	571,145	635,669	717,711	807,220
Residential	247,526	242,035	244,122	269,784	323,381	370,045	418,261
New single family	105,336	112,569	108,178	132,015	170,768	193,600	218,523
New multi-family	28,538	14,686	15,037	22,510	31,500	41,806	51,899
Improvements[2]	113,652	114,780	120,907	115,260	121,113	134,639	147,838
Nonresidential	344,121	263,255	257,803	301,360	312,288	347,666	388,959
Lodging	25,388	11,201	8,395	10,197	13,028	15,698	20,571
Office	37,282	24,368	23,738	27,448	30,133	38,403	48,064
General	33,346	22,203	21,231	25,111	28,419	36,500	46,358
Financial	3,720	2,122	2,284	1,945	1,597	1,863	1,620
Commercial	51,128	37,154	39,153	44,312	50,947	60,761	64,867
Automotive	4,535	3,546	4,270	4,834	4,639	4,945	6,028
Sales	1,564	1,355	1,816	2,079	1,956	2,075	2,376
Service/parts	2,056	1,679	1,977	2,221	2,313	2,122	2,834
Parking	916	511	478	534	370	748	818
Food/beverage	4,868	4,605	5,268	5,845	6,594	7,426	7,284
Food	1,978	2,027	2,840	2,430	2,937	3,651	3,239
Dining/drinking	2,237	1,911	1,650	2,436	2,141	2,133	2,631
Fast food	653	667	779	979	1,516	1,642	1,414
Multi-retail	18,390	12,486	13,361	14,904	16,686	20,144	20,720
General merchandise	3,976	3,794	3,368	3,773	2,942	2,947	2,675
Shopping center	11,447	6,725	6,748	7,934	9,806	13,339	14,025
Shopping mall	2,171	1,332	2,356	2,288	2,938	2,884	3,111
Other commercial	6,300	4,220	3,720	3,934	4,786	4,619	5,285
Drug store	1,882	1,077	721	765	893	981	584
Building supply store	1,118	772	611	493	506	527	594
Other stores	2,528	1,741	1,743	2,069	2,468	2,362	2,982
Warehouse	9,730	5,661	6,543	7,047	8,766	13,793	16,547
General commercial	8,781	5,229	6,219	6,575	8,107	13,060	15,425
Mini-storage	873	401	241	365	472	584	1,028
Farm	7,305	6,637	5,991	7,748	9,476	9,835	8,996
Health Care	35,309	29,552	28,906	31,429	29,696	28,556	31,010
Hospital	24,723	21,528	20,483	21,223	19,199	17,700	20,111
Medical building	7,531	5,276	5,429	6,415	7,015	6,936	6,608
Special care	3,055	2,748	2,995	3,791	3,482	3,921	4,291
Educational	16,851	13,418	14,081	16,625	16,919	16,699	17,651
Preschool	716	492	349	396	420	482	360
Primary/secondary	3,362	2,585	2,973	2,795	2,997	3,713	3,418
Higher education	10,812	8,322	8,477	10,809	10,868	10,432	12,045
Instructional	6,211	4,993	4,658	5,786	4,928	4,893	5,540
Dormitory	2,496	1,654	2,165	2,902	3,924	3,589	3,968
Sports/recreation	832	790	660	709	981	962	1,030
Other educational	1,643	1,687	1,869	2,206	2,226	1,737	1,582
Gallery/museum	1,395	1,522	1,468	1,407	1,486	1,293	1,123

(Continued on next page)

Value of U.S. Private Construction Put in Place: 2009-2015 (cont.)

(In Millions of US$)

Type of Construction	2009	2010	2011	2012	2013	2014	2015
Religious	6,177	5,237	4,205	3,819	3,565	3,242	3,368
House of worship	5,001	4,214	3,256	3,143	3,055	2,768	2,811
Other religious	1,176	1,023	950	675	510	474	557
Auxiliary building	1,031	795	651	521	419	348	437
Public Safety	471	241	205	103	125	219	251
Amusement & Recreation	8,402	6,483	6,744	6,217	6,916	7,481	10,046
Theme/amusement park	280	353	472	512	670	820	1,041
Sports	2,023	1,596	1,096	1,043	1,336	1,853	3,851
Fitness	1,763	1,150	1,067	1,251	1,310	1,137	1,206
Performance/meeting center	796	565	510	517	673	569	573
Social center	1,134	914	662	614	789	790	911
Movie theater/studio	338	426	510	362	618	731	874
Transportation	9,056	9,894	9,537	10,883	11,029	11,823	13,108
Air	512	259	535	1,044	921	658	732
Land	8,484	9,503	8,944	9,810	9,988	10,785	12,150
Railroad	8,042	8,973	8,546	9,279	9,272	10,025	11,381
Communication	19,712	17,689	17,536	15,952	17,619	16,919	19,341
Power	76,064	66,117	64,262	86,402	81,278	89,389	76,704
Electric	60,394	48,972	50,291	68,967	54,411	62,236	48,948
Gas	10,849	8,297	10,357	10,317	10,897	11,506	12,007
Oil	4,821	8,848	3,614	7,118	15,970	15,647	15,749
Sewage & Waste Disposal	468	439	520	597	356	239	222
Water Supply	319	717	635	373	591	559	516
Manufacturing	57,355	40,607	39,768	46,774	49,863	57,239	82,731
Food/beverage/tobacco	3,342	4,000	4,792	4,380	4,934	6,387	6,202
Textile/apparel/leather/furniture	274	591	301	171	377	804	435
Wood	397	438	369	335	732	636	337
Paper/print/publishing[3]	698	634	684	1,190	1,016	681	941
Print/publishing[3]	173	(S)	(S)	(S)	(S)	(S)	(S)
Petroleum/coal	23,892	11,612	5,453	4,777	4,036	5,606	4,133
Chemical	10,310	7,580	7,345	10,848	15,483	23,200	41,973
Plastic/rubber	583	728	947	1,686	1,537	2,305	3,117
Nonmetallic mineral	1,815	1,150	896	857	916	872	1,333
Primary metal	4,772	4,738	3,130	4,691	4,487	3,522	5,533
Fabricated metal	1,507	1,104	1,593	1,726	1,484	894	884
Machinery	1,075	1,030	1,373	1,631	1,442	1,136	1,809
Computer/electronic/electrical	3,711	4,614	8,815	9,592	7,312	5,014	2,932
Transportation equipment	3,590	1,962	3,260	3,869	5,132	5,074	12,117
Miscellaneous	1,390	428	812	1,022	976	1,110	985

Notes: Details may not add to totals since all types of construction are not shown separately. S = Estimate does not meet publication standards.

[1] Total private construction includes the following categories of construction not shown: highway and street, and conservation and development.

[2] Private residential improvements does not include expenditures to rental, vacant, or seasonal properties.

[3] As of 2010, print/publishing is in paper/print/publishing.

Source: U.S. Census Bureau
Plunkett Research,® Ltd.
www.plunkettresearch.com

Value of U.S. Public Construction Put in Place: 2009-2015

(In Millions of US$)

Type of Construction	2009	2010	2011	2012	2013	2014	2015
Total Public Construction[1]	**314,895**	**303,966**	**286,407**	**279,311**	**270,682**	**275,698**	**291,278**
Residential	8,015	10,294	8,524	6,272	5,836	5,054	6,476
Nonresidential	306,880	293,672	277,883	273,038	264,846	270,644	284,802
Office	14,626	13,482	12,273	10,352	7,847	7,653	7,985
Commercial	3,609	2,947	3,663	3,023	2,212	1,947	2,537
Health care	9,536	9,792	11,297	11,116	10,993	9,854	8,952
Educational	86,351	74,986	70,903	68,047	62,141	63,001	67,229
Public safety	13,316	10,913	10,202	10,328	9,382	9,160	8,646
Amusement & recreation	11,002	10,461	9,251	9,263	8,291	9,150	10,635
Transportation	27,646	28,446	25,200	26,979	28,430	29,963	31,678
Power	12,797	11,828	10,923	11,032	12,040	11,826	10,508
Highway and street	82,056	82,417	79,285	80,393	81,250	83,961	89,523
Sewage & waste disposal	24,362	25,552	22,190	21,664	22,069	23,082	24,802
Water supply	15,152	14,605	13,528	12,846	13,006	12,775	13,034
Conservation & development	5,720	7,146	7,459	6,151	5,856	7,142	7,957
Total State & Local Construction[1]	**286,456**	**272,833**	**254,753**	**252,378**	**246,955**	**252,963**	**268,420**
Residential	5,772	7,576	5,962	4,672	4,537	4,136	5,601
Nonresidential	280,684	265,256	248,791	247,705	242,418	248,827	262,819
Office	9,170	8,317	7,440	6,055	5,220	5,379	5,733
Commercial	2,003	1,456	1,682	1,412	1,119	870	1,416
Health care	6,837	6,193	7,009	7,040	6,928	6,249	5,713
Educational	83,703	71,948	67,965	65,622	60,027	61,033	65,321
Public safety	9,432	7,586	7,245	7,641	6,592	6,224	5,901
Amusement & recreation	10,585	9,668	8,542	8,876	8,081	8,799	10,130
Transportation	25,461	26,493	23,340	24,797	25,871	27,795	29,487
Power	11,775	10,770	9,492	9,794	10,951	10,694	9,281
Highway & street	81,254	81,309	78,369	79,722	80,610	83,449	88,971
Sewage & waste disposal	23,455	24,555	21,196	20,946	21,037	22,028	23,999
Water supply	14,838	14,420	13,393	12,746	12,919	12,700	12,989
Conservation & development	2,016	2,035	2,289	2,225	2,500	3,019	3,127
Total Federal Construction[2]	**28,439**	**31,133**	**31,654**	**26,933**	**23,727**	**22,735**	**22,858**
Residential	2,243	2,717	2,562	1,600	1,299	918	876
Nonresidential	26,196	28,416	29,092	25,333	22,428	21,817	21,983
Office	5,456	5,165	4,833	4,297	2,627	2,275	2,253
Commercial	1,606	1,491	1,982	1,610	1,093	1,076	1,121
Health care	2,699	3,599	4,289	4,076	4,064	3,605	3,240
Educational	2,649	3,039	2,938	2,425	2,114	1,968	1,908
Public safety	3,884	3,327	2,957	2,687	2,789	2,936	2,745
Amusement & recreation	417	792	710	387	210	350	506
Transportation	2,185	1,953	1,861	2,182	2,559	2,168	2,190
Power	1,022	1,058	1,431	1,238	1,089	1,132	1,227
Highway & street	802	1,109	916	671	640	513	553
Conservation & development	3,704	5,111	5,170	3,927	3,357	4,123	4,830

Note: Details may not add to totals due to rounding.

[1] Includes the following categories of construction not shown separately: lodging, religious, communication and manufacturing.

[2] Includes the following categories of federal construction not shown separately: lodging, religious, communication, sewage and waste disposal, water supply, and manufacturing.

Source: U.S. Census Bureau
Plunkett Research,® Ltd.
www.plunkettresearch.com

Estimates of the Total Housing Inventory for the U.S.: 2014-2015

(In Thousands of Units)

Type of Housing	Fourth Quarter 2014 Estimate	Fourth Quarter 2015 Estimate	90% Confidence Interval (+/-)*		2015 % of Total
			of 2015 Estimate	of Difference	
All housing units	134,215	134,991	na	na	100
Occupied	117,313	117,775	192	176	87.2
Owner occupied	75,032	75,194	629	430	55.7
Renter occupied	42,281	42,581	569	439	31.5
Vacant	16,902	17,216	365	323	12.8
Year-round vacant	12,756	13,014	358	308	9.6
For rent	3,233	3,233	160	179	2.4
For sale only	1,458	1,446	91	115	1.1
Rented or sold, awaiting occupancy	964	1,014	70	97	0.8
Held off market	7,104	7,321	275	238	5.4
For occasional use	1,936	2,026	148	127	1.5
Temporarily occupied by persons with usual residence elsewhere	1,371	1,415	124	107	1.0
For other reasons	3,797	3,880	203	176	2.9
Seasonal vacant	4,147	4,202	235	205	3.1

Estimates may not add to total due to rounding.

* A 90-percent confidence interval is a measure of an estimate's reliability. The larger the confidence interval is, in relation to the size of the estimate, the less reliable the estimate.

na = Not applicable. Because the number of housing units is set equal to an independent national measure, there is no sampling error, and hence no confidence interval.

Source: U.S. Census Bureau
Plunkett Research,® Ltd.
www.plunkettresearch.com

Assets & Liabilities, U.S. Agency- & Government Sponsored Enterprise (GSE)-Backed Securities by Holder: 2007-2015

(In Billions of US$; Amounts Outstanding End of Period, Not Seasonally Adjusted)

	2007	2008	2009	2010	2011	2012	2013	2014	2015
Total liabilities	**7,397.7**	**8,166.7**	**8,106.8**	**7,598.2**	**7,584.5**	**7,560.3**	**7,798.2**	**7,948.4**	**8,152.9**
Budget agencies	23.1	23.3	23.5	24.2	25.3	24.9	24.5	24.4	24.6
Government-sponsored enterprises	2,910.2	3,181.9	2,706.6	6,434.5	6,247.3	6,092.7	6,200.2	6,275.5	6,352.9
Agency- & GSE-backed mortgage pools	4,464.4	4,961.4	5,376.7	1,139.5	1,311.9	1,442.7	1,573.5	1,648.5	1,775.5
Total assets	**7,397.7**	**8,166.7**	**8,106.8**	**7,598.2**	**7,584.5**	**7,560.3**	**7,798.2**	**7,948.4**	**8,152.9**
Household sector	872.7	1,058.9	359.3	355.6	363.4	302.3	298.4	99.7	134.6
Nonfinancial corporate business	12.8	10.2	14.3	16.0	14.3	13.1	8.9	11.7	9.7
Federal government	0.0	54.4	196.4	149.2	31.1	0.0	0.0	0.0	0.0
State and local governments	496.6	478.5	488.2	510.6	502.8	482.4	452.9	438.7	427.4
Monetary authority	0.0	19.7	1,068.3	1,139.6	1,003.4	1,547.4	1,775.5	1,780.4	1,736.4
U.S.-chartered depository institutions	1,106.3	1,242.5	1,417.4	1,527.2	1,634.1	1,669.6	1,717.3	1,760.3	1,924.7
Foreign banking offices in the U.S.	57.1	50.6	31.3	26.5	30.6	32.1	25.4	20.5	23.1
Banks in U.S.-affiliated areas	26.3	24.4	20.5	12.8	4.8	2.6	1.9	1.7	1.8
Credit unions	68.4	82.9	110.7	151.5	182.1	197.0	199.2	187.5	175.1
Property-casualty insurance companies	125.8	114.3	116.2	115.8	122.7	114.3	109.5	105.2	105.9
Life insurance companies	382.9	366.2	371.9	376.0	374.4	360.9	354.1	339.1	339.1
Private pension funds	190.6	190.3	184.0	190.4	174.6	176.7	171.0	160.8	154.3
Federal government retirement funds	5.4	5.2	5.2	5.8	6.5	8.1	6.4	5.9	6.0
State and local govt. retirement funds	191.2	160.6	157.6	159.4	137.3	112.0	114.6	109.8	101.4
Money market mutual funds	235.9	756.2	543.0	402.8	403.7	343.5	361.3	384.6	460.1
Mutual funds	576.6	553.3	603.2	680.1	794.0	879.8	854.2	929.3	943.9
GSEs	702.9	910.0	924.5	377.0	358.9	310.6	287.9	295.7	272.6
ABS Issuers	349.8	323.7	99.6	3.6	0.3	0.3	0.1	0.1	0.0
REITs	88.9	89.6	105.1	143.3	248.1	357.6	261.9	269.8	229.2
Brokers & dealers	315.2	242.6	110.9	149.8	147.7	169.6	114.2	121.9	109.7
Holding companies	9.9	25.8	24.3	21.1	33.1	22.9	28.0	28.0	23.7
Rest of the world	1,582.4	1,406.9	1,154.9	1,084.0	1,078.2	1,001.2	883.8	902.6	930.2

Notes: Agency- and GSE-backed securities include: issues of federal budget agencies such as those for the TVA; issues of GSEs such as FNMA and FHLB; and agency- and GSE-backed mortgage pool securities issued by GNMA, FNMA, FHLMC, and the Farmers Home Administration. Only the budget agency issues are considered officially to be part of the total debt of the federal government.

ABS = Asset-Backed Security. REIT = Real Estate Investment Trust.

Source: U.S. Federal Reserve

Plunkett Research, Ltd.

www.plunkettresearch.com

Commercial, Residential & Farm Mortgages by Holder, U.S.: 2007-2015

(In Billions of US$; Amounts Outstanding End of Period, Not Seasonally Adjusted)

Holder	2007	2008	2009	2010	2011	2012	2013	2014	2015
Total mortgages	**14,621.3**	**14,713.7**	**14,420.8**	**13,795.3**	**13,499.5**	**13,270.0**	**13,262.0**	**13,442.8**	**13,795.8**
Home	11,237.8	11,150.9	10,937.1	10,446.7	10,206.9	9,976.5	9,879.8	9,884.3	9,986.0
Multi-family residential	797.1	847.9	854.4	852.2	857.5	888.1	924.3	993.2	1,098.8
Commercial	2,473.7	2,580.2	2,483.3	2,342.3	2,267.9	2,232.2	2,273.1	2,369.0	2,505.9
Farm	112.7	134.7	146.0	154.1	167.2	173.2	184.8	196.3	205.1
Total liabilities	**14,621.3**	**14,713.7**	**14,420.8**	**13,795.3**	**13,499.5**	**13,270.0**	**13,262.0**	**13,442.8**	**13,795.8**
Household sector	10,814.4	10,778.2	10,611.4	10,123.3	9,903.1	9,689.7	9,603.3	9,610.2	9,711.1
Nonfinancial business	3,650.1	3,772.0	3,642.5	3,495.9	3,406.9	3,385.9	3,456.5	3,608.9	3,848.0
Corporate	967.2	902.2	774.0	643.2	592.6	477.1	481.5	498.0	528.4
Noncorporate	2,683.0	2,869.9	2,868.5	2,852.7	2,814.3	2,908.8	2,975.0	3,110.9	3,319.6
Federal govt.	0.0	0.0	0.0	0.0	0.0	0.0	0.0	0.0	0.0
REITs	156.7	163.5	166.9	176.1	189.5	194.4	202.2	223.8	236.7
Total assets	**14,621.3**	**14,713.7**	**14,420.8**	**13,795.3**	**13,499.5**	**13,270.0**	**13,262.0**	**13,442.8**	**13,795.8**
Household sector	109.9	110.9	108.0	100.1	100.9	82.8	77.1	82.5	79.2
Nonfinancial corporate business	41.4	33.6	29.5	28.0	27.4	25.5	27.1	28.7	30.2
Nonfinancial noncorporate business	42.1	39.1	37.9	42.1	36.3	33.9	34.8	36.5	39.3
Federal govt.	82.4	95.8	108.4	106.6	109.9	111.5	115.5	117.2	116.5
State & local govt.	193.2	187.3	192.3	204.6	202.4	205.0	203.8	209.9	227.3
U.S.-chartered depository institutions	4,658.6	4,615.2	4,371.9	4,194.9	4,049.6	4,028.8	3,983.5	4,090.7	4,296.3
Foreign banking offices in the U.S.	39.0	44.2	37.8	35.4	32.8	30.9	30.3	37.6	51.3
Banks in U.S.-affiliated areas	40.8	42.6	42.3	35.8	33.4	34.9	32.4	30.4	26.6
Credit unions	280.2	312.2	316.9	317.0	320.5	327.8	345.9	372.5	402.6
Property-casualty insurance companies	4.8	5.0	4.4	4.1	4.9	5.6	7.9	9.9	12.0
Life insurance companies	326.2	342.4	326.1	317.5	332.5	344.4	363.2	384.8	415.4
Private pension funds	34.2	26.1	23.8	26.5	21.9	23.1	22.4	22.5	21.1
State & local govt. retirement funds	17.9	15.1	11.4	11.9	11.0	10.0	9.4	9.2	6.5
GSEs*	643.1	705.3	707.7	5,021.0	4,924.0	4,823.5	4,877.8	4,870.5	4,918.9
Agency- & GSE-backed mortgage pools	4,464.4	4,961.4	5,376.7	1,139.5	1,311.9	1,442.7	1,573.5	1,648.5	1,775.5
ABS issuers	2,970.2	2,618.2	2,247.2	1,922.3	1,713.0	1,495.5	1,201.0	1,099.0	1,020.6
Finance companies	549.9	482.9	430.3	243.5	211.2	179.5	157.0	148.1	119.5
REITs	123.1	76.3	48.3	44.6	56.1	64.5	199.3	244.6	237.0

* FHLB loans to savings institutions are included in other loans and advances.

REIT = Real Estate Investment Trust. GSE = Government Sponsored Enterprise. ABS = Asset Backed Security.

Source: U.S. Federal Reserve

Plunkett Research,® Ltd.

www.plunkettresearch.com

Home Mortgages by Holder, U.S.: 2007-2015

(In Billions of US$; Amounts Outstanding End of Period; Not Seasonally Adjusted)

	2007	2008	2009	2010	2011	2012	2013	2014	2015
Total Liabilities	11,237.8	11,150.9	10,937.1	10,446.7	10,206.9	9,976.5	9,879.8	9,884.3	9,986.0
Household sector	10,610.5	10,577.3	10,417.3	9,915.5	9,702.0	9,490.8	9,401.2	9,400.6	9,490.6
Nonfinancial corporate business	42.2	32.7	20.3	13.6	10.2	9.7	10.0	11.5	13.4
Nonfinancial noncorporate business	585.1	540.9	499.4	517.6	494.8	476.0	468.6	472.3	482.0
Total Assets	11,237.8	11,150.9	10,937.1	10,446.7	10,206.9	9,976.5	9,879.8	9,884.3	9,986.0
Household sector	90.8	91.2	83.2	75.2	67.2	59.2	51.2	43.2	35.2
Nonfinancial corporate business	25.0	20.2	17.7	16.8	16.4	15.3	16.2	17.2	18.1
Nonfinancial noncorporate business	15.4	14.3	13.9	15.4	13.3	12.4	12.8	13.4	14.4
Federal government	13.7	16.4	22.1	23.9	24.0	25.2	25.6	26.8	28.4
State & local governments	99.5	96.5	99.0	105.4	104.2	105.6	105.0	108.1	117.1
U.S.-chartered depository institutions	3,068.0	2,883.6	2,692.5	2,615.6	2,538.0	2,503.6	2,385.7	2,401.8	2,449.3
Foreign banking offices in U.S.	0.0	7.0	0.9	1.1	1.3	1.8	1.4	2.9	1.5
Banks in U.S.-affiliated areas	21.5	22.9	22.6	20.1	17.8	20.0	18.9	17.4	15.8
Credit unions	280.2	312.2	316.9	317.0	320.5	327.8	345.9	372.5	402.6
Life insurance companies	9.4	8.6	8.7	9.0	9.0	11.0	11.3	13.7	15.4
Private pension funds	1.2	1.3	2.0	1.9	1.6	1.6	1.4	1.4	1.2
State & local govt. retirement funds	6.4	5.4	4.1	4.3	4.0	3.6	3.4	3.3	2.1
Government-sponsored enterprises	447.9	456.6	431.3	4,690.7	4,587.7	4,475.6	4,546.4	4,537.9	4,567.7
Agency- and GSE-backed mortgage pools	4,371.8	4,864.0	5,266.5	1,068.8	1,222.9	1,326.6	1,423.9	1,474.8	1,568.8
ABS issuers	2,212.7	1,900.5	1,576.9	1,303.0	1,109.5	928.3	793.0	704.5	605.9
Finance companies	493.7	416.1	366.0	169.5	149.7	132.6	114.7	104.6	89.8
REITs	80.7	34.3	12.6	9.0	19.8	26.5	23.1	41.1	52.6
Memo:									
Home equity loans included above*	1,133.2	1,115.9	1,033.3	928.5	853.5	769.6	703.3	673.0	635.5
U.S.-chartered depository institutions	872.7	894.7	841.4	783.3	723.1	652.7	596.2	568.2	532.9
Foreign banking offices in U.S.	0.0	0.8	0.3	0.3	0.3	0.2	0.4	0.4	0.4
Credit unions	94.1	98.7	94.6	88.2	82.2	75.7	72.0	73.6	76.3
ABS issuers	71.8	46.6	31.5	22.8	18.0	14.5	11.8	9.9	7.9
Finance companies	94.5	75.1	65.5	33.9	29.9	26.5	22.9	20.9	18.0

Note: Home mortgages are mortgages on 1-4 family properties including mortgages on farm houses.

* Loans made under home equity lines of credit and home equity loans secured by junior liens. Excludes home equity loans held by individuals.
REIT = Real Estate Investment Trust. GSE = Government-Sponsored Enterprise. ABS = Asset Backed Security.

Source: U.S. Federal Reserve
Plunkett Research, Ltd.
www.plunkettresearch.com

Mortgage Loans Outstanding, U.S.:
1980-2015

(In Billions of US$, Latest Year Available)

Year	Total	1- to 4-family	Multifamily	Commercial	Farm
1980	1,457.8	957.9	142.5	259.9	97.5
1981	1,579.5	1,030.2	142.4	299.7	107.2
1982	1,661.3	1,070.2	146.1	333.7	111.3
1983	1,850.5	1,186.1	161.2	389.4	113.7
1984	2,091.5	1,321.1	186.1	471.9	112.4
1985	2,367.9	1,526.2	205.9	541.7	94.1
1986	2,654.9	1,729.4	239.4	602.0	84.1
1987	2,955.9	1,927.7	258.4	694.1	75.8
1988	3,273.4	2,162.1	274.5	765.9	70.8
1989	3,526.0	2,368.9	287.0	801.3	68.8
1990	3,781.1	2,606.3	287.4	819.8	67.6
1991	3,932.3	2,774.3	284.1	806.4	67.5
1992	4,043.3	2,941.7	270.9	762.7	67.9
1993	4,174.6	3,100.7	267.7	737.8	68.4
1994	4,338.8	3,277.9	268.2	722.7	69.9
1995	4,524.8	3,445.4	273.9	733.8	71.7
1996	4,805.6	3,681.6	286.1	763.4	74.4
1997	5,119.0	3,916.2	298.0	826.4	78.5
1998	5,603.6	4,272.6	332.3	915.6	83.1
1999	6,210.3	4,697.8	372.9	1,052.4	87.2
2000	6,768.3	5,121.5	402.2	1,159.9	84.7
2001	7,477.3	5,674.6	444.4	1,269.7	88.5
2002	8,379.5	6,430.8	483.4	1,369.9	95.4
2003	9,390.6	7,254.7	555.9	1,496.7	83.2
2004	10,665.9	8,291.7	606.1	1,672.4	95.7
2005	12,109.6	9,420.2	668.8	1,915.9	104.8
2006	13,525.9	10,499.8	710.4	2,207.7	108.0
2007	14,621.3	11,237.8	797.1	2,473.7	112.7
2008	14,713.7	11,150.9	847.9	2,580.2	134.7
2009	14,420.8	10,937.1	854.4	2,483.3	146.0
2010	13,795.3	10,446.7	852.2	2,342.3	154.1
2011	13,499.5	10,206.9	857.5	2,267.9	167.2
2012	13,270.0	9,976.5	888.1	2,232.2	173.2
2013	13,262.0	9,879.8	924.3	2,273.1	184.8
2014	13,442.8	9,884.3	993.2	2,369.0	196.3
2015	13,795.8	9,986.0	1,098.8	2,505.9	205.1

Source: U.S. Federal Reserve
Plunkett Research, Ltd.
www.plunkettresearch.com

New Privately-Owned Housing Units Started, U.S.: 1980-2015

(In Thousands; Latest Year Available)

Year	Single Unit	2-4 Units	5+ Units	Total
1980	852.2	109.5	330.5	1,292.2
1981	705.4	91.2	287.7	1,084.2
1982	662.6	80.1	319.6	1,062.2
1983	1,067.6	113.5	522.0	1,703.0
1984	1,084.2	121.4	543.9	1,749.5
1985	1,072.4	93.5	576.0	1,741.8
1986	1,179.4	84.0	542.0	1,805.4
1987	1,146.4	65.1	408.7	1,620.5
1988	1,081.3	58.7	348.0	1,488.1
1989	1,003.3	55.3	317.6	1,376.1
1990	894.8	37.6	260.4	1,192.7
1991	840.4	35.6	137.9	1,013.9
1992	1,029.9	30.9	139.0	1,199.7
1993	1,125.7	29.4	132.6	1,287.6
1994	1,198.4	35.2	223.5	1,457.0
1995	1,076.2	33.8	244.1	1,354.1
1996	1,160.9	45.3	270.8	1,476.8
1997	1,133.7	44.5	295.8	1,474.0
1998	1,271.4	42.6	302.9	1,616.9
1999	1,302.4	31.9	306.6	1,640.9
2000	1,230.9	38.7	299.1	1,568.7
2001	1,273.3	36.6	292.8	1,602.7
2002	1,358.6	38.5	307.9	1,704.9
2003	1,499.0	33.5	315.2	1,847.7
2004	1,610.5	42.3	303.0	1,955.8
2005	1,715.8	41.1	311.4	2,068.3
2006	1,465.4	42.7	292.8	1,800.9
2007	1,046.0	31.7	277.3	1,355.0
2008	622.0	17.5	266.0	905.5
2009	445.1	11.6	97.3	554.0
2010	471.2	11.4	104.3	586.9
2011	430.6	10.9	167.3	608.8
2012	535.3	11.4	233.9	780.6
2013	617.6	13.6	293.7	924.9
2014	647.9	13.7	341.7	1,003.3
2015	714.5	11.5	385.8	1,111.8

Note: Single-family estimates prior to 1999 include an upward adjustment of 3.3 percent made to account for structures started in permit-issuing areas without permit authorization. Components may not add up due to rounding.

Source: U.S. Census Bureau
Plunkett Research, Ltd.
www.plunkettresearch.com

Homeownership Rates by Region, U.S.:
Selected Years, 1965-2015

(Percent of All Households; Latest Year Available)

Area	1965	1970	1975	1980	1985	1990	1995	2000	2005
United States	**63.3**	**64.2**	**64.6**	**65.6**	**63.9**	**63.9**	**64.7**	**67.4**	**68.9**
Inside Metropolitan Areas[1]	59.7	60.3	61.1	61.6	60.2	61.3	62.7	65.5	67.4
In Principal Cities[1]	48.0	48.9	49.5	49.6	47.7	48.7	49.5	51.4	54.2
Not in Principal Cities[1] (suburbs)	72.0	70.6	70.9	70.8	69.1	70.1	71.2	74.0	76.4
Outside Metropolitan Areas[1]	69.1	71.6	72.1	74.1	71.9	73.2	72.7	75.2	76.3
Northeast	57.7	58.1	59.2	60.8	60.8	62.6	62.0	63.4	65.2
Midwest	68.6	69.5	70.2	69.8	66.9	67.5	69.2	72.6	73.1
South	64.4	66.0	66.0	68.7	66.4	65.7	66.7	69.6	70.8
West	59.3	60.0	60.8	60.0	59.0	58.0	59.2	61.7	64.4
Area	**2007**	**2008**	**2009**	**2010**	**2011**	**2012**	**2013**	**2014**	**2015**
United States	**68.1**	**67.8**	**67.4**	**66.9**	**66.1**	**65.4**	**65.1**	**64.5**	**63.7**
Inside Metropolitan Areas[1]	66.8	66.4	65.9	65.4	64.6	63.9	63.4	62.9	62.2
In Principal Cities[1]	53.6	53.2	52.8	52.1	51.3	50.8	50.3	49.7	48.9
Not in Principal Cities[1] (suburbs)	75.5	75.1	74.6	74.0	73.5	72.6	72.2	71.5	70.7
Outside Metropolitan Areas[1]	75.1	75.2	74.7	74.5	73.9	73.4	74.0	72.7	71.6
Northeast	65.0	64.6	64.0	64.1	63.6	63.5	63.0	62.2	60.9
Midwest	71.9	71.7	71.0	70.8	70.2	69.6	69.7	69.0	68.3
South	70.1	69.9	69.6	69.0	68.3	67.2	66.7	65.9	65.1
West	63.5	63.0	62.6	61.4	60.5	59.8	59.4	59.2	58.7

[1] Approximately every 10 years, metropolitan/nonmetropolitan area definitions are redefined by the Office of Management and Budget. Therefore metropolitan/nonmetropolitan data for 1986 to 1994, 1995 to 2004, and 2005 to 2014 and 2015 and later are not comparable to each other.

Source: U.S. Census Bureau

Plunkett Research, Ltd.

www.plunkettresearch.com

Homeownership Rates by Race & Ethnicity of Householder, U.S.: 1994-2015

(Percent of All Households; Latest Year Available)

Year (4th Quarter)	U.S.	Non-Hispanic White Alone	Black Alone	All Other Races*	Hispanic (of any race)
1994	64.2	70.2	42.6	47.6	42.2
1995	65.1	70.9	42.2	46.7	42.8
1996	65.4	71.8	44.4	51.4	42.3
1997	65.7	71.9	45.1	52.5	44.0
1998	66.4	72.6	45.9	52.7	45.7
1999	66.9	73.3	46.8	54.3	45.5
2000	67.5	73.9	47.8	52.4	47.5
2001	68.0	74.4	48.1	53.2	48.8
2002	68.3	75.0	47.7	55.2	48.3
2003	68.6	75.5	49.4	56.6	47.7
2004	69.2	76.2	49.1	58.9	48.9
2005	69.0	76.0	48.0	60.1	50.0
2006	68.9	76.0	48.2	60.0	49.5
2007	67.8	74.9	47.7	58.6	48.5
2008	67.5	74.8	46.8	58.3	48.6
2009	67.2	74.5	46.0	58.4	48.4
2010	66.5	74.2	44.9	57.7	46.8
2011	66.0	73.7	45.1	56.5	46.6
2012	65.4	73.6	44.5	55.2	45.0
2013	65.2	73.4	43.2	56.0	45.5
2014	64.0	72.3	42.1	55.3	44.5
2015	63.8	72.2	41.9	53.3	46.7

* Includes people who reported Asian, Native Hawaiian or Other Pacific Islander, or American Indian or Alaska Native regardless of whether they reported any other race, as well as all other combinations of two or more races.

Source: U.S. Census Bureau
Plunkett Research, Ltd.
www.plunkettresearch.com

Homeownership Rates by Age of Householder, U.S.: Selected Years, 1990-2015

(In Percentage of Households)

Age Range	1990	1995	2000	2010	2011	2012	2013	2014	2015
United States, Total	**63.9**	**64.7**	**67.4**	**66.9**	**66.1**	**65.4**	**65.1**	**64.0**	**63.8**
Less than 25 years	15.7	15.9	21.7	22.8	22.6	21.7	22.2	20.4	20.8
25 to 29 years	35.2	34.4	38.1	36.8	34.6	34.3	34.1	32.0	31.8
30 to 34 years	51.8	53.1	54.6	51.6	49.8	47.9	48.1	47.4	45.9
35 to 39 years	63.0	62.1	65.0	61.9	59.8	56.9	55.8	55.2	56.2
40 to 44 years	69.8	68.6	70.6	67.9	66.9	65.5	65.0	62.3	62.4
45 to 49 years	73.9	73.7	74.7	72.0	71.0	69.9	69.6	68.2	67.9
50 to 54 years	76.8	77.0	78.5	75.0	74.4	73.4	72.6	72.7	72.1
55 to 59years	78.8	78.8	80.4	77.7	77.3	76.2	75.8	74.9	74.3
60 to 64 years	79.8	80.3	80.3	80.4	79.8	78.6	77.6	76.8	76.2
65 to 69 years	80.0	81.0	83.0	81.6	81.7	81.1	80.5	80.1	80.0
70 to 74 years	78.4	80.9	82.6	82.4	83.3	83.5	82.8	81.7	82.0
75 years and over	72.3	74.6	77.7	78.9	79.3	80.0	80.0	77.8	77.4
Less than 35 years	38.5	38.6	40.8	39.1	37.7	36.7	36.8	35.3	34.7
35 to 44 years	66.3	65.2	67.9	65.0	63.5	61.4	60.6	58.8	59.3
45 to 54 years	75.2	75.2	76.5	73.5	72.7	71.7	71.2	70.5	70.1
55 to 64 years	79.3	79.5	80.3	79.0	78.5	77.3	76.6	75.8	75.2
65 years and over	76.3	78.1	80.4	80.5	80.9	81.1	80.8	79.5	79.3

Source: U.S. Census Bureau
Plunkett Research,® Ltd.
www.plunkettresearch.com

Average Sales Prices of New Homes Sold in U.S.: 1963-2015
(In Constant 2009 Chained US$)

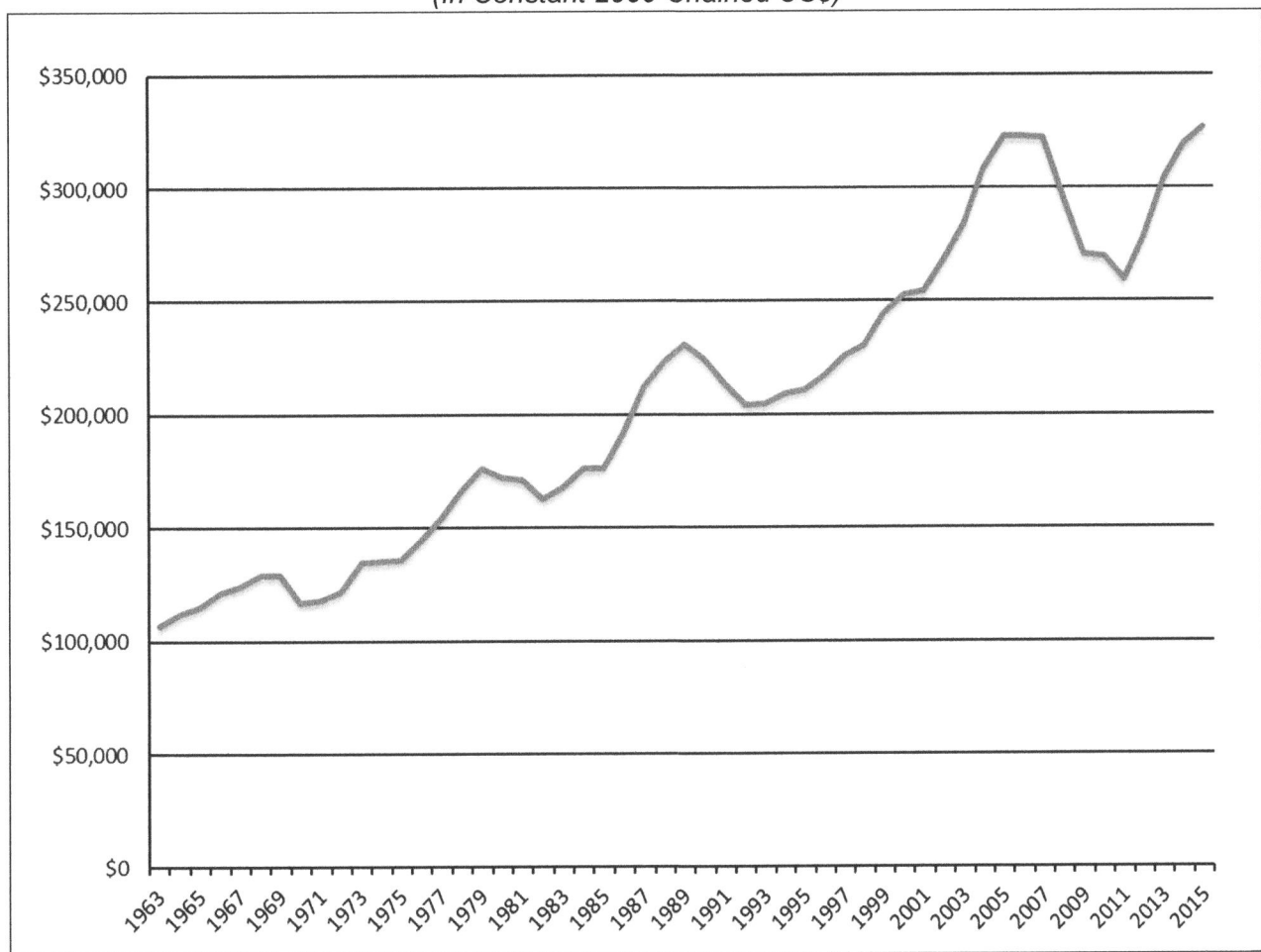

Source: U.S. Census Bureau
Plunkett Research,® Ltd.
www.plunkettresearch.com

Median & Average Sales Price of New Single Family Homes by Region, U.S.: 1985-2015

(In US$)

Period	Median sales price					Average sales price				
	U.S.	Northeast	Midwest	South	West	U.S.	Northeast	Midwest	South	West
1985	84,300	103,300	80,300	75,000	92,600	100,800	121,900	95,400	88,900	111,800
1986	92,000	125,000	88,300	80,200	95,700	111,900	151,300	102,600	95,300	116,100
1987	104,500	140,000	95,000	88,000	111,000	127,200	170,900	115,500	106,600	134,600
1988	112,500	149,000	101,600	92,000	126,500	138,300	179,300	123,700	114,800	155,700
1989	120,000	159,600	108,800	96,400	139,000	148,800	188,600	130,600	123,100	173,900
1990	122,900	159,000	107,900	99,000	147,500	149,800	190,500	133,000	123,500	180,600
1991	120,000	155,900	110,000	100,000	141,100	147,200	188,800	134,500	123,000	176,400
1992	121,500	169,000	115,600	105,500	130,400	144,100	194,900	136,400	126,900	157,800
1993	126,500	162,600	125,000	115,000	135,000	147,700	183,600	143,100	133,600	161,900
1994	130,000	169,000	132,900	116,900	140,400	154,500	200,500	152,700	136,800	168,900
1995	133,900	180,000	134,000	124,500	141,000	158,700	216,600	157,200	142,000	169,800
1996	140,000	186,900	137,500	125,000	153,900	166,400	226,800	158,100	143,100	185,900
1997	146,000	190,000	149,900	129,600	160,000	176,200	234,100	173,000	151,400	198,200
1998	152,500	200,000	157,500	135,800	163,500	181,900	240,100	179,200	159,700	200,500
1999	161,000	210,500	164,000	145,900	173,700	195,600	247,900	186,800	173,000	221,700
2000	169,000	227,400	169,700	148,000	196,400	207,000	274,800	199,300	179,000	238,900
2001	175,200	246,400	172,600	155,400	213,600	213,200	294,300	201,300	185,700	250,000
2002	187,600	264,300	178,000	163,400	238,500	228,700	301,300	209,800	197,500	276,500
2003	195,000	264,500	184,300	168,100	260,900	246,300	315,700	218,200	208,900	306,800
2004	221,000	315,800	205,000	181,100	283,100	274,500	366,100	240,800	232,800	340,000
2005	240,900	343,800	216,900	197,300	332,600	297,000	397,000	249,800	249,200	388,700
2006	246,500	346,000	213,500	208,200	337,700	305,900	428,300	257,100	257,700	405,900
2007	247,900	320,200	208,600	217,700	330,900	313,600	437,700	256,800	269,800	403,700
2008	232,100	343,600	198,900	203,700	294,800	292,600	475,500	250,000	253,400	361,500
2009	216,700	302,500	189,200	194,800	263,700	270,900	411,300	227,700	241,200	321,600
2010	221,800	329,900	197,700	196,800	259,300	272,900	415,800	232,800	244,900	316,600
2011	227,200	322,800	203,300	211,400	256,000	267,900	389,900	241,700	248,900	301,800
2012	245,200	368,800	230,600	227,000	270,000	292,200	418,600	270,300	266,100	321,300
2013	268,900	371,200	255,300	246,600	310,500	324,500	469,900	301,500	292,600	368,900
2014	282,800	402,800	269,700	257,700	333,900	345,800	545,200	316,700	306,900	397,500
2015	293,600	444,500	276,000	272,400	345,400	355,500	573,100	326,000	319,400	413,300

Source: U.S. Census Bureau
Plunkett Research,® Ltd.
www.plunkettresearch.com

New Single Family Homes Sold by Region, U.S.: 1980-2015

(Not Seasonally Adjusted; In Thousands)

Period	U.S.	Northeast	Midwest	South	West
1980	545	50	81	267	145
1981	436	46	60	219	112
1982	412	47	48	219	99
1983	623	76	71	323	152
1984	639	94	76	309	160
1985	688	112	82	323	171
1986	750	136	96	322	196
1987	671	117	97	271	186
1988	676	101	97	276	202
1989	650	86	102	260	202
1990	534	71	89	225	149
1991	509	57	93	215	144
1992	610	65	116	259	170
1993	666	60	123	295	188
1994	670	61	123	295	191
1995	667	55	125	300	187
1996	757	74	137	337	209
1997	804	78	140	363	223
1998	886	81	164	398	243
1999	880	76	168	395	242
2000	877	71	155	406	244
2001	908	66	164	439	239
2002	973	65	185	450	273
2003	1,086	79	189	511	307
2004	1,203	83	210	562	348
2005	1,283	81	205	638	358
2006	1,051	63	161	559	267
2007	776	65	118	411	181
2008	485	35	70	266	114
2009	375	31	54	202	87
2010	323	31	45	174	74
2011	306	21	45	168	72
2012	368	29	47	195	97
2013	429	31	61	233	105
2014	437	28	59	243	108
2015	501	24	61	286	131

Note: Components may not add to total because of rounding. Estimates prior to 1999 include an upward adjustment of 3.3 percent made to account for houses sold in permit-issuing areas that will never have a permit authorization.

Source: U.S. Census Bureau

Plunkett Research,® Ltd.

www.plunkettresearch.com

New Single Family Homes Sold by Type of Financing, U.S.: 1988-2015

(In Thousands)

Period	Total Sold	Type of Financing				
		Conventional*	FHA Insured	VA Guaranteed	Rural Housing Service	Cash
1988	676	437	127	44	6	62
1989	650	416	118	44	14	58
1990	534	337	105	33	10	50
1991	509	329	92	36	9	43
1992	610	428	86	48	7	41
1993	666	476	92	55	6	37
1994	670	490	78	51	9	41
1995	667	490	79	50	9	39
1996	757	570	89	51	9	38
1997	804	616	90	47	6	46
1998	886	693	90	46	9	48
1999	880	689	106	37	6	41
2000	877	695	108	30	4	40
2001	908	726	106	35	2	39
2002	973	788	106	34	4	42
2003	1,086	911	94	36	4	41
2004	1,203	1,047	77	28	6	46
2005	1,283	1,151	51	28	1	52
2006	1,051	949	38	25	1	38
2007	776	695	28	24	2	30
2008	485	358	77	27	NA	23
2009	375	234	92	32	NA	17
2010	323	189	81	35	NA	19
2011	306	190	61	35	NA	20
2012	368	234	75	35	NA	24
2013	429	296	67	36	NA	31
2014	437	311	51	38	NA	37
2015	501	347	81	43	NA	29

* Includes houses reporting other types of financing. Beginning in 2008, also includes Rural Housing Service.

Note: Estimates prior to 1999 include an upward adjustment of 3.3 percent made to account for houses sold in permit-issuing areas that will never have a permit authorization.

Source: U.S. Census Bureau

Plunkett Research,® Ltd.

www.plunkettresearch.com

New Single Family Homes for Sale at End of Period, by Region, U.S.: 1980-2015

(Not Seasonally Adjusted; In Thousands)

Period	U.S.	Northeast	Midwest	South	West
1980	342	40	55	149	97
1981	278	41	34	127	76
1982	255	39	27	129	60
1983	304	42	33	149	79
1984	358	55	41	177	85
1985	350	66	34	172	79
1986	361	88	32	153	87
1987	370	103	39	149	79
1988	371	112	43	133	82
1989	366	108	41	123	93
1990	321	77	42	105	97
1991	284	62	41	97	83
1992	267	48	41	104	74
1993	295	53	48	121	73
1994	340	55	63	140	82
1995	374	62	69	158	86
1996	326	38	67	146	74
1997	287	26	65	127	69
1998	300	28	63	142	68
1999	315	28	64	153	70
2000	301	28	65	146	62
2001	310	28	70	142	69
2002	344	36	77	161	70
2003	377	29	97	172	79
2004	431	30	111	200	91
2005	515	47	109	249	109
2006	537	54	97	267	119
2007	496	48	79	248	121
2008	352	37	57	175	83
2009	232	27	38	118	48
2010	188	22	27	98	41
2011	150	19	20	79	32
2012	148	14	24	79	31
2013	186	16	29	100	40
2014	212	18	29	118	47
2015	234	24	31	124	55

Note: Components may not add to total because of rounding. Estimates prior to 1999 include an upward adjustment of 3.3 percent made to account for houses for sale in permit-issuing areas that will never have a permit authorization.

Source: U.S. Census Bureau

Plunkett Research,® Ltd.

www.plunkettresearch.com

Price Deflator (Fisher) Index of New One-Family Houses Under Construction, U.S.: 1980-February 2016

(2005 = 100)

Year	Annual		Monthly											
	Ann. Index	% Chg.	Jan	Feb	Mar	Apr	May	Jun	Jul	Aug	Sep	Oct	Nov	Dec
1980	37.9	10.5%	36.5	36.7	37.0	37.3	37.4	37.7	38.0	38.1	38.5	38.9	39.3	39.5
1981	40.5	6.9%	39.8	40.1	40.1	40.3	40.4	40.5	40.7	40.7	40.8	40.9	41.1	41.0
1982	41.7	3.0%	41.3	41.3	41.5	41.5	41.7	41.7	41.7	42.0	42.3	42.1	42.1	42.2
1983	42.9	2.9%	42.5	42.7	42.8	42.6	42.5	42.6	42.7	42.8	42.9	43.3	43.3	43.4
1984	44.5	3.7%	43.8	44.1	44.2	44.1	44.1	44.2	44.4	44.8	45.0	45.0	45.0	45.0
1985	45.4	2.0%	45.2	45.3	45.5	45.2	45.0	45.0	45.0	45.2	45.4	45.7	45.9	46.1
1986	47.4	4.4%	46.5	46.6	46.7	46.8	47.0	47.3	47.6	47.6	47.8	48.2	48.6	48.8
1987	49.6	4.6%	49.2	49.2	49.2	49.3	49.4	49.5	49.7	49.8	49.8	50.0	50.2	50.6
1988	51.6	4.0%	51.2	51.4	51.4	51.3	51.4	51.4	51.4	51.6	51.8	51.9	52.2	52.3
1989	53.7	4.1%	52.6	53.0	53.0	53.4	53.7	53.7	53.9	53.8	53.9	54.1	54.1	54.5
1990	55.4	3.2%	55.0	55.0	55.1	55.1	55.1	55.4	55.8	55.8	55.7	55.7	55.8	55.7
1991	55.9	0.9%	55.8	55.7	55.4	55.4	55.5	55.8	56.0	56.2	56.3	56.2	56.1	56.1
1992	57.0	2.0%	56.4	56.5	56.5	56.6	56.5	56.7	56.9	57.0	57.2	57.7	58.0	58.3
1993	59.8	4.9%	58.7	58.9	59.0	59.1	59.3	59.7	59.9	60.1	60.3	60.5	60.9	61.1
1994	62.5	4.5%	61.3	61.5	61.5	61.5	61.7	62.1	62.3	62.7	63.1	63.4	63.8	64.4
1995	65.2	4.3%	64.6	64.8	64.8	64.9	65.0	65.2	65.1	65.4	65.6	65.6	65.6	65.6
1996	66.4	1.8%	65.6	65.8	65.7	65.6	65.8	66.2	66.6	66.8	66.8	67.0	67.1	67.3
1997	68.4	3.0%	67.6	67.6	67.6	67.8	67.8	68.1	68.4	68.6	69.0	69.2	69.3	69.3
1998	70.2	2.6%	69.4	69.2	69.2	69.4	69.5	70.0	70.3	70.4	70.4	70.8	71.2	71.6
1999	73.3	4.4%	72.0	72.1	72.3	72.7	73.0	73.3	73.6	73.6	73.7	73.9	74.3	74.6
2000	76.7	4.6%	75.5	75.8	76.0	76.2	76.4	76.7	76.9	76.9	77.1	77.2	77.3	77.7
2001	80.2	4.6%	78.2	78.4	78.6	79.1	79.7	80.2	80.8	81.6	81.6	82.1	82.1	81.8
2002	82.1	2.4%	81.2	81.4	81.7	81.8	82.1	82.2	82.1	81.8	82.3	82.9	82.6	83.2
2003	86.1	4.9%	84.1	84.6	85.1	85.3	85.3	85.4	85.4	86.1	86.8	87.4	88.1	88.5
2004	93.0	8.0%	89.1	89.4	90.3	91.1	92.2	92.8	93.4	93.9	94.4	94.9	95.6	96.3
2005	100.0	7.5%	96.5	96.3	96.9	97.2	98.3	99.3	100.7	100.8	101.3	101.9	102.8	104.1
2006	106.2	6.2%	104.5	104.7	105.7	105.9	106.1	106.1	105.6	105.9	107.0	107.4	107.5	107.8
2007	107.2	0.9%	107.8	107.9	108.2	107.9	106.9	106.5	106.7	107.0	107.2	107.2	106.7	106.4
2008	104.1	-2.9%	105.8	105.5	105.2	104.8	105.2	104.0	104.1	103.4	102.6	102.2	102.2	102.6
2009	99.5	-4.4%	102.4	101.8	101.8	100.8	99.3	98.4	98.0	97.9	98.3	98.6	99.0	99.5
2010	98.0	-1.5%	99.5	98.8	98.0	97.3	97.1	97.4	97.8	97.7	97.5	98.1	98.4	98.8
2011	98.7	0.7%	98.7	98.2	98.5	98.6	98.5	98.4	98.4	99.0	98.8	99.1	99.2	99.4
2012	99.7	1.0%	98.6	98.1	98.9	99.1	98.8	98.7	99.4	100.2	100.3	100.4	100.7	101.0
2013	105.2	5.5%	101.9	102.7	103.4	104.0	104.2	104.2	104.5	105.0	106.0	107.4	108.2	109.1
2014	113.0	7.4%	110.2	111.0	111.5	110.7	110.9	110.4	112.9	114.0	114.4	115.2	116.0	116.8
2015	115.7	2.4%	117.2	115.7	115.2	114.7	114.6	115.3	115.6	115.1	115.5	116.2	115.8	117.2
2016	--	--	117.2	117.3[P]	--	--	--	--	--	--	--	--	--	--

Fisher Ideal Index (Price Deflator): This index helps answer the question, "What is the (unbiased) value of today's homes being constructed in constant dollars?" In doing this it attempts to eliminate two kinds of problems associated with two other indexes: the tendency to overstate inflation (Laspeyres); and the tendency to understate inflation (Paasche). The Fisher Ideal index is the geometric average of Laspeyres and Paasche indexes for the same time period. The geometric average is calculated by multiplying the Laspeyres index by the Paasche index and then taking the square root of the result. The biases associated with each component index are minimized by calculating the geometric average.

[P] Preliminary Estimate.

Source: U.S. Census Bureau
Plunkett Research,® Ltd.
www.plunkettresearch.com

Constant Quality (Laspeyres) Price Index of New Single-Family Houses Under Construction, U.S.: 1980-February 2016

(2005 = 100)

Year	Annual		Monthly											
	Ann. Index	% Chg.	Jan	Feb	Mar	Apr	May	Jun	Jul	Aug	Sep	Oct	Nov	Dec
1980	39.8	11.3%	38.1	38.4	38.6	39.0	39.2	39.6	40.1	40.3	40.5	40.7	41.0	41.3
1981	42.6	7.1%	41.6	41.9	42.0	42.3	42.6	42.7	42.7	42.8	43.0	43.2	43.5	43.5
1982	43.4	1.9%	43.4	43.4	43.5	43.6	43.5	43.4	43.2	43.4	43.6	43.5	43.7	43.7
1983	44.7	2.8%	44.0	44.1	44.0	44.0	44.3	44.4	44.6	44.7	44.9	45.2	45.4	45.6
1984	46.7	4.6%	45.8	46.0	46.1	46.1	46.2	46.5	46.5	46.9	47.3	47.5	47.6	47.5
1985	47.9	2.5%	47.7	47.5	47.6	47.5	47.4	47.5	47.6	47.8	48.0	48.2	48.4	48.8
1986	50.4	5.2%	49.1	49.2	49.3	49.5	49.8	50.2	50.4	50.6	50.8	51.1	51.5	51.8
1987	52.7	4.6%	52.1	52.1	52.2	52.4	52.4	52.6	52.7	52.8	52.8	53.0	53.3	53.7
1988	54.5	3.5%	54.1	54.3	54.1	54.1	54.3	54.3	54.3	54.5	54.8	54.9	55.2	55.2
1989	56.4	3.4%	55.4	55.6	55.8	56.1	56.3	56.3	56.5	56.6	56.8	56.8	56.8	57.0
1990	58.0	2.9%	57.2	57.4	57.5	57.6	57.6	57.9	58.2	58.2	58.2	58.1	58.5	58.5
1991	58.2	0.2%	58.5	58.4	58.0	57.9	58.0	58.3	58.4	58.5	58.6	58.3	58.2	58.1
1992	58.9	1.2%	58.4	58.4	58.4	58.5	58.4	58.5	58.5	58.7	58.9	59.4	59.7	60.0
1993	61.8	4.9%	60.4	60.7	60.8	60.9	61.1	61.4	61.8	61.9	62.1	62.5	63.0	63.2
1994	64.6	4.6%	63.3	63.5	63.5	63.5	63.8	64.2	64.4	64.7	65.1	65.5	66.0	66.7
1995	67.3	4.3%	67.0	67.0	67.0	67.0	67.2	67.3	67.2	67.4	67.6	67.7	67.6	67.6
1996	68.6	1.9%	67.9	67.9	67.9	67.7	67.9	68.3	68.8	69.0	69.0	69.2	69.3	69.4
1997	70.6	2.9%	69.8	69.8	69.8	70.0	70.1	70.3	70.6	70.7	71.2	71.4	71.6	71.6
1998	72.5	2.6%	71.6	71.4	71.4	71.7	71.8	72.4	72.7	72.7	72.7	73.0	73.5	74.0
1999	72.7	0.3%	71.3	71.4	71.7	72.0	72.4	72.7	73.0	73.2	73.2	73.4	73.5	73.8
2000	75.9	4.4%	74.7	75.0	75.2	75.3	75.5	75.9	76.0	76.1	76.4	76.5	76.7	77.1
2001	79.7	5.0%	77.7	78.0	78.2	78.6	79.2	79.6	80.0	80.8	81.0	81.5	81.5	81.2
2002	81.7	2.5%	80.6	80.9	81.2	81.4	81.7	81.9	81.7	81.4	81.8	82.3	81.9	82.6
2003	85.9	5.1%	83.7	84.5	85.0	85.3	85.2	85.3	85.1	85.9	86.5	87.1	87.8	88.4
2004	93.1	8.4%	88.9	89.4	90.5	91.4	92.5	93.0	93.6	94.1	94.5	95.1	95.7	96.5
2005	100.0	7.4%	96.7	96.4	96.9	97.1	98.3	99.3	100.7	100.8	101.2	101.8	102.8	104.0
2006	106.0	6.0%	104.4	104.7	105.6	105.7	105.9	105.8	105.4	105.9	106.9	107.3	107.5	107.8
2007	107.0	0.9%	107.9	108.1	108.4	108.0	106.8	106.3	106.4	106.6	106.7	106.7	106.2	105.8
2008	103.3	-3.5%	105.1	104.9	104.5	103.8	104.1	103.0	103.3	102.7	101.8	101.4	101.6	102.0
2009	98.1	-5.0%	101.7	100.8	100.5	99.2	97.7	96.9	96.7	96.6	96.9	97.2	97.5	97.9
2010	96.4	-1.7%	97.8	97.2	96.4	95.7	95.6	95.9	96.4	96.4	96.0	96.6	96.7	97.0
2011	97.4	1.0%	97.0	96.6	96.8	97.5	97.4	97.3	97.2	97.9	97.4	97.8	97.8	97.8
2012	98.4	1.0%	96.8	96.2	96.9	97.4	97.4	97.4	98.4	99.3	99.5	99.4	99.5	99.9
2013	104.8	6.5%	100.9	101.9	102.8	103.5	103.8	103.9	104	104.5	105.8	107.5	108.4	108.9
2014	111.8	6.7%	109.9	110.7	111.0	109.8	109.6	108.8	111.3	112.8	113.1	113.8	114.7	115.5
2015	114.3	2.2%	116.0	114.7	114.0	114.0	114.0	114.6	114.9	113.7	113.5	114.4	113.9	115.0
2016	--	--	115.2	115.7P	--	--	--	--	--	--	--	--	--	--

Notes: Laspeyres Price Index (Constant Quality). This index answers the question, "How much is the sales price today for the same quality house as in the base year?" The base year we are now using is 2005; its index value is set to 100.0. Quality includes not only the physical size and amenities of the house, but also its geographic location. A hypothetical calculation is made in which the base year kind of house is held constant over time while its selling price is calculated in current dollars.

P Preliminary Estimate.

Source: U.S. Census Bureau

Plunkett Research,® Ltd.

www.plunkettresearch.com

Rental & Homeowner Vacancy Rates by Area, U.S.:
Selected Years, 1990-2015

(Percentage of Rental Units; End of Year)

Area	1990	1995	2000	2005	2010	2011	2012	2013	2014	2015
United States	**7.2**	**7.6**	**8.0**	**9.8**	**10.2**	**9.5**	**8.7**	**8.3**	**7.6**	**7.1**
Inside Metropolitan Areas[1]	7.1	7.6	7.7	9.7	10.3	9.5	8.6	8.0	7.4	6.8
In Principal Cities[1]	7.8	8.4	8.2	10.0	10.7	9.9	8.8	8.4	7.5	7.1
Not in Principal Cities[1] (suburbs)	6.3	6.6	7.2	9.4	9.8	9.0	8.4	7.7	7.2	6.4
Outside Metropolitan Areas[1]	7.6	7.9	9.5	10.5	9.9	9.5	9.5	10.2	8.9	9.0
Northeast	6.1	7.2	5.6	6.5	7.6	7.3	7.3	7.1	6.0	5.5
Midwest	6.4	7.2	8.8	12.6	10.8	10.2	9.3	9.1	8.0	7.6
South	8.8	8.3	10.5	11.8	12.7	12.0	10.8	10.0	9.5	8.9
West	6.6	7.5	5.8	7.3	8.2	7.0	6.4	6.2	5.6	5.1

(Percentage of Homeowner Units; End of Year)

Area	1990	1995	2000	2005	2010	2011	2012	2013	2014	2015
United States	**1.7**	**1.5**	**1.6**	**1.9**	**2.6**	**2.5**	**2.0**	**2.0**	**1.9**	**1.8**
Inside Metropolitan Areas[1]	1.7	1.5	1.4	1.9	2.6	2.5	2.0	2.0	1.8	1.7
In Principal Cities[1]	2.2	2.1	1.9	2.4	3.1	2.8	2.4	2.3	2.1	2.0
Not in Principal Cities[1] (suburbs)	1.5	1.3	1.3	1.7	2.4	2.3	1.8	1.8	1.7	1.6
Outside Metropolitan Areas[1]	1.8	1.6	2.1	1.9	2.4	2.4	2.3	2.2	2.1	2.4
Northeast	1.6	1.5	1.2	1.5	1.7	2.2	1.9	1.8	1.7	1.9
Midwest	1.3	1.3	1.3	2.2	2.6	2.6	2.0	2.1	1.8	1.7
South	2.1	1.7	1.9	2.1	2.8	2.6	2.2	2.2	2.2	2.1
West	1.8	1.7	1.5	1.4	2.7	2.3	1.9	1.6	1.6	1.4

[1] Approximately every 10 years, metropolitan/nonmetropolitan area definitions are redefined by the Office of Management and Budget. Therefore metropolitan/nonmetropolitan data 1990, 1995 to 2000, 2005 to 2014 and 2015 and later are not comparable to each other.

Source: U.S. Census Bureau
Plunkett Research,® Ltd.
www.plunkettresearch.com

Construction Industry Employment, U.S.: 2011- February 2016

(In Thousands; Not Seasonally Adjusted)

NAICS	Industry Type	2011	2012	2013	2014	2015	2016*
23	**Construction**	**5,533.0**	**5,646.0**	**5,856.0**	**6,151.0**	**6,446.0**	**6,215.0**
236	**Construction of buildings**	**1,222.1**	**1,240.2**	**1,285.9**	**1,358.0**	**1,415.9**	**1,392.9**
2361	Residential building	565.6	580.8	611.6	659.5	691.3	680.7
236115	New single-family general contractors	281.1	283.6	294.9	319.2	330.9	325.6
236116	New multifamily general contractors	21.0	22.2	24.9	27.7	29.9	30.7
236118	Residential remodelers	240.5	252.0	265.9	282.9	298.9	292.6
2362	Nonresidential building	656.5	659.4	674.4	698.4	724.6	712.2
23621	Industrial building	151.8	153.0	154.5	156.7	159.9	154.6
23622	Commercial building	504.6	506.4	519.9	541.7	564.7	557.6
237	**Heavy & civil engineering construction**	**836.8**	**868.3**	**885.0**	**912.3**	**935.0**	**836.3**
2371	Utility system construction	406.5	430.7	447.2	469.1	472.8	444.3
23711	Water & sewer system construction	151.6	151.8	154.0	160.4	166.5	153.3
23712	Oil & gas pipeline construction	111.5	127.2	132.7	141.5	137.2	122.4
23713	Power & communication system construction	143.4	151.8	160.6	167.2	169.2	168.6
2372	Land subdivision	45.2	41.5	41.4	40.7	40.4	40.2
2373	Highway, street & bridge construction	286.3	294.4	292.7	294.4	312.5	255.2
2379	Other heavy construction	98.8	101.7	103.6	108.0	109.3	96.6
238	**Specialty trade contractors**	**3,474.4**	**3,537.1**	**3,684.4**	**3,880.6**	**4,095.1**	**3,986.0**
238 pt.1	Residential specialty trade contractors	1,450.0	1,475.2	1,565.0	1,669.9	1,775.7	1,750.3
238 pt.2	Nonresidential specialty trade contractors	2,024.4	2,061.8	2,119.5	2,210.7	2,319.4	2,235.7
2381	Building foundation & exterior contractors	668.4	688.3	714.0	756.0	805.2	773.4
2381 pt.1	Res. building foundation & ext. contractors	303.1	311.8	330.2	352.6	372.6	363.2
2381 pt.2	Nonres. building foundation & ext. contractors	365.3	376.5	383.9	403.4	432.6	410.2
23811	Poured concrete structure contractors	141.0	152.9	165.6	176.5	189.4	174.8
23812	Steel and precast concrete contractors	64.1	70.4	70.7	73.5	75.5	77.0
23813	Framing contractors	51.7	53.4	63.0	70.6	72.8	71.7
23814	Masonry contractors	128.0	129.5	129.7	136.0	144.3	136.2
23815	Glass and glazing contractors	47.9	48.9	50.0	54.3	58.9	61.1
23816	Roofing contractors	169.9	164.6	165.9	169.9	186.6	176.5
23817	Siding contractors	31.1	30.2	30.6	33.6	33.9	30.9
23819	Other building exterior contractors	34.7	38.4	38.6	41.6	43.8	45.2
2382	Building equipment contractors	1,652.0	1,680.0	1,748.8	1,833.3	1,930.2	1,941.5
2382 pt.1	Residential building equipment contractors	611.6	623.8	660.7	701.3	751.0	762.0
2382 pt.2	Nonresidential building equipment contractors	1,040.4	1,056.2	1,088.1	1,132.0	1,179.2	1,179.5
23821	Electrical contractors	718.0	734.4	764.2	791.8	829.7	839.9
23822	Plumbing and HVAC contractors	815.1	825.6	860.8	911.8	968.5	970.6
23829	Other building equipment contractors	119.0	120.0	123.8	129.7	132.1	131.0
2383	Building finishing contractors	624.4	631.9	668.4	705.4	744.5	733.9
2383 pt.1	Residential building finishing contractors	329.3	333.5	361.2	387.4	412.2	414.0
2383 pt.2	Nonresidential building finishing contractors	295.1	298.4	307.2	318.1	332.2	319.9
23831	Drywall & insulation contractors	197.5	195.4	207.4	215.0	227.4	227.8
23832	Painting & wall covering contractors	162.8	167.4	176.9	185.8	191.1	178.4
23833	Flooring contractors	55.7	57.4	60.5	64.3	69.5	70.5
23834	Tile & terrazzo contractors	39.2	41.6	43.3	48.3	52.2	52.5
23835	Finish carpentry contractors	107.1	107.6	113.9	123.2	131.3	133.5
23839	Other building finishing contractors	62.1	62.5	66.5	68.9	73.0	71.2

(Continued on next page)

Construction Industry Employment, U.S.: 2011- February 2016 (cont.)

(In Thousands; Not Seasonally Adjusted)

NAICS	Industry Type	2011	2012	2013	2014	2015	2016*
2389	Other specialty trade contractors	529.6	536.8	553.2	585.8	615.3	537.2
2389 pt.1	Other residential trade contractors	206.0	206.1	212.8	228.6	239.9	211.1
2389 pt.2	Other nonresidential trade contractors	323.6	330.7	340.4	357.2	375.4	326.1
23891	Site preparation contractors	267.5	269.5	277.3	294.7	306.5	278.3
23899	All other specialty trade contractors	262.1	267.3	275.8	291.2	308.8	258.9

*Preliminary estimate as of February 2016.

Source: U.S. Bureau of Labor Statistics
Plunkett Research,® Ltd.
www.plunkettresearch.com

Miscellaneous Real Estate & Construction Industry Employment, U.S.: 2011- February 2016

(In Thousands; Not Seasonally Adjusted)

NAICS	Industry Type	2011	2012	2013	2014	2015	2016*
Durable Goods							
3271	Clay products & refractories	40.8	40.0	39.4	39.1	40.1	40.4
3273	Cement & concrete products	164.9	163.5	168.2	175.2	182.9	169.4
3323	Architectural & structural metals	328.3	341.0	349.5	361.9	368.9	370.2
33231	Plate work & fabricated structural products	151.8	159.8	163.6	168.3	168.7	166.7
332311,3	Prefabricated metal buildings, components and plate work	74.0	75.7	77.1	78.9	78.0	75.9
33232	Ornamental & architectural metal products	176.5	181.2	185.9	193.7	200.3	203.5
332321	Metal windows & doors	50.3	50.5	52.4	53.9	56.8	59.2
332322	Sheet metal work	93.7	96.5	98.5	102.5	105.3	105.1
332323	Ornamental & architectural metal work	32.6	34.2	35.0	37.2	38.2	39.2
33312	Construction machinery	67.9	72.9	71.9	72.1	72.0	68.9
Financial Activities							
522292	Real estate credit	207.0	207.8	223.0	207.9	215.5	217.6
52231	Mortgage & nonmortgage loan brokers	58.1	68.6	75.0	72.7	78.8	79.2
531	Real estate	1,400.8	1,420.0	1,458.7	1,487.1	1,518.4	1,519.5
5311	Lessors of real estate	563.7	568.4	575.0	577.2	582.4	577.3
53111	Lessors of residential buildings	345.8	344.9	345.3	346.3	347.6	345.3
53112	Lessors of nonresidential buildings	134.5	138.7	142.1	141.5	144.6	140.1
53113	Miniwarehouse & self-storage unit operators	43.8	45.0	47.0	48.6	49.3	49.9
53119	Lessors of other real estate property	39.7	39.9	40.6	41.0	40.9	42.0
5312	Offices of real estate agents & brokers	280.5	273.8	281.5	286.8	293.3	292.2
5313	Activities related to real estate	556.6	577.8	602.3	623.0	642.7	650.0
53131	Real estate property managers	480.4	498.3	517.4	537.8	554.7	557.5
531311	Residential property managers	347.7	362.1	378.0	393.7	406.0	406.5
531312	Nonresidential property managers	132.7	136.2	139.4	144.2	148.7	151.0
53132	Offices of real estate appraisers	35.2	35.2	36.3	33.9	33.7	33.8
53139	Other activities related to real estate	41.0	44.3	48.6	51.3	54.3	58.7
Professional & Business Services							
54131	Architectural services	152.9	154.9	157.7	166.3	175.7	180.0
54132	Landscape architectural services	30.2	29.9	30.3	32.2	36.0	33.1
54133,4	Engineering & drafting services	882.9	901.3	915.0	926.7	949.0	951.0
54135,6,7	Building inspection, surveying & mapping svcs.	74.5	76.8	79.5	84.2	84.9	82.0
Leisure & Hospitality							
7211	Traveler accommodation	1,739.3	1,768.7	1,807.3	1,832.5	1,856.1	1,807.9
72111	Hotels & motels, except casino hotels	1,438.1	1,466.0	1,499.3	1,523.6	1,545.5	1,505.8
72112	Casino hotels	263.9	265.3	267.9	267.7	267.8	262.7

* Preliminary estimate as of February 2016.

Source: U.S. Bureau of Labor Statistics

Plunkett Research,® Ltd.

www.plunkettresearch.com

Chapter 3

IMPORTANT REAL ESTATE & CONSULTING INDUSTRY CONTACTS

Addresses, Telephone Numbers and Internet Sites

Contents:
1) Affordable Housing Associations
2) Alternative Energy-General
3) Careers-First Time Jobs/New Grads
4) Careers-General Job Listings
5) Careers-Job Reference Tools
6) Construction Industry Resources & Associations
7) Construction Resources-Energy Efficient Buildings
8) Corporate Information Resources
9) Design & Architectural Associations
10) Economic Data & Research
11) Engineering Industry Associations
12) Engineering, Research & Scientific Associations
13) Environmental Industry Associations
14) Environmental Organizations
15) Financial Industry Resources
16) Forest Products Associations
17) Home Values Online
18) Industry Research/Market Research
19) Long Term Care, Assisted Living Associations
20) MBA Resources
21) Modular Building Associations
22) Mortgage Industry Associations
23) Mortgage Industry Resources
24) Property Tax Professionals Associations
25) Real Estate Industry Associations
26) Real Estate Industry Resources
27) Remodeling Industry Associations
28) RFID Industry Associations
29) Robotics & Automation Industry Associations
30) Science Parks
31) Seniors Housing
32) Shopping Center Directories
33) Shopping Center Resources
34) Spa Industry Associations
35) Telecommunications Industry Associations
36) Travel Business & Professional Associations
37) Travel-Local Transportation, Bus & Car Rental
38) U.S. Government Agencies

1) Affordable Housing Associations

National Community Reinvestment Coalition (NCRC)
727 15th St. NW, Ste. 900
Washington, DC 20005 US
Phone: 202-628-8866
Fax: 202-628-9800
E-mail Address: *jvantol@ncrc.org*
Web Address: www.ncrc.org
The National Community Reinvestment Coalition (NCRC) is a national nonprofit 501(c)3 organization with 640 dues-paying chapters located in every state

in America. NCRC was founded in 1990 to unite efforts around the nation to increase the flow of private capital into traditionally underserved communities, with a focus on affordable mortgages and loans for housing and small businesses in low income neighborhoods.

2)   Alternative Energy-General

Alliance to Save Energy (ASE)
1850 M St. NW, Ste. 600
Washington, DC 20036 USA
Phone: 202-857-0666
E-mail Address: info@ase.org
Web Address: www.ase.org
The Alliance to Save Energy (ASE) promotes energy-efficiency worldwide to achieve a healthier economy, a cleaner environment and energy security.

American Council for an Energy-Efficient Economy (ACEEE)
529 14th St. NW, Ste. 600
Washington, DC 20045-1000 USA
Phone: 202-507-4000
Fax: 202-429-2248
Web Address: www.aceee.org
The American Council for an Energy-Efficient Economy (ACEEE) is a nonprofit organization dedicated to advancing energy-efficiency as a means of promoting both economic prosperity and environmental protection.

Center for Energy Efficiency and Renewable Technologies (CEERT)
1100 11th St., Ste. 311
Sacramento, CA 95814 USA
Phone: 916-442-7785
E-mail Address: info@ceert.org
Web Address: www.ceert.org
The Center for Energy Efficiency and Renewable Technologies (CEERT) provides technical support to environmental advocates and clean technology developers.

3)   Careers-First Time Jobs/New Grads

CollegeGrad.com, Inc.
950 Tower Ln., Fl. 6
Foster City, CA 94404 USA
E-mail Address: info@quinstreet.com
Web Address: www.collegegrad.com

CollegeGrad.com, Inc. offers in-depth resources for college students and recent grads seeking entry-level jobs.

MonsterCollege
799 Market St., Ste. 500
San Francisco, CA 94103 USA
E-mail Address: info@college.monster.com
Web Address: www.college.monster.com
MonsterCollege provides information about internships and entry-level jobs, as well as career advice and resume tips, to recent college graduates.

National Association of Colleges and Employers (NACE)
62 Highland Ave.
Bethlehem, PA 18017-9085 USA
Phone: 610-868-1421
E-mail Address: customer_service@naceweb.org
Web Address: www.naceweb.org
The National Association of Colleges and Employers (NACE) is a premier U.S. organization representing college placement offices and corporate recruiters who focus on hiring new grads.

4)   Careers-General Job Listings

CareerBuilder, Inc.
200 N La Salle St., Ste. 1100
Chicago, IL 60601 USA
Phone: 773-527-3600
Fax: 773-353-2452
Toll Free: 800-891-8880
Web Address: www.careerbuilder.com
CareerBuilder, Inc. focuses on the needs of companies and also provides a database of job openings. The site has over 1 million jobs posted by 300,000 employers, and receives an average 23 million unique visitors monthly. The company also operates online career centers for 140 newspapers and 9,000 online partners. Resumes are sent directly to the company, and applicants can set up a special e-mail account for job-seeking purposes. CareerBuilder is primarily a joint venture between three newspaper giants: The McClatchy Company, Gannett Co., Inc. and Tribune Company.

CareerOneStop
Toll Free: 877-872-5627
E-mail Address: info@careeronestop.org
Web Address: www.careeronestop.org
CareerOneStop is operated by the employment commissions of various state agencies. It contains job

listings in both the private and government sectors, as well as a wide variety of useful career resources and workforce information. CareerOneStop is sponsored by the U.S. Department of Labor.

LaborMarketInfo (LMI)
Employment Development Dept.
800 Capitol Mall, MIC 83
Sacramento, CA 95814 USA
Phone: 916-262-2162
Fax: 916-262-2352
Web Address: www.labormarketinfo.edd.ca.gov
LaborMarketInfo (LMI) provides job seekers and employers a wide range of resources, namely the ability to find, access and use labor market information and services. It provides statistics for employment demographics on both a local and regional level, as well as career searching tools for California residents. The web site is sponsored by California's Employment Development Office.

Recruiters Online Network
E-mail Address: rossi.tony@comcast.net
Web Address: www.recruitersonline.com
The Recruiters Online Network provides job postings from thousands of recruiters, Careers Online Magazine, a resume database, as well as other career resources.

USAJOBS
1900 E St. NW, Ste. 6500
Washington, DC 20415-0001 USA
Web Address: www.usajobs.gov
USAJOBS, a program of the U.S. Office of Personnel Management, is the official job site for the U.S. Federal Government. It provides a comprehensive list of U.S. government jobs, allowing users to search for employment by location; agency; type of work; or by senior executive positions. It also has special employment sections for individuals with disabilities, veterans and recent college graduates; an information center, offering resume and interview tips and other information; and allows users to create a profile and post a resume.

5)   Careers-Job Reference Tools

Vault.com, Inc.
132 W. 31st St., Fl. 17
New York, NY 10001 USA
Fax: 212-366-6117
Toll Free: 800-535-2074
E-mail Address: customerservice@vault.com

Web Address: www.vault.com
Vault.com, Inc. is a comprehensive career web site for employers and employees, with job postings and valuable information on a wide variety of industries. Its features and content are largely geared toward MBA degree holders.

6)   Construction Industry Resources & Associations

American Concrete Institute (ACI)
38800 Country Club Dr.
Farmington Hills, MI 48331 USA
Phone: 248-848-3700
Fax: 248-848-3701
Web Address: www.concrete.org
The American Concrete Institute (ACI) is a professional organization with 20,000 members in 120 countries and provides information on the use of concrete for structures and facilities. It's a leading authority for development of consensus-based standards, technical resources, educational & training programs and certification programs.

American Society of Professional Estimators (ASPE)
2525 Perimeter Place Dr., Ste. 103
Nashville, TN 37214 USA
Phone: 615-316-9200
Fax: 615-316-9800
E-mail Address: psmith@aspenational.org
Web Address: www.aspenational.org
The American Society of Professional Estimators (ASPE) serves construction estimators by providing education, fellowships and opportunities for professional development.

American Subcontractors Association (ASA)
1004 Duke St.
Alexandria, VA 22314-3588 USA
Phone: 703-684-3450
Fax: 703-836-3482
E-mail Address: ASAoffice@asa-hq.com
Web Address: www.asaonline.com
The American Subcontractors Association (ASA) is an association of professional construction contractors and suppliers.

Associated Builders and Contractors (ABC)
440 1st St. NW, Ste. 200
Washington, DC 20001 USA
Phone: 202-595-1505
E-mail Address: gotquestions@abc.org

Web Address: www.abc.org
Associated Builders and Contractors (ABC) is a national trade association representing more than 21,000 merit shop contractors, subcontractors, material suppliers and related firms in 70 chapters across the United States.

Associated General Contractors of America (The, AGC)

2300 Wilson Blvd., Ste. 300
Arlington, VA 22201 USA
Phone: 703-548-3118
Fax: 703-548-3119
Toll Free: 800-242-1767
E-mail Address: info@agc.org
Web Address: www.agc.org
The Associated General Contractors of America (AGC) is a membership organization dedicated to furthering the agenda of commercial construction contractors, improving job site safety, expanding the use of cutting edge technologies and techniques and strengthening the dialogue between contractors and owners.

BNi Building News

990 Park Center Dr., Ste. E
Vista, CA 8352 USA
Toll Free: 888-264-2665
Web Address: www.bnibooks.com
BNi Building News web site offers users the opportunity to browse and purchase from thousands of construction, architecture, estimating, codes and standards, engineering and construction law titles. It is a premier distributor of building codes and references throughout the U.S.

Canadian Construction Association (CCA)

1900-275 Slater St.
Ottawa, ON K1P 5H9 Canada
Phone: 613-236-9455
Fax: 613-236-9526
E-mail Address: cca@cca-acc.com
Web Address: www.cca-acc.com
The Canadian Construction Association (CCA), or, in French, the Association Canadienne de la Construction (ACC), is a Canadian organization devoted to setting standards for construction industry practices and labor issues nationwide.

Canadian Wood Council (The) (CWC)

99 Bank St., Ste. 400
Ottawa, ON K1P 6B9 Canada
Phone: 613-747-5544

Fax: 613-747-6264
Toll Free: 800-844-1275
E-mail Address: info@cwc.ca
Web Address: www.cwc.ca
The CWC is the Canadian national association of manufacturers of wood products that are used in construction.

Certified Aging in Place Specialist (CAPS)

1201 15th St. NW
National Association of Home Builders
Washington, DC 20005 USA
Toll Free: 800-368-5242
E-mail Address: info@nahb.org
Web Address:
http://www.nahb.org/en/learn/designations/certified-aging-in-place-specialist.aspx
The Certified Aging in Place Specialist (CAPS) designation program teaches the technical, business management and customer service skills necessary to make residential environments capable of housing people as they age. It is offered by the National Association of Home Builders (NAHB)

Construction Industry Institute (CII)

3925 W. Braker Ln.
WPR Building, Fl.2
Austin, TX 78759-5316 USA
Phone: 512-232-3000
Fax: 512-499-8101
E-mail Address: wcrew@cii.utexas.edu
Web Address: www.construction-institute.org
The Construction Industry Institute (CII) is a consortium of leading owners, contractors and suppliers who work with academia to find better ways to plan and execute capital construction programs.

Contractors Group (The)

555 NW Fairhaven Dr.
c/o Diane Dennis Enterprises
Oak Harbor, WA 98277 USA
Phone: 760-490-2557
Fax: 866-480-7105
E-mail Address: diane@TheContractorsGroup.com
Web Address: www.thecontractorsgroup.com
The Contractors Group offers free editable construction forms, tutorials, forums and related links. It was founded, and is still run by Diane Dennis Enterprises.

Metal Building Manufacturers Association (MBMA)
1300 Sumner Ave.
Cleveland, OH 44115-2851 USA
Phone: 216-241-7333
Fax: 216-241-0105
E-mail Address: mbma@mbma.com
Web Address: www.mbma.com
The Metal Building Manufacturers Association (MBMA) promotes the design and construction of metal building and metal roofing systems in the low-rise, non-residential building and roofing marketplace.

National Asphalt Pavement Association (NAPA)
5100 Forbes Blvd.
Lanham, MD 20706 USA
Phone: 301-731-4748
Fax: 301-731-4621
Toll Free: 888-468-6499
Web Address: www.asphaltpavement.org
The National Asphalt Pavement Association is the only trade association that exclusively represents the interests of the asphalt material pavement producer and paving contractor on the national level with Congress, government agencies, and other national trade and business organizations. The association provides technical, educational, and marketing materials and information to its members; supplies product information to users and specifiers of paving materials; and conducts training courses. Currently, it has 1,200 member companies.

National Association of Home Builders (NAHB)
1201 15th St. NW
Washington, DC 20005 USA
Toll Free: 800-368-5242
E-mail Address: info@nahb.org
Web Address: www.nahb.org
The National Association of Home Builders (NAHB) exists to represent the building industry by serving its members and affiliated state and local builders associations.

National Association of Women in Construction (NAWIC)
327 S. Adams St.
Fort Worth, TX 76104 USA
Phone: 817-877-5551
Fax: 817-877-0324
Toll Free: 800-552-3506
E-mail Address: nawic@nawic.org
Web Address: www.nawic.org
The National Association of Women in Construction (NAWIC) is an association to promote the advancement of women in the construction industry. It offers networking, professional development, education, leadership training and public service opportunities to its members.

Precast/Prestressed Concrete Institute
200 W. Adams St., Ste. 2100
Chicago, IL 60606-6938 USA
Phone: 312-786-0300
Web Address: www.pci.org
The Precast/Prestressed Concrete Institute (PCI) is an organization dedicated to the precast and prestressed concrete industry and includes a staff of technical and marketing specialists.

Public Works and Government Services Canada (PWGSC)
11 Laurier St., Phase III
Place du Portage
Gatineau, Quebec K1A 0S5 Canada
Toll Free: 800-926-9105
E-mail Address: questions@tpsgc-pwgsc.gc.ca
Web Address: www.tpsgc-pwgsc.gc.ca/comm/index-eng.html
Public Works and Government Services Canada (PWGSC) is a government agency that performs the following functions for all 106 Canadian federal departments: purchases goods and services; provides office accommodations for public servants; manages national heritage properties; oversees public construction; and offers information technology, telecommunications, translation, banking and auditing services to the government.

Singapore Contractors Association Ltd. (SCAL)
Construction House
1 Bukit Merah Ln. 2
Singapore, 159760 Singapore
Phone: 65-6278-9577
Fax: 65-6273-3977
E-mail Address: enquiry@scal.com.sg
Web Address: www.scal.com.sg
The Singapore Contractors Association (SCAL) has a membership of more than 2000 members under her wing, making it a representative of the construction industry in Singapore.

Sweets
Toll Free: 800-393-6343
E-mail Address: support@construction.com
Web Address: http://sweets.construction.com/

Sweets is an on-line building product resource for architects, contractors and engineers. It offers access to product and manufacturer information quickly and easily through a searchable database. Its design facilitates design specifications and document creation with downloadable, insertable product information. Sweets is a brand of Dodge Data & Analytics based in New York.

7) Construction Resources-Energy Efficient Buildings

Efficient Windows Collaborative (EWC)
21629 Zodiac St., N.E.
Wyoming, MN 55092 USA
E-mail Address:
efficientwindowscollaborative@gmail.com
Web Address: www.efficientwindows.org
The Efficient Windows Collaborative (EWC) web site provides unbiased information on the benefits of energy-efficient windows, descriptions of how they work and recommendations for their selection and use. The web site is sponsored by the U.S. Department of Energy's Windows and Glazings Program and the participation of industry members.

Green Building Initiative (GBI)
5410 S.W. Macadam, Ste. 150
Portland, OR 97239 USA
Phone: 503-274-0448
E-mail Address: info@thegbi.org
Web Address: www.thegbi.org
The Green Building Initiative (GBI) is a nonprofit network of building industry leaders committed to bringing green to mainstream residential and commercial construction. The GBI believes in building approaches that are environmentally progressive, but also practical and affordable for builders to implement.

GreenSource
2 Penn Plz., Fl. 9
New York, NY 10121-2298 USA
Phone: 515-237-3681
Fax: 717-399-8900
Toll Free: 866-664-8243
E-mail Address: greensourcemag@McGraw-Hill.com
Web Address: www.greensource.construction.com
GreenSource is McGraw-Hill Construction's on-line directory of information on sustainable design, practice and products. It includes information from GreenSource, Architectural Record and Engineering News-Record magazines for architects, engineers, contractors and consumers. It also offers Internet-only exclusives.

HomeInnovation Research Labs
400 Prince George's Blvd.
Upper Marlboro, MD 20774 USA
Phone: 301-249-4000
Fax: 301-430-6180
Toll Free: 800-638-8556
Web Address: www.homeinnovation.com
Home Innovation Research Labs (formerly the NAHB Research Center) was founded in 1964 as a wholly-owned, independent subsidiary of the National Association of Home Builders (NAHB). Originating as a small product testing laboratory, it has since grown to become a full-service market research, consulting, product testing, and accredited third-party certification agency dedicated solely to issues related to the home building industry.

International Self-Powered Building Council (ISPBC)
Web Address: www.ispbc.org
International Self-Powered Building Council (ISPBC) is a non-profit organization based in Washington, D.C. with worldwide chapters comprised of developers, architects, builders, property owners, renewable energy companies, engineers, designers, green building material manufacturers and suppliers. ISPBC is dedicated to the global deployment of Self-Powered-Building (SPB), a power generating, energy-efficient and economically superior Building 2.0 with a sophisticated approach to urban design. ISPBC seeks to achieve this by combining the most advanced photovoltaic (PV) and other renewable energy technologies with cutting edge design to provide stylish and seamlessly integrated modern buildings.

Passive House Institute US (PHIUS)
116 W. Illinois St., Suite 5E
Chicago, IL 60654 USA
Phone: 312-561-4588
E-mail Address: info@passivehouse.com
Web Address: www.phius.org/home-page
PHIUS is a 501(c)3 organization committed to making high-performance passive building the mainstream market standard. Founded in 2007, PHIUS has trained more than 1,700 architects, engineers, energy consultants, energy raters, and builders. It also is a leading certifier of passive buildings.

Sustainable Buildings Industry Council (SBIC)
1090 Vermont Ave., Ste. 700
Washington, DC 20005 USA
Phone: 202-289-7800
Fax: 202-289-1092
E-mail Address: rcolker@nibs.org
Web Address: www.nibs.org/?page=sbic
The Sustainable Buildings Industry Council (SBIC)
is an independent, nonprofit organization that
concentrates on providing information on energy
conservation in regard to building construction.

U.S. Green Building Council (USGBC)
2101 L St. NW, Ste. 500
Washington, DC 20037 USA
Phone: 202-742-3792
Toll Free: 800-795-1747
Web Address: www.usgbc.org
The United States Green Building Council (USGBC)
is a coalition of building industry leaders working to
promote environmentally responsible commercial
and residential structures.

**Zero Energy Building Designs Database -Dept of
Energy Efficiency**
1000 Independence Ave. SW
Washington, DC 20585 USA
Web Address: http://zeb.buildinggreen.com
The Zero Energy Buildings Database features
profiles of commercial buildings that produce as
much energy as they use over the course of a year.

8) Corporate Information Resources

bizjournals.com
120 W. Morehead St., Ste. 400
Charlotte, NC 28202 USA
Web Address: www.bizjournals.com
Bizjournals.com is the online media division of
American City Business Journals, the publisher of
dozens of leading city business journals nationwide.
It provides access to research into the latest news
regarding companies both small and large. The
organization maintains 42 websites and 64 print
publications and sponsors over 700 annual industry
events.

Business Wire
101 California St., Fl. 20
San Francisco, CA 94111 USA
Phone: 415-986-4422
Fax: 415-788-5335
Toll Free: 800-227-0845

E-mail Address: info@businesswire.com
Web Address: www.businesswire.com
Business Wire offers news releases, industry- and
company-specific news, top headlines, conference
calls, IPOs on the Internet, media services and access
to tradeshownews.com and BW Connect On-line
through its informative and continuously updated
web site.

Edgar Online, Inc.
11200 Rockville Pike, Ste. 310
Rockville, MD 20852 USA
Phone: 301-287-0300
Fax: 301-287-0390
Toll Free: 888-870-2316
Web Address: www.edgar-online.com
Edgar Online, Inc. is a gateway and search tool for
viewing corporate documents, such as annual reports
on Form 10-K, filed with the U.S. Securities and
Exchange Commission.

PR Newswire Association LLC
350 Hudson St., Ste. 300
New York, NY 10014-4504 USA
Fax: 800-793-9313
Toll Free: 800-776-8090
E-mail Address: MediaInquiries@prnewswire.com
Web Address: www.prnewswire.com
PR Newswire Association LLC provides
comprehensive communications services for public
relations and investor relations professionals, ranging
from information distribution and market intelligence
to the creation of online multimedia content and
investor relations web sites. Users can also view
recent corporate press releases from companies
across the globe. The Association is owned by United
Business Media plc.

9) Design & Architectural Associations

American Institute of Architects (The) (AIA)
1735 New York Ave. NW
Washington, DC 20006-5292 USA
Fax: 202-626-7547
Toll Free: 800-242-3837
E-mail Address: infocentral@aia.org
Web Address: www.aia.org
The American Institute of Architects (AIA) is a
professional trade group for licensed architects in the
United States. It serves as the voice of architecture
industry, as well as the interests of its members.

American Institute of Building Design (AIBD)
7059 Blair Rd. NW, Ste. 400
Washington, DC 20012 USA
Fax: 866-204-0293
Toll Free: 800-366-2423
E-mail Address: info@aibd.org
Web Address: www.aibd.org
The American Institute of Building Design (AIBD) is
a nonprofit professional organization dedicated to the
development, recognition and enhancement of the
profession of building design.

Center for Universal Design (The) (CUD)
College of Design, North Carolina State University
Campus Box 8613
Raleigh, NC 27695-8613 USA
Phone: 919-515-3082
Fax: 919-515-8951
E-mail Address: cud@ncsu.edu
Web Address: www.ncsu.edu/ncsu/design/cud/
The Center for Universal Design (CUD) is a national
information, technical assistance and research center
that evaluates, develops and promotes products and
environments so that they can be used by all people,
regardless of physical or mental limitations.

**National Council of Architectural Registration
Board (NCARB)**
1801 K St. NW, Ste. 700K
Washington, DC 20006-1310 USA
Phone: 202-783-6500
Fax: 202-783-0290
Web Address: www.ncarb.org
The National Council of Architectural Registration
Board (NCARB) is a nonprofit federation of
architectural licensing boards in the U.S. that
provides state registration requirements, information
on the Intern Development Program (IDP) and
Architect Registration Examination (ARE)
certification, continuing education and an architect
database.

Royal Architectural Institute of Canada (RAIC)
55 Murray St., Ste. 330
Ottawa, Ontario K1N 5M3 Canada
Phone: 613-241-3600
Fax: 613-241-5750
E-mail Address: info@raic.org
Web Address: www.raic.org
The Royal Architectural Institute of Canada (RAIC)
is a voluntary national association representing
professional architects, and faculty and graduates of
accredited Canadian Schools of Architecture.

Royal Institute of British Architects (RIBA)
66 Portland Pl.
London, W1B 1AD UK
Phone: 44-20-7580-5533
E-mail Address: info@riba.org
Web Address: www.riba.org
The Royal Institute of British Architects (RIBA) is a
professional association of architects in the United
Kingdom, which also offers lectures, exhibitions and
events, community architecture projects and
community architecture schemes.

Singapore Institute of Architects (SIA)
79 Neil Rd.
Singapore, 088904 Singapore
Phone: 65-6226-2668
Fax: 65-6226-2663
E-mail Address: info@sia.org.sg
Web Address: www.sia.org.sg
The Singapore Institute of Architects (SIA) is the
national organization representing architects in
Singapore.

The Architectural Society of China (ASC)
No. 9 Sanlihe Rd.
Beijing, 100835 China
Phone: 86-10-8808-2224
Web Address: www.chinaasc.org
The Architectural Society of China (ASC) represents
professional architects in China.

10) Economic Data & Research

**Centre for European Economic Research (The,
ZEW)**
L 7, 1
Mannheim, 68161 Germany
Phone: 49-621-1235-01
Fax: 49-621-1235-224
E-mail Address: info@zew.de
Web Address: www.zew.de/en
Zentrum fur Europaische Wirtschaftsforschung, The
Centre for European Economic Research (ZEW),
distinguishes itself in the analysis of internationally
comparative data in a European context and in the
creation of databases that serve as a basis for
scientific research. The institute maintains a special
library relevant to economic research and provides
external parties with selected data for the purpose of
scientific research. ZEW also offers public events
and seminars concentrating on banking, business and
other economic-political topics.

Economic and Social Research Council (ESRC)
Polaris House
North Star Ave.
Swindon, SN2 1UJ UK
Phone: 44-01793 413000
E-mail Address: esrcenquiries@esrc.ac.uk
Web Address: www.esrc.ac.uk
The Economic and Social Research Council (ESRC)
funds research and training in social and economic
issues. It is an independent organization, established
by Royal Charter. Current research areas include the
global economy; social diversity; environment and
energy; human behavior; and health and well-being.

Eurostat
5 Rue Alphonse Weicker
Joseph Bech Bldg.
Luxembourg, L-2721 Luxembourg
Phone: 352-4301-33444
Fax: 352-4301-35349
E-mail Address: eurostat-pressoffice@ec.europa.eu
Web Address: www.epp.eurostat.ec.europa.eu
Eurostat is the European Union's service that
publishes a wide variety of comprehensive statistics
on European industries, populations, trade,
agriculture, technology, environment and other
matters.

Federal Statistical Office of Germany
Gustav-Stresemann-Ring 11
Wiesbaden, D-65189 Germany
Phone: 49-611-75-1
Fax: 49-611-72-4000
Web Address: www.destatis.de
Federal Statistical Office of Germany publishes a
wide variety of nation and regional economic data of
interest to anyone who is studying Germany, one of
the world's leading economies. Data available
includes population, consumer prices, labor markets,
health care, industries and output.

India Brand Equity Foundation (IBEF)
Apparel House, Fl. 5
#519-22, Sector 44
Gurgaon, Haryana 122003 India
Phone: 91-124-4499600
Fax: 91-124-4499615
E-mail Address: info.brandindia@ibef.org
Web Address: www.ibef.org
India Brand Equity Foundation (IBEF) is a public-
private partnership between the Ministry of
Commerce and Industry, the Government of India
and the Confederation of Indian Industry. The

foundation's primary objective is to build positive
economic perceptions of India globally. It aims to
effectively present the India business perspective and
leverage business partnerships in a globalizing
marketplace.

National Bureau of Statistics (China)
57, Yuetan Nanjie, Sanlihe
Xicheng District
Beijing, 100826 China
Fax: 86-10-6878-2000
E-mail Address: info@gj.stats.cn
Web Address: www.stats.gov.cn/english
The National Bureau of Statistics (China) provides
statistics and economic data regarding China's
economy and society.

**Organization for Economic Co-operation and
Development (OECD)**
2 rue Andre Pascal
Cedex 16
Paris, 75775 France
Phone: 33-1-45-24-82-00
Fax: 33-1-45-24-85-00
Web Address: www.oecd.org
The Organization for Economic Co-operation and
Development (OECD) publishes detailed economic,
government, population, social and trade statistics on
a country-by-country basis for over 30 nations
representing the world's largest economies. Sectors
covered range from industry, labor, technology and
patents, to health care, environment and
globalization.

**Statistics Bureau, Director-General for Policy
Planning (Japan)**
19-1 Wakamatsu-cho
Shinjuku-ku
Tokyo, 162-8668 Japan
Phone: 81-3-5273-2020
E-mail Address: toukeisoudan@soumu.go.jp
Web Address: www.stat.go.jp/english
The Statistics Bureau, Director-General for Policy
Planning (Japan) and Statistical Research and
Training Institute, a part of the Japanese Ministry of
Internal Affairs and Communications, plays the
central role of producing and disseminating basic
official statistics and coordinating statistical work
under the Statistics Act and other legislation.

Statistics Canada
150 Tunney's Pasture Driveway
Ottawa, ON K1A 0T6 Canada

Phone: 514-283-8300
Fax: 877-287-4369
Toll Free: 800-263-1136
E-mail Address: infostats@statcan.gc.ca
Web Address: www.statcan.gc.ca
Statistics Canada provides a complete portal to
Canadian economic data and statistics. Its conducts
Canada's official census every five years, as well as
hundreds of surveys covering numerous aspects of
Canadian life.

11) Engineering Industry Associations

National Society of Professional Engineers (NSPE)
1420 King St.
Alexandria, VA 22314-2794 USA
Fax: 703-836-4875
Toll Free: 888-285-6773
Web Address: www.nspe.org
The National Society of Professional Engineers
(NSPE) represents individual engineering
professionals and licensed engineers across all
disciplines. NSPE serves approximately 45,000
members and has more than 500 chapters.

12) Engineering, Research & Scientific Associations

American National Standards Institute (ANSI)
1899 L St. NW, Fl. 11
Washington, DC 20036 USA
Phone: 202-293-8020
Fax: 202-293-9287
E-mail Address: info@ansi.org
Web Address: www.ansi.org
The American National Standards Institute (ANSI) is
a private, nonprofit organization that administers and
coordinates the U.S. voluntary standardization and
conformity assessment system. Its mission is to
enhance both the global competitiveness of U.S.
business and the quality of life by promoting and
facilitating voluntary consensus standards and
conformity assessment systems and safeguarding
their integrity.

American Society of Civil Engineers (ASCE)
1801 Alexander Bell Dr.
Reston, VA 20191-4400 USA
Phone: 703-295-6300
Toll Free: 800-548-2723
Web Address: www.asce.org

The American Society of Civil Engineers (ASCE) is
a leading professional organization serving civil
engineers. It ensures safer buildings, water systems
and other civil engineering works by developing
technical codes and standards.

Association for Facilities Engineering (AFE)
8200 Greensboro Dr., Ste. 400
McLean, VA 22102 USA
Phone: 571-395-8777
Fax: 571-766-2142
Web Address: www.afe.org
The Association for Facilities Engineering (AFE)
provides education, certification, technical
information and other relevant resources for plant and
facility engineering, operations and maintenance
professionals worldwide.

China Academy of Building Research (CABR)
30, Bei San Huan Dong Lu
Beijing, 100013 China
Phone: 010-84272233
Fax: 010-84281369
E-mail Address: office@cabr.com.cn
Web Address: www.cabr.cn
CABR is responsible for the development and
management of the major engineering construction
and product standards of China and is also the largest
comprehensive research and development institute in
the building industry in China. Some related
institutes include Institute of Earthquake
Engineering, Institute of Building Fire Research,
Institute of Building Environment and Energy
Efficiency (Building Physics), Institute of Foundation
Engineering as well as many others.

**Earthquake Engineering Research Institute
(EERI)**
499 14th St., Ste. 220
Oakland, CA 94612-1934 USA
Phone: 510-451-0905
Fax: 510-451-5411
E-mail Address: eeri@eeri.org
Web Address: www.eeri.org
The Earthquake Engineering Research Institute
(EERI) is a national nonprofit technical organization
of engineers, geoscientists, architects, planners,
public officials and social scientists aimed at
reducing earthquake risk by advancing the science
and practice of earthquake engineering.

Illuminating Engineering Society (IES)
120 Wall St., Fl. 17

New York, NY 10005-4001 USA
Phone: 212-248-5000
Fax: 212-248-5017
E-mail Address: ies@ies.org
Web Address: www.ies.org
A recognized authority on lighting in North America,
the Illuminating Engineering Society (IES)
establishes scientific lighting recommendations.
Members include engineers, architects, designers,
educators, students, manufacturers and scientists.

Institute of Structural Engineers (IStructE)
International HQ
47-58 Bastwick St.
London, EC1V 3PS UK
Phone: 44-20-7235-4535
Fax: 44-20-7235-4294
E-mail Address: pr@istructe.org
Web Address: www.istructe.org.uk
The Institute of Structural Engineers (IStructE) is a
professional organization, headquartered in the U.K.,
that sets and maintains standards for professional
structural engineers. It has 27,000 members in 105
countries worldwide.

**National Academy of Building Inspection
Engineers (NABIE)**
P.O. Box 860
Shelter Island, NY 11964 USA
Web Address: www.nabie.org
The National Academy of Building Inspection
Engineers (NABIE) is a chartered affinity group of
the National Society of Professional Engineers.
NABIE accepts only state-licensed engineers and
architects as members in its building inspection
association.

World Federation of Engineering Organizations
Maison de l'UNESCO
1, rue Miollis
Paris, 75015 France
Phone: 33-1-45-68-48-46
Fax: 33-1-45-68-48-65
E-mail Address: info@wfeo.net
Web Address: www.wfeo.org
World Federation of Engineering Organizations
(WFEO) is an international non-governmental
organization that represents major engineering
professional societies in over 90 nations. It has
several standing committees including engineering
and the environment, technology, communications,
capacity building, education, energy and women in
engineering.

13) Environmental Industry Associations

Indoor Air Quality Association (IAQA)
1791 Tullie Cir. NE
Atlanta, GA 30329 USA
Phone: 844-802-4103
E-mail Address: info@iaqa.org
Web Address: www.iaqa.org
The Indoor Air Quality Association (IAQA) was
established in 1995 to promote uniform standards,
procedures and protocols in the indoor air quality
industry. In 2005 IAQA's membership was
consolidated with two very similar organizations:
The American Council for Accredited Certifications
(ACAC) and the Indoor Environmental Standards
Organization (IESO). The group publishes research
and holds conferences, among its many other
activities.

**International Association of Certified Indoor Air
Consultants (IAC2)**
1750 30 St.
Boulder, CO 80301 USA
E-mail Address: fastreply@iac2.org
Web Address: www.iac2.org
The International Association of Certified Indoor Air
Consultants (IAC2) is the nonprofit, certifying body
for home and building inspectors who have fulfilled
certain educational and testing requirements
including those in the area of indoor air quality.
Indoor air quality issues include mold, radon,
biologicals, carbon monoxide, formaldehyde,
pesticides, asbestos and lead.

**International Society of Indoor Air Quality and
Climate**
2548 Empire Grade
Santa Cruz, CA 95060 USA
Phone: 831-426-0148
Fax: 831-426-6522
E-mail Address: info@isiaq.org
Web Address: www.isiaq.org
The International Society of Indoor Air Quality and
Climate is an international, independent,
multidisciplinary, scientific, non-profit organization
whose purpose is to support the creation of healthy,
comfortable and productive indoor environments.
The organization advances the science and
technology of indoor air quality; publishes a journal
Indoor Air as well as newsletter; and develops
guideline documents focused on specific issues.

14) Environmental Organizations

Global Footprint Network
312 Clay St., Ste. 300
Oakland, CA 94607-3510 USA
Phone: 510-839-8879
Fax: 510-251-2410
Web Address: www.footprintnetwork.org
Global Footprint Network publishes regional studies
of human demands on the ecology which it calls an
Ecological Footprint. The Footprint takes into
consideration human use of land, water and other
resources to fill needs for housing, agriculture,
energy and more, along with nature's ability to fulfill
those demands. The organization's analysis creates a
scale by which one nation may compare its footprint
against that of others.

15) Financial Industry Resources

SNL Financial
1 SNL Plz.
212 7th St. NE
Charlottesville, VA 22902 USA
Phone: 434-977-1600
Fax: 434-977-4466
Toll Free: 866-296-3743
E-mail Address: SNLInfo@SNL.com
Web Address: www.snl.com
SNL Financial provides industry-specific research
and statistics in the banking, financial services,
insurance, real estate and energy sectors.

16) Forest Products Associations

Ontario Forest Industries Association (OFIA)
8 King St. E., Ste. 1704
Toronto, ON M5C 1B5 Canada
Phone: 416-368-6188
Fax: 416-368-5445
E-mail Address: info@ofia.com
Web Address: www.ofia.com
The Ontario Forest Industries Association (OFIA)
represents the interests of Canadian lumber
manufacturers and various supporting industries as
well.

17) Home Values Online

Trulia
535 Mission St., Ste. 700
San Francisco, CA 94105 USA

Phone: 415-648-4358
Web Address: www.trulia.com
Trulia is an all-in-one real estate web site for finding
price trends for specific properties. The site includes
many tools for buyers, sellers and real estate
professionals and offers insights about the property,
neighborhood and the real estate process. Trulia is a
wholly-owned subsidiary of Zillow Group.

Zillow
1301 Second Ave., Fl. 31
Seattle, WA 98101 USA
Phone: 206-470-7000
E-mail Address: press@zillow.com
Web Address: www.zillow.com
Zillow is an extremely popular site launched in 2005.
It provides online access to aerial home photos, and
enables users to estimate home values based on a
variety of factors, using a proprietary algorithm.

18) Industry Research/Market Research

Forrester Research
60 Acorn Park Dr.
Cambridge, MA 02140 USA
Phone: 617-613-5730
E-mail Address: press@forrester.com
Web Address: www.forrester.com
Forrester Research is a publicly traded company that
identifies and analyzes emerging trends in technology
and their impact on business. Among the firm's
specialties are the financial services, retail, health
care, entertainment, automotive and information
technology industries.

MarketResearch.com
11200 Rockville Pike, Ste. 504
Rockville, MD 20852 USA
Phone: 240-747-3093
Fax: 240-747-3004
Toll Free: 800-298-5699
E-mail Address:
customerservice@marketresearch.com
Web Address: www.marketresearch.com
MarketResearch.com is a leading broker for
professional market research and industry analysis.
Users are able to search the company's database of
research publications including data on global
industries, companies, products and trends.

Plunkett Research, Ltd.
P.O. Drawer 541737
Houston, TX 77254-1737 USA

Phone: 713-932-0000
Fax: 713-932-7080
E-mail Address:
customersupport@plunkettresearch.com
Web Address: www.plunkettresearch.com
Plunkett Research, Ltd. is a leading provider of market research, industry trends analysis and business statistics. Since 1985, it has served clients worldwide, including corporations, universities, libraries, consultants and government agencies. At the firm's web site, visitors can view product information and pricing and access a large amount of basic market information on industries such as financial services, InfoTech, e-commerce, health care and biotech.

STR Global (Smith Travel Research)
735 E. Main St.
Hendersonville, TN 37075 USA
Phone: 615 824 8664
E-mail Address: info@str.com
Web Address: www.strglobal.com
In 2008, STR brought together Deloitte's HotelBenchmark and The Bench to form STR Global. STR Global offers monthly, weekly, and daily STAR benchmarking reports to more than 38,000 hotel clients, representing nearly 5 million rooms worldwide. STR Global and STR are now the world's foremost sources of hotel performance trends and will offer the definitive global hotel database and development pipeline. STR is headquartered in Hendersonville, TN, and STR Global is based in London, with a satellite office in Singapore.

19) Long Term Care, Assisted Living Associations

Argentum
1650 King St., Ste. 602
Alexandria, VA 22314-2747 USA
Phone: 703-894-1805
E-mail Address: jvann@alfa.org
Web Address: www.alfa.org
Argentum, formerly the Assisted Living Federation of America (ALFA) represents for-profit and nonprofit providers of assisted living, continuing care retirement communities, independent living and other forms of housing and services.

20) MBA Resources

MBA Depot
Web Address: www.mbadepot.com
MBA Depot is an online community and information portal for MBAs, potential MBA program applicants and business professionals.

21) Modular Building Associations

Modular Building Institute
944 Glenwood Station Ln., Ste. 204
Charlottesville, VA 22901 USA
Phone: 434-296-3288
Fax: 434-296-3361
Toll Free: 888-811-3288
E-mail Address: info@modular.org
Web Address: www.modular.org
Founded in 1983, the Modular Building Institute (MBI) is an international non-profit trade association serving modular construction. Members are manufacturers, contractors, and dealers in two distinct segments of the industry - permanent modular construction (PMC) and relocatable buildings (RB). Associate members are companies supplying building components, services, and financing.

22) Mortgage Industry Associations

Association of Mortgage Investors (AMI)
900 19th St. NW, Ste. 800
Washington, DC 20006 USA
Phone: 202-327-8100
Fax: 202-327-8101
E-mail Address: info@the-ami.org
Web Address: www.the-ami.org
The Association of Mortgage Investors (AMI) is an agency formed to help homeowners avoid foreclosure. It represents private investors, public and private pension funds and endowments with interests in mortgage securities.

Canadian Association of Accredited Mortgage Professionals
2235 Sheppard Ave. E, Ste. 1401
Toronto, ON M2J 5B5 Canada
Phone: 416.385.2333
Fax: 416-385-1177
Toll Free: 888-442-4625
E-mail Address: info@caamp.org
Web Address: www.caamp.org

The Canadian Association of Accredited Mortgage Professionals, founded in 1994, includes members from all segments of Canada's mortgage industry. It offers professional development and networking through regional and national events as well as sponsoring industry-related research and publications. The group launched the Accredited Mortgage Professional (AMP) designation in Canada in 2004, and it also advocates for the mortgage industry through ongoing government and regulatory lobbying, media outreach and other activities.

Council of Mortgage Lenders (CML)
Bush House, N. W. Wing, Fl. 3
Aldwych
London, WC2B 4PJ UK
Phone: 44-845-373-6771
E-mail Address: enquiries@cml.org.uk
Web Address: www.cml.org.uk
The Council of Mortgage Lenders is a trade organization whose members account for approximately 94% of U.K. mortgage lending industry. The Council provides economic, legal, statistical, research and other market information.

European Mortgage Federation (EMF)
Rue de la Science 14, Fl. 2
Brussels, B-1040 Belgium
Phone: 32-2-285-40-30
Fax: 32-2-285-40-31
E-mail Address: emfinfo@hypo.org
Web Address: www.hypo.org
The European Mortgage Federation (EMF) is the voice of the mortgage industry in the E.U. on the retail side of the business. The EMF's aim is to ensure a sustainable housing environment for EU citizens. To this end, the EMF is a key talking partner of the European Commission, the European Parliament, the 3rd pillar committees and the Basel Committee on Banking Supervision on all industry related questions.

Mortgage Bankers Association (MBA)
1919 M St. NW, Fl. 5
Washington, DC 20036 USA
Phone: 202-557-2700
Toll Free: 800-793-6222
Web Address: www.mbaa.org
The Mortgage Bankers Association (MBA) serves the real estate finance industry by representing its legislative and regulatory interests before Congress and federal agencies; providing educational programs, periodicals and publications; and

supporting its business interests with research initiatives, products and services.

National Association of Professional Mortgage Women (NAPMW)
c/o Agility Resources Group LLC
705 North Mountain Business Center, Ste. E-104
Newington, CT 06111 USA
Toll Free: 800-827-3034
E-mail Address: napmw1@napmw.org
Web Address: www.napmw.org
The National Association of Professional Mortgage Women (NAPMW) is an association for women in the banking and mortgage industry. It offers business, personal and leadership development opportunities to advance women in mortgage-related professions.

National Reverse Mortgage Lenders Association (NRMLA)
1400 16th St. NW, Ste. 420
Washington, DC 20036 USA
Phone: 202-939-1760
Fax: 202-265-4435
E-mail Address: pbell@dworbell.com
Web Address: www.nrmlaonline.org
The National Reverse Mortgage Lenders Association (NRMLA) was established in 1997 to provide consumer education, industry events and information, and public policy initiatives for the reverse mortgage industry.

23) Mortgage Industry Resources

CitiMortgage
MC 2197 BSC
P.O. Box 6205
Sioux Falls, SD 57117-9893 USA
Toll Free: 800-248-4638
Web Address: www.citimortgage.com/Mortgage/Home.do
CitiMortgage provides customers with rate quotes and with mortgages, home buying, refinancing and debt consolidation; it is powered by Citigroup, Inc.

E-Loan
85 Broad St., Fl. 10
New York, NY 10004 USA
Toll Free: 866-576-7283
Web Address: www.eloan.com
E-Loan provides information on a large variety of loan types, debt consolidation, refinancing, home equity and mortgage management. Additionally, the

company's web site offers several loan calculators and free credit reports.

Fannie Mae
3900 Wisconsin Ave. NW
Washington, DC 20016-2892 USA
Phone: 202-752-7115
Toll Free: 800-732-6643
E-mail Address: andrew_j_wilson@fanniemae.com
Web Address: www.fanniemae.com
Fannie Mae is one of the world's largest non-bank financial services companies and one of the nation's largest sources of financing for home mortgages. It operates under special sanction by the U.S. Congress. Its purpose is to provide liquidity in the national mortgage market.

Federal Agricultural Mortgage Corporation
1999 K St. NW, Fl. 4
Washington, DC 20006 USA
Fax: 800-999-1814
Toll Free: 800-879-3276
Web Address: www.farmermac.com
Federal Agricultural Mortgage Corporation, known as Farmer Mac, is America's secondary market for agricultural real estate and rural housing mortgage loans. Congress created Farmer Mac in 1988 to improve the availability of mortgage credit to America's farmers, ranchers and rural homeowners, businesses and communities.

Federal Citizen Information Center (FCIC)
1275 First St., Fl. 11
Washington, DC 20417 USA
Phone: 719-295-2675
Toll Free: 800-333-4636
E-mail Address: webteam@gpo.gov
Web Address: publications.usa.gov/USAPubs.php
The Federal Citizen Information Center (FCIC) offers information and resources for consumers on topics such as cars, children, education, housing, small businesses and more, as well as current consumer news.

Federal Home Loan Mortgage Corporation (Freddie Mac)
8200 Jones Branch Dr.
McLean, VA 22102-3110 USA
Phone: 703-903-2000
Toll Free: 800-424-5401
E-mail Address: corprel@FreddieMac.com
Web Address: www.freddiemac.com

The Federal Home Loan Mortgage Corporation (Freddie Mac) operates under special sanction by the U.S. Congress. Its purpose is to provide liquidity in the national mortgage market.

Government National Mortgage Association (Ginnie Mae)
550 12th St. SW, Fl. 3
Washington, DC 20024 USA
Phone: 202-708-1535
Toll Free: 888-446-6434
Web Address: www.ginniemae.gov
The Government National Mortgage Association (Ginnie Mae) is a Federal Government enterprise that provides mortgages to low-income families.

Homepath
Toll Free: 800-732-6643
Web Address: www.homepath.com
Homepath, powered by Fannie Mae, helps users find a home, get the most out of their current home, find lenders and other services that will assist in the purchase of a home and help customers avoid mortgage fraud.

LendingTree, LLC
11115 Rushmore Dr.
Charlotte, NC 28277 USA
Phone: 704-943-8208
Toll Free: 800-813-4620
E-mail Address: megan.greuling@tree.com
Web Address: www.lendingtree.com
LendingTree, LLC serves as an online loan center that connects borrowers to a nationwide network of lenders with exceptional speed and efficiency. LendingTree's simple online forms allow borrowers to provide required information in a matter of minutes. Additionally, the site's network of some of the most respected financial institutions in the United States allows for quick loan offers, often in less than 48 hours.

Mortgage 101
909 N. Sepulveda Blvd., Fl. 11
El Segundo, CA 90245 USA
Web Address: www.mortgage101.com
Mortgage 101 lets users find information on rates and local companies, as well as apply for financing. It also offers various calculators concerning refinancing, renting or buying and how much home can be afforded based on monthly mortgage payments. Mortgage 101 is a division of Internet Brands, Inc.

Mortgage-calc.com
Web Address: www.mortgage-calc.com
Mortgage-calc.com is a web site with a multitude of
calculators for home mortgage, amortization,
mortgage refinance, home equity loans and debt
consolidation. It is operated by Bankrate, Inc.

ReverseMortgage.org
1400 16th St. NW, Ste. 420
c/o National Reverse Mortgage Lenders Association
Washington, DC 20036 USA
Phone: 202-939-1760
E-mail Address: dhicks@nrmlaonline.org
Web Address: www.reversemortgage.org
ReverseMortgage.org is a web site that is maintained
by the National Reverse Mortgage Lenders
Association. ReverseMortgage.org offers excellent
information to consumers who want to learn more
about how reverse mortgages work.

24) Property Tax Professionals Associations

**International Association of Assessing Officers
(IAAO)**
314 W. 10th St.
IAAO Headquarters
Kansas City, MO 64105-1616 USA
Phone: 816-701-8100
Fax: 816-701-8149
Toll Free: 800-616-4226
E-mail Address: info@iaao.org
Web Address: www.iaao.org
The International Association of Assessing Officers
(IAAO) is an association of professionals in the field
of property assessment and taxation.

25) Real Estate Industry Associations

Affordable Housing Tax Credit Coalition
1909 K St. NW, Fl. 12
Washington, DC 20006 USA
Phone: 202-661-7698
Fax: 202-661-2299
E-mail Address: info@taxcreditcoalition.org
Web Address: www.taxcreditcoalition.org
The Affordable Housing Tax Credit Coalition is a
nonprofit organization of individuals and groups
involved in providing affordable housing under the
low-income housing tax credit program.

American Escrow Association (AEA)
1000 Q St., Ste. 205
Sacramento, CA 95811-6518 USA
Phone: 916-446-5165
Fax: 916-443-6719
E-mail Address: hq@a-e-a.org
Web Address: www.a-e-a.org
The American Escrow Association (AEA) is a
national trade association of settlement and escrow
industry professionals. It is dedicated to enhance
professional education, improve escrow and closing
services, coordinate legislative efforts and increase
public knowledge and understanding of such
services.

American Institute of Inspectors
P.O. Box 7243
S. Lake Tahoe, CA 96185 USA
Fax: 530-577-1407
Toll Free: 800-877-4770
E-mail Address: contact@inspection.org
Web Address: www.inspection.org
The American Institute of Inspectors is a nonprofit
association for residential and commercial building
inspectors across North America.

American Property Tax Counsel (APTC)
77 W. Washington, Ste. 900
Chicago, IL 60602 USA
Toll Free: 844-227-0407
Web Address: www.aptcnet.com
The American Property Tax Counsel (APTC) is a
professional association of property and tax
attorneys, offering services regarding property tax
reporting and tax reduction needs at local and
national level.

American Society of Appraisers
11107 Sunset Hill Rd., Ste. 310
Reston, VA 20190 USA
Phone: 703-478-2228
Fax: 703-742-8471
Toll Free: 800-272-8258
E-mail Address: asainfo@appraisers.org
Web Address: www.appraisers.org
The American Society of Appraisers is an
organization that provides education and
accreditation for appraisers, plus an appraiser locator
and electronic recruitment resource for employers
and job seekers.

Association of Foreign Investors in Real Estate (AFIRE)
1300 Pennsylvania Ave. NW
Ronald Reagan Bldg.
Washington, DC 20004 USA
Phone: 202-312-1400
E-mail Address: afireinfo@afire.org
Web Address: www.afire.org
The Association of Foreign Investors in Real Estate (AFIRE) is the only not-for-profit association for the foreign real estate investment community, with more than 180 members representing 21 countries.

Building Owners and Managers Association (BOMA) International
1101 15th St. NW, Ste. 800
Washington, DC 20005 USA
Phone: 202-408-2662
Fax: 202-326-6377
E-mail Address: info@boma.org
Web Address: www.boma.org
The Building Owners and Managers Association (BOMA) International is a premier network of 91 local associations in the U.S. as well as 18 international affiliates, representing commercial real estate professionals, including building owners, managers, developers, leasing professionals, asset managers, and corporate facility managers.

Canadian Real Estate Association
200 Catherine St., 6th Fl.
Ottawa, ON K2P 2K9 Canada
Phone: 613-237-7111
Fax: 613-234-2567
Toll Free: 800-842-2732
E-mail Address: info@crea.ca
Web Address: www.crea.ca
The Canadian Real Estate Association (CREA) is among Canada's largest single-industry trade organizations. It represents more than 100,000 real estate agents and salespeople. The organization owns the Realtor trademark in Canada and operates several industry websites, including Realtor.ca; HowRealtorsHelp.ca; and ICX.ca, a commercial real estate web site. CREA monitors public policy and maintains active relationships with Canadian government offices in order to represent and promote the work of its members.

Certified Commercial Investment Member Institute (CCIM)
430 N. Michigan Ave., Ste. 800
Chicago, IL 60611-4092 USA
Phone: 312-321-4460
Web Address: www.ccim.com
The Certified Commercial Investment Member Institute (CCIM) is an organization that provides accreditation for certified commercial investment members.

Certified New Home Specialist
2222 Colony Plz.
Dennis Walsh & Associates, Inc.
Newport Beach, CA 92660 USA
Phone: 949-706-3500
Fax: 949-706-3502
Toll Free: 800-428-1122
E-mail Address: contactus@sellnewhomes.com
Web Address: www.sellnewhomes.com
The Certified New Home Specialist web site offers certification as a New Home Specialist, an accreditation endorsed by GMAC, Coldwell Banker, ERA Real Estate and Prudential Real Estate. The web site is offered by Dennis Walsh & Associates, Inc.

China Commercial Real Estate Association
Yangfangdian Rd., Haidian District
Beijing Oriental Plaza, 18 Guangyao N Block 905
Beijing, 100049 China
Phone: 86-10-63940686
Fax: 86-10-63940687
E-mail Address: ccrea_org@163.com
Web Address: www.ccrea.com.cn
The China Commercial Real Estate Association (CCREA) primary members are commercial real estate development enterprises, retailers, chain store companies, commercial consultant companies, designing companies, funds companies, devices suppliers and professionals.

Commercial Real Estate Women Network (CREW Network)
1201 Wakarusa Dr., Ste. D
Lawrence, KS 66049-3803 USA
Phone: 785-832-1808
Fax: 785-832-1551
E-mail Address: laural@crewnetwork.org
Web Address: www.crewnetwork.org
The Commercial Real Estate Women Network (CREW Network) is a national association of women in the commercial real estate industry.

Community Associations Institute (CAI)
6402 Arlington Blvd., Ste. 500
Falls Church, VA 22042 USA
Phone: 703-970-9220

Fax: 703-970-9558
Toll Free: 888-224-4321
E-mail Address: cai-info@caionline.org
Web Address: www.caionline.org
The Community Associations Institute (CAI) is an
association of condominium, cooperative and
homeowner associations. The organization offers
education programs, seminars, workshops,
conferences, published resources as well as conducts
research. Its membership consists of 33,500 members
with 60 chapters.

Confederation of Real Estate Developer's Associations of India

703 Ansal Bhawan
16 Kasturba Gandhi Marg
New Delhi, 110 001 India
Phone: 91-11-43126262
Fax: 91-11-43126211
E-mail Address: info@credai.org
Web Address: www.credai.org
Confederation of Real Estate Developer's
Associations of India (CREDAI) is the apex body for
private real estate developers/builders across India.

Corporate Housing Providers Association (CHPA)

9100 Purdue Rd., Ste. 200
Indianapolis, IN 46268 USA
Phone: 317-328-4631
Web Address: www.chpaonline.org
The Corporate Housing Providers Association
(CHPA) is a trade association dedicated to supporting
those in the corporate housing industry. The
corporate housing industry, a segment of the lodging
industry, provides a furnished apartment,
condominium or house for rent on a short term basis,
typically lasting 30 days or more. Corporate housing
is usually used by those traveling for business, those
relocating to a new area or those temporarily
displaced.

Council of International Restaurant Real Estate Brokers (CIRB)

8350 N. Central Expy., Ste. 1300
Dallas, TX 75206 USA
Fax: 866-247-2329
Toll Free: 866-247-2123
Web Address: www.cirb.net/about.html
The Council of International Restaurant Real Estate
Brokers (CIRB) is an international network of
independent real estate brokers who specialize solely
in restaurant and related real estate.

Council of Residential Specialists (CRS)

430 N. Michigan Ave., Fl. 3
Chicago, IL 60611 USA
Phone: 312-321-4400
Fax: 312-329-8551
Toll Free: 800-462-8841
Web Address: www.crs.com
The Council of Residential Specialists (CRS) is an
association of realtors providing training for the
Certified Residential Specialist designation.

Counselors of Real Estate (CRE)

430 N. Michigan Ave.
Chicago, Il 60611 USA
Phone: 312-329-8427
E-mail Address: info@cre.org
Web Address: www.cre.org
The Counselors of Real Estate (CRE) is an
international group of high-profile professionals,
including members of prominent real estate,
financial, legal and accounting firms as well as
leaders in government and academia, who provide
objective advice on complex real property situations
and land-related matters.

European Public Real Estate Association (EPRA)

Square de Meeus 23, Fl. 9
Brussels, B-1000 Belgium
Phone: 31-2739-1010
Fax: 32-2739-1020
E-mail Address: info@epra.com
Web Address: www.epra.com
The European Public Real Estate Association
(EPRA) is a trade association of European real estate
companies.

Foundation of Real Estate Appraisers (FREA)

2645 Financial Ct., Ste. A
San Diego, CA 92117 USA
Toll Free: 800-820-5700
Web Address: www.frea.com
The Foundation of Real Estate Appraisers (FREA) is
an organization of appraisers, home inspectors and
environmental site assessors offering errors and
omissions insurance, continuing education and other
benefits.

Hong Kong Institute of Surveyors (HKIS)

111 Connaught Rd. Central, Sheung Wan
Fl. 12, Rm. 1205, Wing on Centre
Hong Kong, Hong Kong Hong Kong
Phone: 852-2526-3679
Fax: 852-2868-4612

E-mail Address: info@hkis.org.hk
Web Address: www.hkis.org.hk
The Hong Kong Institute of Surveyors (HKIS) is the
professional organization representing the surveying
industry and surveying professionals in Hong Kong.
It's activities include establishing standards for
professional services, codes of ethics and
requirements for admission as professional surveyors.

India Institute of Real Estate
Orchard Dr. Pai Marg, Nachiket Pk. Baner,
Ste. 400
Pune, 411 045 India
Phone: 9022909557
E-mail Address: dean@iire.co.in
Web Address: www.iire.co.in
The India Institute of Real Estate is a nonprofit
organization supporting the real estate industry. The
organization provides accreditation and certifications
pertaining to the real estate practice in India.

**Institute for Responsible Housing Preservation
(IRHP)**
799 9th St. NW, Ste. 500
Washington, DC 20001 USA
Phone: 202-737-0019
Fax: 202-737-0021
E-mail Address: info@HousingPreservation.org
Web Address: www.housingpreservation.org
The Institute for Responsible Housing Preservation
(IRHP) is a nonprofit association of owners and
managers of Section 221(d)(3) and Section 236
projects participating in HUD preservation programs.

Institute of Housing Management (IHM)
2800-14th Ave, Ste. 210
Markham, ON L3R 0E4 Canada
Phone: 416-493-7382
Fax: 416-491-1670
Toll Free: 866-212-4377
E-mail Address: ihm@associationconcepts.ca
Web Address: www.ihm-canada.com
The Institute of Housing Management (IHM) is a
Canadian organization of property management
professionals. It offers professional accreditation to
professionals from the for-profit and non-profit
management sectors.

Institute of Real Estate Management (IREM)
430 N. Michigan Ave.
Chicago, IL 60611 USA
Fax: 800-338-4736
Toll Free: 800-837-0706

E-mail Address: getinfo@irem.org
Web Address: www.irem.org
The Institute of Real Estate Management (IREM)
seeks to educate real estate managers, certify their
competence and professionalism, serve as an
advocate on issues affecting the real estate
management industry and enhance its members'
professional competence so they can better identify
and meet the needs of those who use their services.

**International Association of Certified Home
Inspectors (InterNACHI)**
1750 30th St., Ste. 301
Boulder, CO 80301 USA
Phone: 303-223-0861
Fax: 650-429-2057
Toll Free: 877-346-3467
E-mail Address: fastreply@nachi.org
Web Address: www.nachi.org
The International Association of Certified Home
Inspectors (InterNACHI) is a nonprofit organization
designed to help home inspectors to achieve financial
success and remain highly educated. The
organizations web site provides free course material
for home inspectors and online tools to help
consumers find a qualified professional, mainly in the
U.S. and Canada.

International Business Broker Association (IBBA)
7100 E. Pleasant Valley Rd., Ste. 160
Independence, OH 44131 USA
Fax: 800-630-2380
Toll Free: 888-686-4222
Web Address: www.ibba.org
The International Business Broker Association
(IBBA) is the largest international nonprofit
association operating exclusively for the benefit of
people and firms engaged in the various aspects of
business brokerage and mergers and acquisitions.

**International Facility Management Association
(IFMA)**
800 Gessner Rd., Ste. 900
Houston, TX 77034-4257 USA
Phone: 713-623-4362
Fax: 713-623-6124
E-mail Address: ifma@ifma.org
Web Address: www.ifma.org
The International Facility Management Association
(IFMA) is a trade association of facilities managers.
IFMA certifies facility managers, provides
educational programs, conducts research, recognizes

facility management degree and certificate programs and produces research reports and white papers.

International Real Estate Institute (IREI)
810 N. Farrell Dr.
Palm Springs, CA 92262 USA
Phone: 760-327-5284 ext. 252
Fax: 760-327-5631
Toll Free: 877-743-6799
E-mail Address: support@assoc-hdqts.org
Web Address: irei-assoc.org
The International Real Estate Institute (IREI) is one of the largest international real estate associations in the world, with members in over 100 countries. It offers most current real estate information, as well as networking opportunities to real estate professionals at national and international level.

International Society of Appraisers (ISA)
225 W. Wacker Dr., Ste. 650
Chicago, IL 60606 USA
Phone: 312-981-6778
Fax: 312-265-2908
E-mail Address: isa@isa-appraisers.org
Web Address: www.isa-appraisers.org
The International Society of Appraisers (ISA) is a nonprofit association that provides education and organizational support to its members and serves the public by producing highly qualified and ethical appraisers who are recognized authorities in professional personal property appraising.

NAREC
6348 N. Milwaukee Ave., Ste. 103
Chicago, IL 60606 USA
Phone: 773-283-6362
E-mail Address: info@narec.org
Web Address: narec.org
NAREC, formerly PeerSpan and, prior to that, the National Association of Real Estate Companies, is composed of representatives of publicly and privately owned real estate companies, significant subsidiaries of publicly owned companies and public accounting firms.

National Apartment Association (NAA)
4300 Wilson Blvd., Ste. 400
Arlington, VA 22203 USA
Phone: 703-518-6141
Fax: 703-248-9440
E-mail Address: webmaster@naahq.org
Web Address: www.naahq.org

The National Apartment Association (NAA) is a national federation of state and local apartment associations designed to serve the interests of multifamily housing owners, managers, developers and suppliers.

National Association of Hispanic Real Estate Professionals (NAHREP)
591 Camino de la Reina, Ste. 720
San Diego, CA 92108 USA
Phone: 858-622-9046
E-mail Address: membership@nahrep.org
Web Address: www.nahrep.org
The National Association of Hispanic Real Estate Professionals (NAHREP) is an organization that serves real estate agents, brokers, loan officers, mortgage brokers, title officers, escrow officers, appraisers and insurance agents from diverse cultural backgrounds, as membership is not limited to professionals of Hispanic descent.

National Association of Home Inspectors (NAHI)
4426 5th St. W.
Bradenton, FL 34207 USA
Phone: 941-462-4265
Fax: 941-896-3187
Toll Free: 800-448-3942
E-mail Address: info@nahi.org
Web Address: www.nahi.org
The National Association of Home Inspectors (NAHI) is a nonprofit association that exists to promote and develop the home inspection industry.

National Association of Housing and Redevelopment Officials (NAHRO)
630 Eye St. NW
Washington, DC 20001 USA
Phone: 202-289-3500
Fax: 202-289-8181
Toll Free: 877-866-2476
E-mail Address: nahro@nahro.org
Web Address: www.nahro.org
The National Association of Housing and Redevelopment Officials (NAHRO) is an association of professionals in affordable housing and community redevelopment. Its members administer Hud programs such as Public Housing, Section 8, CDBG and HOME.

National Association of Independent Fee Appraisers (NAIFA)
330 N. Wabash Ave., Ste. 2000
Chicago, IL 60611 USA

Phone: 312-321-6830
Fax: 312-673-6652
E-mail Address: info@naifa.com
Web Address: www.naifa.com
The National Association of Independent Fee
Appraisers (NAIFA) offers education, professional
designations, an online appraiser directory and other
services for real estate appraisers and the public.

**National Association of Industrial and Office
Properties (NAIOP)**
2201 Cooperative Way, Ste. 300
Herndon, VA 20171-3034 USA
Phone: 703-904-7100
Fax: 703-904-7942
Web Address: www.naiop.org
The National Association of Industrial and Office
Properties (NAIOP) is a trade association of
commercial real estate professionals, including
developers, owners and investors of industrial, office
and mixed-use properties.

**National Association of Mold Professionals
(NAMP)**
3130 Old Farm Ln., Ste. 1
Commerce Twp., MI 48390 USA
Phone: 248-669-5673
E-mail Address: info@moldpro.org
Web Address: www.moldpro.org
The National Association of Mold Professionals
(NAMP) is a nonprofit organization that was
established with the goal of developing and
promoting the mold inspection and remediation
industry.

**National Association of Real Estate Brokers
(NAREB)**
9831 Greenbelt Rd., Ste. 309
Lanham, MD 20706 USA
Phone: 301-552-9340
Fax: 301-552-9216
E-mail Address: info@nareb.com
Web Address: www.nareb.com
The National Association of Real Estate Brokers
(NAREB) is a national trade organization dedicated
to bringing together the nation's minority
professionals in the real estate industry.

**National Association of Real Estate Investment
Trusts (NAREIT)**
1875 I St. NW, Ste. 600
Washington, DC 20006 USA
Phone: 202-739-9400

Fax: 202-739-9401
Toll Free: 800-362-7348
Web Address: www.reit.com
The National Association of Real Estate Investment
Trusts (NAREIT) is the representative to
governmental policymakers for U.S. Real Estate
Investment Trusts (REITs) and publicly traded real
estate companies worldwide.

**National Association of Real-Estate Inspection &
Evaluation Services (NARIES)**
P.O. Box 532
Edmonds, WA 98020 USA
Phone: 425-319-5783
Fax: 425-774-5651
Toll Free: 800-583-5821
E-mail Address: crmsnsky@frontier.com
Web Address: http://naries.org
The National Association of Real-Estate Inspection
& Evaluation Services (NARIES) is an association
dedicated to the education of consumers hiring
building and home inspectors, appraisers, real-estate
professionals, architects, engineers and building
officials.

National Association of Realtors (NAR)
430 N. Michigan Ave.
Chicago, IL 606-4087 USA
Toll Free: 800-874-6500
Web Address: www.realtor.org
The National Association of Realtors (NAR) is
composed of realtors involved in residential and
commercial real estate as brokers, salespeople,
property managers, appraisers and counselors and in
other areas of the industry. NAR also sponsors
Realtor.com, operated by Move, Inc.

**National Association of Residential Property
Managers (NARPM)**
638 Independence Pkwy., Ste. 100
Chesapeake, VA 23320 USA
Fax: 866-466-2776
Toll Free: 800-782-3452
E-mail Address: info@narpm.org
Web Address: www.narpm.org
National Association of Residential Property
Managers (NARPM) is an association of real estate
management professionals that specializes in single-
family and small residential properties.

**National Council of Real Estate Investment
Fiduciaries (NCREIF)**
Aon Center

200 E. Randolph St., Ste. 5135
Chicago, IL 60601 USA
Phone: 312-819-5890
Fax: 312-819-5891
E-mail Address: info@ncreif.org
Web Address: www.ncreif.org
The National Council of Real Estate Investment
Fiduciaries (NCREIF) is an association of
institutional real estate professionals such as
investment managers, plan sponsors, academicians,
consultants, appraisers, CPAs and other service
providers who have a significant involvement in
pension fund real estate investments.

National Multifamily Housing Council (NMHC)
1850 M St. NW, Ste. 540
Washington, DC 20036-5803 USA
Phone: 202-974-2300
Fax: 202-775-0112
E-mail Address: info@nmhc.org
Web Address: www.nmhc.org
The National Multifamily Housing Council (NMHC)
is a trade association representing apartment owners,
managers, developers, lenders and service providers.

Pension Real Estate Association (PREA)
100 Pearl St., Fl. 13
Hartford, CT 06103 USA
Phone: 860-692-6341
Fax: 860-692-6351
E-mail Address: membership@prea.org
Web Address: www.prea.org
The Pension Real Estate Association (PREA) is a
nonprofit organization whose members are engaged
in the investment of tax-exempt pension and
endowment funds into real estate assets.

Property Management Association (PMA)
7508 Wisconsin Ave., Fl. 4
Bethesda, MD 20814 USA
Phone: 301-657-9200
Fax: 301-907-9326
E-mail Address: info@pma-dc.org
Web Address: www.pma-dc.org
The Property Management Association (PMA) is a
real estate management organization that promotes
the knowledge and practical education of the industry
through monthly meetings, seminars and
publications.

**Real Estate Developers' Association of Singapore
(REDAS)**
190 Clemenceau Ave.

07-01 Singapore Shopping Ctr.
Singapore, 239924 Singapore
Phone: 65-6336-6655
Fax: 65-6337-2217
E-mail Address: enquiry@redas.com
Web Address: www.redas.com
Real Estate Developers' Association of Singapore's
(REDAS) website, Redas.com, is an information
center as well as an electronic marketplace that offers
a broad array of services and resources including but
not limited to property search engines, market trends,
policy updates, on-line purchases of products and
services, business message exchange, and links to
other trade associations, professional bodies,
government agencies and statutory boards in
Singapore and the region.

Real Estate Institute of Canada (REIC)
5407 Eglinton Ave. W., Ste. 208
Toronto, ON M9C 5K6 Canada
Phone: 416-695-9000
Fax: 416-695-7230
Toll Free: 800-542-7342
E-mail Address: infocentral@reic.com
Web Address: www.reic.ca
The Real Estate Institute of Canada (REIC) is an
association of professionals dedicated to establishing,
maintaining, promoting and advancing high standards
of practice through education, certification and
accreditation.

Realtors Land Institute (RLI)
430 N. Michigan Ave., Ste. 600
Chicago, IL 60611 USA
Fax: 312-329-8633
Toll Free: 800-441-5263
E-mail Address: rli@realtors.org
Web Address: www.rliland.com
Realtors Land Institute (RLI) serves professionals
specializing in land brokerage. The web site includes
membership benefits, land listings and state chapters.

Research Institute for Housing America (RIHA)
MBA
1919 M St. NW, Fl. 5
Washington, DC 20036 USA
Phone: 202-557-2700
Toll Free: 800-793-6222
E-mail Address: MBAResearch@mba.org.
Web Address: www.mba.org/news-research-and-
resources/forecasts-data-and-reports/research-
institute-for-housing-america

The Research Institute for Housing America of the Mortgage Bankers Association is a 501(c)(3) trust fund. Its chief purpose is to encourage and aid - through grants and sponsored research to distinguished scholars, educational institutions, research facilities, and government organizations - the pursuit of knowledge of mortgage markets and real estate finance. Excellent research papers on various housing issues are available on its web site.

Royal Institution of Chartered Surveyors
Parliament Square
12 Great George St.
London, SW1P 3AD UK
Phone: 44-870-333-1600
Fax: 44-20-7334-3811
E-mail Address: contactrics@rics.org
Web Address: www.rics.org
The Royal Institution of Chartered Surveyors (RICS) is a U.K.-based trade organization representing 160,000 members operating in 140 countries; it has offices around the globe, on every continent.

Society of Industrial and Office Realtors (SIOR)
1201 New York Ave. NW, Ste. 350
Washington, DC 20005-6126 USA
Phone: 202-449-8200
Fax: 202-216-9325
E-mail Address: membership@sior.com
Web Address: www.sior.com
The Society of Industrial and Office Realtors (SIOR) provides support for industrial and office real estate specialists holding the SIOR designation. It has 3,100 members in 34 countries worldwide.

Surveyors and Valuers Accreditation Association (SAVA)
SAVA, National Energy Ctr., Davy Ave.
Milton Keynes, Surrey MK5 8NA UK
Phone: 44-1908-672787
Fax: 44-1908-662296
Web Address: www.nesltd.co.uk
The Surveyors and Valuers Accreditation Association (SAVA) sets surveying standards for the Royal Institution of Chartered Surveyors.

TriState Commercial Alliance
1201 E. Hector St.
Conshohocken, PA 19428 USA
Phone: 610-238-9950
E-mail Address: tristate@tristaterca.com
Web Address: www.tristaterca.com

The TriState Commercial Alliance is a partnership between TriState REALTORS Commercial Alliance and TriState Brokers Commercial Alliance. The association serves the specific needs of commercial real estate professionals in Pennsylvania, New Jersey and Delaware.

Women's Council of Realtors (WCR)
430 N. Michigan Ave.
Chicago, IL 60611 USA
Fax: 312-329-3290
Toll Free: 800-245-8512
E-mail Address: wcr@wcr.org
Web Address: www.wcr.org
The Women's Council of Realtors (WCR) is a community of women real estate professionals. It promotes the professional growth of its members through networking, leadership development, resources, infrastructure and accessibility

26) Real Estate Industry Resources

CoStar Group
1331 L St. NW
Washington, DC 20005 USA
Phone: 202-346-6500
Fax: 800-613-1301
Toll Free: 800-204-5960
E-mail Address: info@costargroup.com
Web Address: www.costar.com
CoStar Group operates a web site with extensive resources for brokers, owners and users of commercial real estate space. The group operates a listings database that allows users to analyze market trends, research property history and compare local rental rates for commercial real estate.

Joint Center for Housing Studies--Harvard University
1 Bow St., Ste. 400
Cambridge, MA 02138 USA
Phone: 617-495-7908
Fax: 617-496-9957
Web Address: www.jchs.harvard.edu
The Joint Center for Housing Studies is Harvard University's unit for information and research on housing in the United States. JCHS analyzes the relationships between housing markets and economic, demographic, and social trends. It publishes several excellent reports each year.

Marcus & Millichap Real Estate Investment Services, Inc.
23975 Park Sorrento, Ste. 400
Calabasas, CA 91302 USA
Phone: 818-212-2250
Fax: 818-212-2260
Web Address: www.marcusmillichap.com
Marcus & Millichap Real Estate Investment Services, Inc., a national investment real estate brokerage firm, offers in-depth reports on various commercial property and apartment sectors, by city and on a nationwide basis. Users can register to receive its e-mail reports.

Metropolitan Institute at Virginia Tech (MI)
1021 Prince St., Ste. 100
Alexandria, VA 22314 USA
Phone: 703-706-8100
Fax: 703-518-8009
E-mail Address: mivt@vt.edu
Web Address: http://spia.vt.edu/mi
The Metropolitan Institute at Virginia Tech (MI) conducts basic and applied research on national and international development patterns, focusing on key forces shaping metropolitan growth such as demographics, environment, technology, design, transportation, and governance. MI publishes several excellent white papers and reports yearly.

MSN Real Estate
1 Microsoft Way
Microsoft Corporation
Redmond, WA 98052-6399 USA
Toll Free: 800-642-7676
Web Address: realestate.msn.com
MSN Real Estate features advice, links and services for buying and selling houses and managing and maintaining residential property.

Real Estate Library
E-mail Address: mschaffer@relibrary.com
Web Address: www.relibrary.com
The Real Estate Library contains resources such as mortgage loan assistance, online real estate courses for certification, real estate web design walkthroughs and other useful tools for buyers, sellers, home owners and real estate professionals.

RealtyTrac
One Venture Plaza, Ste. 300
Irvine, CA 92618 USA
Toll Free: 800-550-4802
E-mail Address: support@realtytrac.com

Web Address: www.realtytrac.com
RealtyTrac offers comprehensive online housing data, analytics and listings regarding properties in the U.S. In addition, the firm provides a listing of historical default, foreclosure auction and bank-owned properties.

RentLaw.com
6 Cypress Ave.
Oakhurst, NJ 07755 USA
Phone: 732-539-2914
Web Address: www.rentlaw.com
RentLaw.com provides landlords, tenants, investors, lawyers, agent associations and homeowners with extensive information regarding landlord-tenant laws and legal advice.

27) Remodeling Industry Associations

National Association of the Remodeling Industry (NARI)
P.O. Box 4250
Des Plaines, IL 60016 USA
Phone: 847-298-9200
Fax: 847-298-9225
E-mail Address: info@nari.org
Web Address: www.nari.org
The National Association of the Remodeling Industry (NARI) provides certifications, education and support to professionals working in the remodeling industry, along with support to consumers.

National Kitchen & Bath Association (NKBA)
687 Willow Grove St.
Hackettstown, NJ 07840 USA
Fax: 908-852-1695
Toll Free: 800-843-6522
E-mail Address: assist@nkba.org
Web Address: www.nkba.org
The National Kitchen & Bath Association (NKBA) is a premier resource for kitchen and bath remodelers, designers, planners and consumers. It offers courses and certifications for professionals in design, installation and education. It is the owner of the Kitchen/Bath Industry Show & Conference each year and it sponsors annual design competitions.

28) RFID Industry Associations

International RFID Business Association (RFIDba)
5 W. 37th St. & 5th Ave., Fl. 9

New York, NY 10018 USA
E-mail Address: info@rfidba.org
Web Address: www.rfidba.org/
The International RFID Business Association (RFIDba) was founded in April 2004 as a not-for-profit, educational, technology and frequency agnostic, trade association dedicated to serving the business needs of the end user community with vendor neutral information on RFID and real time location system (RTLS) technologies along with information on other associated, complimentary technologies. It has branches that are focused on construction, apparel and footwear and health care.

29) Robotics & Automation Industry Associations

Continental Automated Buildings Association (CABA)
1173 Cyrville Rd., Ste. 210
Ottawa, ON K1J 7S6 Canada
Phone: 613-686-1814
Fax: 613-744-7833
Toll Free: 888-798-2222
E-mail Address: caba@caba.org
Web Address: www.caba.org
The Continental Automated Buildings Association (CABA) is an international not-for-profit industry association dedicated to the advancement of intelligent home and intelligent building technologies. The organization is supported by an international membership of nearly 335 companies involved in the design, manufacture, installation and retailing of products relating to home automation and building automation. Public organizations, including utilities and government are also members.

30) Science Parks

International Association of Science Parks (IASP)
Calle Maria Curie 35, Campanillas
Malaga, 29590 Spain
Phone: 34-95-202-83-03
Fax: 34-95-202-04-64
E-mail Address: iasp@iasp.ws
Web Address: www.iasp.ws
The International Association of Science Parks (IASP) is a worldwide network of science and technology parks. It enjoys Special Consultative status with the Economic and Social Council of the United Nations. Its 394 members represent science parks in 75 nations. It is also a founding member of

the World Alliance for Innovation (WAINOVA). Its world headquarters are located in Spain, with an additional office in the Tsinghua University Science Park, Beijing, China.

31) Seniors Housing

American Seniors Housing Association (ASHA)
5225 Wisconsin Ave. NW, Ste. 502
Washington, DC 20015 USA
Phone: 202-237-0900
Fax: 202-237-1616
Web Address: www.seniorshousing.org
The American Seniors Housing Association (ASHA) was originally formed as a committee of the National Multi Housing Council in 1991, and became an independent non-profit organization ten years later on January 1, 2001. The group represents building owners and managers who develop housing options, services and amenities for seniors.

National Investment Center for the Seniors Housing & Care Industry (NIC)
1997 Annapolis Exchange Pkwy., Ste. 480
Annapolis, MD 21401 USA
Phone: 410-267-0504
Fax: 410-268-4620
Web Address: www.nic.org
NIC serves as a resource to lenders, investors, developers/operators, and others interested in meeting the housing and healthcare needs of America's seniors. NIC serves the entire industry as an objective purveyor of information: NIC's research and educational efforts are neither association-driven nor company-oriented. As an impartial observer and unbiased source, NIC has become the primary link between the financial markets and seniors housing developers/operators, connecting each side through relevant research and practical information.

32) Shopping Center Directories

Chain Store Guide (CSG)
10117 Princess Palm Ave., Ste. 375
Tampa, FL 33610 USA
Fax: 813-627-6888
Toll Free: 800-927-9292
E-mail Address: webmaster@chainstoreguide.com
Web Address: www.chainstoreguide.com
The Chain Store Guide (CSG) is a provider of comprehensive retail and foodservice intelligence. The CSG database contains over 700,000 retailers,

foodservice operators, distributors and wholesalers in the U.S. and Canada.

Directory of Major Malls, Inc. (The)
P.O. Box 837
Nyack, NY 10960 USA
Phone: 845-348-7000
Toll Free: 800-898-6255
Web Address: www.shoppingcenters.com
The Directory of Major Malls, Inc. offers information on centers that have above 250,000 square feet of gross leasable area (GLA). Information includes location, GLA, household income, area population and design of various centers.

Value Retail News
1221 Avenue of the Americas, Fl. 41
New York, NY 10020 USA
Phone: 646-728-3800
Fax: 732-694-1755
E-mail Address: icsc@icsc.org
Web Address: www.valueretailnews.com
Value Retail News, a division of the International Council of Shopping Centers, publishes several directories, including the Global Outlet Project Directory, a listing of factory outlet information; and the Value Retail Directory, which provides factory outlet tenant data.

33) Shopping Center Resources

ChainLinks Retail Advisors
300 Galleria Pkwy., Fl. 12
Atlanta, GA 30339 USA
Phone: 805-684-7767
Fax: 770-951-0054
Web Address: www.chainlinks.com
ChainLinks is a major retail-only, full-service real estate broker in the United States and Canada.

CoreNet Global
133 Peachtree St. NE, Ste. 3000
Atlanta, GA 30303 USA
Phone: 404-589-3200
Fax: 404-589-3201
Toll Free: 800-726-8111
E-mail Address: membership@corenetglobal.org
Web Address: www.corenetglobal.org
CoreNet Global is an organization for business leaders engaged in the strategic management of real estate for major corporations worldwide.

International Council of Shopping Centers (ICSC)
1221 Ave. of the Americas, Fl. 41
New York, NY 10020-1099 USA
Phone: 646-728-3800
Fax: 732-694-1690
E-mail Address: membership@icsc.org
Web Address: www.icsc.org
The International Council of Shopping Centers (ICSC) is the global trade association of the shopping center industry, and includes shopping center owners, developers, managers, marketing specialists, investors, lenders and retailers. ICSC's 70,000 members reach to over 100 countries and the organization links with more than 25 national and regional shopping center councils throughout the world.

34) Spa Industry Associations

International Spa Association (ISPA)
2365 Harrodsburg Rd., Ste. A325
Lexington, KY 40504 USA
Phone: 859-226-4326
Fax: 859-226-4445
Toll Free: 888-651-4772
E-mail Address: ispa@ispastaff.com
Web Address: www.experienceispa.com
The International Spa Association (ISPA) is a leading professional organization for the spa industry. It provides educational and networking opportunities, promotes the value of the spa experience and speaks as the authoritative voice to foster professionalism and growth. ISPA represents health and wellness facilities and providers in over 70 countries.

35) Telecommunications Industry Associations

Fiber to the Home (FTTH) Council
6841 Elm St., Ste. 843
McLean, VA 22101-0843 USA
Phone: 202-524-9550
E-mail Address: heather.b.gold@ftthcouncil.org
Web Address: www.ftthcouncil.org
The Fiber-to-the-Home (FTTH) Council is a nonprofit organization established in 2001 to educate the public on the opportunities and benefits of FTTH solutions. Its website is an excellent resource for statistics, general reference and trends in the delivery of fiber optic cable directly to the home and office.

36) Travel Business & Professional Associations

Hospitality Financial and Technology Professionals (HFTP)
11709 Boulder Ln., Ste. 110
Austin, TX 78726 USA
Phone: 512-249-5333
Fax: 512-249-1533
Toll Free: 800-646-4387
E-mail Address: Education@hftp.org
Web Address: www.hftp.org
Hospitality Financial and Technology Professionals (HFTP) is a organization that caters to the needs of the hospitality industry, principally to those individuals who perform accounting, financial, management or information technology activities.

37) Travel-Local Transportation, Bus & Car Rental

American Public Transportation Association (APTA)
1300 I St. NW, Ste. 1200
Washington, DC 20005 USA
Phone: 202-496-4800
Fax: 202-496-4324
E-mail Address: rsheridan@apta.com
Web Address: www.apta.com
APTA is a nonprofit international association of more than 1,500 member organizations including public transportation systems; planning, design, construction and finance firms; product and service providers; academic institutions; and state associations and departments of transportation. APTA members serve more than 90 percent of persons using public transportation in the United States and Canada.

38) U.S. Government Agencies

Bureau of Economic Analysis (BEA)
1441 L St. NW
Washington, DC 20230 USA
Phone: 202-606-9900
E-mail Address: customerservice@bea.gov
Web Address: www.bea.gov
The Bureau of Economic Analysis (BEA), an agency of the U.S. Department of Commerce, is the nation's economic accountant, preparing estimates that illuminate key national, international and regional aspects of the U.S. economy.

Bureau of Labor Statistics (BLS)
2 Massachusetts Ave. NE
Postal Square Building
Washington, DC 20212-0001 USA
Phone: 202-691-5200
Fax: 202-691-7890
Toll Free: 800-877-8339
E-mail Address: blsdata_staff@bls.gov
Web Address: stats.bls.gov
The Bureau of Labor Statistics (BLS) is the principal fact-finding agency for the Federal Government in the field of labor economics and statistics. It is an independent national statistical agency that collects, processes, analyzes and disseminates statistical data to the American public, U.S. Congress, other federal agencies, state and local governments, business and labor. The BLS also serves as a statistical resource to the Department of Labor.

Department of Housing and Urban Development (HUD)
451 7th St. SW
Washington, DC 20410 USA
Phone: 202-708-1112
Web Address: www.hud.gov
The Department of Housing and Urban Development, commonly known as HUD, is a Federal Government agency involved in increasing homeownership, supporting community development and increasing access to affordable housing free from discrimination.

Energy Efficiency and Renewable Energy (EERE)
1000 Independence Ave. SW
Forrestal Building
Washington, DC 20585 USA
Phone: 202-586-5000
Fax: 202-586-4403
E-mail Address: The.Secretary@hq.doe.gov
Web Address: energy.gov/eere/office-energy-efficiency-renewable-energy
The Energy Efficiency and Renewable Energy (EERE), an office of the U.S. Department of Energy, provides information on bioenergy, geothermal, hydrogen, hydropower, tidal, hydropower, solar, wind and energy conservation methods. The Office also works with U.S. industries to advance the development of various alternative energy technologies.

Federal Housing Administration (FHA)
451 7th St. SW
Washington, DC 20410 USA

Phone: 202-708-1112
Toll Free: 800-225-5342
Web Address:
portal.hud.gov/hudportal/HUD?src=/federal_housing
_administration
The Federal Housing Administration, generally
known as FHA, is the largest government insurer of
mortgages in the world, insuring over 35 million
properties since its inception in 1934. A part of the
United States Department of Housing and Urban
Development (HUD), FHA provides mortgage
insurance on single-family, multifamily,
manufactured homes and hospital loans made by
FHA-approved lenders throughout the United States
and its territories.

Federal Housing Finance Agency (FHFA)
400 7th St. SW
Washington, DC 20024 USA
Phone: 202-649-3800
Fax: 202-649-1071
Web Address: www.fhfa.gov
The Federal Housing Finance Agency (FHFA),
created by the Housing and Economic Recovery Act
of 2008 that combined the staffs of the Office of
Federal Housing Enterprise Oversight (OFHEO), the
Federal Housing Finance Board (FHFB), and the
GSE mission office at the Department of Housing
and Urban Development (HUD), oversees vital
components of our country's secondary mortgage
markets, Fannie Mae, Freddie Mac, and the Federal
Home Loan Banks.

U.S. Census Bureau
4600 Silver Hill Rd.
Washington, DC 20233-8800 USA
Phone: 301-763-4636
Toll Free: 800-923-8282
Web Address: www.census.gov
The U.S. Census Bureau is the official collector of
data about the people and economy of the U.S.
Founded in 1790, it provides official social,
demographic and economic information. In addition
to the Population & Housing Census, which it
conducts every 10 years, the U.S. Census Bureau
numerous other surveys annually.

U.S. Department of Commerce (DOC)
1401 Constitution Ave. NW
Washington, DC 20230 USA
Phone: 202-482-2000
E-mail Address: TheSec@doc.gov
Web Address: www.commerce.gov

The U.S. Department of Commerce (DOC) regulates
trade and provides valuable economic analysis of the
economy.

U.S. Department of Labor (DOL)
200 Constitution Ave. NW
Frances Perkins Bldg.
Washington, DC 20210 USA
Toll Free: 866-487-2365
Web Address: www.dol.gov
The U.S. Department of Labor (DOL) is the
government agency responsible for labor regulations.

U.S. Securities and Exchange Commission (SEC)
100 F St. NE
Washington, DC 20549 USA
Phone: 202-942-8088
Web Address: www.sec.gov
The U.S. Securities and Exchange Commission
(SEC) is a nonpartisan, quasi-judicial regulatory
agency responsible for administering federal
securities laws. These laws are designed to protect
investors in securities markets and ensure that they
have access to disclosure of all material information
concerning publicly traded securities. Visitors to the
web site can access the EDGAR database of
corporate financial and business information.

Chapter 4

THE REAL ESTATE 450: WHO THEY ARE AND HOW THEY WERE CHOSEN

Includes Indexes by Company Name, Industry & Location

The companies chosen to be listed in PLUNKETT'S REAL ESTATE & CONSTRUCTION INDUSTRY ALMANAC comprise a unique list. THE REAL ESTATE 450 (the actual count is 445 companies) were chosen specifically for their dominance in the many facets of the real estate and construction industry in which they operate. Complete information about each firm can be found in the "Individual Profiles," beginning at the end of this chapter. These profiles are in alphabetical order by company name.

THE REAL ESTATE 450 companies are from all parts of the United States, Asia, Canada, Europe and beyond. Essentially, THE REAL ESTATE 450 includes companies that are deeply involved in the technologies, services and trends that keep the entire industry forging ahead.

Simply stated, THE REAL ESTATE 450 contains the largest, most successful, fastest growing firms in real estate, construction and related industries in the world. To be included in our list, the firms had to meet the following criteria:

1) Generally, these are corporations based in the U.S., however, the headquarters of 118 firms are located in other nations.

2) Prominence, or a significant presence, in real estate, construction and supporting fields. (See the following Industry Codes section for a

complete list of types of businesses that are covered).

3) The companies in THE REAL ESTATE 450 do not have to be exclusively in the real estate and construction field.

4) Financial data and vital statistics must have been available to the editors of this book, either directly from the company being written about or from outside sources deemed reliable and accurate by the editors. A small number of companies that we would like to have included are not listed because of a lack of sufficient, objective data.

INDEX OF COMPANIES WITHIN INDUSTRY GROUPS

The industry codes shown below are based on the 2012 NAIC code system (NAIC is used by many analysts as a replacement for older SIC codes because NAIC is more specific to today's industry sectors, see www.census.gov/NAICS). Companies are given a primary NAIC code, reflecting the main line of business of each firm.

Industry Group/Company	Industry Code	2015 Sales	2015 Profits
Airport Related Services, Baggage Handling			
Ferrovial SA	488119	11,074,235,392	821,158,528
Architectural Services			
Adrian Smith + Gordon Gill Architecture (AS+GG)	541310	81,199,000	
Aedas	541310		
Callison LLC	541310		
Cambridge Seven Associates Inc	541310		
Dahlin Group Architecture Planning	541310	17,670,000	
Gensler	541310	1,075,000,000	
Goettsch Partners Inc	541310	29,620,000	
Jerde Partnership Inc (The)	541310	18,000,000	
Kohn Pedersen Fox Associates (KPF)	541310	220,000,000	
MulvannyG2 Architecture	541310	57,190,000	
NBBJ	541310		
Perkins Eastman	541310	182,900,000	
Populous Holdings Inc	541310		
Rebel Design+Group	541310	18,792,889	
Robert A.M. Stern Architects LLP	541310	62,860,000	
Smallwood Reynolds Stewart Stewart & Associates Inc	541310	22,740,000	
Steelman Partners LLP	541310	42,000,000	
STUDIOS Architecture	541310		
VOA Associates Inc	541310	70,000,000	
Building Material Dealers			
84 Lumber Company	444190	2,500,000,000	
Building Materials Holding Corp (BMC)	444190	1,576,745,984	-4,831,000
CRH plc	444190	26,955,667,456	825,720,512
Ferguson Enterprises Inc	444190	13,004,000,000	
Five Star Products Inc	444190		
Pacific Coast Building Products Inc	444190		
Stock Building Supply Inc	444190	1,576,745,984	-4,831,000
Wolseley PLC	444190		
Car Reservations (e.g. Uber), Ticket Offices, Time Share and Vacation Club Rentals and Specialty Reservation Services			
Bluegreen Corporation	561599	614,765,000	82,009,000
Diamond Resorts Holdings LLC	561599	954,040,000	149,478,000
ResortQuest International Inc	561599	505,000,000	
Westgate Resorts	561599		
WorldMark by Wyndham Inc	561599		
Wyndham Vacation Ownership	561599	2,660,000,000	
Casino Hotels and Casino Resorts			
American Casino & Entertainment Properties LLC	721120	373,067,000	12,062,000

Industry Group/Company	Industry Code	2015 Sales	2015 Profits
Ameristar Casinos Inc	721120	1,275,000,000	
Boyd Gaming Corp	721120	2,199,431,936	47,234,000
Caesars Entertainment Corporation	721120	4,654,000,128	5,920,000,000
Kerzner International Holdings Limited	721120	400,000,000	
Las Vegas Sands Corp (The Venetian)	721120	11,688,461,312	1,966,236,032
MGM Growth Properties LLC	721120	0	-261,954,000
MGM Resorts International	721120	9,190,068,224	-447,720,000
Red Rock Resorts Inc	721120		
Rio Properties Inc	721120		
Societe Anonyme des Bains de Mer et du Cercle des Etrangers a Monaco	721120	514,150,000	11,373,400
Station Casinos LLC	721120	1,352,135,000	132,504,000
Wynn Resorts Limited	721120	4,075,883,008	195,290,000
Cement and Ready-Mixed Concrete Manufacturing			
HeidelbergCement AG	327310	15,356,462,080	912,512,320
HeidelbergCement North America	327310	4,252,833,800	
LafargeHolcim Ltd	327310	24,682,106,880	-1,537,398,784
Commercial and Institutional Building Construction			
Clark Construction Group LLC	236220	4,151,000,000	
Gilbane Inc	236220	3,850,000,000	
Kumho Industrial Co Ltd	236220	1,357,435,200	-5,416,942
McCarthy Building Companies Inc	236220	3,000,000,000	
Structure Tone Inc	236220	3,500,000,000	
TDIndustries	236220	433,000,000	
Turner Corp (The)	236220	10,797,500,000	
Walsh Group (The)	236220	4,608,100,000	
Whiting-Turner Contracting Co (The)	236220	6,347,100,000	
Commercial Real Estate Investment and Operations, Including Office Buildings, Shopping Centers, Industrial Properties and Related REITs			
CBL & Associates Properties Inc	531120	1,055,017,984	103,371,000
Cousins Properties Inc	531120	381,643,008	125,518,000
DCT Industrial Trust Inc	531120	354,696,992	94,048,000
DDR Corp	531120	1,028,070,976	-72,168,000
Dividend Capital Group LLC	531120		
Far East Organization	531120	3,600,000,000	
Fibra Uno Administracion SA de CV	531120		
Forest Realty Trust Inc	531120		
Hines Interests LP	531120		
Jones Lang LaSalle Inc	531120	5,965,670,912	438,672,000
LBA Realty LLC	531120		
Mapletree Investments Pte Ltd	531120	1,633,923,000	1,826,956,000
Nomura Real Estate Master Fund Inc	531120		
Paramount Group Inc	531120	662,408,000	-4,419,000
Pearlmark Real Estate Partners	531120		
ProLogis Inc	531120	2,197,073,920	869,438,976
Rouse Properties Inc	531120	305,384,000	41,699,000
Strategic Hotels & Resorts Inc	531120	1,381,600,000	
Trammell Crow Company	531120	1,600,000,000	

Industry Group/Company	Industry Code	2015 Sales	2015 Profits
Transwestern Commercial Services	531120		
Trump Entertainment Resorts Inc	531120		
Unibail-Rodamco SE	531120	1,921,738,880	2,661,922,304
Washington Real Estate Investment Trust	531120	306,427,008	89,740,000
Watson Land Company	531120		
Westfield Corporation	531120	1,296,430,592	2,420,844,288
WeWork Companies Inc	531120	251,000,000	
WP Carey & Co LLC	531120	938,382,976	172,258,000
Computer Software, Business Management & ERP			
RealPage Inc	511210H	468,520,000	-9,218,000
Computer Software, Data Base & File Management			
Reis Inc	511210J	50,890,440	10,304,983
Concrete Manufacturing, Ready-Mix			
CEMEX Inc	327320	3,500,000,000	
Oldcastle Inc	327320	11,000,000,000	
Construction of Railways, Marine Facilities and Subways			
Balfour Beatty plc	237990	9,833,028,608	-291,244,288
China Communications Construction Company Ltd	237990	62,498,332,672	2,425,672,448
China Railway Construction Corporation Limited (CRCC)	237990	92,806,062,080	1,954,206,976
Odebrecht SA	237990	45,800,000,000	
Orion Marine Group Inc	237990	466,497,984	-8,060,000
RailWorks Corp	237990	703,000,000	
Samwhan Corporation	237990		
Construction of Telecommunications Lines and Systems & Electric Power Lines and Systems			
Bouygues SA	237130	36,995,483,800	557,826,750
Doosan Heavy Industry & Construction Co	237130	14,059,352,557	-1,519,132,785
Construction of Water & Sewer Lines and Systems			
Aegion Corp	237110	1,333,570,048	-8,067,000
Layne Christensen Company	237110	797,601,024	-110,151,000
MWH Global Inc	237110	1,900,000,000	
Sterling Construction Company Inc	237110	623,595,008	-20,402,000
Electrical Contractors and Other Wiring Installation Contractors			
EMCOR Group Inc	238210	6,718,726,144	172,286,000
Integrated Electrical Services Inc	238210	573,857,024	16,538,000
M.C. Dean Inc	238210	810,000,000	
MMR Group Inc	238210	845,000,000	
Rosendin Electric	238210	1,300,000,000	
Elevator and Moving Stairway Manufacturing			
Hyundai Elevator Co Ltd	333921	1,273,888,381	-3,562,287
Engineered Wood Member (except Truss) Manufacturing			
Boise Cascade Corp	321213	3,633,414,912	52,182,000
Engineering Services, Including Civil, Mechanical, Electronic, Computer and Environmental Engineering			
Amey plc	541330		
Arcadis NV	541330	3,899,698,944	112,612,768
Black & Veatch Holding Company	541330	3,030,000,000	

Industry Group/Company	Industry Code	2015 Sales	2015 Profits
Burns & McDonnell	541330	2,700,000,000	
CH2M HILL Companies Ltd	541330	5,410,000,000	
Chiyoda Corporation	541330	4,384,534,016	102,206,952
ENGlobal Corp	541330	79,605,000	10,536,000
HDR Inc	541330		
Louis Berger Group Inc (The)	541330	1,000,000,000	
McDermott International Inc	541330	3,070,275,072	-17,983,000
Parsons Corp	541330	3,200,000,000	
Skidmore Owings & Merrill LLP	541330	356,070,000	
STV Group Inc	541330	450,000,000	
Trevi-Finanziaria Industriale SpA (Trevi Group)	541330		
YIT Corporation	541330	1,975,570,432	53,831,500
Environmental Control Manufacturing			
Siemens Building Technologies	334512	6,891,651,200	635,173,035
Facilities Support Services			
Abertis Infraestructuras SA	561210	5,539,211,264	2,144,035,712
Fossil Fuel Electric Power Generation			
Sembcorp Industries Ltd	221112	7,090,783,744	407,749,248
Heavy Construction, Including Civil Engineering-Construction, Major Construction Projects, Land Subdivision, Infrastructure, Utilities, Highways and Bridges			
Amec Foster Wheeler plc	237000	7,712,317,952	-361,934,624
Bechtel Group Inc	237000	37,200,000,000	
Bilfinger SE	237000	7,391,909,376	-557,361,344
CDM Smith Inc	237000	1,260,000,000	
Chicago Bridge & Iron Company NV (CB&I)	237000	12,929,504,256	-504,415,008
Daelim Industrial Co Ltd	237000	8,306,102,486	180,553,952
Fluor Corp	237000	18,114,048,000	412,512,000
Fomento de Construcciones y Contratas SA (FCC)	237000	7,385,891,840	-52,794,788
GS Engineering & Construction Corp	237000	7,873,167,914	-28,665,380
HOCHTIEF AG	237000	24,060,649,472	237,550,896
Hyundai Engineering & Construction Company Ltd	237000	16,568,302,821	318,782,537
Jacobs Engineering Group Inc	237000	12,114,832,384	302,971,008
KBR Inc	237000	5,096,000,000	203,000,000
Lend Lease Corporation Limited	237000	10,060,364,800	470,883,776
PCL Construction Group Inc	237000	7,232,900,000	
Salini Impregilo SpA	237000	5,404,678,144	69,105,048
Samsung C&T Corporation	237000		
Skanska USA Inc	237000	7,100,000,000	
Highway, Street, Tunnel & Bridge Construction			
Acciona SA	237310	7,462,875,136	236,452,592
ACS Actividades de Construccion y Servicios SA	237310		
AECOM Technology Corporation	237310	17,989,879,808	-154,844,992
Aecon Group Inc	237310	2,230,813,952	52,502,144
Empresas ICA SAB de CV	237310		
Granite Construction Inc	237310	2,371,028,992	60,485,000
Kiewit Corp	237310	10,380,000,000	
Meadow Valley Corporation	237310		

Industry Group/Company	Industry Code	2015 Sales	2015 Profits
Tutor Perini Corporation	237310	4,920,472,064	45,292,000
VINCI SA	237310	44,663,037,952	2,333,458,688
Home and Apartment Builders (Residential Construction)			
Cyrela Brazil Realty SA	236110	1,193,813,760	123,143,776
Desarrolladora Homex SAB de CV	236110	21,729,133	1,058,139,784
Sekisui House Ltd	236110	17,436,084,224	822,468,736
Home Centers, Building Materials			
Home Depot Inc	444110	83,175,997,440	6,344,999,936
Lowe's Companies Inc	444110	56,222,998,528	2,697,999,872
Hotels, Motels, Inns and Resorts (Lodging and Hospitality)			
Accor North America	721110	635,410,802	255,037,440
Accor SA	721110	6,365,118,976	278,281,504
Amanresorts International Pte Ltd (Aman Resorts)	721110		
Banyan Tree Holdings Limited	721110	275,387,424	-20,444,110
Barcelo Crestline Corporation	721110		
Belmond Ltd	721110	551,385,024	16,265,000
Best Western International Inc	721110	7,000,000,000	
Carlson Companies Inc	721110	3,150,000,000	
Carlson Rezidor Hotel Group	721110	1,135,234,050	
China Lodging Group Ltd	721110		
Choice Hotels International Inc	721110	859,878,016	128,029,000
Club Mediterranee SA (Club Med)	721110	1,450,000,000	
Commune Hotels & Resorts	721110	500,000,000	
Days Inn Worldwide Inc	721110		
Doyle Collection (The)	721110	200,000,000	
Extended Stay America Inc	721110	1,284,753,000	113,040,000
Fairmont Raffles Hotels International Inc	721110	1,000,000,000	
Four Seasons Hotels Inc	721110	4,300,000,000	
Golden Tulip Hospitality Group	721110		
Groupe du Louvre	721110	1,600,000,000	
Hilton Worldwide Inc	721110	11,271,999,488	1,404,000,000
Homeinns Hotel Group	721110	1,168,000,000	
Hongkong and Shanghai Hotels Ltd	721110	740,184,000	128,929,448
Hotel Properties Ltd	721110	429,044,861	60,658,066
Howard Johnson International Inc	721110		
Hyatt Hotels Corporation	721110	4,328,000,000	124,000,000
Indian Hotels Company Limited (The)	721110	629,558,336	-56,828,944
InterContinental Hotels Group plc	721110	1,803,000,064	1,222,000,000
Interstate Hotels & Resorts Inc	721110		
InTown Suites Management Inc	721110		
Jameson Inn Inc	721110		
Janus Hotels and Resorts Inc	721110		
John Q Hammons Hotels & Resorts LLC	721110	461,000,000	
Kimpton Hotel & Restaurant Group LLC	721110	1,200,000,000	
La Quinta Holdings Inc	721110	1,029,974,016	26,365,000
Loews Hotels Holding Corporation	721110	604,000,000	12,000,000
Mandarin Oriental International Ltd	721110	607,299,968	89,300,000

Industry Group/Company	Industry Code	2015 Sales	2015 Profits
Marcus Corporation (The)	721110	488,067,008	23,995,000
Marriott International Inc	721110	14,485,999,616	859,000,000
Maui Land & Pineapple Company Inc	721110	22,786,000	6,813,000
Melia Hotels International SA	721110		
Meritus Hotels & Resorts Inc	721110		
Millennium & Copthorne Hotels plc	721110	1,197,494,656	91,897,464
Morgans Hotel Group Co	721110	219,982,000	22,097,000
NH Hotel Group SA	721110	1,570,048,256	1,069,787
Oakwood Worldwide	721110	860,000,000	
Oberoi Group (EIH Ltd)	721110	250,608,522	9,487,804
Ramada Worldwide Inc	721110		
Red Lion Hotels Corporation	721110	142,920,000	2,719,000
Rezidor Hotel Group AB	721110		
Ritz-Carlton Hotel Company LLC (The)	721110	2,400,000,000	
Rosewood Hotels & Resorts LLC	721110	575,000,000	
Ryman Hospitality Properties Inc	721110	1,092,124,032	111,511,000
Scandic Hotels AB	721110	1,494,083,703	25,843,517
Shangri-La Asia Ltd	721110		
Shun Tak Holdings Limited	721110		
Sonesta International Hotels Corp	721110		
Starwood Capital Group Global LLC	721110		
Starwood Hotels & Resorts Worldwide Inc	721110	5,762,999,808	489,000,000
Sunburst Hospitality Corporation	721110		
Super 8 Worldwide Inc	721110		
TMI Hospitality Inc	721110		
TRT Holdings Inc	721110		
Wyndham Worldwide Corporation	721110	5,536,000,000	612,000,000
Xanterra Parks & Resorts Inc	721110	106,000,000	
Internet Search Engines, Online Publishing, Sharing and Consumer Services, Online Radio, TV and Entertainment Sites and Social Media			
CoStar Group Inc	519130	711,763,968	-3,465,000
Move Inc (Realtor.com)	519130	260,000,000	
Opendoor	519130	18,685,720	
Zillow Inc	519130	644,676,992	-148,874,000
Iron and Steel Mills and Ferroalloy Manufacturing			
Essar Group Ltd	331110	35,000,000,000	
LB Foster Company	331110	624,523,008	-44,445,000
Janitorial Services			
ABM Industries Inc	561720	4,897,800,192	76,300,000
Land Development and Subdivision			
Newhall Land & Farming Company	237210		
St Joe Company (The)	237210	103,871,000	-1,731,000
Lumber, Cement, Roofing and Other Building Materials Distributors (Wholesale Distribution)			
Cemex SAB de CV	423300		
Metal Buildings and Components (Prefabricated) Manufacturing			
Butler Manufacturing Co	332311		

Industry Group/Company	Industry Code	2015 Sales	2015 Profits
NCI Building Systems Inc	332311	1,563,693,056	17,818,000
Mobile, Modular & Prefabricated Homes and Buildings Manufacturing			
Champion Enterprises Holdings LLC	321992		
Clayton Homes Inc	321992		
Fairmont Homes Inc	321992		
Palm Harbor Homes Inc	321992		
Skyline Corporation	321992	186,984,992	-10,414,000
Southern Energy Homes Inc	321992		
Mortgage Brokers and Loan Brokers			
Bank of America Home Loans	522310	8,830,000,000	
CitiMortgage Inc	522310		
E-LOAN Inc	522310		
LendingTree LLC	522310	254,200,000	48,047,000
Mortgage Real Estate Investment Trusts (REITs), Mortgage Underwriting and Investing			
Annaly Capital Management Inc	525990	664,033,024	466,556,000
ARMOUR Residential REIT Inc	525990	365,300,000	-31,205,000
Capstead Mortgage Corporation	525990	122,355,000	108,325,000
Chimera Investment Corporation	525990	0	250,348,992
Colony Capital Inc	525990	841,976,000	149,980,000
Community Development Trust	525990		
Dynex Capital Inc	525990	34,212,000	16,544,000
Impac Mortgage Holdings Inc	525990	154,604,000	80,799,000
iStar Inc	525990	321,787,008	-2,435,000
MFA Financial Inc	525990	365,655,008	313,225,984
MMA Capital Management LLC	525990	20,145,000	17,843,000
Newcastle Investment Corp	525990	349,718,016	21,847,000
PennyMac Mortgage Investment Trust	525990	373,472,992	90,100,000
Redwood Trust Inc	525990	188,803,008	102,088,000
Starwood Property Trust Inc	525990	803,118,016	450,696,992
Movie (Motion Pictures) Theaters			
Dalian Wanda Group Co Ltd	512131	38,800,000,000	
New Home Builders (Production Builders)			
AV Homes Inc	236117	517,766,016	11,950,000
Beazer Homes Usa Inc	236117	1,627,412,992	344,094,016
CalAtlantic Group Inc	236117	3,540,112,896	213,508,992
Capital Pacific Real Estate	236117		
Centex Corp	236117		
Dominion Homes Inc	236117	181,000,000	
DR Horton Inc	236117	10,823,999,488	750,700,032
Hovnanian Enterprises Inc	236117	2,148,480,000	-16,100,000
KB Home	236117	3,032,029,952	84,643,000
Lennar Corporation	236117	9,474,008,064	802,894,016
M.D.C. Holdings Inc	236117	1,860,226,048	65,791,000
M/I Homes Inc	236117	1,418,395,008	51,763,000
Meritage Homes Corp	236117	2,592,556,032	128,738,000
NVR Inc	236117	5,169,562,112	382,927,008
PulteGroup Inc	236117	5,981,963,776	494,089,984

Industry Group/Company	Industry Code	2015 Sales	2015 Profits
Toll Brothers Inc	236117	4,171,248,128	363,167,008
WCI Communities Inc	236117	562,768,000	35,400,000
William Lyon Homes Inc	236117		
Non Bank Lending and Financing (Shadow Banking)			
GE Capital	522200	10,801,000,000	-7,983,000,000
Nursing Care Facilities (Skilled Nursing Facilities)			
Diversicare Healthcare Services Inc	623110	387,595,008	1,624,000
Extendicare Inc	623110	748,890,816	177,418,816
ManorCare Inc (HCR ManorCare)	623110	5,300,000,000	
Ocean Cargo and Deep Sea Shipping			
Orient Overseas (International) Ltd	483111	5,953,443,840	283,851,008
Swire Pacific Ltd	483111	7,849,869,824	1,731,393,664
Online Shopping, B2B and B2C Sales on the Internet (Ecommerce)			
Bluestem Group Inc	454111	1,743,041,000	14,215,000
Plastics Pipe and Pipe Fitting Manufacturing			
Royal Group Inc	326122		
Plumbing Contractors, and Heating and Air Conditioning (HVAC) Contractors			
ACCO Engineered Systems Inc	238220	770,000,000	
APi Group Inc	238220	2,000,000,000	
Brand Energy & Infrastructure Services Inc	238220	3,000,000,000	
Comfort Systems USA Inc	238220	1,580,519,040	49,364,000
McKinstry	238220	500,000,000	
Pottery, Ceramics and Plumbing Fixture Manufacturing			
Kohler Company	327110	6,000,000,000	
Real Estate Agents & Brokers			
CBRE Group Inc	531210	10,855,810,048	547,132,032
Century 21 Real Estate LLC	531210		
Christies International Real Estate	531210	118,000,000	
Coldwell Banker Real Estate LLC	531210		
Colliers International Group Inc	531210	1,721,986,048	23,347,000
CORE Network (The)	531210		
Cushman & Wakefield Inc	531210	5,000,000,000	
HomeServices of America Inc	531210	191,000,000	
NAI Global Inc	531210		
NRT LLC	531210	7,515,000,000	
RE/MAX Holdings Inc	531210	176,868,000	16,655,000
Realogy Holdings Corp	531210	5,705,999,872	184,000,000
Redfin	531210	125,000,000	
ZipRealty Inc	531210	85,000,000	
Real Estate Rental, Leasing, Development and Management, including REITs			
Brookfield Asset Management Inc	531100	21,752,999,936	2,340,999,936
China Overseas Land & Investment Limited	531100	18,066,966,528	4,294,908,672
China State Construction Engineering Corp	531100		
Chinese Estates Holdings Ltd	531100		
Crest Nicholson plc	531100	1,137,831,936	175,453,472

Industry Group/Company	Industry Code	2015 Sales	2015 Profits
DLF Limited	531100	1,149,613,824	81,197,776
Emaar Properties PJSC	531100	3,719,379,272	1,249,535,249
Evergrande Real Estate Group	531100		
FirstService Corporation	531100		
Gafisa SA	531100		
Guocoleisure Ltd	531100	423,200,000	47,900,000
Hang Lung Group Ltd	531100	1,228,439,936	413,992,480
Hang Lung Properties Limited	531100		
Henderson Land Development Co Ltd	531100		
Hongkong Land Holdings Ltd	531100	1,932,099,968	2,011,699,968
Hutchison Whampoa Properties Ltd	531100	4,385,000,000	
Hypo Real Estate Holding AG (Hypo Bank)	531100	815,000,000	
Hysan Development Co Ltd	531100	442,228,032	374,282,208
John Laing plc	531100	245,628,600	153,504,750
MRV Engenharia E Participacoes SA	531100	1,309,822,464	150,583,280
New World Development Company Limited	531100	7,122,707,456	2,464,099,584
PDG Realty SA Empreendimentos e Participacoes	531100	501,677,792	-760,197,504
Rossi Residencial SA	531100		
Saha Pathana Inter-Holding PCL	531100	115,940,467	35,823,135
Shimao Property Holdings Ltd	531100		
Sino Land Company Limited	531100	2,815,628,032	1,208,324,864
Sun Hung Kai Properties Ltd	531100	8,610,295,808	4,007,385,344
Triple Five Group	531100		
Unitech Limited	531100		
Wharf (Holdings) Limited The	531100	5,269,991,424	2,065,965,568
Wheelock and Company Limited	531100	7,404,547,072	1,834,924,032
REITS (Real Estate Investment Trusts) - Nonresidential			
Alexandria Real Estate Equities Inc	531120A	751,737,984	144,216,992
Anderson-Tully Lumber Company	531120A		
ARC Group of Companies (The)	531120A		
Boston Properties Inc	531120A	2,490,821,120	583,105,984
Brandywine Realty Trust	531120A	602,630,976	-30,401,000
Brixmor Property Group Inc	531120A	1,265,980,032	193,720,000
Capital Automotive REIT	531120A		
CNL Lifestyle Properties Inc	531120A	337,664,992	141,155,008
Condor Hospitality Trust Inc	531120A	57,341,000	13,125,000
Corporate Office Properties Trust	531120A	625,465,984	178,300,000
Crescent Real Estate Equities LP	531120A		
Duke Realty Corp	531120A	949,432,000	615,310,016
EastGroup Properties Inc	531120A	235,008,000	47,866,000
EPR Properties	531120A	341,102,016	194,532,000
Equity Commonwealth REIT	531120A	714,891,008	99,857,000
Equity Office Management LLC	531120A		
Equity One Inc	531120A	360,152,992	65,453,000
Federal Realty Investment Trust	531120A	744,012,032	210,219,008
FelCor Lodging Trust Inc	531120A	886,254,016	-7,428,000
First Industrial Realty Trust Inc	531120A	365,761,984	73,802,000
General Growth Properties Inc	531120A	2,403,906,048	1,374,561,024

Industry Group/Company	Industry Code	2015 Sales	2015 Profits
Glenborough LLC	531120A		
HCP Inc	531120A	2,544,312,064	-559,235,008
Healthcare Realty Trust Inc	531120A	351,929,984	69,436,000
Highwoods Properties Inc	531120A	604,670,976	97,078,000
Hospitality Properties Trust	531120A	1,921,904,000	166,418,000
Host Hotels & Resorts LP	531120A	5,386,999,808	558,000,000
IDI Gazeley	531120A		
Inland Real Estate Corporation	531120A	203,900,000	25,531,000
Innkeepers USA Trust	531120A		
InvenTrust Properties Corp	531120A	450,044,000	3,464,000
Kilroy Realty Corporation	531120A	581,275,008	234,080,992
Kimco Realty Corp	531120A	1,166,769,024	894,115,008
LaSalle Hotel Properties	531120A	1,164,358,016	135,552,000
Lexington Realty Trust	531120A	430,839,008	111,703,000
Liberty Property Trust	531120A	808,772,992	238,039,008
Lillibridge Healthcare Real Estate Trust	531120A		
Link Real Estate Investment Trust (The)	531120A	995,722,112	3,510,748,928
Macerich Company (The)	531120A	1,288,148,992	487,561,984
Mack-Cali Realty Corp	531120A	594,883,008	-125,752,000
National Health Investors Inc	531120A	228,988,000	148,862,000
Omega Healthcare Investors Inc	531120A	743,617,024	224,524,000
Parkway Properties Inc	531120A	473,983,008	67,335,000
Pennsylvania REIT	531120A	425,411,008	-116,683,000
PS Business Parks Inc	531120A	373,675,008	130,475,000
Realty Income Corp	531120A	1,023,284,992	283,766,016
Regency Centers Corp	531120A	569,763,008	150,056,000
Riverview Realty Partners	531120A		
RXR Realty	531120A		
Senior Housing Properties Trust	531120A	998,772,992	123,968,000
Simon Property Group Inc	531120A	5,266,102,784	1,827,719,936
SL Green Realty Corp	531120A	1,662,829,056	284,084,000
Sunstone Hotel Investors Inc	531120A	1,249,180,032	347,355,008
Tanger Factory Outlet Centers Inc	531120A	439,368,992	211,200,000
Taubman Centers Inc	531120A	557,171,968	134,127,000
Ventas Inc	531120A	3,286,397,952	417,843,008
Vornado Realty Trust	531120A	2,502,266,880	760,433,984
Weingarten Realty Investors	531120A	512,844,000	174,352,000
Wells Real Estate Funds Inc	531120A		
WP GLIMCHER Inc	531120A	921,656,000	-85,297,000
REITs (Real Estate Investment Trusts) - Residential			
AvalonBay Communities Inc	531110A	1,856,028,032	742,038,016
Boardwalk Real Estate Investment Trust	531110A	364,005,280	22,053,698
CAPREIT Inc	531110A		
Equity Lifestyle Properties Inc	531110A	821,654,016	139,371,008
Equity Residential	531110A	2,744,965,120	870,120,000
Essex Property Trust Inc	531110A	1,194,407,040	232,120,000
Gables Residential Trust	531110A		
Home Properties Inc	531110A	647,984,000	
Investors Real Estate Trust	531110A	283,190,016	24,087,000

Industry Group/Company	Industry Code	2015 Sales	2015 Profits
Mid-America Apartment Communities Inc	531110A	1,042,779,008	332,287,008
Post Properties Inc	531110A	384,006,016	80,623,000
Sawyer Realty Holdings LLC	531110A		
UDR Inc	531110A	894,638,016	340,383,008
Residential Real Estate Investment and Operations, Including Apartment Properties and Related REITs			
American Campus Communities Inc	531110	753,380,992	115,991,000
American Realty Investors Inc	531110	104,188,000	-1,960,000
AMLI Residential Properties Trust	531110		
Apartment Investment and Management Co	531110	981,310,016	248,710,000
Camden Property Trust	531110	913,404,032	249,315,008
Hometown America LLC	531110		
Lightstone Group LLC (The)	531110		
Related Group (The)	531110		
Sun Communities Inc	531110	674,731,008	155,446,000
Transcontinental Realty Investors Inc	531110	102,220,000	-7,636,000
Welltower Inc	531110	3,859,825,920	883,750,016
Residential Remodelers			
Anthony & Sylvan Pools	236118		
True Home Value Inc	236118		
Retirement Communities and Assisted Living Facilities for the Elderly			
Atria Senior Living Group	623310		
Brookdale Senior Living Inc	623310	4,960,608,256	-457,476,992
Capital Senior Living Corp	623310	412,176,992	-14,284,000
Erickson Living	623310	775,000,000	
Sunrise Senior Living	623310	1,600,000,000	
Secondary Market Financing			
Fannie Mae (Federal National Mortgage Association)	522294	22,325,999,616	10,954,000,384
Federal Agricultural Mortgage Corp	522294	145,948,000	68,700,000
Freddie Mac (Federal Home Loan Mortgage Corp, FHLMC)	522294	11,347,000,320	6,376,000,000
Self Storage and Miniwarehouse Real Estate, Including Operations, Investment and REITs			
Public Storage Inc	531130	2,381,696,000	1,311,244,032
Sovran Self Storage Inc	531130	366,601,984	112,524,000
Ship and Boat Building (Shipyards) and Repairing			
Hanjin Heavy Industries and Construction Co Ltd	336600		
Structural Steel and Precast Concrete Contractors			
Schuff International Inc	238120		
Title Insurance Underwriters (Direct Carriers)			
Fidelity National Financial Inc	524127	9,132,000,256	527,000,000
First American Financial Corporation	524127	5,175,455,744	288,086,016
Investors Title Company	524127	127,200,072	12,533,905
Stewart Information Services Corp	524127	2,033,885,056	-6,204,000
Truck, Utility Trailer and RV (Recreational Vehicle) Rental and Leasing			
AMERCO (U-Haul)	532120	3,074,531,072	356,740,992

Industry Group/Company	Industry Code	2015 Sales	2015 Profits
Utilities Construction, including Pipelines, Water & Sewer Systems, Telecommunications Lines and Systems and Electric Power Lines and Systems			
Matrix Service Company	237100	1,343,134,976	17,157,000

ALPHABETICAL INDEX

Dalian Wanda Group Co Ltd
Days Inn Worldwide Inc
DCT Industrial Trust Inc
DDR Corp
Desarrolladora Homex SAB de CV
Diamond Resorts Holdings LLC
Diversicare Healthcare Services Inc
Dividend Capital Group LLC
DLF Limited
Dominion Homes Inc
Doosan Heavy Industry & Construction Co
Doyle Collection (The)
DR Horton Inc
Duke Realty Corp
Dynex Capital Inc
EastGroup Properties Inc
E-LOAN Inc
Emaar Properties PJSC
EMCOR Group Inc
Empresas ICA SAB de CV
ENGlobal Corp
EPR Properties
Equity Commonwealth REIT
Equity Lifestyle Properties Inc
Equity Office Management LLC
Equity One Inc
Equity Residential
Erickson Living
Essar Group Ltd
Essex Property Trust Inc
Evergrande Real Estate Group
Extended Stay America Inc
Extendicare Inc
Fairmont Homes Inc
Fairmont Raffles Hotels International Inc
Fannie Mae (Federal National Mortgage Association)
Far East Organization
Federal Agricultural Mortgage Corp
Federal Realty Investment Trust
FelCor Lodging Trust Inc
Ferguson Enterprises Inc
Ferrovial SA
Fibra Uno Administracion SA de CV
Fidelity National Financial Inc
First American Financial Corporation
First Industrial Realty Trust Inc
FirstService Corporation
Five Star Products Inc
Fluor Corp
Fomento de Construcciones y Contratas SA (FCC)
Forest Realty Trust Inc
Four Seasons Hotels Inc
Freddie Mac (Federal Home Loan Mortgage Corp, FHLMC)
Gables Residential Trust
Gafisa SA
GE Capital
General Growth Properties Inc

Gensler
Gilbane Inc
Glenborough LLC
Goettsch Partners Inc
Golden Tulip Hospitality Group
Granite Construction Inc
Groupe du Louvre
GS Engineering & Construction Corp
Guocoleisure Ltd
Hang Lung Group Ltd
Hang Lung Properties Limited
Hanjin Heavy Industries and Construction Co Ltd
HCP Inc
HDR Inc
Healthcare Realty Trust Inc
HeidelbergCement AG
HeidelbergCement North America
Henderson Land Development Co Ltd
Highwoods Properties Inc
Hilton Worldwide Inc
Hines Interests LP
HOCHTIEF AG
Home Depot Inc
Home Properties Inc
Homeinns Hotel Group
HomeServices of America Inc
Hometown America LLC
Hongkong and Shanghai Hotels Ltd
Hongkong Land Holdings Ltd
Hospitality Properties Trust
Host Hotels & Resorts LP
Hotel Properties Ltd
Hovnanian Enterprises Inc
Howard Johnson International Inc
Hutchison Whampoa Properties Ltd
Hyatt Hotels Corporation
Hypo Real Estate Holding AG (Hypo Bank)
Hysan Development Co Ltd
Hyundai Elevator Co Ltd
Hyundai Engineering & Construction Company Ltd
IDI Gazeley
Impac Mortgage Holdings Inc
Indian Hotels Company Limited (The)
Inland Real Estate Corporation
Innkeepers USA Trust
Integrated Electrical Services Inc
InterContinental Hotels Group plc
Interstate Hotels & Resorts Inc
InTown Suites Management Inc
InvenTrust Properties Corp
Investors Real Estate Trust
Investors Title Company
iStar Inc
Jacobs Engineering Group Inc
Jameson Inn Inc
Janus Hotels and Resorts Inc
Jerde Partnership Inc (The)
John Laing plc

John Q Hammons Hotels & Resorts LLC
Jones Lang LaSalle Inc
KB Home
KBR Inc
Kerzner International Holdings Limited
Kiewit Corp
Kilroy Realty Corporation
Kimco Realty Corp
Kimpton Hotel & Restaurant Group LLC
Kohler Company
Kohn Pedersen Fox Associates (KPF)
Kumho Industrial Co Ltd
La Quinta Holdings Inc
LafargeHolcim Ltd
Las Vegas Sands Corp (The Venetian)
LaSalle Hotel Properties
Layne Christensen Company
LB Foster Company
LBA Realty LLC
Lend Lease Corporation Limited
LendingTree LLC
Lennar Corporation
Lexington Realty Trust
Liberty Property Trust
Lightstone Group LLC (The)
Lillibridge Healthcare Real Estate Trust
Link Real Estate Investment Trust (The)
Loews Hotels Holding Corporation
Louis Berger Group Inc (The)
Lowe's Companies Inc
M.C. Dean Inc
M.D.C. Holdings Inc
M/I Homes Inc
Macerich Company (The)
Mack-Cali Realty Corp
Mandarin Oriental International Ltd
ManorCare Inc (HCR ManorCare)
Mapletree Investments Pte Ltd
Marcus Corporation (The)
Marriott International Inc
Matrix Service Company
Maui Land & Pineapple Company Inc
McCarthy Building Companies Inc
McDermott International Inc
McKinstry
Meadow Valley Corporation
Melia Hotels International SA
Meritage Homes Corp
Meritus Hotels & Resorts Inc
MFA Financial Inc
MGM Growth Properties LLC
MGM Resorts International
Mid-America Apartment Communities Inc
Millennium & Copthorne Hotels plc
MMA Capital Management LLC
MMR Group Inc
Morgans Hotel Group Co
Move Inc (Realtor.com)

MRV Engenharia E Participacoes SA
MulvannyG2 Architecture
MWH Global Inc
NAI Global Inc
National Health Investors Inc
NBBJ
NCI Building Systems Inc
New World Development Company Limited
Newcastle Investment Corp
Newhall Land & Farming Company
NH Hotel Group SA
Nomura Real Estate Master Fund Inc
NRT LLC
NVR Inc
Oakwood Worldwide
Oberoi Group (EIH Ltd)
Odebrecht SA
Oldcastle Inc
Omega Healthcare Investors Inc
Opendoor
Orient Overseas (International) Ltd
Orion Marine Group Inc
Pacific Coast Building Products Inc
Palm Harbor Homes Inc
Paramount Group Inc
Parkway Properties Inc
Parsons Corp
PCL Construction Group Inc
PDG Realty SA Empreendimentos e Participacoes
Pearlmark Real Estate Partners
Pennsylvania REIT
PennyMac Mortgage Investment Trust
Perkins Eastman
Populous Holdings Inc
Post Properties Inc
ProLogis Inc
PS Business Parks Inc
Public Storage Inc
PulteGroup Inc
RailWorks Corp
Ramada Worldwide Inc
RE/MAX Holdings Inc
Realogy Holdings Corp
RealPage Inc
Realty Income Corp
Rebel Design+Group
Red Lion Hotels Corporation
Red Rock Resorts Inc
Redfin
Redwood Trust Inc
Regency Centers Corp
Reis Inc
Related Group (The)
ResortQuest International Inc
Rezidor Hotel Group AB
Rio Properties Inc
Ritz-Carlton Hotel Company LLC (The)
Riverview Realty Partners

Robert A.M. Stern Architects LLP
Rosendin Electric
Rosewood Hotels & Resorts LLC
Rossi Residencial SA
Rouse Properties Inc
Royal Group Inc
RXR Realty
Ryman Hospitality Properties Inc
Saha Pathana Inter-Holding PCL
Salini Impregilo SpA
Samsung C&T Corporation
Samwhan Corporation
Sawyer Realty Holdings LLC
Scandic Hotels AB
Schuff International Inc
Sekisui House Ltd
Sembcorp Industries Ltd
Senior Housing Properties Trust
Shangri-La Asia Ltd
Shimao Property Holdings Ltd
Shun Tak Holdings Limited
Siemens Building Technologies
Simon Property Group Inc
Sino Land Company Limited
Skanska USA Inc
Skidmore Owings & Merrill LLP
Skyline Corporation
SL Green Realty Corp
Smallwood Reynolds Stewart Stewart & Associates Inc
Societe Anonyme des Bains de Mer et du Cercle des
Etrangers a Monaco
Sonesta International Hotels Corp
Southern Energy Homes Inc
Sovran Self Storage Inc
St Joe Company (The)
Starwood Capital Group Global LLC
Starwood Hotels & Resorts Worldwide Inc
Starwood Property Trust Inc
Station Casinos LLC
Steelman Partners LLP
Sterling Construction Company Inc
Stewart Information Services Corp
Stock Building Supply Inc
Strategic Hotels & Resorts Inc
Structure Tone Inc
STUDIOS Architecture
STV Group Inc
Sun Communities Inc
Sun Hung Kai Properties Ltd
Sunburst Hospitality Corporation
Sunrise Senior Living
Sunstone Hotel Investors Inc
Super 8 Worldwide Inc
Swire Pacific Ltd
Tanger Factory Outlet Centers Inc
Taubman Centers Inc
TDIndustries
TMI Hospitality Inc

Toll Brothers Inc
Trammell Crow Company
Transcontinental Realty Investors Inc
Transwestern Commercial Services
Trevi-Finanziaria Industriale SpA (Trevi Group)
Triple Five Group
TRT Holdings Inc
True Home Value Inc
Trump Entertainment Resorts Inc
Turner Corp (The)
Tutor Perini Corporation
UDR Inc
Unibail-Rodamco SE
Unitech Limited
Ventas Inc
VINCI SA
VOA Associates Inc
Vornado Realty Trust
Walsh Group (The)
Washington Real Estate Investment Trust
Watson Land Company
WCI Communities Inc
Weingarten Realty Investors
Wells Real Estate Funds Inc
Welltower Inc
Westfield Corporation
Westgate Resorts
WeWork Companies Inc
Wharf (Holdings) Limited The
Wheelock and Company Limited
Whiting-Turner Contracting Co (The)
William Lyon Homes Inc
Wolseley PLC
WorldMark by Wyndham Inc
WP Carey & Co LLC
WP GLIMCHER Inc
Wyndham Vacation Ownership
Wyndham Worldwide Corporation
Wynn Resorts Limited
Xanterra Parks & Resorts Inc
YIT Corporation
Zillow Inc
ZipRealty Inc

INDEX OF U.S. HEADQUARTERS LOCATION BY STATE

To help you locate members of the firms geographically, the city and state of the headquarters of each company are in the following index.

ALABAMA
Southern Energy Homes Inc; Addison

ARIZONA
AV Homes Inc; Scottsdale
Best Western International Inc; Phoenix
Meadow Valley Corporation; Phoenix
Meritage Homes Corp; Scottsdale
Schuff International Inc; Phoenix

CALIFORNIA
ACCO Engineered Systems Inc; Glendale
AECOM Technology Corporation; Los Angeles
Alexandria Real Estate Equities Inc; Pasadena
Bank of America Home Loans; Calabasas
Bechtel Group Inc; San Francisco
CalAtlantic Group Inc; Irvine
Capital Pacific Real Estate; Newport Beach
CBRE Group Inc; Los Angeles
Colony Capital Inc; Santa Monica
Commune Hotels & Resorts; San Francisco
Dahlin Group Architecture Planning; Pleasanton
Essex Property Trust Inc; San Mateo
First American Financial Corporation; Santa Ana
Gensler; San Francisco
Glenborough LLC; San Mateo
Granite Construction Inc; Watsonville
HCP Inc; Irvine
Impac Mortgage Holdings Inc; Irvine
Jacobs Engineering Group Inc; Pasadena
Jerde Partnership Inc (The); Venice
KB Home; Los Angeles
Kilroy Realty Corporation; Los Angeles
Kimpton Hotel & Restaurant Group LLC; San Francisco
Macerich Company (The); Santa Monica
Move Inc (Realtor.com); Campbell
Newhall Land & Farming Company; Valencia
Oakwood Worldwide; Los Angeles
Opendoor; San Francisco
Pacific Coast Building Products Inc; Rancho Cordova
Parsons Corp; Pasadena
PennyMac Mortgage Investment Trust; Moorpark
ProLogis Inc; San Francisco
PS Business Parks Inc; Glendale
Public Storage Inc; Glendale
Realty Income Corp; San Diego
Rebel Design+Group; Marina Del Rey
Redwood Trust Inc; Mill Valley
Rosendin Electric; San Jose
Sunstone Hotel Investors Inc; Aliso Viejo

Tutor Perini Corporation; Sylmar
Watson Land Company; Carson
Westfield Corporation; Century City
William Lyon Homes Inc; Newport Beach
ZipRealty Inc; Emeryville

COLORADO
Apartment Investment and Management Co; Denver
CH2M HILL Companies Ltd; Englewood
DCT Industrial Trust Inc; Denver
Dividend Capital Group LLC; Denver
M.D.C. Holdings Inc; Denver
MWH Global Inc; Broomfield
RE/MAX Holdings Inc; Denver
UDR Inc; Highlands Ranch
Xanterra Parks & Resorts Inc; Greenwood Village

CONNECTICUT
EMCOR Group Inc; Norwalk
Five Star Products Inc; Fairfield
GE Capital; Norwalk
Starwood Capital Group Global LLC; Greenwich
Starwood Property Trust Inc; Greenwich

DISTRICT OF COLUMBIA
Fannie Mae (Federal National Mortgage Association); Washington
Federal Agricultural Mortgage Corp; Washington
STUDIOS Architecture; Washington

FLORIDA
ARMOUR Residential REIT Inc; Vero Beach
Bluegreen Corporation; Boca Raton
CNL Lifestyle Properties Inc; Orlando
Fidelity National Financial Inc; Jacksonville
Innkeepers USA Trust; Palm Beach
Janus Hotels and Resorts Inc; Boca Raton
Lennar Corporation; Miami
Parkway Properties Inc; Orlando
Regency Centers Corp; Jacksonville
Related Group (The); Miami
ResortQuest International Inc; Fort Walton Beach
St Joe Company (The); WaterSound
WCI Communities Inc; Bonita Springs
Westgate Resorts; Orlando
WorldMark by Wyndham Inc; Orlando
Wyndham Vacation Ownership; Orlando

GEORGIA
Beazer Homes Usa Inc; Atlanta
Brand Energy & Infrastructure Services Inc; Kennesaw
Cousins Properties Inc; Atlanta
Gables Residential Trust; Atlanta
Home Depot Inc; Atlanta
IDI Gazeley; Atlanta
InTown Suites Management Inc; Atlanta
Jameson Inn Inc; Smyrna

Oldcastle Inc; Atlanta
Post Properties Inc; Atlanta
PulteGroup Inc; Atlanta
Smallwood Reynolds Stewart Stewart & Associates Inc;
Atlanta
Wells Real Estate Funds Inc; Peachtree Corners

HAWAII
Maui Land & Pineapple Company Inc; Maui

IDAHO
Boise Cascade Corp; Boise
Building Materials Holding Corp (BMC); Raleigh

ILLINOIS
Adrian Smith + Gordon Gill Architecture (AS+GG);
Chicago
AMLI Residential Properties Trust; Chicago
Equity Lifestyle Properties Inc; Chicago
Equity Office Management LLC; Chicago
Equity Residential; Chicago
First Industrial Realty Trust Inc; Chicago
General Growth Properties Inc; Chicago
Goettsch Partners Inc; Chicago
Hometown America LLC; Chicago
Hyatt Hotels Corporation; Chicago
Inland Real Estate Corporation; Oak Brook
InvenTrust Properties Corp; Oak Brook
Jones Lang LaSalle Inc; Chicago
Lillibridge Healthcare Real Estate Trust; Chicago
Pearlmark Real Estate Partners; Chicago
Riverview Realty Partners; Chicago
Skidmore Owings & Merrill LLP; Chicago
Strategic Hotels & Resorts Inc; Chicago
Ventas Inc; Chicago
VOA Associates Inc; Chicago
Walsh Group (The); Chicago

INDIANA
Duke Realty Corp; Indianapolis
Fairmont Homes Inc; Nappanee
Simon Property Group Inc; Indianapolis
Skyline Corporation; Elkhart

KANSAS
Black & Veatch Holding Company; Overland Park

KENTUCKY
Atria Senior Living Group; Louisville
True Home Value Inc; Louisville

LOUISIANA
MMR Group Inc; Baton Rouge

MARYLAND
CAPREIT Inc; Rockville
Choice Hotels International Inc; Rockville

Clark Construction Group LLC; Bethesda
Corporate Office Properties Trust; Columbia
Erickson Living; Catonsville
Federal Realty Investment Trust; Rockville
Host Hotels & Resorts LP; Bethesda
LaSalle Hotel Properties; Bethesda
Marriott International Inc; Bethesda
MMA Capital Management LLC; Baltimore
Omega Healthcare Investors Inc; Hunt Valley
Ritz-Carlton Hotel Company LLC (The); Chevy Chase
Sunburst Hospitality Corporation; Silver Spring
Washington Real Estate Investment Trust; Washington
D.C.
Whiting-Turner Contracting Co (The); Baltimore

MASSACHUSETTS
Boston Properties Inc; Boston
Cambridge Seven Associates Inc; Cambridge
CDM Smith Inc; Boston
Equity Commonwealth REIT; Newton
Hospitality Properties Trust; Newton
Sawyer Realty Holdings LLC; Newton
Senior Housing Properties Trust; Newton
Sonesta International Hotels Corp; Newton

MICHIGAN
Champion Enterprises Holdings LLC; Troy
Sun Communities Inc; Southfield
Taubman Centers Inc; Bloomfield Hills

MINNESOTA
APi Group Inc; New Broghton
Bluestem Group Inc; Eden Praire
Carlson Companies Inc; Minnetonka
Carlson Rezidor Hotel Group; Minnetonka
HomeServices of America Inc; Minneapolis

MISSISSIPPI
Anderson-Tully Lumber Company; Vicksburg
EastGroup Properties Inc; Jackson

MISSOURI
Aegion Corp; Chesterfield
Burns & McDonnell; Kansas City
Butler Manufacturing Co; Kansas City
CitiMortgage Inc; O'Fallon
EPR Properties; Kansas City
John Q Hammons Hotels & Resorts LLC; Springfield
McCarthy Building Companies Inc; St. Louis
Populous Holdings Inc; Kansas City

NEBRASKA
Condor Hospitality Trust Inc; Norfolk
HDR Inc; Omaha
Kiewit Corp; Omaha

NEVADA
AMERCO (U-Haul); Reno
American Casino & Entertainment Properties LLC; Las Vegas
Ameristar Casinos Inc; Las Vegas
Boyd Gaming Corp; Las Vegas
Caesars Entertainment Corporation; Las Vegas
Diamond Resorts Holdings LLC; Las Vegas
Las Vegas Sands Corp (The Venetian); Las Vegas
MGM Growth Properties LLC; Las Vegas
MGM Resorts International; Las Vegas
Red Rock Resorts Inc; Las Vegas
Rio Properties Inc; Las Vegas
Station Casinos LLC; Las Vegas
Steelman Partners LLP; Las Vegas
Wynn Resorts Limited; Las Vegas

NEW JERSEY
ARC Group of Companies (The); Clifton
Century 21 Real Estate LLC; Madison
Coldwell Banker Real Estate LLC; Parsippany
Days Inn Worldwide Inc; Parsippany
Hovnanian Enterprises Inc; Red Bank
Howard Johnson International Inc; Parsippany
Louis Berger Group Inc (The); Morristown
Mack-Cali Realty Corp; Edison
NRT LLC; Madison
Ramada Worldwide Inc; Parsippany
Realogy Holdings Corp; Madison
Super 8 Worldwide Inc; Parsippany
Wyndham Worldwide Corporation; Parsippany

NEW YORK
ABM Industries Inc; New York
Annaly Capital Management Inc; New York
Brixmor Property Group Inc; New York
Chimera Investment Corporation; New York
Christies International Real Estate; New York
Community Development Trust; New York
Cushman & Wakefield Inc; New York
E-LOAN Inc; New York
Equity One Inc; New York
Home Properties Inc; Rochester
iStar Inc; New York
Kimco Realty Corp; New Hyde Park
Kohn Pedersen Fox Associates (KPF); New York
LBA Realty LLC; New York
Lexington Realty Trust; New York
Lightstone Group LLC (The); New York
Loews Hotels Holding Corporation; New York
MFA Financial Inc; New York
Morgans Hotel Group Co; New York
NAI Global Inc; New York
Newcastle Investment Corp; New York
Paramount Group Inc; New York
Perkins Eastman; New York
RailWorks Corp; New York
Reis Inc; New York

Robert A.M. Stern Architects LLP; New York
Rouse Properties Inc; New York
RXR Realty; Uniondale
Skanska USA Inc; New York
SL Green Realty Corp; New York
Sovran Self Storage Inc; Williamsville
Starwood Hotels & Resorts Worldwide Inc; White Plains
Structure Tone Inc; New York
Trump Entertainment Resorts Inc; Atlantic City
Turner Corp (The); New York
Vornado Realty Trust; New York
WeWork Companies Inc; New York
WP Carey & Co LLC; New York

NORTH CAROLINA
Extended Stay America Inc; Charlotte
Highwoods Properties Inc; Raleigh
Investors Title Company; Chapel Hill
LendingTree LLC; Charlotte
Lowe's Companies Inc; Mooresville
Stock Building Supply Inc; Raleigh
Tanger Factory Outlet Centers Inc; Greensboro

NORTH DAKOTA
Investors Real Estate Trust; Minot
TMI Hospitality Inc; Fargo

OHIO
DDR Corp; Beachwood
Dominion Homes Inc; Dublin
Forest Realty Trust Inc; Cleveland
M/I Homes Inc; Columbus
ManorCare Inc (HCR ManorCare); Toledo
Welltower Inc; Toledo
WP GLIMCHER Inc; Columbus

OKLAHOMA
Matrix Service Company; Tulsa

PENNSYLVANIA
84 Lumber Company; Eighty Four
Anthony & Sylvan Pools; Doylestown
Brandywine Realty Trust; Radnor
LB Foster Company; Pittsburgh
Liberty Property Trust; Malvern
Pennsylvania REIT; Philadelphia
STV Group Inc; Douglassville
Toll Brothers Inc; Horsham

RHODE ISLAND
Gilbane Inc; Providence

TENNESSEE
Brookdale Senior Living Inc; Brentwood
CBL & Associates Properties Inc; Chattanooga
Clayton Homes Inc; Maryville
Diversicare Healthcare Services Inc; Brentwood

Healthcare Realty Trust Inc; Nashville
Mid-America Apartment Communities Inc; Memphis
National Health Investors Inc; Murfreesboro
Ryman Hospitality Properties Inc; Nashville

TEXAS
Accor North America; Carrollton
American Campus Communities Inc; Bee Cave
American Realty Investors Inc; Dallas
Camden Property Trust; Houston
Capital Senior Living Corp; Dallas
Capstead Mortgage Corporation; Dallas
CEMEX Inc; Houston
Centex Corp; Dallas
Comfort Systems USA Inc; Houston
Crescent Real Estate Equities LP; Fort Worth
DR Horton Inc; Fort Worth
ENGlobal Corp; Houston
FelCor Lodging Trust Inc; Irving
Fluor Corp; Irving
HeidelbergCement North America; Houston
Hines Interests LP; Houston
Integrated Electrical Services Inc; Houston
KBR Inc; Houston
La Quinta Holdings Inc; Irving
Layne Christensen Company; The Woodlands
McDermott International Inc; Houston
NCI Building Systems Inc; Houston
Orion Marine Group Inc; Houston
Palm Harbor Homes Inc; Addison
RealPage Inc; Carrollton
Rosewood Hotels & Resorts LLC; Dallas
Sterling Construction Company Inc; The Woodlands
Stewart Information Services Corp; Houston
TDIndustries; Dallas
Trammell Crow Company; Dallas
Transcontinental Realty Investors Inc; Dallas
Transwestern Commercial Services; Houston
TRT Holdings Inc; Dallas
Weingarten Realty Investors; Houston

VIRGINIA
AvalonBay Communities Inc; Arlington
Barcelo Crestline Corporation; Fairfax
Capital Automotive REIT; McLean
CORE Network (The); Virginia Beach
Dynex Capital Inc; Glen Allen
Ferguson Enterprises Inc; Newport News
Freddie Mac (Federal Home Loan Mortgage Corp, FHLMC); McLean
Hilton Worldwide Inc; McLean
Interstate Hotels & Resorts Inc; Arlington
M.C. Dean Inc; Dulles
NVR Inc; Reston
Sunrise Senior Living; McLean

WASHINGTON
Callison LLC; Seattle
CoStar Group Inc; Northwest
McKinstry; Seattle
MulvannyG2 Architecture; Seattle
NBBJ; Seattle
Red Lion Hotels Corporation; Spokane
Redfin; Seattle
Zillow Inc; Seattle

WISCONSIN
Kohler Company; Kohler
Marcus Corporation (The); Milwaukee

INDEX OF NON-U.S. HEADQUARTERS LOCATION BY COUNTRY

AUSTRALIA
Lend Lease Corporation Limited; Millers Point

BAHAMAS
Kerzner International Holdings Limited; Paradise Island

BELGIUM
Rezidor Hotel Group AB; Brussels

BERMUDA
Belmond Ltd; Hamilton

BRAZIL
Cyrela Brazil Realty SA; Sao Paulo
Gafisa SA; Sao Paulo
MRV Engenharia E Participacoes SA; Belo Horizonte
Odebrecht SA; Salvador
PDG Realty SA Empreendimentos e Participacoes; Botafogo
Rossi Residencial SA; Sao Paulo

CANADA
Aecon Group Inc; Toronto
Boardwalk Real Estate Investment Trust; Calgary
Brookfield Asset Management Inc; Toronto
Colliers International Group Inc; Toronto
Extendicare Inc; Markham
Fairmont Raffles Hotels International Inc; Toronto
FirstService Corporation; Toronto
Four Seasons Hotels Inc; Toronto
PCL Construction Group Inc; Edmonton
Royal Group Inc; Woodbridge
Triple Five Group; Edmonton

CHINA
China Communications Construction Company Ltd; Beijing
China Lodging Group Ltd; Shanghai
China Railway Construction Corporation Limited (CRCC); Beijing
China State Construction Engineering Corp; Wanchai
Dalian Wanda Group Co Ltd; Beijing
Evergrande Real Estate Group; Guangzhou City
Homeinns Hotel Group; Shanghai

FINLAND
YIT Corporation; Helsinki

FRANCE
Accor SA; Paris
Bouygues SA; Paris
Club Mediterranee SA (Club Med); Paris
Golden Tulip Hospitality Group; Paris
Groupe du Louvre; Paris
Unibail-Rodamco SE; Paris
VINCI SA; Cedex

GERMANY
Bilfinger SE; Mannheim
HeidelbergCement AG; Heidelberg
HOCHTIEF AG; Essen
Hypo Real Estate Holding AG (Hypo Bank); Unterschleissheim

HONG KONG
China Overseas Land & Investment Limited; Hong Kong
Chinese Estates Holdings Ltd; Wanchai
Hang Lung Group Ltd; Hong Kong
Hang Lung Properties Limited; Hong Kong
Henderson Land Development Co Ltd; Hong Kong
Hongkong and Shanghai Hotels Ltd; Central
Hongkong Land Holdings Ltd; Hamilton
Hutchison Whampoa Properties Ltd; Hong Kong
Hysan Development Co Ltd; Hong Kong
Link Real Estate Investment Trust (The); Kwun Tong
Mandarin Oriental International Ltd; Hong Kong
New World Development Company Limited; Hong Kong
Orient Overseas (International) Ltd; Wanchai
Shangri-La Asia Ltd; Quarry Bay
Shimao Property Holdings Ltd; Hong Kong
Shun Tak Holdings Limited; Hong Kong
Sino Land Company Limited; Kowloon
Sun Hung Kai Properties Ltd; Hong Kong
Swire Pacific Ltd; Hong Kong
Wharf (Holdings) Limited The; Kowloon
Wheelock and Company Limited; Hong Kong

INDIA
DLF Limited; New Delhi
Essar Group Ltd; Mumbai
Indian Hotels Company Limited (The); Mumbai
Oberoi Group (EIH Ltd); Delhi
Unitech Limited; Gurgaon

ITALY
Salini Impregilo SpA; Milan
Trevi-Finanziaria Industriale SpA (Trevi Group); Cesena

JAPAN
Chiyoda Corporation; Yokohama
Nomura Real Estate Master Fund Inc; Tokyo
Sekisui House Ltd; Osaka

KOREA
Daelim Industrial Co Ltd; Seoul
Doosan Heavy Industry & Construction Co; Changwon
GS Engineering & Construction Corp; Seoul
Hanjin Heavy Industries and Construction Co Ltd; Busan
Hyundai Elevator Co Ltd; Gyeonggi-do
Hyundai Engineering & Construction Company Ltd; Seoul

Kumho Industrial Co Ltd; Jeollanam-do
Samsung C&T Corporation; Seoul
Samwhan Corporation; Seoul

MEXICO
Cemex SAB de CV; San Pedro Garza Garcia
Desarrolladora Homex SAB de CV; Culiacán , Sinaloa
Empresas ICA SAB de CV; Mexico City
Fibra Uno Administracion SA de CV; Zedec Santa Fe

MONACO
Societe Anonyme des Bains de Mer et du Cercle des
Etrangers a Monaco; Monte Carlo

SINGAPORE
Amanresorts International Pte Ltd (Aman Resorts);
Singapore
Banyan Tree Holdings Limited; Singapore
Far East Organization; Singapore
Guocoleisure Ltd; Singapore
Hotel Properties Ltd; Singapore
Mapletree Investments Pte Ltd; Singapore
Meritus Hotels & Resorts Inc; Singapore
Sembcorp Industries Ltd; Singapore

SPAIN
Abertis Infraestructuras SA; Barcelona
Acciona SA; Alcobendas
ACS Actividades de Construccion y Servicios SA; Madrid
Ferrovial SA; Madrid
Fomento de Construcciones y Contratas SA (FCC);
Madrid
Melia Hotels International SA; Palma de Mallorca
NH Hotel Group SA; Madrid

SWEDEN
Scandic Hotels AB; Stockholm

SWITZERLAND
LafargeHolcim Ltd; Jona
Siemens Building Technologies; Zug
Wolseley PLC; Zug

THAILAND
Saha Pathana Inter-Holding PCL; Bangkok

THE NETHERLANDS
Arcadis NV; Amsterdam
Chicago Bridge & Iron Company NV (CB&I); The Hague

UNITED ARAB EMIRATES
Emaar Properties PJSC; Dubai

UNITED KINGDOM
Aedas; London
Amec Foster Wheeler plc; London
Amey plc; Oxford
Balfour Beatty plc; London
Crest Nicholson plc; Chertsey
CRH plc; Dublin
Doyle Collection (The); Dublin
InterContinental Hotels Group plc; Denham
John Laing plc; London
Millennium & Copthorne Hotels plc; London

Individual Profiles
On Each Of
THE REAL ESTATE 450

84 Lumber Company

NAIC Code: 444190

www.84lumber.com

TYPES OF BUSINESS:

Hardware Stores
Building Materials
Construction Financing
Builder's Insurance
Travel Agency

BRANDS/DIVISIONS/AFFILIATES:

84 Lumber Travel
84 National Sales
Affordable Collection
Oaks Collection
Maggie's Management LLC

CONTACTS: *Note: Officers with more than one job title may be intentionally listed here more than once.*

Joe Hardy, CEO
Maggie Hardy Magerko, Pres.

GROWTH PLANS/SPECIAL FEATURES:

84 Lumber Company is a supplier of building materials, equipment and expertise to professional homebuilders, commercial contractors, remodelers and individuals. The company operates over 250 locations in 30 states including door shops, installation centers, engineered wood product shops and component manufacturing facilities, which offer lumber, plywood, insulation, trim, molding, flooring, siding, drywall, trusses, roofing, skylights, engineered lumber, hardware, doors and windows. 84 Lumber also offers a variety of services, such as turn-key installation and onsite management services. The company's manufacturing division builds metal plate connected roof and floor trusses and wall panels. 84 Lumber Travel is a full-service accredited travel agency offering no-fee service to professional contractors and other 84 Lumber customers. Through a team of professionals, the firm's 84 National Sales provides national homebuilders, who construct multi-family and single-family units for commercial sales, with geographic information and quotes. The firm offers builder's risk, general liability, workers compensation, commercial auto and personal insurance through Maggie's Management, LLC. 84 Lumber's installation services include framing, roofing, insulation, windows, doors, trip and siding. The firm sells entire home packages such as the Affordable Collection, which contains plans for easy-to-build homes from 500 square feet up to 2,851 square feet; and the Oaks Collection, which includes more elaborate designs in sizes up to 4,963 square feet. In addition, 84 Lumber has an affiliation with Nelson Design Group, an online provider of over 900 house plans.

The company offers its employees life, disability, medical, mental and dental insurance; a 401(k) plan; a profit sharing plan; and training and development programs.

FINANCIAL DATA: *Note: Data for latest year may not have been available at press time.*

In U.S. $	2015	2014	2013	2012	2011	2010
Revenue	2,500,000,000	2,300,000,000	2,100,000,000	1,875,000,000	1,400,000,000	1,350,000,000
R&D Expense						
Operating Income						
Operating Margin %						
SGA Expense						
Net Income		110,000,000	101,000,000	90,000,000		
Operating Cash Flow						
Capital Expenditure						
EBITDA						
Return on Assets %						
Return on Equity %						
Debt to Equity						

CONTACT INFORMATION:

Phone: 724-228-8820   Fax: 724-228-8058
Toll-Free:
Address: 1019 Rte. 519, Eighty Four, PA 15330 United States

STOCK TICKER/OTHER:

Stock Ticker: Private   Exchange:
Employees: 4,200   Fiscal Year Ends: 12/31
Parent Company:

SALARIES/BONUSES:

Top Exec. Salary: $ Bonus: $
Second Exec. Salary: $ Bonus: $

OTHER THOUGHTS:

Estimated Female Officers or Directors: 1
Hot Spot for Advancement for Women/Minorities:

Abertis Infraestructuras SA

www.abertis.com

NAIC Code: 561210

TYPES OF BUSINESS:

Transport & Communications
Airports

BRANDS/DIVISIONS/AFFILIATES:

Sanef
Abertis Telecom
Retevision
Tradia
Grupo Hispastat

CONTACTS: *Note: Officers with more than one job title may be intentionally listed here more than once.*

Francisco Reynes Massanet, CEO
Salvador Alemany Mas, Pres.
Francisco Jose Aljaro Navarro, CFO
Joan Rafel Herrero, Human Resources
Jose Maria Coronas Guinart, Sec.
Salvador Alemany Mas, Chmn.

GROWTH PLANS/SPECIAL FEATURES:

Abertis Infraestructuras SA is a transport and communications infrastructure management company based in Barcelona and is a leading operator of motorways and car parks. The company is active in 12 countries in Europe and the Americas. Two-thirds of the firm's revenue is generated outside Spain, especially in France, Brazil and Chile. Abertis operates through two business divisions: toll roads, deriving 87% of revenue, and telecommunications, 13%. The toll roads division, which includes its French operating subsidiary Sanef, operates 4,536 miles of toll roads in Spain, other parts of Europe, Chile, Brazil and Puerto Rico. Abertis Telecom, the company's telecommunications division, consists of two companies, Tradia and Retevision. Tradia's activities include provision of radio and TV signal broadcasting services and the renting of space for telecommunications operators. Retevision, which provides national coverage with its analog and digital network, focuses on audiovisual signal transportation and broadcasting. In addition, the company also owns 57.05% stake in Grupo Hispastat, a leading satellite company in Latin America.

FINANCIAL DATA: *Note: Data for latest year may not have been available at press time.*

In U.S. $	2015	2014	2013	2012	2011	2010
Revenue	5,539,211,000	6,360,844,000	5,746,327,000	4,826,302,000	4,443,551,000	4,519,456,000
R&D Expense						
Operating Income	-73,828,990	2,130,271,000	1,962,249,000	1,471,468,000	1,730,544,000	1,732,887,000
Operating Margin %	-1.33%	33.49%	34.14%	30.48%	38.94%	38.34%
SGA Expense						
Net Income	2,144,036,000	747,116,200	703,488,800	1,168,360,000	821,265,700	754,570,500
Operating Cash Flow	1,743,986,000	2,184,683,000	2,018,945,000	1,746,373,000	1,130,123,000	1,574,181,000
Capital Expenditure	1,014,109,000	1,273,468,000	1,030,968,000	775,721,100	300,225,800	837,763,000
EBITDA	935,828,700	3,556,070,000	3,333,933,000	2,577,474,000	2,796,579,000	3,187,603,000
Return on Assets %	7.02%	2.34%	2.15%	3.95%	2.99%	2.65%
Return on Equity %	58.58%	19.60%	16.74%	29.67%	20.32%	15.91%
Debt to Equity	4.08	4.66	4.19	4.03	4.39	3.54

CONTACT INFORMATION:

Phone: 34 932305000   Fax: 34 932305001
Toll-Free:
Address: Avinguda del Parc Logistic, 12-20, Barcelona, 08040 Spain

STOCK TICKER/OTHER:

Stock Ticker: ABRTY
Employees: 16,749
Parent Company:

Exchange: PINX
Fiscal Year Ends: 12/31

SALARIES/BONUSES:

Top Exec. Salary: $ Bonus: $
Second Exec. Salary: $ Bonus: $

OTHER THOUGHTS:

Estimated Female Officers or Directors:
Hot Spot for Advancement for Women/Minorities:

ABM Industries Inc

NAIC Code: 561720

www.abm.com

TYPES OF BUSINESS:

Janitorial Services
Parking Facilities
Maintenance Personnel
Security Services
Lighting Services
Billing & Accounting Services
Supplier Management
Energy Efficiency Technology

BRANDS/DIVISIONS/AFFILIATES:

ABM Janitorial Services
Westway Services Holdings Ltd

CONTACTS: Note: Officers with more than one job title may be intentionally listed here more than once.

Diego Scaglione, CFO
Maryellen Herringer, Chairman of the Board
Dean Chin, Chief Accounting Officer
James Mcclure, COO
Scott Salmirs, Director
Sarah McConnell, Executive VP
David Farwell, Senior VP, Divisional
Angelique Carbo, Senior VP, Divisional

GROWTH PLANS/SPECIAL FEATURES:

ABM Industries, Inc. is one of the country's largest facility services providers. ABM provides janitorial, parking, engineering, security and mechanical services to commercial, industrial, institutional and retail facilities throughout the U.S. and 20 additional countries. The company operates through a number of subsidiaries, which are grouped into four segments: janitorial, parking, facility services and building & energy solutions. Janitorial services, operated through ABM Janitorial Services, include floor cleaning and finishing, window washing, furniture polishing and carpet cleaning and dusting. Parking provides parking and transportation services for clients such as commercial office buildings, airports and other transportation centers, education institutions, health facilities, hotels, municipalities, retail centers and stadiums. Facility services provides onsite mechanical engineering and technical services and solutions for facilities and infrastructure systems for a variety of client facilities, including transportation centers, commercial infrastructure, corporate office buildings, data centers, educational institutions, high technology manufacturing facilities, museums, resorts, and shopping centers. Building & energy solutions provides heating, ventilation, air-conditioning, electrical, lighting and other general maintenance and repair services. In 2015, the firm sold its security business to Universal Protection Service, a division of Universal Services of America; and acquired Westway Services Holdings Ltd.

The firm offers employees dental, medical, vision, disability, life, AD&D and business travel accident insurance; flexible spending accounts; workers compensation insurance; 401(k); employee stock plans; employee assistance; tuition reimbursement; and credit union membership.

FINANCIAL DATA: Note: Data for latest year may not have been available at press time.

In U.S. $	2015	2014	2013	2012	2011	2010
Revenue	4,897,800,000	5,032,800,000	4,809,281,000	4,300,265,000	4,246,842,000	3,495,747,000
R&D Expense						
Operating Income	73,600,000	128,600,000	119,025,000	96,566,000	117,568,000	108,839,000
Operating Margin %	1.50%	2.55%	2.47%	2.24%	2.76%	3.11%
SGA Expense	390,000,000	363,900,000	348,274,000	327,855,000	324,762,000	241,526,000
Net Income	76,300,000	75,600,000	72,900,000	62,582,000	68,504,000	64,121,000
Operating Cash Flow	145,300,000	120,700,000	135,313,000	150,612,000	159,990,000	149,864,000
Capital Expenditure	26,500,000	37,400,000	32,593,000	28,052,000	22,124,000	23,942,000
EBITDA	139,600,000	192,400,000	185,734,000	153,545,000	174,141,000	145,027,000
Return on Assets %	3.51%	3.50%	3.65%	3.33%	3.99%	4.17%
Return on Equity %	7.72%	8.01%	8.24%	7.60%	8.92%	8.99%
Debt to Equity	0.15	0.33	0.34	0.25	0.37	0.19

CONTACT INFORMATION:

Phone: 212 297-0200   Fax: 212 297-0375
Toll-Free:
Address: 551 5th Ave., Ste. 300, New York, NY 10176 United States

STOCK TICKER/OTHER:

Stock Ticker: ABM
Employees: 118,000
Parent Company:

Exchange: NYS
Fiscal Year Ends: 10/31

SALARIES/BONUSES:

Top Exec. Salary: $688,931   Bonus: $
Second Exec. Salary: Bonus: $
$688,931

OTHER THOUGHTS:

Estimated Female Officers or Directors: 4
Hot Spot for Advancement for Women/Minorities: Y

Acciona SA

www.acciona.es

NAIC Code: 237310

TYPES OF BUSINESS:

Heavy Construction
Infrastructure Services
Road Concessions
Logistics Services
Airport Services
Passenger Ferries
Urban & Environmental Services
Alternative Energy Installation

BRANDS/DIVISIONS/AFFILIATES:

Acciona Agua
Acciona TrasMediterranea
Bestinver SA
Hijos De Antonio Barcelo SA
ACCIONA Real Estate

CONTACTS: Note: Officers with more than one job title may be intentionally listed here more than once.

Carlos Arilla de Juana, CFO
Pio Cabanillas, Chief Global Brand & Mktg. Officer
Macarena Carrion, Chief of Staff
Jorge Vega-Penichet, General Counsel
Juan Muro-Lara, Chief Corp. Dev. Officer
Joaquin Mollinedo, Chief Institutional Rel. Officer
Juan Muro-Lara, Chief Investor Rel. Officer
Alfonso Callejo, Chief Corp. Resources Officer
Pedro Martinez, Pres., Acciona Infrastructure
Luis Castilla, Pres., Acciona Agua
Rafael Mateo, CEO-Acciona Energy
Jose Manuel Entrecanales, Chmn.
Carmen Becerril, Chief Intl Officer

GROWTH PLANS/SPECIAL FEATURES:

Acciona SA develops and manages infrastructure and related projects in Spain and internationally. The company operates two main business lines: Energy and Infrastructure. The Energy business line is focused on the development of renewable energy facilities, primarily through the installation of wind farms and solar arrays. It is also active in the design and manufacture of wind turbine generators, the production of biofuels and the development of charging infrastructure for electric vehicles. The Infrastructure business line includes four divisions: water, services, industrial and construction. The water division, through Acciona Agua, is involved in the engineering, construction and management of drinking water plants, sewage treatment plants and reverse-osmosis desalination plants. The division has built more than 70 desalination plants across the globe. The services division handles the needs of a variety of customers from many sectors through a broad range of services, such as cleaning, ancillary services, technical, energy, environmental maintenance, urban management, handling, production and design, catering and security. The industrial division focuses on range of industrial processes, such as engineering, acquisition of equipment and operations and management, as well as installation of projects. The construction division builds roads, bridges, railways, dams, canals, sewer systems, hospitals, seaports and airports. Additional operations not falling into the above divisions include Bestinver SA, an asset management firm; Hijos de Antonio Barcelo SA, an international wine producer; Acciona TrasMediterranea, which transports cargo and passengers by sea; and ACCIONA Real Estate, which promotes general real estate and housing developments.

FINANCIAL DATA: Note: Data for latest year may not have been available at press time.

In U.S. $	2015	2014	2013	2012	2011	2010
Revenue	7,462,875,000	7,411,527,000	7,535,280,000	8,001,688,000	7,579,744,000	7,142,969,000
R&D Expense						
Operating Income	714,999,800	652,323,800	-2,019,749,000	736,989,800	720,442,300	600,537,200
Operating Margin %	9.58%	8.80%	-26.80%	9.21%	9.50%	8.40%
SGA Expense						
Net Income	236,452,600	210,933,900	-2,249,485,000	216,011,400	230,451,300	210,091,100
Operating Cash Flow	778,906,500	923,795,400	902,261,600	858,707,100	1,012,047,000	1,414,333,000
Capital Expenditure	235,078,300	479,171,100	482,300,600	857,875,600	1,120,104,000	1,313,067,000
EBITDA	1,514,240,000	1,276,311,000	1,417,127,000	1,633,187,000	1,832,033,000	1,564,346,000
Return on Assets %	1.29%	1.12%	-10.77%	.94%	.98%	.81%
Return on Equity %	6.03%	5.59%	-46.63%	3.58%	3.64%	2.91%
Debt to Equity	1.68	1.75	1.93	1.32	1.25	0.87

CONTACT INFORMATION:

Phone: 34 916632850   Fax: 34 916632851
Toll-Free:
Address: Ave. De Europa, 18, Parque Empresarial La Moreleja, Alcobendas, Madrid 28108 Spain

STOCK TICKER/OTHER:

Stock Ticker: ACXIF
Employees: 33,559
Parent Company:

Exchange: PINX
Fiscal Year Ends: 12/31

SALARIES/BONUSES:

Top Exec. Salary: $ Bonus: $
Second Exec. Salary: $ Bonus: $

OTHER THOUGHTS:

Estimated Female Officers or Directors: 2
Hot Spot for Advancement for Women/Minorities:

Sales, profits and employees may be estimates. Financial information, benefits and other data can change quickly and may vary from those stated here.

ACCO Engineered Systems Inc

www.accoes.com

NAIC Code: 238220

TYPES OF BUSINESS:
Mechanical Contractors
HVAC Contractors
Plumbing Contractors

BRANDS/DIVISIONS/AFFILIATES:

CONTACTS: Note: Officers with more than one job title may be intentionally listed here more than once.
John Aversano, Pres.

GROWTH PLANS/SPECIAL FEATURES:

ACCO Engineered Systems, Inc. is an employee-owned mechanical contractor that specializes in providing heating, ventilating, air conditioning, refrigeration, plumbing, process piping, industrial construction and building automation services to the new construction and existing building markets. The construction work undertaken by ACCO has included office buildings, biopharmaceutical labs, semi-conductor, medical centers and hospitals, micro-electronics, manufacturing plants, entertainment, retail, telecommunications, educational facilities, data centers, and high rise residential. Headquartered in Glendale, California, the firm has offices in San Leandro, Sacramento, Orange County, Inland Empire, San Diego, Bakersfield, Fresno, Petaluma, Redding and Vacaville, California; Boise and Twin Falls, Idaho; Las Vegas and Reno/Tahoe, Nevada; and Seattle, Washington. To aid in its work, the firm has fully automated computer-aided fabrication facilities in Los Angeles, the San Francisco Bay Area, Sacramento, and Vacaville. These facilities produce over 12 million pounds of sheet metal and pre-fabricate over 160,000 diameter inches of welded and grooved process, plumbing and HVAC piping every year.

FINANCIAL DATA: Note: Data for latest year may not have been available at press time.

In U.S. $	2015	2014	2013	2012	2011	2010
Revenue	770,000,000	765,000,000	712,000,000	634,000,000	501,000,000	408,500,000
R&D Expense						
Operating Income						
Operating Margin %						
SGA Expense						
Net Income						
Operating Cash Flow						
Capital Expenditure						
EBITDA						
Return on Assets %						
Return on Equity %						
Debt to Equity						

CONTACT INFORMATION:
Phone: 818-244-6571   Fax:
Toll-Free: 800-998-2226
Address: 6265 San Fernando Rd., Glendale, CA 91201 United States

STOCK TICKER/OTHER:
Stock Ticker: Private
Employees: 1,029
Parent Company:
Exchange:
Fiscal Year Ends:

SALARIES/BONUSES:
Top Exec. Salary: $ Bonus: $
Second Exec. Salary: $ Bonus: $

OTHER THOUGHTS:
Estimated Female Officers or Directors:
Hot Spot for Advancement for Women/Minorities:

Sales, profits and employees may be estimates. Financial information, benefits and other data can change quickly and may vary from those stated here.

Accor North America
NAIC Code: 721110

www.accorhotels.com/gb/usa/index.shtml

TYPES OF BUSINESS:
Hotels

BRANDS/DIVISIONS/AFFILIATES:
Accor SA
Sofitel
Novotel
Pullman

CONTACTS: *Note: Officers with more than one job title may be intentionally listed here more than once.*
Roland de Bonadona, CEO-Americas
Didier Bosc, CFO
Jeff Winslow, CIO
Didier Bosc, Chief Admin. Officer
Alan Rabinowitz, Exec. VP
Jim Amorosia, Pres.
Jeff Winslow, Chief Investment Officer
Robert Moore, Sr. VP-Technical Service

GROWTH PLANS/SPECIAL FEATURES:
Accor North America, a subsidiary of French hotel and human resources conglomerate Accor SA, operates 16 hotels with approximately 4,715 rooms across the U.S. and Canada. The firm's North American hotel chains include seven Sofitel hotels, a French luxury brand that incorporates local culture into its decor. The brand offers visitors first-rate accommodations with upscale restaurants, complete business facilities, fitness centers, fine art and antiques. The Novotel chain is another more upscale offering, which consists of a relaxed modern decor that makes it accessible to both business and leisure travelers. Novotel properties offer rooms with sitting/working areas, mid-scale restaurants and pools and golf course privileges, with eight locations in the U.S. and Canada. In addition, the firm operates one Pullman hotel, which offers upscale, executive lodging for business and leisure stays. Accor incorporates a green policy into all of its chains that consists of water-saving shower heads and faucet aerators, Energy Star program participation, power-reducing heating and cooling systems, recycled paper and soy ink for its directories, energy efficient fluorescent lighting and the use of green Ecolab products for laundry and cleaning.

FINANCIAL DATA: *Note: Data for latest year may not have been available at press time.*

In U.S. $	2015	2014	2013	2012	2011	2010
Revenue	635,410,802	525,348,522	302,399,520	308,574,820	304,171,460	
R&D Expense						
Operating Income						
Operating Margin %						
SGA Expense						
Net Income	255,037,440	269,380,549	153,772,184			
Operating Cash Flow						
Capital Expenditure						
EBITDA						
Return on Assets %						
Return on Equity %						
Debt to Equity						

CONTACT INFORMATION:
Phone: 972-360-9000   Fax: 972-716-6590
Toll-Free: 800-557-3435
Address: 4001 International Pkwy., Carrollton, TX 75007 United States

STOCK TICKER/OTHER:
Stock Ticker: Subsidiary
Employees: 21,563
Parent Company: Accor SA
Exchange:
Fiscal Year Ends: 12/31

SALARIES/BONUSES:
Top Exec. Salary: $ Bonus: $
Second Exec. Salary: $ Bonus: $

OTHER THOUGHTS:
Estimated Female Officers or Directors: 1
Hot Spot for Advancement for Women/Minorities:

Accor SA
NAIC Code: 721110   **www.accor.com**

TYPES OF BUSINESS:
Hotels

BRANDS/DIVISIONS/AFFILIATES:
Ibis Budget
Ibis
Formule 1
Novotel
Sofitel
Grand Mercure Apartments
MGallery
Adagio

CONTACTS: Note: Officers with more than one job title may be intentionally listed here more than once.
Sebastien Bazin, CEO
Sophie Stabile, CFO
Vivek Badrinath, Deputy CEO-Mktg. & Info. Systems
Sven Boinet, Deputy CEO-Human Resources
Sven Boinet, Deputy CEO-Legal
Christophe Alaux, CEO-Hotel Svcs. France
Roland de Bonadona, CEO-Hotel Svcs. Americas
Jean-Jacques Dessors, CEO-Hotel Svcs. Mediterranean, Middle East &Africa
Michael Issenberg, CEO-Hotel Svcs. Asia Pacific
Sebastien Bazin, Chmn.
Peter Verhoeven, CEO-Hotel Svcs. Northern, Central & Eastern Europe

GROWTH PLANS/SPECIAL FEATURES:
Accor SA is a leading hotel operator, with locations around the world. Accor has over 3,900 hotels under 17 brands in 92 countries worldwide. Ibis Budget offers budget accommodations throughout Europe with services such as wireless Internet access, snacks and an all-you-can-eat breakfast. The HotelF1 and Formule 1 brands are similar to Ibis Budget, but are offered in South Africa, Australia, Brazil, Indonesia and Japan. The firm's economy offerings include Ibis and Adagio Access, which provide higher quality lodgings at modest prices. These brands can be found in North America, Europe, Australia, Asia and Africa. The company's midscale hotels include Novotel, Suite Novotel, Mama Shelter and Mercure and primarily cater to travelers in international cities or vacation locations. Extended stay capabilities are available through the Adagio brand. Accor's luxury and upscale hotels include Pullman, MGallery, The Sebel, Adagio Premium, Grand Mercure Apartments and Thalassa Sea & Spa. Brands in this category are primarily used by tourists in destination cities. Accor's premier luxury brand, Sofitel, provides guests with gourmet cuisine, specialized treatment and high quality sleeping amenities. The company derives its profits primarily from France, which alone accounts for 28% of business, with 25% coming from the rest of Europe, 26% from the Asia Pacific, 11% from the Mediterranean/Middle East/Africa and 10% from the Americas. In 2015, Accor launched its Accorhotels app for Apple Watch, available in 10 languages, which promotes top hotels and destinations and allows users to manage current bookings on Accorhotels; and sold Accor franchisee Zurich MGallery, but will continue to manage it. In February 2016, the firm acquired a 49% stake in Squarebreak, a French start-up and acquired 30% stake in Oasis Collections, a curated marketplace for private rentals.

FINANCIAL DATA: Note: Data for latest year may not have been available at press time.

In U.S. $	2015	2014	2013	2012	2011	2010
Revenue	6,365,119,000	6,220,276,000	6,313,797,000	6,442,673,000	6,957,037,000	6,783,682,000
R&D Expense						
Operating Income	758,431,100	686,579,800	611,306,900	599,901,900	604,463,900	149,405,200
Operating Margin %	11.91%	11.03%	9.68%	9.31%	8.68%	2.20%
SGA Expense					675,174,800	424,265,200
Net Income	278,281,500	254,331,000	143,702,700	-683,158,300	57,024,900	4,117,197,000
Operating Cash Flow	895,290,900	785,803,100	597,620,900	143,702,700	732,199,700	758,431,100
Capital Expenditure				341,008,900	441,372,700	387,769,300
EBITDA	1,124,531,000	1,052,680,000	986,530,700	969,423,200	936,348,800	634,116,900
Return on Assets %	2.75%	2.82%	1.72%	-7.69%	.32%	35.25%
Return on Equity %	6.58%	7.20%	4.75%	-19.02%	.75%	108.31%
Debt to Equity	0.72	0.76	0.67	0.54	0.45	0.48

CONTACT INFORMATION:
Phone: 33 169368080   Fax: 33 169367900
Toll-Free:
Address: 2 rue de la Mare-Neuve, Paris, 75013 France

STOCK TICKER/OTHER:
Stock Ticker: ACC N   Exchange: MEX
Employees: 190,000   Fiscal Year Ends: 12/31
Parent Company:

SALARIES/BONUSES:
Top Exec. Salary: $ Bonus: $
Second Exec. Salary: $ Bonus: $

OTHER THOUGHTS:
Estimated Female Officers or Directors: 1
Hot Spot for Advancement for Women/Minorities: Y

ACS Actividades de Construccion Y Servicios SA

www.grupoacs.com
NAIC Code: 237310

TYPES OF BUSINESS:

Heavy Construction
Engineering Services
Civic Construction & Infrastructure
Industrial Services
Facility Maintenance
Passenger Transportation
Transportation Concessions
Water Treatment & Desalination Plants

BRANDS/DIVISIONS/AFFILIATES:

Grupo Dragados SA
Vias Y Construcciones SA
Tecsa Empresa Constructora SA
Pulice Construction Inc
John P Picone Inc
Geocisa SA
Urbaser SA
ACS Servicios Y Concesiones SL

CONTACTS: Note: Officers with more than one job title may be intentionally listed here more than once.

Florentino Perez Rodriguez, CEO
Antonio Garcia Ferrer, Co-Vice Chmn.
Pablo Vallhona Vadell, Co-Vice Chmn.
Antonio Garcia Ferrer, Exec. Vice Chmn.

GROWTH PLANS/SPECIAL FEATURES:

ACS Actividades de Construccion Y Servicios SA (ACS) is a leading Spanish construction and engineering group that services a wide variety of sectors, including transportation infrastructure, real estate, offshore activities, public facilities, energy and environment. The group is present in over 60 countries, operating through numerous subsidiaries, the most significant being Grupo Dragados SA, a leading Spanish construction company. ACS operates primarily in three major areas: construction, environmental services and industrial services. The construction business builds a variety of civil works projects as well as commercial and residential structures. In addition to Dragados, this segment includes Vias Y Construcciones SA, a general contractor in railway projects; Drace Infraestructuras, which specializes in water treatment and desalination plants; Tecsa Empresa Constructora SA, a leading contractor in railway projects; Geocisa SA, involved in land engineering and structural inspection activities; Pulice Construction, Inc., an Arizona-based construction and engineering firm; and John P. Picone, Inc., a New York-based firm active in civil construction projects. The environmental services segment specializes in waste management and recycling and treatment. This segment operates through Urbaser SA, which manages solid urban waste treatment plants in Spain, and ACS Servicios Y Concesiones, SL. The industrial services division serves the energy and communications sectors through its applied design, installation and maintenance of industrial infrastructure projects. Its operations are carried out through several subsidiaries managed by two parent companies: Cobra Gestion de Infraestructuras SA (Grupo Cobra) and Dragados Industrial.

FINANCIAL DATA: Note: Data for latest year may not have been available at press time.

In U.S. $	2015	2014	2013	2012	2011	2010
Revenue		39,781,550,000	43,763,780,000	43,790,760,000	32,472,120,000	17,540,470,000
R&D Expense						
Operating Income		1,094,469,000	1,875,735,000	1,815,099,000	1,567,216,000	1,228,266,000
Operating Margin %		2.75%	4.28%	4.14%	4.82%	7.00%
SGA Expense						
Net Income		817,839,600	800,106,000	-2,197,098,000	1,097,091,000	1,545,234,000
Operating Cash Flow		939,796,500	1,153,971,000	1,482,134,000	1,467,420,000	1,569,861,000
Capital Expenditure		961,450,000	1,459,020,000	1,994,984,000	2,642,973,000	1,691,363,000
EBITDA		2,034,244,000	3,253,351,000	3,490,344,000	2,655,196,000	3,145,069,000
Return on Assets %		1.81%	1.72%	-4.30%	2.34%	3.98%
Return on Equity %		22.75%	23.68%	-64.47%	25.65%	30.94%
Debt to Equity		1.95	2.20	2.16	1.08	2.30

CONTACT INFORMATION:

Phone: 34 913439200   Fax: 34 913439456
Toll-Free:
Address: Avda. Pio XII, No. 102, Madrid, 28036 Spain

STOCK TICKER/OTHER:

Stock Ticker: ACSAF
Employees: 210,345
Parent Company:

Exchange: PINX
Fiscal Year Ends: 12/31

SALARIES/BONUSES:

Top Exec. Salary: $ Bonus: $
Second Exec. Salary: $ Bonus: $

OTHER THOUGHTS:

Estimated Female Officers or Directors:
Hot Spot for Advancement for Women/Minorities:

Sales, profits and employees may be estimates. Financial information, benefits and other data can change quickly and may vary from those stated here.

Adrian Smith + Gordon Gill Architecture (AS+GG) smithgill.com
NAIC Code: 541310

TYPES OF BUSINESS:
Architectural Services
Architecture
Renewable Energy
Skyscraper

BRANDS/DIVISIONS/AFFILIATES:

CONTACTS: *Note: Officers with more than one job title may be intentionally listed here more than once.*
Adrian Smith, Partner
Gordon Gill, Partner
Robert Forest, Partner

GROWTH PLANS/SPECIAL FEATURES:
Adrian Smith + Gordon Gill Architecture (AS+GG) is an architectural and design firm. It focuses on the design and development of energy-efficient and sustainable architecture. AS+GG's services consist of architecture, urban design and interiors. Within architecture, it designs high-performance, energy-efficient, aesthetically striking architecture on an international scale as well as a wide range of typology and scale that includes low- and mid-rise residential, commercial and cultural buildings to mixed-use supertall towers. This division is currently working on projects in the UAE, Saudi Arabia, China, India, South Korea, Malaysia, Canada and the U.S. Its urban design services offer urban design, city planning, infrastructure design and landscape architectural services to private and public clients. Its urban design team develops sustainable master plans and urban infill projects that create outcomes for municipalities, developers and institutions. Its major projects include master plans for Canary Wharf in London, King Abdullah Economic City in Jeddah, Beijing Central Business District, the Nation of Bahrain, and Za'abeel Energy City in Dubai. Its interiors services works closely with architecture and engineering teams to promote new ideas to maximize natural light and to utilize sustainable materials. Designs include interior spaces for residential towers, mixed-use developments and corporate headquarters buildings, including the world's first large-scale positive-energy building Masdar Headquarters in Abu Dubai and the recently established IRENA (International Renewable Energy Agency), an intergovernmental organization to promote adoption and sustainable use of renewable energy. AS+GG's current supertall skyscraper project, the Kingdom Tower, will overtake the Burj Khalifa as the world's tallest building when completed in 2020.

FINANCIAL DATA: *Note: Data for latest year may not have been available at press time.*

In U.S. $	2015	2014	2013	2012	2011	2010
Revenue	81,199,000					
R&D Expense						
Operating Income						
Operating Margin %						
SGA Expense						
Net Income						
Operating Cash Flow						
Capital Expenditure						
EBITDA						
Return on Assets %						
Return on Equity %						
Debt to Equity						

CONTACT INFORMATION:
Phone: 312-920-1888   Fax: 312-920-1775
Toll-Free:
Address: 111 W. Monroe, Ste. 2300, Chicago, IL 60603 United States

STOCK TICKER/OTHER:
Stock Ticker: Private   Exchange:
Employees: 125   Fiscal Year Ends:
Parent Company:

SALARIES/BONUSES:
Top Exec. Salary: $ Bonus: $
Second Exec. Salary: $ Bonus: $

OTHER THOUGHTS:
Estimated Female Officers or Directors:
Hot Spot for Advancement for Women/Minorities:

AECOM Technology Corporation

www.aecom.com

NAIC Code: 237310

TYPES OF BUSINESS:

Engineering & Design Services
Transportation Projects
Environmental Projects
Power & Mining Support
Consulting
Economic Development Consulting

BRANDS/DIVISIONS/AFFILIATES:

URS Corp.

CONTACTS: Note: Officers with more than one job title may be intentionally listed here more than once.

David Gan, Assistant General Counsel
Ed Mayer, VP, Divisional
Daniel Tishman, CEO, Subsidiary
Michael Burke, CEO
Ronald Osborne, Chief Accounting Officer
W. Rudd, Executive VP
Stephen Kadenacy, President
Randall Wotring, President, Divisional
Michael Donnelly, President, Divisional
Daniel McQuade, President, Divisional
Fredrick Werner, President, Divisional
Keenan Driscoll, Treasurer
Will Gabrielski, Vice President, Divisional
Christina Ching, Vice President

GROWTH PLANS/SPECIAL FEATURES:

AECOM Technology Corporation is a global engineering and design company engaged in facility, transportation, environment and specialty engineering projects for corporate, institutional and government clients. Its operations are divided into three segments: design and consulting services (DCS); management services (MS); and construction services (CS). The DCS segment provides planning, consulting, architectural and engineering design services to commercial and government clients in markets such as transportation, facilities, environmental, energy, water and government. The MS segment focuses on program and facilities management and maintenance, training, logistics, consulting, technical assistance, systems integration and information technology services, primarily for agencies of the U.S. government and other national governments globally. The CS segment offers construction services, including building construction and energy, infrastructure and industrial construction mainly in the U.S. The firm operates in 150 countries worldwide and serves end-markets including oil, gas and chemicals, healthcare, power, sports and venues, commercial and residential, education, leisure and hospitality and government. In 2015, the firm acquired URS Corp. an engineering and construction management company.

The company offers its employees medical, dental and life insurance; a stock purchase program with company match; a 401(k) and Roth IRA; and a merit based scholarship program for the children of employees.

FINANCIAL DATA: Note: Data for latest year may not have been available at press time.

In U.S. $	2015	2014	2013	2012	2011	2010
Revenue	17,989,880,000	8,356,783,000	8,153,495,000	8,218,180,000	8,037,374,000	6,545,791,000
R&D Expense						
Operating Income	129,018,000	352,882,000	376,989,000	53,606,000	421,223,000	340,795,000
Operating Margin %	.71%	4.22%	4.62%	.65%	5.24%	5.20%
SGA Expense	113,975,000	80,908,000	97,318,000	80,903,000	90,298,000	110,463,000
Net Income	-154,845,000	229,854,000	239,243,000	-58,567,000	275,800,000	236,887,000
Operating Cash Flow	764,433,000	360,625,000	408,598,000	433,352,000	132,012,000	158,635,000
Capital Expenditure	69,426,000	62,852,000	52,117,000	62,874,000	77,991,000	68,490,000
EBITDA	747,422,000	451,024,000	474,917,000	165,553,000	534,897,000	419,694,000
Return on Assets %	-1.53%	3.89%	4.22%	-1.02%	4.99%	5.24%
Return on Equity %	-5.53%	10.92%	11.41%	-2.59%	12.45%	12.40%
Debt to Equity	1.30	0.42	0.53	0.41	0.48	0.43

CONTACT INFORMATION:

Phone: 213 593-8000   Fax: 213 593-8730
Toll-Free:
Address: 555 S. Flower St., Ste. 3700, Los Angeles, CA 90071 United States

STOCK TICKER/OTHER:

Stock Ticker: ACM
Employees: 92,000
Parent Company:

Exchange: NYS
Fiscal Year Ends: 09/30

SALARIES/BONUSES:

Top Exec. Salary: $1,153,858   Bonus: $
Second Exec. Salary: $680,060   Bonus: $120,000

OTHER THOUGHTS:

Estimated Female Officers or Directors: 6

Hot Spot for Advancement for Women/Minorities: Y

Aecon Group Inc

NAIC Code: 237310

www.aecon.com

TYPES OF BUSINESS:

Construction
Infrastructure Development
Utility Systems
Steam Power Generation
Renovation

BRANDS/DIVISIONS/AFFILIATES:

Aecon Construction and Materials Limited
Yellowline Asphalt Products Ltd
Aecon Constructors
AGI Traffic Technology Inc
Aecon Atlantic Industrial Inc
Tristar
Aecon Mining
Aecon Mining Construction Management Inc

CONTACTS: Note: Officers with more than one job title may be intentionally listed here more than once.

Terrance McKibbon, CEO
David Smales, CFO
John Beck, Chairman of the Board
Paula Palma, Chief Information Officer
Brian Tobin, Director
L. Brian Swartz, Executive VP, Divisional
Steven Nackan, President, Divisional

GROWTH PLANS/SPECIAL FEATURES:

Aecon Group, Inc. is one of Canada's largest construction and infrastructure development companies. The firm operates in five principal segments through its network of subsidiaries: infrastructure, energy, mining, concessions and multi-sector services. The infrastructure segment, through its various business units, provides all aspects of the construction of both public and private infrastructure, including roads, highways, bridges, airport facilities, dams, tunnels, marine facilities, transit systems and power projects. This segment also offers design, project management and construction management services; and utility infrastructure services for gas projects, hydro distribution networks, telecommunication networks, water mains and sewers. Subsidiaries in this segment include Aecon Construction and Materials Limited (ACML), Yellowline Asphalt Products, Ltd. and Aecon Constructors. The energy segment offers a range of services for designing, building, and commissioning power generation plants, which include oil and gas facilities, cogeneration plants, hydroelectric facilities, natural gas power plants, nuclear plants and waste heat recovery systems. Some of the subsidiaries in this segment are AGI Traffic Technology, Inc., Aecon Atlantic Industrial, Inc. and Tristar. The mining segment consists of Aecon Mining and Aecon Mining Construction Management, Inc., which is involved in extraction and refinement of natural resources ranging from oil and gas reserves, potash and uranium and high-grade deposits of base and precious metals. The concession segment includes the development, financing and operation of infrastructure projects by way of build-operate-transfer and other public-private partnership contract structures. The multi-sector services segment offers distinctive services such as specialty pipe and custom steel fabrication; mechanical and electrical contracting; and environmental and geotechnical services.

Aecon Group offers its employees medical and dental insurance, an employee stock purchase program, a pension plan and tuition reimbursement.

FINANCIAL DATA: Note: Data for latest year may not have been available at press time.

In U.S. $	2015	2014	2013	2012	2011	2010
Revenue	2,230,814,000	1,998,408,000	2,345,887,000	2,252,764,000	2,214,060,000	2,099,446,000
R&D Expense						
Operating Income	109,043,000	48,703,440	74,350,200	83,912,300	76,854,630	47,301,390
Operating Margin %	4.88%	2.43%	3.16%	3.72%	3.47%	2.25%
SGA Expense	129,844,500	114,619,900	113,146,000	120,169,300	106,134,200	90,366,040
Net Income	52,502,140	22,966,490	31,036,330	59,612,560	43,998,080	20,876,400
Operating Cash Flow	44,420,070	57,184,580	102,456,300	52,654,280	150,970,900	-79,894,200
Capital Expenditure	40,333,160	45,227,360	45,072,170	39,149,750	114,930,300	51,350,840
EBITDA	162,047,400	108,072,900	116,570,100	137,393,000	132,276,300	86,762,280
Return on Assets %	3.70%	1.57%	1.96%	3.79%	2.82%	1.44%
Return on Equity %	9.99%	4.83%	7.18%	15.14%	12.06%	5.90%
Debt to Equity	0.37	0.41	0.63	1.02	1.09	1.04

CONTACT INFORMATION:

Phone: 416 293-7004   Fax: 416 293-0271
Toll-Free: 877-232-2677
Address: 20 Carlson Ct., Ste. 800, Toronto, ON M9W 7K6 Canada

STOCK TICKER/OTHER:

Stock Ticker: ARE   Exchange: TSE
Employees: 8,373   Fiscal Year Ends: 12/31
Parent Company:

SALARIES/BONUSES:

Top Exec. Salary: $ Bonus: $
Second Exec. Salary: $ Bonus: $

OTHER THOUGHTS:

Estimated Female Officers or Directors: 2
Hot Spot for Advancement for Women/Minorities:

Sales, profits and employees may be estimates. Financial information, benefits and other data can change quickly and may vary from those stated here.

Aedas

www.aedas.com

NAIC Code: 541310

TYPES OF BUSINESS:

Architectural Services

BRANDS/DIVISIONS/AFFILIATES:

Aedas Arts Team
Aedas RHWL
Renton Howard Hood Levin

CONTACTS: Note: Officers with more than one job title may be intentionally listed here more than once.

Peter Shaw, Managing Dir.
Keith Griffiths, Chmn.

GROWTH PLANS/SPECIAL FEATURES:

Aedas is a leading global architecture and design practice. The company believes great design can only be delivered with a deep social and cultural understanding of the communities in which it is designing for. Aedas' specialties include architecture, graphics, interiors, landscape, urban design and master planning. Recent featured projects include The Forum at Exchange Square in Hong Kong, an office building wholly occupied by Standard Chartered, an international banking group; THR350, a nine-story private residence located at a hillside on the Hong Kong Island, with a sculptural staircase that resembles three stacking ice cubes in order to create a sense of drama for its art-loving residents; Hotel Indigo on Hong Kong Island, consisting of 29 floors, 138 rooms and six suites as well as a glass-bottomed rooftop infinity pool adjacent to its Skybar; and Sandcrawler, a regional headquarters for Lucasfilm Singapore, which has a horseshoe form. In 2015, the firm acquired Renton Howard Hood Levin (RHWL), a U.K. architectural company known for its projects such as the Crucible Theatre in Sheffield, hotels including the St. Pancras Renaissance London Hotel as well as major office complexes. The merger formed two business units, Aedas Arts Team, which works on arts and cultural commissions in the Chinese markets, and Aedas RHWL, which supports the firm's Asian clients as well as mixed commercial and residential work in London.

FINANCIAL DATA: Note: Data for latest year may not have been available at press time.

In U.S. $	2015	2014	2013	2012	2011	2010
Revenue						
R&D Expense						
Operating Income						
Operating Margin %						
SGA Expense						
Net Income						
Operating Cash Flow						
Capital Expenditure						
EBITDA						
Return on Assets %						
Return on Equity %						
Debt to Equity						

CONTACT INFORMATION:

Phone: 44-20-7837-9789   Fax: 44-20-7837-9678
Toll-Free:
Address: 5-8 Hardwick St., London, EC1R 4RG United Kingdom

STOCK TICKER/OTHER:

Stock Ticker: Private
Employees: 1,450
Parent Company:

Exchange:
Fiscal Year Ends:

SALARIES/BONUSES:

Top Exec. Salary: $ Bonus: $
Second Exec. Salary: $ Bonus: $

OTHER THOUGHTS:

Estimated Female Officers or Directors:
Hot Spot for Advancement for Women/Minorities:

Aegion Corp

NAIC Code: 237110

www.aegion.com

TYPES OF BUSINESS:

Construction, Heavy & Civil Engineering
Sewer & Pipe Rehabilitation Services

BRANDS/DIVISIONS/AFFILIATES:

Insituform Technologies Inc
Brinderson

CONTACTS: *Note: Officers with more than one job title may be intentionally listed here more than once.*

David Martin, CFO
Alfred Woods, Chairman of the Board
Michael White, Controller
Charles Gordon, Director
David Morris, Secretary
John Huhn, Senior VP, Divisional
Stephen Callahan, Senior VP, Divisional

GROWTH PLANS/SPECIAL FEATURES:

Aegion Corp., formerly Insituform Technologies, Inc., is a holding company that provides proprietary technologies and pipeline rehabilitation services to the sewer, water, energy and mining infrastructure markets through its various subsidiaries. Most of Aegion's installation operations are project-oriented contracts for municipal entities, concerning the maintenance of municipal sewers, commercial pipes and water mains. The company and its subsidiaries/joint ventures operate principally in the U.S., the U.K., Portugal, Chile, Canada, Argentina, Brazil, the UAE, Mexico, Singapore, Saudi Arabia, Oman and Morocco. The firm has three operating segments: infrastructure solutions, corrosion protection and energy services. The infrastructure solutions segment consist of infrastructure, sewer and water rehabilitation activities including the installation and construction operations performed directly through subsidiaries. The firm's proprietary Insituform Cured-In-Place Pipe (CIPP) process, a trenchless technology, allows pipelines to be repaired without digging and major service disruptions. The operations also include strengthening of pipelines, buildings, bridges, tunnels, industrial developments and waterfront structures. The corrosion protection segment undertakes maintenance rehabilitation and corrosion protection services for oil and gas, industrial and mineral piping systems and structures and products for gas release and leak detection systems. Through its subsidiaries, the firm offers fully-integrated corrosion prevention services including engineering; product and material sales; construction and installation; inspection, monitoring and maintenance; and coatings. The energy services segment consists solely of Brinderson and its affiliated entities. Brinderson is based in Costa Mesa, California and performs engineering, procurement, construction, maintenance and turnaround services supporting all segments of the oil and gas industry from upstream to downstream. Additionally, Brinderson provides project management and engineering professionals across various disciplines, including chemical, civil, structural, mechanical, electrical, instrumentation, project controls, estimating, procurement and safety.

Employees are offered medical, dental, vision and prescription coverage; retirement plans; a 401(k) plan; disability coverage; life insurance; flexible spending; and tution reimbursement.

FINANCIAL DATA: *Note: Data for latest year may not have been available at press time.*

In U.S. $	2015	2014	2013	2012	2011	2010
Revenue	1,333,570,000	1,331,421,000	1,091,420,000	1,027,963,000	938,585,000	914,975,000
R&D Expense						
Operating Income	19,946,000	-19,812,000	66,882,000	79,122,000	44,537,000	87,035,000
Operating Margin %	1.49%	-1.48%	6.12%	7.69%	4.74%	9.51%
SGA Expense	209,477,000	234,105,000	178,483,000	170,882,000	151,764,000	144,245,000
Net Income	-8,067,000	-37,167,000	44,351,000	52,661,000	26,547,000	60,462,000
Operating Cash Flow	132,023,000	80,823,000	84,304,000	110,721,000	22,884,000	53,029,000
Capital Expenditure	30,957,000	34,822,000	28,117,000	46,446,000	22,684,000	39,461,000
EBITDA	61,050,000	21,280,000	112,500,000	116,288,000	82,878,000	117,152,000
Return on Assets %	-.63%	-2.79%	3.43%	4.49%	2.59%	6.77%
Return on Equity %	-1.33%	-5.56%	6.29%	7.85%	4.25%	10.57%
Debt to Equity	0.58	0.56	0.51	0.31	0.34	0.15

CONTACT INFORMATION:

Phone: 636 530-8000   Fax: 636 519-8010
Toll-Free: 800-234-2992
Address: 17988 Edison Ave., Chesterfield, MO 63005 United States

STOCK TICKER/OTHER:

Stock Ticker: AEGN   Exchange: NAS
Employees: 6,200   Fiscal Year Ends: 12/31
Parent Company:

SALARIES/BONUSES:

Top Exec. Salary: $650,000   Bonus: $
Second Exec. Salary: Bonus: $
$392,000

OTHER THOUGHTS:

Estimated Female Officers or Directors: 2
Hot Spot for Advancement for Women/Minorities: Y

Alexandria Real Estate Equities Inc

www.are.com

NAIC Code: 0

TYPES OF BUSINESS:

Real Estate Investment Trust
Scientific Properties
Office & Laboratory Space
Property Development

BRANDS/DIVISIONS/AFFILIATES:

CONTACTS: *Note: Officers with more than one job title may be intentionally listed here more than once.*

Joel Marcus, CEO
Dean Shigenaga, CFO
Stephen Richardson, COO
Jennifer Banks, Executive VP
Thomas Andrews, Executive VP
Daniel Ryan, Executive VP
Peter Moglia, Other Executive Officer

GROWTH PLANS/SPECIAL FEATURES:

Alexandria Real Estate Equities, Inc. is a real estate investment trust (REIT) that owns, acquires, operates, manages, expands and redevelops scientific properties containing a mixture of office and laboratory space. Alexandria's portfolio consists of 199 properties in the U.S. and Canada, comprising an asset base of 32.0 million square feet, including approximately 20.1 million rentable square feet of office/laboratory space. The company's portfolio also includes four development/redevelopment projects in Asia. The firm focuses its activities principally on the life science markets of California (in the San Diego and San Francisco Bay areas), Seattle, suburban Washington, D.C. (both in Maryland and Virginia), greater Boston, New Jersey, New York City, suburban Philadelphia and the Southeast as well as certain parts of Canada. The company's goal is to acquire high-quality life science facilities, such as existing offices, warehouses or vacant space, which it can then expand or redevelop into generic laboratory space. Its properties are leased principally to tenants in a broad spectrum of sectors within the life science industry, such as institutional (universities and independent not-for-profit institutions), pharmaceutical, biopharmaceutical, medical device, life science product, service and translational entities, as well as government agencies. The company's three largest client tenants are Novartis AG, Illumina, Inc. and ARIAD Pharmaceuticals, Inc., collectively accounting for 13.9% of rented space.

Alexandria offers its employees medical, dental, vision, disability and life insurance; financial, retirement and 401(k) options; travel; and an employee assistance program.

FINANCIAL DATA: *Note: Data for latest year may not have been available at press time.*

In U.S. $	2015	2014	2013	2012	2011	2010
Revenue	751,738,000	647,578,000	563,199,000	516,889,000	503,551,000	417,661,000
R&D Expense						
Operating Income	146,157,000	99,142,000	134,525,000	101,446,000	136,235,000	124,454,000
Operating Margin %	19.44%	15.30%	23.88%	19.62%	27.05%	29.79%
SGA Expense	320,853,000	272,694,000	237,559,000	222,318,000	209,790,000	166,668,000
Net Income	144,217,000	101,574,000	136,217,000	102,126,000	131,418,000	135,293,000
Operating Cash Flow	342,611,000	334,325,000	312,727,000	305,533,000	246,960,000	219,346,000
Capital Expenditure	813,139,000	625,660,000	715,458,000	591,201,000	735,068,000	717,103,000
EBITDA						
Return on Assets %	1.37%	.92%	1.48%	.98%	1.63%	1.86%
Return on Equity %	3.19%	1.99%	3.29%	2.23%	3.67%	4.71%
Debt to Equity	0.49	0.43	0.80	0.67	0.92	1.01

CONTACT INFORMATION:

Phone: 626-578-0777   Fax: 626 578-0896
Toll-Free:
Address: 385 E. Colorado Blvd., Ste. 299, Pasadena, CA 91101 United States

STOCK TICKER/OTHER:

Stock Ticker: ARE
Employees: 278
Parent Company:

Exchange: NYS
Fiscal Year Ends: 12/31

SALARIES/BONUSES:

Top Exec. Salary: $450,000   Bonus: $650,000
Second Exec. Salary: $425,000   Bonus: $650,000

OTHER THOUGHTS:

Estimated Female Officers or Directors: 3
Hot Spot for Advancement for Women/Minorities: Y

Amanresorts International Pte Ltd (Aman Resorts)

www.amanresorts.com

NAIC Code: 721110

TYPES OF BUSINESS:

Hotels & Resorts

BRANDS/DIVISIONS/AFFILIATES:

Aman Resorts Group Ltd.
Aman-i-Khas
Amandari
Amankora
Amansara
Le Melezin
Amangani
Amangalla

CONTACTS: Note: Officers with more than one job title may be intentionally listed here more than once.

Olivier Jolivet, CEO
Michel Checoury, CFO
Lisa Bovio, CMO
T.C. Goyle, CEO
Vladislav Doronin, Chmn.

GROWTH PLANS/SPECIAL FEATURES:

Amanresorts International Pte. Ltd. (Aman Resorts) owns and operates luxury resorts. The company name derives from the Sanskrit word aman, meaning peace. Since opening its first location, the Amanpuri (meaning place of peace) in Phuket, Thailand in 1988, the firm has expanded to operating 30 resorts in 20 countries worldwide, mostly in and around Southeast Asia. These resorts are small, ranging from 10 luxurious, air-conditioned tents at the Aman-i-Khas, located near a wildlife sanctuary in Ranthambhore, India, to the 70 pavilions and villas in Amanpuri. Its other resorts include Amandari, Amankila, Amanusa, Amanwana and Amanjiwo, all located in Indonesia. Amankora is spread across five towns in Bhutan, Amansara is in Cambodia, Amantaka is in Laos and Amanpulo is in the Philippines. Hotel Bora Bora in French Polynesia is recently renovated. Amanbagh and Aman New Delhi comprise its other Indian locations. Amanfayun and Aman at the Summer Palace are its Chinese locations. Amangalla and Amanwella are in Sri Lanka. Amanjena is in Marrakech, Morocco. Amanyara is located in Turks and Caicos, in the Caribbean. Aman Sveti Stefen, opening in phases after renovation, is in Montenegro. Le Melezin is in Courcheval, France, while Amangani is in Wyoming, near the Grand Tetons and Yellowstone. Another U.S. property, Amangiri, is a small 34 suite resort in Southern Utah. The firm also recently opened an Aman Spa in The Connaught Hotel in Mayfair, London. The company sources the decor for its resorts locally, in order to reflect the natural surroundings and local cultures. Formerly owned by DLF Global Hospitality, Aman Resorts was purchased by Aman Resorts Group Ltd., a joint venture between Peak Hotels and Resorts Group and Adrian Zecha, the original Aman Resorts founder.

FINANCIAL DATA: Note: Data for latest year may not have been available at press time.

In U.S. $	2015	2014	2013	2012	2011	2010
Revenue						
R&D Expense						
Operating Income						
Operating Margin %						
SGA Expense						
Net Income						
Operating Cash Flow						
Capital Expenditure						
EBITDA						
Return on Assets %						
Return on Equity %						
Debt to Equity						

CONTACT INFORMATION:

Phone: 65-6883-2555   Fax: 65-6883-0555

Toll-Free:

Address: 1 Orchard Spring Ln., #05-01 Tourism Ct., Singapore, 247729 Singapore

STOCK TICKER/OTHER:

Stock Ticker: Subsidiary   Exchange:

Employees: 1,252   Fiscal Year Ends:

Parent Company: Aman Resorts Group Ltd

SALARIES/BONUSES:

Top Exec. Salary: $ Bonus: $

Second Exec. Salary: $ Bonus: $

OTHER THOUGHTS:

Estimated Female Officers or Directors:

Hot Spot for Advancement for Women/Minorities:

Amec Foster Wheeler plc

www.amecfw.com

NAIC Code: 237000

TYPES OF BUSINESS:

Heavy Construction and Engineering
Design
Environmental Services
Nuclear Power Services
Renewable Energy Services
Support & Maintenance Services
Consulting
Mining

BRANDS/DIVISIONS/AFFILIATES:

CONTACTS: Note: Officers with more than one job title may be intentionally listed here more than once.

Samir Brikho, CEO
Jeff Reilly, Pres., Strategy & Bus. Dev.
Ian McHoul, CFO
Will Serle, Group Dir.-Human Resources
Alison Yapp, General Counsel
Sue Scholes, Dir.-Comm.
Nicola-Jane Brooks, Head-Investor Rel.
Hisham Mahmoud, Pres., Growth Regions
Simon Naylor, Pres., Americas
John Connolly, Chmn.
John Pearson, Pres., Europe

GROWTH PLANS/SPECIAL FEATURES:

Amec Foster Wheeler plc is a global engineering and construction management firm with operations in about 50 countries worldwide. Its operations are divided into three geographic segments: Americas; Northern Europe & Commonwealth of Independent States; and Africa, Middle East & Southern Europe. Within those geographic segments, the firm operates in four key markets: oil & gas, mining, clean energy and environment & infrastructure. The oil & gas market services include both the conventional and unconventional. Conventional oil & gas services focus on delivering project execution, asset support, pipeline services, specialist services and petrochemicals. Unconventional oil & gas services provide a full range of services to the oil sands industry. The mining market services provide front-end consulting, mineral processing and design, project delivery, mine water treatment and geo-environmental (such as water management and regulatory/permitting services). The clean energy segment is concerned with power transmission and distribution, nuclear power, renewables, bio-process energy and conventional power delivery. The company's nuclear service areas include new build, licensing, regulatory and project support; reactor support, life time extension and asset performance improvement; clean up, specialist decommissioning and waste management services; and nuclear defense, engineering and technical services. The environmental & infrastructure segment is organized into federal government, mining, oil and gas, transportation and water units. Environmental services include consultancy and engineering as well as specialty services such as environmental assessment, materials testing, specialty water services and clean-up. Moreover, Amec has a global power group that operates four manufacturing facilities, two co-generation plants, 11 engineering & service centers in eight countries. This unit designs, supplies and erects boilers, auxiliary steam & air pollution control equipment, as well as a wide range of aftermarket products and services.

FINANCIAL DATA: Note: Data for latest year may not have been available at press time.

In U.S. $	2015	2014	2013	2012	2011	2010
Revenue	7,712,318,000	5,645,332,000	5,618,470,000	5,878,610,000	4,610,425,000	4,171,579,000
R&D Expense						
Operating Income	-288,416,700	231,864,400	353,451,800	339,313,700	322,348,000	335,072,300
Operating Margin %	-3.73%	4.10%	6.29%	5.77%	6.99%	8.03%
SGA Expense	1,232,840,000	500,487,700	414,245,500	414,245,500	361,934,600	329,982,600
Net Income	-361,934,600	115,932,200	253,071,500	305,382,300	328,003,300	326,589,400
Operating Cash Flow	199,346,800	206,415,800	339,313,700	342,141,300	244,588,600	189,308,800
Capital Expenditure	53,724,670	43,828,020	32,517,560	48,069,440	32,517,560	19,086,400
EBITDA	-213,484,900	330,830,900	460,901,100	459,487,300	438,280,200	418,486,900
Return on Assets %	-4.43%	1.96%	7.30%	8.68%	9.83%	11.00%
Return on Equity %	-14.24%	5.26%	16.26%	17.61%	17.54%	20.12%
Debt to Equity	0.40	0.30				

CONTACT INFORMATION:

Phone: 44-20-7429-7500   Fax: 44-20-7429-7550
Toll-Free:
Address: 128 Queen Victoria St., Old Change House, 4/Fl, London, EC4V 4BJ United Kingdom

STOCK TICKER/OTHER:

Stock Ticker: AMFW
Employees: 24,225
Parent Company:

Exchange: NYS
Fiscal Year Ends: 12/31

SALARIES/BONUSES:

Top Exec. Salary: $1,348,772
Bonus: $

Second Exec. Salary: $749,318
Bonus: $

OTHER THOUGHTS:

Estimated Female Officers or Directors: 4

Hot Spot for Advancement for Women/Minorities: Y

Sales, profits and employees may be estimates. Financial information, benefits and other data can change quickly and may vary from those stated here.

AMERCO (U-Haul)

NAIC Code: 532120

www.amerco.com

TYPES OF BUSINESS:

Truck Rental & Leasing Services
Moving & Storage Services & Supplies
Property & Casualty Insurance
Life Insurance
Annuities
Self-Storage Properties
Propane Tank Refilling
Car Sharing Services

BRANDS/DIVISIONS/AFFILIATES:

U-Haul International Inc
Amerco Real Estate Company
Repwest Insurance Company
Oxford Life Insurance Company
Uhaul.com

CONTACTS: Note: Officers with more than one job title may be intentionally listed here more than once.

Rocky Wardrip, Assistant Treasurer
Edward Shoen, CEO
Jason Berg, CFO
Samuel Shoen, Director
Laurence Derespino, General Counsel
Mark Haydukovich, President, Subsidiary
John Taylor, President, Subsidiary
Douglas Bell, President, Subsidiary
Carlos Vizcarra, President, Subsidiary
Gary Horton, Treasurer
James Shoen, Vice President, Subsidiary

GROWTH PLANS/SPECIAL FEATURES:

AMERCO is a holding company operating through four primary subsidiaries: U-Haul International, Inc.; Amerco Real Estate Company; Repwest Insurance Company; and Oxford Life Insurance Company. Accordingly, the firm has three reportable business segments: moving & storage, property & casualty insurance and life insurance. Moving & storage consists of U-Haul, with its rental equipment fleet of trucks, trailers and tow dollies offered at 1,600 company operated locations and 18,200 independent dealer outlets. It also provides furniture pads, utility dollies and hand trucks; sells a wide selection of moving supplies; and offers protection packages for moving and storage. U-Haul owns more than 135,000 trucks, 107,000 trailers and 38,000 towing devices. The firm's Uhaul.com online reservation portal allows its self-storage customers to make reservations, access all U-Haul storage centers and affiliate partners. This segment is also operated by Amerco Real Estate Company, which manages 491,000 rentable rooms comprising 44.2 million square feet of rentable storage space located in the U.S. and Canada. The property & casualty insurance segment, operated by Repwest Insurance Company, provides loss adjusting and claims handling for U-Haul through regional offices across North America. This segment also underwrites components of the Safemove, Safetow and Safestor protection packages to U-Haul customers. The life insurance segment, operated by Oxford Life Insurance Company, provides life and health insurance products primarily to the senior market through the direct writing and reinsuring of life insurance, Medicare supplement and annuity policies.

FINANCIAL DATA: Note: Data for latest year may not have been available at press time.

In U.S. $	2015	2014	2013	2012	2011	2010
Revenue	3,074,531,000	2,835,252,000	2,558,587,000	2,502,675,000	2,241,275,000	2,002,005,000
R&D Expense						
Operating Income	663,024,000	630,214,000	499,183,000	416,007,000	377,695,000	193,537,000
Operating Margin %	21.56%	22.22%	19.51%	16.62%	16.85%	9.66%
SGA Expense	488,200,000	257,168,000	180,676,000	310,839,000	341,238,000	121,105,000
Net Income	356,741,000	342,391,000	264,708,000	205,367,000	183,575,000	65,623,000
Operating Cash Flow	808,190,000	709,504,000	661,530,000	664,605,000	572,794,000	399,872,000
Capital Expenditure	1,111,899,000	999,365,000	655,984,000	589,799,000	480,418,000	259,491,000
EBITDA	1,011,739,000	923,383,000	755,542,000	645,796,000	590,019,000	423,126,000
Return on Assets %	5.54%	6.05%	5.31%	4.45%	4.30%	1.40%
Return on Equity %	20.91%	24.84%	23.37%	19.37%	18.93%	6.94%
Debt to Equity	1.16	1.27	1.12	1.43	1.40	1.65

CONTACT INFORMATION:

Phone: 775-688-6300   Fax: 775 688-6338
Toll-Free:
Address: 1325 Airmotive Way, Ste. 100, Reno, NV 89502 United States

STOCK TICKER/OTHER:

Stock Ticker: UHAL   Exchange: NAS
Employees: 25,400   Fiscal Year Ends: 03/31
Parent Company:

SALARIES/BONUSES:

Top Exec. Salary: $675,104   Bonus: $200,000
Second Exec. Salary: Bonus: $
$565,962

OTHER THOUGHTS:

Estimated Female Officers or Directors: 1
Hot Spot for Advancement for Women/Minorities:

American Campus Communities Inc

americancampus.com

NAIC Code: 531110

TYPES OF BUSINESS:
Student Housing Development

BRANDS/DIVISIONS/AFFILIATES:
American Campus Communities Operating Partnership

CONTACTS: *Note: Officers with more than one job title may be intentionally listed here more than once.*
William Bayless, CEO
Jonathan Graf, CFO
Jorge De Cardenas, Chief Technology Officer
Kim Voss, Controller
Edward Lowenthal, Director
James Wilhelm, Executive VP, Divisional
Daniel Perry, Executive VP, Divisional
Jennifer Beese, Executive VP, Divisional
James Hopke, Executive VP
William Talbot, Executive VP

GROWTH PLANS/SPECIAL FEATURES:

American Campus Communities, Inc. (ACC) develops, owns and manages high-quality student housing communities, primarily operating via subsidiary American Campus Communities Operating Partnership LP. ACC has developed more than 350 student housing properties, and have acquired over $6.8 billion in student housing assets. It is also a leader in third-party development of more than 78 on-campus projects. The company operates through four business segments: wholly-owned properties, on-campus participating properties, development services and property management services. The wholly-owned properties segment consists of off-campus properties generally located in close proximity to the school campus, as well as premier on-campus modern, well-amenitized locations. The on-campus participating properties segment includes five on-campus properties that are operated under long-term ground/facility leases with three university systems. The development services segment focuses on providing development and construction management services to colleges and universities seeking to modernize on-campus student housing properties. Services include pre-development, feasibility studies, design, financial structuring for short-term consulting projects to long-term full-scale development and construction projects. The property management segment includes revenues generated from third-party management contracts, under which the firm is responsible for all aspects of operations, including marketing, leasing administration, facilities maintenance, business administration, accounts payable, accounts receivable, financial reporting, capital projects and residence life student development. In 2015, the firm's owned and third-party managed portfolio included 201 properties with approximately 128,900 beds in approximately 43,400 units.

FINANCIAL DATA: *Note: Data for latest year may not have been available at press time.*

In U.S. $	2015	2014	2013	2012	2011	2010
Revenue	753,381,000	733,915,000	657,462,000	491,290,000	390,317,000	344,991,000
R&D Expense						
Operating Income	156,904,000	154,986,000	131,753,000	115,915,000	99,893,000	86,942,000
Operating Margin %	20.82%	21.11%	20.03%	23.59%	25.59%	25.20%
SGA Expense	20,838,000	18,935,000	16,666,000	22,965,000	16,360,000	14,505,000
Net Income	115,991,000	62,839,000	104,644,000	56,636,000	56,629,000	16,210,000
Operating Cash Flow	260,986,000	259,898,000	246,678,000	203,038,000	131,033,000	118,084,000
Capital Expenditure	444,675,000	188,418,000	332,815,000	1,181,227,000	285,271,000	250,688,000
EBITDA	415,880,000	350,596,000	313,959,000	230,538,000	183,987,000	163,883,000
Return on Assets %	1.95%	1.09%	1.95%	1.39%	1.98%	.65%
Return on Equity %	4.31%	2.40%	3.96%	2.81%	4.37%	1.53%
Debt to Equity	1.07	1.13	1.04	0.83	1.05	1.10

CONTACT INFORMATION:
Phone: 512 732-1000   Fax: 512 732-2450
Toll-Free:
Address: 12700 Hill Country Blvd., Ste. T-200, Bee Cave, TX 78738 United States

STOCK TICKER/OTHER:
Stock Ticker: ACC
Employees: 3,227
Parent Company:

Exchange: NYS
Fiscal Year Ends: 12/31

SALARIES/BONUSES:
Top Exec. Salary: $750,000   Bonus: $
Second Exec. Salary: $400,000   Bonus: $

OTHER THOUGHTS:
Estimated Female Officers or Directors: 4
Hot Spot for Advancement for Women/Minorities: Y

American Casino & Entertainment Properties LLC

aceplc.com/InvestorRelations/Investor
NAIC Code: 721120

TYPES OF BUSINESS:

Casino Hotel Properties
Retail Outlets
Restaurants
Recreational Vehicle Park

BRANDS/DIVISIONS/AFFILIATES:

Stratosphere Casino Hotel & Tower
Arizona Charlie's Decatur
Arizona Charlie's Boulder
Aquarius Casino Resort
Whitehall Street Real Estate Fund
Goldman Sachs Group

CONTACTS: Note: Officers with more than one job title may be intentionally listed here more than once.

Frank V. Riolo, CEO
Ned Martin, CFO
Phyllis A. Gilland, General Counsel
Paul Hobson, Gen. Mgr.-Stratosphere Casino, Hotel & Tower
Ronald P. Lurie, Exec. VP
Mark Majetich, Sr. VP
Sean Hammond, Gen. Mgr.-Aquarius Casino Resort

GROWTH PLANS/SPECIAL FEATURES:

American Casino & Entertainment Properties LLC (ACEP), a wholly-owned subsidiary of the Goldman Sachs Group affiliate Whitehall Street Real Estate Fund, is a holdings company that operates four gaming and entertainment properties across Nevada. The firm's entertainment properties consist of the Stratosphere Tower Hotel & Casino, a mixed-use resort destination on the Las Vegas Strip; Arizona Charlie's Decatur and Arizona Charlie's Boulder, two casinos also located in Las Vegas; and the Aquarius Casino Resort in Laughlin, Nevada, south of Las Vegas. The Stratosphere is the firm's flagship property, featuring 2,427 hotel rooms, 80,000 square feet of casino space and 110,000 square feet of retail space as well as a revolving restaurant, a roller coaster and the Stratosphere Tower, the tallest freestanding observation tower in the U.S. The Arizona Charlie's Decatur and Arizona Charlie's Boulder are off-Strip full-service casinos and restaurants featuring slot machines, game tables, poker lounges and various other gaming features. Arizona Charlie's Boulder also features a recreational vehicle (RV) park offering amenities such as game and exercise rooms, a swimming pool and shower and laundry facilities. The Aquarius Casino Resort features a 57,070-square-foot casino and 1,907 hotel rooms, including 90 suites and seven restaurants. The Aquarius also maintains 35,000 square feet of rentable meeting space, indoor and outdoor entertainment facilities and a wedding chapel.

FINANCIAL DATA: Note: Data for latest year may not have been available at press time.

In U.S. $	2015	2014	2013	2012	2011	2010
Revenue	373,067,000	351,100,000	337,400,000	339,729,000	341,890,000	336,838,000
R&D Expense						
Operating Income						
Operating Margin %						
SGA Expense						
Net Income	12,062,000	6,980,000	-15,100,000	-15,798,000	-20,264,000	-36,747,000
Operating Cash Flow						
Capital Expenditure						
EBITDA						
Return on Assets %						
Return on Equity %						
Debt to Equity						

CONTACT INFORMATION:

Phone: 702-380-7777   Fax: 702-383-4734
Toll-Free:
Address: 2000 S. Las Vegas Blvd., Las Vegas, NV 89104 United States

STOCK TICKER/OTHER:

Stock Ticker: Private   Exchange:
Employees: 4,300   Fiscal Year Ends: 12/31
Parent Company: Whitehall Street Real Estate Funds

SALARIES/BONUSES:

Top Exec. Salary: $ Bonus: $
Second Exec. Salary: $ Bonus: $

OTHER THOUGHTS:

Estimated Female Officers or Directors: 1
Hot Spot for Advancement for Women/Minorities: Y

American Realty Investors Inc

www.amrealtytrust.com

NAIC Code: 531110

TYPES OF BUSINESS:

Real Estate Development & Management, Apartments
Real Estate Investments
Apartments
Land
Commercial/Office Property
Hotels

BRANDS/DIVISIONS/AFFILIATES:

Pillar Income Asset Management Inc
Transcontinental Realty Investors Inc

CONTACTS: Note: Officers with more than one job title may be intentionally listed here more than once.

Daniel Moos, CEO
Gene Bertcher, CFO
Henry Butler, Chairman of the Board
Alfred Crozier, Executive VP, Divisional
Louis Corna, Executive VP

GROWTH PLANS/SPECIAL FEATURES:

American Realty Investors, Inc. (ARI) owns, develops and manages residential and commercial real estate throughout the U.S. Its investments include apartments, commercial/office buildings, hotels, parcels of land and other equity ownership interests. The company develops residential complexes through third-party agreements, in which it contributes land and in some cases cash support. ARI owns 46 residential apartment communities with a total of 6,344 units. The firm's commercial portfolio consists of nine properties, totaling more than 2.1 million square feet of rentable space, that include four office buildings, one industrial warehouse, a golf course and three retail centers. ARI generates revenue through its income producing properties, third-party development strategies and land acquisitions in high growth suburban markets. The firm also buys large tracts of underdeveloped or partially developed land in areas that are or will be likely growth communities; these tracts of land are located primarily in suburban communities. It uses the purchased land for its own developments or sells it for a profit to other developers. Through majority-owned subsidiary Transcontinental Realty Investors, Inc. (TRI), ARI invests in under-valued properties. ARI is advised and managed by Pillar Income Asset Management, Inc.

FINANCIAL DATA: Note: Data for latest year may not have been available at press time.

In U.S. $	2015	2014	2013	2012	2011	2010
Revenue	104,188,000	79,412,000	89,636,000	119,521,000	118,357,000	157,030,000
R&D Expense						
Operating Income	6,308,000	-3,199,000	-12,416,000	17,207,000	-45,316,000	-57,915,000
Operating Margin %	6.05%	-4.02%	-13.85%	14.39%	-38.28%	-36.88%
SGA Expense	16,668,000	19,225,000	18,100,000	21,300,000	76,090,000	28,410,000
Net Income	-1,960,000	30,885,000	41,276,000	-5,585,000	290,000	-94,747,000
Operating Cash Flow	-34,509,000	-37,968,000	-42,162,000	-23,111,000	23,553,000	-9,627,000
Capital Expenditure	239,143,000	95,665,000	9,316,000	16,984,000	54,503,000	62,896,000
EBITDA	65,264,000	35,612,000	9,550,000	67,818,000	26,697,000	-19,651,000
Return on Assets %	-.30%	3.02%	3.73%	-.67%	-.15%	-5.78%
Return on Equity %	-2.62%	27.98%	59.51%	-16.67%	-5.11%	-115.59%
Debt to Equity	6.66	5.43	7.76	19.09	17.63	36.99

CONTACT INFORMATION:

Phone: 469-522-4200   Fax: 469 522-4299
Toll-Free:
Address: 1603 Lyndon B Johnson Fwy, Ste. 800, Dallas, TX 75234
United States

STOCK TICKER/OTHER:

Stock Ticker: ARL
Employees: 895
Parent Company:

Exchange: NYS
Fiscal Year Ends: 12/31

SALARIES/BONUSES:

Top Exec. Salary: $ Bonus: $
Second Exec. Salary: $ Bonus: $

OTHER THOUGHTS:

Estimated Female Officers or Directors: 1
Hot Spot for Advancement for Women/Minorities:

Ameristar Casinos Inc

www.ameristar.com

NAIC Code: 721120

TYPES OF BUSINESS:

Casino Resorts
Casino Management

BRANDS/DIVISIONS/AFFILIATES:

Pinnacle Entertainment Inc
Ameristar Kansas City
Ameristar St. Charles
Ameristar Council Bluffs
Ameristar Vicksburg
Horseshu Hotel & Casino
Ameristar Black Hawk
Ameristar Casino Hotel East Chicago

CONTACTS: *Note: Officers with more than one job title may be intentionally listed here more than once.*

Anthony Sanfilippo, CEO-Pinnacle Entertainment
Neil Walkoff, Exec. VP-Oper.-Pinnacle Entertainment
Carlos Ruisanchez, CFO
Christina Donelson, Sr.VP-Human Resources-Pinnacle Entertainment
Jim Frank, Sr. VP

GROWTH PLANS/SPECIAL FEATURES:

Ameristar Casinos, Inc., a subsidiary of Pinnacle Entertainment, is a gaming and entertainment company that develops, owns and operates eight casino facilities. Its properties include Ameristar St. Charles and Ameristar Kansas City in Missouri; Ameristar Vicksburg in Mississippi; Ameristar Council Bluffs in southwestern Iowa; Ameristar Black Hawk in Denver, Colorado; Ameristar Casino Hotel East Chicago; and Cactus Pete's Resort Casino and Horseshu Hotel & Casino, both near the Idaho border in Jackpot, Nevada. The casinos typically offer slot machines and a variety of table games, including blackjack, craps, roulette, baccarat and numerous live poker variations such as Texas Hold 'Em and Pai Gow. In addition, some locations offer sports book wagering. The casinos also offer a variety of casual dining and upscale restaurants, sports bars and private clubs for Star Awards members. Ameristar St. Charles offers two ballrooms for its guests, five meeting rooms and an executive board room. Ameristar Kansas City features an 18-screen movie theater, a 4,280 square foot arcade and an activity center called Kids Quest. Ameristar Vicksburg is a permanently docked riverboat casino located on the Mississippi River, while Ameristar Council Bluffs offers a cruising riverboat casino that travels down the Missouri River. The East Chicago property is a 56,000 square foot complex that features a 550-seat ballroom. Cactus Pete's features an outdoor amphitheater, arcades, an 18-hole golf course and tennis courts. Horseshu Hotel & Casino, located across the street from Cactus Pete's, is an old-western style casino, and includes onsite amenities such as a grocery store, gas station and garage.

The firm offers employees medical, dental and vision insurance; prescription drug coverage; a 401(k); life & disability insurance; flexible spending accounts; an employee assistance program; and tuition reimbursement.

FINANCIAL DATA: *Note: Data for latest year may not have been available at press time.*

In U.S. $	2015	2014	2013	2012	2011	2010
Revenue	1,275,000,000	1,250,000,000	1,200,000,000	1,195,220,992	1,214,505,984	1,189,282,048
R&D Expense						
Operating Income						
Operating Margin %						
SGA Expense						
Net Income						
Operating Cash Flow						
Capital Expenditure						
EBITDA						
Return on Assets %						
Return on Equity %						
Debt to Equity						

CONTACT INFORMATION:

Phone: 702-567-7000   Fax:
Toll-Free:
Address: 3773 Howard Hughes Pkwy., Ste. 490S, Las Vegas, NV 89169 United States

STOCK TICKER/OTHER:

Stock Ticker: Subsidiary   Exchange:
Employees: 7,115   Fiscal Year Ends: 12/31
Parent Company: Pinnacle Entertainment Inc

SALARIES/BONUSES:

Top Exec. Salary: $ Bonus: $
Second Exec. Salary: $ Bonus: $

OTHER THOUGHTS:

Estimated Female Officers or Directors: 1
Hot Spot for Advancement for Women/Minorities: Y

Amey plc

NAIC Code: 541330

TYPES OF BUSINESS:

Engineering Services
Facilities Management
Design Services
Consulting Services
Asset Management
Infrastructure Services
Highway Design Services
IT Services

BRANDS/DIVISIONS/AFFILIATES:

Grupo Ferrovial SA
Travel Point Trading Ltd

GROWTH PLANS/SPECIAL FEATURES:

Amey plc, a subsidiary of the Spanish company Grupo Ferrovial SA, provides business and infrastructure support services through more than 150 locations across the U.K. Amey works with public and regulated sector customers, operating more than 320 contracts. These services involve industries such as consulting, water, gas, strategic highways, rail, local highways, environmental and waste, social housing, defense, total facilities management, social justice and transportation. In January 2016, the firm acquired Travel Point Trading Ltd., a strategic asset management consultancy.

The firm offers its employees life, dental, travel and personal accident insurance; parental leave; paid time off; and a pension plan.

CONTACTS: Note: Officers with more than one job title may be intentionally listed here more than once.

Mel Ewell, CEO
Andrew Nelson, CFO
Ian Deninson, Dir.-Human Resources
Darryl Salmons, CIO
Wayne Robertson, Head-Legal
John Faulkner, Dir.-Strategy & Development
Valerie Hughes-D'Aeth, Dir.-Communications
Andy Milner, Managing Dir.-Consulting, Rail & Strategic Highway
Gillian Duggan, Managing Dir.-Built Environment
Dan Holland, Managing Dir.-Utilities & Defense
Nick Gregg, Managing Dir.-Gov't
Richard Mottram, Chmn.

FINANCIAL DATA: Note: Data for latest year may not have been available at press time.

In U.S. $	2015	2014	2013	2012	2011	2010
Revenue		3,501,722,776	3,189,383,510	1,879,789,127	1,801,696,276	1,540,810,713
R&D Expense						
Operating Income						
Operating Margin %						
SGA Expense						
Net Income		138,775,832	101,431,558			
Operating Cash Flow						
Capital Expenditure						
EBITDA						
Return on Assets %						
Return on Equity %						
Debt to Equity						

CONTACT INFORMATION:

Phone: 44-1865-713-100   Fax: 44-1865-713-357
Toll-Free:
Address: Edmund Halley Rd., The Sherard Bldg., Oxford, OX4 4DQ United Kingdom

STOCK TICKER/OTHER:

Stock Ticker: Subsidiary
Employees: 21,000
Parent Company: Grupo Ferrovial SA

Exchange:
Fiscal Year Ends: 12/31

SALARIES/BONUSES:

Top Exec. Salary: $ Bonus: $
Second Exec. Salary: $ Bonus: $

OTHER THOUGHTS:

Estimated Female Officers or Directors: 2
Hot Spot for Advancement for Women/Minorities:

AMLI Residential Properties Trust
www.amli.com

NAIC Code: 531110

TYPES OF BUSINESS:
Real Estate Investment Trust, Luxury Apartments
Apartment Communities
Construction & Landscape Services
Management & Leasing Services
Investment Services

BRANDS/DIVISIONS/AFFILIATES:
AMLI Management Company
AMLI Institutional Advisors Inc
AMLI Residential Construction LLC
Morgan Stanley
AMLI Nest Egg
FAMLI

CONTACTS: Note: Officers with more than one job title may be intentionally listed here more than once.
Gregory T. Mutz, CEO
Allan J. Sweet, Pres.
Leslie Silverman, Sr. VP-Human Resources
Rick Fox, CIO
Charlotte Sparrow, Sr. VP-Legal & Risk Mgmt.
Dustin Fjeld, Sr. VP-Strategic Bus. Svcs. Dept.
Rosita Lin, Sr. VP-Acct.
Ken Veltri, Sr. VP-Asset Mgmt.
James McCormick, Sr. VP-Acquisitions
Fred Shapiro, Sr. VP-Acquisitions & Dispositions
Gregory T. Mutz, Chmn.

GROWTH PLANS/SPECIAL FEATURES:
AMLI Residential Properties Trust, part of Morgan Stanley's real estate division, is a self-administered and self-managed real estate investment trust (REIT) engaged in the development, acquisition and management of luxury apartment communities. The company manages apartment communities in Atlanta; Austin; Chicago; Dallas; Denver; Houston; Seattle; Miami, Doral, Fort Lauderdale, Miramar, Sunrise and Davie, Florida; and Glendale, Marina Del Ray, Camarillo, Orange, Woodland Hills, Rancho Cucamonga, California. Overall, AMLI manages 74 apartment communities, comprised of more than 25,000 units. The company also provides property management, institutional advisory and construction management services through its subsidiary companies, AMLI Management Company; AMLI Institutional Advisors, Inc.; and AMLI Residential Construction, LLC. AMLI's growth strategies employ a combination of brand and market promotions, acquisition, development and co-investments with institutional investors in new apartment communities. The company offers its tenants the AMLI Nest Egg program, an equity credit plan that allows 20% of base apartment rent to be applied as non-monetary credit towards the purchase price of a new home from KB Home, Pulte Homes or JEH Homes. AMLI also created the FAMLI volunteer program as a means of connecting AMLI residents and employees with local non-profit organizations.

AMLI offers its employees a benefits package that includes health and dependent care flexible spending accounts; a 401(k) plan; bonus programs; adoption assistance; medical, dental and vision coverage; life insurance; tuition assistance; and housing discounts.

FINANCIAL DATA: Note: Data for latest year may not have been available at press time.

In U.S. $	2015	2014	2013	2012	2011	2010
Revenue						
R&D Expense						
Operating Income						
Operating Margin %						
SGA Expense						
Net Income						
Operating Cash Flow						
Capital Expenditure						
EBITDA						
Return on Assets %						
Return on Equity %						
Debt to Equity						

CONTACT INFORMATION:
Phone: 312-283-4700   Fax: 312-283-4720
Toll-Free:
Address: 200 W. Monroe St., Ste. 2200, Chicago, IL 60606 United States

STOCK TICKER/OTHER:
Stock Ticker: Subsidiary   Exchange:
Employees: Fiscal Year Ends: 12/31
Parent Company: Morgan Stanley

SALARIES/BONUSES:
Top Exec. Salary: $ Bonus: $
Second Exec. Salary: $ Bonus: $

OTHER THOUGHTS:
Estimated Female Officers or Directors: 5
Hot Spot for Advancement for Women/Minorities: Y

Anderson-Tully Lumber Company

www.andersontully.com

NAIC Code: 0

TYPES OF BUSINESS:

Real Estate Investment Trust
Timber Tract Operations
Hardwood Products
Hardwood Flooring

BRANDS/DIVISIONS/AFFILIATES:

Anderson-Tully Worldwide
Anderson-Tully China
Anderson-Tully de Mexico
Anderson-Tully Vietnam
Louisiana Hardwood Products
International Hardwood Resource
Patton Tully Transportation LLC
Forestland Group LLC (The)

CONTACTS: Note: Officers with more than one job title may be intentionally listed here more than once.

Richard Wilkerson, Pres.
Kimberly Opiela, General Acct. Mgr.

GROWTH PLANS/SPECIAL FEATURES:

Anderson-Tully Lumber Company (ATCO) is a REIT (real estate investment trust) focusing entirely on timberland investments. It is one of the few REITs that manage timberlands and one of the only ones to focus on hardwood timberlands. Originally a crate manufacturing company, founded in 1889, the company moved into the timberland ownership business to provide itself with sustainable operations, eventually shifting its focus away from crates and boxes to exclusively operating its timberlands. Today, it owns hardwood timber tracts along the Mississippi River in Mississippi, Louisiana and Arkansas. The firm was one of the first major Southeastern timber companies certified by the Forest Stewardship Council, an organization devoted to the promotion of responsible forestry. ATCO operates its own sawmills, lumber inspectors, holding facilities and transport networks. Customers may choose between a variety of hardwoods, including cottonwood, hackberry, pecan, poplar, red oak, sweetgum, sycamore, white ash, white oak and willow. Its operating companies include Anderson-Tully Worldwide, the firm's sales and marketing arm; Anderson-Tully China, with offices in Shanghai and Guangzhou; Anderson-Tully de Mexico, based in Mexico City; Anderson-Tully Vietnam, based in Ho Chi Minh City; and Louisiana Hardwood Products. Patton Tully Transportation, a sister company, provides shipping for the firm's products along the Mississippi river systems. Additionally, ATCO created a network of producers called the International Hardwood Resource, designed to insure the smooth flow of hardwood to its customers. The firm is owned by the Heartwood Forestland Fund V, an affiliate of The Forestland Group LLC.

FINANCIAL DATA: Note: Data for latest year may not have been available at press time.

In U.S. $	2015	2014	2013	2012	2011	2010
Revenue						
R&D Expense						
Operating Income						
Operating Margin %						
SGA Expense						
Net Income						
Operating Cash Flow						
Capital Expenditure						
EBITDA						
Return on Assets %						
Return on Equity %						
Debt to Equity						

CONTACT INFORMATION:

Phone: 601-629-3283   Fax: 601-629-3284
Toll-Free:
Address: 1725 N. Washington St., Vicksburg, MS 39181 United States

STOCK TICKER/OTHER:

Stock Ticker: Private   Exchange:
Employees: Fiscal Year Ends: 12/31
Parent Company: Forestland Group LLC (The)

SALARIES/BONUSES:

Top Exec. Salary: $ Bonus: $
Second Exec. Salary: $ Bonus: $

OTHER THOUGHTS:

Estimated Female Officers or Directors: 2
Hot Spot for Advancement for Women/Minorities:

Annaly Capital Management Inc

www.annaly.com

NAIC Code: 525990

TYPES OF BUSINESS:
Mortgage REIT
Mortgage-Backed Securities

BRANDS/DIVISIONS/AFFILIATES:
Fixed Income Discount Advisory Company
Rcap Securities Inc
Annaly Commercial Real Estate Group Inc
Annaly Middle Market Lending LLC

CONTACTS: Note: Officers with more than one job title may be intentionally listed here more than once.
Glenn Votek, CFO
Wellington Denahan, Chairman of the Board
Kevin Keyes, Director
R. Singh, Other Executive Officer

GROWTH PLANS/SPECIAL FEATURES:
Annaly Capital Management, Inc. is a self-managed, self-administered real estate investment trust (REIT) that focuses entirely on owning and managing a portfolio of mortgage-backed securities. To date, over 93.4% of its total assets have consisted of agency mortgage-backed securities and debentures. The company maintains several wholly-owned subsidiaries, including Fixed Income Discount Advisory Company (FIDAC); Rcap Securities, Inc. (Rcap); Annaly Commercial Real Estate Group, Inc. (ACREG); and Annaly Middle Market Lending LLC (MML). FIDAC manages an affiliated REIT for which it earns fee income. Rcap operates as a broker-dealer and is a member of the Financial Industry Regulatory Authority (or FINRA). ACREG specializes in acquiring, financing and managing commercial mortgage loans and other commercial real estate debt, commercial mortgage-backed securities and other commercial real estate-related assets. Finally, MML engages in corporate middle market lending transactions.

FINANCIAL DATA: Note: Data for latest year may not have been available at press time.

In U.S. $	2015	2014	2013	2012	2011	2010
Revenue	664,033,000	-627,616,000	3,969,992,000	2,067,308,000	639,716,000	1,471,616,000
R&D Expense						
Operating Income	463,793,000	-836,954,000	3,737,911,000	1,831,749,000	402,372,000	1,299,769,000
Operating Margin %	69.84%		94.15%	88.60%	62.89%	88.32%
SGA Expense	200,240,000	209,338,000	232,081,000	235,559,000	237,344,000	171,847,000
Net Income	466,556,000	-842,083,000	3,729,698,000	1,735,900,000	344,461,000	1,267,280,000
Operating Cash Flow	-3,167,019,000	6,128,468,000	-12,892,720,000	7,639,507,000	2,420,063,000	10,863,110,000
Capital Expenditure					3,555,000	
EBITDA						
Return on Assets %	.48%	-1.07%	3.39%	1.39%	.34%	1.63%
Return on Equity %	3.37%	-7.64%	27.60%	11.08%	2.59%	13.10%
Debt to Equity	0.26	0.10	0.07	0.05	0.03	0.06

CONTACT INFORMATION:
Phone: 212 696-0100 Fax:
Toll-Free: 888-826-6259
Address: 1211 Ave. of the Americas, Ste. 2902, New York, NY 10036 United States

STOCK TICKER/OTHER:
Stock Ticker: NLY
Employees: 8
Parent Company:

Exchange: NYS
Fiscal Year Ends: 12/31

SALARIES/BONUSES:
Top Exec. Salary: $1,500,000 Bonus: $
Second Exec. Salary: $91,346 Bonus: $

OTHER THOUGHTS:
Estimated Female Officers or Directors:

Hot Spot for Advancement for Women/Minorities:

Anthony & Sylvan Pools

www.anthonysylvan.com

NAIC Code: 236118

TYPES OF BUSINESS:

Swimming Pools
Pool-Related Products & Services

BRANDS/DIVISIONS/AFFILIATES:

iSWIM
Sylvan Fence

CONTACTS: *Note: Officers with more than one job title may be intentionally listed here more than once.*

Mark Koide, CEO
Alan Walker, VP-Operations
Francis Babik, Dir-Human Resources
Alan Walker, VP
Mark Koide, VP-Mktg.
Mark Koide, Chmn.

GROWTH PLANS/SPECIAL FEATURES:

Anthony & Sylvan Pools (ASP) is an installer of custom-designed, in-ground concrete and fiberglass residential swimming pools and spas. Its operations span a network of sales offices serving markets in Connecticut, Delaware, New Jersey, New York, Maryland, Pennsylvania, North Carolina, South Carolina, Virginia, West Virginia, Texas, Nevada and Washington, D.C. In addition to designing and offering residential swimming pools, the company provides pool renovation services in select markets; operates retail stores that sell spa- and pool-related products such as chemicals, replacement parts, toys, accessories and equipment; and operates service centers that offer post-installation services such as weekly maintenance as well as parts, equipment and accessories. ASP also offers eco-friendly swimming pools with filtration systems that are more energy efficient and do not require chlorine. ASP sells its products primarily to residential homeowners and homebuilders. The company offers its customers financing options and warranty packages. On ASP's web site, customers can make payments through the iSWIM online payment system. The firm also provides fence building through its Sylvan Fence division, serving Maryland, Northern Virginia, Delaware, Eastern Pennsylvania, New Jersey and Washington, D.C.

The firm offers employee benefits that include health insurance, dental coverage, company-paid life insurance, discounted prescription drugs, short- and long-term disability insurance, a 401(k) savings plan and career development training.

FINANCIAL DATA: *Note: Data for latest year may not have been available at press time.*

In U.S. $	2015	2014	2013	2012	2011	2010
Revenue						
R&D Expense						
Operating Income						
Operating Margin %						
SGA Expense						
Net Income						
Operating Cash Flow						
Capital Expenditure						
EBITDA						
Return on Assets %						
Return on Equity %						
Debt to Equity						

CONTACT INFORMATION:

Phone: 215-489-5600   Fax:
Toll-Free: 877-729-7946
Address: 3739 Easton Rd., Route 611, Doylestown, PA 18901 United States

STOCK TICKER/OTHER:

Stock Ticker: Private
Employees: 460
Parent Company:

Exchange:
Fiscal Year Ends: 12/31

SALARIES/BONUSES:

Top Exec. Salary: $ Bonus: $
Second Exec. Salary: $ Bonus: $

OTHER THOUGHTS:

Estimated Female Officers or Directors:
Hot Spot for Advancement for Women/Minorities:

Apartment Investment and Management Co   www.aimco.com

NAIC Code: 531110

TYPES OF BUSINESS:

Real Estate Investment Trust
Apartment Communities
Property Management Services

BRANDS/DIVISIONS/AFFILIATES:

AIMCO-LP Trust
AIMCO Properties LP
AIMCO Operating Partnership
AIMCO-GP Inc

CONTACTS: Note: Officers with more than one job title may be intentionally listed here more than once.

Terry Considine, CEO
Paul Beldin, CFO
Keith Kimmel, Executive VP, Divisional
Miles Cortez, Executive VP
John Bezzant, Executive VP
Lisa Cohn, Executive VP
Patti Fielding, Treasurer

GROWTH PLANS/SPECIAL FEATURES:

Apartment Investment and Management Co. (AIMCO) is a self-administered and self-managed real estate investment trust (REIT) engaged in the acquisition, ownership, management and redevelopment of apartment properties. The company operates in three segments: property operations, portfolio management and redevelopment & development. Property operations, which focuses on the owning and operating of apartments, includes both conventional apartments and affordable apartments, which AIMCO owns and operates throughout 22 U.S. states and Washington D.C. Conventional properties, which consist of market-rate apartments with rents that are paid for by the resident, include 140 properties with 40,464 units and an average ownership of 98%. Affordable properties are those paid either in part or fully by the government and include 56 properties with 8,685 units. Portfolio management involves the ongoing allocation of investment capital to meet geographic and product type goals. AIMCO targets geographic balance in its diversified portfolio in order to optimize risk-adjusted returns and to avoid the risk of undue concentration in any particular market. The portfolio is also balanced by product type, with both high quality apartment communities in excellent locations and also high land value apartment communities that support redevelopment activities. The redevelopment & development segment includes investments in certain apartment communities in superior locations. Redevelopment projects are executed using a phased approach, where the firm renovates portions of an apartment community in stages. The company's operations are almost completely conducted through AIMCO Properties, L.P., also known as the AIMCO Operating Partnership, which is majority-held through its wholly-owned subsidiaries AIMCO-GP, Inc. and AIMCO-LP Trust.

The firm offers employees medical, dental and vision insurance; life insurance; disability coverage; a 401(k) plan; an employee assistance program; flexible spending accounts; fitness club discounts; professional certification reimbursement; a tuition aid program; and an apartment discount.

FINANCIAL DATA: Note: Data for latest year may not have been available at press time.

In U.S. $	2015	2014	2013	2012	2011	2010
Revenue	981,310,000	984,363,000	974,053,000	1,033,197,000	1,079,584,000	1,144,934,000
R&D Expense						
Operating Income	256,215,000	262,350,000	249,019,000	200,160,000	166,287,000	130,509,000
Operating Margin %	26.10%	26.65%	25.56%	19.37%	15.40%	11.39%
SGA Expense	43,178,000	44,195,000	45,708,000	49,602,000	50,950,000	53,365,000
Net Income	248,710,000	309,249,000	207,290,000	132,456,000	-57,087,000	-71,728,000
Operating Cash Flow	359,891,000	321,424,000	325,596,000	316,827,000	258,819,000	257,500,000
Capital Expenditure	536,627,000	651,365,000	401,629,000	359,926,000	265,348,000	178,929,000
EBITDA	569,852,000	551,007,000	577,967,000	597,677,000	556,847,000	565,087,000
Return on Assets %	3.85%	4.93%	3.26%	1.23%	-1.44%	-1.63%
Return on Equity %	18.84%	30.93%	23.32%	14.95%	-32.25%	-25.90%
Debt to Equity	2.64	3.96	4.87	5.53	20.58	14.17

CONTACT INFORMATION:

Phone: 303-757-8101   Fax:
Toll-Free: 888-789-8600
Address: 4582 S. Ulster St. Pkwy., Ste. 1100, Denver, CO 80237 United States

STOCK TICKER/OTHER:

Stock Ticker: AIV   Exchange: NYS
Employees: 1,693   Fiscal Year Ends: 12/31
Parent Company:

SALARIES/BONUSES:

Top Exec. Salary: $600,000   Bonus: $
Second Exec. Salary: Bonus: $150,000
$400,000

OTHER THOUGHTS:

Estimated Female Officers or Directors: 3
Hot Spot for Advancement for Women/Minorities: Y

Sales, profits and employees may be estimates. Financial information, benefits and other data can change quickly and may vary from those stated here.

APi Group Inc

NAIC Code: 238220

TYPES OF BUSINESS:

Mechanical Contractors
Fire Protection Installation
Safety Systems Contractors
Piping Installation
Communications Contractors
Specialty Construction

BRANDS/DIVISIONS/AFFILIATES:

3S Incorporated
Reliance Fire Protection Inc
Vipond
J. Koski Company
M&N Construction
Summit Pipeline Services
Api Construction Co
Industrial Fabricators Inc

CONTACTS: Note: Officers with more than one job title may be intentionally listed here more than once.

Russell Becker, CEO
Steven Ulmer, COO
Thomas Lydon, CFO
Julius Chepey, CIO
Lee R. Anderson, Sr., Chmn.

GROWTH PLANS/SPECIAL FEATURES:

APi Group, Inc. is the parent company to over 40 companies working in the fields of fire protection, industrial and specialty construction in North America and the U.K. The firm divides its group companies into three divisions: life safety, energy and specialty. Life safety companies include 3S Incorporated, Reliance Fire Protection, Inc. and Vipond; energy companies include J. Koski Company, M&N Construction and Summit Pipeline Services; and specialty companies include APi Construction Co., APi Distribution and Industrial Fabricators, Inc. The services offered by APi Group include architectural metals, audio systems, boilers, communication systems, excavation and right of way, fire alarm systems, fire protection systems, garage doors, industrial & commercial building material, industrial silencers, insulation contractors and piping systems. The industries served by the firm include commercial, higher education, industrial, medical, oil & gas energy, residential and security & defense.

FINANCIAL DATA: Note: Data for latest year may not have been available at press time.

In U.S. $	2015	2014	2013	2012	2011	2010
Revenue	2,000,000,000	1,800,000,000	1,777,000,000	1,594,000,000	374,500,000	266,300,000
R&D Expense						
Operating Income						
Operating Margin %						
SGA Expense						
Net Income						
Operating Cash Flow						
Capital Expenditure						
EBITDA						
Return on Assets %						
Return on Equity %						
Debt to Equity						

CONTACT INFORMATION:

Phone: 651-636-4320   Fax: 651-636-0312
Toll-Free: 800-223-4922
Address: 1100 Old Hgwy. 8 NW, New Broghton, MN 55112 United States

STOCK TICKER/OTHER:

Stock Ticker: Private
Employees:
Parent Company:

Exchange:
Fiscal Year Ends:

SALARIES/BONUSES:

Top Exec. Salary: $ Bonus: $
Second Exec. Salary: $ Bonus: $

OTHER THOUGHTS:

Estimated Female Officers or Directors: 1
Hot Spot for Advancement for Women/Minorities:

ARC Group of Companies (The)

www.arcproperties.com

NAIC Code: 0

TYPES OF BUSINESS:

Real Estate Investment Trust
Credit Lease Properties

GROWTH PLANS/SPECIAL FEATURES:

The ARC Group of Companies consists of three operating companies: ARC International Fund; ARC Property Trust (the firm's publicly listed REIT); and ARC Properties, Inc., a development and management wing. Together, the companies specialize in the acquisition and development of credit lease properties throughout the U.S., with special emphasis on metropolitan areas east of the Mississippi. Retail and net lease properties are general retail, office or industrial buildings leased long-term to credit-worthy clients, who in turn pay the majority or entirety of operating expenses and insurance. Major retail tenants have included Home Depot, Lowe's, Staples, Walgreen's and Bed, Bath & Beyond. The company seeks undeveloped land from 1-100 acres in size, generally along the Eastern seaboard from New York City to Washington, D.C. and in Miami. Its preferences are less narrow concerning the location of its developed properties, but it tends to focus on the metropolitan areas of Miami, Florida; Boston, Massachusetts; Washington, D.C.; Philadelphia, Pennsylvania; Chicago, Illinois; and Charlotte, North Carolina. These properties tend to range anywhere from $1 million to $50 million. Additionally, ARC has leased residential properties near greater New York, Philadelphia and Baltimore.

BRANDS/DIVISIONS/AFFILIATES:

ARC International Fund
ARC Property Trust
ARC Properties Inc

CONTACTS: Note: Officers with more than one job title may be intentionally listed here more than once.

Robert J. Ambrosi, CEO
James M. Steuterman, COO
Bruce Nelson, CFO
Michael R. Ambrosi, Mktg. & Leasing
Gary S. Baumann, General Counsel
Gary Sheehan, Managing Dir.-Capital Resources
James M. Steuterman, Exec. VP-Capital Markets
Joseph Morena, Exec. VP
Steven L. Maloy, Sr. VP
Robert J. Martone, VP-Acquisitions
Robert J. Ambrosi, Chmn.

FINANCIAL DATA: Note: Data for latest year may not have been available at press time.

In U.S. $	2015	2014	2013	2012	2011	2010
Revenue						
R&D Expense						
Operating Income						
Operating Margin %						
SGA Expense						
Net Income						
Operating Cash Flow						
Capital Expenditure						
EBITDA						
Return on Assets %						
Return on Equity %						
Debt to Equity						

CONTACT INFORMATION:

Phone: 973-249-1000   Fax: 973-249-1001
Toll-Free:
Address: 1401 Broad St., Clifton, NJ 07013 United States

STOCK TICKER/OTHER:

Stock Ticker: Private   Exchange:
Employees: Fiscal Year Ends: 12/31
Parent Company:

SALARIES/BONUSES:

Top Exec. Salary: $ Bonus: $
Second Exec. Salary: $ Bonus: $

OTHER THOUGHTS:

Estimated Female Officers or Directors: 1
Hot Spot for Advancement for Women/Minorities:

Arcadis NV

NAIC Code: 541330

www.arcadis.com/index.aspx

TYPES OF BUSINESS:

Engineering Services
Consulting Services
Project Management
Environmental Services
Infrastructure Construction Management
Power Generation Facilities
Industrial & Residential Development

BRANDS/DIVISIONS/AFFILIATES:

Malcolm Pirnie Inc

CONTACTS: Note: Officers with more than one job title may be intentionally listed here more than once.

Neil McArthur, CEO
Renier Vree, CFO
Lia Belilos, Dir.-Human Resources
Bartheke Weerstra, General Counsel
Roland van Dijk, Global Dir.-Corp. Dev.
Rob Mooren, Global Dir.-Infrastructure Solutions
Stephanie Hottenhuis, Dir.-Europe
Eleanor Allen, Global Dir.-Water Solutions
Manoel Antonio Amarante Avelino da Silva, CEO-ARCADIS Latin America
Gary Coates, CEO-Arcadis U.S.

GROWTH PLANS/SPECIAL FEATURES:

Arcadis NV, based in the Netherlands, is an international provider of consulting, engineering and project management services for the infrastructure, environment and facilities sector. The company develops, designs, implements, maintains and operates projects for private and public sector clients through four divisions: environment, infrastructure, water and buildings. The environment division provides consulting on environmental policy for companies and governments, conducts environmental impact assessments and supports environmental management and environmentally conscious engineering practices. The division provides soil and groundwater contamination testing and develops cost-effective solutions for the remediation of contaminated soil and water. The infrastructure division consults, designs and manages the construction of infrastructure projects, including railroads, highways, harbors, waterways, dikes and retention ponds. The company also develops utilities for rail signaling, safety, communications and energy supply; constructs bridges and tunnels; and develops small power plants, wind farms and hydroelectric facilities. The water division oversees projects through the entire water cycle, from the supply of clean drinking water to wastewater treatment and water management services. It operates primarily through Malcolm Pirnie, Inc. Lastly, the buildings division develops and maintains buildings, including offices, stores, commercial properties, schools, museums, prisons, stadiums and railway stations. To carry out the activities of these divisions, Arcadis works from offices in Europe, the U.S. and South America. The firm conducts international business through its Multinational Client Program.

Arcadis offers its U.S. employees medical, dental and vision coverage; flexible spending accounts; an employee assistance program; 401(k); and a discount stock purchase plan.

FINANCIAL DATA: Note: Data for latest year may not have been available at press time.

In U.S. $	2015	2014	2013	2012	2011	2010
Revenue	3,899,699,000	3,005,118,000	2,869,401,000	2,901,954,000	2,300,801,000	2,284,197,000
R&D Expense						
Operating Income	184,148,200	171,428,200	172,317,800	172,762,600	143,519,100	147,874,700
Operating Margin %	4.72%	5.70%	6.00%	5.95%	6.23%	6.47%
SGA Expense			319,072,600			
Net Income	112,612,800	104,450,200	110,175,500	101,474,700	90,677,570	84,317,010
Operating Cash Flow	194,881,400	159,065,200	159,772,400	180,236,300	90,814,430	104,690,900
Capital Expenditure	60,866,090	43,434,720	36,766,230	39,690,470	40,221,940	40,712,360
EBITDA	287,595,900	235,755,700	230,678,200	226,880,400	201,266,000	191,072,200
Return on Assets %	3.63%	4.27%	5.59%	5.34%	5.32%	5.39%
Return on Equity %	10.39%	12.32%	17.09%	17.95%	18.74%	19.86%
Debt to Equity	0.68	0.56	0.54	0.56	0.81	0.81

CONTACT INFORMATION:

Phone: 31 202011011   Fax: 31 202011002
Toll-Free:
Address: Gustav Mahlerplein 97-103, Amsterdam, 1082 MS Netherlands

SALARIES/BONUSES:

Top Exec. Salary: $ Bonus: $
Second Exec. Salary: $ Bonus: $

STOCK TICKER/OTHER:

Stock Ticker: ARCVF   Exchange: GREY
Employees: 25,792   Fiscal Year Ends: 12/31
Parent Company:

OTHER THOUGHTS:

Estimated Female Officers or Directors: 4
Hot Spot for Advancement for Women/Minorities: Y

ARMOUR Residential REIT Inc

www.armourreit.com

NAIC Code: 525990

TYPES OF BUSINESS:
Mortgage REIT

BRANDS/DIVISIONS/AFFILIATES:
ARMOUR Capital Management LP

GROWTH PLANS/SPECIAL FEATURES:
Armour Residential REIT, Inc., managed by ARMOUR Capital Management LP (ACM), invests mainly in hybrid adjustable rate, adjustable rate and fixed rate residential mortgage backed securities issued or guaranteed by a U.S. Government-chartered entity, such as the Federal National Mortgage Association (Fannie Mae) and the Federal Home Loan Mortgage Corporation (Freddie Mac), or guaranteed by the Government National Mortgage Administration (Ginnie Mae). From time to time, a portion of its portfolio may be invested in unsecured notes and bonds issued by U.S. Government-chartered entities, U.S. treasuries and money market instruments, subject to certain income tests it must satisfy for its qualification as a real estate investment trust (REIT). The company seeks attractive long-term investment returns by investing its equity capital and borrowed funds in its targeted asset class.

CONTACTS:
Note: Officers with more than one job title may be intentionally listed here more than once.

James Mountain, CFO
Daniel Staton, Chairman of the Board
Scott Ulm, Co-CEO
Jeffrey Zimmer, Co-CEO
Mark Gruber, COO

FINANCIAL DATA:
Note: Data for latest year may not have been available at press time.

In U.S. $	2015	2014	2013	2012	2011	2010
Revenue	365,300,000	450,927,000	505,443,000	308,851,000	12,176,400	9,388,286
R&D Expense						
Operating Income	269,073,000	348,216,000	385,074,000	222,282,000	-9,390,741	6,385,490
Operating Margin %	73.65%	77.22%	76.18%	71.97%	-77.12%	68.01%
SGA Expense	33,390,000	35,114,000	35,042,000	23,597,000	8,458,185	1,265,954
Net Income	-31,205,000	-179,048,000	-187,044,000	222,306,000	-9,442,054	6,536,857
Operating Cash Flow	238,255,000	315,100,000	370,438,000	343,681,000	118,071,600	9,164,650
Capital Expenditure						
EBITDA	269,073,000	348,216,000	385,074,000	222,282,000	-9,390,741	6,385,490
Return on Assets %	-.31%	-1.21%	-1.09%	1.62%	-.25%	.97%
Return on Equity %	-3.14%	-10.66%	-9.56%	15.01%	-2.56%	10.04%
Debt to Equity						

CONTACT INFORMATION:
Phone: 772 617-4340   Fax:
Toll-Free:
Address: 3001 Ocean Drive, Vero Beach, FL 32963 United States

STOCK TICKER/OTHER:
Stock Ticker: ARR
Employees: 19
Parent Company:

Exchange: NYS
Fiscal Year Ends: 12/31

SALARIES/BONUSES:
Top Exec. Salary: $ Bonus: $
Second Exec. Salary: $ Bonus: $

OTHER THOUGHTS:
Estimated Female Officers or Directors:
Hot Spot for Advancement for Women/Minorities:

Atria Senior Living Group

www.atriaseniorliving.com

NAIC Code: 623310

TYPES OF BUSINESS:

Assisted Living Facilities
Assisted Living Centers
Alzheimer Care
Short-Term Health Care

BRANDS/DIVISIONS/AFFILIATES:

A Dash and A Dollop
Engage Life Directors
Life Guidance

CONTACTS: Note: Officers with more than one job title may be intentionally listed here more than once.

John A. Moore, CEO
Mark Jessee, Pres.

GROWTH PLANS/SPECIAL FEATURES:

Atria Senior Living Group is one of the nation's largest operators of facilities providing assisted living services for the country's burgeoning senior population. The company currently operates more than 180 communities across 28 states and seven Canadian provinces, which provide housing and support services for over 21,000 seniors. Each program that Atria offers is tailored to the individual, with residents free to bring their own furnishings and pets. Atria provides seniors with the following programs: The Independent Living program, which is a retirement lifestyle that frees seniors from the worries of home maintenance, encourages them to engage in activities and hobbies and allows them to choose their own degree of privacy or sociability; the Assisted Living program, which is available to help seniors with daily activities such as bathing, eating, dressing and medication management; the Life Guidance program, which is available at some communities to provide a separate and secure environment for seniors with Alzheimer's disease and other forms of memory impairment; and short-term retreat programs, which are tailored for seniors on a temporary basis for seasonal stays, hospital recovery or trial-period stays. The firm's Engage Life program works to provide residents with other interesting and meaningful activities, such as book clubs, exercise classes, gardening, arts and crafts, bingo and card games which are coordinated by Engage Life Directors that spend individual time with each resident to help them plan activities according to their interests. Atria also released a cookbook, titled A Dash and A Dollop, which includes dishes and stories from Atria residents. It can be purchased online or at any Barnes and Noble store.

FINANCIAL DATA: Note: Data for latest year may not have been available at press time.

In U.S. $	2015	2014	2013	2012	2011	2010
Revenue						
R&D Expense						
Operating Income						
Operating Margin %						
SGA Expense						
Net Income						
Operating Cash Flow						
Capital Expenditure						
EBITDA						
Return on Assets %						
Return on Equity %						
Debt to Equity						

CONTACT INFORMATION:

Phone: 502-779-4700   Fax: 502-779-4701
Toll-Free: 877-719-1600
Address: 300 East Market Street, Ste 100, Louisville, KY 40202 United States

STOCK TICKER/OTHER:

Stock Ticker: Subsidiary   Exchange:
Employees: 13,500   Fiscal Year Ends: 12/31
Parent Company: Ventas Inc

SALARIES/BONUSES:

Top Exec. Salary: $ Bonus: $
Second Exec. Salary: $ Bonus: $

OTHER THOUGHTS:

Estimated Female Officers or Directors: 1
Hot Spot for Advancement for Women/Minorities:

AV Homes Inc

NAIC Code: 236117   www.avhomesinc.com

TYPES OF BUSINESS:

Construction, Home Building and Residential
Retirement Communities
Residential Properties
Commercial Properties
Home Building
Property Management
Utility Operations
Title Insurance Agency

BRANDS/DIVISIONS/AFFILIATES:

Vitalia at Tradition
Solivita
CantaMia

CONTACTS: Note: Officers with more than one job title may be intentionally listed here more than once.

Roger Cregg, CEO
Michael Burnett, CFO
Joshua Nash, Chairman of the Board
Joseph Mulac, Executive VP
S. Shullaw, Executive VP

GROWTH PLANS/SPECIAL FEATURES:

AV Homes, Inc. is a diversified real estate company engaged primarily in the development of lifestyle communities, including active adult and primary residential communities in Florida, Arizona and North Carolina. Overall, the firm owns 3,505 developed residential lots, 2,853 partially developed residential lots, 9,572 undeveloped residential lots and 7,220 acres of mixed use, commercial and industrial land. Solivita, Vitalia at Tradition and CantaMia serve as the company's flagship communities. Solivita comprises approximately 10,387 lots on 7.193 acres in Central and south of Orlando, Florida areas. It offers its residents numerous activities through the community's Lifestyles program and approximately 148,000 square feet of recreation facilities. These facilities include two fitness centers, 14 heated swimming pools, restaurants, arts and crafts rooms, a cafe and other meeting and ballroom facilities in a 452-acre active adult community in St. Lucie County, Florida, located between Vero Beach and West Palm Beach on Florida's east coast. CantaMia is a 1,696-unit active adult community located on 541 acres in the Estrella Mountain Ranch master planned community in Goodyear, Arizona. Amenities include an exercise facility and swimming pools, a demonstration kitchen, library, technology center, rooms for arts/crafts and games, a movement studio for yoga and aerobics and a cafe. Residents have exclusive use of the 30,000 square foot recreation and lifestyle facility situated on the focal point 10-acre man-made lake system. CantaMia also has space for outdoor sporting venues including swimming, softball, pickle ball, bocce ball, tennis and horseshoes. Projects under development include Encore at Eastmark in Arizona and Creekside at Bethpage, located in Virginia. In 2015, the firm agreed to acquire the home building assets of privately held, North-Carolina based Bonterra Builders LLC.

FINANCIAL DATA: Note: Data for latest year may not have been available at press time.

In U.S. $	2015	2014	2013	2012	2011	2010
Revenue	517,766,000	285,913,000	143,700,000	107,487,000	88,982,000	59,138,000
R&D Expense						
Operating Income	21,117,000	3,771,000	-5,341,000	-79,969,000	-30,020,000	-29,590,000
Operating Margin %	4.07%	1.31%	-3.71%	-74.39%	-33.73%	-50.03%
SGA Expense	16,900,000	15,941,000	15,975,000	16,148,000	17,502,000	20,508,000
Net Income	11,950,000	-1,932,000	-9,477,000	-90,235,000	-165,881,000	-35,108,000
Operating Cash Flow	-58,977,000	-81,409,000	-62,437,000	-48,313,000	-17,568,000	-10,438,000
Capital Expenditure	1,209,000	1,815,000	1,023,000	4,421,000	831,000	53,000
EBITDA	25,118,000	7,469,000	-2,638,000	-75,305,000	-152,426,000	-25,732,000
Return on Assets %	1.69%	-.34%	-5.31%	-24.16%	-34.75%	-6.15%
Return on Equity %	4.06%	-.67%	-9.45%	-42.93%	-48.48%	-8.03%
Debt to Equity	1.08	1.04	0.36	0.63	0.41	0.17

CONTACT INFORMATION:

Phone: 480-214-7400   Fax:
Toll-Free:
Address: 8601 N. Scottsdale Rd., Suite 225, Scottsdale, AZ 85253 United States

STOCK TICKER/OTHER:

Stock Ticker: AVHI
Employees: 214
Parent Company:

Exchange: NAS
Fiscal Year Ends: 12/31

SALARIES/BONUSES:

Top Exec. Salary: $400,000   Bonus: $50,000
Second Exec. Salary: $300,000   Bonus: $45,000

OTHER THOUGHTS:

Estimated Female Officers or Directors: 1
Hot Spot for Advancement for Women/Minorities: Y

AvalonBay Communities Inc

www.avalonbay.com

NAIC Code: 0

TYPES OF BUSINESS:

Real Estate Investment Trust
Apartment Complexes
Residential Construction

BRANDS/DIVISIONS/AFFILIATES:

CONTACTS: Note: Officers with more than one job title may be intentionally listed here more than once.

Timothy Naughton, CEO
Kevin OShea, CFO
Keri Shea, Chief Accounting Officer
Leo Horey, Chief Administrative Officer
Sean Breslin, COO
Stephen Wilson, Executive VP, Divisional
William Mclaughlin, Executive VP, Divisional
Edward Schulman, Executive VP
Matthew Birenbaum, Other Executive Officer
Michael Feigin, Other Executive Officer

GROWTH PLANS/SPECIAL FEATURES:

AvalonBay Communities, Inc. is a real estate investment trust (REIT) engaged in the development, redevelopment, acquisition, ownership and operation of multifamily communities in high barrier-to-entry markets in the U.S. The company owns or operates 257 apartment communities containing over 75,549 apartment homes in 10 states and Washington, D.C. AvalonBay also has 26 wholly-owned communities currently under construction and holds development rights for 32 additional communities. Its investment markets are located in the following regions of the U.S.: New England, the New York/New Jersey Metro area, the Mid-Atlantic, the Pacific Northwest and Northern and Southern California. These investments are divided into three classes: established communities, other stabilized communities and development/redevelopment communities. Established communities are generally operating communities that were owned and had stabilized occupancy and operating expenses at the beginning of the prior year. Other stabilized communities are generally all other operating communities that have stabilized occupancy and operating expenses during the current year, but had not achieved stabilization as of the beginning of the prior year. Development/redevelopment communities consist of communities that are under construction, communities where substantial redevelopment is in progress or is planned to begin during the current year and communities under lease-up. AvalonBay's primary markets are Boston, Massachusetts; Chicago, Illinois; San Jose and Los Angeles, California; Shelton, Connecticut; Virginia Beach, Virginia; Seattle, Washington; New York City; and Washington, D.C.

The firm offers employees medical, dental and vision insurance; disability benefits; AD&D insurance; flexible spending accounts; a dependent care account; an employee assistance program; an employee stock purchase plan; a 401(k); relocation assistance; and housing discounts for employees who live in an AvalonBay community.

FINANCIAL DATA: Note: Data for latest year may not have been available at press time.

In U.S. $	2015	2014	2013	2012	2011	2010
Revenue	1,856,028,000	1,685,061,000	1,462,921,000	1,038,660,000	968,711,000	895,266,000
R&D Expense						
Operating Income	548,304,000	615,365,000	352,114,000	368,916,000	331,273,000	272,515,000
Operating Margin %	29.54%	36.51%	24.06%	35.51%	34.19%	30.43%
SGA Expense	42,396,000	41,425,000	39,573,000	34,101,000	29,371,000	26,846,000
Net Income	742,038,000	683,567,000	353,141,000	423,869,000	441,622,000	175,331,000
Operating Cash Flow	1,056,754,000	886,641,000	724,501,000	540,819,000	429,391,000	332,106,000
Capital Expenditure	1,625,191,000	1,294,657,000	1,277,551,000	765,059,000	688,568,000	462,235,000
EBITDA	1,026,227,000	1,058,047,000	925,829,000	629,010,000	581,542,000	505,457,000
Return on Assets %	4.47%	4.33%	2.66%	4.31%	5.41%	2.29%
Return on Equity %	7.83%	7.74%	4.57%	7.54%	11.46%	5.51%
Debt to Equity	0.65	0.72	0.72	0.29	0.83	1.23

CONTACT INFORMATION:

Phone: 703-329-6300   Fax: 703 329-9130
Toll-Free:
Address: 671 N. Glebe Rd., Ste. 800, Arlington, VA 22203 United States

STOCK TICKER/OTHER:

Stock Ticker: AVB   Exchange: NYS
Employees: 2,981   Fiscal Year Ends: 12/31
Parent Company:

SALARIES/BONUSES:

Top Exec. Salary: $950,000   Bonus: $
Second Exec. Salary: Bonus: $
$490,385

OTHER THOUGHTS:

Estimated Female Officers or Directors: 14
Hot Spot for Advancement for Women/Minorities: Y

Balfour Beatty plc
NAIC Code: 237990

www.balfourbeatty.com

TYPES OF BUSINESS:
Subway Construction
Engineering Services
Railway Services
Property Management
Utility & Roadway Infrastructure Management

BRANDS/DIVISIONS/AFFILIATES:
Gwynt Y Mor Offshore Transmission Project

CONTACTS: Note: Officers with more than one job title may be intentionally listed here more than once.
Leo Quinn, CEO
Philip Harrison, CFO
Paul Raby, Dir.-Human Resources
Chris Vaughan, General Counsel
Peter Zinkin, Dir.-Planning & Dev.
Mark Layman, CEO-Construction
George Pierson, CEO-Professional Services
Ian Rylatt, CEO-Infrastructure Investments
Kevin Craven, CEO-Services
Philip Aiken, Chmn.

GROWTH PLANS/SPECIAL FEATURES:
Balfour Beatty plc provides engineering, construction and financial services for rail, road, power and building projects worldwide. It is one of the largest fixed rail infrastructure contracting companies in the world. Balfour Beatty divides its business into three categories: construction services, support services and infrastructure investments. Construction services include building design, civil and ground engineering, rail engineering, refurbishment and fit-out and mechanical and electrical services. The firm has an established presence in the U.K. and U.S. through its subsidiaries and is expanding its business into Hong Kong and the Middle East. Balfour Beatty provides ongoing operation and maintenance of assets after construction and offers business services outsourcing through its support services division. The division encompasses the company's utilities, facilities management, rail renewals and highway management activities. The infrastructure investments segment is a leader in the public private partnership (PPP) contracts sector. It maintains concessions in the U.K., primarily in the education, health and roads/street lighting sectors; in the U.S., primarily involved in the military housing market; and in Singapore. In 2015, the firm acquired Gwynt Y Mor Offshore Transmission Project (OFTO) in the U.K.; and sold its 50% stake in Signalling Solutions Ltd. to Alstom.

FINANCIAL DATA: Note: Data for latest year may not have been available at press time.

In U.S. $	2015	2014	2013	2012	2011	2010
Revenue	9,833,029,000	10,269,890,000	12,363,740,000	13,407,130,000	13,422,680,000	13,057,920,000
R&D Expense					8,482,843	5,655,229
Operating Income	-257,312,900	-397,279,800	67,862,740	104,621,700	343,555,100	311,037,600
Operating Margin %	-2.61%	-3.86%	.54%	.78%	2.55%	2.38%
SGA Expense						12,724,260
Net Income	-291,244,300	-84,828,430	-49,483,250	62,207,520	262,968,100	202,174,400
Operating Cash Flow	-182,381,100	-525,936,300	-247,416,300	-336,486,100	-53,724,670	149,863,600
Capital Expenditure	117,346,000	182,381,100	169,656,900	104,621,700	135,725,500	139,966,900
EBITDA	-182,381,100	-308,210,000	219,140,100	319,520,400	588,143,700	549,971,000
Return on Assets %		-1.09%	-.60%	.76%	3.33%	2.64%
Return on Equity %		-5.30%	-2.99%	3.47%	15.83%	13.70%
Debt to Equity		0.76	0.81		0.27	0.25

CONTACT INFORMATION:
Phone: 44 2072166800   Fax: 44 2072166950
Toll-Free:
Address: 130 Wilton Rd., London, SW1V 1LQ United Kingdom

STOCK TICKER/OTHER:
Stock Ticker: BAFBF
Employees: 36,000
Parent Company:

Exchange: GREY
Fiscal Year Ends: 12/31

SALARIES/BONUSES:
Top Exec. Salary: $ Bonus: $
Second Exec. Salary: $ Bonus: $

OTHER THOUGHTS:
Estimated Female Officers or Directors: 2
Hot Spot for Advancement for Women/Minorities:

Bank of America Home Loans

www.bankofamerica.com/home-loans/overview.go

NAIC Code: 522310

TYPES OF BUSINESS:

Mortgage Banking
Banking

BRANDS/DIVISIONS/AFFILIATES:

Bank of America Corp

CONTACTS: *Note: Officers with more than one job title may be intentionally listed here more than once.*

Brian T. Moynihan, CEO
Thomas K. Montag, COO
Bruce R. Thompson, CFO
Brian Moynihan, CEO-Bank of America
Paul Morrison, Chief Risk Officer-Home Loans & Insurance

GROWTH PLANS/SPECIAL FEATURES:

Bank of America Home Loans (BOAHL), a wholly-owned subsidiary of Bank of America Corp., is engaged in mortgage lending and other real estate finance-related businesses, including mortgage banking; banking and mortgage warehouse lending; and securities. Home loan products include home loans, refinancing, home equity lines of credit, construction loans and reverse mortgages. Through the banking division, BOAHL takes deposits and invests in mortgage loans and home equity lines of credit, sourced primarily through its mortgage banking operation as well as through purchases from non-affiliates. The segment also offers short-term secured financing to mortgage lenders. Banking products include certificates of deposit, money market accounts, savings accounts and credit cards.

The company's parent offers employees benefits including medical, dental, vision, life and disability coverage; tuition reimbursement; an employee assistance program; a 401(k); auto & home insurance; adoption reimbursement; dependent care assistance; and flexible work arrangements.

FINANCIAL DATA: *Note: Data for latest year may not have been available at press time.*

In U.S. $	2015	2014	2013	2012	2011	2010
Revenue	8,830,000,000	8,511,600,000	7,716,000,000	8,751,000,000	-3,154,000,000	10,329,000,000
R&D Expense						
Operating Income						
Operating Margin %						
SGA Expense						
Net Income		-4,833,000,000	-5,155,000,000	-6,439,000,000	-19,465,000,000	-8,947,000,000
Operating Cash Flow						
Capital Expenditure						
EBITDA						
Return on Assets %						
Return on Equity %						
Debt to Equity						

CONTACT INFORMATION:

Phone: 818-225-3000   Fax:
Toll-Free: 800-283-8875
Address: 4500 Park Granada, Calabasas, CA 91302 United States

SALARIES/BONUSES:

Top Exec. Salary: $ Bonus: $
Second Exec. Salary: $ Bonus: $

STOCK TICKER/OTHER:

Stock Ticker: Subsidiary   Exchange:
Employees: Fiscal Year Ends: 12/31
Parent Company: BANK OF AMERICA CORP

OTHER THOUGHTS:

Estimated Female Officers or Directors: 1
Hot Spot for Advancement for Women/Minorities: Y

Banyan Tree Holdings Limited

www.banyantree.com

NAIC Code: 721110

TYPES OF BUSINESS:

Hotels & Resorts
Spas
Fine Art Galleries
Design Services

BRANDS/DIVISIONS/AFFILIATES:

Architrave
Angsana
Cassia
Banyan Tree
Banyan Tree Management Academy
Canopy Marketing Group Ptd Ltd
Banyan Tree Residences
Banyan Tree Hotels & Resorts

CONTACTS: *Note: Officers with more than one job title may be intentionally listed here more than once.*

Chiang See Ngoh Claire, Managing Dir.
Eddy See Hock Lye, CFO
Claire Chiang, Managing Dir.-Retail Oper.
Ho KwonCjan, Chief Group Designer
Shankar Chandran, Managing Dir.-Spa Oper.
Dharmali Kusumadi, Managing Dir.-Design Svcs.
Ho KwonPing, Exec. Chmn.

GROWTH PLANS/SPECIAL FEATURES:

Banyan Tree Holdings Limited, operating as Banyan Tree Hotels & Resorts, is a Singapore-based hospitality company that develops, manages, operates and invests in resorts, hotels, spas and other properties across 27 countries. The company's portfolio includes more than 30 hotels and resorts, 60 spas, 80 retail galleries and two golf course clubs. All employees are trained at the Banyan Tree Management Academy, which is centrally located in Phuket, Thailand. Banyan Tree's operations are divided into several segments: hotel management & investment, in which it owns and manages luxury hotels under its Angsana, Cassia, Dhawa and Banyan Tree brands as well as hotels that are managed by other operators; Canopy Marketing Group Ptd Ltd., a wholly-owned subsidiary that provides marketing insights on global niche markets such as hotels, resorts, spas, and gallery as well as residential ownership; spa operations, which oversees its more than 60 spa outlets worldwide; gallery operations, which oversees its more than 80 stores worldwide; hotel residences, which are sold under the Banyan Tree Residences brand name; property sales, which are properties not part of hotel operations and sold by subsidiary Laguna Resorts and Hotels; design and other services, which comprises subsidiary Architrave, Bayan's in-house architectural arm, and other services comprise the firm's golf course clubs; and Real Estate Hospitality Funds, which was set up to tap private equity and other sources of investments in order to provide a cost effective structure to fund the group's future developments. The majority of Banyan Tree's properties and operations are located in the Asia Pacific region and Mexico, with destinations in Bahrain, China, Indonesia, Korea, Maldives, Seychelles, Thailand and the UAE.

FINANCIAL DATA: *Note: Data for latest year may not have been available at press time.*

In U.S. $	2015	2014	2013	2012	2011	2010
Revenue	275,387,400	243,203,100	264,584,800	251,412,300	244,782,500	226,812,300
R&D Expense						
Operating Income	4,681,812	21,678,820	38,364,560	34,533,380	15,375,240	46,855,260
Operating Margin %		8.91%	14.49%	13.73%	6.28%	20.65%
SGA Expense	66,532,690	57,060,600	83,848,420	50,423,460	73,309,500	51,220,600
Net Income	-20,444,110	761,482	13,480,830	11,041,860	2,232,441	11,658,470
Operating Cash Flow	-82,259,330	-24,368,150	-4,221,952	23,002,690	-5,723,370	8,339,896
Capital Expenditure	17,589,850	16,445,770	13,157,660	19,911,440	20,334,900	13,494,200
EBITDA	24,758,180	40,512,310	57,105,920	57,926,090	31,758,610	82,027,540
Return on Assets %		.07%	1.30%	1.08%	.11%	1.12%
Return on Equity %		.18%	3.31%	2.74%	.29%	3.04%
Debt to Equity		0.78	0.60	0.56	0.57	0.33

CONTACT INFORMATION:

Phone: 65 6849-5888   Fax: 65 6462-2463
Toll-Free: 800-591-0439
Address: 211 Upper Bukit Timah Rd., Singapore, 588182 Singapore

STOCK TICKER/OTHER:

Stock Ticker: BYNEF   Exchange: GREY
Employees: Fiscal Year Ends: 12/31
Parent Company:

SALARIES/BONUSES:

Top Exec. Salary: $ Bonus: $
Second Exec. Salary: $ Bonus: $

OTHER THOUGHTS:

Estimated Female Officers or Directors: 3
Hot Spot for Advancement for Women/Minorities: Y

Barcelo Crestline Corporation

www.crestlinehotels.com

NAIC Code: 721110

TYPES OF BUSINESS:

Hotels

BRANDS/DIVISIONS/AFFILIATES:

Barcelo Corporacion Empresarial SA
Crestline Hotels & Resorts LLC

GROWTH PLANS/SPECIAL FEATURES:

Barcelo Crestline Corporation is a leading independent hospitality management company. It is a privately-owned subsidiary of Barcelo Corporacion Empresarial SA (based in Palma de Mallorca, Spain). Through its operating company and wholly-owned subsidiary Crestline Hotels & Resorts, LLC (Crestline), the firm manages properties including 106 hotels, resorts and conference and convention centers with nearly 15,700 rooms in 28 states and Washington, D.C. Crestline operates five private-label hotels, with the remainder falling under brand lines such as Starwood, Marriott, InterContinental Hotels Group, Hyatt and Hilton. It generally acquires boutique or premium branded limited-service hotels, or first class, full-service hotels located in either suburban or urban markets.

CONTACTS:
Note: Officers with more than one job title may be intentionally listed here more than once.

James Carroll, CEO
James Carroll, Pres.
Pam Siegler, CFO
Vicki Denfeld, Exec. VP-Sales & Mktg.
Jerry Galindo, VP-IT
Pierre Donahue, General Counsel
Ed Hoganson, Exec. VP-Bus. Dev.
Ed Hoganson, Exec. VP-Finance
Carolee Moore, VP-Sales & Mktg.
Pan Siegler, VP-Hotel Accounting & Finance
Bruce Wardinski, Chmn.

FINANCIAL DATA:
Note: Data for latest year may not have been available at press time.

In U.S. $	2015	2014	2013	2012	2011	2010
Revenue						
R&D Expense						
Operating Income						
Operating Margin %						
SGA Expense						
Net Income						
Operating Cash Flow						
Capital Expenditure						
EBITDA						
Return on Assets %						
Return on Equity %						
Debt to Equity						

CONTACT INFORMATION:

Phone: 571-529-6000   Fax: 571-529-6050
Toll-Free:
Address: 3950 University Dr., Ste. 301, Fairfax, VA 22030 United States

STOCK TICKER/OTHER:

Stock Ticker: Subsidiary   Exchange:
Employees: 3,677   Fiscal Year Ends: 12/31
Parent Company: Barcelo Corporacion Empresarial SA

SALARIES/BONUSES:

Top Exec. Salary: $ Bonus: $
Second Exec. Salary: $ Bonus: $

OTHER THOUGHTS:

Estimated Female Officers or Directors: 4
Hot Spot for Advancement for Women/Minorities: Y

Beazer Homes Usa Inc

www.beazer.com

NAIC Code: 236117

TYPES OF BUSINESS:

Construction, Home Building and Residential
Home Design

BRANDS/DIVISIONS/AFFILIATES:

CONTACTS: Note: Officers with more than one job title may be intentionally listed here more than once.

Allan Merrill, CEO
Robert Salomon, CFO
Kenneth Khoury, Chief Administrative Officer
Brian Beazer, Director
Stephen Zelnak, Director

GROWTH PLANS/SPECIAL FEATURES:

Beazer Homes USA, Inc., a top national homebuilder, designs, sells and builds single-family and multi-family homes in 13 states. Its operations are divided into three geographic regions: West, which covers markets in Arizona, California, Nevada and Texas; East, covering markets in Maryland/Delaware, Virginia, Indiana and Tennessee; and Southeast, which consists of Florida, Georgia, North Carolina and South Carolina. Beazer's homes are designed primarily for first-time buyers and are designed at various price points to appeal to a wide demographic. They are also generally offered for sale in advance of their construction. Most homes are built from standardized plans, which help facilitate a greater profit margin through faster construction times and standard materials. However, the company does offer some measure of customization in the form of flooring, cabinetry, countertop and wall covering options. Beazer's business strategy involves geographically diverse project development to minimize losses caused by local economic downturns and a localized management scheme based on local market knowledge and expertise. The company also seeks to make energy saving, water conservation and improved air quality standard components in all homes. The company's home warranties are issued, administered and insured by independent third parties. The firm is a national sponsor of HomeAid America, a nonprofit national program that provides housing for the temporarily homeless. In 2014, the firm sold its assets in Beazer Pre-Owned Rental Homes, Inc.

FINANCIAL DATA: Note: Data for latest year may not have been available at press time.

In U.S. $	2015	2014	2013	2012	2011	2010
Revenue	1,627,413,000	1,463,767,000	1,287,577,000	1,005,677,000	742,405,000	1,009,841,000
R&D Expense						
Operating Income	51,587,000	55,689,000	27,261,000	-62,058,000	-132,245,000	-113,822,000
Operating Margin %	3.16%	3.80%	2.11%	-6.17%	-17.81%	-11.27%
SGA Expense	207,519,000	194,491,000	174,085,000	153,636,000	170,087,000	186,556,000
Net Income	344,094,000	34,383,000	-33,868,000	-145,326,000	-204,859,000	-34,049,000
Operating Cash Flow	-81,049,000	-160,469,000	-174,642,000	-20,845,000	-178,936,000	69,685,000
Capital Expenditure	15,964,000	14,553,000	10,761,000	17,363,000	20,514,000	10,849,000
EBITDA	64,925,000	68,968,000	40,045,000	-90,951,000	-112,710,000	-100,417,000
Return on Assets %	15.33%	1.69%	-1.70%	-7.34%	-10.55%	-1.73%
Return on Equity %	75.66%	13.23%	-13.47%	-63.09%	-68.80%	-11.47%
Debt to Equity	2.42	5.50	6.28	5.71	7.50	3.12

CONTACT INFORMATION:

Phone: 770-829-3700   Fax: 770-481-2808
Toll-Free:
Address: 1000 Abernathy Rd., Ste. 1200, Atlanta, GA 30328 United States

STOCK TICKER/OTHER:

Stock Ticker: BZH   Exchange: NYS
Employees: 1,063   Fiscal Year Ends: 09/30
Parent Company:

SALARIES/BONUSES:

Top Exec. Salary: $900,000   Bonus: $
Second Exec. Salary: $525,000   Bonus: $

OTHER THOUGHTS:

Estimated Female Officers or Directors:
Hot Spot for Advancement for Women/Minorities: Y

Sales, profits and employees may be estimates. Financial information, benefits and other data can change quickly and may vary from those stated here.

Bechtel Group Inc

www.bechtel.com

NAIC Code: 237000

TYPES OF BUSINESS:

Heavy Construction
Civic Engineering
Outsourcing
Financial Services
Atomic Propulsion Systems Engineering
Airport Construction
Electric Power Plant Construction
Nuclear Power Plant Construction

BRANDS/DIVISIONS/AFFILIATES:

CONTACTS: *Note: Officers with more than one job title may be intentionally listed here more than once.*

Bill Dudley, CEO
Brendan Bechtel, COO
Peter Dawson, CFO
Andy Greig, Head-Human Resources
Carol Zierhoffer, Head-Info. Systems
Michael Bailey, General Counsel
Charlene Wheeless, Head-Corp. Affairs
Anette Sparks, Controller
Steve Katzman, Pres., Asia
Jose Ivo, Pres., Americas
Charlene Wheeless, Head-Sustainability Svcs.
Michael Wilkinson, Head-Risk Mgmt.
Riley P. Bechtel, Chmn.
David Welch, Pres., EMEA

GROWTH PLANS/SPECIAL FEATURES:

Bechtel Group, Inc., founded in 1906 by Warren A. Bechtel, is one of the world's largest engineering companies. The privately-owned firm offers engineering, construction and project management services, with a broad project portfolio including road and rail systems, airports and seaports, nuclear power plants, petrochemical facilities, mines, defense and aerospace facilities, environmental cleanup projects, telecommunication networks, pipelines and oil fields development. Bechtel has four main business units, infrastructure; nuclear, security & environmental; oil, gas & chemicals; and mining & metals. The infrastructure segment oversees projects pertaining to hydroelectric power plants, ports, harbors, bridges, airports and airport systems, commercial and light-industrial buildings, wireless sites, railroads, rapid-transit and rail systems. The nuclear, security & environmental includes missile defense infrastructure, scientific and national security facility operations, environmental restoration and recovery, commercial and U.S. navy nuclear reactor services and chemical weapons dematerialization projects. The oil, gas & chemicals segment offers integrated design, procurement, construction and project management of oil, gas and natural gas facilities. The mining and metal segment encompasses mining and metal projects across six continents including procurement, construction, engineering and solutions for mining of coal, ferrous, industrial and nonferrous metals. The firm has participated in such notable endeavors as the construction of the Hoover Dam, the creation of the Bay Area Rapid Transit system in San Francisco, the massive James Bay Hydroelectric Project in Quebec and the quelling of oil field fires in Kuwait following the Persian Gulf War. Bechtel also constructed the Trans-Alaska Oil Pipeline, covering 800 miles between the Prudhoe Bay oil field and Valdez. Bechtel has also been contracted to develop the New Doha International Airport in Qatar.

The firm offers employees benefits including medical, dental and vision coverage; short- and long-term disability; flexible spending accounts; an employee assistance program; life insurance; and a 401(k) plan.

FINANCIAL DATA: *Note: Data for latest year may not have been available at press time.*

In U.S. $	2015	2014	2013	2012	2011	2010
Revenue	37,200,000,000	37,200,000,000	39,400,000,000	37,900,000,000	32,900,000,000	27,900,000,000
R&D Expense						
Operating Income						
Operating Margin %						
SGA Expense						
Net Income						
Operating Cash Flow						
Capital Expenditure						
EBITDA						
Return on Assets %						
Return on Equity %						
Debt to Equity						

CONTACT INFORMATION:

Phone: 415-768-1234   Fax: 415-768-9038
Toll-Free:
Address: 50 Beale St., San Francisco, CA 94105-1895 United States

STOCK TICKER/OTHER:

Stock Ticker: Private
Employees: 58,000
Parent Company:

Exchange:
Fiscal Year Ends: 12/31

SALARIES/BONUSES:

Top Exec. Salary: $ Bonus: $
Second Exec. Salary: $ Bonus: $

OTHER THOUGHTS:

Estimated Female Officers or Directors: 4
Hot Spot for Advancement for Women/Minorities: Y

Sales, profits and employees may be estimates. Financial information, benefits and other data can change quickly and may vary from those stated here.

Belmond Ltd

www.belmond.com

NAIC Code: 721110

TYPES OF BUSINESS:

Hotels, Luxury
Tourist Railroads
Cruise Lines
Safari Tours
Restaurants

BRANDS/DIVISIONS/AFFILIATES:

Belmond Le Manoir aux Quat'Saisons
Belmond Grand Hotel Europe
Belmond Copacabana Palace
Belmond Mount Nelson Hotel
Eastern & Oriental Express Railway
Belmond Royal Scotsman Railway
21 Club

CONTACTS: Note: Officers with more than one job title may be intentionally listed here more than once.

H. Roeland Vos, CEO
Philippe Cassis, COO
John M. Scott III, Pres.
Martin O'Grady, CFO
Ralph Aruzza, Chief Mktg. Officer
Katherine Blaisdell, VP-Tech. Svcs.
Maurizio Saccani, Chief-Prod. Dev.
Edwin S. Hetherington, General Counsel
Amy Brandt, VP-Corp. Finance & Investor Rel.
Neil Gribben, VP-Acct. & Control
Peter Massey-Cook, VP-Compliance
Philip Calvert, VP-Legal & Commercial Affairs
Raymond Blanc, VP-Orient-Express Hotels Ltd.
Richard M. Levine, Chief Legal Officer
Roland Hernandez, Chmn.
Maurizio Saccani, VP-Italy

GROWTH PLANS/SPECIAL FEATURES:

Belmond Ltd. manages and owns complete or partial interests in a portfolio of 46 properties in 23 countries. Its properties, operating under the name Belmond, include 35 deluxe hotels (with 3,510 individual guestrooms and multi-room suites), six tourist trains, three river or canal cruises and one stand-alone restaurant. The hotels and restaurants segment generates approximately 86% of revenues, while tourist trains and cruises together accounts for 14%, with property development activities accounting for the remainder. Belmond's properties include hotels such as Belmond Le Manoir aux Quat'Saisons, a 16th Century manor house in England; Belmond Grand Hotel Europe, which occupies an entire city block in St. Petersburg, Russia; Belmond Copacabana Palace in Rio de Janeiro, Brazil, complete with a casino and a 500-seat theater; and Belmond Mount Nelson Hotel, a famous historic property in Cape Town, South Africa, with gardens and pools just below Table Mountain. The firm's trains include the Belmond Royal Scotsman, which visits clan castles, historic battlegrounds and famous Scotch whiskey distilleries in Scotland, and the Eastern & Oriental Express, which travels weekly between Singapore, Kuala Lumpur and Bangkok. The company also owns the historic 21 Club restaurant in New York City, which originated as a Prohibition-era speakeasy. In 2015, the firm sold its 50% interest in the 167-room Hotel Ritz by Belmond in Madrid, Spain, to a joint venture between Mandarin Oriental International Limited and The Olayan Group.

FINANCIAL DATA: Note: Data for latest year may not have been available at press time.

In U.S. $	2015	2014	2013	2012	2011	2010
Revenue	551,385,000	585,715,000	594,081,000	545,418,000	588,559,000	571,942,000
R&D Expense						
Operating Income	61,545,000	54,441,000	13,750,000	40,371,000	-4,187,000	-12,400,000
Operating Margin %	11.16%	9.29%	2.31%	7.40%	-.71%	-2.16%
SGA Expense	206,197,000	219,584,000	227,270,000	210,010,000	226,675,000	190,594,000
Net Income	16,265,000	-1,880,000	-31,559,000	-7,061,000	-87,780,000	-62,759,000
Operating Cash Flow	67,924,000	51,336,000	67,124,000	45,612,000	45,720,000	33,968,000
Capital Expenditure	57,109,000	63,723,000	66,646,000	97,131,000	60,610,000	59,866,000
EBITDA	107,968,000	96,876,000	68,074,000	84,042,000	2,418,549,000	40,084,000
Return on Assets %	1.02%	-.10%	-1.67%	-.36%	-4.31%	-2.98%
Return on Equity %	2.29%	-.22%	-3.42%	-.74%	-8.68%	-6.43%
Debt to Equity	0.87	0.80	0.62	0.56	0.58	0.56

CONTACT INFORMATION:

Phone: 441-2952244   Fax: 441 2928666
Toll-Free:
Address: 22 Victoria St., Hamilton, BU HM 12 Bermuda

STOCK TICKER/OTHER:

Stock Ticker: BEL   Exchange: NYS
Employees: 8,000   Fiscal Year Ends: 12/31
Parent Company:

SALARIES/BONUSES:

Top Exec. Salary: Bonus: $
$4,040,170
Second Exec. Salary: Bonus: $
$260,100

OTHER THOUGHTS:

Estimated Female Officers or Directors: 5

Hot Spot for Advancement for Women/Minorities: Y

Sales, profits and employees may be estimates. Financial information, benefits and other data can change quickly and may vary from those stated here.

Best Western International Inc

www.bestwestern.com

NAIC Code: 721110

TYPES OF BUSINESS:

Hotels

BRANDS/DIVISIONS/AFFILIATES:

Best Western
Best Western Plus
Best Western Premier

CONTACTS: Note: Officers with more than one job title may be intentionally listed here more than once.

David T. Kong, CEO
David T. Kong, Pres.
Mark Straszynski, CFO
Dorothy Dowling, Sr. VP-Mktg. & Sales
Barbara Bras, VP-Human Resources
Scott Gibson, CIO
Larry Cuculic, General Counsel
David Velasquez, VP-Info. Systems Oper.
Mark Williams, VP-North American Dev.
Greg Adams, VP-e-commerce
Terry Porter, Sec.
Glenn De Souza, VP-Int'l Oper., Asia
Ron Pohl, Sr. VP-Brand Mgmt. & Member Svcs.
Michael Morton, VP-Member Svcs.
Wendy Ferrill, VP-Worldwide Sales
Dilipkumar (Danny) Patel, Chmn.
Suzi MacDonald Yoder, VP-Int'l Oper.
Scott Gibson, Sr. VP-Dist. & Strategic Svcs.

GROWTH PLANS/SPECIAL FEATURES:

Best Western International, Inc. is one of the largest lodging companies in the world, providing over 400,000 rooms in 100 countries and territories worldwide through more than 4,100 independently owned and operated hotels. Developers can select from three different memberships based on amenities and pricing: the Best Western Premier, which includes upgraded fitness centers, contemporary business centers, onsite lounges with HD Televisions and receives an AAA 3-Diamond rating; Best Western Plus, which is considered an upper midscale AAA 3-Diamond hotel; and the most basic Best Western, which is a midscale hotel and includes free high-speed Internet access. The company is organized as a nonprofit association of member hotels, allowing hotels independence to maintain local control and autonomy while reaping the benefits of a large international lodging firm. The company offers benefits and rewards programs through exclusive partnerships with major airlines, MasterCard, Avis, Budget, Teleflora and Vinesse Wine Clubs. Best Western also offers a prepaid Travel Card that can be filled with U.S. or Canadian dollars, euros, British pounds or several other international currencies and used for hotel accommodations and other services provided by Best Western hotels. Regular amenities include breakfast services, data port connections, free local calls and a variety of other services that are already offered in many locations and are currently being extended to other facilities.

FINANCIAL DATA: Note: Data for latest year may not have been available at press time.

In U.S. $	2015	2014	2013	2012	2011	2010
Revenue	7,000,000,000					
R&D Expense						
Operating Income						
Operating Margin %						
SGA Expense						
Net Income						
Operating Cash Flow						
Capital Expenditure						
EBITDA						
Return on Assets %						
Return on Equity %						
Debt to Equity						

CONTACT INFORMATION:

Phone: 602-957-4200   Fax: 602-957-5641
Toll-Free:
Address: 6201 N. 24th Pkwy., Phoenix, AZ 85016 United States

STOCK TICKER/OTHER:

Stock Ticker: Nonprofit   Exchange:
Employees: 1,000   Fiscal Year Ends: 11/30
Parent Company:

SALARIES/BONUSES:

Top Exec. Salary: $ Bonus: $
Second Exec. Salary: $ Bonus: $

OTHER THOUGHTS:

Estimated Female Officers or Directors: 6
Hot Spot for Advancement for Women/Minorities: Y

Bilfinger SE
NAIC Code: 237000

www.bilfingerberger.com

TYPES OF BUSINESS:
Heavy Construction

BRANDS/DIVISIONS/AFFILIATES:

CONTACTS: Note: Officers with more than one job title may be intentionally listed here more than once.
Herbert Bodner, Chmn-Exec. Board
Joachim Muller, Dir.-Finance, Acct. & Controlling
Jochen Keysberg, Dir.-Human Resources
Joachim Muller, Dir.-IT

GROWTH PLANS/SPECIAL FEATURES:
Bilfinger SE provides industrial services, power services, building/facility services, construction and concessions. The firm offers services in three categories: industry, energy and real estate. Within industry services, Bilfinger provides integrated technical services for the process industry The firm provides services for the design, construction, maintenance and modernization of plants, primarily in the sectors oil & gas, refineries, petrochemicals, chemicals & agro-chemicals, pharmaceuticals, food & beverages, power generation, steel and aluminum. Within the energy category, Bilfinger's services comprise maintenance, repair, efficiency enhancements, service life extensions and demolition of existing plants as well as the design, manufacture and assembly of components for power plant construction with a focus on boilers and high-pressure piping systems. In real estate, the firm provides technical, commercial and infrastructural real-estate services in Europe, USA and MENA (Middle East and North Africa) countries. This division manages all kinds of facilities and offers development, design, management and construction services for real estate, as well as the organization of construction logistics. In early 2016, the firm sold its water technologies division to Chengdu Techcent Environment Group; and received offers for the acquisition of its building and facility services businesses.

FINANCIAL DATA: Note: Data for latest year may not have been available at press time.

In U.S. $	2015	2014	2013	2012	2011	2010
Revenue	7,391,909,000	8,778,640,000	9,596,834,000	9,704,269,000	9,361,778,000	9,131,624,000
R&D Expense						
Operating Income	152,484,600	-33,530,640	327,437,000	473,192,600	379,329,600	390,962,700
Operating Margin %	2.06%	-.38%	3.41%	4.87%	4.05%	4.28%
SGA Expense	759,799,700	920,039,700	954,482,700	971,476,200	879,666,000	893,694,100
Net Income	-557,361,300	-81,431,550	197,078,000	313,522,900	451,523,100	326,638,600
Operating Cash Flow	140,965,500	74,132,370	184,304,500	119,638,200	320,822,100	460,647,100
Capital Expenditure	92,152,230	158,073,000	194,226,800	162,749,100	144,843,200	160,468,100
EBITDA	307,250,100	362,792,400	549,948,100	694,905,300	605,262,300	583,364,700
Return on Assets %	-8.75%	-1.14%	2.58%	3.77%	5.03%	3.57%
Return on Equity %	-28.62%	-3.49%	8.27%	14.40%	21.95%	16.99%
Debt to Equity	0.35	0.27	0.24	0.48	0.29	1.00

CONTACT INFORMATION:
Phone: 49 6214590   Fax: 49 6214592366
Toll-Free:
Address: Carl-ReiÃŸ-Platz 1-5, Mannheim, 68165 Germany

STOCK TICKER/OTHER:
Stock Ticker: BFLBF   Exchange: GREY
Employees: 37,945   Fiscal Year Ends: 12/31
Parent Company:

SALARIES/BONUSES:
Top Exec. Salary: $ Bonus: $
Second Exec. Salary: $ Bonus: $

OTHER THOUGHTS:
Estimated Female Officers or Directors:
Hot Spot for Advancement for Women/Minorities:

Black & Veatch Holding Company   www.bv.com

NAIC Code: 541330

TYPES OF BUSINESS:

Heavy & Civil Engineering, Construction
Infrastructure & Energy Services
Environmental & Hydrologic Engineering
Consulting Services
IT Services
Power Plant Engineering and Construction
LNG and Gas Processing Plant Engineering
Climate Change Services

BRANDS/DIVISIONS/AFFILIATES:

RCC Consultants

CONTACTS: Note: Officers with more than one job title may be intentionally listed here more than once.

Steven L. Edwards, CEO
Steven L. Edwards, Pres.
Karen L. Daniel, CFO
Lori Kelleher, Chief Human Resources Officer
James R. Lewis, Chief Admin. Officer
Timothy W. Triplett, General Counsel
Cindy Wallis-Lage, Pres., Water
O.H. Oskvig, CEO-Energy Business
William R. Van Dyke, Pres., Federal Svcs.
Steven L. Edwards, Chmn.
Hoe Wai Cheong, Sr. VP-Water-Asia Pacific
John E. Murphy, Pres., Construction & Procurement

GROWTH PLANS/SPECIAL FEATURES:

Black & Veatch Holding Company (B&V) is an engineering, consulting and construction company specializing in infrastructure development for the energy, water, telecommunications, federal, management consulting and environmental markets. The company is employee-owned and operates over 100 offices worldwide. B&V divides its service offerings into 13 categories. The firm's asset management services include power and water asset optimization solutions. Its environmental services encompass energy optimization, water planning, greenhouse gas services, climate economics and government services. The company provides construction services for energy facilities, water and wastewater treatment facilities, water distribution systems, desalination facilities, wireless sites and aerospace and defense sites. The construction management service segment offers tailored services aimed at reducing costs while improving quality. Design-build services cover engineering, procurement and construction. Engineering and design services include power delivery, siting, new generation engineering, power plant upgrade and bulk materials handling services. The nexus of water and energy segment addresses the interdependency of water and energy, reducing costs for both. Management consulting services comprise enterprise management and utility efficiency solutions. Procurement services include project procurement strategy, transportation logistics and inspection services. The program management segment manages energy, water, telecom and federal programs. Smart grid/smart utility services include utilities planning, system modeling, distribution automation and field construction. The security services segment offers cyber, electronic and physical security to the communications and utility industries. B&V also offers consulting and planning services. In August 2015, the firm acquired RCC Consultants, a global engineering and consulting firm.

The company offers employees medical, dental, vision and prescription drug coverage; flexible spending accounts; employee assistance programs; tuition reimbursement; adoption assistance; and credit union membership.

FINANCIAL DATA: Note: Data for latest year may not have been available at press time.

In U.S. $	2015	2014	2013	2012	2011	2010
Revenue	3,030,000,000	3,600,000,000	3,560,000,000	3,279,000,000	2,600,000,000	2,265,000,000
R&D Expense						
Operating Income						
Operating Margin %						
SGA Expense						
Net Income			15,200,000	134,000,000	104,000,000	35,000,000
Operating Cash Flow						
Capital Expenditure						
EBITDA						
Return on Assets %						
Return on Equity %						
Debt to Equity						

CONTACT INFORMATION:

Phone: 913-458-2000   Fax: 913-458-2934
Toll-Free:
Address: 11401 Lamar Ave., Overland Park, KS 66211 United States

SALARIES/BONUSES:

Top Exec. Salary: $ Bonus: $
Second Exec. Salary: $ Bonus: $

STOCK TICKER/OTHER:

Stock Ticker: Private   Exchange:
Employees: 10,285   Fiscal Year Ends: 12/31
Parent Company:

OTHER THOUGHTS:

Estimated Female Officers or Directors: 4
Hot Spot for Advancement for Women/Minorities: Y

Bluegreen Corporation

www.bluegreencorp.com

NAIC Code: 561599

TYPES OF BUSINESS:

Timeshare Resorts
Time shares
Residential Properties
Property Development & Subdivisions
Resort Management Services

BRANDS/DIVISIONS/AFFILIATES:

Woodbridge Holdings LLC
BFC Financial Corporation
Bluegreen Vacations Unlimited Inc
Bluegreen Vacation Club

CONTACTS: *Note: Officers with more than one job title may be intentionally listed here more than once.*

David L. Pontius, Chief Strategy Officer
John M. Maloney, Jr., Pres.
Anthony M. Puleo, CFO
David A. Bidgood, Pres., Sales & Mktg
Susan J. Saturday, Chief Human Resources Officer
Chanse Rivera, CIO
Allan J. Herz, Sr. VP-Mortgage Oper.
Raymond S. Lopez, Chief Acct. Officer

GROWTH PLANS/SPECIAL FEATURES:

Bluegreen Corporation markets, sells and manages vacation ownership interests (VOIs) in resorts. The firm markets, sells and manages real estate-based vacation ownership interests in resorts generally located in popular high-volume vacation destinations. It also provides property association management services, mortgage servicing, VOI title services, reservation services and construction design and development services. The company's Bluegreen Vacation Club is a flexible, points-based, deeded VOI program that connects more than 195,000 owners with over 60 Bluegreen resorts in more than 40 popular destinations in the continental U.S. and the Caribbean. Resort locations include Alabama, Florida, Georgia, Louisiana, North Carolina, South Carolina, Tennessee, Virginia and the Caribbean. Bluegreen Vacations Unlimited, Inc., a wholly-owned subsidiary of the firm, manages and markets the Bluegreen Vacation Club. Bluegreen Corporation is a wholly-owned subsidiary of Woodbridge Holdings LLC, which itself is an indirect subsidiary of BFC Financial Corporation.

The company offers employees medical, dental and vision insurance; a health care flexible spending account; life and accidental death & dismemberment coverage; short- and long-term disability insurance; educational assistance; and an employee assistance program.

FINANCIAL DATA: *Note: Data for latest year may not have been available at press time.*

In U.S. $	2015	2014	2013	2012	2011	2010
Revenue	614,765,000	580,328,000	445,000,000	429,000,000	403,446,016	365,676,992
R&D Expense						
Operating Income						
Operating Margin %						
SGA Expense						
Net Income	82,009,000	68,957,000				
Operating Cash Flow						
Capital Expenditure						
EBITDA						
Return on Assets %						
Return on Equity %						
Debt to Equity						

CONTACT INFORMATION:

Phone: 561 912-8000   Fax: 561 912-8100
Toll-Free:
Address: 4960 Conference Way N., Ste. 100, Boca Raton, FL 33431
United States

STOCK TICKER/OTHER:

Stock Ticker: Subsidiary   Exchange:
Employees: 4,600   Fiscal Year Ends: 12/31
Parent Company: Woodbridge Holdings LLC

SALARIES/BONUSES:

Top Exec. Salary: $ Bonus: $
Second Exec. Salary: $ Bonus: $

OTHER THOUGHTS:

Estimated Female Officers or Directors: 1
Hot Spot for Advancement for Women/Minorities:

Bluestem Group Inc

www.bluestem.com

NAIC Code: 454111

TYPES OF BUSINESS:
Electronic Shopping
Online Retail of Women's Apparel & Accessories
Online Retail of Men's Apparel & Accessories

BRANDS/DIVISIONS/AFFILIATES:
Appleseed's
Bedford Blair
Tog Shop (The)
WinterSilks
Norm Thompson
Haband
Orchard Brands Corporation
Capmark Financial Group Inc

CONTACTS: Note: Officers with more than one job title may be intentionally listed here more than once.
Steve Nave, CEO
Vince Jones, COO
William Gallagher, Pres.
Mark Wagener, CFO
Chidam Chidambaram, CMO
Shawn Moren, Sr.. VP-Human Resources
Linda A. Pickles, Chief Admin. Officer
Thomas L. Fairfield, General Counsel
Paul W. Kopsky, Jr., Principal Acct. Officer
Michael Lipson, Exec. VP

GROWTH PLANS/SPECIAL FEATURES:
Bluestem Group Inc., formerly Capmark Financial Group, Inc., is a holding company whose businesses include Bluestem Brands, Inc., a multi-brand, online retailer of a broad selection of name€ brand and private label general merchandise serving low€ to middle€ income consumers in the U.S. Bluestem Brands is the parent to 16 e-commerce brands including: Appleseed's, providing apparel and modern updates to classic styles for women; Bedford Blair, an apparel value brand for women; Blair, offering men's and women's apparel and accessories; The Tog Shop, offering an assortment of stylish, classical apparel for mature women; WinterSilks, providing apparel, sleepwear & intimates and underlayers in luxurious fabrics; Norm Thompson, featuring high-quality men's and women's apparel, shoes and accessories, as well as one-of-a-kind gifts and gourmet foods; and Sahalie, offering apparel and accessories in high-performance fabrics, organically produced materials and sustainable product alternatives. Other brands Bluestem Brands is parent to include: Draper's & Damon's, Fingerhut, Gettington.com, Gold Violin, Haband, LinenSource, Old Pueblo Traders, PayCheck Direct and Solutions. Each brand is complemented by a large selection of merchandise with a variety of payment options to provide customers with the flexibility of paying over time. In July 2015, the Bluestem Group acquired Orchard Brands Corporation, a multi-brand family of 13 catalog and e-commerce brands serving the boomer and senior demographics.

FINANCIAL DATA: Note: Data for latest year may not have been available at press time.

In U.S. $	2015	2014	2013	2012	2011	2010
Revenue	1,743,041,000	478,484,000	151,445,000	245,016,000	22,800,000	
R&D Expense						
Operating Income						
Operating Margin %						
SGA Expense						
Net Income	14,215,000	103,006,000	83,716,000			
Operating Cash Flow						
Capital Expenditure						
EBITDA						
Return on Assets %						
Return on Equity %						
Debt to Equity						

CONTACT INFORMATION:
Phone: 952-656-3700   Fax:
Toll-Free:
Address: 6509 Flying Cloud Dr., Eden Praire, MN 55344 United States

STOCK TICKER/OTHER:
Stock Ticker: BGRP   Exchange: PINX
Employees: 35   Fiscal Year Ends: 12/31
Parent Company:

SALARIES/BONUSES:
Top Exec. Salary: $ Bonus: $
Second Exec. Salary: $ Bonus: $

OTHER THOUGHTS:
Estimated Female Officers or Directors: 1
Hot Spot for Advancement for Women/Minorities:

Boardwalk Real Estate Investment Trust   www.boardwalkreit.com

NAIC Code: 0

TYPES OF BUSINESS:

Real Estate Investment Trust
Apartment Development & Management

BRANDS/DIVISIONS/AFFILIATES:

CONTACTS: *Note: Officers with more than one job title may be intentionally listed here more than once.*

Sam Kolias, CEO
William Wong, CFO
Michael Guyette, Chief Information Officer
P. Burns, General Counsel
Roberto Geremia, President
Van Kolias, Senior VP, Divisional
William Chidley, Senior VP, Divisional
Al Mawani, Trustee
James Dewald, Trustee
Arthur Havener, Trustee
Gary Goodman, Trustee
Andrea Stephen, Trustee
Samantha Kolias-Gunn, Trustee
Lisa Russell, Vice President, Divisional
Ian Dingle, Vice President, Divisional
Kelly Mahajan, Vice President, Divisional
Jonathan Brimmell, Vice President, Divisional
William Zigomanis, Vice President, Divisional
Helen Mix, Vice President, Divisional

GROWTH PLANS/SPECIAL FEATURES:

Boardwalk Real Estate Investment Trust is one of Canada's largest owners and operators of multi-family rental communities. The company currently owns and operates over 220 properties with more than 32,000 units, totaling approximately 28 million rentable square feet. Boardwalk's portfolio includes properties in the provinces of Alberta, Saskatchewan, Ontario and Quebec, with approximately 57% of its rentable portfolio and nearly 65% of its net operating income concentrated in Alberta. The firm is not involved in construction; instead it focuses on the purchase of existing structures within its acquisition program, especially mid-sized suburban and downtown apartment buildings and mid-sized community and neighborhood residential centers located in urban markets, to ensure a diversified portfolio. The company targets well-located properties that it believes have been poorly managed, creating a higher-than-average vacancy rate, and which are located in areas characterized by limited new housing supply. Boardwalk's strength is in its operating, leasing and financing capabilities and its ability to enhance the long-term value of a property. The firm attempts to acquire properties at a substantial discount to replacement cost. It then restores the apartment buildings and focuses on adding high-tech amenities, such as fiber-optic cable. Boardwalk sometimes sells its properties after renovation, either entirely to other companies or as individual condos to residents. In 2015, the firm agreed to sell its Windsor, Ontario, property portfolio.

FINANCIAL DATA: *Note: Data for latest year may not have been available at press time.*

In U.S. $	2015	2014	2013	2012	2011	2010
Revenue	364,005,300	361,786,000	353,206,300	336,295,200	323,166,000	318,107,500
R&D Expense						
Operating Income	127,114,600	118,310,800	118,609,000	105,628,900	91,971,450	30,951,470
Operating Margin %	34.92%	32.70%	33.58%	31.40%	28.45%	9.72%
SGA Expense	25,538,960	25,787,420	24,623,110	22,100,330	20,078,280	19,720,510
Net Income	22,053,700	188,666,600	258,187,600	526,354,700	936,552,800	54,835,340
Operating Cash Flow	131,658,600	132,689,100	130,043,300	114,088,600	103,136,700	106,492,700
Capital Expenditure	14,612,260	67,923,220	5,465,263	4,722,189	56,362,000	55,722,130
EBITDA	84,455,850	259,587,300	331,804,600	609,501,000	459,286,900	176,171,200
Return on Assets %	.48%	4.14%	5.61%	12.79%	33.03%	3.04%
Return on Equity %	.87%	7.41%	10.25%	25.06%	104.89%	
Debt to Equity	0.61	0.50	0.54	0.63	0.70	

CONTACT INFORMATION:

Phone: 403-531-9255   Fax: 403 261-9267
Toll-Free:
Address: 1501 1st St. SW, Ste. 200, Calgary, AB T2R 0W1 Canada

STOCK TICKER/OTHER:

Stock Ticker: BEI.UN   Exchange: TSE
Employees: 1,500   Fiscal Year Ends: 12/31
Parent Company:

SALARIES/BONUSES:

Top Exec. Salary: $ Bonus: $
Second Exec. Salary: $ Bonus: $

OTHER THOUGHTS:

Estimated Female Officers or Directors:
Hot Spot for Advancement for Women/Minorities: Y

Sales, profits and employees may be estimates. Financial information, benefits and other data can change quickly and may vary from those stated here.

Boise Cascade Corp

www.bc.com

NAIC Code: 321213

TYPES OF BUSINESS:

Lumber Products, Manufacturing
Engineered Wood Products
Wholesale Building Materials Distribution
Packaging & Newsprint
Cottonwood Fiber Farming

BRANDS/DIVISIONS/AFFILIATES:

Madison Dearborn Partners IV
OfficeMax

CONTACTS: Note: Officers with more than one job title may be intentionally listed here more than once.

Thomas Corrick, CEO
Wayne Rancourt, CFO
Thomas Carlile, Chairman of the Board
Kelly Hibbs, Chief Accounting Officer
Nick Stokes, Executive VP, Divisional
Daniel Hutchinson, Executive VP, Divisional
John Sahlberg, General Counsel
Frank Elfering, Vice President, Divisional
Rich Viola, Vice President, Divisional
Dennis Huston, Vice President, Divisional

GROWTH PLANS/SPECIAL FEATURES:

Boise Cascade Corp. is a major wholesale distributor of wood products and building materials. The firm operates through three divisions: wood products, building materials distribution and corporate & other. The wood products segment manufactures structural, appearance and industrial plywood panels, and also manufactures engineered wood products consisting of laminated veneer lumber (LVL), I-joists and laminated beams. This division also produces studs, particleboard and ponderosa pine lumber. Its wood products are primarily used in new residential construction, residential repair-and-remodeling markets as well as light commercial construction. The building materials distribution segment is a national stocking wholesale distributor of building materials. It distributes a broad line of building materials including engineered wood products, oriented strand board, plywood, lumber and general line items such as siding, metal products, insulation, roofing and composite decking. The corporate & other segment provides corporate staff services, related assets and liabilities as well as foreign currency exchange gains and losses. These support services include finance, accounting, legal, information technology and human resource functions. The company is majority owned by private equity investors Madison Dearborn Partners IV (with a 68.7% interest) and OfficeMax (with a 20% interest). In December 2015, the firm agreed to acquire engineered lumber production facilities at Thorsby, Alabama and Roxboro, North Carolina from Georgia-Pacific LLC.

The firm offers its employees health, medical, prescription drug, dental and vision coverage; flexible spending accounts; 16-31 days paid-time off; life, AD&D and long-term disability insurance; a 401(k) plan with a 4% employer match; a wellness program; and an employee assistance program.

FINANCIAL DATA: Note: Data for latest year may not have been available at press time.

In U.S. $	2015	2014	2013	2012	2011	2010
Revenue	3,633,415,000	3,573,732,000	3,273,496,000	2,779,062,000	2,248,088,000	2,240,591,000
R&D Expense						
Operating Income	103,189,000	145,549,000	98,757,000	63,131,000	-27,046,000	-13,162,000
Operating Margin %	2.84%	4.07%	3.01%	2.27%	-1.20%	-.58%
SGA Expense	322,733,000	312,662,000	290,772,000	278,177,000	242,240,000	242,503,000
Net Income	52,182,000	80,009,000	116,936,000	41,496,000	-46,363,000	-33,297,000
Operating Cash Flow	80,331,000	101,843,000	33,427,000	80,136,000	-42,981,000	10,287,000
Capital Expenditure	87,526,000	61,217,000	45,751,000	27,386,000	33,537,000	35,751,000
EBITDA	160,411,000	198,406,000	138,384,000	100,771,000	12,096,000	25,682,000
Return on Assets %	4.22%	6.88%	12.05%	4.77%	-5.13%	
Return on Equity %	10.16%	16.93%	42.50%	21.81%	-16.40%	
Debt to Equity	0.64	0.61	0.66	2.81	0.77	

CONTACT INFORMATION:

Phone: 208-384-6161   Fax: 208-384-7189
Toll-Free:
Address: 1111 W. Jefferson St., Ste. 300, Boise, ID 83702-5389 United States

STOCK TICKER/OTHER:

Stock Ticker: BCC
Employees: 5,290
Parent Company:

Exchange: NYS
Fiscal Year Ends: 12/31

SALARIES/BONUSES:

Top Exec. Salary: $701,923   Bonus: $
Second Exec. Salary: $452,308   Bonus: $

OTHER THOUGHTS:

Estimated Female Officers or Directors: 1
Hot Spot for Advancement for Women/Minorities:

Boston Properties Inc

www.bostonproperties.com

NAIC Code: 0

TYPES OF BUSINESS:

Real Estate Investment Trust
Offices
Hotel
Industrial Space

BRANDS/DIVISIONS/AFFILIATES:

Boston Properties Office Value-Added Fund LP

CONTACTS: *Note: Officers with more than one job title may be intentionally listed here more than once.*

Owen Thomas, CEO
Michael LaBelle, CFO
Mortimer Zuckerman, Chairman Emeritus
Bruce Duncan, Chairman of the Board
Lori Silverstein, Chief Accounting Officer
Douglas Linde, Director
Bryan Koop, Executive VP, Divisional
Peter Johnston, Executive VP, Divisional
Robert Pester, Executive VP, Geographical
John Powers, Executive VP, Geographical
Frank Burt, General Counsel
Raymond Ritchey, Senior Executive VP

GROWTH PLANS/SPECIAL FEATURES:

Boston Properties, Inc. is a fully integrated, self-administered and self-managed real estate investment trust (REIT) and one of the largest owners and developers of office properties in the U.S. The trust develops, redevelops, acquires, manages, operates and owns a diverse portfolio of Class A offices. It holds properties in the U.S., with holdings concentrated in four core markets: Boston, Massachusetts; Washington, D.C.; midtown Manhattan; and San Francisco, California. The company's subsidiary, Boston Properties Limited Partnership, conducts the vast majority of its business. Boston Properties owns or has interests in 168 properties, including 158 office properties, five retail properties, four residential properties (including one under construction) and one hotel. The sum of this property is 46.5 million net rentable square feet and 15 million square feet of structured parking. In addition, the trust owns or controls approximately 457.1 acres of undeveloped land. The company has a minority interest in the Boston Properties Office Value-Added Fund, L.P., which acquires deficient properties and remodels/refurbishes them for resale. Some of the company's properties include the John Hancock Tower and Garage, Prudential Tower and Prudential Center in Boston; Times Square Tower and General Motors Building (60% owned) in New York; Metropolitan Square (51% owned) in Washington, D.C.; One, Two, Three and Four Embarcadero Center in San Francisco; and 17 office buildings in Carnegie Center in Princeton.

Boston Properties offers its employees health, dental, life, business travel accident and disability insurance; domestic and same sex partner coverage; an employee assistance program; a 401(k); a stock purchase plan; paid time off; legal referral services; tuition reimbursement; commuter benefits; and a scholarship program.

FINANCIAL DATA: *Note: Data for latest year may not have been available at press time.*

In U.S. $	2015	2014	2013	2012	2011	2010
Revenue	2,490,821,000	2,396,998,000	2,135,539,000	1,876,267,000	1,759,526,000	1,550,804,000
R&D Expense						
Operating Income	849,365,000	801,822,000	678,120,000	651,681,000	618,640,000	610,514,000
Operating Margin %	34.09%	33.45%	31.75%	34.73%	35.15%	39.36%
SGA Expense	97,578,000	102,077,000	115,329,000	82,382,000	81,442,000	79,658,000
Net Income	583,106,000	443,611,000	749,811,000	289,650,000	272,679,000	159,072,000
Operating Cash Flow	799,411,000	695,553,000	777,926,000	642,949,000	606,328,000	375,893,000
Capital Expenditure	631,991,000	599,094,000	993,556,000	1,194,392,000	445,997,000	860,519,000
EBITDA	1,495,761,000	1,442,334,000	1,715,925,000	1,161,830,000	1,147,141,000	948,885,000
Return on Assets %	2.99%	2.16%	4.16%	1.91%	1.93%	1.23%
Return on Equity %	10.40%	7.84%	13.94%	5.81%	5.90%	3.60%
Debt to Equity	1.64	1.83	2.04	1.74	1.78	1.78

CONTACT INFORMATION:

Phone: 617 236-3300   Fax: 617 536-3128
Toll-Free:
Address: 800 Boylston St., The Prudential Ctr., Ste. 1900, Boston, MA 02199-8103 United States

STOCK TICKER/OTHER:

Stock Ticker: BXP   Exchange: NYS
Employees: 750   Fiscal Year Ends: 12/31
Parent Company:

SALARIES/BONUSES:

Top Exec. Salary: $773,077   Bonus: $2,558,333
Second Exec. Salary: Bonus: $1,805,000
$713,462

OTHER THOUGHTS:

Estimated Female Officers or Directors: 1
Hot Spot for Advancement for Women/Minorities: Y

Bouygues SA

NAIC Code: 237130

TYPES OF BUSINESS:

Construction & Telecommunications
Construction
Road Building
Property Development
Precasting
Cellular Phone Service
Media Operation
Research & Development

BRANDS/DIVISIONS/AFFILIATES:

Bouygues Construction
Bouygues Immobilier
Colas
Buoygues Telecom
Bbox Miami
TF1
TMC
Alstom

CONTACTS: Note: Officers with more than one job title may be intentionally listed here more than once.

Martin Bouygues, CEO
Olivier Bouygues, Co-CEO
Philippe Marien, CFO
Alain Pouyat, Exec. VP-New Tech.
Jean-Claude Tostivin, Sr. VP-Admin.
Jean Francois Guillemin, Corp. Sec.
Olivier Bouygues, Deputy CEO
Philippe Marien, Chmn.-Bourgues Telecom
Martin Bouygues, Chmn.

GROWTH PLANS/SPECIAL FEATURES:

Buoygues SA, based in Paris, France, was founded in 1952 as a construction and industrial works company. Operating through its subsidiaries, the firm provides services to three sectors: construction, telecoms and media. The construction sector operates through three subsidiaries: Bouygues Construction, which is dedicated to electrical contracting, building, energy services and civil works; Bouygues Immobilier, committed to property development; and Colas, dedicated to building roads and other complementary activities, such as safety, road marking and signaling services. The telecom sector operates through subsidiary Buoygues Telecom, a mobile telephone operator and full-service player with a subscriber base of 14 million customers. Its Bbox Miami offering is a TV box operating under the Android operating system, combines traditional TV content with apps and content from the web within a highly seamless interface. The media sector operates through TF1, a leading private media group in freeview television within France. TF1 produces four complementary freeview TV channels TF1, TMC, NT1 and HD1 that together claimed an average 27.8% audience share in the first half of 2015. Buoygues maintains operations in Europe, Central and South America, North America, Asia, the Middle East and Africa. In January 2016, the firm acquired an additional minority stake in Alstom, a French company that operates in the worldwide rail transport markets, now holding 28.3% of Alstom's capital.

FINANCIAL DATA: Note: Data for latest year may not have been available at press time.

In U.S. $	2015	2014	2013	2012	2011	2010
Revenue	36,995,483,800	37,234,898,478	46,193,335,211	44,296,705,342	43,488,384,000	45,123,600,000
R&D Expense						
Operating Income						
Operating Margin %						
SGA Expense						
Net Income	557,826,750	561,322,800	-1,048,710,129	853,836,721	1,422,753,000	1,547,710,000
Operating Cash Flow						
Capital Expenditure						
EBITDA						
Return on Assets %						
Return on Equity %						
Debt to Equity						

CONTACT INFORMATION:

Phone: 33-1-44-20-10-00   Fax: 33-1-30-60-4861
Toll-Free:
Address: 32 Ave. Hoche, Paris, 75008 France

STOCK TICKER/OTHER:

Stock Ticker: EN   Exchange: Paris
Employees: 128,070   Fiscal Year Ends: 12/31
Parent Company:

SALARIES/BONUSES:

Top Exec. Salary: $ Bonus: $
Second Exec. Salary: $ Bonus: $

OTHER THOUGHTS:

Estimated Female Officers or Directors: 5
Hot Spot for Advancement for Women/Minorities: Y

Boyd Gaming Corp

www.boydgaming.com

NAIC Code: 721120

TYPES OF BUSINESS:

Casinos & Hotels
Casino Management

BRANDS/DIVISIONS/AFFILIATES:

California Hotel & Casino
Borgata Hotel Casino & Spa
Blue Chip Hotel and Casino
Delta Downs Racetrack and Casino
Gold Coast Hotel and Casino
Fremont Hotel & Casino
IP Casino Resort Spa
Eldorado Casino

CONTACTS: *Note: Officers with more than one job title may be intentionally listed here more than once.*

William Boyd, Chairman of the Board
Anthony McDuffie, Chief Accounting Officer
Robert Boughner, Director
Marianne Johnson, Director
William Boyd, Director
Theodore Bogich, Executive VP, Divisional
Stephen Thompson, Executive VP, Divisional
Josh Hirsberg, Executive VP
Keith Smith, President
Brian Larson, Secretary

GROWTH PLANS/SPECIAL FEATURES:

Boyd Gaming Corp. is a multi-jurisdictional gaming company and one of the country's leading casino operators. It currently owns and operates 22 casinos (21 wholly-owned) totaling over 1.2 million square feet and housing 29,736 slot machines, 757 table games and 11,391 hotel rooms. The firm divides its properties into five segments: Las Vegas local, downtown Las Vegas, Midwest & South, peninsula and Borgata. The Las Vegas local properties include The Orleans Hotel and Casino, Sam's Town Hotel & Gambling Hall, Gold Coast Hotel and Casino, Eldorado Casino, Jokers Wild Casino and Suncoast Hotel and Casino. Downtown Las Vegas facilities consist of the Fremont Hotel & Casino; California Hotel & Casino; and Main Street Casino, Brewery and Hotel. The Midwest & South properties include IP Casino Resort Spa, Sam's Town Hotel and Gambling Hall, Delta Downs Racetrack and Casino & Hotel, Sam's Town Hotel and Casino, Treasure Chest Casino, Par-a-Dice Hotel and Casino and Blue Chip Hotel and Casino. The peninsula properties include Diamond Jo Dubuque, Diamond Jo Worth, Evangeline Downs Racetrack and Casino, Amelia Belle Casino and Kansas Star Casino. The Borgata property includes Borgata Hotel Casino & Spa. Additionally, Boyd owns and operates a travel agency and an insurance company that underwrites travel-related insurance, each located in Hawaii.

Boyd Gaming offers its employees pharmacy, life, short-term disability, medical, dental and vision coverage; flexible spending accounts; and a 401(k).

FINANCIAL DATA: *Note: Data for latest year may not have been available at press time.*

In U.S. $	2015	2014	2013	2012	2011	2010
Revenue	2,199,432,000	2,701,319,000	2,894,438,000	2,487,426,000	2,336,238,000	2,140,899,000
R&D Expense						
Operating Income	344,623,000	251,516,000	278,301,000	-854,875,000	233,104,000	183,938,000
Operating Margin %	15.66%	9.31%	9.61%	-34.36%	9.97%	8.59%
SGA Expense	406,268,000	509,904,000	562,507,000	670,202,000	604,099,000	571,680,000
Net Income	47,234,000	-53,041,000	-80,264,000	-908,865,000	-3,854,000	10,310,000
Operating Cash Flow	339,846,000	322,859,000	277,035,000	142,445,000	253,510,000	285,070,000
Capital Expenditure	131,170,000	149,374,000	144,520,000	125,974,000	87,224,000	87,477,000
EBITDA	471,768,000	493,546,000	506,749,000	-639,511,000	439,796,000	394,898,000
Return on Assets %	1.06%	-1.03%	-1.32%	-14.88%	-.06%	.20%
Return on Equity %	9.98%	-11.68%	-20.74%	-120.70%	-.32%	.87%
Debt to Equity	6.37	7.83	9.26	15.89	2.78	2.68

CONTACT INFORMATION:

Phone: 702 792-7200   Fax: 702 792-7266
Toll-Free:
Address: 3883 Howard Hughes Pkwy., 9th Fl., Las Vegas, NV 89169 United States

STOCK TICKER/OTHER:

Stock Ticker: BYD
Employees: 18,290
Parent Company:

Exchange: NYS
Fiscal Year Ends: 12/31

SALARIES/BONUSES:

Top Exec. Salary: $1,040,000   Bonus: $250,000
Second Exec. Salary: $1,275,000   Bonus: $

OTHER THOUGHTS:

Estimated Female Officers or Directors: 3

Hot Spot for Advancement for Women/Minorities: Y

Brand Energy & Infrastructure Services Inc

www.beis.com

NAIC Code: 238220

TYPES OF BUSINESS:

Mechanical Contractors
Industrial Coatings and Insulation
Contracting and Construction for the Oil and Gas Industry

BRANDS/DIVISIONS/AFFILIATES:

Aluma Systems Concrete Construction
Brand Energy Solutions LLC
MATCOR Inc
Industrial Specialists LLC
Industrial Specialty Services
CP Masters Inc

CONTACTS: Note: Officers with more than one job title may be intentionally listed here more than once.

Paul T. Wood, CEO
Chi Nguyen, CFO
Paul T. Wood, Chmn.

GROWTH PLANS/SPECIAL FEATURES:

Brand Energy & Infrastructure Services Inc. (BEIS) is a provider of specialty contracting services to the energy markets in North America. The firm's services include scaffolding, coatings, insulation, refractory, cathodic protection and mechanical services. BEIS offers its services to those in the upstream oil & gas industry, downstream oil & gas industry, midstream oil & gas industry, power generation industry and infrastructure Industry. The firm operates through various subsidiary companies: Aluma Systems Concrete Construction offers both concrete construction services, such as formwork and engineering solutions, and industrial services, such as equipment rental, erection and dismantling services; Brand Energy Solutions, LLC offers industrial services, such as industrial coatings, industrial insulation and fireproofing; MATCOR, Inc. offers cathodic protection and corrosion control services; Industrial Specialists, LLC offers refractory installation services; and Industrial Specialty Services offers integrated services and solutions, such as line freezing and hot tapping. In 2015, the firm acquired MATCOR and merged CP Masters, Inc. with and into it.

FINANCIAL DATA: Note: Data for latest year may not have been available at press time.

In U.S. $	2015	2014	2013	2012	2011	2010
Revenue	3,000,000,000	2,335,000,000	1,975,000,000	1,730,000,000	1,500,000,000	
R&D Expense						
Operating Income						
Operating Margin %						
SGA Expense						
Net Income						
Operating Cash Flow						
Capital Expenditure						
EBITDA						
Return on Assets %						
Return on Equity %						
Debt to Equity						

CONTACT INFORMATION:

Phone: 678-285-1400   Fax: 770-514-0285
Toll-Free:
Address: 1325 Cobb Int'l. Dr., Ste. A-1, Kennesaw, GA 30152 United States

STOCK TICKER/OTHER:

Stock Ticker: Private   Exchange:
Employees: 12,806   Fiscal Year Ends:
Parent Company: Clayton, Dubilier & Rice

SALARIES/BONUSES:

Top Exec. Salary: $ Bonus: $
Second Exec. Salary: $ Bonus: $

OTHER THOUGHTS:

Estimated Female Officers or Directors: 2
Hot Spot for Advancement for Women/Minorities:

Sales, profits and employees may be estimates. Financial information, benefits and other data can change quickly and may vary from those stated here.

Brandywine Realty Trust
NAIC Code: 0

www.brandywinerealty.com

TYPES OF BUSINESS:
Real Estate Investment Trust
Office & Industrial Properties
Property Management Services
Development Support Services
Consulting

BRANDS/DIVISIONS/AFFILIATES:
Brandywine Realty Services Corporation
BTRS Inc
Brandywine Properties I Limited Inc
BDN Brokerage LLC
Brandywine Properties Management LP
Brandywine Brokerage Services LLC

CONTACTS: Note: Officers with more than one job title may be intentionally listed here more than once.
Gerard Sweeney, CEO
Thomas Wirth, CFO
Anthony Nichols, Chairman Emeritus
Walter DAlessio, Chairman of the Board
George Johnstone, Executive VP, Divisional
Brad Molotsky, Executive VP
William Redd, Executive VP
George Sowa, Executive VP
H. Devuono, Executive VP
Michael Joyce, Trustee
Wyche Fowler, Trustee
Charles Pizzi, Trustee
James Diggs, Trustee
Daniel Palazzo, Vice President

GROWTH PLANS/SPECIAL FEATURES:
Brandywine Realty Trust is a real estate investment trust (REIT) that acquires, develops, redevelops, manages and leases office and industrial properties across the U.S. The firm owns 179 properties, which consist of 106 office properties, six industrial properties, three mixed-use properties, one retail, two development properties, one redevelopment property, one re-entitlement property and 59 properties held for sale. It also owns412 acres of undeveloped land and interests in 16 unconsolidated real estate ventures. Its portfolio is divided into seven geographic segments: Pennsylvania suburbs; Philadelphia central business district; metropolitan Washington, D.C; New Jersey/Delaware; Richmond, Virginia; Austin, Texas; and California, which includes properties in Oakland, Concord and Carlsbad. Additionally, the firm provides a number of third-party real estate management services, which it offers through six management companies: Brandywine Realty Services Corporation; BTRS, Inc.; Brandywine Properties I Limited, Inc.; BDN Brokerage, LLC; Brandywine Properties Management, L.P.; and Brandywine Brokerage Services, LLC. Combined, these organizations manage properties containing an aggregate of 33.9 million net rentable square feet, including Brandywine's 25.1 million net rentable square feet and 8.9 million net rentable square feet related to properties owned by third parties and the firm's real estate ventures. In December 2015, the firm announced that it has agreed upon a series of disposition transactions for five office properties containing an aggregate of 1,216,800 rentable square feet.

The company offers employees medical, dental, prescription drug and vision insurance; short- and long-term disability coverage; a 401(k); reimbursement accounts for health, dependent care and public transportation; an employee stock purchase plan; military leave with pay for training or reserve duty; tuition reimbursement; scholarships for employees' children; and business casual environments in the corporate and property management offices.

FINANCIAL DATA: Note: Data for latest year may not have been available at press time.

In U.S. $	2015	2014	2013	2012	2011	2010
Revenue	602,631,000	596,982,000	562,210,000	559,833,000	581,805,000	566,897,000
R&D Expense						
Operating Income	33,901,000	125,669,000	115,792,000	118,187,000	105,712,000	100,355,000
Operating Margin %	5.62%	21.05%	20.59%	21.11%	18.16%	17.70%
SGA Expense	29,406,000	26,779,000	27,628,000	25,413,000	24,602,000	23,306,000
Net Income	-30,401,000	6,975,000	42,777,000	6,595,000	-4,499,000	-17,074,000
Operating Cash Flow	195,099,000	188,999,000	183,484,000	157,283,000	179,015,000	185,127,000
Capital Expenditure	539,429,000	255,373,000	314,113,000	176,667,000	171,830,000	231,624,000
EBITDA	304,800,000	340,066,000	365,298,000	308,422,000	342,274,000	319,547,000
Return on Assets %	-.79%		.76%	-.18%	-.28%	-.54%
Return on Equity %	-1.84%	-.01%	1.94%	-.45%	-.70%	-1.37%
Debt to Equity	1.23	1.14	1.36	1.40	1.28	1.31

CONTACT INFORMATION:
Phone: 610 325-5600   Fax: 610 325-5622
Toll-Free:
Address: 555 E. Lancaster Ave., Ste. 100, Radnor, PA 19087 United States

STOCK TICKER/OTHER:
Stock Ticker: BDN
Employees: 424
Parent Company:

Exchange: NYS
Fiscal Year Ends: 12/31

SALARIES/BONUSES:
Top Exec. Salary: $683,333   Bonus: $
Second Exec. Salary: Bonus: $
$370,833

OTHER THOUGHTS:
Estimated Female Officers or Directors:
Hot Spot for Advancement for Women/Minorities: Y

Sales, profits and employees may be estimates. Financial information, benefits and other data can change quickly and may vary from those stated here.

Brixmor Property Group Inc

www.brixmor.com

NAIC Code: 0

TYPES OF BUSINESS:

Real Estate Investment Trust
Community Shopping Centers

BRANDS/DIVISIONS/AFFILIATES:

Blackstone Real Estate Partners VI LP
Blackstone Group LP (The)

CONTACTS: Note: Officers with more than one job title may be intentionally listed here more than once.

Daniel Hurwitz, CEO
Barry Lefkowitz, CFO
John Schreiber, Chairman of the Board
Michael Cathers, Chief Accounting Officer
Carolyn Singh, Executive VP, Divisional
Brian Finnegan, Executive VP, Divisional
Michael Moss, Executive VP, Divisional
Steven Siegel, Executive VP
Matthew Berger, Senior VP, Divisional
David Gerstenhaber, Senior VP, Divisional
Stacy Slater, Senior VP, Divisional

GROWTH PLANS/SPECIAL FEATURES:

Brixmor Property Group, Inc. is a U.S. shopping center real estate investment trust (REIT). The firm is controlled by Blackstone Real Estate Partners VI LP, an affiliate of The Blackstone Group L.P. The company engages in the ownership, management and development of community and neighborhood shopping centers, malls and lifestyle centers, with more than 520 properties in 38 states comprising 87 million square feet of gross leasable space. The firm has a mix of convenience shopping and general merchandise properties, with both grocery and non-grocery anchors. In regards to new development, the firm has a policy of having a lease executed with an anchor tenant prior to making an investment. The firm's model for growth is both internal and external, focusing on aggressive management to maintain high occupancy rates and recognized anchor tenants as well as on selective acquisitions of income-producing shopping centers or of centers that the firm can add value to through redevelopment or management. Major tenants such as Bed, Bath & Beyond, Dollar Tree, Kohl's, Kroger, Publix, TJ Maxx, Wal-Mart, Office Depot, Ross, Big Lots, PETCO, Sears, PetSmart and Staples serve as its anchor stores.

FINANCIAL DATA: Note: Data for latest year may not have been available at press time.

In U.S. $	2015	2014	2013	2012	2011	2010
Revenue	1,265,980,000	1,236,599,000	1,174,697,000	1,125,797,000	1,113,380,000	1,117,453,000
R&D Expense						
Operating Income	428,658,000	394,605,000	274,572,000	232,369,000	183,907,000	77,668,000
Operating Margin %	33.85%	31.91%	23.37%	20.64%	16.51%	6.95%
SGA Expense	98,454,000	80,175,000	121,093,000	88,870,000	107,871,000	94,634,000
Net Income	193,720,000	89,002,000	-93,534,000	-122,567,000	67,508,000	-321,387,000
Operating Cash Flow	534,025,000	479,210,000	331,990,000	268,847,000	173,839,000	308,895,000
Capital Expenditure	242,142,000	214,678,000	156,838,000	183,213,000	115,928,000	78,216,000
EBITDA	860,024,000	817,213,000	693,916,000	738,084,000	978,764,000	473,914,000
Return on Assets %	2.01%	.89%	-.94%	-1.25%	.67%	
Return on Equity %	6.70%	3.38%	-4.61%	-6.86%	3.62%	
Debt to Equity	2.08	2.08	2.55	3.77	3.60	

CONTACT INFORMATION:

Phone: 212-869-3000   Fax: 212-869-8969
Toll-Free:
Address: 420 Lexington Ave., New York, NY 10170 United States

STOCK TICKER/OTHER:

Stock Ticker: BRX   Exchange: NYS
Employees: 443   Fiscal Year Ends: 12/31
Parent Company:

SALARIES/BONUSES:

Top Exec. Salary: $800,000   Bonus: $1,000,000
Second Exec. Salary: Bonus: $601,388
$408,154

OTHER THOUGHTS:

Estimated Female Officers or Directors: 2
Hot Spot for Advancement for Women/Minorities: Y

Brookdale Senior Living Inc

www.brookdaleliving.com

NAIC Code: 623310

TYPES OF BUSINESS:

Assisted Living Facilities
Retirement Communities
Assisted Living Communities
Continued Care Retirement Communities (CCRCs)
Managed Facilities

BRANDS/DIVISIONS/AFFILIATES:

CONTACTS: Note: Officers with more than one job title may be intentionally listed here more than once.

T. Smith, CEO
Lucinda Baier, CFO
Dawn Kussow, Chief Accounting Officer
Bryan Richardson, Chief Administrative Officer
Tricia Conahan, Chief Marketing Officer
Labeed Diab, COO
Daniel Decker, Director
Kristin Ferge, Executive VP
H. Kaestner, Executive VP, Divisional
Mary Patchett, Executive VP, Divisional
George Hicks, Executive VP, Divisional
Glenn Maul, Executive VP
Chad White, General Counsel
Mark Ohlendorf, President
Ross Roadman, Senior VP, Divisional
Julie Davis, Vice President, Divisional

GROWTH PLANS/SPECIAL FEATURES:

Brookdale Senior Living, Inc. (BSL) is one of the largest senior living facility operators in the U.S. It operates 1,123 owned, leased or managed senior living facilities in 47 states that can serve approximately 108,000 residents. BSL operates five business segments: retirement centers, assisted living, continuing care retirement communities (CCRCs) rentals, Brookdale ancillary services and management services. It has 130 retirement center communities with 24,486 units, 915 assisted living communities with 62,567 units and 78 rental CCRC communities with 21,367 units. The facilities strive to offer residents a home-like setting and typically feature assistance with daily living, multiple forms of therapy and various home health services. BSL offers a full spectrum of care options, including independent living, personalized assisted living, rehabilitation and skilled nursing. It operates memory care communities, which are freestanding assisted living communities designed for residents with Alzheimer's disease and other dementias. The company maintains its own culinary arts institute, which offers a training ground for chefs and dining staff. Leased communities generated the largest share (48.7%) of 2015 revenues, followed by owned communities (38.8%). BSL generated 81.9% of its 2015 revenue from private pay customers, with the remainder generated by Medicare, Medicaid and other various third-party payer programs. In June 2015, Brookdale acquired a portfolio of 35 private pay senior housing communities from Chartwell Retirement Residences, in which Brookdale owns 90% and HCP, Inc. 10%, respectively.

FINANCIAL DATA: Note: Data for latest year may not have been available at press time.

In U.S. $	2015	2014	2013	2012	2011	2010
Revenue	4,960,608,000	3,831,706,000	2,891,966,000	2,770,085,000	2,457,918,000	2,213,264,000
R&D Expense						
Operating Income	-165,206,000	-84,905,000	131,288,000	82,286,000	90,180,000	65,994,000
Operating Margin %	-3.33%	-2.21%	4.53%	2.97%	3.66%	2.98%
SGA Expense	822,548,000	671,046,000	461,277,000	451,270,000	423,185,000	407,222,000
Net Income	-457,477,000	-148,990,000	-3,584,000	-65,645,000	-68,175,000	-48,901,000
Operating Cash Flow	292,366,000	242,652,000	366,121,000	290,969,000	268,427,000	228,244,000
Capital Expenditure	411,051,000	304,245,000	257,527,000	208,412,000	160,131,000	93,681,000
EBITDA	570,768,000	461,969,000	421,382,000	353,544,000	340,971,000	353,612,000
Return on Assets %	-4.44%	-1.95%	-.07%	-1.43%	-1.51%	-1.06%
Return on Equity %	-17.13%	-7.63%	-.35%	-6.42%	-6.49%	-4.55%
Debt to Equity	2.52	2.07	2.38	2.16	2.26	2.35

CONTACT INFORMATION:

Phone: 615 221-2250   Fax: 615 221-2289
Toll-Free: 866-785-9025
Address: 111 Westwood Place, Ste. 400, Brentwood, TN 37027 United States

STOCK TICKER/OTHER:

Stock Ticker: BKD   Exchange: NYS
Employees: 82,000   Fiscal Year Ends: 12/31
Parent Company:

SALARIES/BONUSES:

Top Exec. Salary: $841,216   Bonus: $
Second Exec. Salary: Bonus: $
$499,538

OTHER THOUGHTS:

Estimated Female Officers or Directors: 1
Hot Spot for Advancement for Women/Minorities:

Sales, profits and employees may be estimates. Financial information, benefits and other data can change quickly and may vary from those stated here.

Brookfield Asset Management Inc

www.brookfield.com

NAIC Code: 531100

TYPES OF BUSINESS:

Real Estate and Industrial Investments
Asset Management
Hydroelectric Generation & Transmission
Paper Production
Agriculture
Financial Services
Timber Development
Wind Power Development

BRANDS/DIVISIONS/AFFILIATES:

Brookfield Renewable Energy Partners LP
Brookfield Office Properties Inc
Brookfield Infrastructure Partners LP
Brookfield Property Partners LP
Bridge Lending Funds I II & III
Brookfield Capital Partners I II III

CONTACTS: Note: Officers with more than one job title may be intentionally listed here more than once.

J. Flatt, CEO
J. Flat, CEO
Brian Lawson, CFO
Frank McKenna, Chairman of the Board
Jeffrey Blidner, Director
George Myhal, Other Corporate Officer
Samuel Pollock, Other Corporate Officer
A. Silber, Secretary

GROWTH PLANS/SPECIAL FEATURES:

Brookfield Asset Management, Inc. is a Canadian holding company that owns and manages assets in the areas of real estate and power generation. The company maintains more than $200 billion in assets under management. Its operations are split into four categories: renewable energy, property, infrastructure and private equity. The renewable energy division, through Brookfield Renewable Energy Partners LP, owns and operates 250 renewable assets, diversified across 73 river systems and eight power markets in the U.S., Canada, Brazil and Europe, with a combined production capacity of over 7,300 megawatts. The property segment, which includes Brookfield Property Partners L.P. invests in the office, retail, multifamily, industrial, hotel and triple net leased sectors; and Brookfield Office Properties, Inc., owns and manages a portfolio of 261 commercial office properties located in major cities in North America, Australia and Europe as well as retail shopping centers in North America, Brazil, the U.K., India and Australia, as well as 172 retail assets primarily in the U.S. and Brazil. Brookfield Infrastructure Partners L.P., which comprises the infrastructure division, manages five key sectors: communication infrastructure (including towers, rooftop sites and services to media and telecom sectors), energy (including energy transmission, distribution and storage), transport (including ports, toll roads and rail lines), utilities (electrical, water and natural gas transmission) and timberlands (consisting of 3.7 million acres). The private equity segment consists of its special situations group, focusing on restructuring and turnaround financing; Brookfield Capital Partners I, II, III, which Invests in financially distressed and/or underperforming companies located primarily in Canada and the U.S.; and Bridge Lending Funds I, II & III, which provides event-driven financing to Canadian and U.S. mid-market companies.

FINANCIAL DATA: Note: Data for latest year may not have been available at press time.

In U.S. $	2015	2014	2013	2012	2011	2010
Revenue	21,753,000,000	18,554,000,000	22,114,000,000	18,697,000,000	15,921,000,000	13,988,000,000
R&D Expense						
Operating Income	2,699,000,000	1,264,000,000	4,026,000,000	870,000,000	266,000,000	383,000,000
Operating Margin %	12.40%	6.81%	18.20%	4.65%	1.67%	2.73%
SGA Expense	106,000,000	123,000,000	152,000,000	158,000,000	481,000,000	417,000,000
Net Income	2,341,000,000	3,110,000,000	2,120,000,000	1,380,000,000	1,957,000,000	1,454,000,000
Operating Cash Flow	2,788,000,000	2,574,000,000	2,278,000,000	1,497,000,000	676,000,000	1,454,000,000
Capital Expenditure	1,114,000,000	1,098,000,000	1,566,000,000	3,544,000,000	2,000,000,000	546,000,000
EBITDA	9,380,000,000	10,581,000,000	8,697,000,000	7,023,000,000	7,341,000,000	5,959,000,000
Return on Assets %	1.74%	2.56%	1.78%	1.25%	2.18%	2.07%
Return on Equity %	11.22%	16.39%	10.99%	7.16%	12.52%	15.14%
Debt to Equity	2.51	2.41	2.41	2.27		2.14

CONTACT INFORMATION:

Phone: 416 363-9491   Fax: 416 365-9642
Toll-Free:
Address: 181 Bay St., Brookfield Pl., Ste. 300, Toronto, ON M5J 2T3 Canada

STOCK TICKER/OTHER:

Stock Ticker: BAM
Employees: 28,000
Parent Company:

Exchange: NYS
Fiscal Year Ends: 12/31

SALARIES/BONUSES:

Top Exec. Salary: $600,000   Bonus: $
Second Exec. Salary: $543,420   Bonus: $

OTHER THOUGHTS:

Estimated Female Officers or Directors: 3
Hot Spot for Advancement for Women/Minorities: Y

Sales, profits and employees may be estimates. Financial information, benefits and other data can change quickly and may vary from those stated here.

Building Materials Holding Corp (BMC)   www.buildwithbmc.com
NAIC Code: 444190

TYPES OF BUSINESS:
Building Materials & Hardware Stores, Retail
Building & Construction Services

BRANDS/DIVISIONS/AFFILIATES:
BMC Design

CONTACTS: Note: Officers with more than one job title may be intentionally listed here more than once.
Peter Alexander, CEO
James Major, CFO
William Gay, Chief Accounting Officer
Paul Street, Chief Administrative Officer
David Bullock, Director
Lisa Hamblet, Executive VP, Divisional
Mark Necaise, Other Corporate Officer
Walter Randolph, President, Divisional
Duff Wakefield, President, Divisional
Steven Wilson, President, Divisional
C. Ball, Senior VP
Michael Farmer, Vice President, Divisional

GROWTH PLANS/SPECIAL FEATURES:
Building Materials Holding Corp. (BMHC) is a leading provider of building materials and services to residential, commercial and industrial contractors as well as professional repair and remodeling contractors and builders. The firm operates 38 lumber yards, 18 truss manufacturing facilities, 23 millwork operations and 8 showroom and design centers. Its principal products include lumber, panel products, engineered wood products, roofing materials, pre-hung doors and millwork, roof and floor trusses, pre-assembled windows, cabinets, hardware, paint and tools. BMHC also provides services such as pre-cutting lumber and pre-assembling windows to meet customer specifications. The firm's BMC Design subsidiary offers expert advice in millwork, windows and cabinetry selections. The company targets primarily professional contractors and builders engaged in residential construction and, to a lesser extent, light commercial and industrial construction. Additionally, the firm provides construction services to high-volume production homebuilders. Services include framing, concrete, plumbing, other construction trades, managing labor and construction schedules as well as sourcing materials. BMHC's primary growth plans consist of expansion through acquisitions, focusing on those that complement existing operations in growing building markets, and those that provide entry into fast-growing, attractive new markets. In June 2015, BMHC agreed to merge with Stock Building Supply Holdings, Inc. in order to expand the geographic reach of both companies across the U.S. That August, the firm agreed to acquire Robert Bowden Inc., a supplier of windows, doors, trim, millwork, siding and decking to professional customers.

BMHC offers its employees medical, dental and health insurance as well as a 401(k).

FINANCIAL DATA: Note: Data for latest year may not have been available at press time.

In U.S. $	2015	2014	2013	2012	2011	2010
Revenue	1,576,746,000	1,295,716,000	1,197,037,000	942,398,000	759,982,000	751,706,000
R&D Expense						
Operating Income	12,248,000	18,324,000	761,000	-18,907,000	-59,301,000	-122,834,000
Operating Margin %	.77%	1.41%	.06%	-2.00%	-7.80%	-16.34%
SGA Expense	306,843,000	279,717,000	254,935,000	221,192,000	213,036,000	246,338,000
Net Income	-4,831,000	10,419,000	-4,635,000	-14,533,000	-42,133,000	-69,994,000
Operating Cash Flow	743,000	16,941,000	-40,264,000	-12,243,000	-7,001,000	-57,999,000
Capital Expenditure	31,319,000	43,306,000	7,448,000	2,741,000	1,339,000	2,506,000
EBITDA	37,621,000	32,454,000	13,694,000	-6,860,000	-44,707,000	-71,587,000
Return on Assets %	-.55%	3.02%	-2.14%	-8.88%	-18.19%	
Return on Equity %	-1.25%	7.75%	-7.99%	-56.11%	-90.07%	
Debt to Equity	0.67	0.68	0.50	0.16	0.01	

CONTACT INFORMATION:
Phone: 208-331-4300   Fax: 208-331-4366
Toll-Free:
Address: 720 Park Blvd., Ste. 200, Raleigh, ID 27617 United States

STOCK TICKER/OTHER:
Stock Ticker: STCK   Exchange: NAS
Employees: 9,600   Fiscal Year Ends: 12/31
Parent Company:

SALARIES/BONUSES:
Top Exec. Salary: $600,000   Bonus: $
Second Exec. Salary: $310,000   Bonus: $200,000

OTHER THOUGHTS:
Estimated Female Officers or Directors: 2
Hot Spot for Advancement for Women/Minorities:

Burns & McDonnell

www.burnsmcd.com

NAIC Code: 541330

TYPES OF BUSINESS:

Engineering Services
Construction
Consulting
Environmental Consulting
Architecture & Design
Energy Transmission

BRANDS/DIVISIONS/AFFILIATES:

CONTACTS: Note: Officers with more than one job title may be intentionally listed here more than once.

Greg Graves, CEO
Greg Graves, Pres.
Dennis W. (Denny) Scott, CFO
Rich Miller, Dir-IT
Mark Taylor, Treas.
Don Greenwood, Pres., Construction
Ray Kowalik, VP
John Nobles, Pres., Process & Industrial
Greg Graves, Chmn.

GROWTH PLANS/SPECIAL FEATURES:

Burns & McDonnell provides engineering, architectural, construction, environmental and consulting services across the U.S. and worldwide. The company divides its businesses into several global practice units. The architecture unit designs & builds plants, facilities and buildings such as federal/military facilities, food/consumer factories, health care facilities, information technology facilities and laboratories. The commissioning unit provides verification of new construction, its subsystems for mechanical, plumbing, electrical, fire/life safety, building envelopes, interior systems such as laboratory units, co-generation utility plants, sustainable systems and more. The construction unit, through a general contractor, designs & builds infrastructure for air quality control, aviation, chemicals, oil & gas, electric power generation, electrical transmission/distribution, manufacturing/industrial, environmental, federal/military, food & consumer products and water sectors. The consulting unit provides insight and understanding concerning the engineering, architecture, construction and environmental features of facilities and buildings. The engineering unit provides integrated engineering services across a variety of disciplines and industries. The operations & maintenance unit helps to streamline and optimize the process of operating and maintaining facilities. The planning unit takes the vision and helps put the facilities, operations and systems in place to make it happen. The studies unit provides critical support for business decision, including financial, environmental and functionality. Burns & McDonnell is a 100% employee-owned company, with more than 35 offices worldwide.

The firm offers employees health and life insurance, short- and long-term disability, flexible spending accounts, personal time off, eight paid holidays, educational seminars and tuition assistance. The company is 100% employee-owned through its employee stock ownership plan.

FINANCIAL DATA: Note: Data for latest year may not have been available at press time.

In U.S. $	2015	2014	2013	2012	2011	2010
Revenue	2,700,000,000	2,645,000,000	2,300,000,000	1,500,000,000	1,174,000,000	1,089,000,000
R&D Expense						
Operating Income						
Operating Margin %						
SGA Expense						
Net Income						
Operating Cash Flow						
Capital Expenditure						
EBITDA						
Return on Assets %						
Return on Equity %						
Debt to Equity						

CONTACT INFORMATION:

Phone: 816-333-9400   Fax: 816-822-3028
Toll-Free:
Address: 9400 Ward Pkwy., Kansas City, MO 64114 United States

STOCK TICKER/OTHER:

Stock Ticker: Private   Exchange:
Employees: 4,900   Fiscal Year Ends: 12/31
Parent Company:

SALARIES/BONUSES:

Top Exec. Salary: $ Bonus: $
Second Exec. Salary: $ Bonus: $

OTHER THOUGHTS:

Estimated Female Officers or Directors:
Hot Spot for Advancement for Women/Minorities:

Butler Manufacturing Co   www.butlermfg.com

NAIC Code: 332311

TYPES OF BUSINESS:

Construction Materials
Prefabricated Metal Building & Component Manufacturing
Pre-Engineered Building Systems & Components
Architectural Aluminum Systems & Components
Construction & Real Estate Services

BRANDS/DIVISIONS/AFFILIATES:

BlueScope Steel Limited
SunLite Strip
Butler GSA
BlueScope Construction
Butler Research Center

CONTACTS: Note: Officers with more than one job title may be intentionally listed here more than once.

John J. Holland, CEO
Tom Gilligan, Pres.
Larry C. Miller, CFO
Robert M. Adams, Pres., Butler Real Estate, Inc.
Moufid Alossi, Pres., Butler (Shanghai), Inc.
John J. Holland, Chmn.

GROWTH PLANS/SPECIAL FEATURES:

Butler Manufacturing Co., a subsidiary of BlueScope Steel Limited, markets, designs and manufactures steel building systems for nonresidential buildings. It also executes construction and real estate operations. The firm manufactures three main types of product: structural systems, consisting of support systems (trusses, joists and anything else that keeps a building standing up); wall systems, which come in a variety of textures but are generally utilitarian in design; and roofs systems. Butler's largest division designs, manufactures, markets and sells steel building systems. This division's products are mostly custom-designed and engineered one- to eight-story steel buildings for use as offices, manufacturing facilities, warehouses, schools, shopping centers, agricultural buildings and other applications. The firm offers a variety of wall forms including fluted and flat panel walls. Roofing systems include flat warehouse style roofing, thermal insulation, SunLite Strip skylight systems and slanted roofing. Butler offers specialized construction services through subsidiary BlueScope Construction, a construction company specializing in large, multiple-site projects. The company offers specialized government construction services through Butler GSA, which has experience in military base construction, Air Force hangars, Coast Guard marine storage and maintenance facilities and more. Additionally the Butler Research Center simulates worst case scenario weather conditions designed to test Butler products and ensure their durability. The company has subsidiaries and joint-ventures located throughout the U.S., China, Europe, Japan, Saudi Arabia and Latin America.

FINANCIAL DATA: Note: Data for latest year may not have been available at press time.

In U.S. $	2015	2014	2013	2012	2011	2010
Revenue						
R&D Expense						
Operating Income						
Operating Margin %						
SGA Expense						
Net Income						
Operating Cash Flow						
Capital Expenditure						
EBITDA						
Return on Assets %						
Return on Equity %						
Debt to Equity						

CONTACT INFORMATION:

Phone: 816-968-3000   Fax: 816-968-3279
Toll-Free:
Address: 1540 Genessee St., Kansas City, MO 64102 United States

STOCK TICKER/OTHER:

Stock Ticker: Subsidiary   Exchange:
Employees: 4,298   Fiscal Year Ends: 06/30
Parent Company: BlueScope Steel Limited

SALARIES/BONUSES:

Top Exec. Salary: $ Bonus: $
Second Exec. Salary: $ Bonus: $

OTHER THOUGHTS:

Estimated Female Officers or Directors:
Hot Spot for Advancement for Women/Minorities:

Caesars Entertainment Corporation

www.caesars.com

NAIC Code: 721120

TYPES OF BUSINESS:
Casino Hotels
Dockside & Riverboat Casinos
Racing Venues
Casino Management
Online Games

BRANDS/DIVISIONS/AFFILIATES:
Harrah's
Caesers
Caesars Entertainment UK
Horseshoe
World Series of Poker
Caesars Interactive Entertainment Inc
Bally's Las Vegas
Flamingo Las Vegas

CONTACTS: Note: Officers with more than one job title may be intentionally listed here more than once.
Mark Frissora, CEO
Eric Hession, CFO
Gary Loveman, Chairman of the Board
Mary Thomas, Executive VP, Divisional
Gregory Miller, Executive VP, Divisional
Janis Jones, Executive VP, Divisional
Timothy Donovan, Executive VP
Tariq Shaukat, Executive VP
Scott Wiegand, Other Corporate Officer
Robert Morse, President, Divisional
Steven Tight, President, Divisional
Thomas Jenkin, President, Divisional
Keith Causey, Senior VP

GROWTH PLANS/SPECIAL FEATURES:
Caesars Entertainment Corporation is one of the largest gaming companies in the world. The firm owns or manages 50 casinos in 14 U.S. states and five countries throughout the world. It operates casino entertainment facilities primarily under the Harrah's, Caesars, and Horseshoe brands in the U.S., including land-based casinos, riverboat or dockside casinos, casino clubs and several racing venues. In Las Vegas alone, the firm owns Caesars Palace, Harrah's Las Vegas, Rio All-Suite Hotel & Casino, Bally's Las Vegas, Flamingo Las Vegas, Paris Las Vegas, Planet Hollywood Resort and Casino, The Quad Resort & Casino and Bill's Gamblin' Hall & Saloon. In addition, the firm owns Caesars Entertainment UK family of casinos. Besides casinos, the firm's properties generally include hotel and convention space, restaurants and non-gaming entertainment facilities. Its facilities contain an aggregate 1.48 million square feet of gaming space and over 15,700 hotel rooms. Through its subsidiary Caesars Interactive Entertainment, Inc., the company owns and operates the World Series of Poker tournament and develops online games, both real money wagered and play-for-fun games. Other subsidiaries are Caesars Entertainment Resort Properties and Caesars Growth Partners, LLC. Caesars was purchased in 2008 for about $30 billion by two private equity companies. Much of the purchase was funded by debt. In January 2015, the company filed a complex bankruptcy for its main operating company in attempt to eliminate much of its vast debts and reduce interest expenses.

FINANCIAL DATA: Note: Data for latest year may not have been available at press time.

In U.S. $	2015	2014	2013	2012	2011	2010
Revenue	4,654,000,000	8,516,000,000	8,559,700,000	8,586,700,000	8,834,500,000	8,818,600,000
R&D Expense						
Operating Income	573,000,000	-452,000,000	-2,234,600,000	-313,400,000	875,500,000	532,300,000
Operating Margin %	12.31%	-5.30%	-26.10%	-3.64%	9.91%	6.03%
SGA Expense	1,485,000,000	2,588,000,000	2,347,600,000	2,316,300,000	2,271,300,000	2,204,700,000
Net Income	5,920,000,000	-2,783,000,000	-2,948,200,000	-1,497,500,000	-687,600,000	-831,100,000
Operating Cash Flow	120,000,000	-735,000,000	-109,400,000	26,500,000	123,100,000	170,800,000
Capital Expenditure	350,000,000	998,000,000	726,300,000	507,700,000	305,900,000	177,200,000
EBITDA	7,091,000,000	132,000,000	-1,464,100,000	751,200,000	1,830,100,000	1,873,800,000
Return on Assets %	33.13%	-11.54%	-11.19%	-5.29%	-2.40%	-2.88%
Return on Equity %				-503.36%	-52.10%	-234.14%
Debt to Equity	6.86				19.62	11.50

CONTACT INFORMATION:
Phone: 702 407-6000   Fax: 702 407-6037
Toll-Free: 800-318-0047
Address: 1 Caesars Palace Dr., Las Vegas, NV 89109 United States

STOCK TICKER/OTHER:
Stock Ticker: CZR
Employees: 68,000
Parent Company:

Exchange: NAS
Fiscal Year Ends: 12/31

SALARIES/BONUSES:
Top Exec. Salary: $1,900,000   Bonus: $
Second Exec. Salary: $1,200,000   Bonus: $

OTHER THOUGHTS:
Estimated Female Officers or Directors: 2

Hot Spot for Advancement for Women/Minorities: Y

Sales, profits and employees may be estimates. Financial information, benefits and other data can change quickly and may vary from those stated here.

CalAtlantic Group Inc

NAIC Code: 236117   www.calatlantichomes.com

TYPES OF BUSINESS:

Construction, Home Building and Residential
Mortgage Financing
Title Services

BRANDS/DIVISIONS/AFFILIATES:

Standard Pacific Corp
Ryland Group (The)
CalAtlantic Homes
HouseWorks

CONTACTS: Note: Officers with more than one job title may be intentionally listed here more than once.

Larry Nicholson, CEO
Jeffrey McCall, CFO
Scott Stowell, Chairman of the Board
Wendy Marlett, Chief Marketing Officer
Peter Skelly, COO
John Babel, Secretary

GROWTH PLANS/SPECIAL FEATURES:

CalAtlantic Group, Inc., formed by the late-2015 merger of Standard Pacific Corp. and The Ryland Group, Inc., is the fourth largest homebuilder in the U.S. CalAtlantic operates under the CalAtlantic Homes label in 41 major markets across the country. The company strives to provide innovative home design, outstanding quality and a commitment to customer satisfaction. CalAtlantic Home design centers are located in Arizona, California, Colorado, Florida, Georgia, Illinois, Indiana, Maryland, Minnesota, Nevada, North Carolina, South Carolina, Texas and Washington, D.C. CalAtlantic Homes' HouseWorks program ensures that new homes are built to leading industry standards in energy efficiency. It also offers a home warranty program that includes wall-to-wall coverage, electrical, plumbing, heating, cooling and ventilation system coverage, as well as major structural defect coverage.

CalAtlantic Group provides medical, vision, dental and prescription plans; flexible spending & health savings accounts; short- and long-term disability plans; life insurance; 401(k) matching; as well as fitness & wellness plans and home-buying discounts.

FINANCIAL DATA: Note: Data for latest year may not have been available at press time.

In U.S. $	2015	2014	2013	2012	2011	2010
Revenue	3,540,113,000	2,435,297,000	1,939,519,000	1,258,258,000	893,900,000	924,874,000
R&D Expense						
Operating Income	429,463,000	351,396,000	249,256,000	81,705,000	9,328,000	52,880,000
Operating Margin %	12.13%	14.42%	12.85%	6.49%	1.04%	5.71%
SGA Expense	390,710,000	275,861,000	230,691,000	172,207,000	154,375,000	150,542,000
Net Income	213,509,000	215,865,000	188,715,000	531,421,000	-16,417,000	-11,724,000
Operating Cash Flow	-271,361,000	-362,397,000	-154,216,000	-283,116,000	-322,613,000	-80,958,000
Capital Expenditure						
EBITDA	470,614,000	354,892,000	261,274,000	87,063,000	11,950,000	30,895,000
Return on Assets %	2.87%	4.18%	3.86%	11.54%	-.43%	-.24%
Return on Equity %	6.50%	10.42%	9.62%	32.62%	-1.49%	-.92%
Debt to Equity	0.98	1.32	1.32	1.30	2.19	2.17

CONTACT INFORMATION:

Phone: 949 789-1600   Fax: 949 789-1609
Toll-Free:
Address: 15360 Barranca Pkwy., Irvine, CA 92618 United States

STOCK TICKER/OTHER:

Stock Ticker: CAA   Exchange: NYS
Employees: 1,250   Fiscal Year Ends: 12/31
Parent Company:

SALARIES/BONUSES:

Top Exec. Salary: $951,667   Bonus: $
Second Exec. Salary: Bonus: $
$665,000

OTHER THOUGHTS:

Estimated Female Officers or Directors: 1
Hot Spot for Advancement for Women/Minorities: Y

Callison LLC

NAIC Code: 541310

www.callison.com

TYPES OF BUSINESS:

Architectural Services

BRANDS/DIVISIONS/AFFILIATES:

Arcadis NV

CONTACTS: Note: Officers with more than one job title may be intentionally listed here more than once.

Lance Josal, CEO
Randy Pace, CFO

GROWTH PLANS/SPECIAL FEATURES:

Callison, LLC, a subsidiary of Arcadis NV, is a global architecture firm with a position in retail stores and mixed-use development, corporate, hospitality, health care, multi-family residential, mission critical and high-rise markets worldwide. The firm has 17 offices worldwide as well as being one the U.S.' largest design firms. Callison has divided its architectural services into three main categories: retail, commercial and specialty. Retail services design facilities for the automotive industry; commercial interiors; offices; and department, luxury and specialty retail stores. Commercial services provides design services for facilities in the hospitality, mixed use, offices residential, health care, retail centers, and mission critical markets. Specialty services reflects the work of the firm in the areas of graphic design, master planning, program management, sustainable design and entertainment. Services offered by the company include architecture, brand building, change management, commercial interiors, commissioning, energy management, health facility planning/transitioning, historic preservation, landscape architecture, protective planning and technology design. Regions Callison serves include Australia, China, Latin America, the Middle East, North America, South Asia, the U.K. and Europe.

FINANCIAL DATA: Note: Data for latest year may not have been available at press time.

In U.S. $	2015	2014	2013	2012	2011	2010
Revenue						
R&D Expense						
Operating Income						
Operating Margin %						
SGA Expense						
Net Income						
Operating Cash Flow						
Capital Expenditure						
EBITDA						
Return on Assets %						
Return on Equity %						
Debt to Equity						

CONTACT INFORMATION:

Phone: 206-632-4646   Fax: 206-623-4625
Toll-Free:
Address: 1420 Fifth Ave., Ste. 2400, Seattle, WA 98101-2343 United States

STOCK TICKER/OTHER:

Stock Ticker: Subsidiary   Exchange:
Employees: 1,000   Fiscal Year Ends:
Parent Company: Arcadis NV

SALARIES/BONUSES:

Top Exec. Salary: $ Bonus: $
Second Exec. Salary: $ Bonus: $

OTHER THOUGHTS:

Estimated Female Officers or Directors:
Hot Spot for Advancement for Women/Minorities:

Cambridge Seven Associates Inc

www.c7a.com

NAIC Code: 541310

TYPES OF BUSINESS:

Architectural Services
Architecture

BRANDS/DIVISIONS/AFFILIATES:

CONTACTS: Note: Officers with more than one job title may be intentionally listed here more than once.

Peter Kuttner, Pres.

GROWTH PLANS/SPECIAL FEATURES:

Cambridge Seven Associates, Inc. (C7A) is an architectural firm established in 1962 by seven designers. C7A's diverse skills in architecture, planning, exhibit design, graphics, industrial design and filmmaking create a single studio with fresh and collaborative design solutions. The company takes on projects of any scale, with a portfolio of building types that include academic, museum, exhibit, hospitality, transportation, retail, office and aquarium facilities. On the exhibit side, the firm designs habitats, outdoor and traveling exhibits, changing exhibit galleries and environmental graphics. Its award-winning exhibits are interactive, immersive and hands-on, with engaging multimedia and technology components that enhance the user experience. C7A has completed projects throughout North America, Europe, the Middle East and Asia, including the Hohhot Children's Museum in China, the KAFD Geo-Science Centre in Saudi Arabia, the College for Life Sciences in Kuwait, the Pomeroy Sports Centre Exhibits in Canada and the Gyeonggi Children's Museum in South Korea.

FINANCIAL DATA: Note: Data for latest year may not have been available at press time.

In U.S. $	2015	2014	2013	2012	2011	2010
Revenue						
R&D Expense						
Operating Income						
Operating Margin %						
SGA Expense						
Net Income						
Operating Cash Flow						
Capital Expenditure						
EBITDA						
Return on Assets %						
Return on Equity %						
Debt to Equity						

CONTACT INFORMATION:

Phone: 617-492-7000   Fax: 617-492-7007
Toll-Free:
Address: 1050 Massachusetts Ave., Cambridge, MA 02138 United States

STOCK TICKER/OTHER:

Stock Ticker: Private   Exchange:
Employees: 57   Fiscal Year Ends:
Parent Company:

SALARIES/BONUSES:

Top Exec. Salary: $ Bonus: $
Second Exec. Salary: $ Bonus: $

OTHER THOUGHTS:

Estimated Female Officers or Directors:
Hot Spot for Advancement for Women/Minorities:

Camden Property Trust

www.camdenliving.com

NAIC Code: 531110

TYPES OF BUSINESS:

Real Estate Investment Trust
Apartment Communities
Construction Services
Consulting Services

BRANDS/DIVISIONS/AFFILIATES:

CONTACTS: Note: Officers with more than one job title may be intentionally listed here more than once.

Richard Campo, CEO
Alexander Jessett, CFO
Michael Gallagher, Chief Accounting Officer
H. Stewart, COO
William Sengelmann, Executive VP, Divisional
Josh Lebar, General Counsel
D. Oden, President
F. Parker, Trustee
Lewis Levey, Trustee
William Paulsen, Trustee
Kelvin Westbrook, Trustee
Scott Ingraham, Trustee
William Mcguire, Trustee
Frances Sevilla-Sacasa, Trustee
Steven Webster, Trustee

GROWTH PLANS/SPECIAL FEATURES:

Camden Property Trust is one of the largest real estate investment trusts (REIT) in the U.S, specializing in several disciplines within the residential real estate industry. The company acquires, develops, manages, disposes and redevelops apartment home communities and offers consulting, building and construction services for third-party clients. Camden currently owns interests in and operates 181 communities consisting of approximately 63,163 apartment homes in the U.S. The firm has approximately 13 projects under construction, comprised of 4,215 units, scheduled to be complete by the end of 2015. Its largest markets by total number of operating properties are Las Vegas, Houston, Metro Washington D.C., Dallas and Florida. Camden maintains a diversified portfolio of residential properties ranging from upscale urban residences to middle-class housing in established suburban neighborhoods.

The firm offers employees medical, vision, prescription, disability and dental insurance; flexible spending accounts; life & AD&D insurance; adoption assistance; a 401(k) plan; an employee stock purchase plan; an apartment discount; education assistance; scholarship funds; and an employee assistance program.

FINANCIAL DATA: Note: Data for latest year may not have been available at press time.

In U.S. $	2015	2014	2013	2012	2011	2010
Revenue	913,404,000	931,638,000	737,482,000	717,626,000	535,810,000	512,848,000
R&D Expense						
Operating Income	260,134,000	303,217,000	157,442,000	170,545,000	30,472,000	-158,413,000
Operating Margin %	28.47%	32.54%	21.34%	23.76%	5.68%	-30.88%
SGA Expense	294,303,000	294,225,000	276,056,000	340,396,000	250,502,000	235,229,000
Net Income	249,315,000	292,089,000	336,364,000	286,241,000	56,379,000	30,216,000
Operating Cash Flow	423,238,000	418,528,000	404,291,000	324,267,000	244,834,000	224,036,000
Capital Expenditure	425,574,000	503,328,000	580,924,000	290,728,000	227,755,000	63,739,000
EBITDA						
Return on Assets %	4.12%	4.99%	6.10%	5.66%	1.05%	.49%
Return on Equity %	8.85%	10.60%	12.80%	13.11%	2.86%	1.44%
Debt to Equity	0.96	0.97	0.94	0.97	1.38	1.52

CONTACT INFORMATION:

Phone: 713-354-2500   Fax: 713 354-2710
Toll-Free: 800-922-6336
Address: 11 Greenway Plz., Ste. 2400, Houston, TX 77046 United States

STOCK TICKER/OTHER:

Stock Ticker: CPT
Employees: 1,780
Parent Company:

Exchange: NYS
Fiscal Year Ends: 12/31

SALARIES/BONUSES:

Top Exec. Salary: $502,655   Bonus: $
Second Exec. Salary: $502,655   Bonus: $

OTHER THOUGHTS:

Estimated Female Officers or Directors: 4
Hot Spot for Advancement for Women/Minorities: Y

Capital Automotive REIT
NAIC Code: 0

www.capitalautomotive.com

TYPES OF BUSINESS:
Real Estate Investment Trust
Automotive Retail Real Estate

BRANDS/DIVISIONS/AFFILIATES:
Brookfield Property Partners

CONTACTS: *Note: Officers with more than one job title may be intentionally listed here more than once.*
Jay M. Ferriero, Pres.
John M. Weaver, General Counsel

GROWTH PLANS/SPECIAL FEATURES:
Capital Automotive REIT (CARS) provides tailored sale-leaseback capital to the automotive retail industry. Through custom tailored real estate finance, CARS assists dealer groups in growing their organizations, acquiring new locations, upgrading existing facilities, constructing new stores and facilitating estate planning and partner buyouts. Through sale-leaseback transactions, automotive dealers sell their real estate to CARS and then lease it back. This arrangement allows dealers to tap into 100% of the equity in their real estate while maintaining long-term control of their property. CARS can also play an integral role during the due diligence of a pending merger or acquisition. It does this by managing the entire real estate portion of a transaction, allowing the dealer to focus on underwriting the operating business. By using CARS, dealers can significantly lower their capital investment in an acquisition, increasing its return on investment. CARS can also provide capital for additional investments to a property during, as well as well after, the initial investment, allowing dealers to refresh or remodel their facilities as necessary while preserving dealer capital and access to credit. The firm is owned by Brookfield Property Partners, a leading real estate owner, operator and investor.

The company offers employees medical, dental and life insurance; a prescription drug plan; 401(k) savings and matching plans; flexible spending accounts; commuter subsidies; an employee assistance program; short- and long-term disability; a discount program; and gym memberships.

FINANCIAL DATA: *Note: Data for latest year may not have been available at press time.*

In U.S. $	2015	2014	2013	2012	2011	2010
Revenue						
R&D Expense						
Operating Income						
Operating Margin %						
SGA Expense						
Net Income						
Operating Cash Flow						
Capital Expenditure						
EBITDA						
Return on Assets %						
Return on Equity %						
Debt to Equity						

CONTACT INFORMATION:
Phone: 703-288-3075   Fax: 703-288-3375
Toll-Free: 877-422-7288
Address: 8270 Greensboro Dr., Ste. 950, McLean, VA 22102 United States

STOCK TICKER/OTHER:
Stock Ticker: Private   Exchange:
Employees: 36   Fiscal Year Ends: 12/31
Parent Company: Brookfield Property Partners

SALARIES/BONUSES:
Top Exec. Salary: $ Bonus: $
Second Exec. Salary: $ Bonus: $

OTHER THOUGHTS:
Estimated Female Officers or Directors:
Hot Spot for Advancement for Women/Minorities:

Capital Pacific Real Estate

www.capitalpacificrealestate.com

NAIC Code: 236117

TYPES OF BUSINESS:

Homebuilding
Land Development
Mortgage Brokerage
Commercial Properties Management

BRANDS/DIVISIONS/AFFILIATES:

Capital Pacific Homes
Capital Pacific Holdings LLC
Makar Properties

CONTACTS: *Note: Officers with more than one job title may be intentionally listed here more than once.*

Scott N. Coler, Pres.
Jason Watson, VP-Mktg. & Sales
Danny Cardenas, Mgr.-Warranties
Mark Mullin, VP

GROWTH PLANS/SPECIAL FEATURES:

Capital Pacific Real Estate (CPRE), based in Newport Beach, California, is a real estate firm that offers services to third-party vendors, banks, investors, landowners and developers. The company operates through three divisions: general contracting, asset management and realty & sales. As the exclusive builder of Capital Pacific Homes (CPH), the company provides new home building, construction and contracting services across California and in parts of Colorado through its general contracting division. The asset management division manages assets for Bank of America, Pacific Western Bank, Institutional Housing Partners and Saybrook Capital. The realty & sales division sells and markets newly-constructed CPH homes, provides maintenance services, brokers the acquisition and disposition of land and mediates institutional receiverships. The firm is financially aligned with Makar Properties, which provides land entitlement, commercial real estate development, mixed-use development, hospitality and real estate investment management services; and Capital Pacific Holdings LLC, which was previously the exclusive builder of CPH homes throughout Colorado, California, Texas and Arizona. CPRE also develops communities for buyers with varying lifestyles and budgets. The company specializes in the construction of homes, ranging from entry level prices starting at $100,000 to estates valued in excess of millions of dollars. Through the CPH brand, the firm has constructed residential homes for more than 30,000 customers and closes over 1,400 homes and lots annually. In addition, the company develops new home communities for third-party owners and provides mortgage brokerage operations and residential design services. Currently, CPRE is developing and selling homes in three new properties: Calder Ranch, located in Menifee, California; Vera Lane, located in Los Angeles, California; and Greenbrier Lane, located in Riverside, California. The company also manages properties in Ventura, Los Angeles and Orange County.

FINANCIAL DATA: *Note: Data for latest year may not have been available at press time.*

In U.S. $	2015	2014	2013	2012	2011	2010
Revenue						
R&D Expense						
Operating Income						
Operating Margin %						
SGA Expense						
Net Income						
Operating Cash Flow						
Capital Expenditure						
EBITDA						
Return on Assets %						
Return on Equity %						
Debt to Equity						

CONTACT INFORMATION:

Phone: 949-622-8400   Fax: 949-622-8404
Toll-Free:
Address: 4100 MacArthur Blvd., Ste. 120, Newport Beach, CA 92660
United States

STOCK TICKER/OTHER:

Stock Ticker: Subsidiary   Exchange:
Employees: Fiscal Year Ends: 02/28
Parent Company: Pacific Continental Corporation

SALARIES/BONUSES:

Top Exec. Salary: $ Bonus: $
Second Exec. Salary: $ Bonus: $

OTHER THOUGHTS:

Estimated Female Officers or Directors:
Hot Spot for Advancement for Women/Minorities:

Sales, profits and employees may be estimates. Financial information, benefits and other data can change quickly and may vary from those stated here.

Capital Senior Living Corp

www.capitalsenior.com

NAIC Code: 623310

TYPES OF BUSINESS:

Assisted Living Facilities
Nursing Homes
Assisted Living Services
Home Care Services

BRANDS/DIVISIONS/AFFILIATES:

CONTACTS: Note: Officers with more than one job title may be intentionally listed here more than once.

Lawrence Cohen, CEO
Carey Hendrickson, CFO
James Moore, Chairman of the Board
Kimberly Herman, Independent Director
Robert Hollister, Other Corporate Officer
Keith Johannessen, President
David Beathard, Senior VP, Divisional
David Brickman, Senior VP
Gloria Holland, Vice President, Divisional
Greg Boemer, Vice President, Divisional
Gary Fernandez, Vice President, Divisional
Christopher Lane, Vice President, Divisional
Joseph Solari, Vice President, Divisional
Glen Campbell, Vice President, Divisional

GROWTH PLANS/SPECIAL FEATURES:

Capital Senior Living Corp. (CSL) is one of the nation's largest operators and developers of residential communities for seniors. The firm operates 126 communities in 26 states, including 76 senior living communities that it either owns or has an ownership interest in and 50 senior living communities that are leased facilities. Its combined facilities can support 15,800 residents. Approximately 96% of the company's annual revenue is generated through private pay parties at these communities. The firm provides senior living services to the elderly in four categories of assistance: independent living, assisted living, continuing care retirement communities and home care services. Its independent living communities provide residents with daily meals, transportation, social and recreational activities, laundry, housekeeping and 24-hour staffing. The firm's assisted living communities, with residents that require additional assistance over independent residents, provide personal care services, such as walking, eating, personal hygiene and medication assistance; and special care services for residents with certain forms of dementia. The continuing care retirement communities provide traditional long-term care through 24-hour-per-day skilled nursing care by registered nurses. The company provides home care services to residents at one senior living community through its home care agency and through third-party providers at a majority of its senior living communities. Many of CSL's communities offer a continuum of care to meet its residents' needs as they change over time. This continuum of care, which integrates independent living, assisted living and home care, sustains residents' autonomy and independence based on their physical and mental abilities. In 2015, it acquired two communities comprised of 127 assisted living units, and sold four non-core communities.

The firm offers employees medical and dental insurance, life insurance and a 401(k) plan.

FINANCIAL DATA: Note: Data for latest year may not have been available at press time.

In U.S. $	2015	2014	2013	2012	2011	2010
Revenue	412,177,000	383,925,000	350,362,000	310,536,000	263,502,000	211,929,000
R&D Expense						
Operating Income	18,835,000	13,900,000	11,250,000	13,655,000	17,911,000	18,515,000
Operating Margin %	4.56%	3.62%	3.21%	4.39%	6.79%	8.73%
SGA Expense	90,397,000	89,326,000	87,633,000	78,716,000	66,928,000	46,707,000
Net Income	-14,284,000	-24,126,000	-16,504,000	-3,119,000	3,025,000	4,254,000
Operating Cash Flow	48,895,000	46,312,000	42,644,000	46,395,000	14,084,000	15,550,000
Capital Expenditure	42,430,000	18,742,000	13,562,000	12,302,000	10,472,000	10,447,000
EBITDA	75,365,000	57,341,000	56,260,000	49,002,000	35,723,000	32,947,000
Return on Assets %	-1.49%	-2.93%	-2.38%	-.56%	.71%	1.11%
Return on Equity %	-10.31%	-16.13%	-10.10%	-1.84%	1.81%	2.64%
Debt to Equity	5.84	4.51	3.21	2.28	1.33	1.03

CONTACT INFORMATION:

Phone: 972-770-5600   Fax: 972 770-5666
Toll-Free:
Address: 14160 Dallas Pkwy., Ste. 300, Dallas, TX 75254 United States

STOCK TICKER/OTHER:

Stock Ticker: CSU   Exchange: NYS
Employees: 6,147   Fiscal Year Ends: 12/31
Parent Company:

SALARIES/BONUSES:

Top Exec. Salary: $725,232   Bonus: $
Second Exec. Salary: Bonus: $
$427,847

OTHER THOUGHTS:

Estimated Female Officers or Directors: 2
Hot Spot for Advancement for Women/Minorities:

CAPREIT Inc

www.capreit.com

NAIC Code: 0

TYPES OF BUSINESS:

Real Estate Investment Trust
Apartments
Property Management

BRANDS/DIVISIONS/AFFILIATES:

CONTACTS: Note: Officers with more than one job title may be intentionally listed here more than once.

Andrew Kadish, Pres.
Miguel J. Gutierrez, COO
Dick Kadish, Pres.
Terence J. Collins, CFO
Ernest L. Heymann, Dir.-IT
Eugene Goodsell, Controller
Sandra Becker, Sr. VP
Kevin Smith, Sr. VP
Wes Pontius, VP
Carol Yates, VP

GROWTH PLANS/SPECIAL FEATURES:

CAPREIT, Inc. is a real estate investment trust (REIT) and property manager. Its focus is on the acquisition of multi-family residential properties. Since its founding in 1993, CAPREIT has owned or operated over 200 rental and condominium apartment homes consisting of over 40,000 units and housing over 100,000 families. The company's current holdings include apartment units spread over a broad market area of 21 states. CAPREIT also manages a portfolio of properties for third parties. Most of its properties are spread over the Northeast and Southeast, with additional properties in the Midwest and in California. In acquiring properties, the firm stresses both a viable knowledge of the local market and establishing consistent relationships with its brokers, many of whom have made the company a preferred customer. It targets markets that have both high economic growth characteristics and identifiable barriers to entry. CAPREIT specializes in financing its acquisitions with tax-exempt multifamily housing revenue bonds and, to date, has acquired over $5 billion. Its acquisitions team works directly with municipal and state issuing authorities, bond trustees and monitoring agents in closing these financing transactions. CAPREIT has a team of real estate professionals that handles day-to-day onsite management of its multifamily communities. In 2015, the firm acquired two Kansas-based communities, known as Aberdeen Apartments and Alvadora Apartments.

CAPREIT offers its employees medical, dental, life, vision and short-term disability insurance; a 401(k); education reimbursement; and an apartment allowance.

FINANCIAL DATA: Note: Data for latest year may not have been available at press time.

In U.S. $	2015	2014	2013	2012	2011	2010
Revenue						
R&D Expense						
Operating Income						
Operating Margin %						
SGA Expense						
Net Income						
Operating Cash Flow						
Capital Expenditure						
EBITDA						
Return on Assets %						
Return on Equity %						
Debt to Equity						

CONTACT INFORMATION:

Phone: 301-231-8700   Fax:
Toll-Free:
Address: 11200 Rockville Pk., Ste. 100, Rockville, MD 20852 United States

STOCK TICKER/OTHER:

Stock Ticker: Private
Employees: 400
Parent Company:

Exchange:
Fiscal Year Ends: 12/31

SALARIES/BONUSES:

Top Exec. Salary: $ Bonus: $
Second Exec. Salary: $ Bonus: $

OTHER THOUGHTS:

Estimated Female Officers or Directors: 3
Hot Spot for Advancement for Women/Minorities: Y

Capstead Mortgage Corporation

NAIC Code: 525990

www.capstead.com

TYPES OF BUSINESS:

Mortgage REIT
Mortgage-Backed Securities
Commercial Real Estate

BRANDS/DIVISIONS/AFFILIATES:

GROWTH PLANS/SPECIAL FEATURES:

Capstead Mortgage Corporation, headquartered in Dallas, Texas, operates as a real estate investment trust, earning income from investing in real estate-related assets on a leveraged basis and from other investment strategies. For the most part, these investments are limited to single-family residential adjustable-rate mortgage securities issued by such government agencies and government-sponsored entities as Fannie Mae, Freddie Mac and Ginnie Mae. The firm also invests in credit-sensitive commercial mortgage assets. Capstead tends to focus on ARM (adjustable-rate mortgages) securities that reset annually or semi-annually and are particularly liquid.

Capstead offers its employees benefits such as basic life and AD&D insurance as well as a 401(k) plan.

CONTACTS: Note: Officers with more than one job title may be intentionally listed here more than once.

Andrew Jacobs, CEO
Phillip Reinsch, CFO
Jack Biegler, Chairman of the Board
Robert Spears, Executive VP
Roy Kim, Senior VP, Divisional

FINANCIAL DATA: Note: Data for latest year may not have been available at press time.

In U.S. $	2015	2014	2013	2012	2011	2010
Revenue	122,355,000	153,421,000	140,604,000	179,040,000	177,557,000	143,786,000
R&D Expense						
Operating Income	108,325,000	140,820,000	126,487,000	163,626,000	160,204,000	126,896,000
Operating Margin %	88.53%	91.78%	89.95%	91.39%	90.22%	88.25%
SGA Expense	14,998,000	12,459,000	13,817,000	15,243,000	16,330,000	15,986,000
Net Income	108,325,000	140,820,000	126,487,000	163,626,000	160,204,000	126,896,000
Operating Cash Flow	244,245,000	248,812,000	261,212,000	263,930,000	220,235,000	202,010,000
Capital Expenditure						
EBITDA						
Return on Assets %	.64%	.89%	.62%	1.04%	1.28%	1.21%
Return on Equity %	8.07%	10.55%	7.09%	11.80%	14.29%	12.67%
Debt to Equity	11.85	0.08	0.08	0.07	0.09	0.12

CONTACT INFORMATION:

Phone: 214-874-2323   Fax: 214 874-2398
Toll-Free: 800-358-2323
Address: 8401 N. Central Expressway, Ste. 800, Dallas, TX 75225 United States

STOCK TICKER/OTHER:

Stock Ticker: CMO
Employees: 14
Parent Company:

Exchange: NYS
Fiscal Year Ends: 12/31

SALARIES/BONUSES:

Top Exec. Salary: $750,000   Bonus: $
Second Exec. Salary: $525,000   Bonus: $

OTHER THOUGHTS:

Estimated Female Officers or Directors: 1
Hot Spot for Advancement for Women/Minorities:

Carlson Companies Inc

www.carlson.com

NAIC Code: 721110

TYPES OF BUSINESS:

Hotels & Resorts
Travel Agencies & Services
Marketing Services
Restaurant Chains
Online Travel Services
Business & Government Travel Management

BRANDS/DIVISIONS/AFFILIATES:

Radisson
Park Plaza
Carlson Wagonlit Travel
Park Inn by Radisson
Country Inns & Suites by Carlson
Rezidor Hotel Group
Carlson Rezidor Hotel Group
Radisson Blu

CONTACTS: *Note: Officers with more than one job title may be intentionally listed here more than once.*

David P. Berg, CEO
Trey Hall, CFO
Cindy Rodahl, VP-Human Resources & Comm.
William A. Van Brunt, General Counsel
Brad Hall, Treasurer
Thorsten Kirschke, Pres., Carlson Rezidor Hotel Group, Americas
Gordon McKinnon, Chief Branding Officer
Diana L. Nelson, Chmn.
Simon Barlow, Pres., Carlson Rezidor Hotel Group, Asia Pacific

GROWTH PLANS/SPECIAL FEATURES:

Carlson Companies, Inc. is an integrated travel and hospitality company with operations in hotels, restaurants and corporate travel management. Carlson and Rezidor Hotel Group established Carlson Rezidor Hotel Group to better align their global activities. Through this partnership, the firm operates over 1,370 hotels in more than 150 countries. Its brand portfolio spans mid-priced properties to full-service premium and luxury hotels and resorts; these include Radisson and Radisson Blu (currently 287 locations with over 68,000 rooms), Park Plaza (48 operating locations and 13 in development), Park Inn by Radisson (211 locations with more than 39,000 rooms), Country Inns & Suites by Carlson (449 and 33 under development locations) and Hotel Quorvus (two properties in operation). Carlson also owns a majority share in Rezidor Hotel Group (50.3%). The firm's travel management division, which is comprised of Carlson Wagonlit Travel (CWT), offers corporate clients business travel management programs with the objective of maximizing productivity while minimizing cost. Through CWT Meetings & Events, it provides numerous events and meetings management services, annually delivering over 40,000 events for its clients. In 2014, the firm announced the sale of T.G.I. Fridays to Sentinel Capital Partners and TriArtisan Capital Partners; and acquired the remaining interest of CWT from JP Morgan Chase & Co, and is now its wholly-owned subsidiary.

FINANCIAL DATA: *Note: Data for latest year may not have been available at press time.*

In U.S. $	2015	2014	2013	2012	2011	2010
Revenue	3,150,000,000	4,400,000,000	4,400,000,000	4,500,000,000	4,130,000,000	3,900,000,000
R&D Expense						
Operating Income						
Operating Margin %						
SGA Expense						
Net Income						
Operating Cash Flow						
Capital Expenditure						
EBITDA						
Return on Assets %						
Return on Equity %						
Debt to Equity						

CONTACT INFORMATION:

Phone: 763-212-5000   Fax: 763-212-2219
Toll-Free:
Address: 701 Carlson Pkwy., Minnetonka, MN 55305 United States

STOCK TICKER/OTHER:

Stock Ticker: Private
Employees: 110,000
Parent Company:

Exchange:
Fiscal Year Ends: 12/31

SALARIES/BONUSES:

Top Exec. Salary: $ Bonus: $
Second Exec. Salary: $ Bonus: $

OTHER THOUGHTS:

Estimated Female Officers or Directors: 6
Hot Spot for Advancement for Women/Minorities: Y

Carlson Rezidor Hotel Group

carlsonrezidor.com

NAIC Code: 721110

TYPES OF BUSINESS:

Hotels & Resorts

BRANDS/DIVISIONS/AFFILIATES:

HNA Tourism Holding (Group) Co Ltd
Radisson Hotels & Resorts
Park Plaza Hotels & Resorts
Country Inns & Suites By Carlson
Park Inn Hotels
Quorvus
Carlson Wagonlit Travel
HNA Group Co Ltd

CONTACTS: Note: Officers with more than one job title may be intentionally listed here more than once.

David P. Berg, CEO
Trudy Rautio, Pres.
William A. Van Brunt, General Counsel
Tony Pellegrin, Sr. VP-Corp. Dev.
Cindy Rodahl, Exec. VP-Comm.
Brad Hall, Treas.
Diana L. Nelson, Chmn.

GROWTH PLANS/SPECIAL FEATURES:

Carlson Rezidor Hotel Group, a subsidiary of HNA Tourism Holding (Group) Co., Ltd., is one of the world's leading hotel franchisors. The firm includes more than 1,400 locations in over 110 countries and territories. Specific brands include Radisson Hotels & Resorts, Park Plaza Hotels & Resorts, Park Inn Hotels, Country Inns & Suites By Carlson and Quorvus. The company's Radisson chain owns 500 full-service hotels at locations throughout North America, Latin America, Asia Pacific, Europe, the Middle East and Africa. Radisson Hotels offer pre-arrival online check-in, a fitness center and free high-speed Internet access. The firm's 48 Park Plaza Hotels include restaurants, meeting rooms, catering, suites and recreational facilities. The Park Inn is the company's economy brand with approximately 211 hotels in operation and under development worldwide. Country Inns & Suites By Carlson is a mid-tier lodging chain with more than 470 locations in the Americas, Europe and India. Specialty services include an in-hotel Read It and Return Lending Library in which guests can borrow a book and return it on their next stay. Quorvus has two hotels in operation in Edinburgh and Kuwait. It is a stylish and contemporary hotel that utilizes the intimate and eclectic creative vision of Rosita Missoni. Quorvus has a hotel in London currently under development. The firm offers a Club Carlson rewards program, which allow clients to accumulate award points towards flights, hotel stays and other bonuses. Other business operations of the company include Carlson Wagonlit Travel (CWT), a business travel management service. In April 2016, Carlson Companies, Inc. sold the Carlson Rezidor Hotel Group to China's HNA Tourism Holding (Group) Co., Ltd., itself a subsidiary of HNA Group Co., Ltd.

FINANCIAL DATA: Note: Data for latest year may not have been available at press time.

In U.S. $	2015	2014	2013	2012	2011	2010
Revenue	1,135,234,050	1,067,256,645	7,500,000,000	7,200,000,000	7,000,000,000	6,500,000,000
R&D Expense						
Operating Income						
Operating Margin %						
SGA Expense						
Net Income						
Operating Cash Flow						
Capital Expenditure						
EBITDA						
Return on Assets %						
Return on Equity %						
Debt to Equity						

CONTACT INFORMATION:

Phone: 763-212-5000   Fax:
Toll-Free:
Address: 701 Carlson Pkwy., Minnetonka, MN 55305 United States

SALARIES/BONUSES:

Top Exec. Salary: $ Bonus: $
Second Exec. Salary: $ Bonus: $

STOCK TICKER/OTHER:

Stock Ticker: Subsidiary   Exchange:
Employees: 90,000   Fiscal Year Ends: 12/31
Parent Company: HNA Group Co Ltd

OTHER THOUGHTS:

Estimated Female Officers or Directors: 6
Hot Spot for Advancement for Women/Minorities: Y

CBL & Associates Properties Inc

www.cblproperties.com

NAIC Code: 531120

TYPES OF BUSINESS:

Malls & Shopping Centers
Retail Property Management
Retail Property Development
Real Estate Investment Trust

BRANDS/DIVISIONS/AFFILIATES:

CBL Holdings I Inc
CBL Holdings II Inc
CBL & Associates Management Inc
Mayfaire Town Center and Community Center

CONTACTS: Note: Officers with more than one job title may be intentionally listed here more than once.

Stephen Lebovitz, CEO
Farzana Mitchell, CFO
Charles Lebovitz, Chairman of the Board
Augustus Stephas, COO
Michael Lebovitz, Executive VP, Divisional
Ben Landress, Executive VP, Divisional
Andrew Cobb, Other Corporate Officer
Jeffery Curry, Other Executive Officer
Stuart Smith, Senior VP, Divisional
Katie Reinsmidt, Senior VP, Divisional
Mike Harrison, Senior VP, Divisional
Howard Grody, Senior VP, Divisional
Jerry Sink, Senior VP, Divisional
Alan Lebovitz, Senior VP, Divisional

GROWTH PLANS/SPECIAL FEATURES:

CBL & Associates Properties, Inc. (CBL) is a self-managed, self-administered real estate investment trust (REIT) engaged in the ownership, development, acquisition, leasing, management and operation of regional shopping malls, open-air centers, community centers and office properties. The firm's properties are located primarily in the southeastern and Midwestern U.S., with its top markets in St. Louis, Missouri; Chattanooga, Tennessee; Madison, Wisconsin; Lexington, Kentucky; and Winston-Salem, North Carolina. CBL owns, holds interests in or manages 81 properties, including regional malls/open-air centers; associated centers, which are retail properties adjacent to regional mall complexes; community centers, which are typically anchored by a combination of supermarkets or value-priced stores; and office buildings. The properties are located in 27 states. The company conducts substantially all of its business through two REIT subsidiaries, CBL Holdings I, Inc. and CBL Holdings II, Inc. Additionally, the company conducts its property management operations through subsidiary CBL & Associates Management, Inc. The firm's onsite property management functions include leasing, management, data processing, rent collection, budgeting and promotions. In 2015, the firm acquired Mayfaire Town Center and Community Center, the open-air center in the coastal market of Wilmington, North Carolina.

FINANCIAL DATA: Note: Data for latest year may not have been available at press time.

In U.S. $	2015	2014	2013	2012	2011	2010
Revenue	1,055,018,000	1,060,739,000	1,053,625,000	1,034,640,000	1,067,340,000	1,071,804,000
R&D Expense						
Operating Income	277,584,000	375,143,000	330,765,000	379,169,000	357,667,000	370,502,000
Operating Margin %	26.31%	35.36%	31.39%	36.64%	33.51%	34.56%
SGA Expense	62,118,000	50,271,000	105,246,000	103,828,000	101,849,000	100,676,000
Net Income	103,371,000	219,150,000	85,204,000	131,600,000	133,936,000	62,151,000
Operating Cash Flow	495,015,000	468,061,000	464,751,000	481,515,000	441,836,000	429,792,000
Capital Expenditure	410,879,000	277,624,000	355,743,000	313,926,000	216,879,000	143,586,000
EBITDA	650,368,000	788,943,000	634,090,000	706,431,000	702,283,000	668,350,000
Return on Assets %	.89%	2.60%	.58%	1.21%	1.28%	.38%
Return on Equity %	4.34%	12.39%	2.94%	6.48%	7.14%	2.44%
Debt to Equity	3.66	3.34	3.45	3.57	3.55	4.00

CONTACT INFORMATION:

Phone: 423-855-0001   Fax: 423 855-8662
Toll-Free: 800-333-7310
Address: 2030 Hamilton Place Blvd., Ste. 500, Chattanooga, TN 37421-6000 United States

STOCK TICKER/OTHER:

Stock Ticker: CBL
Employees: 744
Parent Company:

Exchange: NYS
Fiscal Year Ends: 12/31

SALARIES/BONUSES:

Top Exec. Salary: $675,000   Bonus: $270,000
Second Exec. Salary: $700,000   Bonus: $229,688

OTHER THOUGHTS:

Estimated Female Officers or Directors: 8
Hot Spot for Advancement for Women/Minorities: Y

CBRE Group Inc

NAIC Code: 531210   **www.cbre.com**

TYPES OF BUSINESS:

Real Estate Brokerage
Real Estate Management Services
Mortgage Banking
Investment Management
Consulting Services

BRANDS/DIVISIONS/AFFILIATES:

CBRE Inc
CBRE Capital Markets
CBRE Ltd
CBRE Global Investors LLC
Trammell Crow Company LLC
Sitehawk Retail Real Estate
Forum Analytics LLC
Tax Credit Group

CONTACTS: *Note: Officers with more than one job title may be intentionally listed here more than once.*

William Concannon, CEO, Divisional
T. Ferguson, CEO, Divisional
Calvin Frese, CEO, Geographical
Robert Sulentic, CEO
James Groch, CFO
Ray Wirta, Chairman of the Board
Gil Borok, Chief Accounting Officer
J. Kirk, Chief Administrative Officer
Michael Lafitte, COO
Laurence Midler, Executive VP
Steve Iaco, Other Corporate Officer

GROWTH PLANS/SPECIAL FEATURES:

CBRE Group, Inc. is one of the world's largest commercial real estate services companies, with over 400 offices in more than 50 countries. It offers services to occupiers, owners, lenders/investors in office, retail, industrial, multi-family and other commercial real estate assets. The firm's core services include commercial property and corporate facilities management, tenant representation, property/agency leasing, property sales, valuation, real estate investment management, commercial mortgage origination and servicing, capital markets (equity and debt), development services and proprietary research. CBRE operates in five segments: Americas, which accounted for 57% of its 2015 revenue; Europe, Middle East and Africa (EMEA), 27.7%; Asia Pacific, 10.5%; global investment management, 4.2%; and development services, 0.6%. Americas operates primarily through CBRE, Inc.; CBRE Capital Markets, Inc.; and CBRE Ltd. EMEA has offices in 43 countries, with its largest located in the France, Germany, Italy, the Netherlands, Russia, Spain and the U.K. Asia Pacific operates in 14 countries, including Australia and New Zealand, China, Hong Kong, India, Japan, Korea and Singapore. In addition, the company has agreements with affiliated offices in Cambodia, Malaysia, the Philippines and Thailand that generate royalty fees and support cross-referral arrangements. Global investment management is handled by subsidiary CBRE Global Investors, LLC. Through the Trammell Crow Company LLC, the firm provides development services primarily in the U.S. to users of and investors in commercial real estate. During 2015, the firm acquired Global Workplace Solutions business from Johnson Controls, Inc.; Sitehawk Retail Real Estate, a retail real estate services firm; Forum Analytics, LLC, provider of sophisticated modeling and mapping solutions; and Tax Credit Group, provider of advisory, investment sales, debt and structured finance services.

The company offers its employees medical, dental and vision insurance; a 401(k); flexible spending accounts; an employee assistance program; and paid vacation and holidays.

FINANCIAL DATA: *Note: Data for latest year may not have been available at press time.*

In U.S. $	2015	2014	2013	2012	2011	2010
Revenue	10,855,810,000	9,049,918,000	7,184,794,000	6,514,099,000	5,905,411,000	5,115,316,000
R&D Expense						
Operating Income	835,944,000	792,254,000	616,128,000	585,081,000	462,862,000	446,379,000
Operating Margin %	7.70%	8.75%	8.57%	8.98%	7.83%	8.72%
SGA Expense	2,633,609,000	2,438,960,000	2,104,310,000	2,002,914,000	1,882,666,000	1,607,682,000
Net Income	547,132,000	484,503,000	316,538,000	315,555,000	239,162,000	200,345,000
Operating Cash Flow	651,897,000	661,780,000	745,108,000	291,081,000	361,219,000	616,587,000
Capital Expenditure	139,464,000	171,242,000	156,358,000	150,232,000	147,980,000	68,464,000
EBITDA	1,312,706,000	1,154,398,000	835,337,000	835,451,000	696,717,000	572,170,000
Return on Assets %	5.86%	6.61%	4.27%	4.19%	3.87%	3.94%
Return on Equity %	22.00%	23.31%	18.43%	23.45%	23.22%	26.06%
Debt to Equity	0.97	0.81	0.98	1.65	2.26	2.03

CONTACT INFORMATION:

Phone: 213-613-3333   Fax: 213 438-4820
Toll-Free:
Address: 400 South Hope Street, 25th Fl, Los Angeles, CA 90071 United States

STOCK TICKER/OTHER:

Stock Ticker: CBG   Exchange: NYS
Employees: 52,000   Fiscal Year Ends: 12/31
Parent Company:

SALARIES/BONUSES:

Top Exec. Salary: $967,500   Bonus: $607,130
Second Exec. Salary: $752,500   Bonus: $500,000

OTHER THOUGHTS:

Estimated Female Officers or Directors: 2
Hot Spot for Advancement for Women/Minorities:

Sales, profits and employees may be estimates. Financial information, benefits and other data can change quickly and may vary from those stated here.

CDM Smith Inc

www.cdm.com

NAIC Code: 237000

TYPES OF BUSINESS:

Water and Sewer Line and Related Structures Construction
Water Management
Environmental Services
Design Services
Information Management & Technology
Consulting
Facilities Design
Geotechnical Services

BRANDS/DIVISIONS/AFFILIATES:

CDM Federal Programs Corporation
CDM International Inc
CDM Consult GmbH
CDM Constructors Inc

CONTACTS: Note: Officers with more than one job title may be intentionally listed here more than once.

Stephen J. Hickox, CEO
Timothy B. Wall, COO
Timothy B. Wall, Pres.
Thierry Desmaris, Exec. VP-Finance
Stephen J. Hickox, Chmn.

GROWTH PLANS/SPECIAL FEATURES:

CDM Smith, Inc. (CDM) provides services in engineering, consulting and construction. The company has three operating units: the North America unit, Federal Services unit and Industrial Services unit. Supporting these units are the consulting services, engineering services, construction services and the operations services units. CDM offers engineering services such as 3D design; automation and instrumentation; and civil, electrical, geotechnical, mechanical, process and structural engineering. Construction services include constructability and value engineering reviews, cost estimating, design-build and alternative delivery methods, engineering services during construction, general contracting, procurement and project controls. Operations services include contract management, contract operations, operations and maintenance and operations optimization. CDM's work involves solid waste and wastewater purification facilities; municipal data management systems; airports, dams, harbors and bridges; a wildlife refuge; major universities; municipal railways; and sports facilities. The firm has a number of subsidiaries, including CDM International, Inc., offering CDM's full range of services in Europe, Latin America, the Middle East and Asia; CDM Constructors, Inc., providing design, construction, remediation, general contracting and equipment fabrication services worldwide; and CDM Federal Programs Corporation, offering environmental management services for the U.S. Environmental Protection Agency (EPA), Department of Energy, Department of Defense and other U.S. government agencies. Additionally, CDM owns a majority interest in geotechnical and environmental consulting and design firm CDM Consult GmbH in Germany.

The firm offers its employees medical, dental and vision coverage; short- and long-term disability; life and AD&D insurance; commuter benefits; flexible spending accounts; a 401(k) program; profit sharing; an employee assistance program; tuition assistance; and business travel insurance.

FINANCIAL DATA: Note: Data for latest year may not have been available at press time.

In U.S. $	2015	2014	2013	2012	2011	2010
Revenue	1,260,000,000	1,100,000,000	1,000,000,000	912,700,000	960,800,000	
R&D Expense						
Operating Income						
Operating Margin %						
SGA Expense						
Net Income						
Operating Cash Flow						
Capital Expenditure						
EBITDA						
Return on Assets %						
Return on Equity %						
Debt to Equity						

CONTACT INFORMATION:

Phone: 617-452-6000 Fax: 617-345-3901
Toll-Free:
Address: 75 State St., Ste. 701, Boston, MA 02109 United States

SALARIES/BONUSES:

Top Exec. Salary: $ Bonus: $
Second Exec. Salary: $ Bonus: $

STOCK TICKER/OTHER:

Stock Ticker: Private
Employees: 5,000
Parent Company:

Exchange:
Fiscal Year Ends: 12/31

OTHER THOUGHTS:

Estimated Female Officers or Directors:
Hot Spot for Advancement for Women/Minorities:

CEMEX Inc

NAIC Code: 327320

www.cemexusa.com

TYPES OF BUSINESS:

Cement Materials & Production
Cement Materials & Production
Ready-Mix Cement

GROWTH PLANS/SPECIAL FEATURES:

CEMEX, Inc., a subsidiary of Mexico-based CEMEX SAB de CV, is among the largest cement and ready-mix companies in the U.S. Headquartered in Houston, Texas, the firm's network includes 13 cement plants, 46 distribution channels, more than 74 aggregate quarries and over 350 ready-mix concrete plants. CEMEX divides its operations into the following six categories: cement, aggregates, ready-mix concrete, fly ash, pipe/precast/stormwater and related products. Cement products, which vary by region in terms of availability, include portland cement, masonry cement, hydraulic cement, mortar cement and well cement. Aggregate products include sand, gravel and crushed stone, all of which are used in concrete products as well as in landscaping settings. The firm's ready-mix concrete, which consists of cement, water and aggregates, is a cost-effective and versatile building material used in almost all types of construction. Special ready-mix products include abrasion resistant, corrosion inhibited, self-consolidating and underwater concretes. The fly ash unit sells the coal-fired power plant byproduct fly ash that, when added to concrete, improves strength and reduces water demand. The pipe/precast/stormwater unit deals with concrete pipes, culvert boxes and other pipeline products. The related products unit manufactures concrete block, architectural products (such as concrete pavers, concrete brick and segmental retaining wall block), asphalt, building materials, gypsum and landscaping materials. Products are marketed principally to industrial, commercial, residential and municipal construction customers. Subsidiaries of the company include Ready Mix USA LLC and CEMEX Southeast LLC. In Fall 2015, the firm broke ground on a 240,000-square-foot office building in Memorial City, Texas, on an 18-acre tract, which will be the home of CEMEX headquarters in the U.S. by Fall 2016.

Employees of the firm receive benefits including health and welfare coverage, tuition reimbursement and professional and technical training.

BRANDS/DIVISIONS/AFFILIATES:

CEMEX SAB de CV
Ready Mix USA LLC
CEMEX Southeast LLC

CONTACTS:
Note: Officers with more than one job title may be intentionally listed here more than once.

Ignacio Madridejos, Pres.
Gullermo Martinez, Exec. VP-Human Resources & Communications
Mike Egan, General Counsel
Luis Oropeza, Exec. VP-Cement Oper.
Juan Carlos Herrera, Exec. VP-Strategic Planning
Frank Craddock, Exec. VP-Commercial & Public Affairs
Scott Ducoff, Pres., Texas & New Mexico
Terry Brasher, Pers., Mid-South Region
Bob Capasso, Exec. VP-Bus. Svcs. Organization
Robert Cutter, Pres., Arizona & Nevada
Kirk Light, Exec. VP-Logistics

FINANCIAL DATA:
Note: Data for latest year may not have been available at press time.

In U.S. $	2015	2014	2013	2012	2011	2010
Revenue	3,500,000,000	3,570,800,000	3,349,940,000	3,296,480,000	3,150,500,000	
R&D Expense						
Operating Income						
Operating Margin %						
SGA Expense						
Net Income						
Operating Cash Flow						
Capital Expenditure						
EBITDA						
Return on Assets %						
Return on Equity %						
Debt to Equity						

CONTACT INFORMATION:

Phone: 713-650-6200   Fax: 713-653-6815
Toll-Free: 800-992-3639
Address: 929 Gessner Rd., Ste. 1900, Houston, TX 77024 United States

STOCK TICKER/OTHER:

Stock Ticker: Subsidiary   Exchange:
Employees: 10,000   Fiscal Year Ends: 12/31
Parent Company: CEMEX SAB de CV

SALARIES/BONUSES:

Top Exec. Salary: $ Bonus: $
Second Exec. Salary: $ Bonus: $

OTHER THOUGHTS:

Estimated Female Officers or Directors:
Hot Spot for Advancement for Women/Minorities:

Cemex SAB de CV

www.cemex.com

NAIC Code: 423300

TYPES OF BUSINESS:

Cement & Concrete Production
Construction Materials Production

GROWTH PLANS/SPECIAL FEATURES:

Cemex SAB de CV founded in Mexico in 1906, is a holding company that produces, distributes, markets and sells cement, ready-mix concrete, aggregates, clinker and other construction materials through its operating subsidiaries. The company has operations in over 50 countries across five continents, with annual production levels of around 94 million tons of cement. Cemex wholly owns 55 cement plants and 1,736 ready-mix concrete facilities and has a minority participation in 12 cement plants. The firm also operates 341 aggregates quarries, 233 land-distribution centers and 63 marine terminals. The company produces 56 million cubic meters of ready-mix concrete and over 168 million tons of aggregates per year. The firm's operations in Mexico are run by subsidiaries CEMEX Mexico and Empresas Tolteca de Mexico. In the U.S., the company owns CEMEX, Inc., which manages companies such as Ready Mix USA LLC and CEMEX Southeast LLC. The company operates through a number of subsidiaries throughout Europe: CEMEX France Gestion SAS, CEMEX U.K., CEMEX Espana SA and CEMEX Deutschland, with additional operations in Austria, Croatia, the Czech Republic, Finland, Hungary, Ireland, Latvia, Lithuania, Norway, Poland and Sweden. In South and Central America, the company operates facilities in Colombia, Argentina, Costa Rica, the Dominican Republic, Panama, Guatemala, Nicaragua, Puerto Rico and Jamaica. CEMEX also has operations in Egypt, the UAE, Israel, the Philippines, China, Thailand, Malaysia and Bangladesh. Sales of cement represent 45% of its total sales; ready-mix concrete, 39%; and aggregates, 16%.

BRANDS/DIVISIONS/AFFILIATES:

CEMEX Deutschland
CEMEX Mexico
CEMEX Espana SA
Empresas Tolteca de Mexico
CEMEX France Gestion SAS
CEMEX Southeast LLC
CEMEX U.K.
CEMEX Inc

CONTACTS: *Note: Officers with more than one job title may be intentionally listed here more than once.*

Fernando A. Gonzalez, CEO
Karl Watson, Pres., CEMEX USA
Jose Antonio Gonzalez, CFO
Juan Romero, Head-Global Tech.
Fernando A. Gonzalez, Exec. VP-Admin.
Juan Pablo San Agustin, Exec. VP-Strategic Planning & New Bus. Dev.
Jaime Elizondo, Pres., CEMEX South, Central America & Caribbean
Ignacio Madridejos, Pres., CEMEX Northern Europe
Juan Romero, Pres., CEMEX Mexico
Jaime Muguiro, Pres., CEMEX Mediterranean
Rogelio Zambrano, Chmn.
Joaquin Estrada, Pres., CEMEX Asia
Jaime Elizondo, Head-Global Procurement

FINANCIAL DATA: *Note: Data for latest year may not have been available at press time.*

In U.S. $	2015	2014	2013	2012	2011	2010
Revenue		11,915,910,000	11,101,060,000	11,179,080,000	10,773,470,000	10,113,800,000
R&D Expense						
Operating Income		967,636,000	828,405,400	652,920,300	684,465,600	615,190,700
Operating Margin %		8.12%	7.46%	5.84%	6.35%	6.08%
SGA Expense		2,558,462,000	2,347,914,000	2,333,276,000	2,363,404,000	2,218,953,000
Net Income		-384,841,700	-614,680,100	-674,082,900	-1,406,377,000	-937,055,200
Operating Cash Flow		680,380,700	72,054,980	319,084,400	367,991,000	1,239,005,000
Capital Expenditure		399,196,000	384,274,300	317,552,600	181,442,400	268,135,300
EBITDA		1,946,279,000	1,758,539,000	1,724,611,000	1,225,672,000	1,293,359,000
Return on Assets %		-1.34%	-2.22%	-2.32%	-4.69%	-3.01%
Return on Equity %		-5.12%	-7.89%	-8.02%	-14.19%	-8.09%
Debt to Equity		1.64	1.40	1.49	1.31	1.01

CONTACT INFORMATION:

Phone: 52-81-8888-8888   Fax: 52-81-8888-4417
Toll-Free:
Address: Ave. Ricardo Margain Zozaya 325, San Pedro Garza Garcia, 66265 Mexico

STOCK TICKER/OTHER:

Stock Ticker: CX
Employees: 44,241
Parent Company:

Exchange: NYS
Fiscal Year Ends: 12/31

SALARIES/BONUSES:

Top Exec. Salary: $ Bonus: $
Second Exec. Salary: $ Bonus: $

OTHER THOUGHTS:

Estimated Female Officers or Directors:
Hot Spot for Advancement for Women/Minorities:

Centex Corp
NAIC Code: 236117

www.centex.com

TYPES OF BUSINESS:
Residential Construction
Mortgages
Real Estate Development
Commercial Construction
Construction Supply Services

BRANDS/DIVISIONS/AFFILIATES:
PulteGroup Inc
Pulte Mortgage LLC
PGP Title
PCIC Insurance

CONTACTS: *Note: Officers with more than one job title may be intentionally listed here more than once.*
Richard J. Dugas, Jr., CEO

GROWTH PLANS/SPECIAL FEATURES:

Centex Corp., a subsidiary of PulteGroup, Inc., focuses principally on residential construction and related activities, including mortgage financing. Operating in 13 states, the firm's home building operations involve the purchase and development of land or lots and the construction and sale of detached and attached single-family homes, including resort and second home properties and lots as well as land or lots. Centex's financial services include mortgage lending, conducted through Pulte Mortgage LLC; title insurance, handled by PGP Title; and insurance with PCIC Insurance. In partnership with HomeTeam Pest Defense, the company installs pest control systems in its homes as they are being built, including Taexx built-in pest control systems and Tubes Under the Slab termite control systems. All new Centex homes include energy saving features such as radiant-barrier roof decking, enhanced insulation, low-emissivity windows and Energy Star appliances.

Parent company PulteGroup offers employees medical, dental and vision insurance; life and AD&D insurance; short- and long-term disability insurance; a 401(k) plan; a college savings plan; tuition reimbursement; and an employee assistance program.

FINANCIAL DATA: *Note: Data for latest year may not have been available at press time.*

In U.S. $	2015	2014	2013	2012	2011	2010
Revenue						
R&D Expense						
Operating Income						
Operating Margin %						
SGA Expense						
Net Income						
Operating Cash Flow						
Capital Expenditure						
EBITDA						
Return on Assets %						
Return on Equity %						
Debt to Equity						

CONTACT INFORMATION:
Phone: 214-981-5000   Fax: 214-981-6859
Toll-Free:
Address: 2728 N. Harwood, Dallas, TX 75201 United States

STOCK TICKER/OTHER:
Stock Ticker: Subsidiary
Employees:
Parent Company: PulteGroup Inc

Exchange:
Fiscal Year Ends: 03/31

SALARIES/BONUSES:
Top Exec. Salary: $ Bonus: $
Second Exec. Salary: $ Bonus: $

OTHER THOUGHTS:
Estimated Female Officers or Directors:
Hot Spot for Advancement for Women/Minorities:

Century 21 Real Estate LLC

www.century21.com

NAIC Code: 531210

TYPES OF BUSINESS:

Real Estate Brokerage
Commercial Brokerage
Residential & Specialty Brokerage
Mortgage Services

BRANDS/DIVISIONS/AFFILIATES:

Realogy Corporation
Century 21 Mortgage
Century 21 Commercial
Century 21 Fine Homes and Estates
Century 21 International
At Home with Century 21
Golden Ruler (The)

CONTACTS: Note: Officers with more than one job title may be intentionally listed here more than once.

Richard W. Davidson, CEO
Greg Sexton, COO
Richard W. Davidson, Pres.

GROWTH PLANS/SPECIAL FEATURES:

Century 21 Real Estate LLC, a subsidiary of Realogy Corporation, is the franchiser of one of the world's largest residential real estate sales organizations. It has approximately 6,900 independently owned and operated franchised broker offices in 78 countries and territories. The firm helps clients find, buy, sell and finance their residential property or new home as well as find, buy, sell and sometimes lease commercial property and vacation properties. Century 21's web site gives the public access to its vast database of listed homes as well as allowing clients to list their homes on said database. The company has offices in every state, providing customers with brokers familiar with their area and current market conditions. Century 21 Mortgage offers mortgage services to its clients in person, over the phone or online, and advertises same-day loan decisions for clients. Other subsidiaries include Century 21 Commercial, helping clients sell, buy or lease commercial real estate; Century 21 International, active in Africa, the Americas, Australia and New Zealand, the Caribbean, Europe, Asia and the Middle East; and Century 21 Fine Homes and Estates, for those looking for higher-priced luxury homes. Its bi-monthly magazine, At Home with Century 21, provides information on home improvement, decor, remodeling, organization and storage, gardening and cooking. The firm also has an online tool called The Golden Ruler, which allows customers to view web traffic statistics for their homes listed online.

FINANCIAL DATA: Note: Data for latest year may not have been available at press time.

In U.S. $	2015	2014	2013	2012	2011	2010
Revenue						
R&D Expense						
Operating Income						
Operating Margin %						
SGA Expense						
Net Income						
Operating Cash Flow						
Capital Expenditure						
EBITDA						
Return on Assets %						
Return on Equity %						
Debt to Equity						

CONTACT INFORMATION:

Phone: 877-221-2765   Fax:
Toll-Free: 866-732-6139
Address: 175 Park Avenue, Madison, NJ 07940 United States

STOCK TICKER/OTHER:

Stock Ticker: Subsidiary   Exchange:
Employees: 101,000   Fiscal Year Ends: 12/31
Parent Company: Realogy Corporation

SALARIES/BONUSES:

Top Exec. Salary: $ Bonus: $
Second Exec. Salary: $ Bonus: $

OTHER THOUGHTS:

Estimated Female Officers or Directors: 1
Hot Spot for Advancement for Women/Minorities:

CH2M HILL Companies Ltd

NAIC Code: 541330

www.ch2m.com

TYPES OF BUSINESS:

Engineering Services-Consultation
Environmental Engineering & Consulting
Nuclear Management Services
Water & Electrical Utility Services
Decommissioning & Decontamination
Facilities Design & Construction
Project Financing & Procurement
Nanotechnology Research

BRANDS/DIVISIONS/AFFILIATES:

CH2M HILL Operations Management International
CH2M HILL Canada Ltd
CH2M HILL Lockwood Greene
CH2M HILL Industrial Design and Construction Inc
CH2M-IDC China

CONTACTS: Note: Officers with more than one job title may be intentionally listed here more than once.

Jacqueline Hinman, CEO
Jacqueline Hinman, Pres.
Gary L. McArthur, CFO
Lisa Glatch, Exec. VP-Sales & Client Solutions
Shelie Gustafson, Chief Human Resources Officer
Randall L. Smith, Pres., Industrial & Advanced Tech. Bus. Group
Elisa M. Speranza, Pres., Oper. & Maintenance Bus. Group
Patrick O'Keefe, Sr. VP-Corp. Affairs
John Corsi, VP-Global Media & Public Rel.
Michael E McKelvy, Pres., Gov't, Environment & Infrastructure Div.
Michael A. Szomjassy, Pres., Energy, Water & Facilities Div.
Robert W. Bailey, Pres., Water Bus. Group
Terry A. Ruhl, Pres., Transportation Bus. Group
Jacqueline Hinman, Chmn.
Jacqueline C. Rast, Pres., Int'l Div.

GROWTH PLANS/SPECIAL FEATURES:

CH2M HILL Companies, Ltd. is an employee-owned firm that offers engineering, consulting, design, construction, procurement, operations, maintenance and program and project management services to clients in the public and private sectors. CH2M conducts business in several countries worldwide. The firm operates in the energy, environmental, facilities, resources, transportation and water markets. Throughout these markets the company offers a variety of services. The company's environmental services division offers its clients ecological and natural resource damage assessments, environmental consulting for remediation projects and treatment systems for properties that have been contaminated by toxic or radioactive waste. The nuclear services segment manages the decontamination and demolition of weapons production facilities and designs nuclear waste treatment and handling facilities. CH2M HILL Operations Management International subsidiary provides water, wastewater and electrical utility services to private and public clients. CH2M HILL Canada, Ltd. is the Canadian division of the company. CH2M HILL Lockwood Greene is a major engineering and construction firm focused on national and multinational industrial and power clients worldwide. CH2M HILL Industrial Design and Construction, Inc. (IDC) is a high-technology facilities design, construction, maintenance and operations company serving process-intensive technology clients. IDC also has interests in nanotechnology research and manufacturing. CH2M-IDC China provides full-service solutions to manufacturing companies that are building or have plants in China.

FINANCIAL DATA: Note: Data for latest year may not have been available at press time.

In U.S. $	2015	2014	2013	2012	2011	2010
Revenue	5,410,000,000	5,880,000,000	6,475,000,000	6,160,553,000	5,555,200,000	5,422,801,000
R&D Expense						
Operating Income						
Operating Margin %						
SGA Expense						
Net Income		118,000,000	100,000,000	93,000,000	113,300,000	93,695,000
Operating Cash Flow						
Capital Expenditure						
EBITDA						
Return on Assets %						
Return on Equity %						
Debt to Equity						

CONTACT INFORMATION:

Phone: 720-286-2000   Fax:
Toll-Free: 888-242-6445
Address: 9191 S. Jamaica St., Englewood, CO 80112 United States

STOCK TICKER/OTHER:

Stock Ticker: Private   Exchange:
Employees: 26,000   Fiscal Year Ends: 12/31
Parent Company:

SALARIES/BONUSES:

Top Exec. Salary: $ Bonus: $
Second Exec. Salary: $ Bonus: $

OTHER THOUGHTS:

Estimated Female Officers or Directors: 3
Hot Spot for Advancement for Women/Minorities: Y

Champion Enterprises Holdings LLC

www.championhomes.com

NAIC Code: 321992

TYPES OF BUSINESS:

Manufactured Housing
Housing Retailing
Steel-Frame Building

BRANDS/DIVISIONS/AFFILIATES:

Centerbridge Partners LP
MAK Capital Fund LP
Sankaty Advisors LLC
Champion Enterprises Inc
Moduline Industries
SRI Homes
Champion
Carolina Building Solutions

CONTACTS: *Note: Officers with more than one job title may be intentionally listed here more than once.*

Keith Anderson, CEO
John Lawless, Pres.
Phyllis Knight, CFO
Roger K. Scholten, General Counsel
Timothy J. Bernlohr, Chmn.

GROWTH PLANS/SPECIAL FEATURES:

Champion Enterprises Holdings LLC is the holding company for a family of companies that build and retail manufactured and modular housing. The primary owners of the privately held firm are Centerbridge Partners LP, MAK Capital Fund LP and Sankaty Advisors LLC. Founded in 1953, the company has sold over 1.7 million factory-built homes since its inception. The company is one of the largest housing manufacturers and retailers in the U.S., manufacturing homes through 27 manufacturing facilities in North America and the U.K. The Champion family of homebuilders includes Carolina Building Solutions, Commander, Dutch, Fortune, Highland, Homes of Merit, New Era, New Image, North American, Redman, Silvercrest and Titan. Most of the homes built are multi-section, ranch-style units, but one and two-story homes, colonial-style homes, Cape Cod style homes and multi-family units are also offered. Homes generally range in size from 600 to 3,000 square feet and typically include 2-4 bedrooms, a living room or family room, dining room, kitchen and two full bathrooms. Additional floor plan options include vaulted ceilings, entertainment centers, spa-style bathrooms, fireplaces, custom cabinetry and various floor covering options. In Canada, the company produces and sells factory-built homes through two divisions: Moduline Industries, operating in western Canada; and SRI Homes, operating in western and central Canada. Additionally, Champion produces pre-engineered and steel-framed modular buildings in the U.K. through Caledonian Building Systems Limited. In 2015, the firm acquired the domestic and international operations of newly-created subsidiary Champion Enterprises, Inc. In January 2016, the firm announced that its Eatontown, New Jersey site will feature all net zero energy manufactured residential homes.

FINANCIAL DATA: *Note: Data for latest year may not have been available at press time.*

In U.S. $	2015	2014	2013	2012	2011	2010
Revenue						
R&D Expense						
Operating Income						
Operating Margin %						
SGA Expense						
Net Income						
Operating Cash Flow						
Capital Expenditure						
EBITDA						
Return on Assets %						
Return on Equity %						
Debt to Equity						

CONTACT INFORMATION:

Phone: 248-614-8200   Fax:
Toll-Free:
Address: 755 W. Big Beaver, Ste. 1000, Troy, MI 48084 United States

STOCK TICKER/OTHER:

Stock Ticker: Private   Exchange:
Employees: Fiscal Year Ends: 12/31
Parent Company: Centerbridge Partners LP

SALARIES/BONUSES:

Top Exec. Salary: $ Bonus: $
Second Exec. Salary: $ Bonus: $

OTHER THOUGHTS:

Estimated Female Officers or Directors: 1
Hot Spot for Advancement for Women/Minorities:

Chicago Bridge & Iron Company NV (CB&I)

NAIC Code: 237000

www.cbi.com

TYPES OF BUSINESS:

Heavy Construction & Civil Engineering
Specialty Engineering & Procurement Services
Liquid & Gas Storage Facilities
Maintenance & Support Services

BRANDS/DIVISIONS/AFFILIATES:

CONTACTS: Note: Officers with more than one job title may be intentionally listed here more than once.

Phillip K. Asheman, CEO
Lasse Petterson, COO
Phillip K. Asheman, Pres.
Ronald A. Ballschmiede, CFO
Daniel M. McCarthy, Exec. VP
Beth A. Bailey, Chief Admin. Officer
Richard E. Chandler, Jr., Chief Legal Officer
Patrick K. Mullen, Exec. VP-Corp. Dev.
E. Chip Ray, Exec. VP
Luke V. Scorsone, Exec. VP
L. Richard Flury, Chmn.

GROWTH PLANS/SPECIAL FEATURES:

Chicago Bridge & Iron Company NV (CB&I), a global engineering, procurement and construction (EPC) company, provides specialty construction for liquid and gas storage facilities. The company has active projects in more than 70 countries. CB&I has four business units, operating both independently and on an integrated basis. The engineering and construction unit offers engineering, procurement and construction services for major energy infrastructure facilities. Fabrication services provide fabrication of piping systems, process and nuclear modules and fabrication and erection of storage tanks and pressure vessels for the oil and gas, petrochemicals, water and wastewater, mining, mineral processing and power generation industries. CB&I technology provides licensed process technologies, catalysts, specialized equipment and engineered products for use in petrochemical facilities, oil refineries, coal gasification plants and gas processing plants, and offers process planning and project development services and a comprehensive program of aftermarket support. The capital services segment provides maintenance, environmental, infrastructure and program management services for government and private sector customers. Services in this segment include decommissioning and decontamination; environmental engineering and consulting; energy efficiency and sustainability; and program management. Additionally, it offers numerous complementary products and services, including low temperature or cryogenic tanks and systems, primarily used by petroleum, chemical, petrochemical and other companies to store, transport and handle liquefied gases and specialty structures including iron and aluminum processing facilities and hydroelectric structures.

The firm offers its U.S. employees medical, dental and vision plans; employee and dependent life insurance options; a 401(k) plan; profit sharing; a stock purchase program; and an education assistance plan.

FINANCIAL DATA: Note: Data for latest year may not have been available at press time.

In U.S. $	2015	2014	2013	2012	2011	2010
Revenue	12,929,500,000	12,974,930,000	11,094,530,000	5,485,206,000	4,550,542,000	3,642,318,000
R&D Expense						
Operating Income	-425,117,000	982,608,000	684,508,000	455,643,000	355,197,000	303,260,000
Operating Margin %	-3.28%	7.57%	6.16%	8.30%	7.80%	8.32%
SGA Expense	387,027,000	405,208,000	379,485,000	227,948,000	205,550,000	185,213,000
Net Income	-504,415,000	543,607,000	454,120,000	301,655,000	255,032,000	204,559,000
Operating Cash Flow	-56,214,000	264,047,000	-112,836,000	202,504,000	413,155,000	288,406,000
Capital Expenditure	78,852,000	117,624,000	90,492,000	72,279,000	40,945,000	24,089,000
EBITDA	-255,697,000	1,172,530,000	871,464,000	530,093,000	433,177,000	381,100,000
Return on Assets %	-5.42%	5.79%	6.62%	7.91%	8.22%	6.90%
Return on Equity %	-21.22%	21.39%	24.46%	23.70%	22.84%	21.20%
Debt to Equity	0.89	0.57	0.69	0.58		0.03

CONTACT INFORMATION:

Phone: 31 703732010   Fax: 31 703732750
Toll-Free:
Address: Oostduinlaan 75, Hoofddorp, The Hague, 2596JJ Netherlands

STOCK TICKER/OTHER:

Stock Ticker: CBI
Employees: 54,400
Parent Company:

Exchange: NYS
Fiscal Year Ends: 12/31

SALARIES/BONUSES:

Top Exec. Salary:
$1,311,848
Second Exec. Salary:
$676,769

Bonus: $

Bonus: $

OTHER THOUGHTS:

Estimated Female Officers or Directors: 2

Hot Spot for Advancement for Women/Minorities: Y

Sales, profits and employees may be estimates. Financial information, benefits and other data can change quickly and may vary from those stated here.

Chimera Investment Corporation

www.chimerareit.com

NAIC Code: 525990

TYPES OF BUSINESS:
Mortgage REIT

BRANDS/DIVISIONS/AFFILIATES:
Chimera Mortgage Securities LLC

CONTACTS: Note: Officers with more than one job title may be intentionally listed here more than once.
Matthew Lambiase, CEO
Robert Colligan, CFO
Paul Donlin, Chairman of the Board
Choudhary Yarlagadda, COO
Phillip Kardis, General Counsel
William Dyer, Other Corporate Officer
Mohit Marria, Other Executive Officer

GROWTH PLANS/SPECIAL FEATURES:
Chimera Investment Corporation operates as a specialty finance company that invests in residential mortgage backed securities (RMBS), commercial mortgage loans, residential mortgage loans, real estate-related securities and various other asset classes. It has elected to be taxed as a real estate investment trust (REIT) for federal income tax purposes. The company's objective is to provide attractive risk-adjusted returns to its investors over the long-term, mainly through dividends and secondarily through capital appreciation. It intends to achieve this objective by investing in a class of financial assets to construct an investment portfolio that is designed to achieve attractive risk-adjusted returns and that is structured to comply with the various federal income tax requirements for REIT status. Since it commenced operations in 2007, the firm has focused its investment activities on acquiring non-Agency RMBS and on purchasing residential mortgage loans that have been originated by select high-quality originators, including the retail lending operations of leading commercial banks. In March 2015, the firm formed Chimera Mortgage Securities LLC, which is a wholly-owned qualified REIT subsidiary.

FINANCIAL DATA: Note: Data for latest year may not have been available at press time.

In U.S. $	2015	2014	2013	2012	2011	2010
Revenue						
R&D Expense						
Operating Income	-118,005,000	-87,665,000	-34,099,000	-59,167,000	-64,527,000	-49,628,000
Operating Margin %						
SGA Expense	118,005,000	87,897,000	35,898,000	58,799,000	59,236,000	46,939,000
Net Income	250,349,000	589,205,000	362,686,000	327,767,000	137,329,000	532,851,000
Operating Cash Flow	396,302,000	182,777,000	304,820,000	334,915,000	447,705,000	305,582,000
Capital Expenditure						
EBITDA	509,715,000	736,992,000	464,687,000	454,326,000	272,793,000	685,843,000
Return on Assets %	1.45%	4.51%	4.94%	4.23%	1.73%	8.39%
Return on Equity %	7.63%	16.98%	10.55%	9.94%	4.08%	18.34%
Debt to Equity	1.44	1.41	0.48	0.70	0.87	1.10

CONTACT INFORMATION:
Phone: 646 454-3759   Fax:
Toll-Free:
Address: 1211 Avenue of the Americas, New York, NY 10036 United States

STOCK TICKER/OTHER:
Stock Ticker: CIM
Employees:
Parent Company:

Exchange: NYS
Fiscal Year Ends: 12/31

SALARIES/BONUSES:
Top Exec. Salary: $ Bonus: $
Second Exec. Salary: $ Bonus: $

OTHER THOUGHTS:
Estimated Female Officers or Directors:
Hot Spot for Advancement for Women/Minorities:

China Communications Construction Company Ltd en.cccltd.cn

NAIC Code: 237990

TYPES OF BUSINESS:

Port Facility Construction
Port Construction
Dredging
Container Cranes
Heavy Machinery Manufacturing

BRANDS/DIVISIONS/AFFILIATES:

China Harbor Engineering Co Ltd
China Road and Bridge Corporation
CCCC Investment Co
John Holland

CONTACTS: Note: Officers with more than one job title may be intentionally listed here more than once.

Qitao Liu, Managing Dir.
Fenjian Chen, Pres.
Junyuan Fu, CFO
ZiYu Sun, Chief Engineer
Min Lin, Chief Engineer
Xigang Zhang, Chief Engineer
Wensheng Liu, Sec.
Yun Chen, VP
Yusheng Chen, VP
Fenjian Chen, VP
Gixin Zhu, VP
Qitao Liu, Chmn.

GROWTH PLANS/SPECIAL FEATURES:

China Communications Construction Company Ltd. (CCCC) is one of China's largest port construction and design companies as well as one of the world's largest container crane manufacturers. The firm's principal activities include the design and construction of transportation infrastructure, dredging and heavy machinery manufacturing. CCCC's operations are separated into seven business units: infrastructure construction, infrastructure design, dredging, heavy machinery manufacturing, overseas business, investments and others. The infrastructure construction business, operating through 10 direct subsidiaries, offers construction and transportation services for ports, roads, bridges, railways and tunnel works. It currently has the only three top-tier qualification certificates granted by the Ministry of Construction for port construction work in China. The infrastructure design business, comprised of 10 subsidiaries, offers a range of design services, including consulting and planning, feasibility studies, design services, engineering consulting, engineering surveys and technical studies, project management and supervision and construction. The dredging business, one of the largest in the world, made up of three subsidiaries, is involved in major dredging and reclamation operations along the China coast and internationally. The heavy machinery manufacturing business, operating through three major subsidiaries, mainly supplies container cranes and bulk material handling machinery. The overseas segment operates through subsidiaries China Harbor Engineering Co., Ltd. and China Road and Bridge Corporation. The firm's investment activities are conducted through CCCC Investment Co. The other businesses, comprised of five subsidiaries, have completed projects in a variety of fields, including railway, road, bridge, machinery manufacturing, logistics services and the trading of construction materials and equipment. In 2015, the firm acquired Leighton Holding's contractor/builder subsidiary, John Holland.

FINANCIAL DATA: Note: Data for latest year may not have been available at press time.

In U.S. $	2015	2014	2013	2012	2011	2010
Revenue	62,498,330,000	56,664,950,000	51,381,840,000	45,778,390,000	45,645,970,000	42,277,190,000
R&D Expense						
Operating Income	2,862,219,000	2,623,836,000	2,354,953,000	2,302,855,000	2,012,834,000	1,775,501,000
Operating Margin %	4.57%	4.63%	4.58%	5.03%	4.40%	4.19%
SGA Expense	3,652,788,000	2,861,145,000	2,337,752,000	2,159,220,000	2,024,976,000	1,679,966,000
Net Income	2,425,672,000	2,146,146,000	1,875,913,000	1,846,774,000	1,792,847,000	1,458,385,000
Operating Cash Flow	4,931,471,000	680,589,600	1,077,159,000	2,061,229,000	264,592,700	2,392,788,000
Capital Expenditure	8,038,426,000	6,617,025,000	3,884,186,000	3,030,955,000	2,487,247,000	2,225,766,000
EBITDA	5,745,507,000	5,333,170,000	4,548,558,000	4,166,150,000	3,770,055,000	3,048,131,000
Return on Assets %	2.30%	2.41%	2.55%	3.01%	3.45%	3.25%
Return on Equity %	11.87%	13.06%	13.25%	15.04%	17.15%	15.37%
Debt to Equity	1.13	1.15	1.01	0.82	0.71	0.57

CONTACT INFORMATION:

Phone: 86 1082016655   Fax: 86 1082016500
Toll-Free:
Address: 85 Deshengmenwai St., Xicheng District, Beijing, 100088 China

STOCK TICKER/OTHER:

Stock Ticker: CCCGF
Employees: 103,357
Parent Company:

Exchange: PINX
Fiscal Year Ends: 12/31

SALARIES/BONUSES:

Top Exec. Salary: $ Bonus: $
Second Exec. Salary: $ Bonus: $

OTHER THOUGHTS:

Estimated Female Officers or Directors:
Hot Spot for Advancement for Women/Minorities:

Sales, profits and employees may be estimates. Financial information, benefits and other data can change quickly and may vary from those stated here.

China Lodging Group Ltd

www.htinns.com

NAIC Code: 721110

TYPES OF BUSINESS:

Hotels & Resorts

BRANDS/DIVISIONS/AFFILIATES:

HanTing Hotels
JI Hotel
HanTing Hi Inn
HuaZhu Club
Starway Hotel
Joya Hotel
HuaZhu Hotel Group
Elan Hotel

CONTACTS: Note: Officers with more than one job title may be intentionally listed here more than once.

Jenny Zhang, CEO
Yunhang Xie, COO
Hui Chen, CFO
Min (Jenny) Zhang, Chief Strategy Officer
Qi Ji, Exec. Chmn.

GROWTH PLANS/SPECIAL FEATURES:

China Lodging Group Ltd. is a China-based holding group active in China's hospitality sector. The firm conducts operations through HuaZhu Hotel Group, a leading chain of economy hotels with 2,763 hotels in 352 cities. China Lodging employs a lease-and-operate model, which is used to directly operate hotels in prime locations, and a franchise-and-manage model, which is used to expand network coverage. The firm's primary brands are HanTing Hotels, with 1,648 existing hotels and 374 under development; HanTing Hi Inn, with 158 hotels and 109 under development; JI Hotel, with 117 hotels and 76 under development; Starway Hotel, with 55 hotels, and an additional 45 leased; Elan Hotel, with 13 hotels and 65 under development; and three Joya Hotels with two under development. The economy hotels of HanTing Hotel and HanTing Hi Inn are targeted towards value-conscious travelers and workers or to appeal to the younger, budget conscious traveler. The midscale hotels of JI Hotel and Starway Hotel are geared more towards the older, experienced traveler and the middle class who want both a good value and a good location. The Joya Hotel will be the firm's upscale or premium hotel brand and is geared towards very affluent, elite travelers who seek a low-key yet elegant experience. The firm's HuaZhu Club rewards program has over 15 million members, who represent 80% of its room nights sold. The company hopes to harness rapidly growing levels of middle-class leisure travelers with increasing amounts of disposable income in China to help fuel its growth.

FINANCIAL DATA: Note: Data for latest year may not have been available at press time.

In U.S. $	2015	2014	2013	2012	2011	2010
Revenue		767,239,100	644,211,600	498,312,000	347,648,200	268,663,400
R&D Expense						
Operating Income		60,171,540	58,808,510	33,957,100	16,558,130	39,609,160
Operating Margin %		7.84%	9.12%	6.81%	4.76%	14.74%
SGA Expense		110,631,900	98,003,220	85,133,750	67,859,800	46,668,120
Net Income		47,496,950	43,248,700	27,026,690	17,745,910	33,341,800
Operating Cash Flow		224,700,600	165,381,800	110,605,900	70,892,770	72,497,820
Capital Expenditure		145,473,600	166,427,700	154,782,500	121,069,700	62,569,580
EBITDA		152,645,100	131,739,600	89,951,940	58,831,540	70,048,510
Return on Assets %		5.40%	5.88%	4.45%	3.49%	9.32%
Return on Equity %		10.18%	10.59%	7.42%	5.26%	14.34%
Debt to Equity						

CONTACT INFORMATION:

Phone: 86 2161959595   Fax:
Toll-Free:
Address: 2266 Hongqiao Rd., Changning District, Shanghai, 200336 China

STOCK TICKER/OTHER:

Stock Ticker: HTHT
Employees: 15,551
Parent Company:

Exchange: NAS
Fiscal Year Ends: 12/31

SALARIES/BONUSES:

Top Exec. Salary: $ Bonus: $
Second Exec. Salary: $ Bonus: $

OTHER THOUGHTS:

Estimated Female Officers or Directors: 3
Hot Spot for Advancement for Women/Minorities: Y

Sales, profits and employees may be estimates. Financial information, benefits and other data can change quickly and may vary from those stated here.

China Overseas Land & Investment Limited   www.coli.com.hk

NAIC Code: 531100

TYPES OF BUSINESS:
Real Estate Holdings & Development
Property Management Services
Construction Design Services

BRANDS/DIVISIONS/AFFILIATES:
China State Construction Engineering Corp
China Overseas Property Management
Hua Yi Designing Consultants Ltd

CONTACTS: Note: Officers with more than one job title may be intentionally listed here more than once.
Jian Min Hao, CEO
Yi Chen, Pres.
Yun Wing Nip, CFO
Liang Luo, VP
Xiao Xiao, Vice-Chmn.
Zhang Yi, VP
Dapeng Qi, VP
Jian Min Hao, Chmn.

GROWTH PLANS/SPECIAL FEATURES:
China Overseas Land & Investment Limited is a Hong Kong-based real estate investment holding firm and a publicly listed subsidiary of government-owned China State Construction Engineering Corp. The firm operates in three segments: property development, property investment and property related businesses. The property development business, the company's largest business segment, engages in the development and sale of property. The property investment business holds approximately 12.27 million square feet of investment properties and about 40 million square feet of property under development. The firm's portfolio includes properties in over 20 Chinese cities and regions, including Beijing, Shanghai, Guangzhou, Shenzhen, Chengdu, Changchun, Nanjing, Xian, Zhongshan, Foshan, Zhuhai, Suzhou, Ningbo, Chongqing, Hangzhou, Qingdao, Dalian, Shenyang, Tianjin, Jinan, Hong Kong and Macau. Some of the major properties held by the company include luxury apartment communities such as Mt. Riviera, a 1,071 unit housing and high-rise apartment complex in Hangzhou; Blossom Riverine, a 766,670 square-foot residential development in Foshan; Olympic City, a 1.2 million square-foot multistory apartment complex in Shenzhen; The Arch, a 293,800 square-foot luxury high-rise apartment and hotel property in Nanjing; International Community, a 4.8 million square-foot residential development in Xi'an; the Orchid Garden, a 1 million square-foot residential garden-style community in Chengdu; and One South Lake, a 293,400 square-foot upscale apartment complex in Changchun. The firm is also engaged in several property related business areas, including property management, provided by subsidiary China Overseas Property Management; construction design services, offered through subsidiary Hua Yi Designing Consultants Ltd.; and several long-term investments in infrastructure and provincial facilities projects.

FINANCIAL DATA: Note: Data for latest year may not have been available at press time.

In U.S. $	2015	2014	2013	2012	2011	2010
Revenue	18,066,970,000	15,471,140,000	10,632,690,000	8,326,353,000	6,263,776,000	5,713,252,000
R&D Expense						
Operating Income	5,918,835,000	5,468,413,000	3,654,834,000	3,490,163,000	3,015,446,000	2,668,009,000
Operating Margin %	32.76%	35.34%	34.37%	41.91%	48.14%	46.69%
SGA Expense	543,193,300	460,299,800	363,198,200	262,574,500	244,589,900	245,839,100
Net Income	4,294,909,000	3,568,788,000	2,971,013,000	2,413,846,000	1,937,215,000	1,595,263,000
Operating Cash Flow	4,943,668,000	-588,357,200	-1,300,444,000	908,941,500	-1,031,134,000	-328,891,800
Capital Expenditure	14,564,640	13,643,320	62,150,440	15,064,890	23,104,930	220,018,900
EBITDA	6,359,815,000	5,673,633,000	4,343,433,000	3,828,670,000	3,145,835,000	2,718,188,000
Return on Assets %	8.56%	8.55%	8.75%	9.22%	8.88%	8.95%
Return on Equity %	20.50%	22.75%	23.36%	23.76%	24.03%	25.55%
Debt to Equity	0.56	0.53	0.63	0.61	0.46	0.62

CONTACT INFORMATION:
Phone: 852-2823-7888   Fax: 852-2865-5939
Toll-Free:
Address: 1 Queen's Rd. E., 3 Pacific Place, 10th Fl., Hong Kong, Hong Kong

STOCK TICKER/OTHER:
Stock Ticker: CAOVF   Exchange: PINX
Employees: 25,705   Fiscal Year Ends: 12/31
Parent Company: China State Construction Engineering Corp

SALARIES/BONUSES:
Top Exec. Salary: $ Bonus: $
Second Exec. Salary: $ Bonus: $

OTHER THOUGHTS:
Estimated Female Officers or Directors: 3
Hot Spot for Advancement for Women/Minorities: Y

China Railway Construction Corporation Limited (CRCC)

www.crcc.cn

NAIC Code: 237990

TYPES OF BUSINESS:

Transportation Infrastructure Construction
Design and Engineering
Highway and Bridge Construction
Railway Construction

BRANDS/DIVISIONS/AFFILIATES:

Chongqing Tiefa Suiyu Highway

CONTACTS: Note: Officers with more than one job title may be intentionally listed here more than once.

Zhuang Shangbiao, Pres.
Wang Xiuming, CFO
Meng Fengchao, Chmn.

GROWTH PLANS/SPECIAL FEATURES:

China Railway Construction Corporation Limited (CRCC) was solely established by China Railway Construction Corporation in 2007 in Beijing, and is now a major construction corporation under the administration of the State-owned Assets Supervision and Administration Commission of the State Council of China (SASAC). CRCC, one of the world's largest integrated construction groups, is a major engineering contractor in China and ranks among the world's biggest construction firms in terms of revenue and number of employees. It covers contracting, design consultation, industrial manufacturing, real estate development, logistics, trade of goods and materials as well as capital operations. CRCC has developed mainly from construction management into a comprehensive group of transportation infrastructure research, planning, survey, design, construction, supervision, maintenance and operations. This enables the firm to provide its clients one-stop integrated services. CRCC has established its leadership position in project design and construction fields in plateau railways, high-speed railways, highways, bridges, tunnels and urban rail traffic. In 2015, the firm acquired a majority stake in Chongqing Tiefa Suiyu Highway. That same year, CCRC completed an 835-mile railway line in Angola, Africa, which commenced operations; and won a $600 million contract to upgrade the Egypt Track.

FINANCIAL DATA: Note: Data for latest year may not have been available at press time.

In U.S. $	2015	2014	2013	2012	2011	2010
Revenue	92,806,060,000	91,481,630,000	90,681,290,000	74,844,760,000	70,680,440,000	72,657,400,000
R&D Expense						
Operating Income	2,566,728,000	2,222,348,000	1,898,254,000	1,651,731,000	1,534,495,000	902,443,700
Operating Margin %	2.76%	2.42%	2.09%	2.20%	2.17%	1.24%
SGA Expense	4,101,327,000	4,032,955,000	3,898,848,000	3,500,628,000	3,337,607,000	3,197,427,000
Net Income	1,954,207,000	1,752,966,000	1,598,643,000	1,310,311,000	1,213,787,000	656,202,600
Operating Cash Flow	7,784,868,000	1,017,248,000	-1,439,364,000	856,908,700	-1,943,518,000	966,259,700
Capital Expenditure	4,194,797,000	3,289,717,000	2,726,874,000	1,622,696,000	1,862,745,000	2,629,376,000
EBITDA	5,290,930,000	4,983,847,000	4,482,734,000	3,928,376,000	3,492,737,000	2,331,266,000
Return on Assets %	1.92%	1.93%	2.00%	1.87%	2.03%	1.34%
Return on Equity %	12.48%	13.19%	13.52%	12.40%	12.85%	7.67%
Debt to Equity	0.70	0.86	0.88	0.43	0.47	0.39

CONTACT INFORMATION:

Phone: 86 1052688600   Fax: 86 1052688302
Toll-Free:
Address: No. 40 Fuxing Rd., Beijing, 100855 China

SALARIES/BONUSES:

Top Exec. Salary: $ Bonus: $
Second Exec. Salary: $ Bonus: $

STOCK TICKER/OTHER:

Stock Ticker: CWYCY
Employees: 249,624
Parent Company:

Exchange: PINX
Fiscal Year Ends: 12/31

OTHER THOUGHTS:

Estimated Female Officers or Directors:
Hot Spot for Advancement for Women/Minorities:

China State Construction Engineering Corp   www.cscec.com.cn

NAIC Code: 531100

TYPES OF BUSINESS:

Construction & Real Estate Development
Contract Engineering
Property Development
Infrastructure Design and Construction
Real Estate Investments

BRANDS/DIVISIONS/AFFILIATES:

China State Construction International Co
China Construction Development Co Ltd
China Construction Decoration Engineering Co
CSCEC Property Management Co
China Construction American Co
China Construction

CONTACTS: Note: Officers with more than one job title may be intentionally listed here more than once.

Guab Qing, Pres.
Zeng Zhaohe, CFO
Mao Zhibing, Chief Engineer
Zeng Zhaohe, General Counsel
Liu Jie, Leader-Discipline Inspection
Liu Jinzhang, VP
Kong Qingping, VP
Wang Xiangming, VP
Yi Jun, Chmn.

GROWTH PLANS/SPECIAL FEATURES:

China State Construction Engineering Corp. (CSCEC) is a Chinese state-owned enterprise that primarily engages in real estate, construction and contract engineering. The firm is China's leading international contracting agent and one of the largest house builders in the world, conducting business operations in over 20 countries. The company also conducts design and planning, property development, machinery leasing, project supervision, property management and trading activities. CSCEC has worked on office buildings, public facilities, airports, hotels, educational institutions, sports facilities, residential complexes, hospitals and military buildings. Its technologies are used for many applications, such as constructing high-rises, installing large industrial works, complex deep pit support and dewatering activities, concrete manufacturing, project management and general contracting of international projects. Specific projects include the Hada Express Way, Wuhan Railway Station, Shaanxi LanShang Expressway, the Tianjin Cihan Ferris Wheel, the Beijing Subway Line 4, Hongheyan Nuclear Station and China World Trade Center. The company has various domestic affiliated companies, such as China State Construction International Co., China Construction Development Co. Ltd., China Construction Decoration Engineering Co. and CSCEC Property Management Co. The most prominent subsidiary of CSCEC is China Construction, which manages all domestic affiliates. It was developed from joint agreements between CSCEC, China National Petroleum Corporation, Baosteel Group Corp. Ltd. and Sinochem Corporation. The firm also has various international affiliated companies such as China Construction (South Pacific) Development Co. Pte. Ltd. in Singapore and China Construction American Co. in Jersey City, New Jersey.

FINANCIAL DATA: Note: Data for latest year may not have been available at press time.

In U.S. $	2015	2014	2013	2012	2011	2010
Revenue		4,440,275,000	3,505,842,000	2,548,292,000	2,111,776,000	1,544,945,000
R&D Expense						
Operating Income		502,685,600	377,555,700	299,802,200	168,476,200	116,241,400
Operating Margin %		11.32%	10.76%	11.76%	7.97%	7.52%
SGA Expense		133,202,200	131,068,300	91,494,020	71,453,090	60,790,500
Net Income		445,762,000	357,411,100	274,807,800	194,348,900	133,606,800
Operating Cash Flow		-217,753,800	-420,143,300	-430,600,400	-129,509,000	5,289,073
Capital Expenditure		66,332,530	48,755,830	51,856,200	67,326,960	46,999,680
EBITDA		616,787,000	485,509,600	388,771,300	285,817,100	184,400,800
Return on Assets %		5.88%	6.29%	6.66%	6.63%	6.50%
Return on Equity %		19.20%	18.67%	18.53%	21.36%	23.22%
Debt to Equity		0.71	0.76	0.74	0.52	1.15

CONTACT INFORMATION:

Phone: 86-10-880-82888   Fax:
Toll-Free:
Address: 15 Sanlihe Rd., Haidan District, Wanchai, China

STOCK TICKER/OTHER:

Stock Ticker: CCOHY   Exchange: PINX
Employees: 10,781   Fiscal Year Ends: 12/31
Parent Company:

SALARIES/BONUSES:

Top Exec. Salary: $ Bonus: $
Second Exec. Salary: $ Bonus: $

OTHER THOUGHTS:

Estimated Female Officers or Directors:
Hot Spot for Advancement for Women/Minorities:

Chinese Estates Holdings Ltd

www.chineseestates.com

NAIC Code: 531100

TYPES OF BUSINESS:

Real Estate Holdings & Development
Commercial Property Development
Residential Property Development

BRANDS/DIVISIONS/AFFILIATES:

Kwong Sang Hong International Limited (The)
Power Jade Limited
Direct Win Development Limited

CONTACTS: Note: Officers with more than one job title may be intentionally listed here more than once.

Chan Sze-wan, CEO
Mun-yi (Connie) Cheung, Head-Legal Dept.
Veng-va (Matthew) Cheong, Gen. Mgr.-Oper.
Chi-ming (Alec) Kong, Head-China Bus.
Ming-yan (Hazel) Lai, Sr. Mgr.-Contracts, Project Dev.
Yik-hei (Kenneth) Ng, Sr. Mgr.-Project Dev. Dept.
Lau Ming-wai, Chmn.

GROWTH PLANS/SPECIAL FEATURES:

Chinese Estates Holdings Ltd. is a Hong Kong-based property development firm. Its core business is focused on investments in properties for rent and the development of properties for sale. Over the course of its 20-year history, the company's primary work has been in Hong Kong, though in recent years it has been expanding its activities in mainland China and Macau. The firm's property portfolio is comprised of approximately 1 million square feet of retail property and approximately 1.2 million square feet of office property, mostly situated in commercial areas of Hong Kong such as Causeway Bay, Tsim Sha Tsui and Wanchai. The company's Hong Kong rental properties include eight shopping centers, four industrial sites and three commercial properties, including the MassMutual Tower, where the firm has its headquarters. Hong Kong properties currently for sale include 17 residential complexes, primarily consisting of high-rise apartments and condominiums. On the Chinese mainland, Chinese Estates owns a mixed-use office/shopping complex property in Shanghai, commercial buildings and a five-star hotel in Beijing and part of a shopping arcade in Shenzhen. Mainland China properties for sale comprise two residential properties in Chengdu. The firm specializes in developing mid to high-end residential and commercial properties. It currently has a land bank totaling approximately 30 million square feet in various parts of Hong Kong, Macau and Mainland China. Chinese Estates has several partially-owned subsidiaries also engaged in property investment and development: The Kwong Sang Hong International Limited, of which it owns 50%; Power Jade Limited, of which it owns 50%; and Direct Win Development Limited, of which it owns 33.33%. In addition to its Hong Kong headquarters, the company maintains offices in Beijing, Chengdu, Shanghai, Shenzhen and Macau.

FINANCIAL DATA: Note: Data for latest year may not have been available at press time.

In U.S. $	2015	2014	2013	2012	2011	2010
Revenue		338,734,800	831,946,400	313,755,500	68,047,940	344,372,100
R&D Expense						
Operating Income		317,504,100	330,109,400	452,791,300	-40,610,330	-1,124,670,000
Operating Margin %		93.73%	39.67%	144.31%	-59.67%	-326.58%
SGA Expense		38,201,280	39,687,450	36,328,190	35,229,720	29,627,990
Net Income		1,127,479,000	814,542,300	1,262,856,000	450,971,500	-1,142,089,000
Operating Cash Flow		1,135,331,000	557,504,000	-70,428,350	129,360,200	-402,734,300
Capital Expenditure		31,053,950	1,551,408	3,461,498	200,590,100	118,348,900
EBITDA		1,213,544,000	947,063,800	1,350,753,000	562,482,000	-1,111,855,000
Return on Assets %		11.87%	8.40%	14.39%	6.00%	-15.47%
Return on Equity %		19.72%	13.63%	24.12%	10.11%	-24.55%
Debt to Equity		0.23	0.35	0.20	0.35	0.44

CONTACT INFORMATION:

Phone: 852-2866-6999   Fax: 852-2866-2833
Toll-Free:
Address: 38 Gloucester Rd., 26th Fl., Wanchai, Wanchai,   Hong Kong

STOCK TICKER/OTHER:

Stock Ticker: CESTY   Exchange: PINX
Employees: 624   Fiscal Year Ends: 12/31
Parent Company:

SALARIES/BONUSES:

Top Exec. Salary: $ Bonus: $
Second Exec. Salary: $ Bonus: $

OTHER THOUGHTS:

Estimated Female Officers or Directors: 2
Hot Spot for Advancement for Women/Minorities: Y

Chiyoda Corporation

NAIC Code: 541330

www.chiyoda-corp.com

TYPES OF BUSINESS:

Engineering & Construction Services
Plant Lifecycle Engineering
Computer-Aided Engineering
Risk Management
Pollution Prevention Systems
Industrial Equipment-Online Procurement

BRANDS/DIVISIONS/AFFILIATES:

Biofiner
EUREKA
Chiyoda International Corporation
Chiyoda U-Tech Co Ltd
Chiyoda System Technologies Corporation
Arrow Business Consulting
Chiyoda Oceania Pty Ltd
Chiyoda Corporation (Shanghai)

CONTACTS: Note: Officers with more than one job title may be intentionally listed here more than once.

Shogo Shibuya, CEO
Shogo Shibuya, Pres.
Masahito Kawashima, CFO
Ryosuke Shimizu, Sr. VP-Tech. & Eng.
Hiroshi Ogawa, Sr. Managing Exec. Officer-Project Oper.
Keiichi Nakagaki, Sr. Exec. VP-Corp. Planning, Mgmt. & Finance
Hiroshi Ogawa, Sr. Exec. VP-Global Project Mgmt.
Katsuo Nagasaka, Exec. VP-Bus. Dev.
Masahiko Kojima, Sr. VP-Corp. Planning, Mgmt. & Finance
Takashi Kubota, Chmn.

GROWTH PLANS/SPECIAL FEATURES:

Chiyoda Corporation is a Japanese engineering firm that operates in the hydrocarbon and chemical industries, with a focus on the refining, petrochemical, nuclear and green energy, gas processing, pharmaceutical and fine chemicals sectors. Much of the firm's activities involve engineering, procurement and construction (EPC) solutions for facilities that prevent disasters, improve the environment and control pollution. Chiyoda provides comprehensive engineering services in three phases: the planning phase, which involves master planning, feasibility studies, licensing and process development support services; the EPC phase, during which the firm provides FEED (Front End Engineering Design), equipment and systems design, site survey and commissioning services; and the operation and maintenance phase, in which Chiyoda offers upgrade and troubleshooting services, such as process and utility system improvement, equipment reliability improvement, environmental technology consulting and maintenance consulting. The company's EUREKA process provides a means of producing clean fuel from heavy residual materials. Chiyoda has also developed many environmental preservation and pollution prevention systems, such as Biofiner, denitration technology; and the CT-121 Flue Gas Desulfurization Process. Research and development efforts focus on creating greener technology and sustainability products, such as synthetic gas, clean coal technologies, carbon capture and storage technologies and transportation of hydrogen across long distances. The firm has number of affiliate companies and domestic and overseas subsidiaries including Chiyoda System Technologies Corporation; Chiyoda U-Tech Co., Ltd; Chiyoda Oceania Pty. Ltd. Chiyoda Corporation (Shanghai) and Arrow Business Consulting. Chiyoda's U.S. activities are overseen by Chiyoda International Corporation, headquartered in Houston, Texas.

FINANCIAL DATA: Note: Data for latest year may not have been available at press time.

In U.S. $	2015	2014	2013	2012	2011	2010
Revenue	4,384,534,000	4,067,011,000	3,636,478,000	2,321,580,000	2,252,363,000	2,853,125,000
R&D Expense						
Operating Income	195,680,900	192,153,100	228,926,400	220,576,300	159,928,500	15,515,180
Operating Margin %	4.46%	4.72%	6.29%	9.50%	7.10%	.54%
SGA Expense						
Net Income	102,207,000	120,420,400	149,418,000	132,316,600	72,443,690	26,919,110
Operating Cash Flow	-220,102,300	-156,583,000	128,962,000	506,978,200	-47,666,800	78,514,850
Capital Expenditure	35,296,580	48,086,120	64,923,110	27,329,330	14,977,350	16,991,950
EBITDA	248,908,400	244,231,900	269,218,500	240,540,000	130,338,500	64,011,520
Return on Assets %	2.22%	2.95%	4.01%	3.99%	2.34%	.86%
Return on Equity %	5.47%	6.98%	9.01%	8.88%	5.25%	2.00%
Debt to Equity	0.04	0.05	0.05		0.06	0.06

CONTACT INFORMATION:

Phone: 81 452257777   Fax:
Toll-Free:
Address: 4-6-2 minatomirai, Nishi-ku, Yokohama, 220-8765 Japan

STOCK TICKER/OTHER:

Stock Ticker: CHYCF   Exchange: PINX
Employees: 1,519   Fiscal Year Ends: 03/31
Parent Company:

SALARIES/BONUSES:

Top Exec. Salary: $ Bonus: $
Second Exec. Salary: $ Bonus: $

OTHER THOUGHTS:

Estimated Female Officers or Directors:
Hot Spot for Advancement for Women/Minorities:

Choice Hotels International Inc

www.choicehotels.com

NAIC Code: 721110

TYPES OF BUSINESS:

Hotels
Motels
Suites
Franchising

BRANDS/DIVISIONS/AFFILIATES:

Econo Lodge
MainStay Suites
Rodeway Inn
Quality Inn
Clarion Hotels
Comfort Inn
Sleep Inn
Comfort Suites

CONTACTS: *Note: Officers with more than one job title may be intentionally listed here more than once.*

David White, CFO
Stewart Bainum, Chairman of the Board
Scott Oaksmith, Chief Accounting Officer
Patrick Pacious, COO
Simone Wu, General Counsel
Stephen Joyce, President
Patrick Cimerola, Senior VP, Divisional
David Pepper, Senior VP, Divisional

GROWTH PLANS/SPECIAL FEATURES:

Choice Hotels International, Inc. is one of the world's largest franchisers of hotel properties. It has 6,300 hotels open and 638 hotels under development in 50 states, Washington, D.C. and over 35 foreign countries and territories around the globe, with over 500,000 rooms worldwide. The firm's 11 proprietary brand names include Comfort Inn, Comfort Suites, Sleep Inn, Quality Inn, Clarion Hotels, Econo Lodge, Rodeway Inn, MainStay Suites, Suburban Extended Stay Hotel, Cambria Suites and Ascend Collection. Choice Hotel's business is based on franchise revenues that consist of initial fees and ongoing royalty fees. The company also collects marketing and reservation fees to support centralized activities. The firm's largest brand is Comfort Inn, which provides mid-scale rooms without food and beverage service, targeted primarily to business and leisure travelers. The Comfort Inn, Comfort Suites and Sleep Inn brands compete in the limited-service midscale without food and beverage market; the Quality Inn and Clarion Hotel brands compete primarily in the full-service midscale with food and beverage market; the Econo Lodge and Rodeway Inn brands compete in the limited-service economy market; the MainStay brand and the Extended Stay Hotel brands compete in the extended stay market; the Ascend Collection represents individual properties that are historic, boutique and/or unique and desire to retain their independent brand identity but have access to Choice's marketing and distribution channels. To support its hotel operations, Choice Hotels maintains call centers, proprietary web sites and global distribution systems to help deliver customers to franchisees through multiple channels.

The firm offers employees a 401(k) plan; an employee stock purchase plan; medical, dental, vision and prescription coverage; life and AD&D insurance; flexible spending accounts; long-term care coverage; an employee assistance program; paid leave; tuition reimbursement; adoption assistance; legal services; hotel discounts; employee banking; and pet insurance.

FINANCIAL DATA: *Note: Data for latest year may not have been available at press time.*

In U.S. $	2015	2014	2013	2012	2011	2010
Revenue	859,878,000	757,970,000	724,307,000	691,509,000	638,793,000	596,076,000
R&D Expense						
Operating Income	225,319,000	214,568,000	194,494,000	193,142,000	171,863,000	160,762,000
Operating Margin %	26.20%	28.30%	26.85%	27.93%	26.90%	26.97%
SGA Expense	134,254,000	121,418,000	113,567,000	101,852,000	106,404,000	94,540,000
Net Income	128,029,000	123,160,000	112,601,000	120,687,000	110,396,000	107,441,000
Operating Cash Flow	159,872,000	183,891,000	152,040,000	161,020,000	134,844,000	144,935,000
Capital Expenditure	27,765,000	20,946,000	31,524,000	15,443,000	10,924,000	24,368,000
EBITDA	238,360,000	224,609,000	208,924,000	204,583,000	179,020,000	173,233,000
Return on Assets %	18.63%	20.74%	21.43%	25.18%	25.69%	28.58%
Return on Equity %						
Debt to Equity						

CONTACT INFORMATION:

Phone: 301 592-5000   Fax: 301 592-6269
Toll-Free:
Address: 1 Choice Hotels Cir., Ste. 400, Rockville, MD 20850 United States

STOCK TICKER/OTHER:

Stock Ticker: CHH
Employees: 1,331
Parent Company:

Exchange: NYS
Fiscal Year Ends: 12/31

SALARIES/BONUSES:

Top Exec. Salary: $998,462   Bonus: $249,970
Second Exec. Salary: $549,142   Bonus: $67,498

OTHER THOUGHTS:

Estimated Female Officers or Directors: 2
Hot Spot for Advancement for Women/Minorities: Y

Christies International Real Estate   www.christiesrealestate.com
NAIC Code: 531210

TYPES OF BUSINESS:
Real Estate Brokerage
Luxury Home Sales

BRANDS/DIVISIONS/AFFILIATES:
Christie's International plc
Christie's International Real Estate Magazine

CONTACTS: Note: Officers with more than one job title may be intentionally listed here more than once.
Don Conn, CEO
Bill Hamm, COO
David Branch, Sr. VP-Dev. Project Mktg.
Joachim Wrang-Widen, Dir.-EMEA
Zackary Wright, Sr. VP-Americas
Kathleen Coumont, Sr. VP-Americas
Rick Moeser, Sr. VP
Clayton Andrews, Regional Mgr.-Asia Pacific & Central Region

GROWTH PLANS/SPECIAL FEATURES:
Christie's International Real Estate, a wholly-owned subsidiary of Christie's International plc, was formed when Christie's Auction House acquired Great Estates, Inc. It is the largest international network of real estate companies that are dedicated to the marketing and selling of high-end properties in the prime and super-prime market categories. The company brings buyers and sellers of luxury real estate together throughout the world. Through an exclusive system of advertising, marketing and listing tools, the company provides access to a worldwide audience. Properties are showcased in Christie's International Real Estate Magazine, which is published four times a year; in custom-designed property brochures distributed worldwide; on its web site; and in Christie's Magazine and other highly regarded international publications. The firm's magazine is also available worldwide in all of the firm's offices, at newsstands and bookstores in affluent communities and many luxury hotels and inns. Christie's real estate clients include some of the highest-salaried executives and heads of major international corporations, celebrities, sports figures and other high-net-worth individuals.

FINANCIAL DATA: Note: Data for latest year may not have been available at press time.

In U.S. $	2015	2014	2013	2012	2011	2010
Revenue	118,000,000					
R&D Expense						
Operating Income						
Operating Margin %						
SGA Expense						
Net Income						
Operating Cash Flow						
Capital Expenditure						
EBITDA						
Return on Assets %						
Return on Equity %						
Debt to Equity						

CONTACT INFORMATION:
Phone: 212-468-7182   Fax: 212-468-7141
Toll-Free:
Address: 20 Rockefeller Plaza, New York, NY 10020 United States

SALARIES/BONUSES:
Top Exec. Salary: $ Bonus: $
Second Exec. Salary: $ Bonus: $

STOCK TICKER/OTHER:
Stock Ticker: Subsidiary   Exchange:
Employees: 32,000   Fiscal Year Ends: 12/31
Parent Company: Christies International plc

OTHER THOUGHTS:
Estimated Female Officers or Directors: 2
Hot Spot for Advancement for Women/Minorities: Y

Sales, profits and employees may be estimates. Financial information, benefits and other data can change quickly and may vary from those stated here.

CitiMortgage Inc

www.citimortgage.com

NAIC Code: 522310

TYPES OF BUSINESS:

Mortgages
Refinancing
Home Equity Loans

BRANDS/DIVISIONS/AFFILIATES:

Citigroup Inc
Home Affordable Unemployment Program
Home Affordable Modification Program
Morgan Stanley Smith Barney

CONTACTS: Note: Officers with more than one job title may be intentionally listed here more than once.

Michael L. Corbat, CEO
Paul Ince, CFO
Fred Bolstad, Managing Dir.-Wholesale Lending
Daniel P. Hoffman, Sr. VP
Jeffrey R. McGuiness, Managing Dir.-Correspondent Lending Division

GROWTH PLANS/SPECIAL FEATURES:

CitiMortgage, Inc., the mortgage lending unit of Citigroup, Inc., is one of the nation's leading residential mortgage originators. The company offers products to help first-time homebuyers as well as those interested in building a new home, refinancing or salvaging the equity built up in an existing home. CitiMortgage allows customers to apply online, by phone or in person at various locations throughout the country. Its web site provides several tools that help customers choose the right mortgage product, determine how much they can realistically expect to receive, determine what their rate will be and compare home equity versus refinancing and renting versus buying. The firm receives customer referrals from Citigroup and its subsidiaries, which have roughly 200 million customers. CitiMortgage allows the customers of its sister subsidiary, Morgan Stanley Smith Barney, a global private wealth management and equity research unit, to use its accounts as collateral for down payments. Since the mortgage crisis, Citigroup has implemented several assistance programs to aid struggling homeowners, including the Home Affordable Unemployment Program, which lowers monthly mortgage payments to an average of $500 for three months; and the Home Affordable Modification Program, which allows borrowers to modify their existing mortgage in the face of financial hardship.

FINANCIAL DATA: Note: Data for latest year may not have been available at press time.

In U.S. $	2015	2014	2013	2012	2011	2010
Revenue						
R&D Expense						
Operating Income						
Operating Margin %						
SGA Expense						
Net Income						
Operating Cash Flow						
Capital Expenditure						
EBITDA						
Return on Assets %						
Return on Equity %						
Debt to Equity						

CONTACT INFORMATION:

Phone: 636-261-2484   Fax: 636-261-2387
Toll-Free: 800-667-8424
Address: 1000 Technology Dr., O'Fallon, MO 63368 United States

STOCK TICKER/OTHER:

Stock Ticker: Subsidiary   Exchange:
Employees: 12,000   Fiscal Year Ends: 12/31
Parent Company: CITIGROUP INC

SALARIES/BONUSES:

Top Exec. Salary: $ Bonus: $
Second Exec. Salary: $ Bonus: $

OTHER THOUGHTS:

Estimated Female Officers or Directors: 1
Hot Spot for Advancement for Women/Minorities:

Clark Construction Group LLC

www.clarkconstruction.com

NAIC Code: 236220

TYPES OF BUSINESS:

Construction & Engineering Services

BRANDS/DIVISIONS/AFFILIATES:

Atkinson Construction
Atkinson Power
Clark Civil
Clark Concrete
Clark Foundations
Edgemoor Infrastructure & Real Estate
CFSG Energy & Structured Finance
S2N Technology Group

CONTACTS: Note: Officers with more than one job title may be intentionally listed here more than once.

Robby D. Moser, Jr., CEO
Harold K. Roach, Jr., COO
Timothy R. Yost, CFO

GROWTH PLANS/SPECIAL FEATURES:

Clark Construction Group LLC is a building and civil construction firm located in the U.S. Clark's portfolio features projects of all sizes and levels of complexity, from intricate interior renovations to some of the most complex civil operations in the country. The firm represents the aviation, education, government, health care, hospitality, office and science markets in its completed portfolio. Specialized services offered by Clark include Atkinson Construction, a civil contractor known for its bridge and dam work; Atkinson Power, delivering energy infrastructure; Clark Civil, offering expertise in heavy civil construction; Clark Concrete, performing structural concrete work on some of the most technically-sophisticated projects; Clark Foundations, providing sophisticated excavation support systems; Edgemoor Infrastructure & Real Estate, performing public-private delivery and turnkey development solutions; CFSG Energy & Structured Finance, helping clients achieve energy independence and lower annual operating costs via turnkey alternative energy solutions; S2N Technology Group, offering design, installation and support for IT services for new construction and renovations; and Shirley Contracting Company LLC, performing infrastructure work. In January 2016, the firm was sold to its management team, following the death of billionaire owner and patriarch, A. James Clark.

FINANCIAL DATA: Note: Data for latest year may not have been available at press time.

In U.S. $	2015	2014	2013	2012	2011	2010
Revenue	4,151,000,000	4,300,000,000	4,264,500,000			
R&D Expense						
Operating Income						
Operating Margin %						
SGA Expense						
Net Income						
Operating Cash Flow						
Capital Expenditure						
EBITDA						
Return on Assets %						
Return on Equity %						
Debt to Equity						

CONTACT INFORMATION:

Phone: 301-272-8100   Fax:
Toll-Free:
Address: 7500 Old Georgetown Rd., Bethesda, MD 20814 United States

STOCK TICKER/OTHER:

Stock Ticker: Private   Exchange:
Employees: 4,200   Fiscal Year Ends:
Parent Company:

SALARIES/BONUSES:

Top Exec. Salary: $ Bonus: $
Second Exec. Salary: $ Bonus: $

OTHER THOUGHTS:

Estimated Female Officers or Directors:
Hot Spot for Advancement for Women/Minorities:

Clayton Homes Inc

www.claytonhomes.com

NAIC Code: 321992

TYPES OF BUSINESS:

Construction Services
Manufactured Housing
Insurance & Financing

BRANDS/DIVISIONS/AFFILIATES:

Berkshire Hathaway Inc
Vanderbilt Mortgage and Finance Inc
Schult
Clayton
Crest Homes
Marlette
Chafin Communities
G&I Homes

CONTACTS: *Note: Officers with more than one job title may be intentionally listed here more than once.*

Kevin T. Clayton, CEO
Kevin T. Clayton, Pres.
Richard D. Strachan, Pres., Mfg.
Leon Van Tonder, Dir.-Oper.

GROWTH PLANS/SPECIAL FEATURES:

Clayton Homes, Inc., a subsidiary of Berkshire Hathaway, Inc., produces, sells, finances and insures modular and manufactured homes, in addition to commercial and educational relocatable buildings. The company's 35 manufacturing plants produce homes that are marketed in 49 states through over 1,300 independent retailers and approximately 448 company-owned sales centers. Clayton's factory-built manufactured homes are completely finished dwellings that are constructed under federal code in factories and then transported by truck to its targeted location. The homes are designed to be permanent, owner-occupied residential sites with attached utilities. The firm manufactures a variety of single- and multi-sectional homes from 1,000 to 2,000 square feet and larger under brand names such as Clayton, Schult, Buccaneer, Cavalier, Crest Homes, Marlette, Karsten, Norris, Golden West, Giles and SEhomes. Standard features offered in Clayton homes include central heating, flooring systems and wall and floor treatments. Customers can choose predesigned homes or custom-design a home by size, number of bedrooms and other features. Through financial subsidiary Vanderbilt Mortgage and Finance, Inc. (VMF), the firm offers financing to manufactured home customers as well as customers purchasing homes from certain third parties. VMF's financing products include manufactured home loans, Federal Housing Authority (FHA) loans, Land Home financing and more. Clayton and its subsidiaries provide financing to 350,000 customers and insurance to 160,000 customers. Additionally, Clayton acts as a reinsurance agent for physical damage, family protection and homebuyer protection insurance and other policies issued by insurance companies in connection with the firm's homes. In 2015, the firm acquired Georgia homebuilder Chafin Communities. In February 2016, it acquired New York-based G&I Homes.

Clayton offers its employees a 401(k) plan; medical, dental, vision and life insurance; training programs; and tuition and fitness reimbursement.

FINANCIAL DATA: *Note: Data for latest year may not have been available at press time.*

In U.S. $	2015	2014	2013	2012	2011	2010
Revenue						
R&D Expense						
Operating Income						
Operating Margin %						
SGA Expense						
Net Income						
Operating Cash Flow						
Capital Expenditure						
EBITDA						
Return on Assets %						
Return on Equity %						
Debt to Equity						

CONTACT INFORMATION:

Phone: 865-380-3000   Fax: 865-380-3742
Toll-Free: 800-822-0633
Address: 500 Clayton Rd., Maryville, TN 37804 United States

STOCK TICKER/OTHER:

Stock Ticker: Subsidiary
Employees: 13,164
Parent Company: Berkshire Hathaway Inc

Exchange:
Fiscal Year Ends: 06/30

SALARIES/BONUSES:

Top Exec. Salary: $ Bonus: $
Second Exec. Salary: $ Bonus: $

OTHER THOUGHTS:

Estimated Female Officers or Directors:
Hot Spot for Advancement for Women/Minorities:

Club Mediterranee SA (Club Med)   www.clubmed-corporate.com

NAIC Code: 721110

TYPES OF BUSINESS:

Resort Hotels
Tour Marketing/Packaging
Cruises
Seminars, Conventions & Event Planning
Licensed Apparel Products

BRANDS/DIVISIONS/AFFILIATES:

Club Med
Club Med Decouverte
Club Med 2
Club Med Villas & Chalets
Fosun International Ltd

CONTACTS: Note: Officers with more than one job title may be intentionally listed here more than once.

Henri G. dEstaing, CEO
Michel Wolfovski, CFO
Sylvain Rabuel, VP-Sales & Mktg., France, Benelux & Switzerland
Sylvie Brisson, VP-Human Resources
Patrick Calvet, VP-Villages Europe-Africa
Laure Baume, CEO-New Markets Europe-Africa
Olivier Horps, VP-Greater China
Xavier Mufraggi, VP-North America
Henri G. dEstaing, Chmn.
Janyck Daudet, VP-Latin America

GROWTH PLANS/SPECIAL FEATURES:

Club Mediterranee SA (Club Med) is an international operator of resort hotels and tours, with operations in over 40 countries worldwide. Club Med is known for being the originator of all-inclusive vacations packages, which group fees for every meal, drink, activity and amenity into one price. The core business of the company is the operation of vacation villages and resorts organized like small towns. These villages and villas include streets with shops and houses and feature a large number of luxury accommodations in hand-picked locations. The company operates approximately 70 properties, concentrated in equatorial locations, including the Caribbean, Southeast Asia and the Mediterranean, as well as several ski resorts. In addition, the firm offers Club Med Decouverte, a tour guide service that takes small groups to places such as Phuket and Cancun Yucatan, as well as weekend getaways to Venice, Istanbul and other well-known cities. It also operates Club Med 2, a luxury sailing ship in the Mediterranean. Club Med organizes corporate events, weddings and other events at its villages and other venues. Besides vacation packages, the company sells licensed products such as sportswear and children's clothing. Through Club Med Villas & Chalets, it offers freehold real estate at its villas on the island of Mauritius and its chalet apartments in the French Alps. In February 2015, Chinese conglomerate Fosun International, Ltd. gained a controlling interest in Club Med. The firm plans to open seven new resorts between 2015 and 2018.

FINANCIAL DATA: Note: Data for latest year may not have been available at press time.

In U.S. $	2015	2014	2013	2012	2011	2010
Revenue	1,450,000,000	1,489,601,792	1,513,246,208	1,555,161,472	1,514,320,896	1,449,836,032
R&D Expense						
Operating Income						
Operating Margin %						
SGA Expense						
Net Income		-12,896,985	-11,822,236	1,074,749	2,149,498	-15,046,483
Operating Cash Flow						
Capital Expenditure						
EBITDA						
Return on Assets %						
Return on Equity %						
Debt to Equity						

CONTACT INFORMATION:

Phone: 33 153353553   Fax: 33 153353616
Toll-Free:
Address: 11 Rue de Cambrai, Paris, 75957 France

STOCK TICKER/OTHER:

Stock Ticker: Private   Exchange:
Employees: 12,811   Fiscal Year Ends: 10/31
Parent Company: Fosun International Ltd

SALARIES/BONUSES:

Top Exec. Salary: $ Bonus: $
Second Exec. Salary: $ Bonus: $

OTHER THOUGHTS:

Estimated Female Officers or Directors: 6
Hot Spot for Advancement for Women/Minorities: Y

CNL Lifestyle Properties Inc
www.cnllifestylereit.com/index.stml

NAIC Code: 0

TYPES OF BUSINESS:
Real Estate Investment Trust
Retail, Hotel & Entertainment Properties
Commercial Financing
Investment Advisory Services
Real Estate Development & Management
Health Care & Assisted Living Properties
Industrial Properties

BRANDS/DIVISIONS/AFFILIATES:

GROWTH PLANS/SPECIAL FEATURES:
CNL Lifestyle Properties (CNL) is a real estate investment trust (REITs) that invests in income-producing properties with a focus on lifestyle-related industries. It acquires properties and leases them on a long-term, triple-net lease basis or engages qualified third-party managers to operate properties on its behalf. CNL invests in properties with the potential for long-term revenue generation. The company's current portfolio consists of 49 properties. CNL also monitors lifestyle trends driven by demographics, develops relationships with industry leaders and invests in income-producing properties. These properties are diversified by geography, operator and lifestyle sector. CNL's investments include properties such as ski and mountain, attraction sites, senior housing and marinas.

Employees of the firm receive a 401(k) plan; educational assistance; adoption assistance; flexible spending accounts; and medical, prescription, hospitalization, dental, vision, disability and life insurance.

CONTACTS: Note: Officers with more than one job title may be intentionally listed here more than once.
Thomas K. Sittema, CEO
Tracy G. Schmidt, CFO
Lisa A. Schultz, Chief Svcs. Officer
Holly Greer, General Counsel
Lisa A. Schultz, Chief Comm. Officer
Andy Hyltin, Pres., Fund Mgmt.
Jeffrey R. Shafer, Pres., CNL Securities Corp.
Paul Ellis, Pres., CNL Commercial Real Estate
Stephen H. Mauldin, Pres., Fund Mgmt.
James M. Seneff, Jr., Chmn.

FINANCIAL DATA: Note: Data for latest year may not have been available at press time.

In U.S. $	2015	2014	2013	2012	2011	2010
Revenue	337,665,000	373,295,000	512,801,000	481,279,000	420,071,000	300,023,000
R&D Expense						
Operating Income	-109,691,000	-10,378,000	-229,843,000	-12,795,000	4,132,000	-58,491,000
Operating Margin %	-32.48%	-2.78%	-44.82%	-2.65%	.98%	-19.49%
SGA Expense	25,976,000	27,298,000	33,947,000	33,200,000	31,735,000	26,294,000
Net Income	141,155,000	-92,144,000	-252,539,000	-76,073,000	-69,610,000	-81,889,000
Operating Cash Flow	62,643,000	126,934,000	135,480,000	76,726,000	83,064,000	79,776,000
Capital Expenditure	49,272,000	207,505,000	315,015,000	259,825,000	192,037,000	140,530,000
EBITDA	46,896,000	132,195,000	-292,000	138,034,000	135,591,000	104,611,000
Return on Assets %	8.51%	-3.69%	-8.95%	-2.60%	-2.50%	-3.06%
Return on Equity %	14.37%	-7.29%	-16.39%	-4.22%	-3.64%	-4.24%
Debt to Equity	0.23	0.74	0.84	0.60	0.48	0.31

CONTACT INFORMATION:
Phone: 866-650-0650   Fax: 877-694-1116
Toll-Free: 800-522-3863
Address: 450 S. Orange Ave., Orlando, FL 32801-3336 United States

STOCK TICKER/OTHER:
Stock Ticker: CLLY
Employees:
Parent Company:

Exchange: GREY
Fiscal Year Ends: 12/31

SALARIES/BONUSES:
Top Exec. Salary: $ Bonus: $
Second Exec. Salary: $ Bonus: $

OTHER THOUGHTS:
Estimated Female Officers or Directors: 3
Hot Spot for Advancement for Women/Minorities: Y

Coldwell Banker Real Estate LLC
www.coldwellbanker.com

NAIC Code: 531210

TYPES OF BUSINESS:
Real Estate Brokerage
Residential & Resort Brokerage
Commercial Brokerage
Mortgage Services
Online Information & Lending

BRANDS/DIVISIONS/AFFILIATES:
Realogy Corporation
Coldwell Banker Previews International
Coldwell Banker Concierge Service
Coldwell Banker Mortgage
Coldwell Banker Commercial
CBWorks.ColdwellBanker.com
CBNet
Coldwell Banker Mobile

CONTACTS:
Note: Officers with more than one job title may be intentionally listed here more than once.

Budge Huskey, CEO
Michael Fischer, COO
Budge Huskey, Pres.
Sean Blankenship, CMO
Steve Bright, Sr. VP
Nelson Bennett, Sr. VP
Frank Lindsey, Sr. VP
Anthony Puthon, Pres., Canadian Oper.

GROWTH PLANS/SPECIAL FEATURES:
Coldwell Banker Real Estate LLC, a subsidiary of Realogy Corporation, offers a full array of commercial and residential real estate services, including buying, selling, leasing, lending, refinancing and a variety of support services. The Coldwell Banker website provides resources for buyers and sellers of homes, such as local real estate office information, agent locators, general information for first time buyers/sellers and a property locator. The website's Home File feature allows customers to keep track of properties they are interested in by creating a personal online account to which they can save home and real estate agent information as well as receive alerts when new properties are listed that meet their search criteria. The firm's Coldwell Banker Concierge Service provides customers with information on local service providers specializing in everything from car and truck rental to home maintenance. Coldwell's sales associates act as local market experts and can help clients with everything from selling, buying and financing to moving into a new home. The company carries out its operations through its licensed trademark brand, Coldwell Banker Commercial. Coldwell Banker Mortgage is a leading telephone- and web-based lender in the U.S. Coldwell Banker Previews International markets and sells luxury homes in the Americas, Europe, Africa, the Middle East, Asia and Australia. The firm has approximately 3,000 franchised residential and commercial real estate offices in 49 countries. Coldwell provides its brokers and sales associates with exclusive access to its Internet portal, CBNet, which has the most up-to-date information on all of the company's products and services, its listings, selling and prospecting and networking tools. Coldwell also maintains CBWorks.ColdwellBanker.com, an education site with a wide variety of classes and continuing education programs that can be taken in person, by phone or via the Internet. In addition, the firm offers Coldwell Banker Mobile, a mobile app for Apple and Android platforms.

FINANCIAL DATA:
Note: Data for latest year may not have been available at press time.

In U.S. $	2015	2014	2013	2012	2011	2010
Revenue						
R&D Expense						
Operating Income						
Operating Margin %						
SGA Expense						
Net Income						
Operating Cash Flow						
Capital Expenditure						
EBITDA						
Return on Assets %						
Return on Equity %						
Debt to Equity						

CONTACT INFORMATION:
Phone: 973-407-2000   Fax: 973-496-7217
Toll-Free: 888-308-6558
Address: 1 Campus Dr., Parsippany, NJ 07054 United States

STOCK TICKER/OTHER:
Stock Ticker: Subsidiary   Exchange:
Employees: Fiscal Year Ends: 12/31
Parent Company: Realogy Corporation

SALARIES/BONUSES:
Top Exec. Salary: $ Bonus: $
Second Exec. Salary: $ Bonus: $

OTHER THOUGHTS:
Estimated Female Officers or Directors:
Hot Spot for Advancement for Women/Minorities:

Colliers International Group Inc

www.colliers.com

NAIC Code: 531210

TYPES OF BUSINESS:

Real Estate Agents
Facilities Management
Commercial Sales
Consulting Services
Valuation & Advisory Services
Portfolio Management
Development & Project Management
Debt, Mezzanine & Equity Capital Financing

BRANDS/DIVISIONS/AFFILIATES:

FirstService Corporation
Pointe Group Advisors
Summit Realty Group

CONTACTS: Note: Officers with more than one job title may be intentionally listed here more than once.

Jay Hennick, CEO
John Friedrichsen, CFO
Peter Cohen, Chairman of the Board
Douglas Cooke, Controller
D. Patterson, COO
Elias Mulamoottil, Senior VP, Divisional
Jeremy Rakusin, Vice President, Divisional
Neil Chander, Vice President, Divisional
Christian Mayer, Vice President, Divisional

GROWTH PLANS/SPECIAL FEATURES:

Colliers International Group, Inc. is a leading commercial estate services provider with 554 offices in 66 countries worldwide. The firm's core services include corporate solutions, investment services, project management, property marketing, real estate management services, valuation & advisory services, brokerage and research services. Colliers' specialized services include asset resolution, such as leasing, disposition, workout strategies and valuation and appraisal of distresses properties; automotive retail services, which includes important retail insights including traffic count, competitive analysis and retail design for automotive dealerships; education services, which consists of advisory and facilities management for academic institutions; energy group, which offers consultancy services ranging from office space, manufacturing space to tank farms; food advisory services, which includes facility and site selection, leasing and build-to-suit projects; logistics and transportation solutions, which include solutions for shipping and transportation in every major market, intermodal gateway and port hub city; and technology solutions, which offers cost evaluation, cloud services and access to data centers. In addition, the firm's specialized services includes, healthcare services, hotels, investment services, law firm services, private equity solutions, retail services, student and senior housing, multifamily, life sciences and land advisory. In June 2015, the firm was spun-off from FirstService Corporation and began trading as an independent public company. Later that year, Colliers acquired Pointe Group Advisors, a commercial real estate services firm and Summit Realty Group, based in Indianapolis, Indiana.

FINANCIAL DATA: Note: Data for latest year may not have been available at press time.

In U.S. $	2015	2014	2013	2012	2011	2010
Revenue	1,721,986,000	2,714,273,000	2,343,634,000	2,305,537,000	2,224,171,000	1,986,271,000
R&D Expense						
Operating Income	80,384,000	134,427,000	90,234,000	78,397,000	98,061,000	97,532,000
Operating Margin %	4.66%	4.95%	3.85%	3.40%	4.40%	4.91%
SGA Expense	502,480,000	758,436,000	661,025,000	639,278,000	634,321,000	620,401,000
Net Income	23,347,000	43,316,000	-18,039,000	5,850,000	74,110,000	13,564,000
Operating Cash Flow	157,238,000	159,068,000	116,277,000	102,991,000	80,214,000	115,051,000
Capital Expenditure	22,515,000	52,506,000	34,824,000	44,395,000	37,400,000	32,460,000
EBITDA	119,008,000	198,816,000	168,163,000	135,355,000	143,837,000	143,361,000
Return on Assets %	1.70%	2.81%	-1.53%	-.29%	5.42%	.32%
Return on Equity %	12.72%	18.54%	-12.06%	-3.53%	81.18%	9.03%
Debt to Equity	1.81	2.02	1.35	3.43	1.71	5.06

CONTACT INFORMATION:

Phone: 416-960-9500   Fax:
Toll-Free:
Address: 1140 Bay St., Ste. 4000, Toronto, Ontario M5S 2B4 Canada

SALARIES/BONUSES:

Top Exec. Salary: $1,584,000   Bonus: $
Second Exec. Salary: $561,400   Bonus: $

STOCK TICKER/OTHER:

Stock Ticker: CIG   Exchange: TSE
Employees: 10,035   Fiscal Year Ends: 12/31
Parent Company:

OTHER THOUGHTS:

Estimated Female Officers or Directors: 2

Hot Spot for Advancement for Women/Minorities: Y

Sales, profits and employees may be estimates. Financial information, benefits and other data can change quickly and may vary from those stated here.

Colony Capital Inc

NAIC Code: 525990

www.colonyinc.com

TYPES OF BUSINESS:

Mortgage REIT
real estate investment

BRANDS/DIVISIONS/AFFILIATES:

Colony Mortgage Capital LLC
Colony American Finance
Colony American Homes
Colony Capital Industrial Platform (The)
Colony Realty Partners
Colony Financial Inc

CONTACTS: Note: Officers with more than one job title may be intentionally listed here more than once.

Richard Saltzman, CEO
Darren Tangen, CFO
Thomas Barrack, Chairman of the Board
Neale Redington, Chief Accounting Officer
Jonathan Grunzweig, Other Corporate Officer
Mark Hedstrom, Other Corporate Officer
Ronald Sanders, Other Corporate Officer
Kevin Traenkle, Other Executive Officer

GROWTH PLANS/SPECIAL FEATURES:

Colony Capital, Inc., formerly known as Colony Financial, Inc. is a real estate and investment management company that mainly participates in equity or debt real estate investments. The company targets attractive risk-adjusted investment returns across a broad range of predominantly real estate equity and debt oriented investment strategies. The company's operational segments include The Colony Capital Distressed Debt Platform, which acquires and resolves portfolios of performing and non-performing loans sourced from governmental agencies, financial institutions, and insurance companies. The firm also originates mortgage loans and provides rescue capital. This segment mainly operates in Europe and in select distressed investment opportunities in U.S. Colony Mortgage Capital, LLC is Colony Capital's lending platform focused on originating middle market lending opportunities secured by transitional assets. Colony American Finance provides low cost, non-recourse revolving bridge and permanent mortgage financing for owners of single-family rental portfolios. Colony American Homes (CAH) deals in large-scale acquisitions of single-family homes and executes a residential rental strategy across the US. The Colony Capital Industrial Platform invests in light industrial properties in major metropolitan markets throughout the U.S. The property area is generally 250,000 sq. ft. in size or smaller. The Colony Realty Partners acquires and manages commercial real estate properties on behalf of world's leading institutional investors, including pension funds, endowments, family trusts and sovereign wealth funds. The firm has 14 offices in 10 countries. In April 2015, Colony Financial, Inc. completed its merger with Colony Capital LLC, and as a result the company was renamed as Colony Capital, Inc.

FINANCIAL DATA: Note: Data for latest year may not have been available at press time.

In U.S. $	2015	2014	2013	2012	2011	2010
Revenue	841,976,000	300,649,000	180,239,000	107,155,000	65,469,000	27,425,000
R&D Expense						
Operating Income	201,462,000	156,096,000	122,048,000	70,602,000	45,408,000	17,543,000
Operating Margin %	23.92%	51.91%	67.71%	65.88%	69.35%	63.96%
SGA Expense	317,776,000	87,011,000	39,043,000	28,305,000	17,050,000	9,327,000
Net Income	149,980,000	123,149,000	101,765,000	62,011,000	42,260,000	17,731,000
Operating Cash Flow	373,126,000	132,759,000	125,289,000	71,026,000	24,417,000	9,857,000
Capital Expenditure	56,335,000	1,618,069,000	122,750,000			
EBITDA	520,811,000	214,854,000	145,730,000	78,618,000	47,566,000	18,552,000
Return on Assets %	1.35%	2.31%	3.95%	4.44%	7.56%	5.23%
Return on Equity %	4.08%	4.79%	5.52%	5.26%	9.02%	5.84%
Debt to Equity	1.46	1.06	0.28	0.08	0.02	0.04

CONTACT INFORMATION:

Phone: 310 282-8820   Fax: 212 593-5433
Toll-Free:
Address: 2450 Broadway, Santa Monica, CA 90404 United States

STOCK TICKER/OTHER:

Stock Ticker: CLNY   Exchange: NYS
Employees: Fiscal Year Ends: 12/31
Parent Company:

SALARIES/BONUSES:

Top Exec. Salary: $750,000   Bonus: $3,000,000
Second Exec. Salary: Bonus: $1,800,000
$600,000

OTHER THOUGHTS:

Estimated Female Officers or Directors:
Hot Spot for Advancement for Women/Minorities:

Comfort Systems USA Inc

www.comfortsystemsusa.com

NAIC Code: 238220

TYPES OF BUSINESS:

Mechanical Contractors
Industrial & Commercial HVAC Systems & Services
Facility Automation Services

BRANDS/DIVISIONS/AFFILIATES:

ShoffnerKalthoff Mechanical Electrical Service Inc
Shoffner Mechanical Services Inc
SKMES Inc

CONTACTS: Note: Officers with more than one job title may be intentionally listed here more than once.

Franklin Myers, Chairman of the Board
Alfred Giardinelli, Director
Brian Lane, Director
William George, Executive VP
James Mylett, Senior VP, Divisional
Julie Shaeff, Senior VP
Trent McKenna, Senior VP

GROWTH PLANS/SPECIAL FEATURES:

Comfort Systems USA, Inc. is a national heating, ventilation and air conditioning (HVAC) and building automation services company with 36 operating units in 85 cities and 94 locations throughout the U.S. It provides installation, maintenance, repair and replacement services for a large base of national, multi-location clients in both commercial and industrial markets. The company's technical experts design, install and fine-tune HVAC systems for downtown high-rises, hospitals, universities, national hotels, factories and industrial plants. The company provides design-and-build services, in which Comfort determines the needed capacity and energy efficiency of the HVAC system that best suits the proposed facility and then estimates the amount of time, labor, materials and equipment needed to build the specified system. It also provides plan-and-spec services, in which it participates in a bid process to provide labor, equipment, materials and installation based on plans and engineering specifications provided by a client. Additionally, the firm installs process cooling systems and building automation controls and monitoring systems. Maintenance, repair and replacement services include the maintenance, repair, replacement, reconfiguration and monitoring of HVAC systems and industrial process piping. Maintenance and repair services are provided either in response to service calls or under a service agreement. During 2015, the firm's revenue was derived primarily from its HVAC services (77%), with plumbing deriving 14%, building automation control systems deriving 5% and other deriving 4%. The company's primary clients include owners of office buildings; retail centers; apartment complexes; manufacturing plants; health care, education and government facilities; and other commercial, industrial and institutional facilities. In February 2016, the firm acquired the ShoffnerKalthoff family of companies which includes ShoffnerKalthoff Mechanical Electrical Service, Inc., Shoffner Mechanical Services, Inc. and SKMES, Inc.

FINANCIAL DATA: Note: Data for latest year may not have been available at press time.

In U.S. $	2015	2014	2013	2012	2011	2010
Revenue	1,580,519,000	1,410,795,000	1,357,272,000	1,331,185,000	1,240,020,000	1,108,282,000
R&D Expense						
Operating Income	90,044,000	42,222,000	46,258,000	22,303,000	-49,368,000	20,042,000
Operating Margin %	5.69%	2.99%	3.40%	1.67%	-3.98%	1.80%
SGA Expense	228,965,000	207,652,000	194,214,000	185,809,000	172,137,000	163,431,000
Net Income	49,364,000	23,063,000	27,269,000	13,463,000	-36,830,000	14,740,000
Operating Cash Flow	97,867,000	42,552,000	38,423,000	30,510,000	29,680,000	32,149,000
Capital Expenditure	20,808,000	19,183,000	17,403,000	11,782,000	8,666,000	7,089,000
EBITDA	113,833,000	63,422,000	66,703,000	43,764,000	-22,725,000	39,518,000
Return on Assets %	7.27%	3.63%	4.61%	2.29%	-5.98%	2.42%
Return on Equity %	15.11%	7.66%	9.63%	5.03%	-12.75%	4.76%
Debt to Equity	0.03	0.13		0.02	0.05	0.09

CONTACT INFORMATION:

Phone: 713-830-9600   Fax: 713 830-9696
Toll-Free: 800-723-8431
Address: 675 Bering Dr., Ste. 400, Houston, TX 77057 United States

STOCK TICKER/OTHER:

Stock Ticker: FIX
Employees: 6,698
Parent Company:

Exchange: NYS
Fiscal Year Ends: 12/31

SALARIES/BONUSES:

Top Exec. Salary: $515,000   Bonus: $
Second Exec. Salary: $300,000   Bonus: $200,000

OTHER THOUGHTS:

Estimated Female Officers or Directors: 3
Hot Spot for Advancement for Women/Minorities: Y

Sales, profits and employees may be estimates. Financial information, benefits and other data can change quickly and may vary from those stated here.

Commune Hotels & Resorts

www.communehotels.com

NAIC Code: 721110

TYPES OF BUSINESS:
Hotels
Day Spas
Restaurants
Hospitality Consulting
Condominium Management

BRANDS/DIVISIONS/AFFILIATES:
Joie de Vivre Hotels
Thompson Hotels
tommie
Alila
Hotel Rex
Zimzala Restaurant & Bar
Millennium Restaurant
Kabuki Springs & Spa

CONTACTS: *Note: Officers with more than one job title may be intentionally listed here more than once.*
Niki Leondakis, CEO
Rick Colangelo, Exec. VP-Oper.
Stephen Miano, CFO
Jorge E. Trevino, Exec. VP-Brand Operations
Greg Smith, Exec. VP-Human Resources
Jennifer Foley Shields, VP-Comm. & Special Projects
Karolina Kiebowicz, Sr. Dir-Public Rel.
John Pritzker, Chmn.

GROWTH PLANS/SPECIAL FEATURES:
Commune Hotels & Resorts (CHR), operating through the hotel brands Joie de Vivre Hotels (JdV), Thompson Hotels, tommie and Alila. JdV's properties include over 30 hotels throughout California as well as a location in Scottsdale, Arizona, 2 hotels in Honolulu and Chicago's Hotel Lincoln, featuring some 24 restaurants and bars and six spas. JdV's boutique hotels each target a niche audience by embodying a particular lifestyle or theme, such as arts and literature, featured at the Hotel Rex; rock and roll, at the Phoenix Hotel; or Japanese pop-culture, at Hotel Tomo. It also owns and operates the Kabuki Springs & Spa in San Francisco, Spa Vitale at Hotel Vitale, Spa Aiyana in Carmel Valley Ranch and a spa located in the Ventana Inn near Big Sur. JdV's restaurants and bars are usually attached to its hotels, such as Zimzala Restaurant & Bar in the Shorebreak Hotel; however, it also has free-standing restaurants such as Millennium Restaurant in Union Square, San Francisco. Thompson brand hotels offers luxury boutique hotels in 10 domestic and international locations. The brand's domestic hotels include locations in New York, Los Angeles, Nashville, Seattle, Chicago and Miami. The international locations of the brand are represented by the Belgraves in London, U.K.; Thompson Toronto in Toronto, Canada; The Cape in Los Cabos, Mexico; and the Thompson Playa in Playa Del Carmen, Mexico. Thompson brand hotels are designed to appeal to a chic, urban, bohemian clientele. tommie is a youthful, micro-lifestyle hotel brand located in Los Angeles and New York City. tommie brand hotels combine the playfulness of JdV with the sophistication of the Thompson. The Alila brand of hotels is a combination of design luxury in unique locations, providing private space, personalized hospitality along with natural cultural elements. Alila hotels are located in China, Indonesia, India and Oman.

FINANCIAL DATA: *Note: Data for latest year may not have been available at press time.*

In U.S. $	2015	2014	2013	2012	2011	2010
Revenue	500,000,000	490,000,000	475,000,000	450,000,000	328,000,000	240,000,000
R&D Expense						
Operating Income						
Operating Margin %						
SGA Expense						
Net Income						
Operating Cash Flow						
Capital Expenditure						
EBITDA						
Return on Assets %						
Return on Equity %						
Debt to Equity						

CONTACT INFORMATION:
Phone: 415-835-0300   Fax:
Toll-Free:
Address: 530 Bush Street, Ste 501, San Francisco, CA 94108 United States

STOCK TICKER/OTHER:
Stock Ticker: Private
Employees: 1,077
Parent Company:

Exchange:
Fiscal Year Ends: 12/31

SALARIES/BONUSES:
Top Exec. Salary: $ Bonus: $
Second Exec. Salary: $ Bonus: $

OTHER THOUGHTS:
Estimated Female Officers or Directors: 3
Hot Spot for Advancement for Women/Minorities: Y

Sales, profits and employees may be estimates. Financial information, benefits and other data can change quickly and may vary from those stated here.

Community Development Trust

www.cdt.biz

NAIC Code: 525990

TYPES OF BUSINESS:
Mortgage REIT
Equity & Mortgage Investments
Apartments
Offices
Retail Properties

BRANDS/DIVISIONS/AFFILIATES:

CONTACTS: Note: Officers with more than one job title may be intentionally listed here more than once.
Joseph F. Reilly, CEO
John J. Divers, COO
Joseph F. Reilly, Pres.
John J. Divers, CFO
Patricia Tagarello, VP-Oper.
John J. Divers, Treas.
Brian Gallagher, Sr. VP-Debt
Brian Dowling, VP-Community Investment
Shelly Cleary, Sr. VP-Underwriting
Joan Berkowitz, Sr. VP-Asset Mgmt.

GROWTH PLANS/SPECIAL FEATURES:
Community Development Trust (CDT) is a real estate investment trust (REIT) entirely devoted to investment in the community development market through both an equity program and a debt program. The company has committed $900 million in debt and equity capital to affordable housing properties across the country. Through CDT's debt and equity programs, the firm invests long-term debt capital by purchasing smaller, fixed-rate multifamily mortgages from community lenders and equity capital in cash or by providing a tax advantaged transition for existing properties to a new set of owners committed to long-term affordability. All of CDT's investments must satisfy Community Reinvestment Act requirements. The company has invested in over 35,000 housing units in 42 states across the country, representing a broad mix of types ranging from scattered-site developments to in-fill urban family apartments and redeveloped high rises. The firm's average investment per transaction is approximately $5 million. Major investors in CDT include Allstate Insurance; Bank of America, California Bank and Trust, Citibank Community Development, Deutsche Bank, Fannie Mae, HSBC Bank USA, JPMorgan Chase Community Development Group, Key Community Development, Local Initiatives Support Corporation. MetLife, Merrill Lynch Community Development Company LLC, NCB Capital Impact, Prudential Financial, The Reinvestment Fund and Wells Fargo Bank.

FINANCIAL DATA: Note: Data for latest year may not have been available at press time.

In U.S. $	2015	2014	2013	2012	2011	2010
Revenue						
R&D Expense						
Operating Income						
Operating Margin %						
SGA Expense						
Net Income						
Operating Cash Flow						
Capital Expenditure						
EBITDA						
Return on Assets %						
Return on Equity %						
Debt to Equity						

CONTACT INFORMATION:
Phone: 212-271-5080   Fax: 212-271-5079
Toll-Free:
Address: 1350 Broadway, Ste. 700, New York, NY 10018-7702 United States

STOCK TICKER/OTHER:
Stock Ticker: Private
Employees: 30
Parent Company:

Exchange:
Fiscal Year Ends: 12/31

SALARIES/BONUSES:
Top Exec. Salary: $ Bonus: $
Second Exec. Salary: $ Bonus: $

OTHER THOUGHTS:
Estimated Female Officers or Directors: 5
Hot Spot for Advancement for Women/Minorities: Y

Condor Hospitality Trust Inc

www.supertelinc.com

NAIC Code: 0

TYPES OF BUSINESS:

Real Estate Investment Trust
Hotels & Motels

BRANDS/DIVISIONS/AFFILIATES:

Supertel Hospitality REIT Trust
E&P REIT Trust
Supertel Limited Partnership
E&P Financing Limited Partnership
Supertel Hospitality Management Inc
TRS Leasing Inc
Royco Hotels Inc
Supertel Hospitality Inc

CONTACTS: Note: Officers with more than one job title may be intentionally listed here more than once.

J. Blackham, CEO
Arinn Cavey, Chief Accounting Officer
Jeffrey Dougan, COO
James Friend, Director
Jonathan Gantt, Senior VP
Patrick Beans, Senior VP

GROWTH PLANS/SPECIAL FEATURES:

Condor Hospitality Trust Inc., formerly Supertel Hospitality, Inc., is a self-administered REIT (real estate investment trust) primarily engaged in acquiring and owning hotels. Through its subsidiaries, Condor's portfolio includes 56 limited-service hotels in 20 states under the following franchise brands: Comfort Inn/Comfort Suites, Days Inn, Super 8, Sleep Inn, Savannah Suites, Quality Inn, Clarion, Hilton Garden Inn, Key West Inn, Rodeway Inn and Supertel Inn. Standard guestroom amenities include high-speed wireless Internet; free local calls; cable TV, including movie channels; an iron and ironing board; and a coffee maker and hairdryer. Select properties also have complimentary continental breakfast, pools, suites, conference centers and meeting facilities. Through its two wholly-owned subsidiaries, Supertel Hospitality REIT Trust and E&P REIT Trust, the firm owns a 99% interest in Supertel Limited Partnership and 100% of E&P Financing Limited Partnership. Together, these four firms, along with other subsidiaries and partnerships, own the firm's properties. Those other subsidiaries and partnerships include Supertel Hospitality Management, Inc.; SPPR-BMI Holdings, Inc.; and Solomon's Beacon Inn Limited Partnership. To enter into management agreements with independent contractors, in compliance with the REIT Modernization Act of 1999, TRS Leasing, Inc. was created, along with its subsidiaries TRS Subsidiary LLC and SPPR TRS Subsidiary LLC. All hotels are leased to TRS Leasing and are managed by Royco Hotels, Inc. and HLC Hotels, Inc. In July 2015, Supertel Hospitality Inc. changed its name to Condor Hospitality Trust Inc. in order to extend the company's brand from economy- and mid-scale hotels to encompass high-end hotels as well. That same month, the firm agreed to acquire three premium-branded hotels in an off-market transaction worth $42.5 million. The properties consist of SpringHill Suites by Marriott, San Antonio, Texas; Hotel Indigo in Atlanta, Georgia; and Courtyard by Marriott in Jacksonville, Florida.

FINANCIAL DATA: Note: Data for latest year may not have been available at press time.

In U.S. $	2015	2014	2013	2012	2011	2010
Revenue	57,341,000	57,409,000	56,163,000	70,573,000	75,827,000	84,114,000
R&D Expense						
Operating Income	3,332,000	3,524,000	-196,000	6,799,000	4,955,000	5,419,000
Operating Margin %	5.81%	6.13%	-.34%	9.63%	6.53%	6.44%
SGA Expense	5,493,000	4,192,000	3,923,000	3,908,000	4,008,000	3,443,000
Net Income	13,125,000	-16,236,000	-1,351,000	-10,210,000	-17,445,000	-10,585,000
Operating Cash Flow	4,970,000	5,387,000	2,017,000	3,789,000	2,865,000	7,672,000
Capital Expenditure	47,908,000	3,374,000	5,133,000	17,168,000	4,964,000	4,344,000
EBITDA	21,185,000	-5,743,000	15,104,000	12,244,000	7,571,000	15,523,000
Return on Assets %	6.53%	-12.36%	-2.51%	-6.32%	-7.91%	-4.54%
Return on Equity %	36.83%	-76.40%	-13.60%	-36.55%	-41.08%	-19.87%
Debt to Equity	2.15	3.91	2.88	3.07	3.57	2.61

CONTACT INFORMATION:

Phone: 402 371-2520   Fax: 402 371-4229
Toll-Free:
Address: 1800 W. Pasewalk Ave., Ste. 200, Norfolk, NE 68701 United States

STOCK TICKER/OTHER:

Stock Ticker: CDOR   Exchange: NAS
Employees: 15   Fiscal Year Ends: 12/31
Parent Company:

SALARIES/BONUSES:

Top Exec. Salary: $290,000   Bonus: $
Second Exec. Salary: $190,000   Bonus: $19,000

OTHER THOUGHTS:

Estimated Female Officers or Directors: 3
Hot Spot for Advancement for Women/Minorities: Y

Sales, profits and employees may be estimates. Financial information, benefits and other data can change quickly and may vary from those stated here.

CORE Network (The)

www.corenetworkcre.org

NAIC Code: 531210

TYPES OF BUSINESS:

Office & Industrial Consulting
Retail Planning
Real Estate Development
Financial Services & Mortgages
Property Management
Real Estate Brokerage

BRANDS/DIVISIONS/AFFILIATES:

Project Tracking System

CONTACTS: Note: Officers with more than one job title may be intentionally listed here more than once.

Rachel G. Krupnick, Exec. Dir.
William K. Montrose, Chmn.

GROWTH PLANS/SPECIAL FEATURES:

The CORE Network is a consortium of mid-sized commercial real estate firms providing services to office and retail businesses. Through its member firms, CORE offers clients a variety of services, including tenant representation, in which a specialist assists clients in negotiating leases, moving into a new facility and defining goals for surplus property and space requirements. The leasing and management service offers clients basic property management as well as providing an evaluation of occupancy alternatives. The investment advisory services offer investors consultation in advantageous portfolio acquisitions through the firm's knowledge of local real estate opportunities. In addition, it offers finance and investment banking services to large corporate and institutional clients that consists of financing, transaction assistance, providing capital and negotiation tactics. The facility management services helps reduce cost and facilitate tenant services; construction management provides budgeting, general contracting and project and scheduling management; and development management services provides assistance to those clients wanting to occupy a property or to corporate clients entering into a joint venture. CORE's portfolio management offers tracking and administrative services to clients with numerous real estate acquisitions. The network's strategic planning and general advisory services incorporate long term consulting and analytical methods for financial growth. Additionally, CORE offers fundamental research, analysis and planning through specialized departments within each member firm. The company employs the expertise of industry-specific account teams and specialists from various member firms in the network, allowing companies to provide better service collectively. Most clients of network members are major property holders with at least 100 leased or owned properties. CORE supports client projects through the Project Tracking System, which allows network members to access an index of project progress online.

FINANCIAL DATA: Note: Data for latest year may not have been available at press time.

In U.S. $	2015	2014	2013	2012	2011	2010
Revenue						
R&D Expense						
Operating Income						
Operating Margin %						
SGA Expense						
Net Income						
Operating Cash Flow						
Capital Expenditure						
EBITDA						
Return on Assets %						
Return on Equity %						
Debt to Equity						

CONTACT INFORMATION:

Phone: 757-490-7871   Fax: 757-490-7844
Toll-Free:
Address: One Columbus Ctr., Ste. 600, Virginia Beach, VA 23462 United States

STOCK TICKER/OTHER:

Stock Ticker: Private
Employees:
Parent Company:

Exchange:
Fiscal Year Ends:

SALARIES/BONUSES:

Top Exec. Salary: $ Bonus: $
Second Exec. Salary: $ Bonus: $

OTHER THOUGHTS:

Estimated Female Officers or Directors: 2
Hot Spot for Advancement for Women/Minorities:

Corporate Office Properties Trust

www.copt.com

NAIC Code: 0

TYPES OF BUSINESS:

Real Estate Investment Trust
Commercial Development & Construction Services
Office Properties
Property Management

BRANDS/DIVISIONS/AFFILIATES:

Corporate Office Properties LP

CONTACTS: Note: Officers with more than one job title may be intentionally listed here more than once.

Roger Waesche, CEO
Thomas Brady, Chairman of the Board
Gregory Thor, Chief Accounting Officer
Stephen Budorick, COO
Wayne Lingafelter, Executive VP, Divisional
Anthony Mifsud, Executive VP
Karen Singer, General Counsel
Philip Hawkins, Trustee
Richard Szafranski, Trustee
Robert Denton, Trustee
Steven Kesler, Trustee
C. Pickett, Trustee
David Jacobstein, Trustee

GROWTH PLANS/SPECIAL FEATURES:

Corporate Office Properties Trust (COPT) is a real estate investment trust (REIT) focused on strategic customer relationships and specialized tenant requirements in the U.S. government and defense information technology sectors, with property investments in the Washington/Baltimore corridor; Virginia; Washington, D.C.; Maryland; Virginia; Huntsville, Alabama; Maryland; Pennsylvania; and San Antonio, Texas. The firm engages in the acquisition, development, management and leasing of office and data center properties primarily situated in large office parks near government demand centers. The company's consolidated portfolio includes 177 operating office properties containing 18.1 million rentable square feet. Additionally, COPT owns 13 office properties under construction, development or redevelopment and land parcels totaling 1,439 acres. Primarily all of the firm's business operations are performed through the operating partnership Corporate Office Properties, LP, of which COPT owns 95.5% of the outstanding preferred units and 96.3% of the outstanding common stock. COPT's four largest tenants accounted for approximately 41.9% of its 2015 rental revenues. These include the U.S. government, Northrop Grumman Corporation, The Boeing Company and General Dynamics Corporation. In 2015, the firm acquired a 368,200 square foot building in Baltimore, Maryland; acquired 100 Light Street and its 560-space structured parking garage, as well as 30 Light Street (together, form the Transamerica Building); and sold One Dulles Tower in Herndon, Virginia.

The firm offers employees life insurance; adoption assistance; educational assistance; medical, dental and vision coverage; flexible savings accounts; and college savings plans.

FINANCIAL DATA: Note: Data for latest year may not have been available at press time.

In U.S. $	2015	2014	2013	2012	2011	2010
Revenue	625,466,000	586,473,000	523,360,000	528,007,000	556,841,000	564,475,000
R&D Expense	13,507,000	5,573,000	5,436,000			
Operating Income	120,094,000	131,612,000	141,910,000	95,965,000	4,122,000	131,313,000
Operating Margin %	19.20%	22.44%	27.11%	18.17%	.74%	23.26%
SGA Expense	31,361,000	31,794,000	30,869,000	37,611,000	29,038,000	28,205,000
Net Income	178,300,000	40,255,000	93,707,000	20,977,000	-117,675,000	42,760,000
Operating Cash Flow	204,008,000	193,885,000	158,979,000	191,838,000	152,143,000	156,436,000
Capital Expenditure	489,772,000	256,142,000	247,698,000	274,030,000	319,624,000	496,437,000
EBITDA	347,713,000	265,702,000	240,597,000	226,066,000	114,500,000	268,076,000
Return on Assets %	4.32%	.61%	1.94%	-.04%	-3.46%	.73%
Return on Equity %	12.63%	1.85%	6.41%	-.15%	-11.32%	2.36%
Debt to Equity	1.54	1.53	1.63	1.95	2.13	1.89

CONTACT INFORMATION:

Phone: 443-285-5400   Fax:
Toll-Free:
Address: 6711 Columbia Gateway Dr., Ste. 300, Columbia, MD 21046
United States

STOCK TICKER/OTHER:

Stock Ticker: OFC
Employees: 378
Parent Company:

Exchange: NYS
Fiscal Year Ends: 12/31

SALARIES/BONUSES:

Top Exec. Salary: $597,500   Bonus: $
Second Exec. Salary: Bonus: $
$438,700

OTHER THOUGHTS:

Estimated Female Officers or Directors: 3
Hot Spot for Advancement for Women/Minorities: Y

CoStar Group Inc

NAIC Code: 519130

TYPES OF BUSINESS:
Online Commercial Real Estate Information

BRANDS/DIVISIONS/AFFILIATES:
CoStar Property Professional
CoStar Tenant
CoStar COMPS Professional
FOCUS
CoStar U.K. Limited
Resolve Technology Inc
Property and Portfolio Research Inc
Apartments.com

CONTACTS: Note: Officers with more than one job title may be intentionally listed here more than once.
Andrew Florance, CEO
Scott Wheeler, CFO
Michael Klein, Chairman of the Board
Frank Simuro, Chief Technology Officer
Brian Radecki, Executive VP
Frank Carchedi, Executive VP, Divisional
Matthew Linnington, Executive VP, Divisional
Giles Newman, Managing Director, Subsidiary
Hans Nordby, Managing Director, Subsidiary
Wayne Warthen, Other Executive Officer
Andrew Thomas, President, Divisional
Frederick Saint, President, Subsidiary
Jonathan Coleman, Secretary
Mark Klionsky, Senior VP, Divisional
Susan Jeffress, Vice President, Divisional
Donna Tanenbaum, Vice President, Divisional
Scott Yinger, Vice President, Divisional

GROWTH PLANS/SPECIAL FEATURES:
CoStar Group, Inc. is a leading national provider of commercial real estate sales information. The company offers customers online access to a comprehensive, verified database of commercial real estate information, analytics and marketing services in U.S. markets as well as in London, other U.K. markets and France. CoStar has a highly developed data collection system that includes sales prices, income and expenses, capitalization rates, loan data, property photographs, buyers, sellers, brokers and other key details. The firm's database contains information about approximately 4.5 million properties, and 1.5 million sale and lease listings. CoStar's subscription-based information services include four primary tools. CoStar Property Professional, the firm's flagship service, provides a comprehensive inventory of office, industrial, retail, multifamily properties and land in markets throughout the U.S., including for-lease and for-sales listings, historical data, building photographs, maps and floor plans. CoStar Tenant is an online business-to-business (B2B) prospecting and analytical tool providing commercial real estate professionals with commercial U.S. tenant information. CoStar COMPS Professional offers comprehensive coverage of comparable sales information in the U.S. commercial real estate industry. CoStarGo is the firm's mobile application. FOCUS (offered by CoStar U.K. Limited, the firm's U.K. subsidiary) is a digital online service offering information on the U.K. real estate market. Other subsidiaries of the company include Resolve Technology, Inc., a real estate investment management software provider; Property and Portfolio Research, Inc., a commercial real estate market forecaster; and Grecam SAS, a commercial property information provider in France; Apartments, LLC, which operates Apartments.com, a national online resource for individuals looking for an apartment; and LoopNet, which offers an online marketplace for estate agents, buyers and tenants to search for available property listings.

The firm offers employees medical, dental and vision insurance; a 401(k); an employee assistance program; tuition reimbursement; life insurance; and disability coverage.

FINANCIAL DATA: Note: Data for latest year may not have been available at press time.

In U.S. $	2015	2014	2013	2012	2011	2010
Revenue	711,764,000	575,936,000	440,943,000	349,936,000	251,738,000	226,260,000
R&D Expense	65,760,000	55,426,000	46,757,000	32,756,000	20,037,000	17,350,000
Operating Income	11,455,000	80,878,000	54,154,000	27,440,000	21,771,000	22,775,000
Operating Margin %	1.60%	14.04%	12.28%	7.84%	8.64%	10.06%
SGA Expense	417,733,000	254,221,000	195,664,000	161,267,000	119,526,000	100,231,000
Net Income	-3,465,000	44,869,000	29,734,000	9,915,000	14,656,000	13,289,000
Operating Cash Flow	131,245,000	143,909,000	108,298,000	86,126,000	25,685,000	39,269,000
Capital Expenditure	35,061,000	27,444,000	19,042,000	14,834,000	15,013,000	57,358,000
EBITDA	90,524,000	151,766,000	94,538,000	60,718,000	34,623,000	36,424,000
Return on Assets %	-.16%	2.68%	2.45%	1.02%	2.42%	3.14%
Return on Equity %	-.22%	3.67%	3.39%	1.33%	2.81%	3.58%
Debt to Equity	0.21	0.24	0.13	0.18		

CONTACT INFORMATION:
Phone: 202-346-6500   Fax: 202 346-6370
Toll-Free: 800-204-5960
Address: 1331 L St., NW, Northwest, WA 20005 United States

STOCK TICKER/OTHER:
Stock Ticker: CSGP
Employees: 2,444
Parent Company:

Exchange: NAS
Fiscal Year Ends: 12/31

SALARIES/BONUSES:
Top Exec. Salary: $638,142   Bonus: $
Second Exec. Salary: $372,175   Bonus: $

OTHER THOUGHTS:
Estimated Female Officers or Directors:
Hot Spot for Advancement for Women/Minorities: Y

Sales, profits and employees may be estimates. Financial information, benefits and other data can change quickly and may vary from those stated here.

Cousins Properties Inc

www.cousinsproperties.com

NAIC Code: 531120

TYPES OF BUSINESS:

Real Estate Investment Trust, Commercial
Residential Communities
Office Properties
Retail Properties
Property Management
Land Development
Open Air Malls
Industrial

BRANDS/DIVISIONS/AFFILIATES:

CONTACTS: *Note: Officers with more than one job title may be intentionally listed here more than once.*

Lawrence Gellerstedt, CEO
Gregg Adzema, CFO
S. Glover, Chairman of the Board
John Mccoll, Executive VP
Michael Connolly, Executive VP
Pamela Roper, General Counsel
James Ellis, Senior VP
John Harris, Treasurer

GROWTH PLANS/SPECIAL FEATURES:

Cousins Properties, Inc., a real estate investment trust (REIT) located in Atlanta, Georgia, owns, develops and manages its own real estate portfolio. The company controls, either directly or through joint ventures, office and mixed-use properties totaling 16.5 million square feet. The firm is responsible for identifying new development projects among all product types and managing all phases of the development and construction process through project stabilization or sale. This process includes not only construction management but also leasing and tenant coordination for first generation office and retail space. The company plans to exit the residential land business over time to better focus on its office properties and has sold the majority of its interests in CL Realty, L.L.C. and Temco Associates, LLC, but continues to own a 50% stake in Paulding County, and 100% stake each in Blalock Lakes and Callaway Gardens, consisting of residential properties in Atlanta, Georgia, as well as its 50% interest of residential land in Padre Island, Texas. During 2015, the firm formed a joint venture to develop Carolina Square, a mixed-use property in Chapel hill, North Carolina; and opened Colorado Tower, a Class A office tower in downtown Austin.

FINANCIAL DATA: *Note: Data for latest year may not have been available at press time.*

In U.S. $	2015	2014	2013	2012	2011	2010
Revenue	381,643,000	361,383,000	210,741,000	148,278,000	178,464,000	228,506,000
R&D Expense						
Operating Income	37,519,000	8,026,000	-16,595,000	-12,413,000	-120,145,000	-24,696,000
Operating Margin %	9.83%	2.22%	-7.87%	-8.37%	-67.32%	-10.80%
SGA Expense	20,529,000	23,436,000	27,675,000	32,256,000	30,571,000	52,498,000
Net Income	125,518,000	52,004,000	121,761,000	45,728,000	-128,425,000	-14,573,000
Operating Cash Flow	151,661,000	142,400,000	137,340,000	95,322,000	55,581,000	79,696,000
Capital Expenditure	184,988,000	710,743,000	1,526,263,000	105,069,000	181,909,000	33,761,000
EBITDA	203,704,000	190,694,000	142,880,000	67,918,000	-34,880,000	64,551,000
Return on Assets %	4.76%	1.84%	6.42%	2.78%	-10.84%	-1.91%
Return on Equity %	7.47%	2.99%	12.03%	7.41%	-27.58%	-4.54%
Debt to Equity	0.42	0.47	0.46	0.94	1.24	0.86

CONTACT INFORMATION:

Phone: 404-407-1000   Fax: 404 407-1002
Toll-Free:
Address: 191 Peachtree St., NE, Ste. 500, Atlanta, GA 30303 United States

STOCK TICKER/OTHER:

Stock Ticker: CUZ   Exchange: NYS
Employees: 257   Fiscal Year Ends: 12/31
Parent Company:

SALARIES/BONUSES:

Top Exec. Salary: $650,000   Bonus: $
Second Exec. Salary: Bonus: $
$405,000

OTHER THOUGHTS:

Estimated Female Officers or Directors: 1
Hot Spot for Advancement for Women/Minorities:

Crescent Real Estate Equities LP

www.crescent.com

NAIC Code: 0

TYPES OF BUSINESS:

Real Estate Investment Trust
Property Management
Land Development
Leasing
Offices
Hotels & Resorts
Residential Properties
Temperature-Controlled Logistics Facilities

BRANDS/DIVISIONS/AFFILIATES:

Barclays Capital
Goff Capital Inc

GROWTH PLANS/SPECIAL FEATURES:

Crescent Real Estate Equities LP, registered as a real estate investment trust (REIT), conducts property management, leasing and development activities. The company owns and operates 25 premier office buildings with over 9.2 million square feet of rentable space. Crescent's properties are concentrated in Dallas, Houston, Denver and Las Vegas. The firm also holds resort/spa properties in its portfolio, including the Canyon Ranch in Tucson, Arizona; the Canyon Ranch in Lenox, Massachusetts; the Fairmont Sonoma Mission Inn & Spa in Sonoma, California; and The Ritz-Carlton Hotel in Dallas. Crescent also holds investments in resort residential developments in locations such as Scottsdale, Vail Valley and Lake Tahoe; a luxury hotel, The Ritz-Carlton in Dallas; and a wellness lifestyle center in Canyon Ranch. The company is a joint venture owned by Barclays Capital & Goff Capital, Inc.

The company offers employees medical, dental and vision insurance as well as a 401(k).

CONTACTS:
Note: Officers with more than one job title may be intentionally listed here more than once.

John Goff, CEO
Jason Anderson, COO
Suzanne M. Stevens, CFO
Joseph Pitchford, Sr. VP-Dev.
Jason Phinney, Controller
Robert L. Carlen, VP-Property Mgmt.
Anthony B. Click, VP-Leasing
Tom Nezworski, Managing Dir.-Resort & Residential
John L. Zogg, Jr., Managing Dir.-Leasing
John Goff, Chmn.

FINANCIAL DATA:
Note: Data for latest year may not have been available at press time.

In U.S. $	2015	2014	2013	2012	2011	2010
Revenue						
R&D Expense						
Operating Income						
Operating Margin %						
SGA Expense						
Net Income						
Operating Cash Flow						
Capital Expenditure						
EBITDA						
Return on Assets %						
Return on Equity %						
Debt to Equity						

CONTACT INFORMATION:

Phone: 817-321-2100   Fax: 817-321-2000
Toll-Free:
Address: 777 Main St., Ste. 2000, Fort Worth, TX 76102 United States

STOCK TICKER/OTHER:

Stock Ticker: Joint Venture
Employees: 748
Parent Company:

Exchange:
Fiscal Year Ends: 12/31

SALARIES/BONUSES:

Top Exec. Salary: $ Bonus: $
Second Exec. Salary: $ Bonus: $

OTHER THOUGHTS:

Estimated Female Officers or Directors: 2
Hot Spot for Advancement for Women/Minorities: Y

Crest Nicholson plc

NAIC Code: 531100

TYPES OF BUSINESS:

Real Estate Development
Environmental Communities
Mixed-Use Developments

BRANDS/DIVISIONS/AFFILIATES:

GROWTH PLANS/SPECIAL FEATURES:

Crest Nicholson plc is a British real estate developer focusing on the construction of environmentally sound master-planned communities and mixed-use developments. The Crest Nicholson regeneration business unit focuses on large-scale developments with public and private partners, including residential and mixed-use projects; and the Crest strategic business unit sources unallocated sites and secures valuable planning permission for medium to long-term development projects. The company's master planned communities feature distinctive residential areas and a good deal of open space as well as amenities such as schools, shops and sports facilities. The company also focuses on building communities that do not rely on the use of a car, featuring designated cycling routes and integrated bus service that link outlying residential areas to the town center. Crest Nicholson actively supports the Building Research Establishment's EcoHomes rating scheme, which accesses the environmental performance of buildings in the areas of energy, transport, pollution, materials, water, land use, ecology, health and well-being.

CONTACTS: Note: Officers with more than one job title may be intentionally listed here more than once.

Stephen Stone, CEO
Steve Evans, Dir.-Prod.
Kevin Maguire, Corp. Sec.
Robin Hoyles, Dir.-Planning & Land
Patrick Bergin, Dir.-Finance
Chris Tinker, Regeneration Chmn.
William Rucker, Chmn.

FINANCIAL DATA: Note: Data for latest year may not have been available at press time.

In U.S. $	2015	2014	2013	2012	2011	2010
Revenue	1,137,832,000	899,605,600	743,238,400	576,833,300	451,145,900	402,086,800
R&D Expense						
Operating Income	230,874,700	181,108,700	128,939,200	103,632,100	79,738,720	66,873,080
Operating Margin %	20.29%	20.13%	17.34%	17.96%	17.67%	16.63%
SGA Expense	82,000,820	76,204,210	70,690,360			
Net Income	175,453,500	139,684,100	93,311,270	90,342,280	57,259,190	-39,021,080
Operating Cash Flow	33,224,470	-54,855,720	27,569,240	24,883,010	-1,272,427	38,596,940
Capital Expenditure	2,262,092	1,696,569	1,413,807	706,904	141,381	565,523
EBITDA	237,095,500	182,381,100	131,766,800	104,197,600	-9,613,889	-17,107,070
Return on Assets %	10.53%	9.99%	7.88%	8.51%	6.28%	-4.68%
Return on Equity %	21.26%	19.62%	16.14%	20.15%	43.08%	
Debt to Equity	0.33	0.27	0.12	0.50	0.56	

CONTACT INFORMATION:

Phone: 44-1932-580555   Fax: 44-870-336-3990
Toll-Free:
Address: Pyrcroft Rd., Crest House, Chertsey, Surrey KT16 9GN United Kingdom

STOCK TICKER/OTHER:

Stock Ticker: CRNHY   Exchange: GREY
Employees: 840   Fiscal Year Ends: 10/31
Parent Company:

SALARIES/BONUSES:

Top Exec. Salary: $ Bonus: $
Second Exec. Salary: $ Bonus: $

OTHER THOUGHTS:

Estimated Female Officers or Directors: 1
Hot Spot for Advancement for Women/Minorities:

CRH plc

NAIC Code: 444190

www.crh.com

TYPES OF BUSINESS:

Building Materials
Aggregates
Asphalt
Cement & Concrete Products
Distribution
Glass Fabrications
Security Gates & Fencing
Insulation Products

BRANDS/DIVISIONS/AFFILIATES:

My Home Industries Ltd

CONTACTS: Note: Officers with more than one job title may be intentionally listed here more than once.

Albert Manifold, CEO
Maeve Carton, Dir.-Finance
N. Hartery, Chmn.

GROWTH PLANS/SPECIAL FEATURES:

CRH plc is the parent company for an international group of companies engaging in the operation of builders' merchanting and do-it-yourself (DIY) stores as well as in the manufacturing and supplying of building materials. The firm focuses on three core businesses: materials, products and distribution. With operations in 31 countries, CRH is divided into six regionally focused business segments: Europe materials, Europe products, Europe distribution, Americas materials, Americas products and Americas distribution. The materials businesses deal with the firm's operations in the production and sale of primary materials such as ready mixed concrete, cement, aggregates, agricultural chemical lime and asphalt/bitumen. The products businesses include the production and sale of structural and architectural concrete products and associated construction accessories and services. Additionally, the products businesses deal with the firm's operations in the production and sale of exterior products such as fabrication and tempered glass products, clay products and inter-related services and products for the construction market. The distribution business deals with the firm's DIY store operations and builders' merchanting activities. The company owns stakes in several companies, including My Home Industries Ltd. and Jilin Yatai Group's cement operations. In August 2015, the firm acquired certain assets from Lafarge S.A. and Holcim Ltd.

FINANCIAL DATA: Note: Data for latest year may not have been available at press time.

In U.S. $	2015	2014	2013	2012	2011	2010
Revenue	26,955,670,000	21,569,100,000	20,564,320,000	21,280,550,000	20,621,340,000	19,585,770,000
R&D Expense						
Operating Income	1,456,416,000	1,045,837,000	114,049,800	963,720,800	993,373,700	796,067,500
Operating Margin %	5.40%	4.84%	.55%	4.52%	4.81%	4.06%
SGA Expense	6,801,930,000	5,209,795,000	5,265,679,000	4,776,406,000	4,538,041,000	4,520,934,000
Net Income	825,720,500	663,769,800	-337,587,400	629,554,900	672,893,800	492,695,100
Operating Cash Flow	2,562,699,000	1,410,796,000	1,245,424,000	1,169,010,000	1,170,151,000	1,586,433,000
Capital Expenditure	1,005,919,000	496,116,600	566,827,500	655,786,300	656,926,800	531,472,000
EBITDA	2,583,228,000	1,989,028,000	950,034,800	1,991,309,000	2,004,995,000	2,088,252,000
Return on Assets %	2.68%	2.74%	-1.42%	2.59%	2.75%	2.06%
Return on Equity %	6.24%	5.86%	-2.93%	5.24%	5.66%	4.32%
Debt to Equity	0.65	0.53	0.47	0.40	0.42	0.45

CONTACT INFORMATION:

Phone: 353-1-404-1000   Fax: 353-1-404-1007
Toll-Free: 800-899-8455
Address: Belgard Castle, Clondalkin, Dublin, 22 United Kingdom

STOCK TICKER/OTHER:

Stock Ticker: CRH
Employees: 75,706
Parent Company:

Exchange: NYS
Fiscal Year Ends: 12/31

SALARIES/BONUSES:

Top Exec. Salary: $1,471,242   Bonus: $1,654,863
Second Exec. Salary: $1,459,837   Bonus: $1,614,945

OTHER THOUGHTS:

Estimated Female Officers or Directors: 2

Hot Spot for Advancement for Women/Minorities: Y

Cushman & Wakefield Inc

www.cushmanwakefield.com

NAIC Code: 531210

TYPES OF BUSINESS:

Real Estate Brokerage
Property Management
Real Estate Documentation Web Site
Advisory Services
Research Services
Property Valuation

BRANDS/DIVISIONS/AFFILIATES:

DTZ
Cushman & Wakefield Sonnenblick Goldman

CONTACTS: *Note: Officers with more than one job title may be intentionally listed here more than once.*

Brett White, CEO
John Santora, COO
Edward C. Forst, Pres.
Duncan Palmer, CFO
Gene Boxer, Global General Counsel
James M. Underhill, CEO-Americas
Sanjay Verma, CEO-Asia Pacific
John Busi, Exec. VP-Valuation & Advisory
Brett White, Chmn.
Carlo Sant'Albano, CEO-Int'l

GROWTH PLANS/SPECIAL FEATURES:

Cushman & Wakefield, Inc. (C&W) is a global commercial real estate brokerage and services company. C&W provides advisory services related to asset buying, selling, financing and leasing, and also provides strategic planning, portfolio analysis and space location services. Agency and brokerage services assist clients with marketing and positioning properties through landlord representation, lease advisory, offices, retail services, supply chain, tenant representation and industrial solutions. Its capital markets group offers real estate buying, financing and investment clients various advisory services, including agency execution, investment management and corporate disposition practice. Global consulting services comprise market access, office platform, retail consulting and supply chain solutions for business customers and transaction and portfolio consulting for real estate customers. The practice groups division serves clients in select industries, including energy and resources, health care, hospitality, law firms, life sciences, mission critical facilities and Japanese multinational corporations. The research division offers property research, market analysis and forecasting solutions. The firm's valuation and advisory services include appraisal, portfolio valuation, development strategy, occupancy strategy and property tax services for real estate. Subsidiary Cushman & Wakefield Sonnenblick Goldman provides real estate financial services, specializing in debt structuring and debt and equity placement. C&W's corporate occupier and investor services manage real estate portfolios through account, transaction and project management, lease administration and facilities management. In 2015, the company was acquired by DTZ (formerly Cassidy Turley), and is now a global firm with 43,000 employees and revenues of $5 billion. The combined firms use the Cushman & Wakefield name.

C&W provides its employees with benefits such as medical, dental, vision, legal, disability and life insurance; domestic partner benefits; educational assistance; 401(k) plans; commuter program; and discounted gym memberships.

FINANCIAL DATA: *Note: Data for latest year may not have been available at press time.*

In U.S. $	2015	2014	2013	2012	2011	2010
Revenue	5,000,000,000	2,849,000,000	2,498,600,000	2,050,000,000	2,000,000,000	1,750,000,000
R&D Expense						
Operating Income						
Operating Margin %						
SGA Expense						
Net Income		61,600,000	46,200,000	18,202,800	14,900,000	25,700,000
Operating Cash Flow						
Capital Expenditure						
EBITDA						
Return on Assets %						
Return on Equity %						
Debt to Equity						

CONTACT INFORMATION:

Phone: 212-841-7500   Fax: 212-841-5002
Toll-Free:
Address: 1290 Avenue of the Americas, New York, NY 10104 United States

STOCK TICKER/OTHER:

Stock Ticker: Subsidiary
Employees: 43,000
Parent Company: DTZ

Exchange:
Fiscal Year Ends: 12/31

SALARIES/BONUSES:

Top Exec. Salary: $ Bonus: $
Second Exec. Salary: $ Bonus: $

OTHER THOUGHTS:

Estimated Female Officers or Directors: 2
Hot Spot for Advancement for Women/Minorities:

Sales, profits and employees may be estimates. Financial information, benefits and other data can change quickly and may vary from those stated here.

Cyrela Brazil Realty SA

NAIC Code: 236110

http://cyrela.globalri.com.br/en/

TYPES OF BUSINESS:

Residential Construction

BRANDS/DIVISIONS/AFFILIATES:

SKR Engenharia Ltda
MAC Construtora e Incorporadora
Plano & Plano Construcoes e Participacoes
Cury Construtora e Incorporadora
Cyrela
Living Construtora
Seller
Preference Cyrela

CONTACTS: *Note: Officers with more than one job title may be intentionally listed here more than once.*

Elie Horn, CEO
Eric Alencar, CFO
Rogerio J. Zylbersztajn, VP-Admin.
Claudio Carvalho de Lima, Corp. Legal Dept. Officer
Eric Alencar, Investor Rel. Officer
Cassio Mantelmacher, Real Estate Dev. Officer
Elie Horn, Chmn.

GROWTH PLANS/SPECIAL FEATURES:

Cyrela Brazil Realty SA Empreendimentos e Participacoes is one of Brazil's largest developers of residential properties. In its nearly 50 years of operation, the firm has built more than 56,000 units and has presence in 67 cities spread across 16 Brazilian states. The company operates primarily in Brazil, but also in Argentina and Uruguay. Its building segment is comprised of wholly-owned subsidiaries Living Construtora and Cyrela as well as its four primary joint ventures: MAC Construtora e Incorporadora, which builds middle and mid-high housing; SKR Engenharia Ltda, a mid-high and luxury home builder in Sao Paulo; and Cury Construtora e Incorporadora and Plano & Plano Construcoes e Participacoes, both of which build economic and super-economic housing. Living Construtora is the division's economic business that aims to reduce labor costs and improve construction by developing alternative methods of construction in assembly line fashion. Cyrela's sales segment is operated by subsidiary Seller, is the internal sales team for Cyrela and Living, and also provides information about customer desires when developing new projects. Seller is also responsible for the firm's online sales channel, and consists of a marketing team that develops campaigns for its products, projects and offerings via media outlets. The servicing segment is operated through Preference Cyrela and Facilities. Preference Cyrela is a service that allows customers to personalize their living quarters through pre-defined options and modifications, and Facilities offers a number of services to customers through specialized service providers such as concierge, guest welcoming, ordering flowers, arranging a mechanic for vehicle service and more.

FINANCIAL DATA: *Note: Data for latest year may not have been available at press time.*

In U.S. $	2015	2014	2013	2012	2011	2010
Revenue	1,193,814,000	1,599,914,000	1,477,253,000	1,548,971,000	1,684,848,000	1,344,765,000
R&D Expense						
Operating Income	189,859,200	268,509,200	283,216,900	271,798,800	200,962,200	205,908,100
Operating Margin %	15.90%	16.78%	19.17%	17.54%	11.92%	15.31%
SGA Expense	237,890,000	256,832,300	227,064,100	232,629,500	251,699,200	207,415,300
Net Income	123,143,800	181,910,400	197,676,800	181,538,900	162,849,500	188,391,300
Operating Cash Flow	178,469,900	284,921,600	73,553,790	138,899,700	-42,428,230	-67,006,660
Capital Expenditure	10,248,870	31,246,290	26,377,740	30,589,870	53,586,800	21,347,760
EBITDA	280,764,200	292,874,000	311,000,400	307,409,000	309,985,400	301,372,000
Return on Assets %	3.42%	4.75%	5.13%	4.88%	3.91%	5.31%
Return on Equity %	7.75%	11.89%	13.66%	13.74%	11.23%	14.67%
Debt to Equity	0.39	0.39	0.47	0.52	0.69	0.59

CONTACT INFORMATION:

Phone: 55-1121177279   Fax: 55 1121177323
Toll-Free:
Address: 1455 Presidente Juscelino Kubitscheck Ave., 3rd Fl, Sao Paulo, SP 04543-011 Brazil

STOCK TICKER/OTHER:

Stock Ticker: CYRBY
Employees:
Parent Company:

Exchange: PINX
Fiscal Year Ends: 12/31

SALARIES/BONUSES:

Top Exec. Salary: $ Bonus: $
Second Exec. Salary: $ Bonus: $

OTHER THOUGHTS:

Estimated Female Officers or Directors:
Hot Spot for Advancement for Women/Minorities:

Daelim Industrial Co Ltd

www.daelim.co.kr

NAIC Code: 237000

TYPES OF BUSINESS:

Heavy Construction and Engineering
Petrochemicals Distribution

BRANDS/DIVISIONS/AFFILIATES:

CONTACTS: *Note: Officers with more than one job title may be intentionally listed here more than once.*

Lee Chul Kyoon, CEO

GROWTH PLANS/SPECIAL FEATURES:

Daelim Industrial Co., Ltd. is a construction and engineering firm operating in three segments: building & housing, civil works and plants. The building & housing division is engaged in the construction, redevelopment and remodeling of offices and apartments; cultural and assembly facilities; commercial buildings, such as markets and department stores; educational facilities; medical and sports facilities, including the King Abdul Aziz University Hospital in Saudi Arabia; and others, such as the Seoul Court House Complex and the U.S. Embassy in Bangladesh. Some of the division's projects include constructing the main stadium for the 1988 Seoul Olympics, the Sejong Performing Arts Center and the main campus of the Arabian Gulf University in Bahrain as well as work on the Meyongdong Cathedral, the Daelim Contemporary Art Museum and the main building of the Bank of Korea. The civil works division has three broad categories of projects. The first combines expressways, airports, airfields and bridges. The second encompasses railroads, subways and tunnels. The third is engaged in the construction of dams, irrigation infrastructure and harbors and marine facilities. Some of the division's pending projects include the Sorok Bridge and Haeoreum Bridge in Korea. The plants segment constructs chemical plants, such as oil refineries and petrochemical processing facilities, implementing 483 projects in the Middle East, Southeast Asia, Africa and China. Daelem's nine affiliate firms are involved in the fields of engineering & construction, trading & logistics, motorcycle manufacturing, concrete manufacturing, hotel facility operations, systems integration and consulting, education and art.

FINANCIAL DATA: *Note: Data for latest year may not have been available at press time.*

In U.S. $	2015	2014	2013	2012	2011	2010
Revenue	8,306,102,486	7,263,028,818	9,512,626,947	9,767,300,000	7,087,900,000	7,086,410,000
R&D Expense						
Operating Income						
Operating Margin %						
SGA Expense						
Net Income	180,553,952	-437,069,808	-23,994,000	372,600,000	324,500,000	325,610,000
Operating Cash Flow						
Capital Expenditure						
EBITDA						
Return on Assets %						
Return on Equity %						
Debt to Equity						

CONTACT INFORMATION:

Phone: 82-2-2011-7114   Fax: 82-2-2011-8000
Toll-Free:
Address: 146-12 Susong-Dong, Jongno-Gu, Seoul, 110-732 South Korea

STOCK TICKER/OTHER:

Stock Ticker: 210
Employees: 4,158
Parent Company:

Exchange: Seoul
Fiscal Year Ends: 12/31

SALARIES/BONUSES:

Top Exec. Salary: $ Bonus: $
Second Exec. Salary: $ Bonus: $

OTHER THOUGHTS:

Estimated Female Officers or Directors:
Hot Spot for Advancement for Women/Minorities:

Dahlin Group Architecture Planning   www.dahlingroup.com
NAIC Code: 541310

TYPES OF BUSINESS:
Architectural Services
Architecture

BRANDS/DIVISIONS/AFFILIATES:

CONTACTS: *Note: Officers with more than one job title may be intentionally listed here more than once.*
Nancy Keenan, Pres.
Doug Dahlin, Chmn.

GROWTH PLANS/SPECIAL FEATURES:

Dahlin Group Architecture Planning is an architecture and planning practice with offices throughout California and China. It consists of more than 200 professionals that make up its team of architects and planners that craft memorable public and private places for companies, communities and individuals. Dahlin specializes in planning and urban design, architecture, sustainability and design visualization. Its portfolio includes the Marina Plaza (California, USA), a redevelopment plan that converts underutilized commercial/retail into an urban village; Linq at Spencer Square (Washington, USA), 94 modener residential flats and lots above 20,000 square feet of commercial space dedicated to health care services; and Jinshan Town Center (Guangzhou, China), a group of three luxury high-rise residential towers overlooking a commercial, retail and recreational plaza. Other completed projects include The Mountain Retreat, Enso, Dongguan Clubhous, Slater 116, Marion V. Ashley Community Center, Los Altos Gardens, Hangar A, Scenic Oaks and Renaissance ClubSport Aliso Viejo.

Dahlin Group offers its employees a comprehensive benefits package and participation in an employee stock ownership plan.

FINANCIAL DATA: *Note: Data for latest year may not have been available at press time.*

In U.S. $	2015	2014	2013	2012	2011	2010
Revenue	17,670,000					
R&D Expense						
Operating Income						
Operating Margin %						
SGA Expense						
Net Income						
Operating Cash Flow						
Capital Expenditure						
EBITDA						
Return on Assets %						
Return on Equity %						
Debt to Equity						

CONTACT INFORMATION:
Phone: 925-251-7200   Fax: 925-251-7201
Toll-Free:
Address: 5865 Owens Dr., Pleasanton, CA 94588 United States

STOCK TICKER/OTHER:
Stock Ticker: Private   Exchange:
Employees: 200   Fiscal Year Ends:
Parent Company:

SALARIES/BONUSES:
Top Exec. Salary: $ Bonus: $
Second Exec. Salary: $ Bonus: $

OTHER THOUGHTS:
Estimated Female Officers or Directors:
Hot Spot for Advancement for Women/Minorities:

Sales, profits and employees may be estimates. Financial information, benefits and other data can change quickly and may vary from those stated here.

Dalian Wanda Group Co Ltd

www.wanda-group.com

NAIC Code: 512131

TYPES OF BUSINESS:

Motion Picture Theaters (except Drive-Ins)
Motion Picture and Video Production
Motion Picture and Video Distribution
Performing Arts Companies
Amusement and Theme Parks
Hotels (except Casino Hotels) and Motels
Department Stores
Internet Publishing and Broadcasting and Web Search Portals

BRANDS/DIVISIONS/AFFILIATES:

Wanda Plazas
Qingdao Oriental Movie Metropolis
Continental Film Distribution
China Times
Popular Cinema
Wanda E-commerce
O2O e-commerce
99bill

CONTACTS: Note: Officers with more than one job title may be intentionally listed here more than once.

Wang Jianlin, Chmn.

GROWTH PLANS/SPECIAL FEATURES:

Dalian Wanda Group Co. Ltd. is a private property developer with registered capital totaling $7 billion. The company operates in four major industries: commercial property, culture & tourism, e-commerce and department stores. Dalian Wanda's commercial property division is the largest commercial real estate company in the world, comprising 125 Wanda Plazas and 81 hotels with a total gross floor area of 70 million square feet (21.57 million square meters). Properties under construction include 70 Wanda Plazas and 69 hotels, with a combined gross floor area of 57 million square feet (17.47 million square meters). The culture & tourism division comprises cinemas, film production, film industry parks, performing arts, film technology entertainment, theme parks, entertainment franchises, print media, art investment and travel. This segment operates 187 theatres, with a total 1,657 screens (including 117 IMAX screens); the Qingdao Oriental Movie Metropolis film and television industrial park, which plans to open itself to the public in 2017; film technology entertainment includes Continental Film Distribution, a film distribution, marketing and planning company that plays a key role in Wanda's film industry operations; and print media operates weekly magazines China Times and Popular Cinema. The e-commerce division comprises Wanda E-commerce, a joint venture with Tencent and Baidu, which is developing its O2O e-commerce platform, spanning the areas of film, parenting, dining, shopping, entertainment, leisure, tourism, lifestyle and finance; and 99bill, an independent third-party payment platform. The department store division operates 99 department stores in major cities such as Beijing, Shanghai, Chengdu and Wuhan.

FINANCIAL DATA: Note: Data for latest year may not have been available at press time.

In U.S. $	2015	2014	2013	2012	2011	2010
Revenue	38,800,000,000	37,339,946,212				
R&D Expense						
Operating Income						
Operating Margin %						
SGA Expense						
Net Income						
Operating Cash Flow						
Capital Expenditure						
EBITDA						
Return on Assets %						
Return on Equity %						
Debt to Equity						

CONTACT INFORMATION:

Phone: 86-10-85853888   Fax: 86-10-85853222
Toll-Free:
Address: Tower B, Wanda Plaza, No. 93, Jianguo Rd., Chaoyang District, Beijing, 100022 China

STOCK TICKER/OTHER:

Stock Ticker: Private   Exchange:
Employees: Fiscal Year Ends:
Parent Company:

SALARIES/BONUSES:

Top Exec. Salary: $ Bonus: $
Second Exec. Salary: $ Bonus: $

OTHER THOUGHTS:

Estimated Female Officers or Directors:
Hot Spot for Advancement for Women/Minorities:

Days Inn Worldwide Inc

www.daysinn.com

NAIC Code: 721110

TYPES OF BUSINESS:
Motels

BRANDS/DIVISIONS/AFFILIATES:
Wyndham Worldwide
Days Hotel
Days Inn & Suites
Days Inn Business Place
Days Inn
Wyndham Rewards Card

CONTACTS: Note: Officers with more than one job title may be intentionally listed here more than once.
Stephen P. Holmes, CEO
Clyde Guinn, Pres.
Stephen P. Holmes, Chmn.

GROWTH PLANS/SPECIAL FEATURES:
Days Inn Worldwide, Inc., a subsidiary of Wyndham Worldwide, operates one of the world's largest franchised hotel networks. Its operations span over 1,800 low-cost hotels located throughout the U.S. and around the world, including Canada, China, India, Jordan, Bahrain, Azerbaijan, Germany, Guam, Latvia, Philippines, Saudi Arabia, Singapore and the U.K. The franchised hotels come in four varieties: Days Inn, a standard roadside motel; Days Hotel, a slightly larger property, usually in an urban center or near an airport, that features restaurants, lounges and meeting and banquet rooms in addition to guest rooms; Days Inn & Suites, similar to Days Inn but providing larger rooms, often for guests who plan to stay for a longer duration; and Days Inn Business Place, inns offering specialized services catering to the needs of business travelers. Some universal amenities that the chain offers its guests include free high-speed internet, complimentary Daybreak breakfast, hairdryers, alarm clocks and complimentary copies of USA Today. On the company's web site, customers can make reservations online and access special promotions and other information. Additionally, Wyndham's hotel group offers a best-available-rate guarantee, providing its online customers with the lowest rates available for all its hotel brands, including Days Inn. It also offers the Wyndham Rewards Card, which allows customers to earn points at any of Wyndham's hotel chains, a total of more than 7,800 hotels.

FINANCIAL DATA: Note: Data for latest year may not have been available at press time.

In U.S. $	2015	2014	2013	2012	2011	2010
Revenue						
R&D Expense						
Operating Income						
Operating Margin %						
SGA Expense						
Net Income						
Operating Cash Flow						
Capital Expenditure						
EBITDA						
Return on Assets %						
Return on Equity %						
Debt to Equity						

CONTACT INFORMATION:
Phone: 973-428-9700   Fax: 973-496-7658
Toll-Free: 800-329-7466
Address: 1 Sylvan Way, Parsippany, NJ 07054 United States

STOCK TICKER/OTHER:
Stock Ticker: Subsidiary   Exchange:
Employees: Fiscal Year Ends: 12/31
Parent Company: Wyndham Worldwide

SALARIES/BONUSES:
Top Exec. Salary: $ Bonus: $
Second Exec. Salary: $ Bonus: $

OTHER THOUGHTS:
Estimated Female Officers or Directors:
Hot Spot for Advancement for Women/Minorities:

DCT Industrial Trust Inc

NAIC Code: 531120

www.dctindustrial.com

TYPES OF BUSINESS:

Industrial Property Leasing
REIT
Property Management Services

BRANDS/DIVISIONS/AFFILIATES:

Airport Distribution Center

CONTACTS: *Note: Officers with more than one job title may be intentionally listed here more than once.*

Philip Hawkins, CEO
Matthew Murphy, CFO
Thomas Wattles, Chairman of the Board
Mark Skomal, Chief Accounting Officer
Teresa Corral, Executive VP, Divisional
Charla Rios, Executive VP, Divisional
John Spiegleman, Executive VP
Neil Doyle, Managing Director, Geographical
John Pharris, Managing Director, Geographical
Michael Ruen, Managing Director, Geographical

GROWTH PLANS/SPECIAL FEATURES:

DCT Industrial Trust, Inc. is a real estate investment trust (REIT) specializing in the ownership, acquisition, development and management of bulk distribution and light industrial properties in the U.S. and Mexico. The company owns, manages or has under development 402 properties, with an average size of 158,000 square feet, leasing them to approximately 900 corporate customers. DCT seeks to acquire properties that have convenient access to major transportation arteries, proximity to densely populated markets and quality design standards that allow for easy reconfiguration of space. The firm maintains large numbers of properties in Southern California (46), Houston (39), Atlanta (34) and Chicago (37). It additionally has a smaller amount of properties in major metropolitan areas including Washington, D.C., Nashville, Dallas, Seattle and Chicago. DCT also manages and owns interests in properties through its institutional capital management program. In 2015, the firm acquired the Airport Distribution Center, a five-building industrial park totaling 691,000 square feet in the I-70/Northeast submarket of Denver.

The firm offers employees medical, vision and dental coverage; life insurance; an employee assistance plan; an employee wellness program; and flexible spending accounts for dependent care, parking, transportation and medical and dental expenses.

FINANCIAL DATA: *Note: Data for latest year may not have been available at press time.*

In U.S. $	2015	2014	2013	2012	2011	2010
Revenue	354,697,000	336,526,000	289,005,000	260,779,000	253,449,000	239,417,000
R&D Expense						
Operating Income	70,025,000	58,838,000	51,264,000	45,396,000	32,448,000	20,886,000
Operating Margin %	19.74%	17.48%	17.73%	17.40%	12.80%	8.72%
SGA Expense	34,577,000	29,079,000	28,010,000	26,064,000	25,925,000	25,262,000
Net Income	94,048,000	49,164,000	15,870,000	-15,086,000	-25,250,000	-37,830,000
Operating Cash Flow	200,508,000	169,994,000	152,893,000	119,683,000	106,966,000	91,002,000
Capital Expenditure	436,999,000	565,621,000	555,645,000	456,143,000	276,944,000	146,481,000
EBITDA	313,766,000	258,542,000	191,331,000	173,768,000	156,618,000	133,549,000
Return on Assets %	2.63%	1.44%	.48%	-.53%	-.93%	-1.40%
Return on Equity %	5.33%	2.94%	1.05%	-1.23%	-2.13%	-3.13%
Debt to Equity	0.84	0.78	0.91	1.00	1.03	0.97

CONTACT INFORMATION:

Phone: 303 597-2400   Fax: 303 228-2201
Toll-Free:
Address: 518 17th St., Ste. 800, Denver, CO 80202 United States

STOCK TICKER/OTHER:

Stock Ticker: DCT   Exchange: NYS
Employees: 143   Fiscal Year Ends: 12/31
Parent Company:

SALARIES/BONUSES:

Top Exec. Salary: $750,000   Bonus: $
Second Exec. Salary: Bonus: $
$425,000

OTHER THOUGHTS:

Estimated Female Officers or Directors: 3
Hot Spot for Advancement for Women/Minorities: Y

DDR Corp

www.ddr.com

NAIC Code: 531120

TYPES OF BUSINESS:

Real Estate Investment Trust
Shopping Centers
Business Centers
Property Management
Commercial Construction & Land Development

BRANDS/DIVISIONS/AFFILIATES:

CONTACTS: *Note: Officers with more than one job title may be intentionally listed here more than once.*

David Oakes, CEO
Luke Petherbridge, CFO
Terrance Ahern, Chairman of the Board
Christa Vesy, Executive VP
David Weiss, Secretary
Paul Freddo, Senior Executive VP, Divisional

GROWTH PLANS/SPECIAL FEATURES:

DDR Corp. is a self-administered and self-managed real estate investment trust (REIT). The company is in the business of acquiring, developing, owning, leasing and managing shopping centers and business centers. The firm currently owns and manages approximately 367 shopping centers in 37 states and Puerto Ricol. The company has 169 shopping centers that are owned through joint ventures. DDR also owns over 1,000 acres of undeveloped land and manages all of the properties in its portfolio. In total, the firm owns and manages more than 115 million square feet of gross leasable area. The company focuses on earning rent payments in properties already owned as well as the acquisition, development, redevelopment, renovation and expansion of income-producing real estate properties (primarily shopping centers). DDR's primary interest is in open-air shopping centers and lifestyle centers between 250,000 and 1.5 million square feet in size. The firm prefers properties with strong national tenant anchor stores such as Wal-Mart, PetSmart, Bed Bath & Beyond, TJX Companies and Kohl's. The Otto family is DDR's biggest individual shareholder.

Employees of the company receive benefits including an employee assistance program; health care coverage; disability, life and AD&D coverage; supplemental insurance; a 401(k); and flexible spending accounts.

FINANCIAL DATA: *Note: Data for latest year may not have been available at press time.*

In U.S. $	2015	2014	2013	2012	2011	2010
Revenue	1,028,071,000	985,675,000	888,788,000	800,375,000	771,018,000	803,069,000
R&D Expense						
Operating Income	-20,070,000	188,084,000	190,814,000	136,678,000	125,530,000	132,011,000
Operating Margin %	-1.95%	19.08%	21.46%	17.07%	16.28%	16.43%
SGA Expense	73,382,000	84,484,000	79,556,000	76,444,000	85,221,000	85,573,000
Net Income	-72,168,000	117,282,000	-10,175,000	-25,822,000	-15,854,000	-209,358,000
Operating Cash Flow	434,587,000	420,282,000	373,974,000	304,196,000	273,195,000	278,124,000
Capital Expenditure	481,745,000	260,897,000	210,709,000	586,904,000	217,861,000	164,391,000
EBITDA	412,177,000	679,161,000	557,598,000	483,887,000	433,249,000	320,857,000
Return on Assets %	-1.01%	.94%	- .48%	- .77%	- .70%	-3.10%
Return on Equity %	-2.89%	2.63%	-1.34%	-2.14%	-2.06%	-10.37%
Debt to Equity	1.65	1.53	1.51	1.47	1.53	1.69

CONTACT INFORMATION:

Phone: 216 755-5500   Fax: 216 755-1500
Toll-Free: 877-225-5337
Address: 3300 Enterprise Pkwy., Beachwood, OH 44122 United States

STOCK TICKER/OTHER:

Stock Ticker: DDR   Exchange: NYS
Employees: 576   Fiscal Year Ends: 12/31
Parent Company:

SALARIES/BONUSES:

Top Exec. Salary: $591,875   Bonus: $
Second Exec. Salary: $440,000   Bonus: $

OTHER THOUGHTS:

Estimated Female Officers or Directors: 3
Hot Spot for Advancement for Women/Minorities: Y

Desarrolladora Homex SAB de CV
www.homex.com.mx
NAIC Code: 236110

| TYPES OF BUSINESS: |
Home Building
Construction
Resort Communities

| BRANDS/DIVISIONS/AFFILIATES: |

GROWTH PLANS/SPECIAL FEATURES:

Desarrolladora Homex SAB de CV (Homex) is a homebuilding and construction company based in Culiacan, Mexico. The firm focuses primarily on building entry-level, middle-income and tourism housing throughout Mexico and entry-level affordable housing in Brazil. The firm's real estate operations include purchase of the plot of land, securing of permits and licenses, design, construction and trading of dwellings, as well as assistance for its clients to secure mortgaged loans. In June 2015, Homex filed the duly executed bankruptcy procedures in the Federal court, which was in accordance with the bankruptcy procedures accepted by the court in June 2014.

CONTACTS: *Note: Officers with more than one job title may be intentionally listed here more than once.*

Gerardo de Nicolas Gutierrez, CEO
Carlos J. Moctezuma Velasco, CFO
Ana Cristina Herrera Lasso Espinosa, VP-Human Resources & Social Responsibility
Ruben Izabal Gonzalez, VP-Construction
Alberto Menchaca Valenzuela, VP-Mexico Div.
Carolina Silva Sanchez, VP-Tourism Div.
Eustaquio Tomas De Nicolas Gutierrez, Chmn.

FINANCIAL DATA: *Note: Data for latest year may not have been available at press time.*

In U.S. $	2015	2014	2013	2012	2011	2010
Revenue	21,729,133	97,013,491	1,029,857,856	1,631,132,032	1,239,872,128	1,114,997,504
R&D Expense						
Operating Income						
Operating Margin %						
SGA Expense						
Net Income	1,058,139,784	-212,868,637	110,312,072	90,504,528	74,007,504	89,639,920
Operating Cash Flow						
Capital Expenditure						
EBITDA						
Return on Assets %						
Return on Equity %						
Debt to Equity						

CONTACT INFORMATION:
Phone: 52 6677585800   Fax:
Toll-Free: 800-224-6639
Address: Blvd. Alfonso Zaragoza, M. 2204 Norte, CuliacÃ¡n , Sinaloa, 80200 Mexico

STOCK TICKER/OTHER:
Stock Ticker: HOMEX   Exchange: MEX
Employees: 1,097   Fiscal Year Ends: 12/31
Parent Company:

SALARIES/BONUSES:
Top Exec. Salary: $ Bonus: $
Second Exec. Salary: $ Bonus: $

OTHER THOUGHTS:
Estimated Female Officers or Directors: 2
Hot Spot for Advancement for Women/Minorities:

Diamond Resorts Holdings LLC

www.diamondresorts.com

NAIC Code: 561599

TYPES OF BUSINESS:

Time-Share Resorts

BRANDS/DIVISIONS/AFFILIATES:

DiamondResorts.com
THE Club
Diamond Resorts Corporation

CONTACTS: Note: Officers with more than one job title may be intentionally listed here more than once.

David Palmer, CEO
C. Bentley, CFO
Stephen Cloobeck, Chairman of the Board
Lisa Gann, Chief Accounting Officer
Howard Lanznar, Chief Administrative Officer
Lowell Kraff, Director
Steven Bell, Executive VP, Divisional
Brian Garavuso, Executive VP
Michael Flaskey, Executive VP
Jared Finkelstein, Secretary
Ronan OGorman, Senior VP, Divisional

GROWTH PLANS/SPECIAL FEATURES:

Diamond Resorts Holdings LLC, operating through Diamond Resorts Corporation (DRC), is a hospitality and vacation ownership firm. It has a worldwide network of 379 vacation destinations located in 35 countries throughout the U.S., Hawaii, Canada, Mexico, the Caribbean, Central America, South America, Europe, Asia, Australia, New Zealand and Africa. This network consists of 109 resort properties with approximately 12,000 units managed by DRC; and 250 affiliated resorts and hotels, as well as 20 cruise itineraries which are not managed by the firm, nor do they carry its brand, but are part of the DRC network through THE Club. THE Club, a points-based vacation system with over 490,000 member-owners. Points, which are renewed annually, may be spent as currency on resort vacations, cruises, airline tickets and other travel purchases. Members may search for a resort by an activity or interest they wish to pursue, or by geographical area. Amenities vary by resort, but may include a full bath, kitchen, dishwasher, washer/dryer, satellite TV, fireplace, deck, pool, sauna, spa, beauty salon, fitness center, movie rentals, high-speed Internet, childcare, 24-hour reception and safe deposit boxes. Besides resort reservations, DRC directly offers flight, cruise and car rental reservations on its DiamondResorts.com site. In October 2015, the firm acquired vacation ownership business of Gold Key Resorts; and in January 2016, it acquired the vacation ownership business of Intrawest resort Club Group.

FINANCIAL DATA: Note: Data for latest year may not have been available at press time.

In U.S. $	2015	2014	2013	2012	2011	2010
Revenue	954,040,000	844,566,000	729,788,000	523,668,000	391,021,000	370,825,000
R&D Expense						
Operating Income	300,128,000	213,416,000	105,208,000	-667,000	69,331,000	-20,433,000
Operating Margin %	31.45%	25.26%	14.41%	-.12%	17.73%	-5.51%
SGA Expense	462,912,000	400,088,000	404,376,000	277,380,000	209,129,000	181,934,000
Net Income	149,478,000	59,457,000	-2,525,000	13,643,000	10,303,000	-19,159,000
Operating Cash Flow	175,894,000	118,058,000	-5,409,000	24,600,000	9,292,000	66,001,000
Capital Expenditure	35,318,000	17,950,000	15,204,000	14,335,000	6,276,000	5,553,000
EBITDA	334,645,000	199,163,000	120,063,000	114,347,000	96,762,000	58,668,000
Return on Assets %	8.37%	4.13%	-.22%	1.49%	1.23%	
Return on Equity %	55.53%	25.02%	-4.62%			
Debt to Equity	4.49	3.57	3.76			

CONTACT INFORMATION:

Phone: 702-804-8600   Fax: 702-304-7066
Toll-Free:
Address: 10600 W. Charleston Blvd., Las Vegas, NV 89135 United States

STOCK TICKER/OTHER:

Stock Ticker: DRII
Employees: 7,100
Parent Company:

Exchange: NYS
Fiscal Year Ends: 09/30

SALARIES/BONUSES:

Top Exec. Salary: $2,000,000   Bonus: $
Second Exec. Salary: $500,000   Bonus: $

OTHER THOUGHTS:

Estimated Female Officers or Directors:

Hot Spot for Advancement for Women/Minorities:

Diversicare Healthcare Services Inc

www.dvcr.com

NAIC Code: 623110

TYPES OF BUSINESS:

Nursing Care Facilities
Assisted Living Facilities

BRANDS/DIVISIONS/AFFILIATES:

Life Steps
Lighthouse

CONTACTS: *Note: Officers with more than one job title may be intentionally listed here more than once.*

Kelly Gill, CEO
James McKnight, CFO
Leslie Campbell, COO
Chad McCurdy, Director
Wallace Olson, Vice Chairman of the Board

GROWTH PLANS/SPECIAL FEATURES:

Diversicare Healthcare Services, Inc. provides long-term care services to nursing home patients and residents of assisted living facilities primarily in the Southeast. The company's operations consist of 55 nursing homes containing 6,556 licensed nursing beds. Facilities are located in the states of Alabama, Florida, Indiana, Kansas, Kentucky, Ohio, Tennessee, Texas and Missouri. The firm's nursing centers offer traditional nursing and social services typically provided in long-term care facilities, including skilled nursing health care services such as medication dispensing and plan of care development for each resident, nutrition services, recreational therapy, social services and housekeeping and laundry services. Additionally, Diversicare offers a variety of rehabilitative, nutritional, respiratory and other specialized ancillary services. These specialty services include rehabilitation therapy services, such as audiology, speech, occupational and physical therapies, which are provided through licensed therapists and registered nurses; and the provision of medical supplies, nutritional support, infusion therapies and related clinical services. The majority of these services are provided using internal resources and clinicians. Many of its nursing centers include specialty sub-units designed to meet the needs of special care patients, including facilities equipped with Life Steps units for patients requiring short-term rehabilitation following an incident such as a stroke, joint replacement or bone fracture; and facilities with Lighthouse units for advanced care for patients suffering from dementia related disorders. Other specialty care services include one adult day care facility and specialty programming for bariatric patients (generally, patients weighing more than 350 pounds). In 2015, the firm acquired two nursing centers in Kentucky and one in Kansas.

FINANCIAL DATA: *Note: Data for latest year may not have been available at press time.*

In U.S. $	2015	2014	2013	2012	2011	2010
Revenue	387,595,000	344,192,000	281,919,000	308,072,000	314,715,000	290,130,000
R&D Expense						
Operating Income	7,431,000	6,009,000	-4,659,000	-1,875,000	4,360,000	7,580,000
Operating Margin %	1.91%	1.74%	-1.65%	-.60%	1.38%	2.61%
SGA Expense	61,605,000	55,500,000	49,818,000	60,313,000	59,490,000	19,680,000
Net Income	1,624,000	4,733,000	-8,534,000	-3,046,000	1,367,000	3,849,000
Operating Cash Flow	3,277,000	2,981,000	-1,226,000	3,639,000	10,041,000	10,056,000
Capital Expenditure	4,646,000	5,494,000	5,293,000	4,850,000	13,357,000	6,363,000
EBITDA	15,294,000	13,082,000	2,313,000	4,888,000	10,851,000	13,280,000
Return on Assets %	1.21%	3.38%	-7.02%	-2.92%	.92%	3.32%
Return on Equity %	12.98%	45.39%	-70.16%	-17.61%	4.70%	16.73%
Debt to Equity	4.09	3.62	6.00	1.63	1.34	1.07

CONTACT INFORMATION:

Phone: 615-771-7575   Fax:
Toll-Free:
Address: 1621 Galleria Blvd., Brentwood, TN 37027 United States

STOCK TICKER/OTHER:

Stock Ticker: DVCR   Exchange: NAS
Employees: 6,100   Fiscal Year Ends: 12/31
Parent Company:

SALARIES/BONUSES:

Top Exec. Salary: $450,000   Bonus: $
Second Exec. Salary: Bonus: $
$300,000

OTHER THOUGHTS:

Estimated Female Officers or Directors: 2
Hot Spot for Advancement for Women/Minorities:

Dividend Capital Group LLC

www.dividendcapital.com

NAIC Code: 531120

TYPES OF BUSINESS:

Real Estate Investment
REIT Holding
Mutual Funds
Investment Advisory Services

BRANDS/DIVISIONS/AFFILIATES:

DCT Industrial Trust Inc
Dividend Capital Diversified Property Fund

CONTACTS: *Note: Officers with more than one job title may be intentionally listed here more than once.*

Jeffrey William Taylor, Pres.
Matthew Holberton, VP-Real Estate Finance

GROWTH PLANS/SPECIAL FEATURES:

Dividend Capital Group LLC (DCG) is a real estate investment holding company. Its subsidiaries are DCT Industrial Trust, Inc. (DCT); and Dividend Capital Diversified Property Fund (DCD). DCT is a public real estate investment trust (REIT) that owns, operates and develops primarily bulk distribution warehouses and light industrial properties in the U.S. and Mexico. DCT invests in industrial properties through active leasing and management, which allows the company to more easily control operating expenses and maintain its properties; and an ongoing strategy of recycling capital by disposing of underperforming assets. DCD is a private REIT with a diverse portfolio offering a sampling of multiple high-quality real estate assets that offers investors pooled access to investing and a daily net asset valuation. IIT is a private REIT with investments in industrial and distribution properties, which it leases to corporate customers. In 2015, the firm acquired South Cape Village, a grocery-anchored shopping center in Massachusetts; Bank of America Tower, located in Boca Raton, Florida; a retail center in Chester Springs, New Jersey; and a grocery-anchored shopping center in Tulsa, Oklahoma. It sold Deguigne, an office property located in Sunnyvale, California; and sold subsidiary Industrial Income Trust, Inc. to Global Logistic Properties Limited.

DCG offers its employees medical, dental and vision coverage; life, AD&D and short- and long-term disability insurance; parking and transportation plans; a health savings account; and a 401(k) plan.

FINANCIAL DATA: *Note: Data for latest year may not have been available at press time.*

In U.S. $	2015	2014	2013	2012	2011	2010
Revenue						
R&D Expense						
Operating Income						
Operating Margin %						
SGA Expense						
Net Income						
Operating Cash Flow						
Capital Expenditure						
EBITDA						
Return on Assets %						
Return on Equity %						
Debt to Equity						

CONTACT INFORMATION:

Phone: 303-228-2200   Fax: 303-228-0128
Toll-Free: 866-324-7348
Address: 518 17th St., Fl. 17, Denver, CO 80202 United States

STOCK TICKER/OTHER:

Stock Ticker: Private   Exchange:
Employees: 126   Fiscal Year Ends: 12/31
Parent Company:

SALARIES/BONUSES:

Top Exec. Salary: $ Bonus: $
Second Exec. Salary: $ Bonus: $

OTHER THOUGHTS:

Estimated Female Officers or Directors:
Hot Spot for Advancement for Women/Minorities:

DLF Limited

NAIC Code: 531100

TYPES OF BUSINESS:

Real Estate Development
Commercial Real Estate Management & Leasing
Hotels
Luxury Resorts

BRANDS/DIVISIONS/AFFILIATES:

DLF Utilities Limited

CONTACTS: Note: Officers with more than one job title may be intentionally listed here more than once.

T.C. Goyal, Managing Dir.
Ashok Kumar Tyagi, CFO
Kushal P. Singh, Chmn.
Pua Seck Guan, CEO-Int'l

GROWTH PLANS/SPECIAL FEATURES:

DLF Limited is an India-based real estate development and construction company. The firm has over 294 msf (million square feet) of planned projects and 47 msf of projects under construction. DLF is primarily engaged in the development of residential, commercial and retail properties. It is divided into two business units: development and annuity. The development unit develops commercial complexes and mid-income, luxury and super luxury residential communities. Products offered include condominiums, duplexes, row houses and apartments. The annuity business is engaged in the rental of office and retail properties. The firm's business model allows it to earn revenues from both development activities and the leasing of completed properties. The company hopes to continue to grow its business through projects in India's special economic zones (SEZs), which are specially legislated areas designed to encourage foreign investment and exports from the country. DLF is also one of the largest owners of wind power plants in India, managing plants with a combined installed capacity of 228.7 megawatts. The firm's clients have included GE, IBM, Microsoft, Canon, Citibank, Vertex, Hewitt, Fidelity Investments, WNS, Bank of America, Cognizant, Infosys, CSC, Symantec and Sapient. In 2015, the company, through its subsidiary, DLF Utilities Limited, agreed to sell its cinema exhibition business operated under the brand name of DT Cinemas to PVR Limited.

FINANCIAL DATA: Note: Data for latest year may not have been available at press time.

In U.S. $	2015	2014	2013	2012	2011	2010
Revenue	1,149,614,000	1,247,207,000	1,367,103,000	1,447,308,000	1,524,724,000	1,180,001,000
R&D Expense						
Operating Income	94,186,100	78,268,560	121,122,600	500,488,800	300,633,300	376,448,700
Operating Margin %	8.19%	6.27%	8.85%	34.58%	19.71%	31.90%
SGA Expense				201,368,300	226,649,200	70,156,840
Net Income	81,197,780	97,125,600	107,002,400	180,484,800	231,583,200	270,899,700
Operating Cash Flow	306,136,500	220,576,900	301,929,400	378,721,900	414,377,200	1,296,772,000
Capital Expenditure	130,638,200	142,684,600	198,296,300	86,548,420	165,523,200	2,090,327,000
EBITDA	477,653,600	180,662,700	245,682,400	673,772,500	403,005,900	425,286,600
Return on Assets %	.82%	1.00%	1.11%	1.88%	2.45%	3.27%
Return on Equity %	2.50%	2.99%	2.60%	4.64%	5.61%	6.64%
Debt to Equity	0.64	0.86	0.56	0.61	0.97	0.71

CONTACT INFORMATION:

Phone: 91-11-4210-2030   Fax:
Toll-Free:
Address: DLF Centre, Sansad Marg, New Delhi, 110 001 India

STOCK TICKER/OTHER:

Stock Ticker: DSFQY   Exchange: GREY
Employees: 2,180   Fiscal Year Ends: 03/31
Parent Company:

SALARIES/BONUSES:

Top Exec. Salary: $ Bonus: $
Second Exec. Salary: $ Bonus: $

OTHER THOUGHTS:

Estimated Female Officers or Directors:
Hot Spot for Advancement for Women/Minorities:

Dominion Homes Inc

www.dominionhomes.com

NAIC Code: 236117

TYPES OF BUSINESS:

Construction, Home Building and Residential
Financial Services
Residential Sales

BRANDS/DIVISIONS/AFFILIATES:

GROWTH PLANS/SPECIAL FEATURES:

Dominion Homes, Inc. provides and services home warranties to new home buyers. During a customer's new home orientation and walk-through, Dominion provides a packet of warranty information, and at closing, provides the Home Care How To homeowner manual. These are resources that contain detailed information about the care and keeping of the new home including recommended seasonal maintenance schedules and how-tos as well as information regarding what the home warranty covers and how to request service. These services exist for those who purchased a Dominion home between January 2010 and August 2014 in Central Ohio, as well as Louisville and Lexington, Kentucky. The company is owned by PulteGroup, Inc.

Dominion offers its employees benefits including life, accidental death, medical, dental and vision insurance; flexible spending accounts; a 401(k); training/development programs; tuition assistance; and home purchase discounts.

CONTACTS: Note: Officers with more than one job title may be intentionally listed here more than once.

Keith Tomlinson, CEO
William G. Cornely, COO

FINANCIAL DATA: Note: Data for latest year may not have been available at press time.

In U.S. $	2015	2014	2013	2012	2011	2010
Revenue	181,000,000	170,000,000	168,000,000	111,000,000		
R&D Expense						
Operating Income						
Operating Margin %						
SGA Expense						
Net Income						
Operating Cash Flow						
Capital Expenditure						
EBITDA						
Return on Assets %						
Return on Equity %						
Debt to Equity						

CONTACT INFORMATION:

Phone: 614-356-5000   Fax: 614-356-6010
Toll-Free:
Address: 4900 Tuttle Crossing Blvd., Dublin, OH 43016 United States

STOCK TICKER/OTHER:

Stock Ticker: Private   Exchange:
Employees: 75   Fiscal Year Ends: 12/31
Parent Company: PulteGroup Inc

SALARIES/BONUSES:

Top Exec. Salary: $ Bonus: $
Second Exec. Salary: $ Bonus: $

OTHER THOUGHTS:

Estimated Female Officers or Directors:
Hot Spot for Advancement for Women/Minorities:

Doosan Heavy Industry & Construction Co www.doosanheavy.com

NAIC Code: 237130

TYPES OF BUSINESS:

Power and Communication Line and Related Structures Construction
Power Plant Construction
Desalination Plant Construction
Civil Works Projects
Architecture Works Projects
Casting & Forging
Material Handling Systems

BRANDS/DIVISIONS/AFFILIATES:

Doosan Power Systems
Doosan Power Systems Europe
Doosan Skoda Power
Doosan IMGB
Doosan Babcock

CONTACTS: Note: Officers with more than one job title may be intentionally listed here more than once.

Geewon Park, CEO
Ji Taik Chung, COO
Ji Taik Chung, Vice Chmn.
Yong-Sung Park, Chmn.

GROWTH PLANS/SPECIAL FEATURES:

Doosan Heavy Industry & Construction Co., established in 1962, has built over 300 thermal, coal-fired, nuclear and combined cycle power plants. It is also developing water treatment plants, wind power systems, fuel cells and other environmentally friendly energy sources. Doosan built the Shuaibah Desalination Plant Phase 3 in Saudi Arabia, one of the world's largest desalination facilities, with a daily capacity of 880,000 tons. The company also built one of the world's first hybrid desalination plants, located in Fujairah, UAE. The plant combines two desalination technologies: Multi-Stage Flash (MSF), which desalinates through a repeated cycle of evaporation and condensation; and Reverse Osmosis (RO), which forces the water across a filtering membrane. The firm is one of a few with proprietary technologies in all three areas of desalination: MSF, RO and MED (Multi-Effect Distillation, where the heat from boiling one chamber is used to heat another). In addition to constructing plants, Doosan produces material handling equipment; supplies forging and casting products; and constructs highways, ports, high-speed railroads, tunnels, subways, apartments and power transmission lines. Its casting and forging products include rotor shafts and turbine casings for nuclear and thermal power plants; shells for reactors in nuclear power plants; crank shafts, marine shafts and stern frame casting for ships; components for steel mills, such as mill housing and rolls; die steel, used in the manufacture of plastic products; tool steel, used to manufacture machine tools; and shells for petrochemical applications. Material handling equipment includes container handling cranes and bulk material handling systems. Firm's subsidiaries include Doosan Power Systems; Doosan Power Systems Europe; Doosan Skoda Power; Doosan IMGB; and Doosan Babcock.

FINANCIAL DATA: Note: Data for latest year may not have been available at press time.

In U.S. $	2015	2014	2013	2012	2011	2010
Revenue	14,059,352,557	15,594,355,153	18,752,304,000	7,432,379,904	6,143,038,976	5,690,646,528
R&D Expense						
Operating Income						
Operating Margin %						
SGA Expense						
Net Income	-1,519,132,785	-44,669,102	67,587,000	37,045,000	158,038,368	102,428,968
Operating Cash Flow						
Capital Expenditure						
EBITDA						
Return on Assets %						
Return on Equity %						
Debt to Equity						

CONTACT INFORMATION:

Phone: 82 552786114   Fax: 82 552645551
Toll-Free:
Address: 22 DoosanVolvo-ro, Seongsan-Gu, Changwon, Gyeongnam 642-792 South Korea

STOCK TICKER/OTHER:

Stock Ticker: 34020   Exchange: Korea
Employees: 8,400   Fiscal Year Ends:
Parent Company:

SALARIES/BONUSES:

Top Exec. Salary: $ Bonus: $
Second Exec. Salary: $ Bonus: $

OTHER THOUGHTS:

Estimated Female Officers or Directors:
Hot Spot for Advancement for Women/Minorities:

Doyle Collection (The)
NAIC Code: 721110

www.doylecollection.com

TYPES OF BUSINESS:
Hotels
Hotel Management

BRANDS/DIVISIONS/AFFILIATES:
Dupont Circle Hotel (The)
Westbury Hotel (The)
Croke Park Hotel (The)
River Lee Hotel (The)
Marylebone Hotel (The)
Bloomsbury Hotel (The)
Kensington Hotel (The)
Bristol Hotel (The)

CONTACTS: Note: Officers with more than one job title may be intentionally listed here more than once.
Patrick King, CEO
Seamus Daly, Corp. Sec.
Bernadette C. Gallagher, Chmn.

GROWTH PLANS/SPECIAL FEATURES:
The Doyle Collection (TDC) is an international hotel operator headquartered in Dublin, Ireland. It has properties in Ireland, the U.K. and the U.S. The company offers business and leisure travelers a portfolio of eight luxury hotels in five cities. The firm's Irish properties are located in Dublin (The Westbury Hotel and The Croke Park Hotel) and Cork (The River Lee Hotel), its U.K. properties are located in London (The Bloomsbury Hotel, The Kensington Hotel and the Marylebone Hotel) and Bristol (The Bristol Hotel), and its U.S. property is located in Washington, D.C. (The Dupont Circle Hotel). Hotels are generally located centrally in their respective cities, giving corporate clients easy access to local business and financial districts. Additionally, the firm's hotels offer corporate hospitality, conference, event and meeting suites to accommodate meetings and events of various sizes. Several of the group's hotels have professional wedding planners on staff to help coordinate wedding and reception activities.

The Doyle Collection offers its U.S. employees discounted health insurance, employee discounts and educational support.

FINANCIAL DATA: Note: Data for latest year may not have been available at press time.
In U.S. $	2015	2014	2013	2012	2011	2010
Revenue	200,000,000	195,000,000	191,000,000	180,000,000	169,523,000	157,197,000
R&D Expense						
Operating Income						
Operating Margin %						
SGA Expense						
Net Income						
Operating Cash Flow						
Capital Expenditure						
EBITDA						
Return on Assets %						
Return on Equity %						
Debt to Equity						

CONTACT INFORMATION:
Phone: 353-1-607-0070   Fax: 353-1-667-2370
Toll-Free:
Address: 146 Pembroke Rd., Ballsbridge, Dublin, 4 United Kingdom

STOCK TICKER/OTHER:
Stock Ticker: Private
Employees: 1,400
Parent Company:
Exchange:
Fiscal Year Ends: 12/31

SALARIES/BONUSES:
Top Exec. Salary: $ Bonus: $
Second Exec. Salary: $ Bonus: $

OTHER THOUGHTS:
Estimated Female Officers or Directors: 1
Hot Spot for Advancement for Women/Minorities:

DR Horton Inc

NAIC Code: 236117

www.drhorton.com

TYPES OF BUSINESS:

Construction, Home Building and Residential
Mortgages
Title Insurance

BRANDS/DIVISIONS/AFFILIATES:

DHI Mortgage

CONTACTS: Note: Officers with more than one job title may be intentionally listed here more than once.

Thomas Montano, Assistant Secretary
Bill Wheat, CFO
Donald Horton, Chairman of the Board
Michael Murray, COO
David Auld, President

GROWTH PLANS/SPECIAL FEATURES:

D.R. Horton, Inc. is a leading national builder of single-family homes with a diversified set of holdings and operating divisions in 27 states and 79 metropolitan markets. The firm generally builds homes between 1,000 to 4,000 square feet, ranging in price from $100,000 to over 1 million. In 2015, the company closed approximately 36,736 homes, with an average closing sales price approximating $285,700. The company is divided into six regional homebuilding segments and one financial services segment. The homebuilding segments are East, operating in eight states; Midwest, four states; Southeast, five states; South Central, three states; Southwest, two states; and West, six states. The building services section constructs residences, tailored to the particular community where they are being built, including single-family residential homes, townhouses, condominiums, duplexes and triplexes. Detached homes sales accounted for 90% of the firm's 2015 revenues. Subcontractors under the supervision of D. R. Horton do substantially all of the actual building. The financial services segment of the company provides mortgage financing and title insurance through its wholly-owned subsidiary, DHI Mortgage. The home builder's current business strategy is to enter into new lot option contracts to purchase finished lots in selected communities to potentially increase sales volumes and profitability. The firm plans to renegotiate existing lot option contracts as necessary to reduce lot costs and better match the scheduled lot purchases with new home demand in each community. The company also manages inventory of homes under construction by selectively starting construction on unsold homes to capture new home demand while monitoring the number and aging of unsold homes and aggressively marketing its unsold, completed homes in inventory. In 2015, the company acquired the homebuilding operations of Pacific Ridge Homes in Seattle, Washington.

FINANCIAL DATA: Note: Data for latest year may not have been available at press time.

In U.S. $	2015	2014	2013	2012	2011	2010
Revenue	10,824,000,000	8,024,900,000	6,259,300,000	4,354,000,000	3,636,800,000	4,400,200,000
R&D Expense						
Operating Income	1,092,500,000	790,900,000	654,400,000	242,900,000	12,100,000	91,400,000
Operating Margin %	10.09%	9.85%	10.45%	5.57%	.33%	2.07%
SGA Expense	1,186,000,000	965,400,000	766,300,000	614,100,000	556,300,000	599,200,000
Net Income	750,700,000	533,500,000	462,700,000	956,300,000	71,800,000	245,100,000
Operating Cash Flow	700,400,000	-661,400,000	-1,231,100,000	-298,100,000	14,900,000	709,400,000
Capital Expenditure	56,100,000	100,200,000	58,000,000	33,600,000	16,300,000	19,200,000
EBITDA	1,177,500,000	852,600,000	685,600,000	288,600,000	83,900,000	204,900,000
Return on Assets %	7.03%	5.59%	5.74%	15.17%	1.27%	3.86%
Return on Equity %	13.63%	11.63%	12.09%	30.78%	2.74%	10.05%
Debt to Equity	0.64	0.71	0.86	0.69	0.65	0.83

CONTACT INFORMATION:

Phone: 817-390-8200   Fax: 817 856-8429
Toll-Free:
Address: 301 Commerce St., Ste. 500, Fort Worth, TX 76102 United States

STOCK TICKER/OTHER:

Stock Ticker: DHI
Employees: 6,230
Parent Company:

Exchange: NYS
Fiscal Year Ends: 09/30

SALARIES/BONUSES:

Top Exec. Salary: $500,000   Bonus: $950,000
Second Exec. Salary: $1,000,000   Bonus: $

OTHER THOUGHTS:

Estimated Female Officers or Directors: 1
Hot Spot for Advancement for Women/Minorities:

Duke Realty Corp

www.dukerealty.com

NAIC Code: 0

TYPES OF BUSINESS:
Real Estate Investment Trust
Real Estate Development
Commercial Real Estate
Construction & Design Services
Commercial Construction
Warehouses

BRANDS/DIVISIONS/AFFILIATES:
Duke Realty LP
Duke Realty Services LLC
Duke Construction LP
Duke Realty Services LP

CONTACTS: Note: Officers with more than one job title may be intentionally listed here more than once.
Mark Denien, CFO
Dennis Oklak, Chairman of the Board
Steven Kennedy, Executive VP, Divisional
Ann Dee, General Counsel
James Connor, President

GROWTH PLANS/SPECIAL FEATURES:
Duke Realty Corp. is a self-administered and self-managed real estate investment trust (REIT) that operates in 21 major U.S. cities, including Atlanta, Chicago, Dallas, Nashville, Orlando and Phoenix. The firm provides in-house leasing, management, development and construction services. Its diversified portfolio consists of 587 rental properties including office, industrial, health care and retail rental properties totaling 142.6 million square feet, including 70 jointly controlled in-service properties. The firm's portfolio of industrial properties totals 130.5 million square feet and consists of bulk warehouses and service center properties also known as flex buildings or light industrial properties. Office properties owned by the firm total 5.5 million square feet and are located primarily in suburban locations. Duke also owns interests in approximately 6.6 million square feet of health care and retail buildings. Additionally, the firm owns and controls through purchase options approximately 3,200 acres of land. The company conducts rental operations through subsidiary Duke Realty LP. It conducts service operations through Duke Realty Services LLC, Duke Construction LP and Duke Realty Services LP.

The firm offers employees a 401 (k) plan; short- and long-term disability; medical, dental and vision insurance; flexible spending accounts; an employee assistance program; professional certification programs; adoption assistance plans; adoption and paternity leave; matching gifts program; and scholarship programs.

FINANCIAL DATA: Note: Data for latest year may not have been available at press time.

In U.S. $	2015	2014	2013	2012	2011	2010
Revenue	949,432,000	1,164,704,000	1,081,790,000	1,109,440,000	1,274,274,000	1,393,603,000
R&D Expense						
Operating Income	448,396,000	444,874,000	296,000,000	160,959,000	219,352,000	227,728,000
Operating Margin %	47.22%	38.19%	27.36%	14.50%	17.21%	16.34%
SGA Expense	58,565,000	49,362,000	42,673,000	55,253,000	43,107,000	50,532,000
Net Income	615,310,000	243,588,000	190,592,000	-73,977,000	95,565,000	65,798,000
Operating Cash Flow	379,381,000	444,487,000	435,676,000	299,157,000	337,537,000	391,156,000
Capital Expenditure	501,096,000	738,926,000	1,041,322,000	393,583,000	275,424,000	711,070,000
EBITDA	676,180,000	829,150,000	678,811,000	533,246,000	549,272,000	616,797,000
Return on Assets %	8.38%	2.64%	1.99%	-1.73%	.42%	-.18%
Return on Equity %	20.36%	7.55%	6.75%	-6.49%	1.58%	-.71%
Debt to Equity	1.05	1.55	1.65	2.26	1.98	2.06

CONTACT INFORMATION:
Phone: 317-808-6000   Fax: 317 808-6794
Toll-Free: 800-875-3366
Address: 600 E. 96th St., Ste. 100, Indianapolis, IN 46240 United States

STOCK TICKER/OTHER:
Stock Ticker: DRE
Employees: 750
Parent Company:

Exchange: NYS
Fiscal Year Ends: 12/31

SALARIES/BONUSES:
Top Exec. Salary: $844,615   Bonus: $
Second Exec. Salary: $490,385   Bonus: $

OTHER THOUGHTS:
Estimated Female Officers or Directors: 5
Hot Spot for Advancement for Women/Minorities: Y

Dynex Capital Inc

www.dynexcapital.com

NAIC Code: 525990

TYPES OF BUSINESS:
Mortgage REIT
Real Estate Investment Trust
Mortgage-Backed Securities
Commercial Mortgage Loans

BRANDS/DIVISIONS/AFFILIATES:

GROWTH PLANS/SPECIAL FEATURES:

Dynex Capital, Inc. is a real estate investment trust (REIT). The firm invests in loans and mortgage-backed securities (MBS) issued or guaranteed by a federally-chartered corporation or an agency of the U.S. government. These securities are referred to as Agency MBS. The company also invests in commercial mortgage-backed securities (CMBS) and non-agency residential mortgage-backed securities, including CMBS interest-only (IO) securities. Dynex's investments are typically financed through a combination of repurchase agreements, securitization financing and equity capital. The firm's investment policy allocates its capital among short-duration, high-grade Agency MBS with less exposure to credit risk, interest rate risk and liquidity risk as well as CMBS rated A or better by one of the nationally recognized rating services. Dynex's primary source of income is net interest income, which is the excess of the interest income earned on its investments over the cost of financing them.

CONTACTS:
Note: Officers with more than one job title may be intentionally listed here more than once.

Byron Boston, CEO
Stephen Benedetti, CFO
Thomas Akin, Chairman of the Board
Jeffrey Childress, Controller
Smriti Popenoe, Executive VP

FINANCIAL DATA:
Note: Data for latest year may not have been available at press time.

In U.S. $	2015	2014	2013	2012	2011	2010
Revenue	34,212,000	43,813,000	81,388,000	86,970,000	58,879,000	37,610,000
R&D Expense						
Operating Income	16,544,000	27,806,000	68,069,000	74,042,000	39,812,000	27,414,000
Operating Margin %	48.35%	63.46%	83.63%	85.13%	67.61%	72.89%
SGA Expense	17,668,000	16,007,000	13,058,000	12,736,000	9,956,000	8,817,000
Net Income	16,544,000	27,806,000	68,069,000	74,042,000	39,812,000	29,472,000
Operating Cash Flow	216,988,000	214,551,000	208,760,000	149,388,000	70,641,000	30,071,000
Capital Expenditure						
EBITDA						
Return on Assets %	.20%	.47%	1.41%	2.09%	1.88%	2.02%
Return on Equity %	1.67%	3.82%	11.59%	15.44%	11.99%	12.59%
Debt to Equity	1.38	0.02	0.02		0.19	0.36

CONTACT INFORMATION:
Phone: 804 217-5800   Fax: 804 217-5860
Toll-Free:
Address: 4991 Lake Brook Dr., Ste. 100, Glen Allen, VA 23060 United States

STOCK TICKER/OTHER:
Stock Ticker: DX
Employees: 17
Parent Company:

Exchange: NYS
Fiscal Year Ends: 12/31

SALARIES/BONUSES:
Top Exec. Salary: $375,000   Bonus: $471,558
Second Exec. Salary: $ Bonus: $

OTHER THOUGHTS:
Estimated Female Officers or Directors: 1
Hot Spot for Advancement for Women/Minorities:

EastGroup Properties Inc

www.eastgroup.net

NAIC Code: 0

TYPES OF BUSINESS:

Real Estate Investment Trust
Industrial Properties
Property Management

BRANDS/DIVISIONS/AFFILIATES:

CONTACTS: *Note: Officers with more than one job title may be intentionally listed here more than once.*

Bruce Corkern, Assistant Secretary
Leland Speed, Chairman Emeritus
David Hoster, Director
N. Mckey, Executive VP
Marshall Loeb, President
Brent Wood, Senior VP
John Coleman, Senior VP
William Petsas, Senior VP

GROWTH PLANS/SPECIAL FEATURES:

EastGroup Properties, Inc. is an equity real estate investment trust (REIT) focused on the acquisition, operation and development of industrial properties in major Sunbelt markets throughout the U.S., with concentration in the states of Florida, Texas, Arizona, North Carolina, Mississippi and California. The company acquires distribution facilities in the 5,000 to 50,000 square foot range, generally clustered near major transportation routes in supply constrained markets. Business distribution space properties are typically multi-tenant buildings with a building depth of 200 feet or less, clear height of 20-24 feet, office finish of 10%-25% and truck courts with a depth of 100-120 feet. Approximately 81% of the firm's properties are distribution centers. In addition to direct property acquisitions and developments, the firm seeks to expand its portfolio through the acquisition of other public and private real estate companies and REITs. During 2015, EastGroup increased its holdings in real estate properties through its acquisition and development programs. The company purchased two warehouse distribution complexes (totalling 335,000 square feet) and 112.6 acres of development land for a total of $50.9 million. Also during 2015, the company began construction of 11 development projects containing 1,283,000 square feet and transferred 17 properties (1,419,000 square feet) from its development program to real estate properties with costs of $96.8 million at the date of transfer. Over 99% of the firm's revenue consists of rental income from real estate properties.

FINANCIAL DATA: *Note: Data for latest year may not have been available at press time.*

In U.S. $	2015	2014	2013	2012	2011	2010
Revenue	235,008,000	219,829,000	202,171,000	186,178,000	174,631,000	173,126,000
R&D Expense						
Operating Income	79,061,000	73,782,000	66,581,000	60,813,000	56,826,000	53,302,000
Operating Margin %	33.64%	33.56%	32.93%	32.66%	32.54%	30.78%
SGA Expense	15,091,000	12,726,000	11,725,000	10,488,000	10,691,000	10,332,000
Net Income	47,866,000	47,941,000	32,615,000	32,384,000	22,359,000	18,325,000
Operating Cash Flow	131,385,000	117,401,000	109,750,000	91,808,000	86,547,000	76,858,000
Capital Expenditure	151,120,000	167,025,000	169,444,000	125,289,000	149,788,000	56,771,000
EBITDA	156,355,000	154,273,000	133,449,000	123,543,000	114,994,000	112,276,000
Return on Assets %	2.95%	3.14%	2.30%	2.45%	1.81%	1.55%
Return on Equity %	8.50%	8.83%	6.52%	7.28%	5.50%	4.29%
Debt to Equity	1.86	1.63	1.73	1.67	2.06	1.79

CONTACT INFORMATION:

Phone: 601 354-3555   Fax: 601 352-1441
Toll-Free:
Address: 190 E. Capital St., Ste. 400, Jackson, MS 39201 United States

STOCK TICKER/OTHER:

Stock Ticker: EGP   Exchange: NYS
Employees: 73   Fiscal Year Ends: 12/31
Parent Company:

SALARIES/BONUSES:

Top Exec. Salary: $605,000   Bonus: $
Second Exec. Salary: Bonus: $
$360,000

OTHER THOUGHTS:

Estimated Female Officers or Directors: 4
Hot Spot for Advancement for Women/Minorities: Y

E-LOAN Inc

NAIC Code: 522310

www.eloan.com

TYPES OF BUSINESS:

Online Mortgage Broker
Debt Consolidation
Small Business Loans
Credit Cards
Education Loans

BRANDS/DIVISIONS/AFFILIATES:

Popular Inc
Banco Popular North America

CONTACTS: Note: Officers with more than one job title may be intentionally listed here more than once.

MarkE. Lefanowicz, CEO
Mark Lefanowicz, Pres.
Darren Nelson, CFO
Etienne Handman, CTO

GROWTH PLANS/SPECIAL FEATURES:

E-LOAN, Inc. is an online provider of a full range of mortgages; credit cards; and home equity, auto, education and personal loans. It is the wholly-owned subsidiary of Banco Popular North America, a New-York state-chartered bank, which is itself a subsidiary of Popular, Inc. The E-LOAN web site offers a variety of services for borrowers, including comparisons of loans from the nation's leading lenders, tools for managing debt, credit reports and scores, exclusive 24-hour loan status access, automatic notification regarding new products that meet specific customer needs and a variety of other services. E-LOAN originates loans through its web site, funds the loans using warehouse and other lines of credit and then sells closed loans. The gain on the sale of these loans is the company's primary source of income. The firm showcases a low-cost guarantee, assuring customers that its combination of low rate, points and lender or broker fees is the lowest-cost loan available. Products offered by E-LOAN include fixed-rate mortgage products, including 30 year and 15 year terms; adjustable rate mortgage products, including 10, 7, 5 and 3 year terms; stated income mortgage products, for individuals who are self-employed or write off a large portion of their income; and home equity lines of credit, home equity loans and installment vehicle loans. In addition, E-LOAN offers federally insured Federal Housing Administration (FHA) loans to assist consumers facing a tight credit market. The company has partnerships with lenders and financial service providers, including lendingtree, LendingClub, RoadLoans and Consolidated Credit Counseling Services, Inc.

FINANCIAL DATA: Note: Data for latest year may not have been available at press time.

In U.S. $	2015	2014	2013	2012	2011	2010
Revenue						
R&D Expense						
Operating Income						
Operating Margin %						
SGA Expense						
Net Income						
Operating Cash Flow						
Capital Expenditure						
EBITDA						
Return on Assets %						
Return on Equity %						
Debt to Equity						

CONTACT INFORMATION:

Phone: 847-994-5800   Fax:
Toll-Free:
Address: 85 Borad Street, 10/Fl, New York, NY 10004 United States

STOCK TICKER/OTHER:

Stock Ticker: Subsidiary   Exchange:
Employees: 930   Fiscal Year Ends: 12/31
Parent Company: POPULAR INC

SALARIES/BONUSES:

Top Exec. Salary: $ Bonus: $
Second Exec. Salary: $ Bonus: $

OTHER THOUGHTS:

Estimated Female Officers or Directors:
Hot Spot for Advancement for Women/Minorities:

Emaar Properties PJSC

www.emaar.com

NAIC Code: 531100

TYPES OF BUSINESS:

Real Estate Development
Retail Construction
Residential Construction
Education & Health Care Construction
Investments

BRANDS/DIVISIONS/AFFILIATES:

Emaar Square
Emaar Malls Group LLC
Emaar Hotels and Resorts LLC
Emaar Hospitality Group LLC
Emaar International
Emaar United Arab Emirates

CONTACTS: *Note: Officers with more than one job title may be intentionally listed here more than once.*

Ahmed Thani Al Matrooshi, Managing Dir.
Amit Jain, COO
Rasha Hassan, Chief Commercial Officer-Sales
Clare Elliott, Human Resources
Ayman Hamdy, Company Sec.
Abdulla Lahej, Group CEO
Nasser Rafi, CEO-Emaar Malls Group LLC
Maitha Al Dossari, CEO-Emaar Retail LLC
Robert Booth, CEO-Emaar Dubai Real Estate
H.E. Mohamed Ali Rashed Alabbar, Chmn.
Fred Durie, CEO-Emaar Int'l

GROWTH PLANS/SPECIAL FEATURES:

Emaar Properties PJSC is a Dubai-based real estate developer whose projects are primarily located in Dubai with operations spanning the Middle East, North Africa, Asia, Europe and North America. Emaar Properties' operating divisions include Emaar United Arab Emirates, principally active in Dubai; Emaar International, with focus on partnerships in the Middle East, North Africa and India; Emaar Malls Group LLC, which manages over 150 malls in South Asia, the Middle East and North Africa; Emaar Hotels and Resorts LLC, a joint venture with Giorgio Armani SpA; and Emaar Hospitality Group LLC, which manages the company's hospitality and leisure projects. Its most ambitious project to date is the Downtown Burj Dubai or Burj Khalifa, the world's tallest tower. Other projects include the Dubai Mall, one of the world's largest shopping malls; and the Old Town, a residence community designed to reflect traditional Arabian architecture including the Old Town Island, Emaar Square, the Residences and Burj Dubai Boulevard, and featuring a man-made lake and fountain as well as parks and gardens.

FINANCIAL DATA: *Note: Data for latest year may not have been available at press time.*

In U.S. $	2015	2014	2013	2012	2011	2010
Revenue	3,719,379,272	2,693,438,606	2,812,000,000	2,243,000,000	2,208,700,000	3,307,550,000
R&D Expense						
Operating Income						
Operating Margin %						
SGA Expense						
Net Income	1,249,535,249	896,542,335	699,000,000	577,000,000	488,300,000	666,450,000
Operating Cash Flow						
Capital Expenditure						
EBITDA						
Return on Assets %						
Return on Equity %						
Debt to Equity						

CONTACT INFORMATION:

Phone: 971-4-367-3333   Fax: 971-4-367-3000
Toll-Free:
Address: 28th Fl-Al Attar Bus Tower, Sheikh Zayed Rd., Dubai, 9267 United Arab Emirates

STOCK TICKER/OTHER:

Stock Ticker: EMAAR
Employees: 4,786
Parent Company:

Exchange: Abu Dhabi
Fiscal Year Ends: 12/31

SALARIES/BONUSES:

Top Exec. Salary: $ Bonus: $
Second Exec. Salary: $ Bonus: $

OTHER THOUGHTS:

Estimated Female Officers or Directors: 1
Hot Spot for Advancement for Women/Minorities:

EMCOR Group Inc

NAIC Code: 238210

TYPES OF BUSINESS:

Electric, Heating and AC Contractors
Mechanical Contracting
Technical Consulting Services
Facilities Management

BRANDS/DIVISIONS/AFFILIATES:

CONTACTS: *Note: Officers with more than one job title may be intentionally listed here more than once.*

Anthony Guzzi, CEO
Mark Pompa, CFO
Stephen Bershad, Chairman of the Board
R. Matz, Executive VP, Divisional
Maxine Mauricio, Executive VP
Sheldon Cammaker, Vice Chairman

GROWTH PLANS/SPECIAL FEATURES:

EMCOR Group, Inc. is a global leader in mechanical and electrical contracting and facilities services. The company offers its services through more than 70 subsidiaries and joint ventures and more than 170 offices located throughout the U.S. as well as in Canada, the U.K. and the Middle East. Services provided to its customers include the design, integration, installation, start up, operation and maintenance of systems for generation and distribution of electrical power; lighting systems; low-voltage systems, such as fire alarm, security, communications and process control systems; voice and data communication systems; heating, ventilation, air conditioning, refrigeration and clean-room process ventilation systems; plumbing, process and high-purity piping systems; water and wastewater treatment systems; and central plant heating and cooling systems. In addition to its construction services, EMCOR offers facilities services, such as site-based operations and maintenance, mobile maintenance and service, facilities management, installation and support for building systems, technical consulting and diagnostic services, small modification and retrofit projects and program development and management for energy systems. Most of the firm's business is done with corporations, municipalities and other government agencies, owner/developers and building tenants. Additional services are provided to a range of general and specialty contractors, with EMCOR operating as a subcontractor.

EMCOR offers its employees benefits including medical, vision and dental coverage; life insurance; flexible spending accounts; disability income; employee wellness and assistance programs; and a 401(k) and stock purchase options.

FINANCIAL DATA: *Note: Data for latest year may not have been available at press time.*

In U.S. $	2015	2014	2013	2012	2011	2010
Revenue	6,718,726,000	6,424,965,000	6,417,158,000	6,346,679,000	5,613,459,000	5,121,285,000
R&D Expense						
Operating Income	287,082,000	289,878,000	210,292,000	249,967,000	210,793,000	-28,686,000
Operating Margin %	4.27%	4.51%	3.27%	3.93%	3.75%	-.56%
SGA Expense	656,573,000	626,478,000	591,063,000	556,242,000	518,121,000	497,808,000
Net Income	172,286,000	168,664,000	123,792,000	146,584,000	130,826,000	-86,691,000
Operating Cash Flow	266,666,000	246,657,000	150,069,000	184,408,000	149,425,000	68,652,000
Capital Expenditure	35,460,000	38,035,000	35,497,000	37,875,000	29,581,000	19,359,000
EBITDA	361,944,000	365,210,000	278,758,000	312,489,000	266,389,000	23,830,000
Return on Assets %	4.96%	4.92%	3.76%	4.78%	4.53%	-3.02%
Return on Equity %	11.91%	11.70%	8.80%	11.35%	10.95%	-7.31%
Debt to Equity	0.20	0.22	0.22	0.11	0.12	0.31

CONTACT INFORMATION:

Phone: 203 849-7800   Fax: 203 849-7900
Toll-Free: 866-890-7794
Address: 301 Merritt Seven, Norwalk, CT 06851 United States

STOCK TICKER/OTHER:

Stock Ticker: EME   Exchange: NYS
Employees: 27,000   Fiscal Year Ends: 12/31
Parent Company:

SALARIES/BONUSES:

Top Exec. Salary: Bonus: $
$1,010,000
Second Exec. Salary: Bonus: $
$630,000

OTHER THOUGHTS:

Estimated Female Officers or Directors: 3

Hot Spot for Advancement for Women/Minorities: Y

Empresas ICA SAB de CV

www.ica.com.mx

NAIC Code: 237310

TYPES OF BUSINESS:

Heavy Construction
Civic Construction
Industrial Construction
Transportation Infrastructure Management
Residential Construction
Design & Engineering Services
Airport Operations

BRANDS/DIVISIONS/AFFILIATES:

ICA Fluor
Grupo Aeroportuario del Centro Norte

CONTACTS: Note: Officers with more than one job title may be intentionally listed here more than once.

Alonso Quintana Kawage, CEO
Alonso Quintana Kawage, Pres.
Gabriel De La Concha Guerrero, CFO
Porfirio Gonzalez Alvarez, CEO-GACN
Diego Quintana Kawage, Exec. VP-Industrial Construction & Airports
Bernardo Quintana Isaac, Chmn.

GROWTH PLANS/SPECIAL FEATURES:

Empresas ICA SAB de CV (ICA) is one of Mexico's largest engineering, procurement and construction companies. It operates three primary divisions: construction (divided into civil construction and industrial construction), airports and concessions. The company's civil construction segment builds highways, dams, airports, bridges, tunnels, subways and port facilities primarily in Mexico, with occasional projects in Latin America, the Caribbean, Asia and the U.S. It also participates in sub-soil construction projects in Spain through an affiliate company. The industrial construction segment, through majority-owned subsidiary ICA Fluor, builds industrial factories such as refineries, petrochemical plants, cement factories, automotive factories and electrical generation plants. The airports segment operates 13 airports, located primarily in the central-north region of Mexico, through its 40.5% ownership in Grupo Aeroportuario del Centro Norte (GACN). The concessions segment focuses on the construction, development, maintenance and operation of long-term concessions of tollroads, tunnels, social infrastructure and water projects. ICA participated in the operating concessioned tunnel (the Acapulco tunnel), and in the management and operation of a water treatment plant in Ciudad Acuna and other water supply systems, including the Aqueduct II water supply system. In early 2015, the firm signed a public works contract to build the Santa Maria reservoir dam with the Federal Water Commission, Conagua, scheduled for a January 2018 completion. In December 2015, the firm sold its 49% stake in Proactiva medio Ambiente, an environmental services company.

FINANCIAL DATA: Note: Data for latest year may not have been available at press time.

In U.S. $	2015	2014	2013	2012	2011	2010
Revenue		2,023,089,000	1,676,907,000	2,697,400,000	2,426,539,000	1,983,796,000
R&D Expense						
Operating Income		248,231,800	177,730,200	208,854,000	209,721,100	142,481,100
Operating Margin %		12.26%	10.59%	7.74%	8.64%	7.18%
SGA Expense		173,934,700	170,905,500	204,361,900	192,203,800	141,983,700
Net Income		-171,544,100	80,698,800	64,091,440	83,985,980	51,562,940
Operating Cash Flow		143,091,400	9,558,179	311,309,100	-576,165,400	-233,045,800
Capital Expenditure		39,739,800	20,242,340	26,374,790	22,995,520	27,916,480
EBITDA		153,085,100	232,157,500	351,552,500	189,420,900	249,356,400
Return on Assets %		-2.75%	1.35%	1.08%	1.70%	1.30%
Return on Equity %		-17.38%	8.03%	6.72%	8.64%	5.30%
Debt to Equity		2.91	1.55	2.42	1.80	1.51

CONTACT INFORMATION:

Phone: 52 5552729991   Fax: 52 5252712431
Toll-Free:
Address: Blvd. Manuel Avila Camacho 36, Mexico City, DF 11000 Mexico

STOCK TICKER/OTHER:

Stock Ticker: ICA
Employees: 31,302
Parent Company:

Exchange: NYS
Fiscal Year Ends: 12/31

SALARIES/BONUSES:

Top Exec. Salary: $ Bonus: $
Second Exec. Salary: $ Bonus: $

OTHER THOUGHTS:

Estimated Female Officers or Directors:
Hot Spot for Advancement for Women/Minorities: Y

ENGlobal Corp

NAIC Code: 541330

TYPES OF BUSINESS:

Engineering Services
Petrochemicals Industry Support Services
Control & Instrumentation Systems
Consulting & Inspection Services
Project Management

BRANDS/DIVISIONS/AFFILIATES:

ENGlobal U.S. Inc
ENGlobal Government Services Inc
ENGlobal Emerging Markets Inc
ENGlobal International Inc

CONTACTS: Note: Officers with more than one job title may be intentionally listed here more than once.

William Coskey, CEO
Mark Hess, CFO
Bruce Williams, COO
Tami Walker, General Counsel
J. Harrison, Senior VP, Divisional

GROWTH PLANS/SPECIAL FEATURES:

ENGlobal Corp. is an international provider of engineering services and systems to the energy market. The firm operates in two primary segments: engineering, procurement & construction management (EPCM) and automation. The EPCM segment offers a range of services relating to the development, management and execution of projects requiring professional engineering primarily to the midstream and downstream energy sectors. Its services include feasibility studies, engineering, design, procurement and construction management. This segment also includes the technical services group, which provides engineering, design, installation and operation and maintenance of various government, public sector and international facilities. Its customers include pipeline, refining, utility, chemical, petroleum, petrochemical, oil and gas and power industries throughout the U.S. The automation segment provides services related to the design, fabrication and implementation of process distributed control and analyzer systems, advanced automation and information technology projects. This segment's customers include members of both domestic and foreign energy related industries. ENGlobal caters to the engineering and construction, automation integration, automation engineering and subsea controls and integration markets. Subsidiaries of the firm include ENGlobal U.S., Inc.; ENGlobal Government Services, Inc.; ENGlobal International, Inc.; and ENGlobal Emerging Markets, Inc.

Employees receive medical, dental and vision coverage; life insurance; flexible spending accounts; and educational reimbursement.

FINANCIAL DATA: Note: Data for latest year may not have been available at press time.

In U.S. $	2015	2014	2013	2012	2011	2010
Revenue	79,605,000	107,900,000	168,963,000	227,916,000	312,747,000	320,615,000
R&D Expense						
Operating Income	2,100,000	6,880,000	-778,000	-21,087,000	-3,984,000	-17,256,000
Operating Margin %	2.63%	6.37%	-.46%	-9.25%	-1.27%	-5.38%
SGA Expense	14,168,000	16,568,000	22,080,000	25,239,000	31,263,000	39,975,000
Net Income	10,536,000	6,031,000	-2,989,000	-33,601,000	-7,076,000	-11,752,000
Operating Cash Flow	-2,847,000	3,767,000	10,666,000	-4,874,000	5,579,000	-6,151,000
Capital Expenditure	1,005,000	438,000	836,000	666,000	664,000	1,174,000
EBITDA	3,556,000	9,206,000	1,291,000	-19,268,000	-420,000	-12,915,000
Return on Assets %	20.28%	12.37%	-4.80%	-36.74%	-6.59%	-10.63%
Return on Equity %	30.51%	23.41%	-12.51%	-80.19%	-11.44%	-16.34%
Debt to Equity						

CONTACT INFORMATION:

Phone: 281 878-1000   Fax: 281 821-5488
Toll-Free:
Address: 654 N. Sam Houston Pkwy E., Ste. 400, Houston, TX 77060
United States

STOCK TICKER/OTHER:

Stock Ticker: ENG
Employees: 459
Parent Company:

Exchange: NAS
Fiscal Year Ends: 12/31

SALARIES/BONUSES:

Top Exec. Salary: $236,912   Bonus: $23,690
Second Exec. Salary: Bonus: $21,630
$216,299

OTHER THOUGHTS:

Estimated Female Officers or Directors: 2
Hot Spot for Advancement for Women/Minorities: Y

EPR Properties

NAIC Code: 0

TYPES OF BUSINESS:

REIT-Entertainment Properties
Megaplex Movie Theaters
Entertainment Retail Centers
Ski Resorts
Vineyards & Wineries
Public Charter Schools

BRANDS/DIVISIONS/AFFILIATES:

CONTACTS: Note: Officers with more than one job title may be intentionally listed here more than once.

Mark Peterson, CFO
Robert Druten, Chairman of the Board
Tonya Mater, Chief Accounting Officer
Morgan Earnest, Other Executive Officer
Gregory Silvers, President
Thomas Wright, Senior VP, Divisional
Craig Evans, Senior VP
Robin Sterneck, Trustee
Jack Newman, Trustee
Barrett Brady, Trustee
Peter Brown, Trustee
Thomas Bloch, Trustee
Michael Hirons, Vice President, Divisional

GROWTH PLANS/SPECIAL FEATURES:

EPR Properties is a self-administered REIT (real estate investment trust) with approximately 131 megaplex movie theaters, nine entertainment retail centers (ERCs), seven family entertainment centers, 70 public charter school properties, 18 early childhood centers, two K-12 private schools and other specialty properties. EPR's owned real estate portfolio of megaplex theatre properties consist of approximately 10.0 million square feet, and its remaining owned entertainment real estate portfolio consist of 1.8 million square feet. Its megaplex theaters usually have at least 10 screens with elevated, stadium-style seating and amenities such as digital projection, which allow greater enhancement to audio and visual experiences for theater patrons. EPR's portfolio includes multiplexes in 34 U.S. states and Ontario, Canada, with theaters leased to operators such as American Multi-Cinema, Inc. (AMC); Regal Entertainment Group; Carmike Cinemas, Inc.; Alamo Draft House Cinmas; AmStar Cinemas, LLC; Southern Theatres; and Cinemark. The firm's ERCs are located in Colorado, New York, Virginia, California and Ontario, Canada. Land parcels are often leased to restaurant and retail operators adjacent to EPR's theater properties. Some of its specialty properties include nine metro ski parks, 19 golf entertainment complexes and five water parks, with $59.5 million in construction in progress for three more golf entertainment complexes and one indoor waterpark hotel.

FINANCIAL DATA: Note: Data for latest year may not have been available at press time.

In U.S. $	2015	2014	2013	2012	2011	2010
Revenue	341,102,000	303,781,000	262,008,000	245,130,000	229,980,000	238,262,000
R&D Expense						
Operating Income	170,017,000	177,278,000	152,193,000	132,841,000	99,716,000	106,612,000
Operating Margin %	49.84%	58.35%	58.08%	54.19%	43.35%	44.74%
SGA Expense	57,387,000	30,018,000	27,568,000	23,574,000	21,903,000	18,227,000
Net Income	194,532,000	179,633,000	180,226,000	121,556,000	115,228,000	114,874,000
Operating Cash Flow	278,460,000	250,295,000	234,120,000	207,255,000	195,799,000	180,391,000
Capital Expenditure	179,820,000	85,205,000	123,497,000	73,188,000	53,175,000	247,117,000
EBITDA						
Return on Assets %	4.31%	4.46%	5.03%	3.27%	2.98%	3.02%
Return on Equity %	8.53%	8.62%	9.94%	6.36%	5.48%	5.50%
Debt to Equity	0.95	0.85	0.87	0.93	0.78	0.74

CONTACT INFORMATION:

Phone: 816-472-1700   Fax:
Toll-Free: 888-377-7348
Address: 909 Walnut St., Ste. 200, Kansas City, MO 64106 United States

STOCK TICKER/OTHER:

Stock Ticker: EPR
Employees: 40
Parent Company:

Exchange: NYS
Fiscal Year Ends: 12/31

SALARIES/BONUSES:

Top Exec. Salary: $585,000   Bonus: $1,099,800
Second Exec. Salary: Bonus: $676,800
$400,000

OTHER THOUGHTS:

Estimated Female Officers or Directors: 2
Hot Spot for Advancement for Women/Minorities:

Equity Commonwealth REIT

www.cwhreit.com

NAIC Code: 0

TYPES OF BUSINESS:
Real Estate Investment Trust
Office Buildings

BRANDS/DIVISIONS/AFFILIATES:
CommonWealth REIT

CONTACTS: Note: Officers with more than one job title may be intentionally listed here more than once.
David Helfand, CEO
Adam Markman, Executive VP
David Weinberg, Executive VP
Orrin Shifrin, Executive VP
Jeffrey Brown, Senior VP
Edward Glickman, Trustee
Kenneth Shea, Trustee
Jim Lozier, Trustee
James Star, Trustee
Gerald Spector, Trustee
James Corl, Trustee
Martin Edelman, Trustee
Mary Robertson, Trustee
Peter Linneman, Trustee
Samuel Zell, Trustee

GROWTH PLANS/SPECIAL FEATURES:

Equity Commonwealth REIT, formerly CommonWealth REIT, is an internally-managed and self-advised real estate investment trust (REIT). The company is one of the largest commercial office REITs in the U.S., with a portfolio of 65 properties and a total of over 24 million square feet located in 23 states and Washington D.C. Its properties are predominately located in central business districts and suburban areas in five major market regions: Metro Denver; Metro Chicago; Metro Philadelphia; Metro Washington, D.C.; and Oahu, Hawaii. In evaluating potential investments and asset sales, Equity Commonwealth considers historic and projected rents received and operating expenses; the growth, tax and regulatory environment of the market; the quality, experience and credit worthiness of the property's tenants; the occupancy, demand for and pricing of similar properties in nearby markets; the construction quality, physical condition and design of the property; and the geographic area and type of property. Its leading tenants, as measured by percentage of square feet, include Flextronics International Ltd.; Royal Dutch Shell plc; PNC Financial Services Group; Carmike Cinemas, Inc.; and Office Depot, Inc. In 2015, the firm disposed of around 91 properties totaling 18.9 million square feet, including the assets in Australia.

The firm offers employees dental, health and vision coverage; life insurance; flexible spending accounts for both medical and dependent care expenses; tuition assistance; employee assistance plans; and fitness discounts.

FINANCIAL DATA: Note: Data for latest year may not have been available at press time.

In U.S. $	2015	2014	2013	2012	2011	2010
Revenue	714,891,000	861,857,000	885,536,000	1,013,092,000	911,948,000	793,370,000
R&D Expense						
Operating Income	121,323,000	-51,884,000	218,286,000	290,337,000	233,943,000	190,391,000
Operating Margin %	16.97%	-6.01%	24.65%	28.65%	25.65%	23.99%
SGA Expense	57,457,000	113,155,000	77,209,000	51,697,000	46,758,000	39,646,000
Net Income	99,857,000	24,012,000	-177,060,000	-95,421,000	109,984,000	135,409,000
Operating Cash Flow	181,544,000	200,392,000	234,651,000	276,783,000	263,332,000	252,134,000
Capital Expenditure	70,633,000	99,651,000	268,821,000	762,579,000	868,399,000	972,913,000
EBITDA	403,538,000	370,699,000	470,364,000	554,265,000	476,868,000	467,827,000
Return on Assets %	1.30%	-.39%	-2.98%	-1.94%	.89%	1.28%
Return on Equity %	2.43%	-.85%	-8.52%	-5.77%	2.32%	3.38%
Debt to Equity	0.57	0.75	1.10	1.75	1.28	1.22

CONTACT INFORMATION:
Phone: 617 332-3990   Fax: 617 332-2261
Toll-Free:
Address: 255 Washington St., 2 Newton Pl., Newton, MA 02458 United States

STOCK TICKER/OTHER:
Stock Ticker: EQC
Employees: 66
Parent Company:

Exchange: NYS
Fiscal Year Ends: 12/31

SALARIES/BONUSES:
Top Exec. Salary: $346,154   Bonus: $1,250,000
Second Exec. Salary: $259,616   Bonus: $625,000

OTHER THOUGHTS:
Estimated Female Officers or Directors: 1
Hot Spot for Advancement for Women/Minorities:

Equity Lifestyle Properties Inc

www.equitylifestyle.com

NAIC Code: 0

TYPES OF BUSINESS:

Real Estate Investment Trust
Manufactured Home & RV Communities

BRANDS/DIVISIONS/AFFILIATES:

MHC Operating Limited Partnership

CONTACTS: *Note: Officers with more than one job title may be intentionally listed here more than once.*

Thomas Heneghan, CEO, Subsidiary
Paul Seavey, CFO
Samuel Zell, Chairman of the Board
Patrick Waite, COO
Howard Walker, Director
Roger Maynard, Executive VP, Divisional
David Eldersveld, Executive VP
Marguerite Nader, President
Ann Wallin, Vice President

GROWTH PLANS/SPECIAL FEATURES:

Equity Lifestyle Properties, Inc. (ELS) is an integrated real estate investment trust (REIT) that owns and operates communities of developed residential sites as well as recreational vehicle (RV) resorts. The company primarily operates through MHC Operating Limited Partnership. The firm owns or has an interest in 387 properties throughout the U.S. and Canada, consisting of about 143,887 residential sites. The heaviest concentrations of these properties are located in Florida, California and Arizona, with 122, 49 and 42 properties respectively. Additional U.S. sites are located in Texas, Pennsylvania, Washington, Colorado, Wisconsin, Oregon, North Carolina, Delaware, Indiana, Nevada, New York, Virginia, New Jersey, Illinois, Maine, Massachusetts, Idaho, Michigan, Minnesota, New Hampshire, South Carolina, Utah, Maryland, North Dakota, Ohio, Tennessee, Alabama, Connecticut and Kentucky. These communities are designed and improved for the placement of detached, single-family manufactured homes that are produced offsite, then installed and set on residential sites within the communities. The owner of each home leases the site on which it is located, while the firm handles property infrastructure issues, such as water, sewage and power. Sites typically contain centralized entrances, paved streets, curbs, gutters, parkways, clubhouses for social activities and recreation, swimming pools, shuffleboard courts, tennis courts, laundry facilities and cable television service, among other amenities. Each community is designed to attract, and is marketed to, retirees, empty-nesters, families or first-time homeowners. The company focuses on owning properties in or near large metropolitan markets as well as popular retirement and vacation destinations.

FINANCIAL DATA: *Note: Data for latest year may not have been available at press time.*

In U.S. $	2015	2014	2013	2012	2011	2010
Revenue	821,654,000	776,809,000	728,375,000	709,877,000	580,073,000	511,361,000
R&D Expense						
Operating Income	146,423,000	137,520,000	75,208,000	67,963,000	40,556,000	58,601,000
Operating Margin %	17.82%	17.70%	10.32%	9.57%	6.99%	11.45%
SGA Expense	47,016,000	42,476,000	43,493,000	37,302,000	51,354,000	32,838,000
Net Income	139,371,000	128,005,000	116,199,000	69,391,000	36,598,000	38,354,000
Operating Cash Flow	352,882,000	285,745,000	255,349,000	236,459,000	175,641,000	163,309,000
Capital Expenditure	117,486,000	145,112,000	182,421,000	99,473,000	713,121,000	48,629,000
EBITDA	370,941,000	367,418,000	308,214,000	310,314,000	226,818,000	228,451,000
Return on Assets %	3.79%	3.47%	3.14%	1.58%	.82%	1.82%
Return on Equity %	16.63%	15.48%	14.44%	7.55%	4.77%	17.21%
Debt to Equity	2.71	2.85	2.89	3.13	3.14	6.22

CONTACT INFORMATION:

Phone: 312 279-1400   Fax: 312 279-1710
Toll-Free:
Address: 2 N. Riverside Plz., Ste. 800, Chicago, IL 60606 United States

STOCK TICKER/OTHER:

Stock Ticker: ELS
Employees: 3,900
Parent Company:

Exchange: NYS
Fiscal Year Ends: 12/31

SALARIES/BONUSES:

Top Exec. Salary: $385,000   Bonus: $
Second Exec. Salary: $350,000   Bonus: $

OTHER THOUGHTS:

Estimated Female Officers or Directors: 3
Hot Spot for Advancement for Women/Minorities: Y

Equity Office Management LLC

www.equityoffice.com

NAIC Code: 0

TYPES OF BUSINESS:

Real Estate Investment Trust
Office Properties
Property Management

BRANDS/DIVISIONS/AFFILIATES:

Blackstone Group LP (The)

CONTACTS: *Note: Officers with more than one job title may be intentionally listed here more than once.*

Eli Khouri, CEO
Chris Hendricks, COO
Tom August, Pres.
Kurt. A. Heister, CFO
Matt Koritz, VP-General Counsel
Mike Ernst, Market Managing Dir.-Central Markets
Adam Goldenberg, Market Managing Dir.-New York
John Moe, Market Managing Dir.-Northern California
Frank Campbell, Market Managing Dir.-Southern California

GROWTH PLANS/SPECIAL FEATURES:

Equity Office Management LLC, operating through its various affiliates and subsidiaries, is one of the largest real estate investment trusts (REITs) in the U.S. The company owns 350 buildings comprising approximately 50 million square feet, most of which are Class A office buildings. Equity Office provides a wide range of office options for local, regional and national customers. The firm has a history of strategic acquisitions that have enabled it to quadruple in size since its 1997 initial public offering. These acquisitions, totaling nearly $20 billion in value, include mergers with Beacon Properties, Inc.; Cornerstone Properties, Inc.; and Spieker Properties, Inc. Its current portfolio includes 100 Montgomery Street in San Francisco, 100 Summer Street in Boston, 1000 Town Center Drive in Oxnard, 1111 W. 22nd Street in Oak Brook, 114 West 41st Street in New York City, 1140 Avenue of the Americas in New York City, 1201 Dove Street in Newport Beach, 1221 Brickell Avenue in Miami, 150 East Colorado in Pasadena, 17911 Von Karman in Irvine, 1977 Saturn Street in Monterey Park, 1999 Avenue of the Stars in Century City in Los Angeles, 2850 Premiere Parkway in Duluth and 3330 Cahuenga Blvd. W. in Los Angeles. The Blackstone Group owns the firm.

The company offers employees a benefits package that includes a 401(k) savings program with company match, health and dependent care flexible spending accounts, insurance benefits for domestic partners, adoption assistance, flexible work arrangements, educational assistance, an employee assistance program and a commuter program.

FINANCIAL DATA: *Note: Data for latest year may not have been available at press time.*

In U.S. $	2015	2014	2013	2012	2011	2010
Revenue						
R&D Expense						
Operating Income						
Operating Margin %						
SGA Expense						
Net Income						
Operating Cash Flow						
Capital Expenditure						
EBITDA						
Return on Assets %						
Return on Equity %						
Debt to Equity						

CONTACT INFORMATION:

Phone: 312-466-3300   Fax:
Toll-Free:
Address: Two N. Riverside Plz., Chicago, IL 60606 United States

STOCK TICKER/OTHER:

Stock Ticker: Private   Exchange:
Employees: 2,000   Fiscal Year Ends: 12/31
Parent Company: Blackstone Group LP (The)

SALARIES/BONUSES:

Top Exec. Salary: $ Bonus: $
Second Exec. Salary: $ Bonus: $

OTHER THOUGHTS:

Estimated Female Officers or Directors:
Hot Spot for Advancement for Women/Minorities:

Equity One Inc

NAIC Code: 0

TYPES OF BUSINESS:

Real Estate Investment Trust
Property Management
Shopping Centers

BRANDS/DIVISIONS/AFFILIATES:

DIM Vastgoed NV
C&C (US) No. 1 Inc

CONTACTS: Note: Officers with more than one job title may be intentionally listed here more than once.

David Lukes, CEO
Chaim Katzman, Chairman of the Board
Michael Makinen, COO
Dori Segal, Director
Michael Berfield, Executive VP, Divisional
William Brown, Executive VP, Divisional
Matthew Ostrower, Executive VP
Aaron Kitlowski, General Counsel
Thomas Caputo, President
Angela Valdes, Vice President

GROWTH PLANS/SPECIAL FEATURES:

Equity One, Inc. is a real estate investment trust (REIT) that acquires, renovates, develops and manages neighborhood shopping centers anchored by national and regional supermarket chains and other necessity-oriented retailers. Its consolidated portfolio consists of 126 properties totaling 12.6 million square feet located in Connecticut, California, Florida, Georgia, Louisiana, Maryland, Massachusetts, New York and North Carolina. These properties include 102 retail properties and five non-retail properties. The firm also has 13 development or redevelopment properties with approximately 2.8 million square feet of gross leasable area (GLA) and six land parcels. Additionally, Equity One has joint venture interests in six retail properties and two office buildings, totaling approximately 1.4 million square feet. The company has leased to supermarkets, such as Albertsons, Winn-Dixie, The Fresh Market, Whole Foods, Trader Joe's, Ralph's, Safeway, Von's, Stop & Shop, ShopRite, Food Emporium, Pathmark and Publix Super Markets, as well as national retailers such as Office Depot, Best Buy, CVS Pharmacy, Home Depot, Walgreens and Wal-Mart. The firm's subsidiaries include DIM Vastgoed, NV and C&C (US) No. 1, Inc.

FINANCIAL DATA: Note: Data for latest year may not have been available at press time.

In U.S. $	2015	2014	2013	2012	2011	2010
Revenue	360,153,000	353,185,000	332,511,000	325,611,000	291,925,000	285,224,000
R&D Expense						
Operating Income	137,339,000	121,173,000	116,084,000	109,486,000	73,708,000	96,992,000
Operating Margin %	38.13%	34.30%	34.91%	33.62%	25.24%	34.00%
SGA Expense	36,277,000	41,174,000	39,514,000	42,474,000	51,707,000	42,041,000
Net Income	65,453,000	48,897,000	77,954,000	-3,477,000	33,621,000	25,112,000
Operating Cash Flow	164,765,000	144,095,000	132,742,000	153,219,000	102,626,000	71,562,000
Capital Expenditure	191,090,000	197,358,000	180,115,000	345,541,000	345,727,000	133,965,000
EBITDA	222,930,000	226,856,000	210,542,000	153,813,000	191,956,000	165,244,000
Return on Assets %	1.97%	1.47%	2.27%	-.10%	1.13%	.97%
Return on Equity %	4.29%	3.39%	5.58%	-.24%	2.48%	2.13%
Debt to Equity	0.87	0.89	1.08	1.14	0.92	0.93

CONTACT INFORMATION:

Phone: 212-796-1760   Fax:
Toll-Free:
Address: 410 Park Avenue, Ste. 1220, New York, NY 10022 United States

STOCK TICKER/OTHER:

Stock Ticker: EQY
Employees: 155
Parent Company:

Exchange: NYS
Fiscal Year Ends: 12/31

SALARIES/BONUSES:

Top Exec. Salary: $850,000   Bonus: $
Second Exec. Salary: $750,000   Bonus: $

OTHER THOUGHTS:

Estimated Female Officers or Directors: 5
Hot Spot for Advancement for Women/Minorities: Y

Equity Residential

NAIC Code: 0

www.equityapartments.com

TYPES OF BUSINESS:

Real Estate Investment Trust
Apartment Communities
Property Management

GROWTH PLANS/SPECIAL FEATURES:

Equity Residential is a real estate investment trust (REIT) engaged in the acquisition, development and management of apartment properties across the U.S. The firm conducts all operations through operating partnership ERP Operating Limited Partnership, of which it owns 96.2%. One of the leading publicly-traded owners and operators of multiple-family properties in the U.S., the firm owns all or a portion of 394 properties throughout 12 states and Washington, D.C., totaling 109,652 units. The firm's most important markets include Los Angeles (with 61 apartment properties), San Francisco Bay Area (53), Seattle/Tacoma (44), South Florida (35) and the New York Metro Area (40). Its properties are highly diversified with respect to design and geography, ranging from high-rise to garden styles. It also owns housing units specially designed for corporate and military use. Corporate housing benefits include fully furnished units, free local calling, basic cable, short-term leases and direct billing options. Military housing provides similar benefits in addition to government credit card acceptance and military discounts. In March 2015, the firm began construction on One Henry Adams, a 241-unit apartment community in San Francisco, California. In January 2016, Equity Residential sold 72 properties, consisting of 23,262 apartment units to controlled affiliates of Starwood Capital Group.

BRANDS/DIVISIONS/AFFILIATES:

ERP Operating Limited Partnership

CONTACTS: Note: Officers with more than one job title may be intentionally listed here more than once.

David Neithercut, CEO
Mark Parrell, CFO
Samuel Zell, Chairman of the Board
Ian Kaufman, Chief Accounting Officer
David Santee, COO
John Powers, Executive VP, Divisional
Mark Tennison, Executive VP, Divisional
Barry Altshuler, Executive VP, Divisional
Alexander Brackenridge, Executive VP, Divisional
Bruce Strohm, Executive VP
Alan George, Executive VP
B. White, Trustee
Charles Atwood, Trustee
John Alexander, Trustee
Linda Bynoe, Trustee
Stephen Sterrett, Trustee
Bradley Keywell, Trustee
John Neal, Trustee
Connie Duckworth, Trustee
Mark Shapiro, Trustee
Mary Haben, Trustee
Gerald Spector, Trustee

FINANCIAL DATA: Note: Data for latest year may not have been available at press time.

In U.S. $	2015	2014	2013	2012	2011	2010
Revenue	2,744,965,000	2,614,748,000	2,387,702,000	2,123,715,000	1,989,463,000	1,995,519,000
R&D Expense						
Operating Income	1,008,820,000	921,375,000	512,288,000	667,958,000	573,332,000	442,001,000
Operating Margin %	36.75%	35.23%	21.45%	31.45%	28.81%	22.14%
SGA Expense	65,082,000	50,948,000	62,179,000	47,248,000	43,606,000	39,887,000
Net Income	870,120,000	631,308,000	1,830,613,000	841,719,000	893,585,000	283,610,000
Operating Cash Flow	1,356,499,000	1,324,073,000	868,916,000	1,046,251,000	798,334,000	732,693,000
Capital Expenditure	1,167,346,000	1,186,333,000	673,021,000	1,208,543,000	1,723,010,000	1,481,374,000
EBITDA	2,118,503,000	1,869,270,000	1,479,131,000	1,454,766,000	1,213,362,000	1,098,576,000
Return on Assets %	3.74%	2.73%	9.12%	4.88%	5.35%	1.70%
Return on Equity %	8.32%	6.03%	20.64%	13.00%	16.98%	5.53%
Debt to Equity	1.01	0.52	1.01	1.17	1.77	2.03

CONTACT INFORMATION:

Phone: 312 474-1300   Fax:
Toll-Free:
Address: 2 N. Riverside Plz., Chicago, IL 60606 United States

STOCK TICKER/OTHER:

Stock Ticker: EQR   Exchange: NYS
Employees: 3,500   Fiscal Year Ends: 12/31
Parent Company:

SALARIES/BONUSES:

Top Exec. Salary: $900,000   Bonus: $
Second Exec. Salary: Bonus: $
$600,000

OTHER THOUGHTS:

Estimated Female Officers or Directors: 2
Hot Spot for Advancement for Women/Minorities:

Sales, profits and employees may be estimates. Financial information, benefits and other data can change quickly and may vary from those stated here.

Erickson Living

NAIC Code: 623310

www.ericksonliving.com

TYPES OF BUSINESS:

Assisted Living Facilities
Supplemental Health Insurance

BRANDS/DIVISIONS/AFFILIATES:

Erickson Advantage
Erickson Health

CONTACTS: *Note: Officers with more than one job title may be intentionally listed here more than once.*

R. Alan Butler, CEO
Debra B. Doyle, COO
William J. Butz, Jr., Pres.
Todd A. Matthiesen, CFO
Tom Neubauer, Exec. VP-Sales & Mktg.
Joseph Machicote, Sr. VP-Human Resources
John F. Triscoli, CIO
John F. Triscoli, Sr. VP-Enterprise Tech. & Programs
Gerald F. Doherty, General Counsel
Kerry Jones, Sr. VP-Oper.
Rick Slosson, Sr. VP-Dev.
Adam Kane, Sr. VP-Corp. Affairs
Matthew Narrett, Chief Medical Officer

GROWTH PLANS/SPECIAL FEATURES:

Erickson Living is a real estate firm focused on the development and operation of retirement communities for men and women ages 62 and up. The firm's portfolio consists of 18 properties spread throughout Colorado, Florida, Kansas, Maryland, Massachusetts, Michigan, New Jersey, North Carolina, Pennsylvania, Texas and Virginia, which altogether house more than 20,000 active senior citizens. The firm's communities feature apartment homes with a variety of floor plans, primarily in one- or two-bedroom options. Each apartment home is designed to promote an independent lifestyle, and all Erickson communities provide residents with services including transportation, grounds maintenance and housekeeping. Erickson's retirement communities offer amenities including restaurants, stores and a fitness center with full-time trainers. Additionally, each community features a medical center staffed by geriatricians who practice exclusively in Erickson communities. Residents pay an all-inclusive monthly service fee, which covers all rent and maintenance service costs. In addition to its retirement communities, Erickson provides Erickson Advantage, a supplemental health insurance plan, and Erickson Health, a health and wellness maintenance system that focuses on fitness and preventative programs.

FINANCIAL DATA: *Note: Data for latest year may not have been available at press time.*

In U.S. $	2015	2014	2013	2012	2011	2010
Revenue	775,000,000	772,000,000	761,000,000	750,000,000	729,000,000	
R&D Expense						
Operating Income						
Operating Margin %						
SGA Expense						
Net Income						
Operating Cash Flow						
Capital Expenditure						
EBITDA						
Return on Assets %						
Return on Equity %						
Debt to Equity						

CONTACT INFORMATION:

Phone: 410-242-2880   Fax: 410-737-8854
Toll-Free:
Address: 701 Maiden Choice Ln., Catonsville, MD 21228 United States

STOCK TICKER/OTHER:

Stock Ticker: Private   Exchange:
Employees: 12,000   Fiscal Year Ends:
Parent Company:

SALARIES/BONUSES:

Top Exec. Salary: $ Bonus: $
Second Exec. Salary: $ Bonus: $

OTHER THOUGHTS:

Estimated Female Officers or Directors: 2
Hot Spot for Advancement for Women/Minorities:

Essar Group Ltd

www.essar.com

NAIC Code: 331110

TYPES OF BUSINESS:

Steel Production
Oil & Gas
Electric Generation
Logistics & Shipping
Investments
Construction
Communications Investments
Trucking

BRANDS/DIVISIONS/AFFILIATES:

Essar Steel
Essar Shipping
Essar Oil
Essar Energy
Essar Power
Essar Communications Holdings Ltd
Essar Ports
Aegis Ltd

CONTACTS: *Note: Officers with more than one job title may be intentionally listed here more than once.*

Prashant Ruia, CEO
V. Ashok, CFO
Adil Malia, Pres., Human Resources
Vikash Saraf, Dir.-Strategy & Planning
Sunil Bajaj, Head-Corp. Rel. Group
Dilip Oommen, CEO-Steel Bus. Group
Lalit Kumar Gupta, Managing Dir.
Alwyn Bowden, CEO-Project Bus. Group
Sushil Maroo, CEO-Energy Group
Shashi Ruia, Chmn.

GROWTH PLANS/SPECIAL FEATURES:

Essar Group, Ltd., based in Mumbai, is a multinational conglomerate active in the steel, energy, oil & gas, communications, construction and shipping industries as well as other various activities. Essar Steel is an integrated flat carbon steel manufacturer with a production capacity of nearly 14 million tons per year. The division is involved in all aspects of production and distribution, from mining iron ore to over 375 end user distribution outlets known as Essar Hypermarts. Essar Energy consists of Essar Oil and Essar Power. Essar Oil operations include exploration and production of oil and gas as well as refining and retail distribution through more than 1,400 Essar-branded service stations across India. Essar Power operates seven power plants with a combined capacity of 6,700 megawatts (MW). Essar Communications, through Essar BPO & Telecom, operates a network of nearly 700 outlets in 90 cities and towns in India. It also owns Aegis, Ltd., which has 56 delivery centers in 13 countries. Essar Projects is a global engineering, procurement and construction company. Essar Shipping and Essar Ports operates ports and terminals for crude oil, petroleum and coal; owns 15 sea transportation vessels; and manages an oilfield drilling business with a fleet of 15 onshore rigs and one semi-submersible offshore rig. Other business areas the group is involved in include realty (through Equinox Realty) and agribusiness (through Essar Agrotech). In March 2014, Essar Power acquired the Tokisud North coal block in Jharkhand.

FINANCIAL DATA: *Note: Data for latest year may not have been available at press time.*

In U.S. $	2015	2014	2013	2012	2011	2010
Revenue	35,000,000,000	32,208,207,000	26,000,000,000	17,000,000,000		
R&D Expense						
Operating Income						
Operating Margin %						
SGA Expense						
Net Income						
Operating Cash Flow						
Capital Expenditure						
EBITDA						
Return on Assets %						
Return on Equity %						
Debt to Equity						

CONTACT INFORMATION:

Phone: 91-22-5001-1100   Fax: 91-22-6660-1809
Toll-Free:
Address: Essar House, 11 Kesharao Khadye Marg, Mumbai, 400 034 India

STOCK TICKER/OTHER:

Stock Ticker: Private   Exchange:
Employees: 60,000   Fiscal Year Ends: 03/31
Parent Company:

SALARIES/BONUSES:

Top Exec. Salary: $ Bonus: $
Second Exec. Salary: $ Bonus: $

OTHER THOUGHTS:

Estimated Female Officers or Directors: 2
Hot Spot for Advancement for Women/Minorities:

Essex Property Trust Inc

www.essexpropertytrust.com

NAIC Code: 0

TYPES OF BUSINESS:
Real Estate Investment Trust
Apartments
Residential Brokerage
Property Management

BRANDS/DIVISIONS/AFFILIATES:
Essex Portfolio LP

CONTACTS: Note: Officers with more than one job title may be intentionally listed here more than once.
Michael Schall, CEO
Angela Kleiman, CFO
George Marcus, Chairman of the Board
John Farias, Chief Accounting Officer
Keith Guericke, Director
Craig Zimmerman, Executive VP, Divisional
John Eudy, Executive VP, Divisional
John Burkart, Senior Executive VP, Divisional
Barb Pak, Vice President, Divisional

GROWTH PLANS/SPECIAL FEATURES:

Essex Property Trust, Inc. (EPT) is a self-administered and self-managed real estate investment trust (REIT) that acquires, develops, redevelops and manages multifamily residential properties in West Coast communities. The firm owns all of its interest in properties either directly or through Essex Portfolio, L.P., in which EPT owns a 96.7% general partnership interest and is the sole general partner. The company's holdings encompass 239 apartment communities, aggregating approximately 57,455 units; four commercial buildings, totaling roughly 325,200 square feet; and 12 active development projects, with 2,920 units in various stages of completion. Most of the firm's holdings are concentrated in coastal markets in Southern California, the San Francisco Bay area and the Seattle metropolitan area. EPT's apartment communities are primarily suburban garden-style apartment communities and town homes comprising multiple clusters of two and three-story buildings situated on three to fifteen acres of land. Its portfolio currently includes 163 garden-style, 72 mid-rise and four high-rise apartment communities averaging approximately 240 units each. These communities contain a mixture of studio, one-, two- and three-bedroom units. EPT's investment strategy has various components, including monitoring of current markets as well as the evaluation of new metropolitan markets in search of areas with relatively high rental growth potential. The company typically targets properties with over 100 units and a total value greater than $10 million. In 2015, the firm acquired an apartment community in downtown Los Angeles, California for $200 million; and sold Sharon Green Apartments, located in Menlo Park, California, for $245 million.

EPT employees receive medical, dental and vision coverage; flexible spending accounts; life and disability insurance, including dependent insurance coverage; a 401(k); educational assistance; and a company-paid employee assistance plan, among other benefits.

FINANCIAL DATA: Note: Data for latest year may not have been available at press time.

In U.S. $	2015	2014	2013	2012	2011	2010
Revenue	1,194,407,000	969,305,000	613,703,000	543,425,000	475,558,000	415,732,000
R&D Expense						
Operating Income	331,174,000	201,514,000	188,705,000	168,925,000	137,441,000	112,251,000
Operating Margin %	27.72%	20.78%	30.74%	31.08%	28.90%	27.00%
SGA Expense	40,090,000	40,878,000	32,282,000	29,820,000	25,304,000	25,962,000
Net Income	232,120,000	122,150,000	156,283,000	125,284,000	47,070,000	35,934,000
Operating Cash Flow	617,410,000	492,983,000	304,982,000	267,499,000	216,571,000	175,530,000
Capital Expenditure	830,249,000	700,606,000	428,704,000	514,982,000	231,954,000	486,220,000
EBITDA	906,489,000	659,581,000	451,216,000	412,127,000	305,492,000	268,078,000
Return on Assets %	1.92%	1.39%	3.00%	2.69%	1.03%	.96%
Return on Equity %	3.74%	3.01%	8.61%	7.84%	3.24%	3.13%
Debt to Equity	0.86	0.81	1.67	1.66	1.73	2.00

CONTACT INFORMATION:
Phone: 650-655-7800   Fax: 650 494-8743
Toll-Free:
Address: 1100 Park Place, Ste 200, San Mateo, CA 94403 United States

STOCK TICKER/OTHER:
Stock Ticker: ESS
Employees: 1,173
Parent Company:

Exchange: NYS
Fiscal Year Ends: 12/31

SALARIES/BONUSES:
Top Exec. Salary: $550,000   Bonus: $300,000
Second Exec. Salary: $268,750   Bonus: $410,000

OTHER THOUGHTS:
Estimated Female Officers or Directors: 6
Hot Spot for Advancement for Women/Minorities: Y

Evergrande Real Estate Group

NAIC Code: 531100

www.evergrande.com/en/

TYPES OF BUSINESS:

Real Estate Development
Architectural Engineering Services
Architectural Design Services
Property Management Services

BRANDS/DIVISIONS/AFFILIATES:

Evergrande Real Estate Development Co Ltd
Evergrande Construction Engineering Company
Evergrande Architectural Design Institute
Evergrande Property Management Company
Evergrande Oasis
Evergrande City
Evergrande Metropolis
Evergrande Construction Engineering Company

CONTACTS: Note: Officers with more than one job title may be intentionally listed here more than once.

Xia Haijun, CEO
Tse Wai Wah, CFO
Xu Wen, VP
Li Gang, Vice Chmn.
Siu Shawn, Exec. VP
Wang Chuan, Exec. VP
He Miaoling, VP
Hui Ka Yan, Chmn.

GROWTH PLANS/SPECIAL FEATURES:

Evergrande Real Estate Group, one of the largest real estate developers in China, is primarily engaged in the planning, design, construction, development and management of real estate properties. The firm's business segments include residential building, which is engaged in large-scale residential projects; cultural tourism, which consists of higher-end business type tourism complexes in 20 provinces; agri-husbandry, which includes bottled water, grain & oil and dairy business units; Evergrande Health, which consists of international hospitals, senior care, medical cosmetology and anti-aging services; hotel property, which includes the five-star Hengda Hotel and Evergrande Commercial Group; and sports culture, which consists of a football club and school and a volleyball club. Major projects, consisting of high-end residential, and commercial properties, include the Evergrande Palace, Evergrande Metropolis, Evergrande City, Evergrande Oasis and Evergrande Splendor. The firm has launched more than 400 major projects throughout 170 cities across China, including Guangzhou, Shanghai, Shenzhen, Tianjin, Chongqing and Shenyang. Evergrande's subsidiaries include Evergrande Real Estate Development Co., Ltd., a real estate development company that has developed projects such as the Evergrande Jinbi Garden, Evergrande Royal Scenic Peninsula and Evergrande Palace; Evergrande Construction Engineering Company, an architectural engineering company that performs industrial and civil construction, high-rise construction, equipment installation and municipal works; Evergrande Architectural Design Institute, which offers architectural design services for residential buildings ranging from planning and interior design to garden landscaping; and Evergrande Property Management Company, which manages a variety of properties throughout China, including multi-story residential buildings, high-rise residential buildings, villas and commercial properties.

FINANCIAL DATA: Note: Data for latest year may not have been available at press time.

In U.S. $	2015	2014	2013	2012	2011	2010
Revenue		17,215,240,000	14,475,850,000	10,085,280,000	9,568,713,000	7,078,057,000
R&D Expense						
Operating Income		4,412,457,000	3,870,210,000	2,553,410,000	3,079,363,000	2,135,856,000
Operating Margin %		25.63%	26.73%	25.31%	32.18%	30.17%
SGA Expense		2,038,659,000	1,202,649,000	868,245,200	754,450,500	457,204,500
Net Income		1,947,805,000	1,948,999,000	1,417,243,000	1,758,905,000	1,172,756,000
Operating Cash Flow		-7,029,089,000	-6,007,084,000	-861,336,300	-577,293,800	-1,811,766,000
Capital Expenditure		1,412,680,000	1,924,666,000	1,231,489,000	1,474,463,000	139,156,200
EBITDA		4,998,220,000	4,088,652,000	2,647,454,000	3,133,179,000	2,846,445,000
Return on Assets %		3.06%	4.29%	4.38%	8.03%	9.05%
Return on Equity %		16.52%	29.04%	25.85%	42.69%	45.31%
Debt to Equity		0.73	1.50	1.07	1.26	1.17

CONTACT INFORMATION:

Phone: 020 8888-3333   Fax:
Toll-Free:
Address: Evergrande Center No.78, Huangpu Ave. W., Guangzhou City, 510060 China

STOCK TICKER/OTHER:

Stock Ticker: EGRNF
Employees: 77,057
Parent Company:

Exchange: GREY
Fiscal Year Ends:

SALARIES/BONUSES:

Top Exec. Salary: $ Bonus: $
Second Exec. Salary: $ Bonus: $

OTHER THOUGHTS:

Estimated Female Officers or Directors: 3
Hot Spot for Advancement for Women/Minorities: Y

Extended Stay America Inc

www.extendedstayhotels.com

NAIC Code: 721110

TYPES OF BUSINESS:

Hotels, Extended Stay

BRANDS/DIVISIONS/AFFILIATES:

Extended Stay America
Extended Stay Canada
Crossland Economy Studios
Hometown Inn

CONTACTS: Note: Officers with more than one job title may be intentionally listed here more than once.

Gerardo Lopez, CEO
Jim Donald, Pres.
Jonathan S. Halkyard, CFO

GROWTH PLANS/SPECIAL FEATURES:

Extended Stay America, Inc. (ESA), is the operator of the Extended Stay Hotels network, which has 682 extended-stay properties with 76,000 rooms in Canada and most of the major metro areas in the U.S. The company owns and manages moderately price extended-stay hotels under four brand names: Extended Stay America (632 locations), Extended Stay Canada (three locations) and Crossland Economy Studios & Hometown Inn (47 locations combined). ESA targets large corporate customers with multi-location extended-stay needs and has customized its rooms for this purpose with full kitchens, Internet connections and work/study areas. It offers daily, weekly and monthly rates with discounts for extended stays. ESA hotels all include kitchen, living and dining areas; laundry facilities; and pet accommodations. Certain locations offer pools, fitness rooms and individual DVD players. The company's properties are typically located near business centers, airports and entertainment areas. All Extended Stay Hotels participate in the affiliate program, which allows individuals to insert the company's link on their website and earn commission when individuals use the link to book stays. Other rewards and promotional programs include Kids Stay Free, AAA and Senior discounts, Suite Offer Email Program and a lowest internet price guarantee.

FINANCIAL DATA: Note: Data for latest year may not have been available at press time.

In U.S. $	2015	2014	2013	2012	2011	2010
Revenue	1,284,753,000	1,213,475,000	1,132,800,000	1,011,500,000	942,700,000	87,600,000
R&D Expense						
Operating Income						
Operating Margin %						
SGA Expense						
Net Income	113,040,000	39,596,000	86,200,000	20,700,000	45,600,000	1,767,100,000
Operating Cash Flow						
Capital Expenditure						
EBITDA						
Return on Assets %						
Return on Equity %						
Debt to Equity						

CONTACT INFORMATION:

Phone: 980-345-1600   Fax:
Toll-Free: 800-804-3724
Address: 11525 N. Community House Rd., Ste. 100, Charlotte, NC 28277
United States

STOCK TICKER/OTHER:

Stock Ticker: STAY
Employees: 3,700
Parent Company:

Exchange: NYS
Fiscal Year Ends: 12/31

SALARIES/BONUSES:

Top Exec. Salary: $ Bonus: $
Second Exec. Salary: $ Bonus: $

OTHER THOUGHTS:

Estimated Female Officers or Directors:
Hot Spot for Advancement for Women/Minorities:

Extendicare Inc

NAIC Code: 623110

www.extendicare.com

TYPES OF BUSINESS:

Long-Term Care
Assisted Living Facilities
Sub-Acute Care
Rehabilitative Services

BRANDS/DIVISIONS/AFFILIATES:

Extendicare Health Services Inc
Silver Group Purchasing
Extendicare (Canada) Inc
Revera Home Health
ParaMed
Nutrional Support System
Extendicare Assist

CONTACTS: Note: Officers with more than one job title may be intentionally listed here more than once.

Timothy Lukenda, CEO
Elaine Everson, CFO
Benjamin Hutzel, Chairman of the Board
Frederic Waks, Director
D.C. Keating, General Counsel, Subsidiary
Paul Tuttle, President, Divisional
Jillian Fountain, Secretary
R. Gurka, Senior VP, Subsidiary
Christina McKey, Vice President, Subsidiary
Deborah Bakti, Vice President, Subsidiary
Richard Luneburg, Vice President, Subsidiary
Timothy Detary, Vice President, Subsidiary
Judith Taubenheim, Vice President, Subsidiary

GROWTH PLANS/SPECIAL FEATURES:

Extendicare, Inc. is a leading provider of long-term care and related services in Canada. Through its subsidiaries, the firm operates 112 senior care centers. The company's long-term care provides long-stay services, short-stay services and complex continuing care for patients with complex health issues. This division includes a chronic care unit and 96 long-term care homes providing support to 13,000 residents across Ontario, Maintoba, Alberta and Saskatchewan. The retirement living segment provides services such as daily personal care, medication reminders, housekeeping, meals and planned social opportunities. Through ParaMed, it offers in-home personal care, homemaking and nursing services, wound and palliative care. ParaMed also offers workplace health and wellness programs including health and wellness clinics, mask fit test clinics for healthcare organizations, immunization clinics, seminars and education and consultation. Through Silver Group Purchasing (SGP) the firm provides cost savings and its Nutritional Support System offers assistance with menu development, nutritional analysis and pricing. The Extendicare Assist subsidiary provides managmeent and consulting services. In 2015, the company acquired the Revera Home Health business from Revera, Inc., and sold its U.S. business to a group of investors led by Formation Capital, LLC.

Extendicare offers its employees and their children educational assistance programs as well as scholarship programs for those pursuing health care professions.

FINANCIAL DATA: Note: Data for latest year may not have been available at press time.

In U.S. $	2015	2014	2013	2012	2011	2010
Revenue	748,890,800	623,906,000	1,547,662,000	1,557,560,000	1,600,882,000	1,581,786,000
R&D Expense						
Operating Income	45,230,420	38,852,360	52,116,840	75,314,970	46,681,400	141,124,400
Operating Margin %	6.03%	6.22%	3.36%	4.83%	2.91%	8.92%
SGA Expense	31,964,410	25,500,740	57,606,570	48,280,690	52,867,560	55,519,540
Net Income	177,418,800	-14,336,280	4,015,045	47,899,210	-23,237,110	39,513,640
Operating Cash Flow	40,362,980	65,444,780	74,854,750	83,165,410	81,630,340	114,180,300
Capital Expenditure	27,122,200	23,104,860	42,622,010	64,295,000	49,162,130	53,477,620
EBITDA	72,672,170	84,007,100	114,626,000	137,840,200	110,399,200	193,697,600
Return on Assets %	15.77%	- .99%	.28%	3.44%	-1.72%	3.07%
Return on Equity %	273.63%	-106.06%	11.35%	121.68%	-90.72%	
Debt to Equity	2.49		26.85	19.00	19.49	34.95

CONTACT INFORMATION:

Phone: 905-470-4000   Fax: 905-470-4003
Toll-Free:
Address: 3000 Steeles Ave. E., Ste. 700, Markham, ON L3R 9W2 Canada

STOCK TICKER/OTHER:

Stock Ticker: EXE   Exchange: TSE
Employees: 16,800   Fiscal Year Ends: 12/31
Parent Company:

SALARIES/BONUSES:

Top Exec. Salary: $ Bonus: $
Second Exec. Salary: $ Bonus: $

OTHER THOUGHTS:

Estimated Female Officers or Directors: 7
Hot Spot for Advancement for Women/Minorities: Y

Fairmont Homes Inc

www.fairmonthomes.com

NAIC Code: 321992

TYPES OF BUSINESS:

Manufactured Housing
Recreational Vehicles

BRANDS/DIVISIONS/AFFILIATES:

Cavco Industries Inc
Foxwood
Canadian
Bayview
Gulf Stream Coach Inc
Amerlite
Gulf Breeze
Sedona

CONTACTS: Note: Officers with more than one job title may be intentionally listed here more than once.

James F. Shea, Sr., CEO
James F. Shea, Sr., Pres
James F. Shea, Sr., Chmn.

GROWTH PLANS/SPECIAL FEATURES:

Fairmont Homes, Inc. manufactures single- and multi-section houses with up to 2,330 square feet. It offers several branded floor plan designs for sectional homes, such as Foxwood, Triumph and Canadian; single section homes, Bayview and Canadian; and modular sectional homes, Friendship, Kingsley and Fairmont. Floor plan features consist of 2-4 bedrooms, 1-3 bathrooms and extra features such as a study, retreat room, library, utility room, family room, great room, porch, basement, den, breakfast nook, dining room, living room, sunroom, morning room, sitting room and park trailer. The company also offers luxury features such as Amish-crafted cabinetry and custom designed kitchens. In addition to its floor plan offerings, builders can custom design homes. Through its partnership with Energy Star, the company provides builders with energy efficient feature choices. Fairmont has retailers throughout the Midwestern and Northeastern U.S. The firm's suppliers include Owens Corning, Moen, Pella and Trane. In addition to manufactured housing, Fairmont's Gulf Stream Coach subsidiary manufactures a full line of RVs, offering 18 brands of travel trailers, fifth wheels, light weight and toy hauler trailers. Gulf Stream brands include Amerlite, Gulf Breeze, Kingsport, Canyon Trail, Conquest, Ice Haven, Innsbruck, Matrix, Northern Express, Sedona, Streamlight, Track & Trail, Trailmaster, Vista Cruiser and Wide Open. In 2015, the firm was acquired by Cavco Industries, Inc., a producer of manufactured and modular housing, park model RVs and vacation cabins in the U.S.

Employees of the company receive health and life insurance, an onsite country club membership, a 401(k) plan, credit union membership, wellness programs and a prescription card.

FINANCIAL DATA: Note: Data for latest year may not have been available at press time.

In U.S. $	2015	2014	2013	2012	2011	2010
Revenue						
R&D Expense						
Operating Income						
Operating Margin %						
SGA Expense						
Net Income						
Operating Cash Flow						
Capital Expenditure						
EBITDA						
Return on Assets %						
Return on Equity %						
Debt to Equity						

CONTACT INFORMATION:

Phone: 574-773-7941   Fax: 574-773-2185
Toll-Free:
Address: 502 S. Oakland Ave., Nappanee, IN 46550 United States

STOCK TICKER/OTHER:

Stock Ticker: Subsidiary
Employees:
Parent Company: Cavco Industries Inc

Exchange:
Fiscal Year Ends: 12/31

SALARIES/BONUSES:

Top Exec. Salary: $ Bonus: $
Second Exec. Salary: $ Bonus: $

OTHER THOUGHTS:

Estimated Female Officers or Directors: 1
Hot Spot for Advancement for Women/Minorities:

Fairmont Raffles Hotels International Inc

www.frhi.com

NAIC Code: 721110

TYPES OF BUSINESS:

Hotels, Luxury
Spa Services
Real Estate Holdings

BRANDS/DIVISIONS/AFFILIATES:

Voyager Partners
Swissotel
Raffles
Fairmont Nanjing
Fairmont Chengdu
Fairmont Jakarta
Fairmont Pekin Moscow
Makkah Clock Royal Tower

CONTACTS: *Note: Officers with more than one job title may be intentionally listed here more than once.*

William R. Fatt, CEO
Chris J. Cahill, COO

GROWTH PLANS/SPECIAL FEATURES:

Fairmont Raffles Hotels International, Inc. is one of the world's largest luxury hotel firms. It operates over 130 hotels worldwide under the Raffles, Fairmont and Swissotel brand names. The firm offers many services to its business travel clients, such as high-speed Internet access; a 24 hour technology help desk; work centers with photocopying services, secretarial services, a private lounge and boardroom; printers for in-room use; a 24-hour fax service; and express checkout. Fairmont offers various types of resort accommodations, including spa resorts, golf resorts, ski resorts, Fairmont Gold and Fairmont Residences. Fairmont Spas feature Willow Stream spa facilities in many of its hotels. Fairmont Golf properties are located in cities around the world such as Acapulco, Mexico; St. Andrews, Scotland; Zimbali, South Africa; and Southampton, Bermuda. Fairmont Ski destinations include the Fairmont Chateau Whistler, Fairmont Le Manoir Richelieu, Fairmont Tremblant and many others. Fairmont Gold is an exclusive, private floor of the hotel with its own private check-in and check-out desk. Fairmont Gold offers a private lounge; a healthy continental breakfast; afternoon canapes and honor bar; complimentary newspapers; computer access in lounge; in-room high-speed Internet access; and a selection of DVDs, CDs, books and games. Fairmont Residences are located worldwide in locations such as Dubai and Vancouver. These properties are designed to be utilized as primary dwelling or as getaway retreats. Recently, Fairmont opened the Fairmont Nanjing and the Fairmont Chengdu in China; the Fairmont Jakarta in Indonesia; the Fairmont Pekin Moscow in Russia; the Makkah Clock Royal Tower in Riyadh, Saudi Arabia; and the Fairmont Jaipur and Fairmont Makati, its first luxury brand hotels in India and the Philippines respectively. The majority owner of the firm is Voyager Partners. In December 2014, Abu Dhabi's first Fairmont Residences became available for purchase. These luxury residences come fully furnished and overlook the Arabian Gulf. In March 2015, Raffles Jakarta was launched, the first Raffles Hotels & Resort property in the Indonesian market. In December of same year, the firm announced its plans to join AccorHotels.

FINANCIAL DATA: *Note: Data for latest year may not have been available at press time.*

In U.S. $	2015	2014	2013	2012	2011	2010
Revenue	1,000,000,000	950,000,000	925,000,000	858,000,000	800,000,000	750,000,000
R&D Expense						
Operating Income						
Operating Margin %						
SGA Expense						
Net Income						
Operating Cash Flow						
Capital Expenditure						
EBITDA						
Return on Assets %						
Return on Equity %						
Debt to Equity						

CONTACT INFORMATION:

Phone: 416-874-2600   Fax: 416-874-2601
Toll-Free: 800-257-7544
Address: 155 Wellington St. W., Ste. 3300, Toronto, ON M5K 0C3 Canada

STOCK TICKER/OTHER:

Stock Ticker: Subsidiary
Employees: 31,000
Parent Company: AccorHotels Group

Exchange:
Fiscal Year Ends: 12/31

SALARIES/BONUSES:

Top Exec. Salary: $ Bonus: $
Second Exec. Salary: $ Bonus: $

OTHER THOUGHTS:

Estimated Female Officers or Directors:
Hot Spot for Advancement for Women/Minorities:

Sales, profits and employees may be estimates. Financial information, benefits and other data can change quickly and may vary from those stated here.

Fannie Mae (Federal National Mortgage Association)

www.fanniemae.com
NAIC Code: 522294

TYPES OF BUSINESS:

Mortgages, Secondary Market
Mortgage-Related Securities
Financial Software

BRANDS/DIVISIONS/AFFILIATES:

Federal National Mortgage Association

CONTACTS: *Note: Officers with more than one job title may be intentionally listed here more than once.*

Timothy Mayopoulos, CEO
David Benson, CFO
Kimberly Johnson, Chief Risk Officer
Egbert Perry, Director
Andrew Bon Salle, Executive VP, Divisional
Brian Brooks, Executive VP
Jeffery Hayward, Other Corporate Officer
Zachary Oppenheimer, Other Corporate Officer
Bruce Lee, Other Corporate Officer
Joy Cianci, Senior VP, Divisional

GROWTH PLANS/SPECIAL FEATURES:

Fannie Mae is a government-sponsored enterprise chartered by the U.S. Congress under the name Federal National Mortgage Association (FNMA). Fannie Mae is currently in a conservatorship under the control of the U.S. federal government. The firm's activities include providing funds to mortgage lenders through the purchase of mortgage assets, then issuing and guarantying mortgage-related securities, which facilitates the flow of additional funds into the mortgage market. Fannie Mae does not offer mortgages directly to homebuyers, but operates exclusively in the secondary mortgage market, securitizing mortgages from primary market lenders. These include commercial banks, savings and loan associations, mortgage companies and securities dealers. The company operates in three segments: single-family credit guaranty, multifamily and capital markets. The single-family credit guaranty segment works with lender customers to securitize single-family mortgage loans into Fannie Mae mortgage-backed securities (MBS) and to facilitate the purchase of single-family mortgage loans. The multifamily segment works with lender customers to securitize multifamily mortgage loans into Fannie Mae MBS and to facilitate the purchase of multifamily mortgage loans. The capital markets segment manages the firm's investment activity in mortgage loans and mortgage-related securities. This segment is also responsible for managing the company's assets, liabilities and its liquidity and capital positions. Fannie Mae has mortgage help centers that review borrower's loan, discuss foreclosure alternatives, collect required documents for federal mortgage relief programs and reach a decision on any pending loan workout efforts.

The company offers its employees a 401(k); health insurance; life, long-term care and disability insurance; an employer-assisted housing program; paid vacation; childbirth and adoption leave; an employee assistance program; flexible spending accounts; and education assistance.

FINANCIAL DATA: *Note: Data for latest year may not have been available at press time.*

In U.S. $	2015	2014	2013	2012	2011	2010
Revenue	22,326,000,000	22,024,000,000	30,551,000,000	19,541,000,000	13,789,000,000	15,964,000,000
R&D Expense						
Operating Income	16,208,000,000	21,150,000,000	38,567,000,000	17,220,000,000	-16,945,000,000	-14,100,000,000
Operating Margin %	72.59%	96.03%	126.23%	88.12%	-122.88%	-88.32%
SGA Expense	1,884,000,000	1,498,000,000	1,446,000,000	1,413,000,000	1,455,000,000	1,485,000,000
Net Income	10,954,000,000	14,208,000,000	83,963,000,000	17,224,000,000	-16,855,000,000	-14,014,000,000
Operating Cash Flow	-6,673,000,000	-1,338,000,000	12,903,000,000	37,001,000,000	-15,238,000,000	-27,395,000,000
Capital Expenditure						
EBITDA						
Return on Assets %		-.03%	-.04%	.04%	-.82%	-1.06%
Return on Equity %						
Debt to Equity						

CONTACT INFORMATION:

Phone: 202 752-7000   Fax: 202 752-4934
Toll-Free: 800-732-6643
Address: 3900 Wisconsin Ave. NW, Washington, DC 20016 United States

STOCK TICKER/OTHER:

Stock Ticker: FNMA
Employees: 7,300
Parent Company:

Exchange: PINX
Fiscal Year Ends: 12/31

SALARIES/BONUSES:

Top Exec. Salary: $1,680,000   Bonus: $625,000
Second Exec. Salary: $2,100,000   Bonus: $

OTHER THOUGHTS:

Estimated Female Officers or Directors: 5

Hot Spot for Advancement for Women/Minorities: Y

Far East Organization

www.fareast.com.sg

NAIC Code: 531120

TYPES OF BUSINESS:

Lessors of Nonresidential Buildings (except Miniwarehouses)
Lessors of Residential Buildings and Dwellings
Hotels (except Casino Hotels) and Motels
Tea & Soft drink Manufacturing

BRANDS/DIVISIONS/AFFILIATES:

Far East Orchard Limited
Yeo Hiap Seng Limited
Far East Hospitality Trust
Adina Apartment Hotels
Marque
Oasia
Rendezvous Hotels
Vibe Hotels

CONTACTS: Note: Officers with more than one job title may be intentionally listed here more than once.

Chee Tat (Philip) Ng, CEO

GROWTH PLANS/SPECIAL FEATURES:

Far East Organization is the largest private property developer in Singapore. The firm has developed over 760 properties in the residential, hospitality, retail commercial, and industrial sectors, including 49,000 or 1 in 6 private homes in Singapore. Far East Organization's listed entities comprise Far East Orchard Limited, Yeo Hiap Seng Limited and Far East Hospitality Trust. Far East Orchard Limited develops residential, commercial and hospitality properties as well as operating business lines in hospitality management and health care real estate segments. Its hospitality portfolio covers over 80 properties with more than 13,000 rooms under the brands Adina Apartment Hotels, Medina Service Apartments, Marque, Oasia, Quincy, Rendezvous Hotels, Travelodge Hotels, Vibe Hotels and Village Hotels and Residences. Far East Orchard also owns a portfolio of medical suites in Singapore's premier medical hub of Novena and is co-developing the SBF Center with Far East Organization, a 31-storey commercial building located in the Central Business District that offers a broad spectrum of office and medical spaces. Yeo Hiap Seng Limited develops authentic Asian beverages and Southeast Asian sauces under the brand names of YEO'S, H-TWO-O, Justea and Pink Dolphin. Far East Hospitality Trust is the first and only Singapore-focused hotel and serviced residence hospitality trust listed on the Singapore Exchange Securities Trading Limited (SGX-ST). It is also Singapore's largest diversified hospitality portfolio by asset value.

FINANCIAL DATA: Note: Data for latest year may not have been available at press time.

In U.S. $	2015	2014	2013	2012	2011	2010
Revenue	3,600,000,000	3,600,000,000				
R&D Expense						
Operating Income						
Operating Margin %						
SGA Expense						
Net Income						
Operating Cash Flow						
Capital Expenditure						
EBITDA						
Return on Assets %						
Return on Equity %						
Debt to Equity						

CONTACT INFORMATION:

Phone: 65-6534-8000   Fax: 65-6235-2411
Toll-Free:
Address: 14 Scotts Rd., #06-00 Far East Plaza, Singapore, 228213 Singapore

STOCK TICKER/OTHER:

Stock Ticker: Private   Exchange:
Employees: 4,500   Fiscal Year Ends:
Parent Company:

SALARIES/BONUSES:

Top Exec. Salary: $ Bonus: $
Second Exec. Salary: $ Bonus: $

OTHER THOUGHTS:

Estimated Female Officers or Directors:
Hot Spot for Advancement for Women/Minorities:

Federal Agricultural Mortgage Corp

www.farmermac.com

NAIC Code: 522294

TYPES OF BUSINESS:

Mortgages, Secondary Market
Agricultural Real Estate Loans
Rural Housing Mortgage Loans

BRANDS/DIVISIONS/AFFILIATES:

Farmer Mac II LLC
Farmer Mac Mortgage Securities Corporation
Contour Valuation Services LLC
AgVantage

CONTACTS: Note: Officers with more than one job title may be intentionally listed here more than once.

Timothy Buzby, CEO
Robert Lynch, CFO
Gregory Ramsey, Chief Accounting Officer
Lowell Junkins, Director
Myles Watts, Director
Stephen Mullery, General Counsel
John Covington, Senior VP, Divisional

GROWTH PLANS/SPECIAL FEATURES:

Federal Agricultural Mortgage Corp., commonly known as Farmer Mac, is a secondary market for agricultural real estate and rural housing mortgage loans. Established by Congress in 1987, Farmer Mac seeks to improve the availability of mortgage credit to U.S. farmers, ranchers and rural homeowners, businesses and communities. The organization provides liquidity and lending capacity to agricultural mortgage lenders by purchasing newly originated and pre-existing mortgage loans directly from lenders; guaranteeing securities backed by eligible mortgage loans, referred to as Farmer Mac guaranteed securities; exchanging newly issued securities for newly originated and seasoned eligible mortgage loans that back those securities in so-called swap transactions; and issuing long-term standby purchase commitments for both newly originated and seasoned mortgage loans. The firm's three existing subsidiaries include Farmer Mac II LLC, Farmer Mac Mortgage Securities Corporation and Contour Valuation Services LLC. Farmer Mac II purchases the portions of certain agricultural, rural development, business & industry and community facilities loans guaranteed by the USDA under the Consolidated Farm and Rural Development Act. Farmer Mac Mortgage facilitates the purchase and issuance of Farmer Mac guaranteed securities. Contour Valuation appraises agricultural real estate. AgVantage is a registered trademark of Farmer Mac used to designate Farmer Mac's guarantees of securities related to general obligations of lenders that are secured by pools of eligible loans. Although created by Congress, Farmer Mac is a publicly traded corporation owned by its stockholders.

FINANCIAL DATA: Note: Data for latest year may not have been available at press time.

In U.S. $	2015	2014	2013	2012	2011	2010
Revenue	145,948,000	103,637,000	163,893,000	125,715,000	73,916,000	101,017,000
R&D Expense						
Operating Income	108,078,000	73,106,000	131,267,000	91,116,000	44,647,000	66,497,000
Operating Margin %	74.05%	70.54%	80.09%	72.47%	60.40%	65.82%
SGA Expense	37,571,000	33,587,000	29,380,000	30,309,000	27,616,000	28,039,000
Net Income	68,700,000	48,090,000	75,328,000	46,773,000	16,663,000	31,993,000
Operating Cash Flow	184,355,000	155,052,000	313,990,000	33,303,000	26,820,000	-484,203,000
Capital Expenditure						
EBITDA						
Return on Assets %	.31%	.27%	.55%	.35%	.12%	.28%
Return on Equity %	13.73%	12.43%	25.30%	16.00%	6.34%	13.88%
Debt to Equity	14.24	16.04	19.33	17.85	18.94	23.89

CONTACT INFORMATION:

Phone: 202-872-7700   Fax: 202-872-7713
Toll-Free: 800-879-3276
Address: 1999 K Street, N.W., 4th Fl., Washington, DC 20006 United States

STOCK TICKER/OTHER:

Stock Ticker: AGM
Employees: 71
Parent Company:

Exchange: NYS
Fiscal Year Ends: 12/31

SALARIES/BONUSES:

Top Exec. Salary: $675,000   Bonus: $
Second Exec. Salary: Bonus: $
$425,000

OTHER THOUGHTS:

Estimated Female Officers or Directors: 1
Hot Spot for Advancement for Women/Minorities:

Federal Realty Investment Trust
www.federalrealty.com

NAIC Code: 0

TYPES OF BUSINESS:
Real Estate Investment Trust
Retail Properties
Real Estate Development & Redevelopment

BRANDS/DIVISIONS/AFFILIATES:

CONTACTS: *Note: Officers with more than one job title may be intentionally listed here more than once.*
Donald Wood, CEO
James Taylor, CFO
Joseph Vassalluzzo, Chairman of the Board
Dawn Becker, Executive VP
David Faeder, Trustee
Jon Bortz, Trustee
Kristin Gamble, Trustee
Warren Thompson, Trustee
Gail Steinel, Trustee

GROWTH PLANS/SPECIAL FEATURES:
Federal Realty Investment Trust is an equity real estate investment trust (REIT) engaged in the ownership, management, development and redevelopment of retail and mixed-use properties. Federal Realty's portfolio encompasses approximately 89 community and neighborhood shopping centers and mixed-use properties throughout California, the Northeast and Mid-Atlantic regions, comprising a total of 20.2 million square feet. The company also holds a 30% share in a joint venture that owns seven properties totaling approximately 0.8 million square feet. The company focuses on the acquisition of commercial properties that have expansion, redevelopment and remerchandising potential. The firm particularly targets affluent and densely populated communities, with its largest markets by number of developments being Maryland (with 18 properties), followed by Virginia (15), California (13) and Pennsylvania (10). Federal Realty's portfolio is 100% leased to national, regional and local retailers. Its leasing clients include H&M, Banana Republic, Safeway, Crate & Barrel, TJ Maxx, Giant Food, CVS and Barnes & Noble, among many others. In 2015, the firm acquired a controlling interest in a 376,000 square-foot (sf) shopping center in Mountain View, California; an 80% interest in CocoWalk, a 198,000sf lifestyle center in Miami, Florida; and an 85% interest in The Shops at Sunset Place, a 515,000sf mixed-use center in South Miami, Florida.

The company offers employees medical, dental and vision coverage; educational assistance with tuition reimbursement; flexible spending accounts; life insurance; free parking; discounts on tickets to movies and entertainment events; and annual summer picnics.

FINANCIAL DATA: *Note: Data for latest year may not have been available at press time.*

In U.S. $	2015	2014	2013	2012	2011	2010
Revenue	744,012,000	686,090,000	637,413,000	608,018,000	553,059,000	544,674,000
R&D Expense						
Operating Income	300,154,000	271,037,000	254,161,000	255,262,000	227,697,000	230,474,000
Operating Margin %	40.34%	39.50%	39.87%	41.98%	41.17%	42.31%
SGA Expense	35,645,000	32,316,000	31,970,000	31,158,000	28,985,000	24,189,000
Net Income	210,219,000	164,535,000	162,681,000	151,925,000	143,917,000	122,790,000
Operating Cash Flow	359,835,000	346,130,000	314,498,000	296,633,000	244,711,000	256,735,000
Capital Expenditure	436,846,000	370,112,000	377,418,000	261,536,000	246,019,000	146,228,000
EBITDA	457,443,000	432,643,000	403,887,000	399,747,000	354,265,000	350,291,000
Return on Assets %	4.43%	3.74%	3.99%	4.00%	4.20%	3.83%
Return on Equity %	12.91%	10.81%	11.94%	12.19%	12.22%	10.59%
Debt to Equity	1.59	1.51	1.61	1.72	1.74	0.94

CONTACT INFORMATION:
Phone: 301 998-8100   Fax: 301 998-3700
Toll-Free: 800-658-8980
Address: 1626 E. Jefferson St., Rockville, MD 20852-4041 United States

STOCK TICKER/OTHER:
Stock Ticker: FRT
Employees: 436
Parent Company:

Exchange: NYS
Fiscal Year Ends: 12/31

SALARIES/BONUSES:
Top Exec. Salary: $850,000   Bonus: $
Second Exec. Salary: $450,000   Bonus: $

OTHER THOUGHTS:
Estimated Female Officers or Directors: 8
Hot Spot for Advancement for Women/Minorities: Y

Sales, profits and employees may be estimates. Financial information, benefits and other data can change quickly and may vary from those stated here.

FelCor Lodging Trust Inc

www.felcor.com

NAIC Code: 0

TYPES OF BUSINESS:
Real Estate Investment Trust
Hotel Ownership

BRANDS/DIVISIONS/AFFILIATES:
FelCor Lodging LP

CONTACTS: *Note: Officers with more than one job title may be intentionally listed here more than once.*
Richard Smith, CEO
Michael Hughes, CFO
Thomas Corcoran, Chairman of the Board
Jeffrey Symes, Chief Accounting Officer
Troy Pentecost, Executive VP
Jonathan Yellen, Executive VP
Thomas Hendrick, Executive VP

GROWTH PLANS/SPECIAL FEATURES:
FelCor Lodging Trust, Inc. (FelCor) is a hotel real estate investment trust (REIT) that, through its 99.5% partnership interest in FelCor Lodging LP, holds ownership interests in 41 hotels with a total of 12,443 rooms and suites. The company is among the largest U.S. owners of upscale, all-suite style hotel properties. Its hotels are operated under some of the most recognized and respected hotel brands, such as Doubletree, Embassy Suites, Fairmont, Hilton, Holiday Inn, Marriott, Renaissance, Sheraton and Wyndham as well as premium independent brands (Morgans, Knickerbocker and Royalton). FelCor's properties are located in major business and leisure travel markets, including New York, San Francisco, southern Florida, Atlanta, Los Angeles, Orlando, Dallas, San Diego, Boston and Philadelphia. The firm has alliances with the four brand owners that manage most of FelCor's hotels: Hilton Worldwide, whose brands include Embassy Suites Hotels, Hilton and Doubletree; InterContinental Hotels Group, owner of the Holiday Inn brand; Starwood Hotels & Resorts, whose brands include Sheraton and Westin; and Marriott International, Inc., which owns the Marriott and Renaissance hotel brands. In recent years, FelCor disclosed plans to dispose of its interests in several hotels. In early 2015, the firm sold Embassy Suites San Antonio International Airport Hotel, Embassy Suites Hotel (Raleigh) and Westin Hotel in Dallas Park; it also announced the opening of its Knickerbocker Hotel in New York City.

FINANCIAL DATA: *Note: Data for latest year may not have been available at press time.*

In U.S. $	2015	2014	2013	2012	2011	2010
Revenue	886,254,000	921,587,000	893,436,000	909,525,000	945,992,000	928,311,000
R&D Expense						
Operating Income	79,713,000	67,192,000	15,367,000	8,552,000	27,105,000	-137,540,000
Operating Margin %	8.99%	7.29%	1.72%	.94%	2.86%	-14.81%
SGA Expense	62,855,000	65,652,000	62,731,000	67,943,000	72,235,000	165,634,000
Net Income	-7,428,000	93,318,000	-61,504,000	-128,007,000	-129,854,000	-223,041,000
Operating Cash Flow	144,609,000	104,818,000	68,461,000	47,309,000	45,865,000	58,812,000
Capital Expenditure	81,961,000	170,229,000	161,910,000	146,324,000	346,638,000	38,936,000
EBITDA	194,165,000	234,264,000	144,541,000	66,490,000	165,997,000	10,123,000
Return on Assets %	-2.26%	2.51%	-4.61%	-7.23%	-7.07%	-10.50%
Return on Equity %					-195.35%	-193.06%
Debt to Equity					19.28	17.24

CONTACT INFORMATION:
Phone: 972 444-4900   Fax: 972 444-4949
Toll-Free:
Address: 545 E. John Carpenter Freeway, Ste. 1300, Irving, TX 75062
United States

STOCK TICKER/OTHER:
Stock Ticker: FCH
Employees: 61
Parent Company:

Exchange: NYS
Fiscal Year Ends: 12/31

SALARIES/BONUSES:
Top Exec. Salary: $764,909   Bonus: $
Second Exec. Salary: $442,554   Bonus: $

OTHER THOUGHTS:
Estimated Female Officers or Directors: 1
Hot Spot for Advancement for Women/Minorities:

Ferguson Enterprises Inc

www.ferguson.com

NAIC Code: 444190

TYPES OF BUSINESS:

Plumbing Supplies, Retail
Wholesale Distribution
Construction Supplies, Retail
Waterworks Supplies
HVAC Equipment, Retail
PVF Supplies, Retail

BRANDS/DIVISIONS/AFFILIATES:

Wolseley plc
Ferguson Xpress
HP Products Corporation
Builders Appliance Center LLC
Ar-Jay Building Products Inc
Equarius Inc

CONTACTS: Note: Officers with more than one job title may be intentionally listed here more than once.

Frank W. Roach, CEO
Kevin Murphy, COO
Frank W. Roach, Pres.
Dave Keltner, CFO

GROWTH PLANS/SPECIAL FEATURES:

Ferguson Enterprises, Inc. is one of the largest wholesale distributors of plumbing supplies in the U.S. The company has been a subsidiary of Wolseley plc, one of largest distributors of plumbing and heating products in the world, since 1982. Ferguson has approximately 1,400 locations in all 50 U.S. states, Puerto Rico, the Caribbean and Mexico. Additionally, the firm operates Ferguson Xpress stores, largely self-service locations that market plumbing and light commercial products to contractors. In general, Ferguson's customers include homeowners, builders, contractors, engineers and other trade professionals. Ferguson operates in eight business groups: residential, heating and cooling equipment, industrial, commercial and mechanical, waterworks, hospitality, government and integrated services. The company's product offerings include plumbing supplies; pipes, valves and fittings; heating, ventilation and air conditioning (HVAC); waterworks; lighting; appliances; tools and safety equipment; gas fireplaces; and fire protection products. An internal delivery service moves products from distribution hubs to Ferguson branches, satellites and customers. Through other divisions, the company is involved in nuclear power provision, fire protection supply, valve assembly and testing and geosynthetic product supply to the mining industry. In 2015, the company acquired HP Products Corporation, a plumbing wholesale distributor; Builders Appliance Center, LLC; an appliance dealer; Ar-Jay Building Products, Inc., a cabinet, lighting and fireplace showroom; Equarius, Inc., a Neptune Meter distributor; Redlon & Johnson Supply, the plumbing distribution division of The Gage Company; and eight other companies.

Ferguson offers employees medical, dental and vision coverage; a 401(k) plan; short- and long-term disability; life insurance; flexible spending accounts; educational assistance; paid leave; onsite employee training courses; and a performance awards program.

FINANCIAL DATA: Note: Data for latest year may not have been available at press time.

In U.S. $	2015	2014	2013	2012	2011	2010
Revenue	13,004,000,000	11,660,000,000	10,600,000,000	10,815,862,000	10,927,063,684	10,640,992,102
R&D Expense						
Operating Income						
Operating Margin %						
SGA Expense						
Net Income		418,051,240	360,389,000	62,864,370	218,370,685	-274,026,747
Operating Cash Flow						
Capital Expenditure						
EBITDA						
Return on Assets %						
Return on Equity %						
Debt to Equity						

CONTACT INFORMATION:

Phone: 757-874-7795   Fax: 757-989-2501
Toll-Free:
Address: 12500 Jefferson Ave., Newport News, VA 23602 United States

STOCK TICKER/OTHER:

Stock Ticker: Subsidiary   Exchange:
Employees: 21,000   Fiscal Year Ends: 07/31
Parent Company: WOLSELEY PLC

SALARIES/BONUSES:

Top Exec. Salary: $ Bonus: $
Second Exec. Salary: $ Bonus: $

OTHER THOUGHTS:

Estimated Female Officers or Directors: 3
Hot Spot for Advancement for Women/Minorities: Y

Ferrovial SA

www.ferrovial.com/en

NAIC Code: 488119

TYPES OF BUSINESS:

Airport Operations
Construction
Infrastructure Services
Toll Roads
Civil Engineering

BRANDS/DIVISIONS/AFFILIATES:

Cintra
Amey
Ferroser
Cespa
Cadagua
Ferrovial Agroman
Budimex

CONTACTS: Note: Officers with more than one job title may be intentionally listed here more than once.

Inigo Meiras, CEO
Ernesto Lopez Mozo, CFO
Jaime Aguirre de Carcer, Dir.-Human Resources
Federico Florez, CIO
Santiago Ortiz, Sec.
Santiago Olivares, CEO-Ferrovial Svcs.
Enrique Diaz-Rato, CEO-Cintra
Alejandro de la Joya, CEO-Ferrovial Agroman
Jorge Gil, CEO-Ferrovial Airports
Rafael del Pino Calvo, Chmn.

GROWTH PLANS/SPECIAL FEATURES:

Ferrovial SA is a leading infrastructure and industrial group with a presence in over 25 countries worldwide. The company has four business units: airports, toll roads, services and construction. Ferrovial's airports segment is one of the leading private airport operators in the world, conducting business largely through London Heathrow Airports (LHR). LHR manages four airports in the U.K.: Heathrow, Southampton, Glasgow and Aberdeen. Cintra is the firm's toll road and car parks division. It manages a total of 28 toll roads in Spain, Portugal, Ireland, Colombia, Australia, Greece, Canada, the U.K. and the U.S. The company's services segment consists of Amey, a British infrastructure maintenance subsidiary; Ferroser, an infrastructure management company in Spain; Cespa, a municipal and waste-water treatment subsidiary; and various other infrastructure and maintenance companies, chiefly in Spain and Portugal. Construction, the firm's original business, covers all aspects of civil engineering and building, including roads, railways, hydraulic works, maritime works, hydroelectric and industrial works. This division includes several subsidiaries: Cadagua, a water and waste treatment plant engineering and construction company; Ferrovial Agroman, the group's flagship construction company, engaged in civil engineering; Budimex, one of Poland's largest construction companies; and Webber, a construction group in Texas. In December 2015, the firm offered to acquire Australian services provider Broadspectrum Ltd. for $494 million (U.S.), but Broadspectrum rejected the offer in January 2016.

FINANCIAL DATA: Note: Data for latest year may not have been available at press time.

In U.S. $	2015	2014	2013	2012	2011	2010
Revenue	11,074,240,000	10,038,660,000	9,313,307,000	8,765,867,000	8,492,148,000	13,878,720,000
R&D Expense						
Operating Income	1,027,589,000	847,390,000	943,191,800	866,778,400	877,042,900	3,869,709,000
Operating Margin %	9.27%	8.44%	10.12%	9.88%	10.32%	27.88%
SGA Expense						
Net Income	821,158,500	458,480,200	829,142,000	809,753,500	1,446,151,000	2,070,004,000
Operating Cash Flow	1,288,763,000	1,629,772,000	1,478,085,000	1,345,788,000	812,034,500	2,856,947,000
Capital Expenditure	201,868,100	131,157,300	109,487,800	134,578,800	109,487,800	144,843,200
EBITDA	1,319,556,000	1,125,671,000	1,208,928,000	1,116,547,000	1,613,805,000	5,209,795,000
Return on Assets %	2.83%	1.66%	3.22%	3.14%	3.83%	4.94%
Return on Equity %	12.27%	7.05%	12.79%	12.05%	22.39%	52.87%
Debt to Equity	1.10	1.53	1.34	1.27	1.09	4.14

CONTACT INFORMATION:

Phone: 34 915862500   Fax: 34 915862677
Toll-Free:
Address: Principe de Vergara, 135, Madrid, 28002 Spain

STOCK TICKER/OTHER:

Stock Ticker: FRRVY   Exchange: PINX
Employees: 74,032   Fiscal Year Ends: 12/31
Parent Company:

SALARIES/BONUSES:

Top Exec. Salary: $ Bonus: $
Second Exec. Salary: $ Bonus: $

OTHER THOUGHTS:

Estimated Female Officers or Directors:
Hot Spot for Advancement for Women/Minorities:

Fibra Uno Administracion SA de CV

www.fibra-uno.com

NAIC Code: 531120

TYPES OF BUSINESS:

Lessors of Nonresidential Buildings (except Miniwarehouses)

BRANDS/DIVISIONS/AFFILIATES:

CuautiPark II
Puerta de Hierro

CONTACTS: *Note: Officers with more than one job title may be intentionally listed here more than once.*

Andre El-Mann, CEO
Charles El-Mann, COO
Javier Elizalde, CFO
Isidoro Attie, VP-Finance & Strategy

GROWTH PLANS/SPECIAL FEATURES:

Fibra Uno Administracion SA de CV is a real estate investment trust (REIT) which acquires, develops and operates commercial real estate in Mexico. The company targets properties in large urban markets as well as medium-sized metropolitan cities where it can focus on underserved communities. The firm currently has 488 properties in its portfolio, consisting of industrial, retail, office and mixed-use facilities. The company is directly involved in highest or best use real estate appraisal, project decision, project financing and administration. It also is involved with property acquisition, permits and licenses and commercialization either directly or through third parties. Its indirect operations include pre-project work and budgeting. During 2015, the firm acquired an office building in Florida as well as in Mexico City and agreed to acquire two additional office buildings and three shopping centers, all in Mexico City. It also acquired CuautiPark II, an industrial park, and six office buildings in Mexico City. In February 2016, the company acquired Puerta de Hierro hospital located in Guadalajara's metropolitan area.

FINANCIAL DATA: *Note: Data for latest year may not have been available at press time.*

In U.S. $	2015	2014	2013	2012	2011	2010
Revenue		396,572,000	221,503,900	88,123,750	30,167,940	
R&D Expense						
Operating Income		324,025,300	166,787,000	72,296,570	21,273,670	
Operating Margin %		81.70%	75.29%	82.03%	70.51%	
SGA Expense		32,623,940	33,757,530	11,308,320	3,498,530	
Net Income		324,025,300	514,903,400	72,296,570	23,548,540	
Operating Cash Flow		-19,812,450	-5,220,526	960,941	-1,084,111	
Capital Expenditure						
EBITDA		449,644,700	557,886,000	82,831,230	27,237,580	
Return on Assets %		4.65%	13.45%	5.92%	4.50%	
Return on Equity %		6.99%	22.07%	7.93%	5.12%	
Debt to Equity		0.32	0.46	0.34		

CONTACT INFORMATION:

Phone: 4170-7070   Fax:
Toll-Free:
Address: Antonio Dovali Jaime No 70, 11th Fl., Tower B, Zedec Santa Fe, 01210 Mexico

STOCK TICKER/OTHER:

Stock Ticker: FBASF   Exchange: PINX
Employees: Fiscal Year Ends:
Parent Company:

SALARIES/BONUSES:

Top Exec. Salary: $ Bonus: $
Second Exec. Salary: $ Bonus: $

OTHER THOUGHTS:

Estimated Female Officers or Directors:
Hot Spot for Advancement for Women/Minorities:

Fidelity National Financial Inc

www.fnf.com

NAIC Code: 524127

TYPES OF BUSINESS:

Title Insurance
Escrow Services
Collection & Trust Activities
Electronic Data Interchange Software
Payment Processing
Equipment Lease Services
Insurance Claims Management

BRANDS/DIVISIONS/AFFILIATES:

Fidelity National Title
Chicago Title
Commonwealth Land Title
Alamo Title
Remy International Inc
Ceridian Corporation
J. Alexander's Corporation
Fidelity National Title Group Inc

CONTACTS: Note: Officers with more than one job title may be intentionally listed here more than once.

Raymond Quirk, CEO
Anthony Park, CFO
William Foley, Chairman of the Board
Michael Nolan, Co-COO
Roger Jewkes, Co-COO
Frank Willey, Director
Brent Bickett, Executive VP, Divisional
Michael Gravelle, Executive VP
Peter Sadowski, Executive VP

GROWTH PLANS/SPECIAL FEATURES:

Fidelity National Financial, Inc. (FNF) is a holding company that provides title insurance, technology and transaction services to the real estate and mortgage industries. The firm claims to be the nation's largest title insurance company through the following title insurance underwriters: Fidelity National Title, Chicago Title, Commonwealth Land Title and Alamo Title. FNF also provides mortgage technology solutions and transaction services through its majority-owned subsidiaries: Black Knight Financial Services, LLC and ServiceLink Holdings, LLC. FNF owns majority and minority equity investment stakes in a number of entities, including American Blue Ribbon Holdings LLC; J. Alexander's LLC; Remy International, Inc.; Ceridian HCM, Inc.; Fleetcor Technologies, Inc.; and Digital Insurance, Inc. Black Knight provides integrated technology, services, data and analytics that lenders and servicers use to manage the life cycle of their loans. American Blue Ribbon Holdings operates over 690 company and franchise family and casual dining restaurants in more than 40 states under the Village Inn, Bakers Square, Max and Erma's, O' Charley's, and Ninety Nine brands. J. Alexander's manages 40 upscale dining restaurants in 14 states. Ceridian offers business service and software solutions that help organizations control costs, save time, optimize their workforce, and grow revenue. Digital Insurance is the nation's leading employee benefits agency specializing in insurance for small businesses and mid-sized companies.

FINANCIAL DATA: Note: Data for latest year may not have been available at press time.

In U.S. $	2015	2014	2013	2012	2011	2010
Revenue	9,132,000,000	8,024,000,000	8,565,000,000	7,201,700,000	4,839,600,000	5,740,300,000
R&D Expense						
Operating Income	867,000,000	392,000,000	651,000,000	843,400,000	414,800,000	562,400,000
Operating Margin %	9.49%	4.88%	7.60%	11.71%	8.57%	9.79%
SGA Expense	5,597,000,000	5,231,000,000	6,074,000,000	4,596,000,000	2,988,800,000	3,368,800,000
Net Income	527,000,000	583,000,000	402,000,000	606,500,000	369,500,000	370,100,000
Operating Cash Flow	917,000,000	567,000,000	484,000,000	620,000,000	124,900,000	182,500,000
Capital Expenditure	241,000,000	210,000,000	145,000,000	79,200,000	35,900,000	53,900,000
EBITDA						
Return on Assets %	3.88%	1.75%	3.93%	6.82%	4.69%	4.67%
Return on Equity %	9.19%	3.86%	8.61%	15.35%	10.46%	10.96%
Debt to Equity	0.48	0.47	0.26	0.31	0.25	0.27

CONTACT INFORMATION:

Phone: 904 854-8100   Fax: 904 357-1007
Toll-Free: 888-934-3354
Address: 601 Riverside Ave., Jacksonville, FL 32204 United States

STOCK TICKER/OTHER:

Stock Ticker: FNF
Employees: 54,091
Parent Company:

Exchange: NYS
Fiscal Year Ends: 12/31

SALARIES/BONUSES:

Top Exec. Salary: $850,030   Bonus: $
Second Exec. Salary: $769,133   Bonus: $

OTHER THOUGHTS:

Estimated Female Officers or Directors: 1
Hot Spot for Advancement for Women/Minorities:

First American Financial Corporation
NAIC Code: 524127

www.firstam.com

TYPES OF BUSINESS:
Title Insurance
Real Estate Services
Escrow Services
Screening Services
Credit Reporting
Property & Casualty Insurance
Trust Services
Internet Services

BRANDS/DIVISIONS/AFFILIATES:
TitleVest Holdings Group, LLC

CONTACTS: Note: Officers with more than one job title may be intentionally listed here more than once.
Dennis Gilmore, CEO
Mark Seaton, CFO
Parker Kennedy, Chairman of the Board
Matthew Wajner, Chief Accounting Officer
Christopher Leavell, COO, Subsidiary
Kenneth Degiorgio, Executive VP
Jeffrey Robinson, Vice President

GROWTH PLANS/SPECIAL FEATURES:

First American Financial Corporation (FAFC) provides title insurance, settlement services, as well as specialty insurance coverage and risk solutions for the real estate and mortgage industries. The company conducts its operations through two units: Title Insurance and Services, which accounted for 92 % of the firm's 2014 revenue; and Specialty Insurance. The Title Insurance and Services division issues title insurance policies on residential and commercial property in the U.S. and internationally. This segment also provides closing and/or escrow services; accommodates tax-deferred exchanges of real estate; provides products, services and solutions involving the use of real property related data designed to mitigate risk or otherwise facilitate real estate transactions; maintains, manages and provides access to title plant records and images; and provides banking, trust and investment advisory services. FAFC's Specialty insurance segment includes property and casualty insurance and home warranties. Its property and casualty insurance business provides insurance coverage to residential homeowners and renters for liability losses and typical hazards such as fire, theft, vandalism and other types of property damage. It is licensed to issue policies in all 50 states and Washington, D.C. Its home warranty business provides residential service contracts that cover residential systems such as heating and air conditioning systems, and certain appliances against failures that occur as the result of normal usage during the coverage period. Most of these policies are issued on resale residences, although policies are also available in some instances for new homes. In March 2015, the firm acquired TitleVest Holdings Group, LLC.

The firm offers its employees medical, dental, prescription drug and vision insurance; life, long-term care and AD&D; employee assistance program; a 401(k) plan; retirement plan; employee stock purchase plan; group legal and auto plans; and fitness discounts.

FINANCIAL DATA: Note: Data for latest year may not have been available at press time.

In U.S. $	2015	2014	2013	2012	2011	2010
Revenue	5,175,456,000	4,677,949,000	4,956,077,000	4,541,821,000	3,820,574,000	3,906,612,000
R&D Expense						
Operating Income	432,765,000	350,560,000	310,708,000	467,406,000	130,293,000	212,106,000
Operating Margin %	8.36%	7.49%	6.26%	10.29%	3.41%	5.42%
SGA Expense	3,251,657,000	2,881,647,000	3,082,276,000	2,756,363,000	2,381,761,000	2,438,324,000
Net Income	288,086,000	233,534,000	186,367,000	301,041,000	78,276,000	127,829,000
Operating Cash Flow	551,323,000	360,637,000	378,472,000	429,675,000	133,820,000	155,543,000
Capital Expenditure	123,697,000	97,222,000	87,142,000	83,892,000	69,797,000	88,725,000
EBITDA						
Return on Assets %	3.61%	3.29%	2.96%	5.27%	1.39%	2.25%
Return on Equity %	10.79%	9.29%	7.76%	13.75%	3.90%	6.39%
Debt to Equity	0.21	0.22	0.12	0.09	0.14	0.14

CONTACT INFORMATION:
Phone: 714 250-3000   Fax:
Toll-Free: 800-854-3643
Address: 1 First American Way, Santa Ana, CA 92707 United States

STOCK TICKER/OTHER:
Stock Ticker: FAF   Exchange: NYS
Employees: 17,955   Fiscal Year Ends: 12/31
Parent Company:

SALARIES/BONUSES:
Top Exec. Salary: $900,000   Bonus: $
Second Exec. Salary: Bonus: $
$725,000

OTHER THOUGHTS:
Estimated Female Officers or Directors: 1
Hot Spot for Advancement for Women/Minorities:

First Industrial Realty Trust Inc

www.firstindustrial.com

NAIC Code: 0

TYPES OF BUSINESS:

Real Estate Investment Trust-Industrial Properties

BRANDS/DIVISIONS/AFFILIATES:

CONTACTS: *Note: Officers with more than one job title may be intentionally listed here more than once.*

Bruce Duncan, CEO
Scott Musil, CFO
David Harker, Executive VP, Divisional
Peter Schultz, Executive VP, Divisional
Johannson Yap, Executive VP, Divisional
Daniel Hemmer, General Counsel

GROWTH PLANS/SPECIAL FEATURES:

First Industrial Realty Trust, Inc. owns, manages, acquires, sells, develops and redevelops industrial real estate in the U.S. Organized as a real estate investment trust (REIT), First Industrial owns and operates approximately 580 in-service industrial properties containing approximately 61.6 million square feet of gross leasable area located in 25 states. These properties fall into four primary sub-groups: light industrial (of which the firm's portfolio contains approximately 234 properties), R&D flex (78 properties), bulk warehouse (167 properties) and regional warehouse (101 properties). Most of its properties are located in business parks, with easy access to highways, rail lines or airports. The tenants that occupy its properties do business in the manufacturing, retail, wholesale trade, distribution and professional service industries. First Industrial's properties and land parcels are held through partnerships, corporations and limited liability companies controlled directly or indirectly by the firm.

The firm offers employees medical, dental and vision insurance; retirement planning; employee stock programs; and medical and family leave.

FINANCIAL DATA: *Note: Data for latest year may not have been available at press time.*

In U.S. $	2015	2014	2013	2012	2011	2010
Revenue	365,762,000	344,599,000	328,226,000	327,273,000	317,835,000	288,541,000
R&D Expense						
Operating Income	109,929,000	93,825,000	85,606,000	76,311,000	74,257,000	17,492,000
Operating Margin %	30.05%	27.22%	26.08%	23.31%	23.36%	6.06%
SGA Expense	25,362,000	23,418,000	23,152,000	25,103,000	20,638,000	26,589,000
Net Income	73,802,000	49,110,000	40,307,000	-1,318,000	-7,445,000	-202,821,000
Operating Cash Flow	162,149,000	137,176,000	125,751,000	136,422,000	87,534,000	83,189,000
Capital Expenditure	318,201,000	219,082,000	188,448,000	138,730,000	90,524,000	89,736,000
EBITDA	265,505,000	215,857,000	203,053,000	200,884,000	199,028,000	157,981,000
Return on Assets %	2.78%	1.79%	.99%	- .83%	- .99%	-7.47%
Return on Equity %	6.95%	4.27%	2.32%	-2.07%	-2.88%	-23.97%
Debt to Equity	1.34	1.28	1.15	1.21	1.44	2.05

CONTACT INFORMATION:

Phone: 312 344-4300   Fax: 312 922-6320
Toll-Free:
Address: 311 S. Wacker Dr., Ste. 3900, Chicago, IL 60606 United States

STOCK TICKER/OTHER:

Stock Ticker: FR   Exchange: NYS
Employees: 173   Fiscal Year Ends: 12/31
Parent Company:

SALARIES/BONUSES:

Top Exec. Salary: $832,000   Bonus: $
Second Exec. Salary: Bonus: $
$379,000

OTHER THOUGHTS:

Estimated Female Officers or Directors:
Hot Spot for Advancement for Women/Minorities:

FirstService Corporation

NAIC Code: 531100

www.firstservice.com

TYPES OF BUSINESS:

Real Estate Services
Residential Property Management
Security Services
Maintenance Services
Outsourcing Services
IT & Logistics Support
Security Systems Design

BRANDS/DIVISIONS/AFFILIATES:

FirstService Financial
FS Energy
Paul Davis Restoration
California Closets
CertaPro Painters
Pillar to Post
College Pro Painters
Floorcoverings International

CONTACTS: *Note: Officers with more than one job title may be intentionally listed here more than once.*

Jay Hennick, CEO
John Friedrichsen, CFO
Peter Cohen, Chairman of the Board
Douglas Cooke, Controller
D. Patterson, COO
Elias Mulamoottil, Senior VP, Divisional
Jeremy Rakusin, Vice President, Divisional
Neil Chander, Vice President, Divisional
Christian Mayer, Vice President, Divisional

GROWTH PLANS/SPECIAL FEATURES:

FirstService Corporation is a diversified provider of services to residential and commercial tenants worldwide. The company has two operating divisions: FirstService Residential and FirstService Brands. In addition, the company offers specialized property insurance, multiple financial products and loan services through FirstService Financial as well as energy advisory services through FS Energy. The FirstService Residential segment oversees residential communities and provides rental and leasing management, energy management, standard property management (leasing, maintenance, etc.), financial and insurance products and specialty management (such as pools). In total, this segment oversees roughly 7,400 residential communities containing over 1.6 million housing units in the U.S. and Canada. FirstService Brands offers property reservation services; owns and operates six franchise systems, including Paul Davis Restoration, California Closets, CertaPro Painters, Pillar to Post, College Pro Painters and Floorcoverings International; and franchising services (the reacquisition of profitable franchises).

FINANCIAL DATA: *Note: Data for latest year may not have been available at press time.*

In U.S. $	2015	2014	2013	2012	2011	2010
Revenue		2,714,273,024	2,343,633,920	2,305,537,024	2,224,171,008	1,986,270,976
R&D Expense						
Operating Income						
Operating Margin %						
SGA Expense						
Net Income		43,316,000	-18,039,000	5,850,000	74,110,000	13,564,000
Operating Cash Flow						
Capital Expenditure						
EBITDA						
Return on Assets %						
Return on Equity %						
Debt to Equity						

CONTACT INFORMATION:

Phone: 416-960-9500   Fax: 416 960-5333
Toll-Free:
Address: FirstService Bldg., 1140 Bay St., Ste. 4000, Toronto, ON M5S 2B4 Canada

STOCK TICKER/OTHER:

Stock Ticker: FSRV   Exchange: NAS
Employees: 24,000   Fiscal Year Ends: 12/31
Parent Company:

SALARIES/BONUSES:

Top Exec. Salary: $ Bonus: $
Second Exec. Salary: $ Bonus: $

OTHER THOUGHTS:

Estimated Female Officers or Directors:
Hot Spot for Advancement for Women/Minorities:

Five Star Products Inc

www.fivestarproducts.com

NAIC Code: 444190

TYPES OF BUSINESS:

Building Material Distribution
Decorating & Finishing Products
Hardware Distribution
Painting Supplies
Technical Services

BRANDS/DIVISIONS/AFFILIATES:

Novolac
Five Star Special Grout 400
Five Star Fluid Epoxy
Five Star EZ-Cure Grout

CONTACTS: *Note: Officers with more than one job title may be intentionally listed here more than once.*

Wilfred A. Martinez, CEO
Fred Rubin, CFO
Steven Lupien, VP-Mktg.
Claudino Petruccelli, Regional Sales Mgr.-North East

GROWTH PLANS/SPECIAL FEATURES:

Five Star Products, Inc. (Five Star) is engaged in the wholesale distribution of home decorating, hardware and finishing products. Five Star serves retail dealers in the Northeast and Mid-Atlantic states. The company offers a variety of high performance cement and epoxy based construction solutions for the industrial, infrastructure and marine markets. Specific products produced include concrete restoration, advanced vibration dampening products, waterproofing coatings, adhesives and machinery foundation systems. The firm is a major distributor of paint sundry items, interior and exterior stains, brushes, rollers, caulking compounds and hardware products. Five Star has also developed elastomeric grout, which absorbs vibration in the train and trolley industries; and a line of Novolac products for coatings, structural concrete and grout. The company offers its own line of patented products, which include Five Star Special Grout 400, Five Star Fluid Epoxy and Five Star EZ-Cure Grout. It distributes its products to lumberyards, do-it-yourself centers, hardware stores and paint stores throughout the world.

FINANCIAL DATA: *Note: Data for latest year may not have been available at press time.*

In U.S. $	2015	2014	2013	2012	2011	2010
Revenue						
R&D Expense						
Operating Income						
Operating Margin %						
SGA Expense						
Net Income						
Operating Cash Flow						
Capital Expenditure						
EBITDA						
Return on Assets %						
Return on Equity %						
Debt to Equity						

CONTACT INFORMATION:

Phone: 203-336-7900   Fax: 203-336-7930
Toll-Free:
Address: 750 Commerce Dr., Fairfield, CT 06825 United States

STOCK TICKER/OTHER:

Stock Ticker: Subsidiary   Exchange:
Employees: 47   Fiscal Year Ends: 12/31
Parent Company: Merit Group Inc (The)

SALARIES/BONUSES:

Top Exec. Salary: $ Bonus: $
Second Exec. Salary: $ Bonus: $

OTHER THOUGHTS:

Estimated Female Officers or Directors:
Hot Spot for Advancement for Women/Minorities:

Fluor Corp

NAIC Code: 237000

TYPES OF BUSINESS:

Heavy Construction and Engineering
Power Plant Construction and Management
Facilities Management
Procurement Services
Consulting Services
Project Management
Asset Management
Staffing Services

BRANDS/DIVISIONS/AFFILIATES:

ServiTrade
Ameco
Fluor Constructors International Inc

CONTACTS: Note: Officers with more than one job title may be intentionally listed here more than once.

David Seaton, CEO
Biggs Porter, CFO
Peter Oosterveer, COO
Glenn Gilkey, Executive VP, Divisional
Garry Flowers, Executive VP, Divisional
Ray Barnard, Executive VP, Divisional
Jose Luis Bustamante, Executive VP, Divisional
Carlos Hernandez, Executive VP
Robin Chopra, Other Corporate Officer
Bruce Stanski, President, Divisional

GROWTH PLANS/SPECIAL FEATURES:

Fluor Corp., through its subsidiaries, is a global provider of engineering, procurement, construction and maintenance services, with offices in over 25 countries. The company provides logistics services in both Afghanistan and Iraq. Besides being a primary service provider to the U.S. federal government, Fluor serves a diverse set of industries including oil and gas, chemical and petrochemicals, transportation, mining and metals, power, life sciences and manufacturing. It operates in five business segments: oil and gas, industrial and infrastructure, government, global services and power. The oil and gas segment offers design, engineering, procurement, construction and project management services to energy-related industries. Industrial and infrastructure provides design, engineering and construction services to the transportation, mining, life sciences, telecommunications, manufacturing, microelectronics and health care sectors. The government segment provides project management services, including environmental restoration, engineering, construction, site operations and maintenance, to the U.S. government, particularly to the Department of Energy, the Department of Homeland Security and the Department of Defense. Global services provides operations, maintenance and construction services as well as industrial fleet outsourcing, plant turnaround services, temporary staffing, procurement services and construction-related support. The power segment provides such services as engineering, procurement, construction, program management, start-up, commissioning and maintenance to the gas fueled, solid fueled, renewable and nuclear marketplaces. Separate from the rest of the businesses, Fluor Constructors International, Inc. provides unionized management and construction services in the U.S. and Canada. Subsidiary Ameco provides integrated mobile equipment and tool solutions and includes Mozambique construction equipment company, Servitrade. In December 2015, the firm agreed to acquire Stork Holding B.V., a Netherland-based provider of maintenance, modification and asset integrity services.

Fluor offers its employees health, dental, vision, life and accident insurance; disability coverage; savings and retirement plans; a tax savings account; and educational assistance.

FINANCIAL DATA: Note: Data for latest year may not have been available at press time.

In U.S. $	2015	2014	2013	2012	2011	2010
Revenue	18,114,050,000	21,531,580,000	27,351,570,000	27,577,140,000	23,381,400,000	20,849,350,000
R&D Expense						
Operating Income	926,367,000	1,216,322,000	1,190,043,000	733,987,000	985,456,000	548,982,000
Operating Margin %	5.11%	5.64%	4.35%	2.66%	4.21%	2.63%
SGA Expense	168,329,000	182,711,000	175,148,000	151,010,000	163,460,000	156,268,000
Net Income	412,512,000	510,909,000	667,711,000	456,330,000	593,728,000	357,496,000
Operating Cash Flow	849,132,000	642,574,000	788,906,000	628,378,000	889,769,000	550,914,000
Capital Expenditure	240,220,000	324,704,000	288,487,000	254,747,000	338,167,000	265,410,000
EBITDA	961,060,000	1,427,184,000	1,411,584,000	974,124,000	1,219,356,000	760,796,000
Return on Assets %	5.21%	6.18%	8.04%	5.51%	7.47%	4.83%
Return on Equity %	13.50%	14.87%	18.81%	13.54%	17.22%	10.51%
Debt to Equity	0.33	0.31	0.13	0.15	0.15	

CONTACT INFORMATION:

Phone: 469 398-7000   Fax: 469 398-7255
Toll-Free:
Address: 6700 Las Colinas Blvd., Irving, TX 75039 United States

STOCK TICKER/OTHER:

Stock Ticker: FLR
Employees: 38,758
Parent Company:

Exchange: NYS
Fiscal Year Ends: 12/31

SALARIES/BONUSES:

Top Exec. Salary: $1,333,302   Bonus: $

Second Exec. Salary: $721,180   Bonus: $350,000

OTHER THOUGHTS:

Estimated Female Officers or Directors: 3

Hot Spot for Advancement for Women/Minorities: Y

Fomento de Construcciones Y Contratas SA (FCC)   www.fcc.es

NAIC Code: 237000

TYPES OF BUSINESS:

Heavy & Civil Engineering Construction
Alternative Energy Development
Integrated Water Management
Cement Manufacturing
Logistics Services
Engineering Services
Railway Concessions

BRANDS/DIVISIONS/AFFILIATES:

Cementos Portland Valderrivas
FCC Aqualia
FCC Construction

CONTACTS: Note: Officers with more than one job title may be intentionally listed here more than once.

Carlos M. Jarque, CEO
Victor Pastor Fernandez, CFO
Francisco Martin Monteagudo, Gen. Dir.-Human Resources
Antonio Gomez Ciria, Gen. Dir.-Admin.
Jose Manuel Velasco Guardado, Gen. Dir.-Comm.
Juan Bejar, Exec. Chmn.-Cementos Portland Valderrivas SA
Eric Marotel, Managing Dir.-Cemusa
Eduardo Gonzalez Gomez, Gen Dir.-Energy
Miguel Hernanz Sanjuan, Gen. Dir.-Internal Audit
Esther Alcocer Koplowitz, Chmn.

GROWTH PLANS/SPECIAL FEATURES:

Fomento de Construcciones Y Contratas SA (FCC) is a Spanish construction and renewable energy developer. It divides its business into three major areas: medio ambiente (environment), water and infrastructure. Medio ambiente manages and treats domestic and industrial waste, cleans streets and maintains parks and gardens. The water division operates under FCC Aqualia, and provides water management services ranging from operating infrastructure to supplying households and businesses. The firm's infrastructure business operates under two companies: FCC Construction and Cementos Portland Valderrivas (CPV). FCC Construction is a leading construction, environmental and water service group in Europe. FCC Construction earns 55% of the FCC group's turnover. CPV is a leading cement production company in Spain, with an international location in Tunisia. FCC is involved in renewable energies, waste-to-power technologies and co-generation by maintaining 14 wind farms in Spain with an installed capacity of 421.8 megawatts (MW). It also operates two photovoltaic farms with 10 MW in capacity and two solar thermal energy farms with 50 MW each in capacity. In October 2014, FCC reached an agreement to sell its North American business (used oil and recycling) to Heritage Crystal Clean LLC.

FINANCIAL DATA: Note: Data for latest year may not have been available at press time.

In U.S. $	2015	2014	2013	2012	2011	2010
Revenue	7,385,892,000	7,223,989,000	7,671,546,000	12,719,090,000	13,406,290,000	13,816,180,000
R&D Expense						
Operating Income	369,324,000	-394,102,500	-345,723,700	-459,515,700	457,107,000	882,431,700
Operating Margin %	5.00%	-5.45%	-4.50%	-3.61%	3.40%	6.38%
SGA Expense						
Net Income	-52,794,790	-826,055,800	-1,717,938,000	-1,172,390,000	3,323,411	357,640,800
Operating Cash Flow	684,622,700	694,402,400	872,566,500	800,589,600	1,139,858,000	1,153,118,000
Capital Expenditure	386,512,400	449,319,700	510,071,700	591,929,800	602,218,200	820,499,300
EBITDA	920,373,900	825,544,900	532,305,800	714,543,600	1,368,426,000	1,856,328,000
Return on Assets %	- .34%	-4.88%	-8.53%	-4.87%	.48%	1.39%
Return on Equity %	-16.75%	-527.02%	-238.51%	-56.50%	4.14%	11.13%
Debt to Equity	19.26	19.96	289.08	3.45	1.80	2.66

CONTACT INFORMATION:

Phone: 34 913595400   Fax: 34 913594923
Toll-Free:
Address: Federico Salmon 13, Madrid, 28016 Spain

SALARIES/BONUSES:

Top Exec. Salary: $ Bonus: $
Second Exec. Salary: $ Bonus: $

STOCK TICKER/OTHER:

Stock Ticker: FMOCF   Exchange: PINX
Employees: 79,568   Fiscal Year Ends: 12/31
Parent Company:

OTHER THOUGHTS:

Estimated Female Officers or Directors: 2
Hot Spot for Advancement for Women/Minorities: Y

Forest Realty Trust Inc

NAIC Code: 531120

www.forestcity.net

TYPES OF BUSINESS:

Real Estate Development & Management
Office Buildings
Apartments
Retail Centers
Hotels
Land Development
Industrial Development
Military Housing

BRANDS/DIVISIONS/AFFILIATES:

Forest City Enterprises Inc
FCE Merger Sub Inc

CONTACTS: Note: Officers with more than one job title may be intentionally listed here more than once.

James Ratner, CEO, Subsidiary
Ronald Ratner, CEO, Subsidiary
David Larue, CEO
Robert OBrien, CFO
Bruce Ratner, Chairman of the Board, Subsidiary
Charles Ratner, Chairman of the Board
Charles Obert, Chief Accounting Officer
Samuel Miller, Co-Chairman Emeritus
Albert Ratner, Co-Chairman Emeritus
Duane Bishop, COO, Subsidiary
Brian Ratner, Director
Deborah Salzberg, Director
Andrew Passen, Executive VP, Divisional
Geralyn Presti, Executive VP
Linda Kane, Senior VP

GROWTH PLANS/SPECIAL FEATURES:

Forest City Realty Trust, Inc., formerly, Forest City Enterprises, Inc., has more than $10.0 billion in real estate assets and owns, develops, acquires and manages premier commercial and residential real estate throughout the U.S. Its portfolio consists of hotels, office buildings, apartment communities and retail and lifestyle centers. It operates through three strategic business units. The commercial group owns interests in 78 completed projects, including 37 office buildings with 10.3 million square feet of leasable space, 24 retail centers with 4.2 million square feet, 14 regional malls with 7.8 million square feet and three office buildings and a mall expansion under construction that total 645,000 square feet. The residential group deals with rental properties, senior housing, mixed-use facilities, military housing and new condos and town homes. It has 24,639 units in 74 apartment properties across 21 states and Washington, D.C., 40 senior subsidized housing properties with 6,813 total units, 11 military housing properties with 14,819 total units and 13 properties under construction that total 3,863 units. The land development group acquires and sells raw land and developed lots to commercial, residential and industrial customers; and develops raw land into master-planned communities, mixed-use projects and other residential developments. It currently owns over 403 acres of undeveloped land as well as a purchase option for 566 acres at its Stapleton project in Denver, Colorado over the next three years. In January 2016, the firm completed its conversion to real estate investment trust (REIT) status. It also announced completion of the merger of FCE Merger Sub, Inc., an Ohio corporation, with and into Forest City Enterprises, Inc., subsequently making Forest City Realty Trust, Inc. the parent company of Forest City Enterprises, Inc.

The company offers employees medical, dental, vision and life insurance; disability coverage; tuition reimbursement; and flexible spending accounts.

FINANCIAL DATA: Note: Data for latest year may not have been available at press time.

In U.S. $	2015	2014	2013	2012	2011	2010
Revenue		966,052,000	1,134,687,000	1,089,977,000	1,177,661,000	1,257,222,000
R&D Expense						
Operating Income		-177,457,000	148,571,000	83,093,000	241,228,000	246,717,000
Operating Margin %		-18.36%	13.09%	7.62%	20.48%	19.62%
SGA Expense		51,116,000				
Net Income		-7,595,000	36,425,000	-86,486,000	58,660,000	-30,651,000
Operating Cash Flow		262,022,000	422,742,000	298,763,000	267,247,000	421,535,000
Capital Expenditure		419,869,000	815,857,000	783,288,000	743,545,000	955,762,000
EBITDA		326,735,000	425,290,000	373,590,000	761,369,000	590,725,000
Return on Assets %		-.08%	.04%	-.91%	.39%	-.26%
Return on Equity %		-.44%	.31%	-7.97%	3.81%	-3.12%
Debt to Equity		2.87	4.61	5.35	6.19	7.65

CONTACT INFORMATION:

Phone: 216-621-6060   Fax: 216 362-2692
Toll-Free:
Address: 1100 Terminal Tower, 50 Public Sq., Cleveland, OH 44113
United States

STOCK TICKER/OTHER:

Stock Ticker: FCE.A
Employees: 2,865
Parent Company:

Exchange: NYS
Fiscal Year Ends: 01/31

SALARIES/BONUSES:

Top Exec. Salary: $600,000   Bonus: $450,000
Second Exec. Salary: $550,000   Bonus: $412,500

OTHER THOUGHTS:

Estimated Female Officers or Directors: 22
Hot Spot for Advancement for Women/Minorities: Y

Sales, profits and employees may be estimates. Financial information, benefits and other data can change quickly and may vary from those stated here.

Four Seasons Hotels Inc

www.fourseasons.com

NAIC Code: 721110

TYPES OF BUSINESS:

Hotels, Luxury
Luxury Condominiums
Conference Centers
Resort Time Shares

BRANDS/DIVISIONS/AFFILIATES:

Regent Hotels
Four Seasons Hotels
Triples Holdings Limited
Cascade Investment LLC
Four Seasons Hotel Bahrain Bay
Four Seasons Resort Dubai
Four Seasons Hotel Doha
Four Seasons Hotel Riyadh

CONTACTS: *Note: Officers with more than one job title may be intentionally listed here more than once.*

J. Allen Smith, CEO
J. Allen Amith, Pres.
John Davison, CFO
Peter Nowlan, Exec. VP
Chris Hunsberger, Exec. VP-Human Resources
Chris Hunsberger, Exec. VP-Prod. & Innovation
Nick Mutton, Exec. VP-Admin.
Sarah Cohen, Exec. VP
Christopher Norton, Exec. VP-Global Product & Oper.
Scott Woroch, Exec. VP-Worldwide Dev.
John Davison, Exec. VP-Residential
Chris Hart, Pres., Asia Pacific
Chris Hunsberger, Pres., Americas
Isadore Sharp, Chmn.

GROWTH PLANS/SPECIAL FEATURES:

Four Seasons Hotels, Inc. is a leading operator of luxury hotels and resorts. Headquartered in Toronto, the company manages 96 properties in 41 countries, mostly operated under the Four Seasons and Regent brands, owning roughly half of them. The firm offers its guests amenities such as monogrammed terry-cloth bathrobes, concierge service, in-room fax machines, overnight sandal and golf shoe repair and in-room exercise equipment installation, if requested. The No Luggage Required program provides a variety of crucial loan-items to customers who have lost their belongings. Many hotels provide experienced meeting and conference personnel to help guests plan business events such as award galas and multimedia presentations. Moreover, the firm offers a number of branded vacation ownership properties and private residences in Jackson Hole, Wyoming; Scottsdale, Arizona; San Francisco, California; Miami, Florida; Austin, Texas; and Toronto, Canada, among others. The Four Seasons has dozens of new hotels under development, including sites in Tanzania as well as India. The company is privately owned by investment firms controlled by Cascade Investment LLC and Triples Holdings Limited. In recent years, the firm added hotels to its portfolio in China, with plans to open approximately 10 more over the long term. The Four Seasons also recently announced development plans for a new resort property in Cesme, Turkey, scheduled to open in 2016. In March 2015, the firm opened the Four Seasons Hotel Bahrain Bay in Bahrain, along with the recently opened Four Seasons Resort Dubai, Four Seasons Hotel Doha and Four Seasons Hotel Riyadh in the brand's growing portfolio in the Gulf region.

The Four Seasons offers its employees such benefits as medical and dental coverage, disability and life insurance, a retirement pension plan, complimentary stays at Four Seasons properties with discounted meals, education assistance and paid holidays and vacation. The firm is consistently listed on Fortune magazine's annual list of the 100 Best Companies to Work For.

FINANCIAL DATA: *Note: Data for latest year may not have been available at press time.*

In U.S. $	2015	2014	2013	2012	2011	2010
Revenue	4,300,000,000	4,025,000,000	4,000,000,000	3,710,000,000	3,520,000,000	3,100,000,000
R&D Expense						
Operating Income						
Operating Margin %						
SGA Expense						
Net Income						
Operating Cash Flow						
Capital Expenditure						
EBITDA						
Return on Assets %						
Return on Equity %						
Debt to Equity						

CONTACT INFORMATION:

Phone: 416-449-1750   Fax: 416-441-4374
Toll-Free:
Address: 1165 Leslie St., Toronto, ON M3C 2K8 Canada

STOCK TICKER/OTHER:

Stock Ticker: Private   Exchange:
Employees: 20,990   Fiscal Year Ends: 12/31
Parent Company: Cascade Investment LLC

SALARIES/BONUSES:

Top Exec. Salary: $ Bonus: $
Second Exec. Salary: $ Bonus: $

OTHER THOUGHTS:

Estimated Female Officers or Directors: 3
Hot Spot for Advancement for Women/Minorities: Y

Sales, profits and employees may be estimates. Financial information, benefits and other data can change quickly and may vary from those stated here.

Freddie Mac (Federal Home Loan Mortgage Corp, FHLMC)

www.freddiemac.com
NAIC Code: 522294

TYPES OF BUSINESS:
Mortgages, Secondary Market
Credit Services

BRANDS/DIVISIONS/AFFILIATES:

CONTACTS: Note: Officers with more than one job title may be intentionally listed here more than once.
Donald Layton, CEO
James Mackey, CFO
Christopher Lynch, Chairman of the Board
Jerry Weiss, Chief Administrative Officer
Robert Lux, Chief Information Officer
David Brickman, Executive VP, Divisional
David Lowman, Executive VP, Divisional
Michael Hutchins, Executive VP, Divisional
William Mcdavid, Executive VP
Anil Hinduja, Executive VP

GROWTH PLANS/SPECIAL FEATURES:

Freddie Mac, officially known as Federal Home Loan Mortgage Corp., is a public company chartered by the U.S. Congress to create a continuous flow of funds to mortgage lenders in support of home ownership and rental housing. Freddie Mac is currently in conservatorship under the control of the U.S. federal government through the Federal Housing Finance Agency. The company purchases residential mortgages from lenders and then packages them into securities, which are sold to investors worldwide. The firm purchases 30-year, 20-year, 15-year and 10-year fixed-rate single-family mortgages; adjustable-rate mortgages (ARMs); and balloon/reset mortgages. Freddie Mac's mortgage securitization business receives the mortgage payments from the original lender or loan servicer and deducts a timeliness guarantee fee and other fees, passing the remainder on to the holder or holders of the mortgage-backed securities. The company implements regular public risk-based capital stress tests, initiates public interest-rate risk sensitivity analyses, discloses credit risk sensitivity analyses and obtains annual ratings from statistical rating organizations. Freddie Mac's primary business objectives are to support U.S. homeowners and renters by maintaining mortgage availability with alternatives that allow them to stay in their homes or avoid foreclosure; to reduce taxpayer exposure to losses by increasing the role of private capital in the mortgage market; and to support and improve the secondary mortgage market.

The company provides its employees benefits including employee assistance programs; an educational assistance program; a fitness center; transportation benefits; adoption expense reimbursement; domestic partner coverage; medical, dental, disability and vision coverage; parental, medical and military leave; and a 401(k) plan.

FINANCIAL DATA: Note: Data for latest year may not have been available at press time.

In U.S. $	2015	2014	2013	2012	2011	2010
Revenue	11,347,000,000	14,150,000,000	24,987,000,000	13,528,000,000	7,519,000,000	5,268,000,000
R&D Expense						
Operating Income	9,274,000,000	11,002,000,000	25,363,000,000	9,445,000,000	-5,666,000,000	-14,882,000,000
Operating Margin %	81.73%	77.75%	101.50%	69.81%	-75.35%	-282.49%
SGA Expense	1,374,000,000	1,296,000,000	1,208,000,000	1,143,000,000	1,444,000,000	1,546,000,000
Net Income	6,376,000,000	7,690,000,000	48,668,000,000	10,982,000,000	-5,266,000,000	-14,025,000,000
Operating Cash Flow	-934,000,000	8,885,000,000	18,532,000,000	8,466,000,000	10,320,000,000	9,861,000,000
Capital Expenditure						
EBITDA						
Return on Assets %		-.11%	-.17%	-.10%	-.53%	-1.27%
Return on Equity %						
Debt to Equity						

CONTACT INFORMATION:
Phone: 703 903-2000 Fax: 703 903-2759
Toll-Free: 800-424-5401
Address: 8200 Jones Branch Dr., McLean, VA 22102-3110 United States

STOCK TICKER/OTHER:
Stock Ticker: FMCC Exchange: PINX
Employees: 5,462 Fiscal Year Ends: 12/31
Parent Company:

SALARIES/BONUSES:
Top Exec. Salary: $2,100,000 Bonus: $
Second Exec. Salary: $2,100,000 Bonus: $

OTHER THOUGHTS:
Estimated Female Officers or Directors: 2

Hot Spot for Advancement for Women/Minorities: Y

Gables Residential Trust

www.gables.com

NAIC Code: 0

TYPES OF BUSINESS:

Real Estate Investment Trust
Apartment Communities

BRANDS/DIVISIONS/AFFILIATES:

Gables Corporate Accommodations
Express Luxury
Gables Great Awards

CONTACTS: Note: Officers with more than one job title may be intentionally listed here more than once.

Susan Ansel, CEO
Cris Sullivan, Operations
Susan Ansel, Pres.
Dawn Severt, CFO
Philip Altschuler, Sr. VP-Human Resources
Benjamin Pisklak, CIO
Cristina Sullivan, Exec. VP-Oper.
David Reece, Sr. VP-Finance & Capital Markets
Dennis E. Rainosek, Sr. VP-Portfolio Mgmt.
Benjamin Pisklak, Chief Investment Officer
Joseph Wilber, Sr. VP-Investments, East
Pamela Wade, VP-Corp. Accommodations

GROWTH PLANS/SPECIAL FEATURES:

Gables Residential Trust is a real estate investment trust (REIT) engaged in the business of multi-family apartment community management, development, construction, disposition and acquisition. The company owns upscale apartment communities in Arizona, California, Colorado, Florida, Georgia, Massachusetts, Maryland, North Carolina, New York, Texas, Virginia, Washington and Washington, D.C. Gables manages over 31,500 apartment homes and over 500,000 square feet of retail space. Management solutions offered by Gables include accounting, asset management, construction, design services, development, maintenance, marketing, renovation, resident services and vendor compliance. Gables offers its tenants Gables Great Awards, a program that provides carpet cleaning and home-improvements such as paint, patio/balcony detailing and vinyl replacement. In addition to residential apartments, the company also offers Gables Corporate Accommodations, a temporary housing provider for business travel, insurance claims and entertainment industry projects. Express Luxury packages for corporate accommodations include a studio, 1-, 2- or 3-bedroom apartment with utensils, cookware, dishes, utilities, a washer and dryer and maid service.

The company offers employees medical, dental and vision coverage; life insurance; free short- and long-term disability insurance; free AD&D insurance; flexible spending plans; a rental housing allowance; a 401(k) plan; an employee assistance plan; an associate discount program; a scholarship program for employees and their children; personal and family counseling; and a sabbatical program.

FINANCIAL DATA: Note: Data for latest year may not have been available at press time.

In U.S. $	2015	2014	2013	2012	2011	2010
Revenue						
R&D Expense						
Operating Income						
Operating Margin %						
SGA Expense						
Net Income						
Operating Cash Flow						
Capital Expenditure						
EBITDA						
Return on Assets %						
Return on Equity %						
Debt to Equity						

CONTACT INFORMATION:

Phone: 404-923-5500   Fax:
Toll-Free:
Address: 3399 Peachtree Road NE, Ste. 600, Atlanta, GA 30326 United States

STOCK TICKER/OTHER:

Stock Ticker: Private
Employees:
Parent Company: Clarion Partners LLC

Exchange:
Fiscal Year Ends: 12/31

SALARIES/BONUSES:

Top Exec. Salary: $ Bonus: $
Second Exec. Salary: $ Bonus: $

OTHER THOUGHTS:

Estimated Female Officers or Directors: 14
Hot Spot for Advancement for Women/Minorities: Y

Sales, profits and employees may be estimates. Financial information, benefits and other data can change quickly and may vary from those stated here.

Gafisa SA

NAIC Code: 531100

TYPES OF BUSINESS:

Construction & Real Estate Services
Construction Services
Real Estate Services
Residential Communities
Commercial Buildings
Third Party Construction Services

BRANDS/DIVISIONS/AFFILIATES:

Construtora Tenda SA
Gafisa Vendas Rio
Gafisa Vendas

CONTACTS: Note: Officers with more than one job title may be intentionally listed here more than once.

Sandro Gambia, CEO
Andre Bergstein, CFO
Fernado Cesar Calamita, Exec. Officer-Oper., Control Dept.
Andre Bergstein, Investor Relations Officer
Sandro Rogerio da Silva Gamba, Officer-Real Estate Dev.
Luiz Carlos Siciliano, Officer-Supply Chain

GROWTH PLANS/SPECIAL FEATURES:

Gafisa SA, based in Sao Paulo, Brazil, provides construction and real estate services to Brazil's housing market. Since its founding, Gafisa has completed and sold more than 1,100 developments and constructed over 12 million square feet of residential space. The firm focuses primarily on residential markets, and over half of its construction revenues annually derive from middle and upper-income residential developments. Its homebuilding operations are divided into three segments. Construtora Tenda SA (Tenda) represents its activities within the low-income/entry-level housing market, with regional store fronts and projects developed in 113 municipalities. Gafisa focuses on residential developments within the upper, upper-middle and middle-income segments. Gafisa's brokerage activities are handled through its subsidiaries Gafisa Vendas, which is active in Sao Paulo, and Gafisa Vendas Rio, active in the state of Rio de Janeiro. Gafisa also provides construction services to third parties across Brazil, primarily to developers of residential and commercial projects who do not construct their own designs. The company has a presence in roughly 130 Brazilian cities, including Sao Paulo, Rio de Janeiro, Salvador, Fortaleza, Natal, Curitiba, Belo Horizonte, Manaus, Porto Alegre and Belem, across 23 states. More than half of new projects are typically located in the populous states of Sao Paulo and Rio de Janeiro.

FINANCIAL DATA: Note: Data for latest year may not have been available at press time.

In U.S. $	2015	2014	2013	2012	2011	2010
Revenue		591,518,600	682,326,200	1,087,142,000	808,631,200	935,829,400
R&D Expense						
Operating Income		-5,371,247	110,513,400	47,245,900	-165,802,500	107,999,400
Operating Margin %		- .90%	16.19%	4.34%	-20.50%	11.54%
SGA Expense		130,076,700	145,218,900	206,236,400	177,274,000	141,785,000
Net Income		-11,700,860	238,544,400	-34,238,260	-259,836,100	72,754,650
Operating Cash Flow		11,519,910	81,853,490	179,008,100	-225,343,200	-296,898,900
Capital Expenditure		24,346,060	22,272,850	31,571,060	26,099,440	17,451,330
EBITDA		47,262,130	150,139,700	66,984,930	-136,158,600	152,521,700
Return on Assets %		- .55%	10.05%	-1.34%	-10.18%	3.27%
Return on Equity %		-1.36%	30.24%	-4.79%	-30.38%	9.22%
Debt to Equity		0.50	0.59	1.08	0.27	0.69

CONTACT INFORMATION:

Phone: 5.51130259e+11   Fax: 5.5113025922e+11
Toll-Free:
Address: Ave. Nacoes Unidas 8501, 19th Fl., Sao Paulo, SP 05425-070 Brazil

STOCK TICKER/OTHER:

Stock Ticker: GFA
Employees: 2,162
Parent Company:

Exchange: NYS
Fiscal Year Ends: 12/31

SALARIES/BONUSES:

Top Exec. Salary: $ Bonus: $
Second Exec. Salary: $ Bonus: $

OTHER THOUGHTS:

Estimated Female Officers or Directors:
Hot Spot for Advancement for Women/Minorities:

GE Capital

NAIC Code: 522200

TYPES OF BUSINESS:

Commercial Lending and Finance
Consumer Financial Products & Services
Real Estate Finance
Commercial Financial Products & Services
Leases
Energy Industry Lending and Finance

BRANDS/DIVISIONS/AFFILIATES:

General Electric Co

CONTACTS: Note: Officers with more than one job title may be intentionally listed here more than once.

Keith S. Sherin, CEO
Tom Gentile, COO
Robert Green, CFO
Stacey Hoin, Sr. VP-Human Resources
Julie Stansbury, CIO
Alex Dimitrief, General Counsel
Aris Kekedjian, VP
Russell Wilkerson, Managing Dir.-Comm. & Public Affairs
Kathy Cassidy, Treas.
Michael Pilot, Chief Commercial Officer
Ryan A. Zanin, Chief Risk Officer
Alec Burger, CEO
Michael Silva, Chief Regulatory Officer
Keith S. Sherin, Chmn.
Richard A. Laxer, Pres.

GROWTH PLANS/SPECIAL FEATURES:

GE Capital is the financial services unit of General Electric Co. Its financial solutions cater to the aviation, healthcare and energy sectors. Aviation financing includes more than 245 customers in over 75 countries, with the division offering a selection of aircraft and engine leasing and lending options tailored to each business' needs. Solutions include leasing, lending, trading, consulting and servicing. This segment's industry specializations include aircraft associations, aircraft manufacturers, preferred aircraft manufacturers, preferred aircraft dealers/brokers and helicopter manufacturers, as well as the commercial, marine and military sectors. Healthcare financing provides capital and services to the health care industry with investments in more than 30 sub-sectors, including senior housing, hospitals, pharmaceuticals and medical devices. Energy financing and investing has committed $16 billion for energy assets, diversifying across renewable energy, thermal power, oil and gas reserves and infrastructure. This division offers structured, common and second- and third-stage equity and debt. In April 2015, GE announced plans to sell nearly all of GE Capital's assets and business interests. GE will then focus on its industrial businesses. However, GE will keep its aircraft leasing, healthcare finance and energy finance businesses. The sale will include most of GE's investment real estate assets. In late 2015, GE closed the sale of GE Capital's health care lending business to bank and credit card company Capital One. In March 2016, it sold the North American portions of GE Capital's commercial distribution finance & vendor finance businesses, as well as a portion of its corporate finance business.

FINANCIAL DATA: Note: Data for latest year may not have been available at press time.

In U.S. $	2015	2014	2013	2012	2011	2010
Revenue	10,801,000,000	11,320,000,000	11,267,000,000	11,268,000,000	11,843,000,000	49,163,000,000
R&D Expense						
Operating Income						
Operating Margin %						
SGA Expense						
Net Income	-7,983,000,000	1,209,000,000	401,000,000	1,245,000,000	1,469,000,000	3,083,000,000
Operating Cash Flow						
Capital Expenditure						
EBITDA						
Return on Assets %						
Return on Equity %						
Debt to Equity						

CONTACT INFORMATION:

Phone: 203-840-6300   Fax:
Toll-Free:
Address: 901 Main Ave., Norwalk, CT 06851 United States

STOCK TICKER/OTHER:

Stock Ticker: Subsidiary   Exchange:
Employees: 47,000   Fiscal Year Ends: 12/31
Parent Company: General Electric Co (GE)

SALARIES/BONUSES:

Top Exec. Salary: $ Bonus: $
Second Exec. Salary: $ Bonus: $

OTHER THOUGHTS:

Estimated Female Officers or Directors: 3
Hot Spot for Advancement for Women/Minorities: Y

General Growth Properties Inc

www.ggp.com

NAIC Code: 0

TYPES OF BUSINESS:

Real Estate Investment Trust
Property Management

BRANDS/DIVISIONS/AFFILIATES:

The Crown Building

CONTACTS: Note: Officers with more than one job title may be intentionally listed here more than once.

Sandeep Mathrani, CEO
Michael Berman, CFO
Bruce Flatt, Chairman of the Board
Shobi Khan, COO
Richard Pesin, Executive VP, Divisional
Marvin Levine, Executive VP
Alan Barocas, Senior Executive VP, Divisional
Tara Marszewski, Senior VP

GROWTH PLANS/SPECIAL FEATURES:

General Growth Properties, Inc. (GGP) is a real estate investment trust (REIT) with ownership and management interests in 131 retail properties located throughout the U. S. comprising approximately 128 million square feet of gross leasable area. The firm's operations consist of development and management of retail and other rental properties, primarily regional malls. Its property portfolio predominantly is comprised of Class A malls, which are the core centers of retail, dining and entertainment within their trade areas. The retail properties account for 96.9 % of revenue and also includes festival market places, urban mixed-use centers and strip/community centers. GGP's three largest tenants are Limited Brands, Inc., The Gap, Inc. and Foot Locker, Inc. In April 2015, the firm acquired The Crown Building, a 26-story retail and office property located at 730 5th Avenue in New York City.

The firm offers employees medical, dental and vision insurance; life insurance; disability coverage; same sex domestic partner coverage; a 401(k) savings plan; flexible spending accounts; an employee assistance program; and various employee discounts.

FINANCIAL DATA: Note: Data for latest year may not have been available at press time.

In U.S. $	2015	2014	2013	2012	2011	2010
Revenue	2,403,906,000	2,535,559,000	2,527,387,000	2,511,850,000	2,742,942,000	
R&D Expense						
Operating Income	923,893,000	941,513,000	832,159,000	743,059,000	592,929,000	
Operating Margin %	38.43%	37.13%	32.92%	29.58%	21.61%	
SGA Expense	233,919,000	87,506,000	240,228,000	232,780,000	280,209,000	
Net Income	1,374,561,000	665,850,000	302,528,000	-481,233,000	-313,172,000	
Operating Cash Flow	1,064,888,000	949,724,000	889,531,000	807,103,000	502,802,000	
Capital Expenditure	1,078,891,000	1,162,186,000	982,472,000	702,346,000	253,276,000	
EBITDA	2,268,325,000	1,837,950,000	1,832,789,000	1,167,113,000	1,780,360,000	
Return on Assets %	5.49%	2.54%	1.08%	-1.69%	-1.01%	
Return on Equity %	17.65%	8.53%	3.72%	-5.97%	-3.37%	
Debt to Equity	1.79	2.17	1.99	2.09	2.04	

CONTACT INFORMATION:

Phone: 312 960-5000   Fax: 312 960-5475
Toll-Free:
Address: 110 N. Wacker Dr., Chicago, IL 60606 United States

STOCK TICKER/OTHER:

Stock Ticker: GGP   Exchange: NYS
Employees: 1,800   Fiscal Year Ends: 12/31
Parent Company:

SALARIES/BONUSES:

Top Exec. Salary: Bonus: $
$1,200,000
Second Exec. Salary: Bonus: $
$750,000

OTHER THOUGHTS:

Estimated Female Officers or Directors: 2

Hot Spot for Advancement for Women/Minorities: Y

Gensler

NAIC Code: 541310

TYPES OF BUSINESS:
Architectural Services
Consulting Services

BRANDS/DIVISIONS/AFFILIATES:

CONTACTS: Note: Officers with more than one job title may be intentionally listed here more than once.

Andy Cohen, Co-CEO
Diane Hoskins, Co-CEO
Dianne Hoskins, Exec. Dir.
David Gensler, Exec. Dir.
Andy Cohen, Chmn.

GROWTH PLANS/SPECIAL FEATURES:
Gensler is an architecture and planning firm with 46 locations in North America, Latin America, Asia and Europe. The company utilizes around 5,000 professionals to provide services in the areas of aviation & transportation, brand design, commercial office building development, consulting, entertainment, financial services firms, hospitality, urban design, product design, retail centers and sports & recreation. The firm incorporates sustainable design in its projects, having designed the first large-scale office building in the U.S. to employ underfloor air systems (the U.S. headquarters of Epson America, Inc.) and the first LEED (Leadership in Energy & Environmental Design)-certified car dealership in the U.S. (Pat Lobb Toyota of McKinney, Texas). Other notable projects include the Abu Dhabi Financial Center, Denver International Airport and Columbia College in Chicago. Gensler has more than 2,700 current clients that span practically every market sector possible.

FINANCIAL DATA: Note: Data for latest year may not have been available at press time.

In U.S. $	2015	2014	2013	2012	2011	2010
Revenue	1,075,000,000	807,000,000	751,000,000	722,000,000	651,000,000	
R&D Expense						
Operating Income						
Operating Margin %						
SGA Expense						
Net Income						
Operating Cash Flow						
Capital Expenditure						
EBITDA						
Return on Assets %						
Return on Equity %						
Debt to Equity						

CONTACT INFORMATION:
Phone: 415-433-3700   Fax:
Toll-Free:
Address: 2 Harrison St., Ste. 400, San Francisco, CA 94105 United States

STOCK TICKER/OTHER:
Stock Ticker: Private
Employees: 5,000
Parent Company:

Exchange:
Fiscal Year Ends:

SALARIES/BONUSES:
Top Exec. Salary: $ Bonus: $
Second Exec. Salary: $ Bonus: $

OTHER THOUGHTS:
Estimated Female Officers or Directors: 2
Hot Spot for Advancement for Women/Minorities:

Gilbane Inc

www.gilbaneco.com

NAIC Code: 236220

TYPES OF BUSINESS:
Construction & Real Estate Services

BRANDS/DIVISIONS/AFFILIATES:
Gilbane Building Company
Gilbane Development Company

CONTACTS: *Note: Officers with more than one job title may be intentionally listed here more than once.*
Thomas F. Gilbane Jr., CEO
John T. Ruggieri, CFO
Thomas F. Gilbane Jr., Chmn.

GROWTH PLANS/SPECIAL FEATURES:
Gilbane Inc., based in Providence, Rhode Island, is one of the largest privately held family-owned real estate development and construction firms in the industry. Gilbane is comprised of two operating companies: Gilbane Building Company (GBC) and Gilbane Development Company (GDC). These two often work jointly as one company to provide integrated expertise in finance, property development, planning and commercial construction. GBC provides a full slate of construction and facilities-related services. Under pre-construction services, GBC offers transition planning & management, building information modeling, conceptual cost modeling, high-performance building & energy modeling and interdisciplinary document coordination. The construction services offered consist of Construction Manager (CM) at risk, CM as agent, design-build, integrated project delivery, lump sum general contracting and program management. Some of GBC's consulting services are environmental solutions, sustainability services, building information modeling and facilities management services. GDC offers turnkey solutions to a broad range of real estate development needs, from strategic planning to delivering a comprehensive building and development program. Services offered by GDC include public private partnerships, build-to-suit, turnkey project management, real estate strategic planning and property management.

FINANCIAL DATA: *Note: Data for latest year may not have been available at press time.*

In U.S. $	2015	2014	2013	2012	2011	2010
Revenue	3,850,000,000	4,150,000,000	4,100,000,000	4,200,000,000	4,300,000,000	3,140,000,000
R&D Expense						
Operating Income						
Operating Margin %						
SGA Expense						
Net Income						
Operating Cash Flow						
Capital Expenditure						
EBITDA						
Return on Assets %						
Return on Equity %						
Debt to Equity						

CONTACT INFORMATION:
Phone: 401-456-5800   Fax:
Toll-Free: 800-445-2263
Address: 7 Jackson Walkway, Providence, RI 02903 United States

STOCK TICKER/OTHER:
Stock Ticker: Private   Exchange:
Employees: 2,607   Fiscal Year Ends:
Parent Company:

SALARIES/BONUSES:
Top Exec. Salary: $ Bonus: $
Second Exec. Salary: $ Bonus: $

OTHER THOUGHTS:
Estimated Female Officers or Directors:
Hot Spot for Advancement for Women/Minorities:

Glenborough LLC

www.glenborough.com

NAIC Code: 0

TYPES OF BUSINESS:

Real Estate Investment Trust
Office Properties
Property Management
Property Development

BRANDS/DIVISIONS/AFFILIATES:

Morgan Stanley Real Estate

CONTACTS: *Note: Officers with more than one job title may be intentionally listed here more than once.*

Andrew Batinovich, CEO
Michael A. Steele, COO
Andrew Batinovich, Pres.
Brad Kettmann, VP-Finance
Carlos Santamaria, VP-Eng. Svcs.
G. Lee Burns, General Counsel
Terri Garnick, Sr. VP-Acct.
Alan Shapiro, Sr. VP-Investments

GROWTH PLANS/SPECIAL FEATURES:

Glenborough, LLC is a privately-held realty trust focused on acquiring, managing, leasing and developing high-quality office properties. The company operates in four business segments: acquisitions, property management and leasing, development and investments. The acquisitions division is responsible for property purchases and investments as well as all legal contract negotiations and underwriting. The property management and leasing division assists clients with energy conservation and environmental and information technology solutions. Development services include entitlement analysis and management, design coordination and review and project and construction management of new development and renovation projects. Investment services compliment the acquisition division by providing asset management and co-investment opportunity programs with institutional investors and pension funds. The company holds a portfolio of primarily high-quality, multi-tenant office properties, with some industrial and multi-use buildings. The firm's portfolio encompasses major markets across the U.S., including Southern California, Denver, Washington, D.C., New York and San Francisco. The company performs all property management, leasing, construction supervision, accounting, finance, acquisition and disposition activities for its portfolio of properties. It also performs all portfolio management activities, including onsite property management, lease negotiations and construction supervision of tenant improvements, property renovations and capital expenditures. The firm is owned by Morgan Stanley Real Estate.

FINANCIAL DATA: *Note: Data for latest year may not have been available at press time.*

In U.S. $	2015	2014	2013	2012	2011	2010
Revenue						
R&D Expense						
Operating Income						
Operating Margin %						
SGA Expense						
Net Income						
Operating Cash Flow						
Capital Expenditure						
EBITDA						
Return on Assets %						
Return on Equity %						
Debt to Equity						

CONTACT INFORMATION:

Phone: 650-343-9300   Fax: 650-343-7438
Toll-Free:
Address: 400 S. El Camino Real, San Mateo, CA 94402-1708 United States

STOCK TICKER/OTHER:

Stock Ticker: Private   Exchange:
Employees: 96   Fiscal Year Ends: 12/31
Parent Company: Morgan Stanley Real Estate

SALARIES/BONUSES:

Top Exec. Salary: $ Bonus: $
Second Exec. Salary: $ Bonus: $

OTHER THOUGHTS:

Estimated Female Officers or Directors: 2
Hot Spot for Advancement for Women/Minorities:

Sales, profits and employees may be estimates. Financial information, benefits and other data can change quickly and may vary from those stated here.

Goettsch Partners Inc

www.gpchicago.com

NAIC Code: 541310

TYPES OF BUSINESS:

Architectural Services

BRANDS/DIVISIONS/AFFILIATES:

CONTACTS: Note: Officers with more than one job title may be intentionally listed here more than once.

James Goettsch, CEO
James Goettsch, Chmn.

GROWTH PLANS/SPECIAL FEATURES:

Goettsch Partners, Inc., based out of Chicago, is an architecture firm providing architectural, interior, planning and building enclosure design services. With additional offices in Shanghai and Abu Dhabi, the firm's work spans four continents. Projects by Goettsch include office towers in Beijing, Chicago and Shanghai as well as luxury business hotels in China, India and the U.S.; a 25-story vertical addition to Chicago's 300 East Randolph, the headquarters of Health Care Service Corporation and its Blue Cross Blue Shield of Illinois division; Northwestern University's new music building in Evanston, Illinois; the China Diamond Exchange Center in Shanghai; a 214-room Wyndham hotel in the King Abdullah Financial District in Riyadh, Saudi Arabia; and a 3 million-square-foot office development in Abu Dhabi that features the headquarters building of the Abu Dhabi Securities Exchange. Specialties of the company comprise corporate/commercial: office buildings & headquarter; hospitality/residential: hotels, apartments & condominiums; institutional: cultural, governmental & higher education facilities; renovation: repositioning, adaptive reuse and restoration; and mixed-use: office, hotel, residential and retail complexes.

FINANCIAL DATA: Note: Data for latest year may not have been available at press time.

In U.S. $	2015	2014	2013	2012	2011	2010
Revenue	29,620,000					
R&D Expense						
Operating Income						
Operating Margin %						
SGA Expense						
Net Income						
Operating Cash Flow						
Capital Expenditure						
EBITDA						
Return on Assets %						
Return on Equity %						
Debt to Equity						

CONTACT INFORMATION:

Phone: 312-356-0600   Fax: 312-356-0601
Toll-Free:
Address: 224 S. Michigan Ave., Fl. 17, Chicago, IL 60604 United States

STOCK TICKER/OTHER:

Stock Ticker: Private   Exchange:
Employees: 110   Fiscal Year Ends:
Parent Company:

SALARIES/BONUSES:

Top Exec. Salary: $ Bonus: $
Second Exec. Salary: $ Bonus: $

OTHER THOUGHTS:

Estimated Female Officers or Directors:
Hot Spot for Advancement for Women/Minorities:

Sales, profits and employees may be estimates. Financial information, benefits and other data can change quickly and may vary from those stated here.

Golden Tulip Hospitality Group

www.goldentulip.com

NAIC Code: 721110

TYPES OF BUSINESS:

Hotels & Resorts

BRANDS/DIVISIONS/AFFILIATES:

Golden Tulip
Tulip Inns
Royal Tulip
Branche Restaurant, Bar & Lounge
Campanile
Tulip Residences
Louvre Hotels
Kyriad

CONTACTS: Note: Officers with more than one job title may be intentionally listed here more than once.

Pierre Frederic Roulot, CEO
Pierre Frederic Roulot, Pres.
Victoire Boissier, CFO
Emmanuelle Greth, Chief Human Resources Officer
Haike Blaauw, Sr. VP-Franchise Oper.

GROWTH PLANS/SPECIAL FEATURES:

Golden Tulip Hospitality Group is a hospitality company with over 240 Tulip Inn, Golden Tulip and Royal Tulip branded hotels in more than 46 countries. Golden Tulip also maintains hotels that operate under the brand names Louvre Hotels, Kyriad, Campanile and Premiere Classe. The company's original brands include Royal Tulip, reserved for five-star hotels located in prime areas near city centers and business districts; Golden Tulip Hotels, which are four-star hotels situated in key locations near city centers, airports, conference venues and business districts; and Tulip Inns, which are three-star budget hotels with fewer amenities than the company's other brands. Golden Tulip also owns and operates the Branche Restaurant, Bar & Lounge chain; and The State Room, exclusively in Royal Tulip Hotels. In addition to its facilities, the company offers several services to business customers including group rates and the Golden Tulip Central Meeting Line, a reservation and planning service for businesses or large groups. The Ambassador Club is the company's frequent stay program through which guests have access to special events and discounts. Golden Tulip franchises and manages hotels in Europe, the Middle East and Africa, the Asia-Pacific Region and the Americas. In recent years, the company has added a new brand concept, Tulip Residences, which provides travelers with extended-stay studios and one-bedroom suites. The accommodations are four-star rated and offer amenities such as high-definition televisions, free Wi-Fi connections, en suite bathrooms and fully equipped kitchen facilities.

FINANCIAL DATA: Note: Data for latest year may not have been available at press time.

In U.S. $	2015	2014	2013	2012	2011	2010
Revenue						
R&D Expense						
Operating Income						
Operating Margin %						
SGA Expense						
Net Income						
Operating Cash Flow						
Capital Expenditure						
EBITDA						
Return on Assets %						
Return on Equity %						
Debt to Equity						

CONTACT INFORMATION:

Phone: 33-01-4291-4600   Fax: 33-01-4291-4601
Toll-Free:
Address: 50 Place de L'Ellipse, Village 5, Paris, CS 70050 France

STOCK TICKER/OTHER:

Stock Ticker: Private   Exchange:
Employees: 1,053   Fiscal Year Ends:
Parent Company: Shanghai Jin Jiang International Hotels

SALARIES/BONUSES:

Top Exec. Salary: $ Bonus: $
Second Exec. Salary: $ Bonus: $

OTHER THOUGHTS:

Estimated Female Officers or Directors:
Hot Spot for Advancement for Women/Minorities:

Sales, profits and employees may be estimates. Financial information, benefits and other data can change quickly and may vary from those stated here.

Granite Construction Inc

www.graniteconstruction.com

NAIC Code: 237310

TYPES OF BUSINESS:

Construction, Heavy & Civil Engineering
Infrastructure Projects
Site Preparation Services
Construction Materials Processing
Heavy Construction Equipment

BRANDS/DIVISIONS/AFFILIATES:

Granite Construction Supply
Granite Construction Company
Kenny Construction Company
Intermountain Slurry Seal, Inc
Garco Testing Laboratories
Granite Land Company

CONTACTS: Note: Officers with more than one job title may be intentionally listed here more than once.

James Roberts, CEO
Laurel Krzeminski, CFO
William Powell, Chairman of the Board
Christopher Miller, Executive VP
Richard Watts, General Counsel
Michael Donnino, Other Corporate Officer
Patrick Kenny, Other Corporate Officer
James Richards, Other Corporate Officer
Martin Matheson, Other Corporate Officer

GROWTH PLANS/SPECIAL FEATURES:

Granite Construction, Inc. is one of the largest heavy civil construction contractors in the U.S. The firm operates nationwide, serving both public and private sector clients. Within the public sector, the company primarily concentrates on infrastructure projects, including the construction of roads, highways, bridges, dams, tunnels, canals, mass transit facilities and airports. Within the private sector, it performs site preparation services for commercial/industrial buildings, residential buildings and other facilities. Granite owns and leases substantial aggregate reserves and owns many construction materials processing plants. Granite operates in four segments: construction, large project construction, construction materials and real estate. The construction division performs various heavy civil construction projects with a large portion of the work focused on new construction and improvement of streets, roads, highways, bridges, site work and other infrastructure projects. The large project construction segment focuses on the firm's larger projects, such as mass transit facilities, highways, dams, bridges and airports. The construction materials division (Granite Construction Supply) produces concrete, gravel, ready-mix asphalt, sand and other products. Granite's real estate segment purchases, develops, operates, sells and invests in real estate. Its current portfolio includes residential, retail and office site development projects held for rental income or for sale to home and commercial property developers. The firm's other subsidiaries include, Granite Construction Company, which is engaged in construction materials market; Kenny Construction Company, which is a national contractor and construction manager; Intermountain Slurry Seal, Inc., which provides a range of pavement preservation services and solutions; Garco Testing Laboratories, which provides materials testing, mix design, inspection and engineering services; and Granite Land Company, which manages corporate real estate and mine-alternative end-use planning and development.

The company offers employees life, AD&D, medical, dental and vision coverage; a 401(k); a stock purchase plan; flexible spending accounts; education reimbursement; and domestic partner benefits.

FINANCIAL DATA: Note: Data for latest year may not have been available at press time.

In U.S. $	2015	2014	2013	2012	2011	2010
Revenue	2,371,029,000	2,275,270,000	2,266,901,000	2,083,037,000	2,009,531,000	1,762,965,000
R&D Expense						
Operating Income	110,308,000	65,100,000	-54,692,000	80,835,000	99,269,000	-109,340,000
Operating Margin %	4.65%	2.86%	-2.41%	3.88%	4.93%	-6.20%
SGA Expense	207,339,000	203,821,000	199,946,000	185,099,000	162,302,000	191,593,000
Net Income	60,485,000	25,346,000	-36,423,000	45,283,000	51,161,000	-58,983,000
Operating Cash Flow	66,978,000	43,142,000	5,380,000	91,790,000	92,345,000	29,318,000
Capital Expenditure	44,179,000	43,428,000	43,682,000	37,622,000	45,035,000	37,004,000
EBITDA	181,993,000	138,008,000	23,256,000	147,733,000	160,341,000	-22,201,000
Return on Assets %	3.72%	1.56%	-2.17%	2.76%	3.31%	-3.63%
Return on Equity %	7.40%	3.21%	-4.51%	5.55%	6.55%	-7.41%
Debt to Equity	0.29	0.34	0.35	0.32	0.27	0.31

CONTACT INFORMATION:

Phone: 831 724-1011   Fax: 831 7617871
Toll-Free:
Address: 585 W. Beach St., Watsonville, CA 95076 United States

STOCK TICKER/OTHER:

Stock Ticker: GVA   Exchange: NYS
Employees: 1,700   Fiscal Year Ends: 12/31
Parent Company:

SALARIES/BONUSES:

Top Exec. Salary: $750,000   Bonus: $
Second Exec. Salary: Bonus: $
$475,000

OTHER THOUGHTS:

Estimated Female Officers or Directors: 2
Hot Spot for Advancement for Women/Minorities:

Groupe du Louvre

www.groupedulouvre.com

NAIC Code: 721110

TYPES OF BUSINESS:

Hotels
Luxury Goods
Perfume House

BRANDS/DIVISIONS/AFFILIATES:

Shanghai Jin Jiang International Hotels
Crillon
Concorde Hotel & Resorts
Kyriad Prestige
Campanile
Premiere Classe
Annick Goutal
Baccarat

CONTACTS: Note: Officers with more than one job title may be intentionally listed here more than once.

Steven Goldman, CEO
Jean-Yves Schapiro, Finance Dir.
Pascal Malbequi, General Counsel
Jean-Yves Schapiro, Dir.-Finance
Brigitte Taitttinger, Pres., Annick Goutal
Bernard Granier, COO-Concorde Hotels & Resorts
Pierre Frederic Roulot, Pres., Louvre Hotels
Herve Martin, Pres., Baccarat

GROWTH PLANS/SPECIAL FEATURES:

Groupe du Louvre (Louvre) is a French hotel and luxury goods company. The Louvre chain of hotels includes 97,000 rooms in 1,100 properties. The company operates in four segments: luxury hotel, budget hotel, crystal manufacturer and perfume house. The luxury hotel division encompasses the Crillon and Concorde hotels and resorts. The segment comprises more than 27 prestigious hotels in locations such as Amsterdam, Barcelona, Cannes, Nice, Paris, Prague and Tokyo. The budget hotel division includes more than 800 hotels with a total of roughly 52,000 rooms in nine European countries under four brands: Kyriad, Kyriad Prestige, Campanile and Premiere Classe. The crystal manufacturer division operates through subsidiary Societe du Louvre, which is the majority shareholder of Baccarat, one of the most prestigious crystal manufacturers in the world. Products, which include jewelry, accessories and wristwatches, are sold through a network of 47 owned stores and points of sales worldwide. The perfume house division operates through Annick Goutal, a unique luxury perfume house created in 1980 that distributes products in more than 20 countries. Annick Goutal offers more than 25 fragrances composed mainly of natural elements as well as a skin care line with active rose serum. The perfumery has 11 boutiques and 1,000 points of sale such as Saks Fifth Avenue, Isetan, Bergdorf Goodman, Harrods and Harvey Nichols. In March 2015, the firm was acquired by Shanghai Jin Jiang International Hotels Development Co. Ltd., owner and operator of economy and mid-scale hotels in Asia.

FINANCIAL DATA: Note: Data for latest year may not have been available at press time.

In U.S. $	2015	2014	2013	2012	2011	2010
Revenue	1,600,000,000	1,600,000,000				
R&D Expense						
Operating Income						
Operating Margin %						
SGA Expense						
Net Income						
Operating Cash Flow						
Capital Expenditure						
EBITDA						
Return on Assets %						
Return on Equity %						
Debt to Equity						

CONTACT INFORMATION:

Phone: 33-1-42-91-4500   Fax:
Toll-Free:
Address: Village 5, 50 Place de l'Ellipse, Paris, 92081 France

STOCK TICKER/OTHER:

Stock Ticker: Subsidiary   Exchange:
Employees: 19,000   Fiscal Year Ends: 12/31
Parent Company: Shanghai Jin Jiang International Hotels

SALARIES/BONUSES:

Top Exec. Salary: $ Bonus: $
Second Exec. Salary: $ Bonus: $

OTHER THOUGHTS:

Estimated Female Officers or Directors: 1
Hot Spot for Advancement for Women/Minorities:

GS Engineering & Construction Corp

www.gsconst.co.kr

NAIC Code: 237000

TYPES OF BUSINESS:

Heavy Construction and Civil Engineering

BRANDS/DIVISIONS/AFFILIATES:

Xi
Eclat

CONTACTS: Note: Officers with more than one job title may be intentionally listed here more than once.

Lim Byeong-Yong, CEO
Myung-Soo Huh, Pres.
Huh Chang-Soo, Chmn.

GROWTH PLANS/SPECIAL FEATURES:

GS Engineering & Construction Corp. (GS E&C) has six main construction divisions: civil engineering, plant, architecture, housing, power and development. The civil engineering division focuses on roads, bridges and railroads; underground spaces; and harbors and dredging. Notable projects include the 4.54 mile West Sea Grand Bridge and an LPG (Liquefied Petroleum Gas) offloading terminal in Incheon. The plant division constructs oil refineries and gas processing plants; petrochemical plants; and power plants and other energy related facilities, such as pipelines and storage facilities. Some of its major projects include the Azerpetrochim refinery in Azerbaijan and a VCM/PVC (vinyl chloride monomer, used to make polyvinyl chloride) plant in Saudi Arabia. The architecture division constructs intelligent buildings; research, educational and medical health care facilities; and cultural and sports facilities. Major projects include the COEX Convention Center and the main stadium for Iman University in Saudi Arabia. GS E&C's housing division constructs residential complexes under the Xi brand and studio apartments under the Eclat brand. One of its major accomplishments was being selected as the contractor for the Korea National Housing Corp. project. The power segment focuses on the construction of combined cycle and thermal fired power plants; co-generation plants and district heating; and nuclear power plants. This division is currently constructing Nuclear Power Plant No.1 & 2 in the southeastern part of Korea. In the environmental sector, this segment also handles water and wastewater treatment facilities as well as waste treatment and recycling facilities. Some of its projects include the China Sanghae Songjiang sewage treatment facilities and the Incheon Namdong-gu food waste recycling center. The development segment builds residential and commercial complexes overseas and domestically. Projects have included Midan City and Xi Riverview Palace.

FINANCIAL DATA: Note: Data for latest year may not have been available at press time.

In U.S. $	2015	2014	2013	2012	2011	2010
Revenue	7,873,167,914	7,736,233,642	9,284,889,000	8,095,900,000	8,029,300,000	7,669,380,000
R&D Expense						
Operating Income						
Operating Margin %						
SGA Expense						
Net Income	-28,665,380	-73,994,092	-857,566,503	113,100,000	375,800,000	370,010,000
Operating Cash Flow						
Capital Expenditure						
EBITDA						
Return on Assets %						
Return on Equity %						
Debt to Equity						

CONTACT INFORMATION:

Phone: 82-2-2154 1112   Fax:
Toll-Free:
Address: Gran Seoul, 33 Jongro, Jongno-gu, Seoul, 110-130 South Korea

STOCK TICKER/OTHER:

Stock Ticker: 6360
Employees: 770
Parent Company:

Exchange: Seoul
Fiscal Year Ends: 12/31

SALARIES/BONUSES:

Top Exec. Salary: $ Bonus: $
Second Exec. Salary: $ Bonus: $

OTHER THOUGHTS:

Estimated Female Officers or Directors:
Hot Spot for Advancement for Women/Minorities:

Guocoleisure Ltd

www.guocoleisure.com

NAIC Code: 531100

TYPES OF BUSINESS:
Hotel & Resort Development
Casino Operations
Real Estate Development

BRANDS/DIVISIONS/AFFILIATES:
Guoman Hotel Management (Malaysia) Sdn Bhd
Guoman Hotel Management (UK) Limited
Clermont Leisure (UK) Limited
Molokai Properties Ltd
Tabua Investments Limited
Clermont Club (The)
Guoman
Thistle

CONTACTS: *Note: Officers with more than one job title may be intentionally listed here more than once.*
Michael Bernard DeNoma, CEO
Ho Kah Meng, CFO
Jeanette Ling, Group Legal Counsel
Clay Rumbaoa, COO-Molokai Properties
Michael Bernard DeNoma, CEO-Guoman Hotels Limited
Andy Hughes, CFO-Guoman & Thistle Hotels & Clermont Leisure
Asem Abdin, VP-Intl Mktg.-Clermont Leisure
Quek Leng Chan, Chmn.

GROWTH PLANS/SPECIAL FEATURES:
GuocoLeisure Ltd. (GL) is an investment holding company with principal investments in hotel management, gaming operations, oil & gas and property development. It is a member of conglomerate the Hong Leong Group Malaysia. GL owns, leases and/or manages and operate hotels in U.K. and Malaysia through is wholly-owned subsidiaries Guoman Hotel Management (UK) Limited and Guoman Hotel Management (Malaysia) Sdn Bhd. The UK subsidiary manages and operates two distinct hotel brands: Guoman, a premium collection of five deluxe hotels in central London; and Thistle, a group of 32 full-service hotels in London, as well as regional locations throughout the U.K. The Malaysian subsidiary manages and operates two hotels under the Thistle brand in Malaysia. Gaming operations are owned and operated by wholly-owned subsidiary Clermont Leisure (UK) Limited. Its gaming operations are located throughout the U.K., including The Clermont Club, an exclusive members-only casino in Mayfair, London. As for the company's oil and gas investments, GL receives royalties from the Bass Strait Oil Trust, a unit trust managed by Bass Strait Oil Management Limited, a wholly-owned subsidiary of the GL Group in Australia. BHP Billiton Limited and Esso Australia Resources Pty Ltd. currently produce hydrocarbons from designated areas in the Bass Strait, Australia. GL's property development division is operated through two wholly-owned subsidiaries: Molokai Properties Limited and Tabua Investments Limited. Molokai Properties owns a 54,677-acre property on the island of Molokai, Hawaii, and maintains a small staff that performs routine maintenance of buildings and other assets. Tabua Investments is GL's property investment arm in Denarau, Fiji. The company announced that it intends to divest its property investments and eventually exit from Fiji.

FINANCIAL DATA: *Note: Data for latest year may not have been available at press time.*

In U.S. $	2015	2014	2013	2012	2011	2010
Revenue	423,200,000	448,800,000	424,900,000	423,500,000	391,100,000	369,700,000
R&D Expense						
Operating Income	76,200,000	79,700,000	82,600,000	99,500,000	116,200,000	66,400,000
Operating Margin %	18.00%	17.75%	19.43%	23.49%	29.71%	17.96%
SGA Expense	168,400,000	22,500,000	26,000,000			17,900,000
Net Income	47,900,000	39,000,000	44,000,000	77,000,000	80,600,000	49,500,000
Operating Cash Flow	99,900,000	108,700,000	75,600,000	128,900,000	95,700,000	100,700,000
Capital Expenditure	55,400,000	39,200,000	12,800,000	46,600,000	44,600,000	8,200,000
EBITDA	107,200,000	114,100,000	120,200,000	144,000,000	149,500,000	108,600,000
Return on Assets %	2.86%	2.37%	2.73%	4.69%	5.00%	3.06%
Return on Equity %	4.04%	3.35%	3.95%	6.99%	7.77%	5.01%
Debt to Equity	0.26	0.08	0.27	0.28	0.34	0.38

CONTACT INFORMATION:
Phone: 65-6438-0002   Fax: 65-6435-0040
Toll-Free:
Address: 20 Collyer Quay, #18-05, Singapore, 049319 Singapore

STOCK TICKER/OTHER:
Stock Ticker: GUORY   Exchange: PINX
Employees: Fiscal Year Ends: 06/30
Parent Company: HONG LEONG GROUP MALAYSIA

SALARIES/BONUSES:
Top Exec. Salary: $ Bonus: $
Second Exec. Salary: $ Bonus: $

OTHER THOUGHTS:
Estimated Female Officers or Directors: 2
Hot Spot for Advancement for Women/Minorities:

Hang Lung Group Ltd

NAIC Code: 531100

www.hanglung.com

TYPES OF BUSINESS:
Real Estate Holdings & Development
Carpark Management
Hotel Management
Dry Cleaning

GROWTH PLANS/SPECIAL FEATURES:
Hang Lung Group Ltd., through its subsidiary, Hang Lung Properties Limited, develops property in Hong Kong and China for both sale and lease. The company's Hong Kong portfolio consists of residential, office and commercial properties. Its China portfolio includes two large scale developments in Shanghai: The Grand Gateway 66, a commercial, office and residential structure; and Plaza 66, a commercial and office complex. The firm is currently expanding in other large cities such as Tianjin, Shenyang, Wuxi, Jinan, Dalian and Kunming, where it is developing a variety of mixed use commercial, office and residential buildings. The Hang Lung Group also manages shopping centers, office buildings, apartments and car parks. The company owns a 50% stake in Hang Lung-Hakuyosha Dry Cleaning. The company's other subsidiaries include AP City Limited, AP Properties Limited, AP Joy Limited, AP Star Limited and Antonis Limited.

BRANDS/DIVISIONS/AFFILIATES:
Hang Lung Properties Limited
Hang Lung-Hakuyosha Dry Cleaning
AP City Limited
AP Joy Limited
AP Properties Limited
Antonis Limited
AP Star Limited

CONTACTS: Note: Officers with more than one job title may be intentionally listed here more than once.
Philip Nan Lok Chen, Managing Dir.
Bella Peck Lim Chhoa, General Counsel
Ronnie Chichung Chan, Chmn.

FINANCIAL DATA: Note: Data for latest year may not have been available at press time.

In U.S. $	2015	2014	2013	2012	2011	2010
Revenue	1,228,440,000	2,269,932,000	1,254,999,000	1,029,631,000	736,702,800	
R&D Expense						
Operating Income	896,833,300	1,867,672,000	1,194,918,000	1,087,391,000	1,011,065,000	
Operating Margin %	73.00%	82.27%	95.21%	105.60%	137.24%	
SGA Expense	93,860,640	90,895,260	89,992,760	89,605,970	63,175,430	
Net Income	413,992,500	879,943,500	587,531,500	678,426,800	454,992,000	
Operating Cash Flow	773,447,800	1,992,476,000	666,823,100	663,986,700	632,399,000	
Capital Expenditure	797,815,500	683,970,800	1,275,886,000	691,061,900	815,349,800	
EBITDA	1,068,567,000	2,044,047,000	1,370,778,000	1,522,012,000	1,051,935,000	
Return on Assets %	1.60%	3.37%	2.38%	3.04%	2.50%	
Return on Equity %	4.23%	9.31%	6.71%	8.37%	6.46%	
Debt to Equity	0.40	0.42	0.61	0.56	0.22	

CONTACT INFORMATION:
Phone: 852 2879-0111   Fax: 852-2868-6086
Toll-Free:
Address: 4 Des Voeux Rd., 28th Fl., Hong Kong, Hong Kong

STOCK TICKER/OTHER:
Stock Ticker: HNLGY
Employees:
Parent Company:

Exchange: PINX
Fiscal Year Ends: 06/30

SALARIES/BONUSES:
Top Exec. Salary: $ Bonus: $
Second Exec. Salary: $ Bonus: $

OTHER THOUGHTS:
Estimated Female Officers or Directors: 1
Hot Spot for Advancement for Women/Minorities: Y

Hang Lung Properties Limited

www.hanglung.com

NAIC Code: 531100

TYPES OF BUSINESS:

Real Estate Holdings & Development
Leasing

BRANDS/DIVISIONS/AFFILIATES:

Hang Lung Group Ltd
Summit (The)
HarbourSide
Grand Gateway 66
Plaza 66
Palace 66
Wesley Hotel (The)
Olympia 66

CONTACTS: *Note: Officers with more than one job title may be intentionally listed here more than once.*

Philip N.L. Chen, Managing Dir.
Norman Ka Ngok Chan, Dir.-Sales & Leasing
Bella Peck Lim Chhoa, General Counsel
Chuk Fai Kwan, Assistant Dir.-Corp. Comm.
Raymond Wai Man Mak, Group Financial Controller
Henry Kam Ling Cheung, Assistant Dir.-Leasing & Mgmt.
Moses Woon Tim Leung, Assistant Dir.-Project. Dev.
Gavin Yee Liang Lu, Assistant Dir.-Project Dev.
Paul Wan Po Chan, Sr. Mgr.-Finance
Ronnie C. Chan, Chmn.

GROWTH PLANS/SPECIAL FEATURES:

Hang Lung Properties Limited (HLPL) is a property investment holding company with interests in the retail, commercial, residential and industrial sectors. As a publicly listed subsidiary of Hang Lung Group Ltd., the firm has approximately 54% of its investment properties located in Hong Kong and the remainder in mainland China. In Hong Kong, HLPL has developed shopping malls and mixed use commercial and residential complexes concentrated along the MTR Line, the city's mass transit system. The company also develops luxury residential properties such as The Summit and HarbourSide, which consists of three blocks of 80 story residences. Its mainland China developments include Plaza 66, a five story shopping mall and office building, and Grand Gateway 66, a complex consisting of retail, office and residential facilities in Shanghai. Development projects are also underway in the mainland Chinese cities of Tianjin, Shenyang, Wuxi, Jinan and Dalian. In addition, HLPL is involved in property leasing activities. Its commercial leasing portfolio in Hong Kong includes properties in strategic locations along busy shopping centers in the city, such as Causeway Bay and Mongkok. Its office properties are located along mass transit railway lines and other transportation networks. It also leases its Plaza 66 and Grand Gateway 66 properties in Shanghai and recently completed construction at its 360,000 square foot Palace 66 shopping mall in Shenyang. The firm operates car parks located in densely populated areas and manages The Wesley Hotel in Hong Kong, a 250 room property accommodating both commercial and business travelers. In December 2015, the firm opened Olympia 66, a shopping mall in Dalian, China.

FINANCIAL DATA: *Note: Data for latest year may not have been available at press time.*

In U.S. $	2015	2014	2013	2012	2011	2010
Revenue		2,195,668,000	1,178,157,000	950,468,000	665,791,700	1,554,502,000
R&D Expense						
Operating Income		1,813,651,000	1,123,362,000	1,342,413,000	936,543,600	3,841,453,000
Operating Margin %		82.60%	95.34%	141.23%	140.66%	247.11%
SGA Expense		83,030,570	82,772,700	80,709,840	66,011,880	58,791,830
Net Income		1,508,990,000	929,839,200	1,082,363,000	746,759,400	2,869,454,000
Operating Cash Flow		1,980,614,000	664,760,200	648,901,900	499,343,700	1,150,438,000
Capital Expenditure		683,455,000	1,274,597,000	686,936,100	814,318,500	472,655,400
EBITDA		1,935,876,000	1,242,751,000	1,360,464,000	952,272,900	3,865,821,000
Return on Assets %		6.23%	4.11%	5.27%	4.44%	21.74%
Return on Equity %		9.11%	5.94%	7.31%	5.86%	28.53%
Debt to Equity		0.22	0.26	0.24	0.11	0.05

CONTACT INFORMATION:

Phone: 852 28790111   Fax: 852 28686086
Toll-Free:
Address: 4 Des Voeux Rd., 28th Fl., Hong Kong,   Hong Kong

STOCK TICKER/OTHER:

Stock Ticker: HLPPF   Exchange: PINX
Employees: Fiscal Year Ends: 06/30
Parent Company: Hang Lung Group Ltd

SALARIES/BONUSES:

Top Exec. Salary: $ Bonus: $
Second Exec. Salary: $ Bonus: $

OTHER THOUGHTS:

Estimated Female Officers or Directors: 14
Hot Spot for Advancement for Women/Minorities: Y

Hanjin Heavy Industries and Construction Co Ltd

www.hanjinsc.com/eng/default.aspx

NAIC Code: 336600

TYPES OF BUSINESS:

Construction & Manufacturing
Shipbuilding
Civil Engineering
Tourism Facilities
Gas & Energy Supply

BRANDS/DIVISIONS/AFFILIATES:

Daeryun Power
Byeollae Energy

CONTACTS: Note: Officers with more than one job title may be intentionally listed here more than once.

Sung-Moon Choi, CEO
S.D. Lee, Gen. Mgr.-Investor Rel.
S.Y. Lee, Managing Dir.-Ship Sales Mktg. Team
H.D. Ki, Managing Dir.-Naval Ship Sales Mktg. Team
J.S. Yoo, Managing Dir.-Architectural Mktg. Dept.
O.C. Shin, Managing Dir.-Infrastructure Sales Dept.
N.H. Cho, Chmn.
P.J. Yu, Managing Dir.-Overseas Bus.

GROWTH PLANS/SPECIAL FEATURES:

Hanjin Heavy Industries and Construction Co. Ltd. (HHIC), established in 1937, engages in shipbuilding, construction and plant servicing operations. The firm operates its business through four divisions: shipbuilding, construction, Daeryun Power and Byeollae Energy. Shipbuilding includes ultra container carriers, liquid natural gas (LNG) carriers and tankers. Shipyards and factories within this division consist of the Subic and Yeongdo shipyards and the Yuldo/Ulsan/Dadaepo factories, which manufacture products that support shipbuilding in the shipyards. HHIC has built approximately 1,000 ships of various kinds. Construction involves projects such as dredging and port, airport terminals and runways as well as steel-framed architectural construction projects in both Korea and abroad. It is a construction industry leader in the areas of housing civil engineering and factories. Daeryun Power supplies district heating and cooling service to the Okjeong and Hoecheon district in Yangju. It is an eco-friendly company through its renewable energy projects that include solar energy generation and fuel cell. Byeollae Energy is an eco-friendly cogeneration plant that provides district heating and cooling service to approximately 60,000 households in the Byeollae and neighboring districts. It also sells electricity at the Korea Power Exchange, contributing to reduction of power shortage in the capital area. HHIC has a R&D center in Jungang-Dong, Busan, and includes drawing and IT teams that create high-tech abilities for the design and manufacturing of its shipbuilding and construction businesses. Firm's overseas offices are located in the U.S., Philippines, London, Singapore, Greece, Hong Kong, Shanghai and Dubai.

FINANCIAL DATA: Note: Data for latest year may not have been available at press time.

In U.S. $	2015	2014	2013	2012	2011	2010
Revenue	13,209,900	13,086,760	14,327,710	11,903,920	11,884,840	24,882,280
R&D Expense						
Operating Income	9,526,080	9,591,987	10,431,430	8,227,030	7,815,982	2,836,578
Operating Margin %	72.11%	73.29%	72.80%	69.11%	65.76%	11.40%
SGA Expense	213,329	189,047	186,446	291,376	499,501	373,759
Net Income	-210,277,100	7,884,490	9,608,463	8,799,376	6,071,196	4,061,050
Operating Cash Flow	8,099,554	8,122,968	10,740,150	8,240,038	7,559,294	17,025,540
Capital Expenditure					3,469	12,141
EBITDA	-206,982,600	12,131,120	11,309,020	10,199,890	9,794,909	4,926,506
Return on Assets %	-24.18%	.83%	1.04%	.95%	.66%	.44%
Return on Equity %	-25.89%	.86%	1.05%	.96%	.67%	.45%
Debt to Equity		0.02				

CONTACT INFORMATION:

Phone: 82-2-2006-7114   Fax: 82-2-2006-7054
Toll-Free:
Address: No. 22-1, Jungangro 4-ga, Jung-gu, Busan, 600-751 South Korea

STOCK TICKER/OTHER:

Stock Ticker: HADDF   Exchange: GREY
Employees: 2,758   Fiscal Year Ends: 12/31
Parent Company:

SALARIES/BONUSES:

Top Exec. Salary: $ Bonus: $
Second Exec. Salary: $ Bonus: $

OTHER THOUGHTS:

Estimated Female Officers or Directors:
Hot Spot for Advancement for Women/Minorities:

HCP Inc

www.hcpi.com

NAIC Code: 0

TYPES OF BUSINESS:

Real Estate Investment Trust
Health Care Properties

BRANDS/DIVISIONS/AFFILIATES:

CONTACTS: Note: Officers with more than one job title may be intentionally listed here more than once.

Lauralee Martin, CEO
Timothy Schoen, CFO
Michael Mckee, Chairman of the Board
Scott Anderson, Chief Accounting Officer
Darren Kowalske, Executive VP, Divisional
John Lu, Executive VP, Divisional
John Stasinos, Executive VP, Divisional
Kendall Young, Executive VP, Divisional
Jonathan Bergschneider, Executive VP, Divisional
Thomas Kirby, Executive VP, Divisional
Thomas Klaritch, Executive VP, Divisional
Troy McHenry, Executive VP
J. Hutchens, Executive VP

GROWTH PLANS/SPECIAL FEATURES:

HCP, Inc. invests primarily in real estate serving the health care industry in the U.S. Headquartered in Long Beach, California, the firm has additional offices in Nashville and San Francisco. HCP's portfolio of investments is comprised of 506 senior housing facilities, 311 post-acute/skilled nursing facilities, 227 medical offices, 118 life science facilities and 16 hospitals in the U.S. Senior housing facilities include independent living facilities, assisted living facilities and continuing care retirement communities. HCP's post-acute/skilled nursing facilities offer restorative, rehabilitative and custodial nursing care for people not requiring the more extensive and sophisticated treatment available at hospitals. The firm's medical office facilities typically contain physicians' offices and examination rooms and may also include pharmacies, hospital ancillary service space and outpatient services such as diagnostic centers, rehabilitation clinics and day-surgery operating rooms. HCP's life science properties are primarily configured in business park formats and contain laboratory and office space for scientific research institutions, government agencies and biotechnology and pharmaceutical companies. All of the company's hospitals, which include acute care hospitals, long-term acute care hospitals, specialty hospitals and rehabilitation hospitals, are leased to single tenants under net lease structures.

FINANCIAL DATA: Note: Data for latest year may not have been available at press time.

In U.S. $	2015	2014	2013	2012	2011	2010
Revenue	2,544,312,000	2,266,279,000	2,099,878,000	1,900,722,000	1,725,386,000	1,255,134,000
R&D Expense						
Operating Income	-587,628,000	882,622,000	833,799,000	1,038,015,000	492,493,000	373,108,000
Operating Margin %	-23.09%	38.94%	39.70%	54.61%	28.54%	29.72%
SGA Expense	96,022,000	82,175,000	109,233,000	79,454,000	96,150,000	83,048,000
Net Income	-559,235,000	922,233,000	970,837,000	832,540,000	538,891,000	330,709,000
Operating Cash Flow	1,222,145,000	1,248,621,000	1,148,987,000	1,034,870,000	724,161,000	580,498,000
Capital Expenditure	1,664,594,000	753,717,000	259,552,000	1,896,446,000	198,385,000	304,847,000
EBITDA	423,534,000	1,793,175,000	1,710,579,000	1,538,203,000	1,281,002,000	989,529,000
Return on Assets %	-2.61%	4.43%	4.84%	4.35%	3.35%	2.40%
Return on Equity %	-5.58%	8.57%	9.11%	8.41%	6.27%	4.67%
Debt to Equity	1.18	0.90	0.80	0.82	0.88	0.60

CONTACT INFORMATION:

Phone: 949-407-0700   Fax:
Toll-Free: 888-604-1990
Address: 1920 Main Street, Ste 1200, Irvine, CA 92614 United States

STOCK TICKER/OTHER:

Stock Ticker: HCP   Exchange: NYS
Employees: 187   Fiscal Year Ends: 12/31
Parent Company:

SALARIES/BONUSES:

Top Exec. Salary: $183,333   Bonus: $1,100,000
Second Exec. Salary: Bonus: $
$800,000

OTHER THOUGHTS:

Estimated Female Officers or Directors: 2
Hot Spot for Advancement for Women/Minorities: Y

HDR Inc

NAIC Code: 541330

www.hdrinc.com

TYPES OF BUSINESS:

Engineering Services
Architectural Services
Consulting Services

BRANDS/DIVISIONS/AFFILIATES:

E.T. Archer Corporation
Claunch & Miller Inc
Stetson Engineering Inc
Doherty & Associates
RAP
CRAVE
SROI

CONTACTS: *Note: Officers with more than one job title may be intentionally listed here more than once.*

George A. Little, CEO
Eric L. Keen, Vice Chmn.
Doug S. Wignall, Pres., HDR Architecture
George A. Little, Chmn.

GROWTH PLANS/SPECIAL FEATURES:

HDR, Inc. is an architectural, engineering and consulting firm that specializes in managing complex projects and solving engineering and architectural challenges for its clients. The employee-owned firm has more than 225 locations globally, including operations in all 50 states and within 60 countries worldwide. The company offers design-build services and program management for a variety of markets, including community architecture, hospitals and other health care projects, science and technology construction, transportation, power, justice and wastewater and water resources. The company operates in 10 markets, including architecture, oil & gas, federal, private development, resource management, transportation, power, mining, waste & industrial and water. HDR's economic tools include RAP, a suite of risk analysis tools; Cost Risk Assessment and Value Engineering (CRAVE), a tool for conducting real-time business case evaluations; and Sustainability Return on Investment (SROI), a tool for assessing the costs and benefits related to sustainable design. Repeat clients account for roughly 80% of the company's business. Subsidiaries of the firm include E.T. Archer Corporation; Claunch & Miller, Inc.; Stetson Engineering, Inc.; and Doherty & Associates.

The company offers employees medical, dental and vision coverage; tuition assistance; a 401(k) plan; a flexible spending account; life insurance; and short- & long-term disability.

FINANCIAL DATA: *Note: Data for latest year may not have been available at press time.*

In U.S. $	2015	2014	2013	2012	2011	2010
Revenue						
R&D Expense						
Operating Income						
Operating Margin %						
SGA Expense						
Net Income						
Operating Cash Flow						
Capital Expenditure						
EBITDA						
Return on Assets %						
Return on Equity %						
Debt to Equity						

CONTACT INFORMATION:

Phone: 402-399-1000   Fax: 402-548-5015
Toll-Free: 800-366-4411
Address: 8404 Indian Hills Dr., Omaha, NE 68114 United States

STOCK TICKER/OTHER:

Stock Ticker: Private   Exchange:
Employees: 10,000   Fiscal Year Ends: 12/31
Parent Company:

SALARIES/BONUSES:

Top Exec. Salary: $ Bonus: $
Second Exec. Salary: $ Bonus: $

OTHER THOUGHTS:

Estimated Female Officers or Directors: 1
Hot Spot for Advancement for Women/Minorities:

Healthcare Realty Trust Inc

www.healthcarerealty.com

NAIC Code: 0

TYPES OF BUSINESS:

Real Estate Investment Trust
Health Care Properties
Medical Facility Construction
Property Management

BRANDS/DIVISIONS/AFFILIATES:

CONTACTS: Note: Officers with more than one job title may be intentionally listed here more than once.

David Emery, CEO
Scott Holmes, CFO
Amanda Callaway, Chief Accounting Officer
B. Whitman, Executive VP, Divisional
Todd Meredith, Executive VP, Divisional
John Bryant, Executive VP
James Douglas, Senior VP, Divisional

GROWTH PLANS/SPECIAL FEATURES:

Healthcare Realty Trust, Inc. (HRT) is a self-managed and self-administered real estate investment trust (REIT) that owns, acquires, manages and develops real estate properties in the health care industry. HRT owns a portfolio of 198 real estate properties throughout 30 U.S. states, with approximately 14.3 million square feet. The company focuses its portfolio on outpatient-related facilities located on or near the campuses of large acute care hospitals and associated with leading health systems. Its portfolio consists of six major facility types: medical offices, physician clinics, inpatient rehabilitation facilities, surgical facilities, specialty outpatient centers and other specialty health care fields. HRT has constructed over two dozen medical real estate facilities valued at more than $750 million in an effort to improve acute care facilities, technology and services for health care providers. Although HRT invests in properties nationwide, the majority of its investments are concentrated in the Southern and Western U.S. During 2015, the firm acquired eight medical office buildings and two parcels of land; and divested nine properties.

FINANCIAL DATA: Note: Data for latest year may not have been available at press time.

In U.S. $	2015	2014	2013	2012	2011	2010
Revenue	351,930,000	301,079,000	264,381,000	242,278,000	219,445,000	195,103,000
R&D Expense						
Operating Income	58,836,000	33,979,000	15,878,000	7,812,000	-3,057,000	4,501,000
Operating Margin %	16.71%	11.28%	6.00%	3.22%	-1.39%	2.30%
SGA Expense	32,185,000	22,808,000	23,729,000	20,908,000	20,991,000	16,894,000
Net Income	69,436,000	31,887,000	6,946,000	5,465,000	-214,000	8,200,000
Operating Cash Flow	160,375,000	125,370,000	120,797,000	116,397,000	105,032,000	80,835,000
Capital Expenditure	220,981,000	142,569,000	177,744,000	97,473,000	230,106,000	369,034,000
EBITDA						
Return on Assets %	2.49%	1.16%	.26%	.21%		.38%
Return on Equity %	5.63%	2.58%	.58%	.51%	- .02%	1.00%
Debt to Equity	1.15	1.14	1.08	1.15	1.38	1.67

CONTACT INFORMATION:

Phone: 615 269-8175   Fax: 615 269-8461
Toll-Free:
Address: 3310 West End Ave., 4th Fl., Ste. 700, Nashville, TN 37203
United States

STOCK TICKER/OTHER:

Stock Ticker: HR   Exchange: NYS
Employees: 236   Fiscal Year Ends: 12/31
Parent Company:

SALARIES/BONUSES:

Top Exec. Salary: $975,513   Bonus: $418,077
Second Exec. Salary: Bonus: $188,880
$396,649

OTHER THOUGHTS:

Estimated Female Officers or Directors:
Hot Spot for Advancement for Women/Minorities: Y

HeidelbergCement AG

NAIC Code: 327310

www.heidelbergcement.com

TYPES OF BUSINESS:
Cement Manufacturing

GROWTH PLANS/SPECIAL FEATURES:

HeidelbergCement AG is engaged in the fields of cement, concrete and other related activities. Products include ready-mixed concrete, concrete products and concrete elements; in some countries, asphalt and building products such as bricks and roof tiles, lime, or sand-lime bricks are also manufactured. HeidelbergCement operates in Russia through subsidiary CJSC Construction Materials. Through subsidiary HC Trading, the firm trades and transports cement, coal and petroleum coke via sea routes. The company has more than 2,300 locations in over 40 countries. In December 2015, the firm formed a partnership with Joule's technology in order to explore carbon-neutral fuel application in cement manufacturing.

BRANDS/DIVISIONS/AFFILIATES:
CJSC Construction Materials
HC Trading

CONTACTS: Note: Officers with more than one job title may be intentionally listed here more than once.
Bernd Scheifele, Chmn.

FINANCIAL DATA: Note: Data for latest year may not have been available at press time.

In U.S. $	2015	2014	2013	2012	2011	2010
Revenue	15,356,460,000	14,386,580,000	15,893,860,000	15,990,010,000	14,714,590,000	13,423,430,000
R&D Expense						
Operating Income	2,105,473,000	1,819,208,000	1,832,438,000	1,839,965,000	1,570,009,000	1,631,254,000
Operating Margin %	13.71%	12.64%	11.52%	11.50%	10.66%	12.15%
SGA Expense						
Net Income	912,512,300	553,939,800	850,127,200	343,518,000	609,254,000	582,680,400
Operating Cash Flow	1,652,924,000	1,687,709,000	1,205,962,000	1,726,030,000	1,304,730,000	1,304,730,000
Capital Expenditure	1,035,230,000	1,073,437,000	1,067,962,000	948,210,000	837,125,500	837,125,500
EBITDA	2,946,704,000	2,553,118,000	3,047,525,000	3,123,140,000	2,646,982,000	2,471,003,000
Return on Assets %	2.83%	1.76%	2.71%	1.05%	1.23%	1.29%
Return on Equity %	5.70%	3.92%	6.15%	2.38%	2.82%	3.05%
Debt to Equity	0.32	0.44	0.57	0.05	0.60	0.65

CONTACT INFORMATION:
Phone: 49 62214810   Fax: 49 6221481554
Toll-Free:
Address: Berliner Strasse 6, Heidelberg, 69120 Germany

STOCK TICKER/OTHER:
Stock Ticker: HDELY   Exchange: PINX
Employees: 45,543   Fiscal Year Ends: 12/31
Parent Company:

SALARIES/BONUSES:
Top Exec. Salary: $ Bonus: $
Second Exec. Salary: $ Bonus: $

OTHER THOUGHTS:
Estimated Female Officers or Directors:
Hot Spot for Advancement for Women/Minorities:

HeidelbergCement North America

www.heidelbergcement.com/en/north-america

NAIC Code: 327310

TYPES OF BUSINESS:

Cement Manufacturing
Aggregates
Building Products
Concrete Manufacturing

BRANDS/DIVISIONS/AFFILIATES:

HeidelbergCement AG

GROWTH PLANS/SPECIAL FEATURES:

HeidelbergCement North America is a leading manufacturer of cement, aggregates and ready-mixed concrete. Asphalt is also manufactured in a few U.S. states, and concrete pipes are produced & distributed in Western Canada. Heidelberg Cement NA comprises cement giant Heidelberg Cement AG's operations within U.S. and Canada. Cement derives 32.8% of the company's annual revenues, aggregates derives 33.9%, ready-mixed concrete/asphalt derives 25.7% and other services derives 7.6%. The company owns and operates 14 cement plants, 47 cement terminals, 187 aggregates plants, 18 aggregates terminals, 149 ready-mixed concrete plants, 48 asphalt plants, three concrete product/precast facilities, two grinding facilities and one blast furnace slag.

CONTACTS: Note: Officers with more than one job title may be intentionally listed here more than once.

Dominik von Achten, Managing Dir.

FINANCIAL DATA: Note: Data for latest year may not have been available at press time.

In U.S. $	2015	2014	2013	2012	2011	2010
Revenue	4,252,833,800	3,406,828,501	4,215,840,000	4,150,000,000	4,015,000,000	
R&D Expense						
Operating Income						
Operating Margin %						
SGA Expense						
Net Income		460,422,465				
Operating Cash Flow						
Capital Expenditure						
EBITDA						
Return on Assets %						
Return on Equity %						
Debt to Equity						

CONTACT INFORMATION:

Phone: 281-616-0700   Fax:
Toll-Free:
Address: 16155 Park Row # 120, Houston, TX 77084 United States

STOCK TICKER/OTHER:

Stock Ticker: Subsidiary
Employees: 7,658
Parent Company: HeidelbergCement AG

Exchange:
Fiscal Year Ends:

SALARIES/BONUSES:

Top Exec. Salary: $ Bonus: $
Second Exec. Salary: $ Bonus: $

OTHER THOUGHTS:

Estimated Female Officers or Directors:
Hot Spot for Advancement for Women/Minorities:

Henderson Land Development Co Ltd

www.hld.com

NAIC Code: 531100

TYPES OF BUSINESS:

Real Estate Holdings & Development
Property Management
Construction
Hotel Operation

BRANDS/DIVISIONS/AFFILIATES:

Hong Kong Ferry (Holdings) Company Limited
Henderson Investment Limited
Hong Kong & China Gas Company Limited (The)
Miramar Hotel & Investment Company, Limited
Towngas China Company Limited
Citistore

CONTACTS: Note: Officers with more than one job title may be intentionally listed here more than once.

Shau Kee Lee, Managing Dir.
Wing Hoo (Billy) Wong, Gen. Mgr.-Construction
Cheung Yuen (Timon) Liu, Corp. Sec.
Wing Kee (Christopher) Wong, Gen. Mgr.-Acct.
David Francis Dumigan, Gen. Mgr.-Project Mgmt. Unit
Ho Ming Wong, Gen. Mgr.-Property Dev.
Kam Leung Leung, Gen. Mgr-Property Planning
Pui Man (Margaret) Lee, Gen. Mgr.-Leasing
Shau Kee Lee, Chmn.

GROWTH PLANS/SPECIAL FEATURES:

Henderson Land Development Company Limited is an investment holding firm that focuses on property development and investment, property management, construction and hotel operation in Hong Kong and China. The total land bank of the company and its subsidiaries in Hong Kong equals approximately 23.8 million square feet of total attributable floor space. It also holds 44.5 million square feet of agricultural land. Henderson's investment portfolio is comprised mainly of shopping arcades and offices, which together account for 85% of the company's investment properties in terms of gross floor area. Henderson's landmark investment project in the city is the International Finance Centre. Henderson also participates in hotel development, having established the Mira Hong Kong, Four Seasons Hotel Hong Kong and Newton Hotel Hong Kong. In addition, the company holds a place in the department store market, under the trade name, Citistore, which targets young and fashion forward clientele. The firm's total land bank in China is approximately 126.1 million square feet in developable gross floor area. Henderson has participated in the development and investment of the Henderson Centre in Beijing, Office Tower II of The Grand Gateway in Shanghai, the Shanghai Skycity and the Heng Bao Garden in Guangzhou. The company's subsidiaries include Hong Kong Ferry (Holdings) Company Limited, which focuses on property development and investment; Henderson Investment Limited, which is involved in infrastructure; The Hong Kong and China Gas Company Limited, which produces and distributes gas in Hong Kong and China; Miramar Hotel and Investment Company Limited, which is involved in hotel operation and the travel business; and Towngas China Company Limited, a distributor of liquefied petroleum and natural gas in China.

FINANCIAL DATA: Note: Data for latest year may not have been available at press time.

In U.S. $	2015	2014	2013	2012	2011	2010
Revenue		3,013,210,000	3,002,638,000	2,010,268,000	1,958,180,000	914,367,600
R&D Expense						
Operating Income		896,833,300	789,692,900	683,197,100	1,742,482,000	1,596,791,000
Operating Margin %		29.76%	26.30%	33.98%	88.98%	174.63%
SGA Expense		414,121,400	410,769,200	352,364,200	338,310,900	279,261,200
Net Income		2,159,826,000	2,056,167,000	2,605,406,000	2,215,524,000	2,039,664,000
Operating Cash Flow		457,957,400	-174,054,800	340,631,600	-77,357,670	-2,228,159,000
Capital Expenditure		674,687,800	65,367,230	64,206,870	60,725,770	48,606,400
EBITDA		2,491,820,000	2,443,084,000	2,937,915,000	2,623,714,000	2,370,239,000
Return on Assets %		5.39%	5.44%	7.42%	6.97%	7.50%
Return on Equity %		7.25%	7.44%	10.34%	9.97%	10.82%
Debt to Equity		0.12	0.17	0.18	0.14	0.21

CONTACT INFORMATION:

Phone: 852-2908-8111   Fax: 852-2524-7102
Toll-Free:
Address: 2 Int'l Finance Centre, 8 Finance St., Fl. 72-76, Hong Kong, Hong Kong

STOCK TICKER/OTHER:

Stock Ticker: HLDCY   Exchange: PINX
Employees: 8,560   Fiscal Year Ends: 06/30
Parent Company:

SALARIES/BONUSES:

Top Exec. Salary: $ Bonus: $
Second Exec. Salary: $ Bonus: $

OTHER THOUGHTS:

Estimated Female Officers or Directors: 3
Hot Spot for Advancement for Women/Minorities: Y

Highwoods Properties Inc

www.highwoods.com

NAIC Code: 0

TYPES OF BUSINESS:
Real Estate Investment Trust
Real Estate Development
Property Management & Leasing
Office, Retail & Industrial Properties
Apartments
Construction Services
Asset Management
Corporate Hosting

BRANDS/DIVISIONS/AFFILIATES:
Highwoods Realty LP

CONTACTS: Note: Officers with more than one job title may be intentionally listed here more than once.
Edward Fritsch, CEO
Mark Mulhern, CFO
O. Sloan, Chairman of the Board
Daniel Clemmens, Chief Accounting Officer
Theodore Klinck, COO
Jeffrey Miller, General Counsel
Kevin Penn, Senior VP, Divisional

GROWTH PLANS/SPECIAL FEATURES:
Highwoods Properties, Inc. is a self-administered real estate investment trust (REIT) engaged in the acquisition, development and operation of rental real estate properties, including office, industrial, retail and residential properties in the Southeastern and Midwestern U.S. Highwoods conducts nearly all of its activities through Highwoods Realty LP, in which it is the sole general partner. The firm's property holdings consist of in-service office, industrial and retail properties, encompassing 28.5 million square feet; two development properties; and 465 acres of development land. The company's properties and development land are located in Florida, Georgia, Missouri, North Carolina, Pennsylvania, Tennessee and Virginia. Its three largest tenants, by rental revenues, are the U.S. government, PPG Industries and HCA Corporation. Highwoods also provides services including asset management, tenant services and corporate hosting. In February 2016, the firm agreed to sell Country Club Plaza in Kansas City.

The firm offers employees medical, dental and vision insurance; life insurance; flexible reimbursement accounts; disability coverage; a 401(k) savings plan; an employee stock purchase plan; educational assistance; credit union membership; an interest free computer purchase program; and discounted auto and homeowner's insurance.

FINANCIAL DATA: Note: Data for latest year may not have been available at press time.

In U.S. $	2015	2014	2013	2012	2011	2010
Revenue	604,671,000	608,468,000	556,810,000	516,102,000	482,852,000	463,321,000
R&D Expense						
Operating Income	149,170,000	150,130,000	139,316,000	134,973,000	125,529,000	130,192,000
Operating Margin %	24.66%	24.67%	25.02%	26.15%	25.99%	28.09%
SGA Expense	37,642,000	36,223,000	37,193,000	37,377,000	35,727,000	32,948,000
Net Income	97,078,000	110,964,000	125,457,000	79,595,000	45,125,000	68,498,000
Operating Cash Flow	288,879,000	266,911,000	256,437,000	193,416,000	195,396,000	190,537,000
Capital Expenditure	464,515,000	347,653,000	609,702,000	128,726,000	184,566,000	102,717,000
EBITDA	387,530,000	397,463,000	322,026,000	305,159,000	283,646,000	266,350,000
Return on Assets %	2.22%	2.77%	3.43%	2.36%	1.27%	2.14%
Return on Equity %	6.16%	7.36%	9.53%	7.39%	3.95%	6.03%
Debt to Equity						

CONTACT INFORMATION:
Phone: 919 872-4924   Fax: 919 876-2448
Toll-Free: 866-449-6637
Address: 3100 Smoketree Ct., Ste. 600, Raleigh, NC 27604 United States

STOCK TICKER/OTHER:
Stock Ticker: HIW
Employees: 447
Parent Company:

Exchange: NYS
Fiscal Year Ends: 12/31

SALARIES/BONUSES:
Top Exec. Salary: $648,068   Bonus: $
Second Exec. Salary: $399,000   Bonus: $

OTHER THOUGHTS:
Estimated Female Officers or Directors: 3
Hot Spot for Advancement for Women/Minorities: Y

Hilton Worldwide Inc

www.hiltonworldwide.com

NAIC Code: 721110

TYPES OF BUSINESS:

Hotels & Resorts
Timeshare Properties
Conference Centers
Franchising
Management Services
Online Reservations

BRANDS/DIVISIONS/AFFILIATES:

Blackstone Group LP (The)
Hilton Grand Vacations Co LLC
Hilton
Hhonors
DoubleTree
Embassy Suites
Homewood Suites
Hampton Inn

CONTACTS: Note: Officers with more than one job title may be intentionally listed here more than once.

Christopher Nassetta, CEO
Kevin Jacobs, CFO
Jonathan Gray, Chairman of the Board
Michael Duffy, Chief Accounting Officer
James Holthouser, Executive VP, Divisional
Mark Wang, Executive VP, Divisional
Kristin Campbell, Executive VP
Matthew Schuyler, Executive VP
Christopher Silcock, Executive VP
Ian Carter, Executive VP

GROWTH PLANS/SPECIAL FEATURES:

Hilton Worldwide, Inc. owns, manages and develops hotels, resorts and timeshare properties and franchises lodging properties worldwide. Hilton Worldwide consists of 12 hotel brands and more than 4,610 hotels in 100 countries, ranging from affordable focus-service hotels to luxury extended stay suites. Hotel brands include Hilton, Hilton Garden Inn, DoubleTree, Embassy Suites, Homewood Suites, Home2 Suites, Hampton Inn, Conrad Hotels, Curio A Collection by Hilton, tru by Hilton, Canopy by Hilton and The Waldorf Astoria Collection. Additionally, Hilton Grand Vacations Co., LLC is the firm's timeshare brand in which ownership of a deeded real estate interest with club membership points provides members with a lifetime of vacation advantages. Hhonors, the firm's loyalty enrollment program for returning customers, has over 44 million members and includes partner benefits with several airlines. In addition, Hilton Worldwide offers architecture and construction and management services to individuals interested in developing their own properties. The Blackstone Group LP owns 45.9% of the firm's stock. In 2015, Hilton sold Waldorf Astoria New York, and signed agreements to redeploy the proceeds to add five landmark properties: Hilton Orlando Bonnet Creek, Waldorf Astoria Orlando, The Reach in Key West, Casa Marina in Key West (all in Florida) and Parc 55 in San Francisco, California. In February 2016, the firm announced plans to split into three companies by years' end, spinning off its real estate and vacation/timeshare businesses, while retaining its present hotel business.

FINANCIAL DATA: Note: Data for latest year may not have been available at press time.

In U.S. $	2015	2014	2013	2012	2011	2010
Revenue	11,272,000,000	10,502,000,000	9,735,000,000	9,276,000,000	8,783,000,000	8,068,000,000
R&D Expense						
Operating Income	2,071,000,000	1,673,000,000	1,102,000,000	1,100,000,000	975,000,000	553,000,000
Operating Margin %	18.37%	15.93%	11.32%	11.85%	11.10%	6.85%
SGA Expense	547,000,000	491,000,000	748,000,000	460,000,000	416,000,000	637,000,000
Net Income	1,404,000,000	673,000,000	415,000,000	352,000,000	253,000,000	128,000,000
Operating Cash Flow	1,394,000,000	1,366,000,000	2,101,000,000	1,110,000,000	1,167,000,000	833,000,000
Capital Expenditure	372,000,000	337,000,000	332,000,000	536,000,000	482,000,000	168,000,000
EBITDA	2,763,000,000	2,393,000,000	1,921,000,000	1,692,000,000	1,403,000,000	1,939,000,000
Return on Assets %	5.41%	2.55%	1.54%	1.29%	.92%	
Return on Equity %	26.15%	14.76%	12.45%	16.88%	13.54%	
Debt to Equity	1.72	2.43	2.90	6.77	8.78	

CONTACT INFORMATION:

Phone: 703-883-1000   Fax:
Toll-Free: 800-445-8667
Address: 7930 Jones Branch Dr., Ste. 1100, McLean, VA 22102 United States

STOCK TICKER/OTHER:

Stock Ticker: HLT
Employees: 157,000
Parent Company:

Exchange: NYS
Fiscal Year Ends: 12/31

SALARIES/BONUSES:

Top Exec. Salary: $1,246,154
Bonus: $

Second Exec. Salary: $748,077
Bonus: $

OTHER THOUGHTS:

Estimated Female Officers or Directors: 2

Hot Spot for Advancement for Women/Minorities:

Hines Interests LP

NAIC Code: 531120

TYPES OF BUSINESS:

Real Estate Development
Property Management
Property Development
Engineering Services
Office Buildings
Industrial Properties
Hotels
Residential Properties

BRANDS/DIVISIONS/AFFILIATES:

Hines European Development Fund LP
Emerging Markets Real Estate Fund II LP
HCM Holdings LP
HC Green Development Fund

CONTACTS: Note: Officers with more than one job title may be intentionally listed here more than once.

Jeffrey C. Hines, CEO
Jeffrey C. Hines, Pres.
Charles M. Baughn, CFO
Stephanie Fore, Sr. VP-Human Resources
C. Hastings Johnson, CIO
Clayton Ulrich, Sr. VP-Eng. Svcs.
David LeVrier, Chief Admin. Officer
Ilene Allen, VP-Corp. Oper. Svcs.
George C. Lancaster, Sr. VP-Corp. Comm.
Kay P. Forbes, Treas.
C. Hastings Johnson, Chief Investment Officer
Rick Vance, Sr. VP-Risk Mgmt.
Tom Owens, Chief Risk Officer-Investments
Keith Montgomery, Sr. VP
Gerald D. Hines, Chmn.
E. Staman Ogilvie, CEO-Eurasia

GROWTH PLANS/SPECIAL FEATURES:

Hines Interests, LP is a private commercial real estate development firm. It handles many aspects of development, including site selection, acquisition, zoning, design, construction management, financing and property management. The company has offices in 199 cities and around the world and controls assets valued at approximately $87 billion. Hines operates in three divisions: investment, development and management. The investment division raises a variety of investment funds, such as the HC Green Development Fund, Hines European Development Fund LP, Emerging Markets Real Estate Fund II LP and HCM Holdings LP. The development division portfolio consists of completed, underway, acquired and third-party managed projects, which include properties with 275 million square feet in office, mixed-use, industrial, hotel, biotechnology/medical, skyscrapers and residential properties as well as large, master-planned communities and land developments. Other development services include budget, administration, construction buyout, design coordination, marketing and public relations, maintenance planning and architectural and engineering contract negotiation. The management division offers property insurance and focuses on investment management, tenant relations, parking, energy management, asset and property management, acquisition and disposition and marketing and leasing. In February 2016, the firm announced a major milestone in its U.S. multi-family platform, acquiring a 308-unit, Class A apartment community called The Domain, located in the Las Vegas suburb of Henderson. Hines recently launched its U.S. multi-family platform, investing nearly $3 billion in the product type, has completed 10 apartment communities since, and has an additional 14 projects underway.

Hines offers its employees medical, dental, life, AD&D, disability and vision coverage; adoption assistance; an employee assistance program; access to a credit union; tuition reimbursement; and wellness benefits.

FINANCIAL DATA: Note: Data for latest year may not have been available at press time.

In U.S. $	2015	2014	2013	2012	2011	2010
Revenue						
R&D Expense						
Operating Income						
Operating Margin %						
SGA Expense						
Net Income						
Operating Cash Flow						
Capital Expenditure						
EBITDA						
Return on Assets %						
Return on Equity %						
Debt to Equity						

CONTACT INFORMATION:

Phone: 713-621-8000   Fax: 713-966-2636
Toll-Free:
Address: 2800 Post Oak Blvd., Williams Tower, Houston, TX 77056-6118
United States

STOCK TICKER/OTHER:

Stock Ticker: Private   Exchange:
Employees: 3,700   Fiscal Year Ends: 12/31
Parent Company:

SALARIES/BONUSES:

Top Exec. Salary: $ Bonus: $
Second Exec. Salary: $ Bonus: $

OTHER THOUGHTS:

Estimated Female Officers or Directors: 4
Hot Spot for Advancement for Women/Minorities: Y

HOCHTIEF AG

NAIC Code: 237000

TYPES OF BUSINESS:

Heavy Construction
Airport Management & Consulting Services
Infrastructure Development
Geothermal Plant Construction
Green Building Engineering Services

BRANDS/DIVISIONS/AFFILIATES:

Turner Construction Company
Flatiron
HOCHTIEF Solutions AG
HOCHTIEF AirPort
Leighton Properties Pty
Leighton Contractors Pty
Leighton Asia (Northern) Ltd
Leighton Asia (Southern) Ltd

CONTACTS: Note: Officers with more than one job title may be intentionally listed here more than once.

Marcelino Fernandez Verdes, CEO
Peter Sassenfeld, CFO
Manfred Wennemer, Chmn.

GROWTH PLANS/SPECIAL FEATURES:

HOCHTIEF AG is a construction services provider that designs, builds, finances and operates facilities worldwide. The company operates through three divisions: HOCHTIEF Americas, HOCHTIEF Europe and HOCHTIEF Asia-Pacific. The Americas division includes the activities of subsidiaries in the U.S. and Canada, including Turner Construction Company, a U.S. general construction contractor and green building engineering firm; and Flatiron, which is among the top 10 suppliers in the U.S. transportation infrastructure construction sector. The Europe division's leading company, HOCHTIEF Solutions AG, provides civil and structural engineering as well as building construction services. The core of the Asia-Pacific division is Leighton Holdings, Ltd., which operates the Australian subsidiaries Leighton Contractors Pty., Leighton Properties Pty., John Holland Group Pty. and Thiess Pty.; and Leighton Asia (Northern), Ltd. and Leighton Asia (Southern), Ltd. in Hong Kong. In addition to these primary divisions, subsidiary HOCHTIEF Insurance Broking & Risk Management Solutions handles insurance services for the company's units, while HOCHTIEF AirPort contains a portfolio of approximately six airports. To further streamline HOCHTIEF Europe, in 2014, the company sold its offshore assets, including installation vessels and pontoons; and also sold its residential developer format business. During 2015, it sold its 50% share in the concession company of the Tunnel San Cristobal toll highway in Santiago, Chile.

FINANCIAL DATA: Note: Data for latest year may not have been available at press time.

In U.S. $	2015	2014	2013	2012	2011	2010
Revenue	24,060,650,000	25,203,920,000	29,303,090,000	29,114,310,000	26,553,340,000	22,991,620,000
R&D Expense						
Operating Income	656,185,500	-85,915,990	979,814,300	678,664,600	714,495,700	815,848,300
Operating Margin %	2.72%	- .34%	3.34%	2.33%	2.69%	3.54%
SGA Expense						
Net Income	237,550,900	287,048,500	195,248,700	180,323,000	-191,478,200	623,028,900
Operating Cash Flow	1,294,700,000	862,254,100	235,825,300	1,146,979,000	1,169,943,000	1,181,536,000
Capital Expenditure	325,583,600	656,168,300	1,041,924,000	1,385,283,000	1,717,630,000	1,053,919,000
EBITDA	1,342,916,000	487,827,500	1,889,433,000	1,762,316,000	1,287,204,000	2,034,768,000
Return on Assets %	1.46%	1.67%	1.07%	.96%	-1.04%	2.09%
Return on Equity %	9.63%	11.32%	6.97%	6.03%	-5.76%	11.32%
Debt to Equity	1.09	1.41	1.17		0.88	

CONTACT INFORMATION:

Phone: 49 2018240   Fax: 49 2018242777
Toll-Free:
Address: Opernplatz 2, Essen, 45128 Germany

STOCK TICKER/OTHER:

Stock Ticker: HOCFY   Exchange: PINX
Employees: 68,426   Fiscal Year Ends: 12/31
Parent Company:

SALARIES/BONUSES:

Top Exec. Salary: $ Bonus: $
Second Exec. Salary: $ Bonus: $

OTHER THOUGHTS:

Estimated Female Officers or Directors: 1
Hot Spot for Advancement for Women/Minorities:

Home Depot Inc

NAIC Code: 444110

TYPES OF BUSINESS:

Home Centers, Retail
Home Improvement Products
Building Materials
Lawn & Garden Products
Online & Catalog Sales
Tool & Truck Rental
Installation & Design Services

BRANDS/DIVISIONS/AFFILIATES:

Husky
Hampton Bay
Behr
RIDGID
Ryobi
Glacier Bay
Blinds.com
Interline Brands

CONTACTS: *Note: Officers with more than one job title may be intentionally listed here more than once.*

Carol Tome, CFO
Matthew Carey, Chief Information Officer
Mark Holifield, Executive VP, Divisional
Edward Decker, Executive VP, Divisional
Marc Powers, Executive VP, Divisional
Timothy Crow, Executive VP, Divisional
Teresa Roseborough, Executive VP
Craig Menear, President

GROWTH PLANS/SPECIAL FEATURES:

Home Depot, Inc. is one of the world's largest home improvement retailers. The company operates approximately 2,269 Home Depot stores throughout the U.S., Canada, Guam, Puerto Rico, the Virgin Islands and Mexico. A typical store encompasses 104,000 square feet of enclosed space with a 24,000 square foot outdoor garden center; these locations usually stock between 30,000 and 40,000 items. These stores sell an assortment of building materials, plumbing materials, electrical materials, kitchen products, hardware, seasonal items, paint, flooring and wall coverings. The firm's proprietary brands include Hampton Bay lighting, Husky hand tools, Behr Premium Plus paint, Vigoro lawn care products, RIDGID and Ryobi power tools and Glacier Bay bath fixtures. Home Depot markets its products primarily to three types of customers: professional customers, such as remodelers, contractors, repairmen and small business owners; do-it-for-me shoppers, who are homeowners that personally purchase Home Depot products but hire third-party individuals for installation and/or project completion; and do-it-yourself (DIY) customers, who are homeowners that both shop for and personally install and/or utilize the firm's materials. In 2014, Home Depot acquired Blinds.com. In July 2015, it acquired Interline Brands from P2 Capital Partners for $1.6 billion.

The company offers its employees medical, dental, vision, life, AD&D and disability insurance; a 401(k) plan; a stock purchase plan; adoption, education and relocation assistance; flexible spending accounts; a legal services plan; auto and homeowners insurance; and veterinary coverage.

FINANCIAL DATA: *Note: Data for latest year may not have been available at press time.*

In U.S. $	2015	2014	2013	2012	2011	2010
Revenue	83,176,000,000	78,812,000,000	74,754,000,000	70,395,000,000	67,997,000,000	
R&D Expense						
Operating Income	10,469,000,000	9,166,000,000	7,766,000,000	6,661,000,000	5,839,000,000	
Operating Margin %	12.58%	11.63%	10.38%	9.46%	8.58%	
SGA Expense	16,834,000,000	16,597,000,000	16,508,000,000	16,028,000,000	15,849,000,000	
Net Income	6,345,000,000	5,385,000,000	4,535,000,000	3,883,000,000	3,338,000,000	
Operating Cash Flow	8,242,000,000	7,628,000,000	6,975,000,000	6,651,000,000	4,585,000,000	
Capital Expenditure	1,442,000,000	1,389,000,000	1,312,000,000	1,221,000,000	1,096,000,000	
EBITDA	12,592,000,000	10,935,000,000	9,537,000,000	8,356,000,000	7,521,000,000	
Return on Assets %	15.77%	13.19%	11.11%	9.63%	8.24%	
Return on Equity %	58.09%	35.54%	25.42%	21.11%	17.43%	
Debt to Equity	1.80	1.17	0.53	0.60	0.46	

CONTACT INFORMATION:

Phone: 770 433-8211   Fax: 770 431-2707
Toll-Free: 800-553-3199
Address: 2455 Paces Ferry Rd. N.W., Atlanta, GA 30339 United States

STOCK TICKER/OTHER:

Stock Ticker: HD
Employees: 371,000
Parent Company:

Exchange: NYS
Fiscal Year Ends: 01/31

SALARIES/BONUSES:

Top Exec. Salary:
$1,300,000
Second Exec. Salary:
$1,051,923

Bonus: $

Bonus: $

OTHER THOUGHTS:

Estimated Female Officers or Directors: 7

Hot Spot for Advancement for Women/Minorities: Y

Home Properties Inc

www.homeproperties.com

NAIC Code: 0

TYPES OF BUSINESS:

Real Estate Investment Trust
Apartment Communities
Property Management

BRANDS/DIVISIONS/AFFILIATES:

Home Properties LP
Home Properties Resident Services Inc
Lone Star Funds

CONTACTS: Note: Officers with more than one job title may be intentionally listed here more than once.

Edward Pettinella, CEO
Clifford Smith, Chairman of the Board
Robert Luken, Chief Accounting Officer
David Gardner, Executive VP
Ann McCormick, Executive VP
Donald Hague, Senior VP, Divisional
Lisa Critchley, Senior VP, Divisional
Bernard Quinn, Senior VP, Divisional
John Smith, Senior VP
Kenneth Hall, Vice President

GROWTH PLANS/SPECIAL FEATURES:

Home Properties, Inc. specializes in the acquisition, development, rehabilitation, ownership and operation of apartment complexes. The firm operates in Florida, New Jersey, Pennsylvania, Maryland, New York, Massachusetts, Illinois and Washington, D.C. It conducts its business through Home Properties, L.P., in which it owns an 85.2% interest; and its management subsidiary Home Properties Resident Services, Inc. Home Properties operates 108 communities with 36,341 apartment units. The company typically manages and owns two- to four-story apartment communities with 150 or more units. Other characteristics the firm uses to evaluate markets include acquisition opportunities below replacement costs, a mature housing stock, high average single-family home prices and stable or moderate job growth. The suburban areas the trust has targeted for growth are the suburbs of Baltimore, Boston, New York City, Philadelphia and Washington, D.C. Home Properties' business strategy involves aggressively managing and improving communities to achieve increased net operating income as well as maintaining a conservative capital structure with efficient access to capital markets. In early 2015, it acquired The Lakes of Schaumburg apartment community, The Mansions of Mountshire and The Mansions Apartments (all three in Chicago), as well as Longbrook Apartments in New Jersey. In October 2015, the firm was acquired by an affiliate of Lone Star Funds, a private equity firm.

Home Properties offers employees medical, dental, vision and prescription coverage; short- and long-term disability; life insurance and AD&D; a corporate fitness center; dependent care flex accounts; employee stock purchase plans; discount programs; and a 401(k) plan.

FINANCIAL DATA: Note: Data for latest year may not have been available at press time.

In U.S. $	2015	2014	2013	2012	2011	2010
Revenue	647,984,000	671,915,008	663,600,000	644,348,032	579,972,992	516,579,008
R&D Expense						
Operating Income						
Operating Margin %						
SGA Expense						
Net Income		160,938,000	160,872,992	135,302,000	37,856,000	20,081,000
Operating Cash Flow						
Capital Expenditure						
EBITDA						
Return on Assets %						
Return on Equity %						
Debt to Equity						

CONTACT INFORMATION:

Phone: 585 546-4900   Fax: 585 546-5433
Toll-Free: 866-243-6256
Address: 850 Clinton Sq., Rochester, NY 14604 United States

STOCK TICKER/OTHER:

Stock Ticker: Private
Employees: 1,200
Parent Company: Lone Star Funds

Exchange:
Fiscal Year Ends: 12/31

SALARIES/BONUSES:

Top Exec. Salary: $ Bonus: $
Second Exec. Salary: $ Bonus: $

OTHER THOUGHTS:

Estimated Female Officers or Directors: 3
Hot Spot for Advancement for Women/Minorities: Y

Homeinns Hotel Group

www.homeinns.com

NAIC Code: 721110

TYPES OF BUSINESS:

Economy Hotels
Midscale Hotels

BRANDS/DIVISIONS/AFFILIATES:

Yitel
Motel 168
Home Inns

CONTACTS: *Note: Officers with more than one job title may be intentionally listed here more than once.*

David Jian Sun, CEO
Jason Xiangxin Zong, COO
Jason Zong, Pres.
Huiping Yan, CFO
May Wu, Chief Strategy Officer
Neil Shen, Co-Chmn.
Yi Liu, Co-Chmn.

GROWTH PLANS/SPECIAL FEATURES:

Homeinns Hotel Group is an economy hotel chain in China, with more than 2,500 locations in 335 cities. Home Inns leases real estate properties on which it develops and operates hotels, franchises its brand to hotel owners and manages these hotel properties. For the leased-and-operated hotels, the company is responsible for hotel development and customization to conform to the standards of Home Inns as well as repairs, maintenance and operating expenses. For the franchised and managed hotels, the firm is responsible for managing the hotel, while the franchisee is responsible for cost of development and customization. A typical Homeinns hotel has 80-160 guest rooms. Each hotel has a standardized design, appearance, decor, color scheme, lighting scheme and set of guest amenities in each room, including free in-room broadband Internet access, a work space, air conditioning and a supply of cold and hot drinking water. The firm's hotels are strategically located to provide guests with convenient access to major business districts, ground transportation hubs, major highways, shopping centers, industrial development zones, colleges and universities and large residential neighborhoods. In addition, the company operates under the Yitel brand, which is aimed at China's midscale hotel market, and the Motel 168 brand.

FINANCIAL DATA: *Note: Data for latest year may not have been available at press time.*

In U.S. $	2015	2014	2013	2012	2011	2010
Revenue	1,168,000,000	969,205,184	921,224,256	837,026,816	573,405,824	459,908,992
R&D Expense						
Operating Income						
Operating Margin %						
SGA Expense						
Net Income		79,295,768	30,323,756	-4,137,910	54,324,900	55,556,260
Operating Cash Flow						
Capital Expenditure						
EBITDA						
Return on Assets %						
Return on Equity %						
Debt to Equity						

CONTACT INFORMATION:

Phone: 86 2133373333   Fax: 86 2164835661
Toll-Free:
Address: Caobao Rd., No. 124, Xuhui District, Shanghai, 200235 China

STOCK TICKER/OTHER:

Stock Ticker: Subsidiary
Employees: 25,176
Parent Company: BTG Hotels Group Holdings Co

Exchange:
Fiscal Year Ends: 12/31

SALARIES/BONUSES:

Top Exec. Salary: $ Bonus: $
Second Exec. Salary: $ Bonus: $

OTHER THOUGHTS:

Estimated Female Officers or Directors: 1
Hot Spot for Advancement for Women/Minorities:

HomeServices of America Inc

www.homeservices.com

NAIC Code: 531210

TYPES OF BUSINESS:
Real Estate Brokerage
Settlement Services
Title Services
Mortgage Services

BRANDS/DIVISIONS/AFFILIATES:
Berkshire Hathaway Energy Company
Beth Allman & Associates
Carol Jones Realtors
Edina Realty
FirstWeber Realtors
Realty South
Semonin Realtors
Woods Bros Realty

CONTACTS: *Note: Officers with more than one job title may be intentionally listed here more than once.*
Ronald J. Peltier, CEO
Robert R. Moline, COO
Mike Warmka, CFO
Stuart Lyle, Sr. VP-Human Resources
Patty Smejkal, VP-IT
Dana Strandmo, General Counsel
Mike Warmka, Chief Acct. Officer
Mary Lee Blaylock, Pres., HomeServices of America Relocation LLC
Melissa J. Buscho, Pres., HomeServices Insurance, Inc.
Todd Johnson, Managing Dir.
Ed McGreen, Pres.
Ronald J. Peltier, Chmn.

GROWTH PLANS/SPECIAL FEATURES:
HomeServices of America, Inc., a subsidiary of Berkshire Hathaway Energy Company, is a real estate brokerage firm and provider of brokerage-owned settlement services. The firm is comprised of 28 real estate brands and operates through several offices and more than 15,000 sales associates. HomeServices maintains operations all over the U.S., operating under various brands in the regions it serves. Brands include Allie Beth Allman & Associates, Carol Jones Realtors, CBSHome, Champion Realty Inc., Edina Realty, EWM Realty International, FirstWeber Realtors, Guarantee.com Real Estate, Harry Normal Realtors, Home Real Estate, Huff Realty, Intero Real Estate Services, Iowa Realty, Long Realty Mortgage Title & Insurance, Realty South, Rector Hayden, ReeceNichols, Roberts Brothers, Semonin Realtors and Woods Bros Realty. Companies under the Bershire Hathaway HomeServices banner include California Properties, Carolinas Realty, First Realty, Fox & Roach Realtors, Georgia Properties, Kansas City Realty, Koenig Rubloff Realty Group, New England Properties/Westchester Properties, Northwest Real Estate, York Simpson Underwood Realty and Yost & Little Realty. In 2015, the firm acquired FirstWeber Realtors, based in Wisconsin; and Allie Beth Allman & Associates, which is engaged in the Dallas luxury real estate market.

FINANCIAL DATA: *Note: Data for latest year may not have been available at press time.*

In U.S. $	2015	2014	2013	2012	2011	2010
Revenue	191,000,000	139,000,000	139,000,000			
R&D Expense						
Operating Income						
Operating Margin %						
SGA Expense						
Net Income						
Operating Cash Flow						
Capital Expenditure						
EBITDA						
Return on Assets %						
Return on Equity %						
Debt to Equity						

CONTACT INFORMATION:
Phone: 　　　　Fax: 612-336-5572
Toll-Free: 888-485-0018
Address: 333 S. 7th St., Fl. 27, Minneapolis, MN 55402 United States

STOCK TICKER/OTHER:
Stock Ticker: Subsidiary　　Exchange:
Employees: 4,326　　Fiscal Year Ends: 12/31
Parent Company: Berkshire Hathaway Energy Company

SALARIES/BONUSES:
Top Exec. Salary: $　　Bonus: $
Second Exec. Salary: $　　Bonus: $

OTHER THOUGHTS:
Estimated Female Officers or Directors: 3
Hot Spot for Advancement for Women/Minorities: Y

Hometown America LLC

www.hometownamerica.com

NAIC Code: 531110

TYPES OF BUSINESS:

Manufactured Home Communities
RV Parks
Property Management
Brokerage Services

BRANDS/DIVISIONS/AFFILIATES:

Hometown America Family Communities
Hometown America Foundation
Hometown America Age-Qualified Communities

CONTACTS: Note: Officers with more than one job title may be intentionally listed here more than once.

Richard G. Cline, Jr., CEO
Stephen Braun, COO
Stephen Braun, Pres.
Patrick Crocetta, VP-Mktg. & Sales
Thomas Curatolo, VP-Finance

GROWTH PLANS/SPECIAL FEATURES:

Hometown America, LLC develops, acquires and manages manufactured housing communities and age-qualified communities. Hometown currently owns over 50 communities in 10 states, including Arizona, California, Delaware, Florida, Illinois, Massachusetts, New Jersey, New York, Rhode Island and Virginia. The company operates through two divisions: The Hometown America Family Communities division and the Hometown America Age-Qualified Communities division. The Hometown America Age-Qualified Communities are targeted toward active seniors. All of the firm's communities are provided with onsite management and new home and resale brokerage services. Many of Hometown's communities feature recreational facilities such as swimming pools, clubhouses, golf courses and basketball courts. The Hometown America Foundation provides funding to nonprofit organizations that address the needs of the homeless and assist in disaster relief.

The company offers employees assistance programs; comprehensive medical, dental, vision, life and short- and long-term disability insurance; bonus opportunities; paid vacation; sick and holiday pay; and dependent care flexible spending accounts.

FINANCIAL DATA: Note: Data for latest year may not have been available at press time.

In U.S. $	2015	2014	2013	2012	2011	2010
Revenue						
R&D Expense						
Operating Income						
Operating Margin %						
SGA Expense						
Net Income						
Operating Cash Flow						
Capital Expenditure						
EBITDA						
Return on Assets %						
Return on Equity %						
Debt to Equity						

CONTACT INFORMATION:

Phone: 312-604-7500   Fax: 312-604-7501
Toll-Free: 888.735.4310
Address: 150 N. Wacker Dr., Ste. 2800, Chicago, IL 60606 United States

STOCK TICKER/OTHER:

Stock Ticker: Private   Exchange:
Employees: 206   Fiscal Year Ends: 12/31
Parent Company:

SALARIES/BONUSES:

Top Exec. Salary: $ Bonus: $
Second Exec. Salary: $ Bonus: $

OTHER THOUGHTS:

Estimated Female Officers or Directors:
Hot Spot for Advancement for Women/Minorities:

Hongkong and Shanghai Hotels Ltd

NAIC Code: 721110

www.hshgroup.com

TYPES OF BUSINESS:

Hotels
Commercial Properties
Resorts & Luxury Clubs
Apartments
Golf Courses
Property Management
Laundry Services
Park Attractions & Tramways

BRANDS/DIVISIONS/AFFILIATES:

Peninsula Group (The)
Landmark (The)
Peninsula Merchandising Ltd
Peak Tramways
Peninsula Office Tower
Peninsula Hong Kong
Peninsula Clubs and Consultancy Services Ltd
Tai Pan Laundry & Dry Cleaning Services Ltd

CONTACTS: Note: Officers with more than one job title may be intentionally listed here more than once.

Clement King Man Kwok, CEO
Peter Camille Borer, COO
Ingvar Herland, Gen. Mgr.-R&D
Ingvar Herland, Gen. Mgr.-Tech.
Christobelle Liao, Corp. Counsel
Paul Tchen, Gen. Mgr.-Oper. Planning & Support
Ming Chen, Sr. Mgr.-Bus. Dev.
Sian Griffiths, Dir.-Comm.-The Peninsula Hotels
Ming Chen, Sr. Mgr.-Investor Rel.
Martin Lew, Gen. Mgr.-Oper., Financial Control
Martyn Sawyer, Gen Mgr.-Properties & Clubs
Jonathan Crook, Gen. Mgr.-Peninsula Shanghai
Rainy Chan, VP-Hong Kong & Bangkok
Nicolas Beliard, Gen. Mgr.-Peninsula Bangkok
Michael Kadoorie, Chmn.
Maria Razumich-Zec, Regional VP-U.S. East Coast

GROWTH PLANS/SPECIAL FEATURES:

Hongkong and Shanghai Hotels, Ltd. (HSH), celebrating its 150th Anniversary in 2016, owns and operates luxury hotels, resorts and other commercial properties. The Peninsula Group, HSH's hotel management division, owns hotels located in Hong Kong, Beijing, Shanghai, Chicago, New York, Bangkok, Manila, Beverly Hills, Tokyo and Paris. The company also owns Repulse Bay, a self-contained residential, retail and commercial complex. The firm's Landmark property in Ho Chi Minh City features 65 fully furnished apartments on the upper floors, a restaurant and bar, rooftop swimming pool with a poolside bar, deck, gymnasium, sauna, squash court and office space on the lower levels. HSH owns the Peninsula Office Tower, part of the extension of the Peninsula Hong Kong. The building includes nine floors of high-end office space with office units ranging from 6,200 to 11,200 square feet. Another office property owned by HSH is the St. John's Building in Hong Kong, with 21 floors of office space. Subsidiary Peninsula Clubs and Consultancy Services, Ltd. manages high-end clubs such as The Hong Kong Club, The Hong Kong Bankers Club and Butterfield's. The firm has a 75% ownership stake in the Thai Country Club, a 72-hole golf course in Thailand. The firm's other operations include Peak Tramways, one of the oldest operating funicular railways in the world, operating out of Hong Kong; Peninsula Merchandising, Ltd., a subsidiary that owns and operates boutiques selling Peninsula-brand products; and Tai Pan Laundry & Dry Cleaning Services, Ltd., a commercial laundry service for major hotels, clubs and restaurants. The company recently entered the European market with the opening of its new hotel, The Peninsula Paris, housed in the 100 year old Beaux Arts building. In December 2015, the firm announced plans to construct a 190-room hotel in central London, The Peninsula London, due to open in 2021.

FINANCIAL DATA: Note: Data for latest year may not have been available at press time.

In U.S. $	2015	2014	2013	2012	2011	2010
Revenue	740,184,000	752,690,200	710,143,400	667,596,700	645,807,600	606,870,900
R&D Expense						
Operating Income	130,734,500	142,467,000	117,454,700	105,335,400	107,527,200	102,370,000
Operating Margin %	17.66%	18.92%	16.53%	15.77%	16.65%	16.86%
SGA Expense						
Net Income	128,929,400	147,753,200	220,727,200	200,485,300	291,251,600	387,819,800
Operating Cash Flow	157,293,900	184,627,000	168,639,700	124,288,000	128,800,500	131,379,100
Capital Expenditure	179,469,800	52,732,140	415,281,800	112,813,300	40,225,990	35,584,530
EBITDA	210,670,700	241,355,900	311,364,600	290,220,200	386,272,600	482,325,100
Return on Assets %	2.24%	2.63%	4.12%	3.98%	6.03%	8.66%
Return on Equity %	2.76%	3.22%	5.01%	4.81%	7.46%	11.53%
Debt to Equity	0.16	0.12	0.14	0.09	0.10	0.11

CONTACT INFORMATION:

Phone: 852-2840-7788   Fax: 852-2840-7567
Toll-Free:
Address: 2 Ice House St., St. George's Bldg., 8th Fl., Central,   Hong Kong

STOCK TICKER/OTHER:

Stock Ticker: HKSHY
Employees: 5,772
Parent Company:

Exchange: PINX
Fiscal Year Ends: 12/31

SALARIES/BONUSES:

Top Exec. Salary: $ Bonus: $
Second Exec. Salary: $ Bonus: $

OTHER THOUGHTS:

Estimated Female Officers or Directors: 13
Hot Spot for Advancement for Women/Minorities: Y

Hongkong Land Holdings Ltd

www.hkland.com

NAIC Code: 531100

TYPES OF BUSINESS:

Property Investment, Management & Development
Office Properties
Homes & Apartments
Highway Development
Water Company Development

BRANDS/DIVISIONS/AFFILIATES:

Alexandra House
Jardine House
Landmark Mandarin Oriental
Gaysorn
Central Park
Marina Bay Suites
MCL Land
Jardine Matheson Ltd

CONTACTS: Note: Officers with more than one job title may be intentionally listed here more than once.

Y. K. Pang, CEO
John R. Witt, CFO
Angie Fung, Head-Admin. & Secretarial Svcs.
Cissy Leung, Head-Legal Svcs.
Jennifer Lim, Head-Comm.
Ben Keswick, Managing Dir.
N.M. McNamara, Corp. Sec.
Ben Keswick, Chmn.
Robert Garman, Exec. Dir.-Commercial Property, South Asia

GROWTH PLANS/SPECIAL FEATURES:

Hongkong Land Holdings Ltd., founded in 1889, is a property investment, management and development group with a portfolio encompassing approximately 8.2 million square feet of commercial and residential property in Hong Kong. The firm operates in two main business sectors: commercial and residential. The commercial sector owns several prime office and retail locations in Hong Kong's Central Business District as well as other high-end retail properties and prime office locations in South East Asia. Its commercial Hong Kong properties include Alexandra House, Chater House, One/Two & Three Exchange Square, The Forum, Jardine House, Landmark Mandarin Oriental, Landmark Atrium, Edinburgh Tower, Gloucester Tower, York House and Prince's Building. Its other commercial properties include One Raffles Link, One Raffles Quay and Marina Bay Financial Centre in Singapore; Gaysorn, in Thailand; Central Building and 63 Ly Thai To in Vietnam; One Central in Macau; Jakarta Land in Indonesia; WF Central and CBD in China; and Exchange Square in Cambodia. The residential sector focuses on luxury and high-end properties in Hong Kong and Southeast Asia. Its Hong Kong residential properties consist of the Serenade residential towers. Its other Asian properties include Central Park and Maple Place at Beijing Riviera in Beijing; Bamboo Grove, Landmark Riverside, Central Avenue and Yorkville in Chongqing; One Central Residences in Macau; Marina Bay Suites, Marina Bay Residences and Terrasse Uber 388, Palms, Hallmark Residences, Ripple Bay, J Gateway, LakeVille and Choa Chu Kang Grove in Singapore; Nava Park and Anandamaya Residences in Indonesia; and Roxas Triangle Tower in the Philippines. The company also holds a 77% ownership interest in Singapore-based residential property developer MCL Land. Asia-based trading group Jardine Matheson owns a 50% stake in Hongkong Land.

FINANCIAL DATA: Note: Data for latest year may not have been available at press time.

In U.S. $	2015	2014	2013	2012	2011	2010
Revenue	1,932,100,000	1,876,300,000	1,857,100,000	1,114,800,000	1,223,700,000	1,340,600,000
R&D Expense						
Operating Income	2,007,600,000	1,091,300,000	834,700,000	1,108,300,000	5,214,400,000	4,079,100,000
Operating Margin %	103.90%	58.16%	44.94%	99.41%	426.11%	304.27%
SGA Expense	106,900,000	105,100,000	93,700,000	84,800,000	75,800,000	81,600,000
Net Income	2,011,700,000	1,327,400,000	1,189,600,000	1,438,500,000	5,306,400,000	4,739,400,000
Operating Cash Flow	896,200,000	699,000,000	907,900,000	298,700,000	336,300,000	690,400,000
Capital Expenditure	152,300,000	136,600,000	134,000,000	515,000,000	38,300,000	200,000
EBITDA	2,246,200,000	1,643,600,000	1,451,700,000	1,665,100,000	5,530,000,000	5,010,200,000
Return on Assets %	5.91%	3.98%	3.67%	4.73%	19.96%	21.68%
Return on Equity %	7.15%	4.87%	4.48%	5.65%	24.01%	29.42%
Debt to Equity	0.13	0.14	0.13	0.14	0.13	0.14

CONTACT INFORMATION:

Phone: 852 28428428   Fax: 852 28459226
Toll-Free:
Address: 1 Exchange Sq. Central, 8th Fl., Hamilton,   Hong Kong

STOCK TICKER/OTHER:

Stock Ticker: HNGKY   Exchange: PINX
Employees: 1,435   Fiscal Year Ends: 12/31
Parent Company:

SALARIES/BONUSES:

Top Exec. Salary: $ Bonus: $
Second Exec. Salary: $ Bonus: $

OTHER THOUGHTS:

Estimated Female Officers or Directors: 4
Hot Spot for Advancement for Women/Minorities: Y

Hospitality Properties Trust

NAIC Code: 0

www.hptreit.com

TYPES OF BUSINESS:

Real Estate Investment Trust
Hotels
Travel Centers

BRANDS/DIVISIONS/AFFILIATES:

InterContinental Hotels & Resorts
Courtyard by Marriott
Candlewood Suites
Staybridge Suites
Residence Inn by Marriott
Crowne Plaza Hotels & Resorts
Hyatt Place
SpringHill Suites by Marriott

CONTACTS: Note: Officers with more than one job title may be intentionally listed here more than once.

John Murray, Assistant Secretary
Mark Kleifges, CFO
Jennifer Clark, Executive VP
Ethan Bornstein, Senior VP
Barry Portnoy, Trustee
John Harrington, Trustee
William Lamkin, Trustee
Adam Portnoy, Trustee
Donna Fraiche, Trustee

GROWTH PLANS/SPECIAL FEATURES:

Hospitality Properties Trust (HPT) is a real estate investment trust (REIT) that owns 302 hotels as well as 193 travel centers, located in 45 U.S. states, Puerto Rico and Canada. HPT's hotel properties currently include the following brands: Courtyard by Marriott, Candlewood Suites, Staybridge Suites, Residence Inn by Marriott, Crowne Plaza Hotels & Resorts, Hyatt Place, InterContinental Hotels & Resorts, Marriott Hotels and Resorts, Radisson Hotels & Resorts, TownePlace Suites by Marriott, Country Inns & Suites by Carlson, Holiday Inn Hotels & Resorts, SpringHill Suites by Marriott, Wyndham Hotels & Resorts and Park Plaza Hotels & Resorts. The company's 193 travel centers include 147 operated under the TravelCenters of America brand and 46 operated under the Petro Stopping Centers or Petro brand name. All of HPT's management agreements or leases share nine points: first, managers are required to pay a fixed minimum rent; second, operators must pay percentage returns on gross hotel revenues exceeding a certain threshold; third, all agreements are long-term (15 years or more); fourth, each hotel is part of a combination of hotels and is subject to cross-default obligations with respect to all other hotels in the same combination; fifth, each combination of hotels is geographically diverse; sixth, contract renewals may only be pursued for a combination of hotels, not for each hotel individually; seventh, the firm's agreements require the deposit of 5% to 6% of gross hotel revenues into escrows to fund periodic renovations; eighth, the properties are located near major demand generators, such as urban centers, airports or educational facilities; and finally, each management agreement or lease includes security terms to ensure payments to HPT. In June 2015, the firm acquired combined economic ownership of approximately half of Reit Management & Research LLC.

FINANCIAL DATA: Note: Data for latest year may not have been available at press time.

In U.S. $	2015	2014	2013	2012	2011	2010
Revenue	1,921,904,000	1,736,322,000	1,563,855,000	1,296,982,000	1,210,333,000	1,085,488,000
R&D Expense						
Operating Income	335,935,000	339,170,000	273,583,000	278,460,000	325,843,000	330,843,000
Operating Margin %	17.47%	19.53%	17.49%	21.46%	26.92%	30.47%
SGA Expense	109,837,000	45,897,000	50,087,000	44,032,000	40,963,000	38,961,000
Net Income	166,418,000	197,185,000	133,178,000	151,923,000	190,440,000	21,351,000
Operating Cash Flow	530,893,000	461,745,000	391,089,000	363,908,000	355,102,000	341,444,000
Capital Expenditure	630,585,000	284,621,000	515,872,000	630,940,000	69,345,000	7,091,000
EBITDA	642,637,000	654,494,000	573,027,000	550,477,000	554,394,000	398,790,000
Return on Assets %	2.35%	2.95%	1.74%	1.92%	3.10%	-.15%
Return on Equity %	5.56%	6.39%	3.96%	4.41%	6.58%	-.32%
Debt to Equity	1.29	1.04	0.96	1.18	0.87	0.85

CONTACT INFORMATION:

Phone: 617 964-8389   Fax: 617 969-5730
Toll-Free:
Address: 255 Washington St., 2 Newton Pl., Ste. 300, Newton, MA 02458
United States

STOCK TICKER/OTHER:

Stock Ticker: HPT   Exchange: NYS
Employees: 500   Fiscal Year Ends: 12/31
Parent Company:

SALARIES/BONUSES:

Top Exec. Salary: $ Bonus: $
Second Exec. Salary: $ Bonus: $

OTHER THOUGHTS:

Estimated Female Officers or Directors: 1
Hot Spot for Advancement for Women/Minorities:

Host Hotels & Resorts LP

www.hosthotels.com

NAIC Code: 0

TYPES OF BUSINESS:

Real Estate Investment Trust
Hotels

BRANDS/DIVISIONS/AFFILIATES:

CONTACTS: *Note: Officers with more than one job title may be intentionally listed here more than once.*

W. Walter, CEO
Gregory Larson, CFO
Richard Marriott, Chairman of the Board
Brian Macnamara, Chief Accounting Officer
Joanne Hamilton, Executive VP, Divisional
Minaz Abji, Executive VP, Divisional
James Risoleo, Executive VP
Elizabeth Abdoo, Executive VP
Nathan Tyrrell, Managing Director, Divisional

GROWTH PLANS/SPECIAL FEATURES:

Host Hotels & Resorts, LP (HHR) is a leading hotel real estate investment trust (REIT). HHR operates a portfolio of 105 primarily luxury and upper-upscale hotels containing approximately 57,000 rooms, with the majority located in the U.S., and 12 properties located outside of the U.S. in Australia, Brazil, Canada, Chile, Mexico and New Zealand. Its brands include Marriott, Westin, Ritz-Carlton, Sheraton, W, Hyatt, Hilton/Embassy Suites, Novotel, ibis, St. Regis, The Luxury Collection, Fairmont and Swissotel. Additionally, the firm owns an interest in five international joint ventures that own 10 luxury and upper upscale hotels located in Italy, Spain, Belgium, the U.K., Poland, France and the Netherlands, and a 25% stake in an Asian joint venture which owns one hotel in Australia, minority interests in two operating hotels in India and five additional hotels in India currently under development. Its properties typically include meeting and banquet facilities, a variety of restaurants and lounges, swimming pools, exercise facilities and/or spas, gift shops and parking facilities. During 2015, the firm acquired 643-room Phoenician, a Luxury Collection resort; and completed the sale of the Sheraton Needham.

HHR offers its employees medical, dental, prescription, vision and hearing coverage; free onsite fitness centers; associate assistance programs; long- and short-term disability and life insurance; a 401(k); company paid parking; hotel discounts; health and dependent care spending accounts; long-term care coverage; an employee stock purchase plan; and tuition assistance.

FINANCIAL DATA: *Note: Data for latest year may not have been available at press time.*

In U.S. $	2015	2014	2013	2012	2011	2010
Revenue	5,387,000,000	5,354,000,000	5,166,000,000	5,286,000,000	4,998,000,000	4,437,000,000
R&D Expense						
Operating Income	650,000,000	710,000,000	512,000,000	383,000,000	324,000,000	223,000,000
Operating Margin %	12.06%	13.26%	9.91%	7.24%	6.48%	5.02%
SGA Expense	716,000,000	656,000,000	719,000,000	901,000,000	869,000,000	768,000,000
Net Income	558,000,000	732,000,000	317,000,000	61,000,000	-15,000,000	-130,000,000
Operating Cash Flow	1,171,000,000	1,150,000,000	1,019,000,000	782,000,000	661,000,000	520,000,000
Capital Expenditure	663,000,000	436,000,000	436,000,000	638,000,000	542,000,000	326,000,000
EBITDA	1,530,000,000	1,676,000,000	1,242,000,000	1,171,000,000	1,013,000,000	818,000,000
Return on Assets %	4.65%	5.85%	2.45%	.46%	-.11%	-1.10%
Return on Equity %	7.75%	10.05%	4.51%	.90%	-.23%	-2.22%
Debt to Equity	0.38	0.54	0.65	0.79	0.86	0.86

CONTACT INFORMATION:

Phone: 240 744-1000   Fax: 240 380-6338
Toll-Free:
Address: 6903 Rockledge Dr., Ste. 1500, Bethesda, MD 20817 United States

STOCK TICKER/OTHER:

Stock Ticker: HST
Employees: 240
Parent Company:

Exchange: NYS
Fiscal Year Ends: 12/31

SALARIES/BONUSES:

Top Exec. Salary: $952,750   Bonus: $
Second Exec. Salary: $551,580   Bonus: $

OTHER THOUGHTS:

Estimated Female Officers or Directors: 2
Hot Spot for Advancement for Women/Minorities: Y

Hotel Properties Ltd

www.hotelprop.com

NAIC Code: 721110

TYPES OF BUSINESS:
Hotels
Condominiums
Restaurants
Retail Properties & Operations
Food Distribution

BRANDS/DIVISIONS/AFFILIATES:
HPL Hotels & Resorts Pte Ltd
Concorde Hotels & Resorts (Malaysia) Sdn Bhd

CONTACTS: *Note: Officers with more than one job title may be intentionally listed here more than once.*
Ong Beng Seng, Managing Dir.
Chuang Sheue Ling, Corp. Sec.
Buong Lik Lau, Head-Hotel Div.
Arthur Tan Keng Hock, Chmn.

GROWTH PLANS/SPECIAL FEATURES:
Hotel Properties Ltd. (HPL) is a diversified hotel holding company headquartered in Singapore. It invests in premium commercial and residential properties mainly throughout the Asia Pacific region, including hotels, condominiums, shopping centers, restaurants, a food distribution chain and retail operations. It owns or has interests in 27 hotels in 11 countries including Bhutan, Czech Republic, Indonesia, Maldives, Malaysia, Seychelles, Singapore, South Africa, Thailand, Tanzania and Vanuatu. These include three Four Seasons properties and a Hard Rock Hotel in Indonesia; two Four Seasons Resorts, one Holiday Inn Resort, a Gili Lankanfushi and a Six Senses Laamu in the Maldives; two Concorde Hotels, a Casa Del Mar, a Hard Rock Hotel and The Lakehouse Cameron Highlands in Malaysia; a Four Seasons Hotel, a Hilton and a Concorde Hotel in Singapore; a Hard Rock Hotel, a Point Yamu and a Metropolitan in Thailand; a Four Seasons Resort in Seychelles; and a Holiday Inn in Vanuatu. Furthermore, through its subsidiaries HPL Hotels & Resorts Pte. Ltd. and Concorde Hotels & Resorts (Malaysia) Sdn. Bhd., HPL provides hotel management services. The company owns three shopping centers and a residential tower in Singapore as well as a residential property in Thailand. In its lifestyle division, the company owns Hard Rock Cafes in Indonesia, Malaysia, the Philippines, Singapore and Thailand.

FINANCIAL DATA: *Note: Data for latest year may not have been available at press time.*

In U.S. $	2015	2014	2013	2012	2011	2010
Revenue	429,044,861	461,300,113	552,954,370	428,473,032	391,200,000	357,860,000
R&D Expense						
Operating Income						
Operating Margin %						
SGA Expense						
Net Income	60,658,066	101,261,000	141,962,460	109,796,027	55,800,000	113,570,000
Operating Cash Flow						
Capital Expenditure						
EBITDA						
Return on Assets %						
Return on Equity %						
Debt to Equity						

CONTACT INFORMATION:
Phone: 65-734-5250   Fax: 65-732-0347
Toll-Free:
Address: 50 Cuscaden Rd., 08-01 HPL House, Singapore, 249724 Singapore

STOCK TICKER/OTHER:
Stock Ticker: H15   Exchange: Singapore
Employees: 4,400   Fiscal Year Ends: 12/31
Parent Company:

SALARIES/BONUSES:
Top Exec. Salary: $ Bonus: $
Second Exec. Salary: $ Bonus: $

OTHER THOUGHTS:
Estimated Female Officers or Directors: 1
Hot Spot for Advancement for Women/Minorities:

Hovnanian Enterprises Inc

www.khov.com

NAIC Code: 236117

TYPES OF BUSINESS:

Construction, Home Building and Residential
Mortgages
Title Insurance
Residential Communities
Commercial Construction

BRANDS/DIVISIONS/AFFILIATES:

CONTACTS: Note: Officers with more than one job title may be intentionally listed here more than once.

Ara Hovnanian, CEO
J. Sorsby, CFO
Brad OConnor, Chief Accounting Officer
Thomas Pellerito, COO
Michael Discafani, Other Corporate Officer
David Bachstetter, Treasurer

GROWTH PLANS/SPECIAL FEATURES:

Hovnanian Enterprises, Inc. designs, constructs, markets and sells single-family detached homes, attached town homes and condominiums, mid-rise and high-rise condominiums, urban infill and active adult homes in planned residential communities. The firm organizes its operations into two divisions: homebuilding and financial services. Homebuilding operations are conducted in six geographic segments: Northeast (New Jersey and Pennsylvania), Mid-Atlantic (Delaware, Maryland, Virginia, West Virginia and Washington, D.C.), Midwest (Illinois, Minnesota and Ohio), Southeast (Florida, Georgia, North Carolina and South Carolina), Southwest (Arizona and Texas) and West (California). With more than 318,000 total homes sold since its inception in 1959, including 5,776 homes delivered in 2015, the firm is one of the largest homebuilders in the U.S. Its homes are available in 219 communities in 16 states, and are generally located in suburban areas easily accessible through public and personal transportation. Base home prices range from $116,000 to $1,673,000. During 2015, the average sales price for a Hovnanian home was $379,000. A residential development generally includes single-family detached homes and/or a number of residential buildings containing from 2-24 individual homes per building, together with amenities such as club houses, swimming pools, tennis courts, tot lots and open areas. Hovnanian's financial services operations provide mortgage loans and title services to its homebuyers.

The firm offers employees medical, dental and vision insurance; an employee assistance program; flexible spending accounts; life insurance; short- and long-term disability; a 401(k) plan; a 529 college bound savings plan; tuition assistance; U.S. government bonds; a home purchase discount; entertainment discounts; and discounted home, auto, boater's and renter's insurance.

FINANCIAL DATA: Note: Data for latest year may not have been available at press time.

In U.S. $	2015	2014	2013	2012	2011	2010
Revenue	2,148,480,000	2,063,380,000	1,851,253,000	1,485,353,000	1,134,907,000	1,371,842,000
R&D Expense						
Operating Income	71,904,000	134,079,000	102,789,000	24,517,000	-188,184,000	-213,651,000
Operating Margin %	3.34%	6.49%	5.55%	1.65%	-16.58%	-15.57%
SGA Expense	250,909,000	254,912,000	220,166,000	190,319,000	211,394,000	238,231,000
Net Income	-16,100,000	307,144,000	31,295,000	-66,197,000	-286,087,000	2,588,000
Operating Cash Flow	-320,535,000	-190,585,000	9,268,000	-66,998,000	-207,415,000	32,487,000
Capital Expenditure	2,054,000	3,423,000	1,558,000	5,059,000	826,000	2,456,000
EBITDA	73,458,000	110,975,000	117,991,000	2,870,000	-185,079,000	-184,787,000
Return on Assets %	-.65%	15.17%	1.81%	-4.02%	-16.73%	.13%
Return on Equity %						
Debt to Equity						

CONTACT INFORMATION:

Phone: 732 747-7800   Fax: 732 747-7159
Toll-Free:
Address: 110 W. Front St., Red Bank, NJ 07701 United States

SALARIES/BONUSES:

Top Exec. Salary: Bonus: $1,180,014
$1,120,972
Second Exec. Salary: Bonus: $432,000
$634,044

STOCK TICKER/OTHER:

Stock Ticker: HOV   Exchange: NYS
Employees: 2,078   Fiscal Year Ends: 10/31
Parent Company:

OTHER THOUGHTS:

Estimated Female Officers or Directors:

Hot Spot for Advancement for Women/Minorities:

Howard Johnson International Inc   www.hojo.com

NAIC Code: 721110

TYPES OF BUSINESS:

Hotels
Restaurants
Vacation Specials
Travel Packages

BRANDS/DIVISIONS/AFFILIATES:

Wyndham Worldwide
HoJo.com Booking Advantage
Rise & Dine
HoJo
TripFinder Vacation Packages

CONTACTS: Note: Officers with more than one job title may be intentionally listed here more than once.

Mary K. Mahoney, CEO
Mary K. Mahoney, Pres.

GROWTH PLANS/SPECIAL FEATURES:

Howard Johnson International, Inc., a subsidiary of Wyndham Worldwide, owns and franchises mid-priced hotels. Its portfolio currently includes over 500 hotels and locations in 14 countries. Howard Johnson's hotels feature a variety of amenities, including an airport shuttle, business centers, free high-speed Internet access, free Rise & Dine breakfast, gym/fitness centers, meeting/banquet facilities, pet friendly rooms, pools and restaurants. Customers can search for a hotel on the firm's web site using any or all of these amenities as search criteria. The firm's Hotel Package Deals feature several destinations for a given week, offering various packages viewable on the firm's web site. These packaged specials often include discounted hotel fees and location-specific promotions. TripFinder Vacation Packages offer a travel package program through which customers can rent cars or purchase air fares in addition to securing hotel reservations. The company participates in the Wyndham Rewards program, which is a free program that offers customers redeemable points or airline miles for staying at any one of Wyndham's hotel locations. Additionally, Howard Johnson features the HoJo.com Booking Advantage program, which guarantees customers a 10% lower price than any competitor. The company also has two HoJo restaurants in the U.S., one in Lake George, New York, and one in Bangor, Maine.

FINANCIAL DATA: Note: Data for latest year may not have been available at press time.

In U.S. $	2015	2014	2013	2012	2011	2010
Revenue						
R&D Expense						
Operating Income						
Operating Margin %						
SGA Expense						
Net Income						
Operating Cash Flow						
Capital Expenditure						
EBITDA						
Return on Assets %						
Return on Equity %						
Debt to Equity						

CONTACT INFORMATION:

Phone: 973-428-9700   Fax: 973-496-7658
Toll-Free: 800-544-9881
Address: 1 Sylvan Way, Parsippany, NJ 07054 United States

SALARIES/BONUSES:

Top Exec. Salary: $ Bonus: $
Second Exec. Salary: $ Bonus: $

STOCK TICKER/OTHER:

Stock Ticker: Subsidiary   Exchange:
Employees: 10,000   Fiscal Year Ends: 12/31
Parent Company: Wyndham Worldwide

OTHER THOUGHTS:

Estimated Female Officers or Directors: 1
Hot Spot for Advancement for Women/Minorities:

Hutchison Whampoa Properties Ltd

www.hutchison-whampoa.com/en/businesses/property.php

NAIC Code: 531100

TYPES OF BUSINESS:

Real Estate Operations & Development
Hotels
Office & Industrial Properties
Residential Properties
Marine Docks & Repair Yards

BRANDS/DIVISIONS/AFFILIATES:

Hutchison Whampoa Limited
Hongkong & Whampoa Dock Company Limited
Hutchison Properties Limited
Cavendish International Holdings Limited
Harbour Plaza Hotels & Resorts
Hutchison Premium Services
Pacific Property Net

CONTACTS: *Note: Officers with more than one job title may be intentionally listed here more than once.*

Kin Ning Fok, Managing Dir.
Frank John Sixt, CFO
Ka-shing Li, Chmn.

GROWTH PLANS/SPECIAL FEATURES:

Hutchison Whampoa Properties Ltd. (Hutchison) is the property development and investment subsidiary of Hutchison Whampoa Limited (HWL). Hutchison was established by HWL to hold the property interests of Hongkong & Whampoa Dock Company Limited (HWD), Hutchison Properties Limited (HPL) and Cavendish International Holdings Limited (CIHL). The company operates within HWL's property & hotels group. HWL's other property & hotels group company is joint-venture Harbour Plaza Hotels & Resorts. Hutchison's properties are organized by location into China, Hong Kong and the U.K. Its Chinese properties make up the majority of its portfolio and are divided by city into Beijing, Changchun, Changsha, Changzhou, Chengdu, Chongqing, Dalian, Dongguan, Foshan, Guangzhou, Huizhou, Jianmen, Nanjing, Qingdao, Shanghai, Shenzhen, Tianjin, Wuhan, Xian, Zhongshan and Zhuhai. The firm's Hong Kong properties consist of assets of 28 Barker Road. Finally its U.K. assets include a Lots Toad Power Station; the Albion Riverside, Chelsea Waterfront and Convoys Wharf. The firm's services include property management, project management, marketing and e-business. Property management services offered by Hutchison include the deployment of security guards, caretakers and maintenance workers. The company's project management team supervises and coordinates the design and construction of its developments through external consultants and contractors. Hutchison's in-house marketing team is responsible for the leasing and sale of its properties, and provides property-related services to other HWL subsidiaries, including tenancy negotiations, valuations, feasibility studies and general real estate guidance. The company's e-business operations comprise subsidiaries Hutchison Premium Services and Pacific Property Net, both of which operate real estate web sites.

FINANCIAL DATA: *Note: Data for latest year may not have been available at press time.*

In U.S. $	2015	2014	2013	2012	2011	2010
Revenue	4,385,000,000	3,303,000,000	2,221,520,000	2,066,013,600	2,211,680,000	2,081,080,000
R&D Expense						
Operating Income						
Operating Margin %						
SGA Expense						
Net Income						
Operating Cash Flow						
Capital Expenditure						
EBITDA						
Return on Assets %						
Return on Equity %						
Debt to Equity						

CONTACT INFORMATION:

Phone: 852-2128-7500   Fax: 852-2128-7888
Toll-Free:
Address: 18 Tak Fung St., Hunghom, Fl. 3, One Harbourfront, Hong Kong, Hong Kong

SALARIES/BONUSES:

Top Exec. Salary: $ Bonus: $
Second Exec. Salary: $ Bonus: $

STOCK TICKER/OTHER:

Stock Ticker: Subsidiary   Exchange:
Employees: 260,000   Fiscal Year Ends: 12/31
Parent Company: Hutchison Whampoa Limited

OTHER THOUGHTS:

Estimated Female Officers or Directors:
Hot Spot for Advancement for Women/Minorities:

Hyatt Hotels Corporation

NAIC Code: 721110

www.hyatt.com

TYPES OF BUSINESS:

Hotel Ownership & Management
Timeshares
Golf Courses
Gaming
Retirement Communities
Motels & Inns
Hotel Franchising

BRANDS/DIVISIONS/AFFILIATES:

Hyatt Regency
Grand Hyatt
Hyatt Place
Hyatt Gold Passport
Andaz
Hyatt House
Hyatt Residence Club
Park Hyatt

CONTACTS: Note: Officers with more than one job title may be intentionally listed here more than once.

Mark Hoplamazian, CEO
Thomas Pritzker, Chairman of the Board
Bradley OBryan, Chief Accounting Officer
Maryam Banikarim, Executive VP
Rena Reiss, Executive VP
H. Floyd, Executive VP
Peter Sears, Executive VP
David Udell, Executive VP
Peter Fulton, Executive VP
Stephen Haggerty, Other Corporate Officer
Robert Webb, Other Executive Officer
Atish Shah, Senior VP

GROWTH PLANS/SPECIAL FEATURES:

Hyatt Hotels Corporation (Hyatt) owns, operates, manages and franchises full-service luxury hotels in 52 countries across the globe. The company owns, manages or franchises approximately 599 hotels with approximately 156,336 rooms. Hyatt's operations consist of several brands. Hyatt and Hyatt Regency host business and leisure travelers, although Hyatt Regency caters mainly to larger groups. Grand Hyatt hotels cater to leisure and business travelers and include accommodations for banquets and conferences. Park Hyatt hotels are smaller, full-service luxury hotels featuring world class art and restaurants in a few of the world's most visited cities. The Andaz branded hotels are boutique-style hotels that feature restaurants and bars aimed at local clientele as well as single travelers. The two select service brands, Hyatt House and Hyatt Place, are extended-stay brands designed to feel more like home. Hyatt Residence Club provides vacation ownership and vacation rental opportunities, offering members timeshare or points-based resort vacation opportunities. Hyatt Ziva and Hyatt Zilara are the company's all-inclusive resort brands which are developed, sold and managed as part of the Hyatt Residence club. Hyatt's guest loyalty program, Hyatt Gold Passport, has over 20 million members.

The firm offers employees complementary hotel rooms; medical, dental, vision and prescription drug coverage; and tuition assistance.

FINANCIAL DATA: Note: Data for latest year may not have been available at press time.

In U.S. $	2015	2014	2013	2012	2011	2010
Revenue	4,328,000,000	4,415,000,000	4,184,000,000	3,949,000,000	3,698,000,000	3,527,000,000
R&D Expense						
Operating Income	323,000,000	279,000,000	233,000,000	159,000,000	153,000,000	108,000,000
Operating Margin %	7.46%	6.31%	5.56%	4.02%	4.13%	3.06%
SGA Expense	308,000,000	349,000,000	323,000,000	316,000,000	283,000,000	276,000,000
Net Income	124,000,000	344,000,000	207,000,000	88,000,000	113,000,000	66,000,000
Operating Cash Flow	538,000,000	473,000,000	456,000,000	499,000,000	393,000,000	450,000,000
Capital Expenditure	269,000,000	253,000,000	232,000,000	301,000,000	331,000,000	520,000,000
EBITDA	582,000,000	950,000,000	731,000,000	518,000,000	445,000,000	421,000,000
Return on Assets %	1.57%	4.21%	2.61%	1.16%	1.53%	.91%
Return on Equity %	2.87%	7.32%	4.31%	1.82%	2.27%	1.30%
Debt to Equity	0.26	0.29	0.27	0.25	0.25	0.29

CONTACT INFORMATION:

Phone: 312 750-1234   Fax:
Toll-Free: 800-323-7249
Address: 71 S. Wacker Dr., 12th Fl., Chicago, IL 60606 United States

STOCK TICKER/OTHER:

Stock Ticker: H   Exchange: NYS
Employees: 45,000   Fiscal Year Ends: 12/31
Parent Company:

SALARIES/BONUSES:

Top Exec. Salary: Bonus: $
$1,060,833
Second Exec. Salary: Bonus: $
$727,083

OTHER THOUGHTS:

Estimated Female Officers or Directors: 3

Hot Spot for Advancement for Women/Minorities: Y

Hypo Real Estate Holding AG (Hypo Bank) www.hyporealestate.com

NAIC Code: 531100

TYPES OF BUSINESS:

Commercial Real Estate Financing
Mortgages
Asset Management

BRANDS/DIVISIONS/AFFILIATES:

Deutsche Pfandbriefbank AG
DEPFA Bank Plc
Flint Nominees Ltd
Hypo Real Estate Capital Corp
Hypo Real Estate Capital Japan Corporation
Hypo Real Estate Transactions SAS
Hypo Real Estate Capital India Corp Pte Ltd
Hypo Pfandbrief Bank International SA

CONTACTS: Note: Officers with more than one job title may be intentionally listed here more than once.

Wolfgang Groth, Treas.
Bernhard Scholz, Head-Real Estate Finance
Andreas Schenk, Chief Risk Officer
Gunter Borgel, Chmn.

GROWTH PLANS/SPECIAL FEATURES:

Hypo Real Estate Holding AG (Hypo Bank) is a financial holding company with interests in real estate and related areas such as financing, consulting, brokering and other services. It does not have any direct banking operations. The firm's businesses include credit and financial service institutions Deutsche Pfandbriefbank AG and DEPFA Bank Plc. Deutsche Pfandbriefbank, headquartered in Munich, Germany, was formed by the merger of Hypo Real Estate Bank and DEPFA Deutsche Pfandbriefbank AG. Subsidiaries of Deutsche Pfandbriefbank include Flint Nominees Ltd. in London, Hypo Real Estate Capital Corp. in New York, Hypo Real Estate Capital Japan Corporation in Tokyo, Hypo Real Estate Transactions SAS in Paris, Mumbai-based Hypo Real Estate Capital India Corp. Pte Ltd. and Hypo Real Estate Capital Singapore Corp. Private Ltd. DEPFA Bank, headquartered in Dublin, is the parent of Hypo Pfandbrief Bank International SA of Luxembourg, DEPFA ACS Bank in Dublin and Hypo Public Finance Bank in Dublin. Both Deutsche Pfandbriefbank and DEPFA Bank are currently restructuring following the firm's recent takeover by the German Financial Markets Stabilization Fund, known as SoFFin. As part of the restructuring plan, the banks have become specialist institutions focused on real estate and public finance and will take on a more regional approach, concentrating its activities primarily in Germany and Europe.

FINANCIAL DATA: Note: Data for latest year may not have been available at press time.

In U.S. $	2015	2014	2013	2012	2011	2010
Revenue	815,000,000	810,000,000	805,000,000	800,000,000	895,747,000	120,215,000
R&D Expense						
Operating Income						
Operating Margin %						
SGA Expense						
Net Income						
Operating Cash Flow						
Capital Expenditure						
EBITDA						
Return on Assets %						
Return on Equity %						
Debt to Equity						

CONTACT INFORMATION:

Phone: 49-89-2880-0   Fax: 49-89-2880-10319
Toll-Free:
Address: Freisinger Strasse 5, Unterschleissheim, D-85716 Germany

STOCK TICKER/OTHER:

Stock Ticker: Private   Exchange:
Employees: 844   Fiscal Year Ends: 12/31
Parent Company:

SALARIES/BONUSES:

Top Exec. Salary: $ Bonus: $
Second Exec. Salary: $ Bonus: $

OTHER THOUGHTS:

Estimated Female Officers or Directors: 1
Hot Spot for Advancement for Women/Minorities:

Hysan Development Co Ltd

NAIC Code: 531100

TYPES OF BUSINESS:

Real Estate Development, Management & Investment
Apartments
Shopping Centers
Office Buildings
Car Parks

BRANDS/DIVISIONS/AFFILIATES:

Service-Scan
Lee Gardens (The)
Leighton Centre
Sunning Plaza
Lee Theatre Plaza
Bamboo Grove
One Hysan Avenue
Lee Gardens Two

CONTACTS: *Note: Officers with more than one job title may be intentionally listed here more than once.*

Siu Chuen Lau, CEO
Roger Shu Yan Hao, CFO
Wendy Wen Yee Yung, Corp. Sec.
Lai Kiu Chan, Dir.-Design & Project
Lawrence Wai Leung Lau, Gen. Mgr.-Property Svcs.
Kitty Man Wai Choy, Gen. Mgr.-Retail Leasing
Jessica Mo Ching Yip, Dir.-Office Leasing
Irene Yun Lien Lee, Chmn.

GROWTH PLANS/SPECIAL FEATURES:

Hysan Development Co., Ltd. is a Hong Kong property development and investment firm that derives revenues by leasing space to commercial and residential tenants. Its investment portfolio totals approximately 4 million square feet of space, located primarily in the Causeway Bay district of Hong Kong. In addition, Hysan owns and manages about 1,200 car parks. Most of the firm's commercial properties combine office and retail within a single building. These holdings include: The Lee Gardens (53 floors), Lee Gardens Two (34 floors), Leighton Centre (28 floors), One Hysan Avenue (26 floors), Sunning Plaza (30 floors) and 111 Leighton Road (24 floors). The firm also owns Lee Theatre Plaza, a shopping and dining complex. Hysan's residential holdings encompass the 345-unit residential complex Bamboo Grove, which includes a club house and sports facilities; and the 59-unit Sunning Court. The firm's development investments consist of a 24.7% interest in The Grand Gateway, a mixed-use commercial, dining, office and residential complex in Shanghai. Additionally, the company provides property management services, based on its proprietary model Service-Scan, and through its Satisfaction Management System, aims to monitor tenant satisfaction and respond accordingly to keep occupancy rates high.

FINANCIAL DATA: *Note: Data for latest year may not have been available at press time.*

In U.S. $	2015	2014	2013	2012	2011	2010
Revenue	442,228,000	415,668,500	394,910,900	320,518,600	247,802,400	227,431,600
R&D Expense						
Operating Income	365,643,900	344,757,300	325,675,800	248,962,800	198,680,300	181,532,700
Operating Margin %	82.68%	82.94%	82.46%	77.67%	80.17%	79.81%
SGA Expense	30,169,490	27,590,900	26,817,330	24,109,810	22,304,800	18,050,120
Net Income	374,282,200	632,012,200	793,947,600	1,283,493,000	1,101,702,000	495,604,800
Operating Cash Flow	327,480,800	308,786,000	293,056,600	221,887,600	181,403,700	167,479,400
Capital Expenditure	773,577	3,094,307	1,031,436	3,996,813	1,031,436	902,506
EBITDA	488,900,500	759,007,700	957,816,900	1,396,306,000	1,208,198,000	568,707,800
Return on Assets %	3.67%	6.32%	8.52%	15.58%	15.83%	8.32%
Return on Equity %	4.29%	7.52%	10.14%	18.62%	19.10%	10.34%
Debt to Equity	0.06	0.07	0.10	0.09	0.10	0.09

CONTACT INFORMATION:

Phone: 852 28955777   Fax: 852 25775153
Toll-Free:
Address: The Lee Gardens, 33 Hysan Ave., 49th Fl.,, Hong Kong, Hong Kong

STOCK TICKER/OTHER:

Stock Ticker: HYSNF   Exchange: GREY
Employees: 638   Fiscal Year Ends: 12/31
Parent Company:

SALARIES/BONUSES:

Top Exec. Salary: $ Bonus: $
Second Exec. Salary: $ Bonus: $

OTHER THOUGHTS:

Estimated Female Officers or Directors: 5
Hot Spot for Advancement for Women/Minorities: Y

Hyundai Elevator Co Ltd

www.hyundaielevator.co.kr

NAIC Code: 333921

TYPES OF BUSINESS:

Elevator Manufacture & Design
Escalator Manufacture & Design
Moving Walk Manufacture & Design
Material Handling System Manufacture & Design
Platform Screen Door Manufacture & Design
Auto Parking System Manufacture & Design

BRANDS/DIVISIONS/AFFILIATES:

H-Series
I-XEL
STVF
LUXEN
YZER
Millenium
Modular
Gap Zero

CONTACTS: *Note: Officers with more than one job title may be intentionally listed here more than once.*

Martin S.H. Han, CEO
Jeong Eun Hyun, Chmn.

GROWTH PLANS/SPECIAL FEATURES:

Hyundai Elevator Co., Ltd. manufactures, sells, installs and maintains elevators, escalators and moving walkways, auto parking systems, material handling systems and platform screen doors. Hyundai Elevator manufactures passenger, observation, hospital bed, automobile, freight and marine elevators. Passenger elevators, sold under the I-XEL, STVF, LUXEN and YZER brands, comprise high-speed traction elevators for heavy traffic high-rise commercial buildings and medium-speed traction elevators for residential or mid-rise commercial buildings. Glass wall observation elevators are designed to provide an aesthetic architectural detail to hotels, shopping centers, office buildings, banks, hospitals and observation towers. Hyundai Elevator's hospital bed elevators are designed for hospitals and other clinics. The firm's automobile elevators transport automobiles from one floor to another in parking garages. Its freight elevators include designs for lightweight cargoes as well as forklifts and heavier loads. Hyundai Elevator's marine elevators are designed for numerous ship types, including tankers, bulk carriers, containerships, Ro-Ro (roll-on roll-off, for wheeled cargo), ferries and navy vessels. The firm sells escalators under such brands as Millennium, designed for hotels, shopping malls, banks and department stores; Modular, designed for heavy-duty use; and H-Series, designed for installation in subway stations, sports complexes, conference halls and airports. Its moving walkways include horizontal, 12-degree inclined and combination designs. Auto parking systems cover vertical, rotating, elevator and horizontal parking solutions. Material handling systems include stacker cranes, automated guided vehicles and robotic transfer vehicles. Hyundai Elevator's platform screen doors, designed to separate subway and light rail transit platforms from railways, include hermetic, semi-hermetic and handrail varieties. The company's newest product categories include mechanical bicycle parking systems, Gap Zero safety floor boards and auto-foldable and auto-sliding canopies. In 2015, the firm opened Hyundai Elevator Technology Training Center, which offers onsite education and training for manufacturing, operating, repairing, installing, and servicing elevators and related technologies and products.

FINANCIAL DATA: *Note: Data for latest year may not have been available at press time.*

In U.S. $	2015	2014	2013	2012	2011	2010
Revenue	1,273,888,381	1,119,391,065	1,038,930,000	893,433,000	771,400,000	629,660,000
R&D Expense						
Operating Income						
Operating Margin %						
SGA Expense						
Net Income	-3,562,287	-203,992,652	-172,475,438	-38,276,480	-229,200,000	27,500,000
Operating Cash Flow						
Capital Expenditure						
EBITDA						
Return on Assets %						
Return on Equity %						
Debt to Equity						

CONTACT INFORMATION:

Phone: 82-31-644-5114   Fax: 82-2-745-4227
Toll-Free:
Address: San 136-1, Ami-ri Boobal-eup, Icheon-si, Gyeonggi-do, 467-734 South Korea

STOCK TICKER/OTHER:

Stock Ticker: 17800
Employees: 1,405
Parent Company:

Exchange: Seoul
Fiscal Year Ends: 12/31

SALARIES/BONUSES:

Top Exec. Salary: $ Bonus: $
Second Exec. Salary: $ Bonus: $

OTHER THOUGHTS:

Estimated Female Officers or Directors:
Hot Spot for Advancement for Women/Minorities:

Hyundai Engineering & Construction Company Ltd

www.hdec.co.kr
NAIC Code: 237000

TYPES OF BUSINESS:

Construction, Heavy & Civil Engineering
Power Plant Construction
Highway & Bridge Construction
Residential Construction
Commercial Construction

BRANDS/DIVISIONS/AFFILIATES:

Hillstate
Hyundai Engineering Co Ltd
Hyperion

CONTACTS: Note: Officers with more than one job title may be intentionally listed here more than once.

Soo-Hyun Jung, CEO

GROWTH PLANS/SPECIAL FEATURES:

Hyundai Engineering & Construction Company Ltd. (HDEC) is an international construction company based in South Korea. HDEC has expertise in a variety of structures, such as civil works, highways and bridges, housing, shipyards, dams, power plants (including nuclear power), airports, stadiums, hotels and retail complexes. The company is organized into four divisions: infra & environment works, building works, plant works and power and energy works. The infra & environment works division is responsible for the engineering and construction of roads, bridges, harbor, railways, highways and other infrastructure developments across Korea and overseas. Its major projects include the Gyeongbu Expressway, the Honam High-speed Railway, the Soyang Multi-purpose Dam, Busan Harbor, Pattani-Narathiwat Highway in Thailand and Jaber Causeway Marine Bridge in Kuwait. The building works division builds skyscrapers, sports/leisure facilities, residential housing and other large-scale architectural projects in both the Korean and international markets. Some of its notable projects include Hamad Medical Center in Qatar, the Hwaseong Stadium in Korea, the Jangbogo Antarctic Research Station, the Four Seasons Hotel in Cairo and the National Center for Korean Traditional Performing Arts. The plant works division constructs hydrocarbon processing and industrial plants, including LNG receiving terminals, LNG supply pipelines, nuclear energy power plant, multi-purpose water gate facilities and integrated steel works. Lastly, the power & energy works division constructs power and desalination plants and overseas operations. This division is also expanding into the renewable energy business with developments in photovoltaic, wind velocity and nuclear energy abroad in Africa, Central America and South America. Subsidiaries of the company include Hyundai Engineering Co., Ltd.; Hillstate, a construction company; and Hyperion, an upscale apartment builder.

FINANCIAL DATA: Note: Data for latest year may not have been available at press time.

In U.S. $	2015	2014	2013	2012	2011	2010
Revenue	16,568,302,821	15,064,930,625	13,490,214,000	12,675,900,000	10,527,900,000	10,274,800,000
R&D Expense						
Operating Income						
Operating Margin %						
SGA Expense						
Net Income	318,782,537	363,618,739	487,526,647	485,000,000	561,100,000	476,880,000
Operating Cash Flow						
Capital Expenditure						
EBITDA						
Return on Assets %						
Return on Equity %						
Debt to Equity						

CONTACT INFORMATION:

Phone: 82-2-746-1114   Fax: 82-2-743-8963
Toll-Free:
Address: Bldg 75, Yulgok-ro, Jongno-gu, Seoul, 110-793 South Korea

STOCK TICKER/OTHER:

Stock Ticker: 720   Exchange: Seoul
Employees: 59,481   Fiscal Year Ends: 12/31
Parent Company:

SALARIES/BONUSES:

Top Exec. Salary: $ Bonus: $
Second Exec. Salary: $ Bonus: $

OTHER THOUGHTS:

Estimated Female Officers or Directors:
Hot Spot for Advancement for Women/Minorities:

IDI Gazeley
NAIC Code: 0

www.brookfieldlogisticsproperties.com/

TYPES OF BUSINESS:
Real Estate Development
Industrial Business Parks
Warehouses
Facility Management & Leasing
Construction Management Services

BRANDS/DIVISIONS/AFFILIATES:
Brookfield Property Partners LP
IDI Investment Management LLC

CONTACTS: Note: Officers with more than one job title may be intentionally listed here more than once.
Jay Cornforth, Managing Dir.
Kent D. Greenawalt, Sr. VP-Oper.
Timothy J. Gunter, Pres.
Matthew Berger, CFO
Debbi Kvietkus, VP-Human Resources
Matt O'Sullivan, Chief Dev. Officer
Rita Skaggs, VP-Comm.
Greg J. Ryan, Sr. VP
Brian T. Mee, Sr. VP-Strategic Dev. Svcs.
S. Michael Parks, Sr. VP-Nat'l Bus. Dev.
G. Bryan Blasingame, Jr., Chief Investment Officer

GROWTH PLANS/SPECIAL FEATURES:
IDI Gazeley (IDI) is a national, full-service industrial real estate developer. It is one of the world's leading investors and developers of logistics warehouses and distribution parks near major markets and transport routes throughout North America, Europe and Asia. IDI has developed over 250 million square feet (sf) of property across the globe for over 900 customers. It develops buildings to suit customer needs, and also owns and operates industrial space. IDI construction managers meet to review procedures and look for ways to improve the building process and manage issues that arise after tenant occupancy. At any given time, IDI has eight to 14 million sf of move-in-ready industrial, office and distribution facilities available. Subsidiary IDI Investment Management LLC is a leading buyer, seller and operator specializing in industrial real estate assets in key locations along major logistics and distribution routes. Acquisition criteria consists of Class A and B quality warehouse and distribution facilities, single and multi-tenant facilities, top industrial markets, no minimum or maximum deal size and a preference for all cash closings with the ability to assume debt on a case-by-case basis. The firm is a subsidiary of Brookfield Logistics Property Partners, an owner, operator and investor of real estate. In October 2015, the firm entered the Netherlands market by acquiring a portfolio of 13 logistics warehouses totaling more than 2 million sf from Eurindustrial NV.

IDI offers its employees medical, dental and vision coverage; life insurance; short- and long-term disability insurance; flexible spending accounts; and a 401(k) plan.

FINANCIAL DATA: Note: Data for latest year may not have been available at press time.

In U.S. $	2015	2014	2013	2012	2011	2010
Revenue						
R&D Expense						
Operating Income						
Operating Margin %						
SGA Expense						
Net Income						
Operating Cash Flow						
Capital Expenditure						
EBITDA						
Return on Assets %						
Return on Equity %						
Debt to Equity						

CONTACT INFORMATION:
Phone: 404-479-4000   Fax: 404-479-4162
Toll-Free:
Address: 1100 Peachtree St., Ste. 1100, Atlanta, GA 30309 United States

STOCK TICKER/OTHER:
Stock Ticker: Subsidiary   Exchange:
Employees: 148   Fiscal Year Ends: 12/31
Parent Company: Brookfield Property Partners LP

SALARIES/BONUSES:
Top Exec. Salary: $ Bonus: $
Second Exec. Salary: $ Bonus: $

OTHER THOUGHTS:
Estimated Female Officers or Directors: 3
Hot Spot for Advancement for Women/Minorities: Y

Impac Mortgage Holdings Inc

NAIC Code: 525990

www.impaccompanies.com

TYPES OF BUSINESS:
Mortgage REIT
Real Estate Investment Trust
Real Estate & Mortgage Services

BRANDS/DIVISIONS/AFFILIATES:
IMH Assets Corp
Integrated Real Estate Service Corporation
Impac Funding Corporation
Impac Mortgage Corp
Excel Mortgage Servicing Inc

CONTACTS: *Note: Officers with more than one job title may be intentionally listed here more than once.*
Joseph Tomkinson, CEO
Todd Taylor, CFO
William Ashmore, Director

GROWTH PLANS/SPECIAL FEATURES:
Impac Mortgage Holdings, Inc. (IMH) is a mortgage real estate investment trust. Following the episode of distress within the mortgage lending and securitization markets, IMH has shifted its focus away from past interests in long-term investments of non-conforming Alt-A residential mortgages. Now, through subsidiary Integrated Real Estate Service Corporation (IRES), the firm's primary goal is to create an integrated services platform to provide solutions to the mortgage and real estate markets. These business activities include mortgage lending activities, portfolio loss mitigation, real estate service activities, surveillance and recovery services and the management of the long-term mortgage portfolio. IMH and subsidiary IMH Assets Corp. maintain master servicing rights on the portfolio of long-term investments that include various adjustable rate, fixed rate and hybrid ARM residential and commercial mortgages acquired and originated by its discontinued mortgage and commercial operations. The company also operates subsidiaries Impac Funding Corporation (IFC) and Impac Mortgage Corp., formerly known as Excel Mortgage Servicing, Inc., among others. The company's correspondent lending division provides services to banks, lenders and credit unions. In January 2015, the firm agreed to acquire the mortgage operations of CashCall, Inc., which will operate as a separate division of Impac Mortgage under the name CashCall Mortgage.

FINANCIAL DATA: *Note: Data for latest year may not have been available at press time.*

In U.S. $	2015	2014	2013	2012	2011	2010
Revenue	154,604,000	53,373,000	82,032,000	91,165,000	73,070,000	69,174,000
R&D Expense						
Operating Income	58,923,000	-2,662,000	-6,042,000	14,289,000	6,923,000	7,804,000
Operating Margin %	38.11%	-4.98%	-7.36%	15.67%	9.47%	11.28%
SGA Expense	133,459,000	50,473,000	77,688,000	68,484,000	58,610,000	54,924,000
Net Income	80,799,000	-6,322,000	-8,184,000	-3,375,000	3,224,000	10,294,000
Operating Cash Flow	30,664,000	29,981,000	174,546,000	183,313,000	173,577,000	271,617,000
Capital Expenditure	-109,000	18,000	362,000	252,000	483,000	1,773,000
EBITDA						
Return on Assets %	1.49%	-.11%	-.13%	-.05%	.05%	.17%
Return on Equity %	115.92%	-24.89%	-29.87%	-11.46%	11.44%	49.26%
Debt to Equity	3.77	212.07	6.01	4.15	2.34	0.44

CONTACT INFORMATION:
Phone: 800 597-4101   Fax:
Toll-Free: 800-597-4101
Address: 19500 Jamboree Rd., Irvine, CA 92612 United States

STOCK TICKER/OTHER:
Stock Ticker: IMH   Exchange: ASE
Employees: 298   Fiscal Year Ends: 12/31
Parent Company:

SALARIES/BONUSES:
Top Exec. Salary: $600,000   Bonus: $200,000
Second Exec. Salary: Bonus: $200,000
$600,000

OTHER THOUGHTS:
Estimated Female Officers or Directors:
Hot Spot for Advancement for Women/Minorities:

Sales, profits and employees may be estimates. Financial information, benefits and other data can change quickly and may vary from those stated here.

Indian Hotels Company Limited (The)   www.tajhotels.com

NAIC Code: 721110

TYPES OF BUSINESS:

Hotels
Spas
Apartments
Private Jet Rental
Air Catering
Travel Agency

BRANDS/DIVISIONS/AFFILIATES:

Taj Wellington Mews
Tata Group
Taj Hotels, Resorts and Places
TajAir Ltd
TajSATS Air Catering Ltd
IndiTravel Private Limited
Roots Corporation Ltd
Jiva Spas

CONTACTS: Note: Officers with more than one job title may be intentionally listed here more than once.

Rakesh Sarna, Managing Dir.
Prabhat Verm, Sr.VP-Operations
Anil P. Goel, CFO
Beejal Desai, VP-Legal
Abhijit Mukerji, Exec. Dir.-Hotel Oper.
Rajiv Gujral, Sr. VP-Mergers, Acquisitions & Dev.
Anil P. Goel, Exec. Dir.-Finance
Veer Vijay Singh, COO-Upper Upscale Hotels
Prabhat Verma, COO-The Gateway Hotels & Resorts
Jyoti Narang, COO-Luxury Div.

GROWTH PLANS/SPECIAL FEATURES:

The Indian Hotels Company Limited (IHCL), which, together with its subsidiaries, does business as Taj Hotels, Resorts and Palaces (Taj), is one of the world's premier operators of luxury hotels. Part of Indian conglomerate Tata Group, IHCL owns more than 93 hotels in 55 locations across India and 16 additional hotels in Sri Lanka, the Maldives, Malaysia, Australia, the U.S., the U.K. Bhutan, Africa and the Middle East. The company operates hotels in the value, mid-market, premium and luxury segments. The firm's hotel brands include Taj, a chain of luxury full-service resorts, hotels and palaces including historic properties, modern business-oriented hotels, beach resorts and safari lodges; Taj Exotica, a resort and spa brand; Taj Safaris, ecotourism lodges in the Indian jungle; The Gateway Hotel, a mid to upscale full-service hotel and resort brand; Ginger, a line of economy hotels operated by wholly-owned subsidiary Roots Corporation Ltd.; and The Gateway Hotel, a pan-India network of hotels and resorts that offers business and leisure travelers a hotel with a nomadic design theme. Taj Wellington Mews provides 80 furnished apartments for short-term and extended-stay travelers, and includes butler service, 24-hour concierge services, baby-sitting services and recreation options. Taj offers a variety of services through Jiva Spas, including yoga, meditation, Indian healing ceremonies and other treatments. Besides hotels, IHCL offers chartered private luxury jets through subsidiary TajAir Ltd. and catering services for domestic and international airlines through TajSATS Air Catering Ltd. Subsidiary Taj Luxury Residences offers apartments in three locations, staffed with personal butlers, a round-the-clock concierge and 24-hour-a-day babysitting service. Finally, IndiTravel Private Limited offers travel services from ticketing to car rentals and passport assistance.

FINANCIAL DATA: Note: Data for latest year may not have been available at press time.

In U.S. $	2015	2014	2013	2012	2011	2010
Revenue	629,558,336	611,153,920	562,632,064	515,940,992	438,113,440	378,912,736
R&D Expense						
Operating Income						
Operating Margin %						
SGA Expense						
Net Income	-56,828,944	-83,244,408	-64,665,660	459,922	-13,115,297	-17,795,682
Operating Cash Flow						
Capital Expenditure						
EBITDA						
Return on Assets %						
Return on Equity %						
Debt to Equity						

CONTACT INFORMATION:

Phone: 91 22 - 61371637   Fax: 91 22 - 61371710
Toll-Free:
Address: 9/Fl, Express Towers, Barrister Rajni Patel Marg, Mumbai, 400021 India

STOCK TICKER/OTHER:

Stock Ticker: INDHOTEL
Employees: 10,020
Parent Company: TATA Group

Exchange: India
Fiscal Year Ends: 03/31

SALARIES/BONUSES:

Top Exec. Salary: $ Bonus: $
Second Exec. Salary: $ Bonus: $

OTHER THOUGHTS:

Estimated Female Officers or Directors: 2
Hot Spot for Advancement for Women/Minorities:

Inland Real Estate Corporation

www.inlandrealestate.com

NAIC Code: 0

TYPES OF BUSINESS:

Real Estate Investment Trust
Property Management
Retail Properties

BRANDS/DIVISIONS/AFFILIATES:

Inland Real Estate Group of Companies Inc (The)
Westbury Square

CONTACTS: *Note: Officers with more than one job title may be intentionally listed here more than once.*

Mark Zalatoris, CEO
Brett Brown, CFO
Thomas DArcy, Chairman of the Board
D. Carr, Executive VP, Divisional
Beth Brooks, General Counsel
Dawn Benchelt, Other Corporate Officer
William Anderson, Senior VP, Divisional

GROWTH PLANS/SPECIAL FEATURES:

Inland Real Estate Corporation (IRC), a member of The Inland Real Estate Group of Companies, Inc., is a self-administered real estate investment trust (REIT) that acquires, owns, operates and develops open-air neighborhood, community and lifestyle shopping centers and single-tenant retail properties located primarily in the Midwestern U.S. IRC owns or has interests in 133 properties comprised of 46 neighborhood retail centers, 40 power centers, 11 single-user retail properties, 35 community centers and one lifestyle center, with nearly 15.4 million square feet of total rentable space. Most of the company's properties are located within a 400-mile radius of its suburban Chicago headquarters and consist of grocery anchored, discount anchored, fashion anchored and convenience retail centers. The firm also purchases freestanding properties leased by single credit tenants, located anywhere in the country. IRC acquires both single assets and entire portfolios. Its emphasis remains on well-tenanted, well-located properties, from 75,000 to 300,000 square feet. Notable property holdings include the real estate in Celebration, Florida where Walt Disney World is located. In March 2015, it acquired Westbury Square, a 115,000 square-foot shopping center space in Alabama. In December 2015, the firm agreed to be acquired by real estate funds managed by DRA Advisors LLC. Upon completion of the merger/acquisition IRC will become a privately-held real estate investment trust.

The firm offers employees medical and dental insurance, a flexible savings plan for health and dependent care needs, a 401(k) plan, life and AD&D insurance and both short- and long-term disability insurance.

FINANCIAL DATA: *Note: Data for latest year may not have been available at press time.*

In U.S. $	2015	2014	2013	2012	2011	2010
Revenue	203,900,000	204,762,000	183,374,000	159,844,000	167,226,000	167,028,992
R&D Expense						
Operating Income						
Operating Margin %						
SGA Expense						
Net Income	25,531,000	39,175,000	111,684,000	17,759,000	-7,184,000	-263,000
Operating Cash Flow						
Capital Expenditure						
EBITDA						
Return on Assets %						
Return on Equity %						
Debt to Equity						

CONTACT INFORMATION:

Phone: 630 218-8000   Fax: 630 218-7350
Toll-Free: 888-331-4732
Address: 2901 Butterfield Rd., Oak Brook, IL 60523 United States

SALARIES/BONUSES:

Top Exec. Salary: $ Bonus: $
Second Exec. Salary: $ Bonus: $

STOCK TICKER/OTHER:

Stock Ticker: Private
Employees: 141
Parent Company: DRA Advisors LLC

Exchange:
Fiscal Year Ends: 12/31

OTHER THOUGHTS:

Estimated Female Officers or Directors: 3
Hot Spot for Advancement for Women/Minorities: Y

Innkeepers USA Trust

www.innkeepersusa.com

NAIC Code: 0

TYPES OF BUSINESS:

Real Estate Investment Trust
Hotels
Hotel Development

BRANDS/DIVISIONS/AFFILIATES:

Island Hospitality Management
Apollo Investment Corporation

CONTACTS: Note: Officers with more than one job title may be intentionally listed here more than once.

Mark A. Murphy, Sec.
Richard A. Mielbye, Sr. VP-Dev.
Roger Pollak, VP-Acct., Innkeepers Hospitality
Marc A. Beilinson, Chief Restructuring Officer
John J. Hannan, Chmn.-Apollo

GROWTH PLANS/SPECIAL FEATURES:

Innkeepers USA Trust, owned by Apollo Investment Corp., is a real estate investment trust (REIT) that specializes in the ownership of multi-brand, upscale, extended-stay hotels in the U.S. The firm seeks to acquire hotel properties in markets with high barriers to entry and with strong underlying demand growth. Innkeepers USA owns interests in 64 hotels with an aggregate of 8,300 rooms/suites in 18 states and Washington, D.C. Sixty-three of these hotels are managed by Island Hospitality Management. The firm's hotels operate under the following brands: Residence Inn, TownePlace Suites and Courtyard by Marriott; Sheraton; Double Tree; Embassy Suites; Hyatt Summerfield Suites; Westin; Best Western; Bulfinch Hotel; Hampton Inn; and Hilton. In addition to its acquisitions of existing hotels in the upscale and extended-stay market, Innkeepers USA also acquires under-performing mid-priced and full service hotels that have the potential for strategic repositioning or re-flagging to a premium franchise brand in the upscale segment. During 2010, burdened with a large debt load, the firm filed for bankruptcy. It is currently in the restructuring process.

FINANCIAL DATA: Note: Data for latest year may not have been available at press time.

In U.S. $	2015	2014	2013	2012	2011	2010
Revenue						
R&D Expense						
Operating Income						
Operating Margin %						
SGA Expense						
Net Income						
Operating Cash Flow						
Capital Expenditure						
EBITDA						
Return on Assets %						
Return on Equity %						
Debt to Equity						

CONTACT INFORMATION:

Phone: 561-835-1800   Fax: 561-835-0457
Toll-Free:
Address: 340 Royal Poinciana Way, Ste. 306, Palm Beach, FL 33480
United States

STOCK TICKER/OTHER:

Stock Ticker: Private   Exchange:
Employees: 33   Fiscal Year Ends: 12/31
Parent Company: Apollo Investment Corporation

SALARIES/BONUSES:

Top Exec. Salary: $ Bonus: $
Second Exec. Salary: $ Bonus: $

OTHER THOUGHTS:

Estimated Female Officers or Directors:
Hot Spot for Advancement for Women/Minorities:

Integrated Electrical Services Inc

NAIC Code: 238210

www.ies-co.com

TYPES OF BUSINESS:

Electric Contractors
Electrical & Communications Installation & Maintenance
Residential Building & Remodeling
Commercial & Industrial Renovations

BRANDS/DIVISIONS/AFFILIATES:

Southern Industrial Sales and Services Inc
Shanahan Mechanica and Electrical Inc
STR Mechanical LLC

CONTACTS: Note: Officers with more than one job title may be intentionally listed here more than once.

Tracy McLauchlin, CFO
Robert Lewey, COO
David Gendell, Director
Gail Makode, General Counsel

GROWTH PLANS/SPECIAL FEATURES:

Integrated Electrical Services (IES) is a provider of electrical contracting services in the U.S., with locations across the country. The company's operations are divided into four business segments: communications, residential, commercial & industrial and infrastructure solutions. The communications segment is a leading provider of network infrastructure products and services for data centers and other mission critical environments. Services offered include the design, installation and maintenance of network infrastructure for the financial, medical, hospitality, government, high-tech manufacturing, educational and information technology industries. The residential business provides electrical installation services for single-family housing and multi-family apartment complexes and CATV (cable television) cabling installations for residential and light commercial applications. In addition to its core electrical construction work, this segment also provides services for the installation of residential solar power, smart meters and electric car charging stations, both for new construction and existing residences. The commercial & industrial segment is one of the largest providers of electrical contracting services in the U.S. The division offers a broad range of electrical design, construction, renovation, engineering and maintenance services to the commercial and industrial markets. The types of projects the firm engages in include high-rise residential and office buildings, power plants, manufacturing facilities, municipal infrastructure, health care facilities and residential developments. The infrastructure solutions segment provides maintenance and repair services to several industries, including electric motor repair and rebuilding for the steel, railroad, marine, petrochemical, pulp and paper, wind energy, mining, automotive and power generation industries. In 2015, the firm acquired Southern Industrial Sales and Services, Inc., a motor repair and related field services company; and Shanahan Mechanical and Electrical, Inc., a provider of mechanical & electrical contracting services. In April 2016, it acquired STR Mechanical LLC, a provider of commercial & industrial mechanical services.

FINANCIAL DATA: Note: Data for latest year may not have been available at press time.

In U.S. $	2015	2014	2013	2012	2011	2010
Revenue	573,857,000	512,395,000	494,593,000	456,115,000	481,607,000	460,633,000
R&D Expense						
Operating Income	18,488,000	7,641,000	426,000	-389,000	-35,348,000	-29,016,000
Operating Margin %	3.22%	1.49%	.08%	-.08%	-7.33%	-6.29%
SGA Expense	81,416,000	75,571,000	66,598,000	58,609,000	69,365,000	84,920,000
Net Income	16,538,000	5,324,000	-3,573,000	-11,802,000	-37,693,000	-32,147,000
Operating Cash Flow	11,506,000	12,598,000	1,954,000	-7,371,000	-11,852,000	-13,166,000
Capital Expenditure	2,779,000	1,982,000	444,000	1,877,000	2,688,000	924,000
EBITDA	21,177,000	9,985,000	2,471,000	1,493,000	-28,914,000	-23,374,000
Return on Assets %	7.69%	2.79%	-2.07%	-6.84%	-19.56%	-13.57%
Return on Equity %	17.37%	7.07%	-6.17%	-20.00%	-45.30%	-27.46%
Debt to Equity	0.10	0.11	0.16		0.15	0.10

CONTACT INFORMATION:

Phone: 713 860-1500   Fax: 713 860-1599
Toll-Free: 877-437-6285
Address: 5433 Westheimer Road, Ste 500, Houston, TX 77056 United States

STOCK TICKER/OTHER:

Stock Ticker: IESC
Employees: 3,106
Parent Company:

Exchange: NAS
Fiscal Year Ends: 09/30

SALARIES/BONUSES:

Top Exec. Salary: $471,875   Bonus: $
Second Exec. Salary: $300,000   Bonus: $

OTHER THOUGHTS:

Estimated Female Officers or Directors: 1
Hot Spot for Advancement for Women/Minorities:

InterContinental Hotels Group plc

www.ihgplc.com

NAIC Code: 721110

TYPES OF BUSINESS:

Hotel & Motel Development & Management
Hotels

BRANDS/DIVISIONS/AFFILIATES:

InterContinental Hotels Group Europe
InterContinental Hotels Group The Americas
InterContinental Hotels Group AMEA
InterContinental Hotels Group Greater China
InterContinental Hotels & Resorts
Hualuxe Hotels and Resorts
Holiday Inn
Kimpton Hotels & Restaurants

CONTACTS: Note: Officers with more than one job title may be intentionally listed here more than once.

Richard Solomons, CEO
Paul Edgecliffe-Johnson, CFO
Tracy Robbins, Exec. VP-Global Human Resources & Oper. Support
George Turner, General Counsel
Kirk Kinsell, Pres., Americas
Kenneth Macpherson, CEO-Greater China
Angela Brav, CEO-Europe
Patrick Cescau, Chmn.
Jan Smits, CEO-Asia, Middle East & Africa

GROWTH PLANS/SPECIAL FEATURES:

InterContinental Hotels Group plc (IHG) is an international hotel and hospitality firm. The company operates under the following 12 brands: InterContinental Hotels & Resorts, Kimpton Hotels & Restaurants, Hualuxe Hotels and Resorts, Hotel Indigo, Even Hotels, Crowne Plaza Hotels & Resorts, Holiday Inn, Holiday Inn Express, Holiday Inn Club Vacations, Staybridge Suites and Candlewood Suites. Hotel use is more prevalent within the U.S., Canada, Latin America and the Caribbean (64%), with the remaining percentage nearly equally-divided between Europe, Asia/Middle East/Africa and Greater China. IHG's hotels are located in nearly 100 countries, with more than 5,000 hotels and 744,000 guest rooms. The company's hotels are primarily franchised (4,219 hotels), 806 hotels are managed by IHG and seven are IHG-owned and -leased. Subsidiaries include InterContinental Hotels Group Europe, with headquarters in Denham, U.K.; InterContinental Hotels Group The Americas, with headquarters in Atlanta, USA; InterContinental Hotels Group Asia, Middle East and Africa with headquarters in Singapore; and InterContinental Hotels Group Greater China, with headquarters in Shanghai. In 2015, the firm acquired Kimpton Hotels & Restaurants, an independent boutique hotel operator; and sold its ownership stake in InterContinental Hong Kong to Supreme Key Limited for $938 million. IHG will retain its 37-year management contract on the Hong Kong hotel, with three 10-year extension rights.

The firm typically offers employees benefits including life, disability, medical and dental insurance; paid time off; employee discounts; a pension or 401(k); incentive programs; and educational assistance.

FINANCIAL DATA: Note: Data for latest year may not have been available at press time.

In U.S. $	2015	2014	2013	2012	2011	2010
Revenue	1,803,000,000	1,858,000,000	1,903,000,000	1,835,000,000	1,768,000,000	1,628,000,000
R&D Expense						
Operating Income	1,499,000,000	680,000,000	673,000,000	610,000,000	616,000,000	437,000,000
Operating Margin %	83.13%	36.59%	35.36%	33.24%	34.84%	26.84%
SGA Expense	420,000,000	483,000,000	541,000,000	379,000,000	359,000,000	366,000,000
Net Income	1,222,000,000	391,000,000	372,000,000	544,000,000	473,000,000	280,000,000
Operating Cash Flow	628,000,000	543,000,000	624,000,000	472,000,000	479,000,000	462,000,000
Capital Expenditure	203,000,000	248,000,000	245,000,000	128,000,000	103,000,000	91,000,000
EBITDA	1,580,000,000	760,000,000	744,000,000	707,000,000	717,000,000	547,000,000
Return on Assets %	37.10%	13.55%	11.96%	17.46%	16.44%	9.86%
Return on Equity %			329.20%	127.25%	115.64%	133.33%
Debt to Equity	4.00			4.03	1.22	2.86

CONTACT INFORMATION:

Phone: 44-1895-512000   Fax: 44-1895-512101
Toll-Free: 800-621-0555
Address: Broadwater Park, Denham, Buckinghamshire UB9 5HR United Kingdom

STOCK TICKER/OTHER:

Stock Ticker: IHG
Employees: 7,311
Parent Company:

Exchange: NYS
Fiscal Year Ends: 12/31

SALARIES/BONUSES:

Top Exec. Salary: $785,000   Bonus: $
Second Exec. Salary: $450,000   Bonus: $

OTHER THOUGHTS:

Estimated Female Officers or Directors: 4
Hot Spot for Advancement for Women/Minorities: Y

Interstate Hotels & Resorts Inc
www.interstatehotels.com

NAIC Code: 721110

TYPES OF BUSINESS:

Hotel Management
Corporate Hotel Management
Engineering & Design Consulting
Construction Management
Procurement Services

BRANDS/DIVISIONS/AFFILIATES:

Thayer Lodging Group
Shanghai Jin Jiang International Hotels
Colony Hotels & Resorts

CONTACTS: Note: Officers with more than one job title may be intentionally listed here more than once.

James R. Abrahamson, CEO
Samuel E. Knighton, COO
Samuel E. Knighton, Pres.
Carrie McIntyre, CFO
George J. Brennan, Exec. VP-Mktg. & Sales
Laura E. FitzRandolph, Exec. VP-Human Resources
Leslie Ng, CIO
Christopher L. Bennett, Chief Admin. Officer
Christopher L. Bennett, General Counsel
Thomas J. Bardenett, Exec. VP-Oper.
Edward J. Blum, Exec. VP-Dev. & Acquisitions
Joseph A. Klam, Exec. VP-Finance
Leslie Ng, Chief Investment Officer
James Rowe, Sr. VP-Investment Strategy & Capital Markets
Jim Biggar, Exec. VP-Hotel Oper., Full Service
Greg Juceam, Exec. VP-Hotel Oper., Full Service
Kenneth W. McLaren, Exec. VP-Int'l Oper.

GROWTH PLANS/SPECIAL FEATURES:

Interstate Hotels & Resorts, Inc. is an independent hotel management companies. The company is wholly-owned subsidiary of a joint venture between Thayer Lodging Group and Shanghai Jin Jiang International Hotels. The firm divides its operations into three primary categories: hotel management services, hotel development services and design and construction services. Interstate manages luxury, full-service, select-service and extended-stay hospitality properties, consisting of more than 430 properties with over 76,000 rooms. These properties are located in 39 U.S. states, Washington, D.C., Canada, the U.K., Ireland, The Netherlands, Belgium, Hungary, Russia, China and India. The company's brand portfolio includes Best Western, Comfort Inn, Courtyard by Marriott, Crowne Plaza, Days Inn, Doubletree, Embassy Suites, Fairfield Inn, Hampton Inn, Hilton, Holiday Inn, Homewood Suites, Hyatt Place, Marriott, Radisson, Renaissance, Sheraton and Westin. The firm offers management services to unbranded hotels through its Colony Hotels & Resorts division. The company's development services comprise hotel brand and site validation, owner and brand relations, joint venture partnerships, financing, acquisitions and conversions and renovations. Finally, its design and construction services consist of pre-design budgeting, construction and design consulting, purchase order placement and reporting, shipment monitoring and technical services.

The firm offers employees medical, dental and vision coverage; health care flexible spending accounts; an employee assistance program; life insurance; short- and long-term disability; a 401(k) plan; discounted hotel rooms; and a corporate discount program.

FINANCIAL DATA: Note: Data for latest year may not have been available at press time.

In U.S. $	2015	2014	2013	2012	2011	2010
Revenue						
R&D Expense						
Operating Income						
Operating Margin %						
SGA Expense						
Net Income						
Operating Cash Flow						
Capital Expenditure						
EBITDA						
Return on Assets %						
Return on Equity %						
Debt to Equity						

CONTACT INFORMATION:

Phone: 703-387-3100   Fax:
Toll-Free:
Address: 4501 N. Fairfax Dr., Ste 500, Arlington, VA 22203 United States

STOCK TICKER/OTHER:

Stock Ticker: Subsidiary   Exchange:
Employees: 19,000   Fiscal Year Ends: 12/31
Parent Company: Thayer Lodging Group

SALARIES/BONUSES:

Top Exec. Salary: $ Bonus: $
Second Exec. Salary: $ Bonus: $

OTHER THOUGHTS:

Estimated Female Officers or Directors: 1
Hot Spot for Advancement for Women/Minorities:

InTown Suites Management Inc

www.intownsuites.com

NAIC Code: 721110

TYPES OF BUSINESS:

Hotels
Extended-Stay Hotels

BRANDS/DIVISIONS/AFFILIATES:

Westmont Hospitality Group
Intown Suites

CONTACTS: Note: Officers with more than one job title may be intentionally listed here more than once.

Jonathan Pertchik, CEO
Dennis Cassel, Pres.
Collier Daily, Dir-Mktg. & Communications

GROWTH PLANS/SPECIAL FEATURES:

InTown Suites Management, Inc. is a private corporation that develops, owns and operates budget extended-stay properties. The company does not offer reservations for less than seven days and prefers long-term commitments or apartment leases. A seven-day stay at InTown Suites is generally cheaper than renting a nightly room at a traditional hotel for a few days. Most facilities are located in predominately retail-oriented locations near shops, restaurants and movie theaters, with proximity to major metropolitan areas, spanning more than 180 locations across 21 states. The firm's properties feature studio suites with full amenities, including complete kitchens, high-speed Internet, a dining area, cable TV, a full size bath, voicemail service, laundry facilities, pool areas and weekly housekeeping. The company also offers specialized services for corporate customers. The firm is currently focused on accelerating the growth of its operations, and it will convert all newly acquired properties to the InTown Suites brand. Certain locations are also undergoing renovation and development, which is in some cases significantly increasing the number of rooms. Westmont Hospitality Group operates InTown Suites' properties. In late 2015, the firm acquired four new locations in Hampton Roads, Virginia: Chesapeake/Battlefield, Hampton, Norfolk and Newport News City Center, which will obtain an InTown Suites makeover before opening.

FINANCIAL DATA: Note: Data for latest year may not have been available at press time.

In U.S. $	2015	2014	2013	2012	2011	2010
Revenue						
R&D Expense						
Operating Income						
Operating Margin %						
SGA Expense						
Net Income						
Operating Cash Flow						
Capital Expenditure						
EBITDA						
Return on Assets %						
Return on Equity %						
Debt to Equity						

CONTACT INFORMATION:

Phone: 770-799-5000   Fax: 770-437-8190
Toll-Free: 800-553-9338
Address: 2727 Paces Ferry Rd., Ste. 2-1200, Atlanta, GA 30339 United States

STOCK TICKER/OTHER:

Stock Ticker: Private
Employees: 1,100
Parent Company:

Exchange:
Fiscal Year Ends: 12/31

SALARIES/BONUSES:

Top Exec. Salary: $ Bonus: $
Second Exec. Salary: $ Bonus: $

OTHER THOUGHTS:

Estimated Female Officers or Directors:
Hot Spot for Advancement for Women/Minorities:

InvenTrust Properties Corp

www.inventrustproperties.com

NAIC Code: 0

TYPES OF BUSINESS:
Real Estate Investment Trust

BRANDS/DIVISIONS/AFFILIATES:
Inland American Real Estate Trust Inc
University House Communities Group Inc
Sonterra Village Shopping Center

CONTACTS: *Note: Officers with more than one job title may be intentionally listed here more than once.*
Thomas P. McGuinness, CEO
Thomas P. McGuinness, Pres.
Michael E. Podboy, CFO
Annie Fitzgerald, Chief Acct. Officer
Robert D. Parks, Chmn.

GROWTH PLANS/SPECIAL FEATURES:
InvenTrust Properties Corp., formerly Inland American Real Estate Trust, Inc. manages commercial real estate assets with a focus on lodging, multi-tenant retail and student housing. InvenTrust consists of 128 multi-tenant properties totaling approximately 19 million square feet of retail space in 24 states. The company's acquisition team seeks to invest in power centers, community centers and grocery-anchored centers with an average deal ranging between $10 million to $100 million. Its geographical focus in divided into Northeast, Mid-Atlantic, Southeast, Midwest, Texas and West/Southwest, with population sizes of 100,000 or greater within a 3-mile radius, and a median household income of $65,000 or more. Inland's University House Communities Group, Inc. is a development, acquisition and management firm that creates communities in university markets across the U.S. Currently, it has 16 properties, including one joint venture asset with 9,600 beds. This subsidiary also provides services such as deal sourcing, underwriting and valuation, legal/physical/financial due diligence, third party coordination, debt financing, closing coordination, rebranding and marketing/advertising. In addition, the firm owns 5.8 million square feet of non-core, office and industrial buildings. In February 2015, the firm spun off Xenia Hotels & Resorts, Inc., which trades publicly on the New York Stock Exchange under the symbol XHR. In April 2015, the firm changed its name form Inland American Real Estate Trust, Inc. to InvenTrust Properties Corp. In January 2016, the firm acquired Sonterra Village Shopping Center in San Antonio, Texas.

The firm offers employees benefits such as life, AD&D, disability, medical and dental insurance; a flexible savings plan for health and dependent care needs; tuition reimbursement; and a 401(k) plan.

FINANCIAL DATA: *Note: Data for latest year may not have been available at press time.*

In U.S. $	2015	2014	2013	2012	2011	2010
Revenue	450,044,000	1,379,141,000	1,321,837,000	1,437,395,000	1,323,151,000	1,094,696,000
R&D Expense						
Operating Income	-15,209,000	132,071,000	-73,128,000	178,097,000	119,865,000	45,196,000
Operating Margin %	-3.37%	9.57%	-5.53%	12.39%	9.05%	4.12%
SGA Expense	78,218,000	85,632,000	93,511,000	76,706,000	71,033,000	72,665,000
Net Income	3,464,000	486,642,000	244,048,000	-69,338,000	-316,253,000	-176,431,000
Operating Cash Flow	194,734,000	340,335,000	422,813,000	456,221,000	397,949,000	356,660,000
Capital Expenditure	449,994,000	465,586,000	77,558,000	113,313,000	71,157,000	109,827,000
EBITDA	255,664,000	769,585,000	382,808,000	636,943,000	466,609,000	516,296,000
Return on Assets %	.05%	5.67%	2.39%	-.63%	-2.83%	-1.54%
Return on Equity %	.11%	11.78%	5.67%	-1.53%	-6.38%	-3.36%
Debt to Equity	0.87	0.79	5.67	1.38	1.26	1.05

CONTACT INFORMATION:
Phone: 630-218-8000   Fax: 630-218-4957
Toll-Free: 800-826-8228
Address: 2901 Butterfield Rd., Oak Brook, IL 60523 United States

STOCK TICKER/OTHER:
Stock Ticker: IARE   Exchange: GREY
Employees: 287   Fiscal Year Ends: 12/31
Parent Company:

SALARIES/BONUSES:
Top Exec. Salary: $ Bonus: $
Second Exec. Salary: $ Bonus: $

OTHER THOUGHTS:
Estimated Female Officers or Directors: 1
Hot Spot for Advancement for Women/Minorities: Y

Investors Real Estate Trust

www.iret.com

NAIC Code: 0

TYPES OF BUSINESS:

Real Estate Investment Trust
Apartment Communities
Retail, Office & Industrial Properties
Medical Properties

BRANDS/DIVISIONS/AFFILIATES:

IRET Properties

CONTACTS: *Note: Officers with more than one job title may be intentionally listed here more than once.*

Joy Newborg, Assistant General Counsel
Timothy Mihalick, CEO
Jeffrey Miller, Chairman of the Board
Diane Bryantt, COO
Ted Holmes, Executive VP
Mark Reiling, Executive VP
Michael Bosh, General Counsel
Cindy Bradehoft, Other Corporate Officer
Charles Greenberg, Senior VP, Divisional
Andrew Martin, Senior VP, Divisional
Jeffrey Woodbury, Trustee
Linda Keller, Trustee
Stephen Stenehjem, Trustee
Jeffrey Caira, Trustee
Terrance Maxwell, Trustee
John Stewart, Trustee
Nancy Andersen, Vice President

GROWTH PLANS/SPECIAL FEATURES:

Investors Real Estate Trust is a self-advised real estate investment trust (REIT) involved in the ownership and operation of multi-family residential, commercial office, medical, industrial and retail properties located primarily in the upper Midwest. The firm conducts its business operations through its operating partnership, IRET Properties. Its portfolio contains approximately 100 multi-family residential properties, with a total of 11,844 apartment units; 53 office properties containing 4.2 million square feet of leasable space; 66 medical properties, including senior housing properties, for a total of approximately 3 million square feet of leasable space; seven industrial properties, with a total of 1.2 million square feet of leasable space; and 23 retail properties, containing approximately 1.2 million square feet of leasable space. The day-to-day management activities of many of the firm's properties are either maintained through the company's employees or by third-party property management firms. The trust's main geographical area of concentration is in the upper Midwest, primarily Minnesota and North Dakota, with these two states together representing approximately 69.5% of annual revenue. The firm also has property management offices in the states where it owns properties. In 2015, the firm acquired the GrandeVille rental townhomes at Cascade Lake in Rochester, Minnesota; and sold 34 commercial office properties, 17 retail assets and four office showroom properties.

FINANCIAL DATA: *Note: Data for latest year may not have been available at press time.*

In U.S. $	2015	2014	2013	2012	2011	2010
Revenue	283,190,000	265,482,000	259,406,000	241,788,000	237,407,000	242,775,000
R&D Expense						
Operating Income	78,650,000	33,116,000	85,116,000	73,967,000	67,960,000	71,062,000
Operating Margin %	27.77%	12.47%	32.81%	30.59%	28.62%	29.27%
SGA Expense	17,663,000	15,908,000	12,421,000	10,931,000	9,526,000	10,123,000
Net Income	24,087,000	-13,174,000	25,530,000	8,212,000	20,082,000	4,001,000
Operating Cash Flow	114,179,000	92,514,000	77,718,000	65,137,000	58,774,000	61,412,000
Capital Expenditure	260,110,000	196,986,000	159,949,000	141,771,000	62,824,000	80,069,000
EBITDA	159,782,000	109,475,000	153,423,000	135,921,000	129,304,000	132,246,000
Return on Assets %	.65%	-1.31%	.90%	.35%	1.08%	.09%
Return on Equity %	2.60%	-5.32%	3.70%	1.47%	4.62%	.47%
Debt to Equity	1.89	2.19	2.21	2.58	2.58	2.76

CONTACT INFORMATION:

Phone: 701 837-4738   Fax: 701 838-7785
Toll-Free: 888-478-4738
Address: 1400 31st Ave. SW, Ste. 60, Minot, ND 58701 United States

STOCK TICKER/OTHER:

Stock Ticker: IRET   Exchange: NYS
Employees: 433   Fiscal Year Ends: 04/30
Parent Company:

SALARIES/BONUSES:

Top Exec. Salary: $442,598   Bonus: $
Second Exec. Salary: Bonus: $
$284,440

OTHER THOUGHTS:

Estimated Female Officers or Directors: 4
Hot Spot for Advancement for Women/Minorities: Y

Investors Title Company

NAIC Code: 524127

www.invtitle.com

TYPES OF BUSINESS:
Title Insurance
Tax-Deferred Exchange Services
Reinsurance

BRANDS/DIVISIONS/AFFILIATES:
Investors Title Insurance Company (ITIC)
National Investors Title Insurance (NITIC)
Investors Title Exchange Corporation (ITEC)
Investors Title Accommodation Corporation (ITAC)
Investors Capital Management Company
Investors Title Management Services Inc
Investors Trust Company

CONTACTS: *Note: Officers with more than one job title may be intentionally listed here more than once.*
James Fine, CEO, Subsidiary
J. Fine, CEO
W. Fine, Chairman of the Board, Subsidiary

GROWTH PLANS/SPECIAL FEATURES:
Investors Title Company is a holding company that, through its subsidiaries, issues and underwrites title insurance policies. Through Investors Title Insurance Company (ITIC) and National Investors Title Insurance (NITIC), the firm underwrites land title insurance for owners and mortgagees as a primary insurer and as a reinsurer for other title insurance companies. ITIC markets title insurance through issuing agents and branch offices in 20 states. NITIC writes title insurance as a primary insurer and as a reinsurer in the states of Texas and New York and is also licensed to write title insurance in 21 states and Washington, D.C. Investors Title also provides tax-deferred real property exchange services through subsidiaries Investors Title Exchange Corporation (ITEC) and Investors Title Accommodation Corporation (ITAC). ITEC acts as an intermediary in tax-deferred exchanges of property held for productive use in a business or for investments. The subsidiary's income is primarily derived from fees for handling exchange transactions. ITAC serves as an exchange accommodation titleholder, offering services for accomplishing reverse exchanges when a taxpayer must acquire replacement property before selling the relinquished property. Through Investors Trust Company and Investors Capital Management Company, the firm offers investment management and trust services to individuals, companies, banks and trusts. Additionally, the company provides management and consulting services to help partners start and successfully operate a title insurance agency through Investors Title Management Services, Inc.

The firm offers employees medical, dental and vision insurance; disability coverage; life and AD&D insurance; flexible spending accounts; an employee assistance program; and a 401(k) plan.

FINANCIAL DATA: *Note: Data for latest year may not have been available at press time.*

In U.S. $	2015	2014	2013	2012	2011	2010
Revenue	127,200,100	123,119,300	126,251,500	115,079,100	90,685,150	71,309,350
R&D Expense						
Operating Income	17,777,050	13,488,500	21,542,740	16,079,910	9,498,936	8,615,626
Operating Margin %	13.97%	10.95%	17.06%	13.97%	10.47%	12.08%
SGA Expense	104,060,100	103,580,200	104,524,900	92,347,820	77,338,060	57,729,070
Net Income	12,533,910	9,648,975	14,708,210	11,102,500	6,933,936	6,372,626
Operating Cash Flow	16,889,630	9,683,980	16,251,090	8,707,514	9,007,159	4,226,501
Capital Expenditure	2,742,619	2,017,379	1,424,108	568,728	361,207	317,530
EBITDA						
Return on Assets %	6.12%	4.99%	8.16%	6.73%	4.45%	4.24%
Return on Equity %	8.94%	7.26%	12.12%	10.04%	6.58%	6.33%
Debt to Equity						

CONTACT INFORMATION:
Phone: 919 968-2200   Fax: 919 968-2223
Toll-Free: 800-326-4842
Address: 121 N. Columbia St., Chapel Hill, NC 27514 United States

STOCK TICKER/OTHER:
Stock Ticker: ITIC
Employees: 263
Parent Company:
Exchange: NAS
Fiscal Year Ends: 12/31

SALARIES/BONUSES:
Top Exec. Salary: $340,000   Bonus: $300,000
Second Exec. Salary: $288,250   Bonus: $350,000

OTHER THOUGHTS:
Estimated Female Officers or Directors: 8
Hot Spot for Advancement for Women/Minorities: Y

iStar Inc

www.istarfinancial.com

NAIC Code: 525990

TYPES OF BUSINESS:

Mortgage REIT
Real Estate Investment Trust
Mortgages & Loans
Asset Services
Corporate Finance
Corporate Leasing

BRANDS/DIVISIONS/AFFILIATES:

iStar Financial Inc

CONTACTS: Note: Officers with more than one job title may be intentionally listed here more than once.

Jay Sugarman, CEO
David DiStaso, CFO
Michelle MacKay, Executive VP
Barclay Jones, Executive VP
Geoffrey Dugan, General Counsel
Nina Matis, Other Executive Officer

GROWTH PLANS/SPECIAL FEATURES:

iStar, Inc., formerly iStar Financial, Inc., is a mortgage finance company focused on the commercial real estate industry and is taxed as a real estate investment trust (REIT). The firm provides custom-tailored financing to high-end private and corporate owners of real estate. iStar operates through four business segments: real estate finance, net lease, operating properties and land. The real estate finance segment primarily consists of senior mortgage loans that are secured by commercial real estate assets where the company is the first lien holder. The net lease division is comprised of properties owned by the company and leased to single creditworthy tenants where the properties are subject to long-term leases. The majority of the leases provide for expenses at the facility to be paid by the tenant on a triple net lease basis. The operating properties segment contains commercial and residential properties which represent a diverse pool of assets across a broad range of geographies and property types. iStar generally seeks to reposition or redevelop these assets with the objective of maximizing their value through the infusion of capital and/or intensive asset management efforts. The land division consists of master planned community projects, urban infill land parcels and waterfront land parcels located throughout the U.S. Master planned communities represent large-scale residential projects that the company will entitle, plan and/or develop and may sell through retail channels to home buildings or in bulk. In August 2015, the firm formally changed its name to iStar, Inc. to reflect its investment focus and real estate capabilities.

iStar's offers its employees health, life and disability insurance plans; a 401(k) plan; and paid time off.

FINANCIAL DATA: Note: Data for latest year may not have been available at press time.

In U.S. $	2015	2014	2013	2012	2011	2010
Revenue	321,787,000	237,545,000	132,895,000	110,568,000	284,558,000	420,097,000
R&D Expense						
Operating Income	-92,334,000	-139,802,000	-214,054,000	-304,249,000	-58,671,000	-207,525,000
Operating Margin %	-28.69%	-58.85%	-161.07%	-275.16%	-20.61%	-49.39%
SGA Expense	295,409,000	265,035,000	249,555,000	232,683,000	200,760,000	189,292,000
Net Income	-2,435,000	16,469,000	-111,951,000	-239,930,000	-22,064,000	79,683,000
Operating Cash Flow	-59,947,000	-10,342,000	-180,465,000	-191,932,000	-31,785,000	-47,396,000
Capital Expenditure	169,744,000	147,453,000	211,767,000	83,070,000	64,169,000	42,863,000
EBITDA						
Return on Assets %	-.95%	-.60%	-2.64%	-3.99%	-.74%	.33%
Return on Equity %	-4.66%	-2.76%	-12.55%	-19.73%	-3.92%	2.29%
Debt to Equity	3.91	3.36	3.34	3.78	3.81	4.45

CONTACT INFORMATION:

Phone: 212 930-9400   Fax: 212 930-9494
Toll-Free:
Address: 1114 Ave. of the Americas, 39th Fl., New York, NY 10036 United States

STOCK TICKER/OTHER:

Stock Ticker: STAR
Employees: 182
Parent Company:

Exchange: NYS
Fiscal Year Ends: 12/31

SALARIES/BONUSES:

Top Exec. Salary: $1,000,000   Bonus: $

Second Exec. Salary: $500,000   Bonus: $

OTHER THOUGHTS:

Estimated Female Officers or Directors: 4

Hot Spot for Advancement for Women/Minorities: Y

Jacobs Engineering Group Inc

www.jacobs.com

NAIC Code: 237000

TYPES OF BUSINESS:

Engineering & Design Services
Facility Management
Construction & Field Services
Technical Consulting Services
Environmental Services

BRANDS/DIVISIONS/AFFILIATES:

J L Patterson & Associates

CONTACTS: Note: Officers with more than one job title may be intentionally listed here more than once.

Noel Watson, Chairman of the Board
Santo Rizzuto, Executive VP, Divisional
Kevin Berryman, Executive VP
Michael Tyler, General Counsel
Andrew Kremer, President, Divisional
Joseph Mandel, President, Divisional
Philip Stassi, President, Divisional
Terence Hagen, President, Divisional
Steven Demetriou, President
Cora Carmody, Senior VP, Divisional
Lori Sundberg, Senior VP, Divisional
Geoffrey Sanders, Senior VP

GROWTH PLANS/SPECIAL FEATURES:

Jacobs Engineering Group, Inc. offers technical, professional and construction services to industrial, commercial and governmental clients throughout North America, Europe, Asia, South America, India, the U.K. and Australia. Jacobs provides project services, which include engineering, design and architecture; process, scientific and systems consulting services; operations and maintenance services; and construction services, which include direct-hire construction and management services. Services are offered to industry groups such as oil & gas exploration, production and refining; pharmaceuticals & biotechnology; chemicals & polymers; buildings, which includes projects in the fields of health care and education as well as civic, governmental and other buildings; infrastructure; technology; energy; consumer & forest products; automotive & industrial; and environmental programs. Jacobs also provides pricing studies, project feasibility reports and automation & control system analysis for U.S. government agencies involved in defense and aerospace programs. In addition, the company is one of the leading providers of environmental engineering and consulting services in the U.S. and abroad, providing support in such areas as underground storage tank removal, contaminated soil and water remediation and long-term groundwater monitoring. Jacobs also designs, builds, installs, operates and maintains various types of soil and groundwater cleanup systems. In December 2015, the firm acquired J. L. Patterson & Associates, a consulting and professional services engineering firm specializing in rail planning, environmental permitting, design and construction management.

Jacobs offers its employees medical, disability, life and AD&D insurance; an employee stock purchase plan; and tuition reimbursement.

FINANCIAL DATA: Note: Data for latest year may not have been available at press time.

In U.S. $	2015	2014	2013	2012	2011	2010
Revenue	12,114,830,000	12,695,160,000	11,818,380,000	10,893,780,000	10,381,660,000	9,915,517,000
R&D Expense						
Operating Income	445,527,000	528,068,000	668,979,000	596,073,000	518,918,000	400,083,000
Operating Margin %	3.67%	4.15%	5.66%	5.47%	4.99%	4.03%
SGA Expense	1,522,811,000	1,545,716,000	1,173,340,000	1,130,916,000	1,040,575,000	932,522,000
Net Income	302,971,000	328,108,000	423,093,000	378,954,000	331,029,000	245,974,000
Operating Cash Flow	484,572,000	721,716,000	448,516,000	299,805,000	236,490,000	196,970,000
Capital Expenditure	88,404,000	132,146,000	127,270,000	102,574,000	98,749,000	49,075,000
EBITDA	598,932,000	699,015,000	773,328,000	705,846,000	620,830,000	490,303,000
Return on Assets %	3.73%	4.17%	5.99%	5.88%	6.16%	5.39%
Return on Equity %	6.91%	7.55%	10.66%	10.77%	10.72%	8.96%
Debt to Equity	0.13	0.17	0.09	0.14		

CONTACT INFORMATION:

Phone: 626 578-3500   Fax: 626 578-6988
Toll-Free:
Address: 155 N. Lake Ave., Pasadena, CA 91101 United States

SALARIES/BONUSES:

Top Exec. Salary: $125,000   Bonus: $5,650,000
Second Exec. Salary: $544,832   Bonus: $1,500,000

STOCK TICKER/OTHER:

Stock Ticker: JEC   Exchange: NYS
Employees: 49,900   Fiscal Year Ends: 09/30
Parent Company:

OTHER THOUGHTS:

Estimated Female Officers or Directors: 1
Hot Spot for Advancement for Women/Minorities: Y

Jameson Inn Inc

www.jamesoninns.com

NAIC Code: 721110

TYPES OF BUSINESS:

Hotels

BRANDS/DIVISIONS/AFFILIATES:

Jameson Inn
America's Best Franchising Inc

GROWTH PLANS/SPECIAL FEATURES:

Jameson Inn, Inc., a subsidiary of America's Best Franchising, operates hotels in nine states in the southeastern and Midwestern U.S. Jameson owns 27 hotels under the Jameson Inn brand. Most of the Jameson Inn hotels are designed with a southern colonial style, although some newer locations are designed in a more contemporary style. All hotels owned by the firm typically offer amenities such as swimming pools, fitness centers, deluxe continental breakfast, free premium channels and wake up service. In addition, the hotel also allows small and medium sized pets for a small fee. In order to attract more business travelers, Jameson offers workstations equipped with data ports, meeting spaces and fax and photocopy machines.

CONTACTS: Note: Officers with more than one job title may be intentionally listed here more than once.

Thomas W. Kitchen, CEO
Sterling F. Stoudenmire, IV, CEO

FINANCIAL DATA: Note: Data for latest year may not have been available at press time.

In U.S. $	2015	2014	2013	2012	2011	2010
Revenue						
R&D Expense						
Operating Income						
Operating Margin %						
SGA Expense						
Net Income						
Operating Cash Flow						
Capital Expenditure						
EBITDA						
Return on Assets %						
Return on Equity %						
Debt to Equity						

CONTACT INFORMATION:

Phone: 404-350-9990   Fax: 404-601-6106
Toll-Free: 800-526-3766
Address: 4770 S. Atlanta Rd., Smyrna, GA 30080 United States

SALARIES/BONUSES:

Top Exec. Salary: $ Bonus: $
Second Exec. Salary: $ Bonus: $

STOCK TICKER/OTHER:

Stock Ticker: Subsidiary   Exchange:
Employees: 1,750   Fiscal Year Ends: 12/31
Parent Company: America's Best Franchising Inc

OTHER THOUGHTS:

Estimated Female Officers or Directors:
Hot Spot for Advancement for Women/Minorities:

Janus Hotels and Resorts Inc

www.janushotels.com

NAIC Code: 721110

TYPES OF BUSINESS:

Hotel Management
Management, Financial & Legal Consulting
Food & Beverage Services
Staffing Services

BRANDS/DIVISIONS/AFFILIATES:

CONTACTS: Note: Officers with more than one job title may be intentionally listed here more than once.

Michael Nanosky, CEO
Michael Nanosky, Pres.
Rick Tonges, CFO
Greg Cappel, VP-Mktg. & Sales
Laura Flannery, Human Resources
Eric Glazer, General Counsel
Scott Wielkiewicz, Corp. Controller
Harry Yeaggy, Vice Chmn.
Barb Soete, Controller
Louis Beck, Chmn.

GROWTH PLANS/SPECIAL FEATURES:

Janus Hotels and Resorts, Inc. is an independently-owned full-service hotel management company. It owns or manages 30 hotels with approximately 8,000 guest rooms, concentrated in Florida. Its locations include nationally recognized brands such as Days Inn, Holiday Inn, Radisson and Best Western. Janus operates each hotel according to a business plan specifically tailored to the characteristics of the hotel and its market, employing centralized management, accounting and purchasing systems to reduce cost and increase operating margins. The firm focuses primarily on continuing sales and marketing to increase revenue. Janus also provides food and beverage services, including lounges, coffee shops and locally and nationally branded restaurants. The company offers support services in the areas of administration, legal issues, brand relationships, maintenance of property, staffing, cash controls, liquor license issues, financial review and management, property development and customer service strategies. The firm also provides receivership services that involve quickly securing the asset, management services designed to stabilize the property and services aimed at restoring staff and customer confidence.

FINANCIAL DATA: Note: Data for latest year may not have been available at press time.

In U.S. $	2015	2014	2013	2012	2011	2010
Revenue						
R&D Expense						
Operating Income						
Operating Margin %						
SGA Expense						
Net Income						
Operating Cash Flow						
Capital Expenditure						
EBITDA						
Return on Assets %						
Return on Equity %						
Debt to Equity						

CONTACT INFORMATION:

Phone: 561-997-2325   Fax: 561-997-5331
Toll-Free:
Address: 2300 Corporate Blvd. NW, Ste. 232, Boca Raton, FL 33431-8596 United States

STOCK TICKER/OTHER:

Stock Ticker: Private   Exchange:
Employees: 30   Fiscal Year Ends: 12/31
Parent Company:

SALARIES/BONUSES:

Top Exec. Salary: $ Bonus: $
Second Exec. Salary: $ Bonus: $

OTHER THOUGHTS:

Estimated Female Officers or Directors: 7
Hot Spot for Advancement for Women/Minorities: Y

Sales, profits and employees may be estimates. Financial information, benefits and other data can change quickly and may vary from those stated here.

Jerde Partnership Inc (The)

www.jerde.com

NAIC Code: 541310

TYPES OF BUSINESS:
Architectural Services
Architecture

BRANDS/DIVISIONS/AFFILIATES:

CONTACTS: *Note: Officers with more than one job title may be intentionally listed here more than once.*
Phil Kim, Co-CEO
John Simones, Co-CEO
Paul Martinkovic, CFO

GROWTH PLANS/SPECIAL FEATURES:

The Jerde Partnership, Inc. is an architecture and urban planning firm that designs unique places people visit every year. These places include public spaces, shops, parks, restaurants, entertainment, housing and nature areas, spanning cities that include Budapest, Hong Kong, Las Vegas, Los Angeles, Osaka, Rotterdam, Seoul, Shanghai, Tokyo, Istanbul, Warsaw and Dubai. Jerde's market expertise includes hotels, casinos and resorts; residential complexes; office and commercial facilities; transit-oriented mixed-use hubs; major urban districts; waterfronts; town centers; community plans; and visionary master plans. The company partners with clients to conceptualize ideas for each venue that will become a powerful economic and social engine. Over 1 billion people visit Jerde-designed places annually; average lease rates for retail is between 97 and 100%; thousands of jobs are generated, as well as profitable annual taxable sales for retailers; and retail sales average $500-800 per square foot. Seventy percent of Jerde projects are mixed-use. Projects the firm has completed include Canal city Hakata in Fukuoka, Universal City Walk in Los Angeles, Bellagio and Fremont Street Experience in Las Vegas, Roppongi Hills in Tokyo, Namba Parks in Osaka, Zlote Tarasy in Poland, Kanyon in Instanbul, D-Cube City and Mecenatpolis in Seoul, Dubai Festival Waterfront Centre in UAE and Langham Place in Hong Kong. Over 110 Jerde places have been realized since 1977. The firm is headquartered in Los Angeles, with offices in Hong Kong, Shanghai and Seoul.

FINANCIAL DATA: *Note: Data for latest year may not have been available at press time.*

In U.S. $	2015	2014	2013	2012	2011	2010
Revenue	18,000,000	18,000,000				
R&D Expense						
Operating Income						
Operating Margin %						
SGA Expense						
Net Income						
Operating Cash Flow						
Capital Expenditure						
EBITDA						
Return on Assets %						
Return on Equity %						
Debt to Equity						

CONTACT INFORMATION:
Phone: 310-399-1987   Fax:
Toll-Free:
Address: 913 Ocean Front Walk, Venice, CA 90291 United States

STOCK TICKER/OTHER:
Stock Ticker: Private
Employees: 103
Parent Company:

Exchange:
Fiscal Year Ends:

SALARIES/BONUSES:
Top Exec. Salary: $ Bonus: $
Second Exec. Salary: $ Bonus: $

OTHER THOUGHTS:
Estimated Female Officers or Directors:
Hot Spot for Advancement for Women/Minorities:

John Laing plc
NAIC Code: 531100

TYPES OF BUSINESS:
Investor & Developer, Infrastructure Projects
Public Infrastructure Management
Rail Operations
Facilities Management
Private Finance Initiative Investor

BRANDS/DIVISIONS/AFFILIATES:
Henderson Group plc
Henderson Infrastructure Holdco Ltd
John Laing Capital Management
John Laing Infrastructure Fund
John Laing Environmental Assets Group

CONTACTS: *Note: Officers with more than one job title may be intentionally listed here more than once.*
Olivier Brousse, CEO
Patrick O'D Bourke, Dir.-Group Finance
Carolyn Cattermole, General Counsel
Derek Potts, Managing Dir.-Bus. Dev.
Phil Nolan, Chmn.

GROWTH PLANS/SPECIAL FEATURES:

John Laing plc, owned by Henderson Infrastructure Holdco Ltd., a subsidiary of Henderson Group plc, operates as a long-term investor, developer and operator of privately financed, public sector infrastructure projects such as roads, railways, hospitals, schools and other major projects in the U.K. and overseas. The company has investments in 38 projects, including 18 projects under construction. The firm's primary investment activities include sourcing and originating bidding for and winning renewable energy projects, while secondary investment activities involve ownership of a substantial portfolio of interests in such projects. Firm's asset management division manages primary and secondary investment portfolios and offers advisory services through subsidiary, John Laing Capital Management to two funds, John Laing Infrastructure Fund (JLIF) and John Laing Environmental Assets Group. The firm's clients include local and education authorities, health trusts and the U.K. Ministry of Defense (MoD). The company serves industry sectors including public safety, defense, education, health care, housing & community regeneration, transportation infrastructure and environmental infrastructure. In the public safety sector, John Laing serves the organizations including the Department for Constitutional Affairs across all of its procurement methods, such as private finance initiative (PFI), public-private partnership (PPP) and private developer schemes (PDS). John Laing's education portfolio includes over 100 development projects, which it delivers through PPP and PFI deals. In health care, the company has helped fund over 40 hospital development projects in the U.K., Canada and Australia. Its transportation infrastructure operations include developing park-and-ride rail lines, roads and bridges throughout Europe and lighting columns and illuminated traffic signs. Its environmental infrastructure projects include renewable energy development and waste management projects. In 2015, the firm made an investment in a biomass plant in North East England and agreed to acquire the Klettwitz wind farm in Brandenburg Schipkau, Germany from Ventotec GmbH, the developer of the site.

FINANCIAL DATA: *Note: Data for latest year may not have been available at press time.*

In U.S. $	2015	2014	2013	2012	2011	2010
Revenue	245,628,600	302,623,650	340,750,850			
R&D Expense						
Operating Income						
Operating Margin %						
SGA Expense						
Net Income	153,504,750	176,895,950	191,531,825			
Operating Cash Flow						
Capital Expenditure						
EBITDA						
Return on Assets %						
Return on Equity %						
Debt to Equity						

CONTACT INFORMATION:
Phone: 44-20-7901-3200   Fax: 44-20-7901-3520
Toll-Free:
Address: 1 Kingsway, London, WC2B 6AN United Kingdom

STOCK TICKER/OTHER:
Stock Ticker: JLG   Exchange: London
Employees: 264   Fiscal Year Ends: 12/31
Parent Company:

SALARIES/BONUSES:
Top Exec. Salary: $ Bonus: $
Second Exec. Salary: $ Bonus: $

OTHER THOUGHTS:
Estimated Female Officers or Directors: 2
Hot Spot for Advancement for Women/Minorities:

John Q Hammons Hotels & Resorts LLC   www.jqhhotels.com
NAIC Code: 721110

TYPES OF BUSINESS:
Hotels

BRANDS/DIVISIONS/AFFILIATES:

GROWTH PLANS/SPECIAL FEATURES:

John Q. Hammons Hotels & Resorts LLC is a leading independent owner, manager and developer of low-cost upscale hotels in the U.S. The company owns and operates 35 hotels with nearly 8,500 rooms in 16 states. These properties operate under the Embassy Suites by Hilton, IHG (Holiday Inn Express & Suites), JQH Hotels & Resorts, Marriott, Renaissance and Sheraton trade names and are marketed to a range of customers, including frequent business travelers, groups and conventions and leisure travelers. John Q. Hammons Hotels Management LLC manages all of the firm's hotels, which are generally located near a state capitol, university, airport, corporate headquarters or other major facility. Most of the hotels contain a multi-storied atrium, extensive meeting space and large rooms or suites, and some hotels feature signature full-service spas. U.S. locations include Alabama, Arizona, Arkansas, Colorado, Illinois, Kansas, Missouri, Nebraska, New Mexico, North Carolina, Oklahoma, South Carolina, South Dakota, Tennessee, Texas and Virginia.

CONTACTS: *Note: Officers with more than one job title may be intentionally listed here more than once.*
Jacqueline Dowdy, CEO
Joe Morrissey, COO
Phill Burgess, VP-Sales & Revenue Mgmt.
Kent Foster, VP-Human Resources
Christopher Smith, Sr. VP-Admin. & Control
Greggory Groves, General Counsel
Joe Morrissey, Sr. VP-Oper.
Rod Dornbusch, VP-Capital Planning & Asset Mgmt.
Kent Foster, VP-Human Resources
Rick Beran, VP-Food & Beverage

FINANCIAL DATA: *Note: Data for latest year may not have been available at press time.*

In U.S. $	2015	2014	2013	2012	2011	2010
Revenue	461,000,000	456,000,000	445,000,000	430,900,000		
R&D Expense						
Operating Income						
Operating Margin %						
SGA Expense						
Net Income						
Operating Cash Flow						
Capital Expenditure						
EBITDA						
Return on Assets %						
Return on Equity %						
Debt to Equity						

CONTACT INFORMATION:
Phone: 417-864-4300   Fax:
Toll-Free: 800-641-4026
Address: 300 John Q. Hammons Pkwy., Ste. 900, Springfield, MO 65806
United States

STOCK TICKER/OTHER:
Stock Ticker: Private
Employees: 5,800
Parent Company:

Exchange:
Fiscal Year Ends: 12/31

SALARIES/BONUSES:
Top Exec. Salary: $ Bonus: $
Second Exec. Salary: $ Bonus: $

OTHER THOUGHTS:
Estimated Female Officers or Directors: 1
Hot Spot for Advancement for Women/Minorities:

Jones Lang LaSalle Inc

NAIC Code: 531120

www.us.jll.com

TYPES OF BUSINESS:

Real Estate Rental, Leasing & Management
Investment Management
Project Management
Consulting Services
Real Estate Investment Banking
Properties Brokerage

BRANDS/DIVISIONS/AFFILIATES:

LaSalle Investment Management
Oak Grove Capital
Corrigo
Tansei Mall Management
Propell National Valuers
Lodgetax
Bill Goold Realty
Washington Partners Inc

CONTACTS: *Note: Officers with more than one job title may be intentionally listed here more than once.*

Alastair Hughes, CEO, Divisional
James Jasionowski, Executive VP
Jeff Jacobson, CEO, Divisional
Gregory OBrien, CEO, Divisional
Guy Grainger, CEO, Geographical
Chris Ireland, CEO, Geographical
Colin Dyer, CEO
Christie Kelly, CFO
David Johnson, Chief Information Officer
Charles Doyle, Chief Marketing Officer
Louis Bowers, Controller
Sheila Penrose, Director
Roger Staubach, Director
Parikshat Suri, Executive VP
Allan Frazier, Executive VP
Patricia Maxson, Executive VP
Mark Ohringer, Executive VP
Joseph Romenesko, Executive VP
Anthony Couse, Managing Director, Subsidiary

GROWTH PLANS/SPECIAL FEATURES:

Jones Lang LaSalle, Inc. (JLL) is a real estate money management firm that provides integrated real estate and investment management expertise on a local, regional and global level to owner, occupier and investor clients. The firm is active in the area of property and corporate facility management services, with a portfolio encompassing roughly 4 billion square feet worldwide. JLL offers its real estate services across three geographically-aligned business segments: the Americas; Europe, the Middle East and Africa (EMEA); and Asia Pacific. The company's range of real estate service areas includes agency leasing, property management, project and development management, valuations, brokerage of properties, capital markets, real estate investment banking and merchant banking, corporate finance, hotel advisory, space acquisition and disposition, facilities management, strategic consulting, energy management and sustainability, value recovery and receivership services as well as money management. These services are offered to for-profit and not-for-profit firms as well as to governmental entities and public-private partnerships across a wide variety of property categories, including offices, hotels, industrial, retail, multi-family residential, hospitals, data centers, sporting facilities, cultural institutions and transportation centers. A fourth business segment encompasses the operations of subsidiary LaSalle Investment Management, a diversified real estate investment management firm with $56.4 billion in assets under management. JLL has operations in over 1,000 locations in 80 countries worldwide, including 230 corporate offices. In 2015 the firm completed 20 acquisitions including Oak Grove Capital; Corrigo; Tansei Mall Management; Lodgetax and Propell National Valuers among others. In early 2016, the firm acquired Bill Goold Realty and Washington Partners, Inc.

FINANCIAL DATA: *Note: Data for latest year may not have been available at press time.*

In U.S. $	2015	2014	2013	2012	2011	2010
Revenue	5,965,671,000	5,429,603,000	4,461,591,000	3,932,830,000	3,584,544,000	2,925,613,000
R&D Expense						
Operating Income	529,798,000	465,664,000	368,819,000	289,403,000	251,205,000	260,658,000
Operating Margin %	8.88%	8.57%	8.26%	7.35%	7.00%	8.90%
SGA Expense	5,293,615,000	4,827,097,000	3,994,604,000	3,519,196,000	3,194,380,000	2,586,996,000
Net Income	438,672,000	386,063,000	269,865,000	208,050,000	164,384,000	153,902,000
Operating Cash Flow	375,769,000	498,861,000	293,167,000	327,698,000	211,338,000	384,270,000
Capital Expenditure	196,703,000	156,927,000	110,684,000	94,758,000	91,538,000	47,609,000
EBITDA	715,415,000	608,266,000	448,672,000	424,346,000	374,424,000	362,082,000
Return on Assets %	7.77%	7.97%	6.02%	5.01%	4.50%	4.76%
Return on Equity %	17.27%	16.89%	13.04%	11.39%	10.06%	10.41%
Debt to Equity	0.19	0.11	0.19	0.22	0.27	0.12

CONTACT INFORMATION:

Phone: 312 782-5800   Fax: 312 782-4339
Toll-Free:
Address: 200 E. Randolph Dr., Chicago, IL 60601 United States

STOCK TICKER/OTHER:

Stock Ticker: JLL   Exchange: NYS
Employees: 58,100   Fiscal Year Ends: 12/31
Parent Company:

SALARIES/BONUSES:

Top Exec. Salary: $750,000   Bonus: $
Second Exec. Salary: Bonus: $
$408,462

OTHER THOUGHTS:

Estimated Female Officers or Directors: 2
Hot Spot for Advancement for Women/Minorities: Y

KB Home

www.kbhome.com

NAIC Code: 236117

<table>
<tr><td>

TYPES OF BUSINESS:

Construction, Home Building and Residential
Mortgage Services
Home Construction

</td></tr>
</table>

BRANDS/DIVISIONS/AFFILIATES:

KBnxt
KB Home Studios
Double ZeroHouse 3.0
Hollister Commons
North Park
Ridge at Bandera (The)

CONTACTS: Note: Officers with more than one job title may be intentionally listed here more than once.

Jeffrey Mezger, CEO
Stephen Bollenbach, Chairman of the Board
William Hollinger, Chief Accounting Officer
Nicholas Franklin, Executive VP, Divisional
Albert Praw, Executive VP, Divisional
Jeff Kaminski, Executive VP
Brian Woram, Executive VP
Thomas Norton, Senior VP, Divisional
William Richelieu, Vice President

GROWTH PLANS/SPECIAL FEATURES:

KB Home is one of the largest homebuilders in the U.S. The firm's homebuilding activities are concentrated in 40 major markets located in 10 states: Arizona, California, Colorado, Florida, Nevada, New Mexico, North Carolina, Maryland, Virginia and Texas. During 2014, the company delivered 7,215 houses at an average selling price of $328,400. KB Home builds innovatively designed homes catering to first-time and first-move-up homebuyers, generally in medium-sized developments close to major metropolitan areas, as well as to luxury home buyers. In addition to homebuilding operations, the firm's financial services segment offers insurance services to its homebuyers, provides title services and has established a marketing services agreement with Nationstar Mortgage LLC to offer mortgage banking services, including mortgage loan originations, to its homebuyers. Through its KBnxt business model, KB Home seeks to keep construction costs and base prices as low as possible while promoting customer choice. Potential buyers may visit one of the firm's large KB Home Studios locations to select options for their home. The firm has established several environmental initiatives to promote its commitment to sustainable building practices. It currently installs exclusively ENERY STAR appliances in all of its new homes and is committed to build all new communities to ENERGY STAR guidelines, meaning that they must be more efficient and use less energy than typical houses. The firm has also introduced net-zero energy design options through its DoubleZeroHouse 3.0 program in several markets. In 2014, the firm acquired land for a new community called Hollister Commons in Houston's Spring Branch submarket, which debuted in 2015. In 2015, it acquired land for a new community in the North Park development of Broomfield, Colorado; and purchased land for a new community called The Ridge at Bandera in the Texas Hill Country.

FINANCIAL DATA: Note: Data for latest year may not have been available at press time.

In U.S. $	2015	2014	2013	2012	2011	2010
Revenue	3,032,030,000	2,400,949,000	2,097,130,000	1,560,115,000	1,315,866,000	1,589,996,000
R&D Expense						
Operating Income	145,953,000	123,829,000	101,194,000	-11,564,000	-96,282,000	-10,931,000
Operating Margin %	4.81%	5.15%	4.82%	-.74%	-7.31%	-.68%
SGA Expense	346,709,000	288,023,000	255,808,000	251,159,000	247,886,000	292,639,000
Net Income	84,643,000	918,349,000	39,963,000	-58,953,000	-178,768,000	-69,368,000
Operating Cash Flow	181,185,000	-630,691,000	-443,486,000	34,617,000	-347,545,000	-133,964,000
Capital Expenditure	4,677,000	5,795,000	2,391,000	1,749,000	242,000	420,000
EBITDA	152,310,000	128,119,000	102,910,000	-7,627,000	-129,933,000	-4,772,000
Return on Assets %	1.72%	23.09%	1.38%	-2.32%	-6.35%	-2.11%
Return on Equity %	5.13%	86.14%	8.75%	-14.38%	-33.27%	-10.36%
Debt to Equity	1.55	1.61	4.01	4.57	3.57	2.80

CONTACT INFORMATION:

Phone: 310 231-4000   Fax: 310 231-4222
Toll-Free: 888-524-6637
Address: 10990 Wilshire Blvd., Los Angeles, CA 90024 United States

STOCK TICKER/OTHER:

Stock Ticker: KBH
Employees: 1,590
Parent Company:

Exchange: NYS
Fiscal Year Ends: 11/30

SALARIES/BONUSES:

Top Exec. Salary: Bonus: $
$1,000,000
Second Exec. Salary: Bonus: $
$656,250

OTHER THOUGHTS:

Estimated Female Officers or Directors: 1

Hot Spot for Advancement for Women/Minorities: Y

KBR Inc
NAIC Code: 237000

TYPES OF BUSINESS:
Heavy Construction and Engineering
Energy & Petrochemical Projects
Program & Project Management
Consulting & Technology Services
Operations & Maintenance Services
Contract Staffing Services

BRANDS/DIVISIONS/AFFILIATES:
Granherne
Energo
GVA
Technical Staffing Resources
Mantenimiento Marino de Mexico
KBR Training Solutions
Weatherly Inc
Plinke GmbH

CONTACTS: Note: Officers with more than one job title may be intentionally listed here more than once.
Adam Kramer, Assistant Secretary
Brian Ferraioli, CFO
Loren Carroll, Chairman of the Board
Nelson Rowe, Chief Accounting Officer
Farhan Mujib, Executive VP, Divisional
Kenneth Hill, Executive VP, Divisional
Jan Egil Braendeland, Executive VP, Divisional
Ian Mackey, Executive VP, Divisional
Eileen Akerson, Executive VP
John Derbyshire, President, Divisional
Andrew Pringle, President, Divisional
David Zelinski, President, Divisional
Roy Oelking, President, Divisional
Jalal Ibrahim, President, Geographical
Stuart Bradie, President

GROWTH PLANS/SPECIAL FEATURES:
KBR, Inc. is a global engineering, construction and services company supporting the hydrocarbons and international government services markets. The company conducts business in over 70 countries and maintains offices throughout the U.S., Australia, Africa, the U.K., Asia and the Middle East. It operates in three segments: technology & consulting, engineering & construction and government services. Technology & consulting combines proprietary KBR technologies, knowledge-based services and its three brands, Granherne, Energo and GVA, under a single global business. It also provides licensed technologies and consulting services to the oil and gas sectors. Engineering & construction leverages the firm's operational and technical abilities as a global provider of engineering, procurement, construction, commissioning and maintenance services for oil and gas, refining, petrochemicals and chemicals customers. Government services focuses on long-term service contracts with annuity streams particularly for the U.K., Australian and U.S. governments. Geographically, U.S. operations account for 37% of total revenue; Australia, 22%; Canada, 12%; Middle East, 11%; Europe, 10%; Africa, 4%; and other countries, 4%. The firm's subsidiaries include Technical Staffing Resources, which caters to engineering, finance, administrative, legal and IT markets; Mantenimiento Marino de Mexico, which provides maintenance and rehabilitation services to PEMEX offshore oil and gas facilities; and KBR Training Solutions, which offers training services to oil and gas and government clients. In January 2016, the firm agreed to sell its Infrastructure Americas business to Stantec Consulting Services, Inc.; and acquired Weatherly, Inc., Plinke GmbH and Chematur Ecoplanning Oy from Chematur Technologies AB, which is owned by Connell Chemical.

FINANCIAL DATA: Note: Data for latest year may not have been available at press time.
In U.S. $	2015	2014	2013	2012	2011	2010
Revenue	5,096,000,000	6,366,000,000	7,283,000,000	7,921,000,000	9,261,000,000	10,099,000,000
R&D Expense						
Operating Income	310,000,000	-794,000,000	471,000,000	299,000,000	587,000,000	609,000,000
Operating Margin %	6.08%	-12.47%	6.46%	3.77%	6.33%	6.03%
SGA Expense	155,000,000	239,000,000	249,000,000	222,000,000	214,000,000	212,000,000
Net Income	203,000,000	-1,262,000,000	229,000,000	144,000,000	480,000,000	327,000,000
Operating Cash Flow	47,000,000	170,000,000	290,000,000	142,000,000	650,000,000	549,000,000
Capital Expenditure	10,000,000	53,000,000	78,000,000	75,000,000	83,000,000	86,000,000
EBITDA	349,000,000	-722,000,000	539,000,000	360,000,000	658,000,000	671,000,000
Return on Assets %	5.33%	-25.98%	4.05%	2.51%	8.65%	6.08%
Return on Equity %	20.22%	-70.91%	8.87%	5.71%	20.24%	14.42%
Debt to Equity	0.04	0.06		0.03		

CONTACT INFORMATION:
Phone: 713 753-3011   Fax: 713 753-5353
Toll-Free:
Address: 601 Jefferson St., Ste. 3400, Houston, TX 77002 United States

STOCK TICKER/OTHER:
Stock Ticker: KBR
Employees: 25,000
Parent Company:
Exchange: NYS
Fiscal Year Ends: 12/31

SALARIES/BONUSES:
Top Exec. Salary: $1,038,485   Bonus: $
Second Exec. Salary: $675,022   Bonus: $

OTHER THOUGHTS:
Estimated Female Officers or Directors: 2

Hot Spot for Advancement for Women/Minorities: Y

Kerzner International Holdings Limited

www.kerzner.com

NAIC Code: 721120

TYPES OF BUSINESS:

Casino Hotels
Luxury Resort Hotels
Resort Development

BRANDS/DIVISIONS/AFFILIATES:

Atlantis
Atlantis, The Palm Dubai
Cove Atlantis (The)
One&Only
Mazagan Beach Resort
Investment Corporation of Dubai

CONTACTS: Note: Officers with more than one job title may be intentionally listed here more than once.

Alan Leibman, CEO
Mark DeCocinis, COO
Bonnie S. Biumi, Pres.
Ali Tabbal, CFO
Helen McCabe-Young, Exec. VP-Sales & Mktg.
Suresh Menon, Exec. VP-Global Human Resources
Monica Digilio, Exec. VP-Admin.
Tim Brown, Sr. VP-Project Planning & Dev.
George Markantonis, Pres., Kerzner Int'l Bahamas
Serge Zaalof, Pres., Atlantis & The Palm
H. E. Mohammed Al Shaibani, Chmn.

GROWTH PLANS/SPECIAL FEATURES:

Kerzner International Holdings Limited (KIHL) is a resort and gaming company that develops, operates and manages premier resorts, casinos and luxury hotels. The company's flagship property is Atlantis Paradise Island in the Bahamas. The over 2,300-room ocean themed resort features three interconnected hotel towers built around 100 acres of pools and marine environments, home to over 50,000 marine animals. Additionally, the resort features the world's largest open-air aquarium. The firm also operates Atlantis The Palm, Dubai, an approximately 1,500-room ocean themed resort located on the manmade island of Palm Jumeirah; The Cove Atlantis in the Bahamas; 11 One&Only Resorts, exclusive luxury resorts located in the Bahamas, Mexico, Montenegro, Jeddah, Hainan, Mauritius, South Africa, the Maldives and Dubai; and Mazagan Beach Resort, an approximately 500-room casino resort in Morocco which features a golf course, lagoon, swimming patio and a 180 degree view of the Atlantic Ocean. In 2014, the Investment Corporation of Dubai acquired a significant equity interest in KIHL from the Kerzner family and several other institutional investors for an undisclosed sum.

FINANCIAL DATA: Note: Data for latest year may not have been available at press time.

In U.S. $	2015	2014	2013	2012	2011	2010
Revenue	400,000,000	380,000,000	395,000,000	375,000,000		
R&D Expense						
Operating Income						
Operating Margin %						
SGA Expense						
Net Income						
Operating Cash Flow						
Capital Expenditure						
EBITDA						
Return on Assets %						
Return on Equity %						
Debt to Equity						

CONTACT INFORMATION:

Phone: 242-363-6000   Fax: 954-809-2337
Toll-Free:
Address: 1 Casino Dr., Paradise Island, C5 Bahamas

STOCK TICKER/OTHER:

Stock Ticker: Private
Employees: 8,574
Parent Company:

Exchange:
Fiscal Year Ends: 12/31

SALARIES/BONUSES:

Top Exec. Salary: $ Bonus: $
Second Exec. Salary: $ Bonus: $

OTHER THOUGHTS:

Estimated Female Officers or Directors: 2
Hot Spot for Advancement for Women/Minorities:

Kiewit Corp

www.kiewit.com

NAIC Code: 237310

TYPES OF BUSINESS:

Construction & Engineering Services

BRANDS/DIVISIONS/AFFILIATES:

Haystack Coal Company
Haystack Coal Company

CONTACTS: Note: Officers with more than one job title may be intentionally listed here more than once.

Bruce E. Grewcock, CEO
Bruce E. Grewcock, Pres.

GROWTH PLANS/SPECIAL FEATURES:

Kiewit Corp., based in Omaha, Nebraska and employee-owned, is one of the largest general contractors in the world. The firm dis organiized into units including building; mining; oil, gas and chemical; power; transportation; and water/wastewater. The building unit focuses on office buildings, industrial complexes, education and sports facilities, hotels, hospitals, transportation terminals, science and technology facilities and manufacturing plants, as well as retail and special-use facilities. The segment is Leadership in Energy and Environmental Design (LEED) certified. The mining division has performed more than 50 mining contracts totaling over $1.3 billion in revenue. This segment specializes in mine management, production, maintenance, contract mining, ore processing and mine infrastructure. Subsidiaries in this division include Haystack Coal Company, working in the Haystack Mine of southwest Wyoming, which is permitted to ship 1.5 million tons of high quality thermal coal per year; and Buckskin Mining Company, working in the Gillette, Wyoming area, with a capacity of 25 million tons per year of coal. The oil, gas and chemical segment engineers, procures, constructs and provides startup services for the energy needs of the U.S. and Canada. This division specializes in offshore, oil sands, gas processing, compressor/pump stations, pipelines, terminals, liquid natural gas and refining services and solutions. Power helps its clients meet the challenge of changing power consumption trends by building run-of-the-river hydroelectric, nuclear and geothermal power plants, as well as cogeneration, combined-cycle and waste-to-energy generation and resource facilities. The transportation division builds highways, bridges, rails and runways in order to connect the world. It has constructed more than 1,100 transportation projects, delivering engineering solutions and constructions for the air, bridge, marine/port, rail, roads and tunnels sectors. Water/wastewater performs water supply projects such as roller-compacted concrete, earth-fill & rock-fill dams, reservoirs, water tunnels and canals across North America.

FINANCIAL DATA: Note: Data for latest year may not have been available at press time.

In U.S. $	2015	2014	2013	2012	2011	2010
Revenue	10,380,000,000	11,000,000,000	10,787,600,000	10,100,000,000	10,000,000,000	9,940,000,000
R&D Expense						
Operating Income						
Operating Margin %						
SGA Expense						
Net Income						
Operating Cash Flow						
Capital Expenditure						
EBITDA						
Return on Assets %						
Return on Equity %						
Debt to Equity						

CONTACT INFORMATION:

Phone: 402-342-2052   Fax: 402-271-2829
Toll-Free:
Address: 1000 Kiewit Plaza, Omaha, NE 68131 United States

STOCK TICKER/OTHER:

Stock Ticker: Private   Exchange:
Employees: 25,700   Fiscal Year Ends:
Parent Company:

SALARIES/BONUSES:

Top Exec. Salary: $ Bonus: $
Second Exec. Salary: $ Bonus: $

OTHER THOUGHTS:

Estimated Female Officers or Directors:
Hot Spot for Advancement for Women/Minorities:

Kilroy Realty Corporation

www.kilroyrealty.com

NAIC Code: 0

TYPES OF BUSINESS:

Real Estate Investment Trust
Office & Industrial Properties
Property Management
Financial Services
Leasing
Construction Management

BRANDS/DIVISIONS/AFFILIATES:

Kilroy Realty Finance Partnership LP
Kilroy Services LLC

CONTACTS: Note: Officers with more than one job title may be intentionally listed here more than once.

John Kilroy, CEO
Tyler Rose, CFO
Heidi Roth, Chief Accounting Officer
Justin Smart, Executive VP, Divisional
A. Robert Paratte, Executive VP, Divisional
David Simon, Executive VP, Geographical
Michael Sanford, Executive VP, Geographical
Jeffrey Hawken, Executive VP
Michelle Ngo, Senior VP

GROWTH PLANS/SPECIAL FEATURES:

Kilroy Realty Corporation (KRC) is a real estate investment trust (REIT) that owns, operates, develops and acquires Class A office and industrial real estate in suburban markets, primarily in Orange County, Los Angeles, San Diego County, the San Francisco Bay area and Seattle. Its portfolio of operating properties consists of 111 office buildings, totaling more than 14 million rentable square feet. The office properties are 94.4% leased to 526 tenants (2014). In addition, the company has six development properties under construction and one leased-up property. The firm's 15 largest tenants in terms of annualized base rental revenues (including DIRECTV LLC; Intuit, Inc.; Bridgepoint Education, Inc.; Delta Dental of California; LinkedIn Corporation and AMN Healthcare, Inc.) represented approximately 24.5% of its total revenue in 2014. KRC owns all of its properties through its subsidiary Kilroy Realty Finance Partnership, L.P. (KRLP) and its operating partnership, of which it owns a 99% interest. KRLP's wholly-owned subsidiary Kilroy Services, LLC conducts substantially all of the firm's development activities. In 2014, the firm acquired a four-story life science property in Seattle for $106.1 million; a four acre parcel in Hollywood, California from The Academy of Motion Pictures Arts and Sciences; a 3.1 acre development site in San Francisco; and proposed to acquire San Francisco Flowers Growers' Association, a 1.9 acre land site in Central SoMa. That same year, the firm sold 13 San Diego office properties for a total of $327 million. In 2015, Kilroy acquired a full city block site, along with three smaller adjacent parcels, which total 2.4 acres, in Seattle.

FINANCIAL DATA: Note: Data for latest year may not have been available at press time.

In U.S. $	2015	2014	2013	2012	2011	2010
Revenue	581,275,000	521,725,000	465,098,000	404,912,000	367,131,000	301,980,000
R&D Expense						
Operating Income	168,977,000	122,833,000	90,072,000	83,713,000	93,897,000	82,478,000
Operating Margin %	29.07%	23.54%	19.36%	20.67%	25.57%	27.31%
SGA Expense	51,361,000	49,227,000	43,164,000	39,356,000	29,927,000	28,947,000
Net Income	234,081,000	180,219,000	43,880,000	270,914,000	66,015,000	19,708,000
Operating Cash Flow	272,008,000	245,253,000	240,576,000	180,724,000	138,256,000	119,827,000
Capital Expenditure	646,599,000	697,046,000	755,465,000	958,470,000	694,557,000	730,551,000
EBITDA	500,580,000	331,362,000	285,417,000	247,762,000	231,065,000	182,687,000
Return on Assets %	3.81%	3.10%	.62%	6.19%	1.62%	.18%
Return on Equity %	8.10%	7.04%	1.43%	15.76%	4.75%	.53%
Debt to Equity	0.75	0.99	0.97	1.02	1.55	1.47

CONTACT INFORMATION:

Phone: 310 481-8400   Fax: 310 481-6580
Toll-Free:
Address: 12200 W. Olympic Blvd., Ste. 200, Los Angeles, CA 90064
United States

STOCK TICKER/OTHER:

Stock Ticker: KRC
Employees: 232
Parent Company:

Exchange: NYS
Fiscal Year Ends: 12/31

SALARIES/BONUSES:

Top Exec. Salary:
$1,225,000
Second Exec. Salary:
$675,000

Bonus: $3,700,000

Bonus: $1,600,000

OTHER THOUGHTS:

Estimated Female Officers or Directors: 1

Hot Spot for Advancement for Women/Minorities:

Kimco Realty Corp

NAIC Code: 0

www.kimcorealty.com

TYPES OF BUSINESS:

Real Estate Investment Trust
Commercial Development
Property Services
Leasing
Consulting Services
Shopping Centers
Regional Malls
Property Management

BRANDS/DIVISIONS/AFFILIATES:

Christown Spectrum

CONTACTS: Note: Officers with more than one job title may be intentionally listed here more than once.

Conor Flynn, CEO
Glenn Cohen, CFO
Milton Cooper, Chairman of the Board
Paul Westbrook, Chief Accounting Officer
Bruce Rubenstein, Executive VP

GROWTH PLANS/SPECIAL FEATURES:

Kimco Realty Corp. is a real estate investment trust (REIT) that primarily owns and operates shopping centers. The firm owns interest in approximately 1,051 properties, including 605 shopping centers, retail store leases, development projects and undeveloped land, totaling 103.3 million square feet of gross leasable area. These properties are located in 38 U.S. states, Puerto Rico and Canada. Kimco's shopping centers are anchored by department store, supermarket or drug store tenants offering day-to-day necessities. Its tenants include major national and regional companies such as TJX Companies, The Home Depot, Bed Bath & Beyond, Royal Ahold and Albertsons which represent 3.2%, 2.4%, 2.1%, 1.9% and 1.9%, of the company's annual base rental revenue. Kimco's property services division provides property management services relating to the management, operation, supervision and maintenance of properties. The leasing department assists potential renters in finding the best locations for their business ventures. Kimco focuses on acquiring properties that have opportunities for redevelopment and renovation as well as other value increasing characteristics. Generally, the company can close a transaction in 45 to 60 days. Kimco also offers retail store owners and developers preferred equity debt. In 2015, the firm agreed with RioCan Real Estate Investment Trust to unwind their Canadian joint venture, which includes 22 properties that will be acquired by RioCan. That same year, Kimco acquired Christown Spectrum, an 850,000-square-foot destination power center in the Phoenix-Mesa-Scottsdale, Arizona.

The firm offers employees medical and dental insurance, disability coverage, life insurance, health care and dependent care reimbursement accounts, a 401(k) savings plan and tuition assistance.

FINANCIAL DATA: Note: Data for latest year may not have been available at press time.

In U.S. $	2015	2014	2013	2012	2011	2010
Revenue	1,166,769,000	993,897,000	946,673,000	922,304,000	916,288,000	898,876,000
R&D Expense						
Operating Income	343,572,000	310,315,000	225,502,000	257,510,000	287,896,000	294,751,000
Operating Margin %	29.44%	31.22%	23.82%	27.92%	31.41%	32.79%
SGA Expense	122,735,000	122,201,000	127,913,000	124,480,000	118,937,000	109,201,000
Net Income	894,115,000	424,001,000	236,281,000	266,073,000	169,051,000	142,868,000
Operating Cash Flow	493,701,000	629,343,000	570,035,000	479,054,000	448,613,000	479,935,000
Capital Expenditure	861,309,000	582,765,000	462,155,000	554,956,000	381,195,000	224,457,000
EBITDA	732,334,000	578,959,000	506,433,000	524,406,000	556,615,000	563,623,000
Return on Assets %	7.68%	3.66%	1.83%	1.78%	1.12%	.91%
Return on Equity %	16.92%	7.77%	3.78%	3.65%	2.28%	1.87%
Debt to Equity	1.06	0.96	0.91	0.88	0.86	0.81

CONTACT INFORMATION:

Phone: 516 869-9000   Fax: 516 869-9001
Toll-Free: 800-285-4626
Address: 3333 New Hyde Park Rd., Ste. 100, New Hyde Park, NY 11042
United States

STOCK TICKER/OTHER:

Stock Ticker: KIM
Employees: 580
Parent Company:

Exchange: NYS
Fiscal Year Ends: 12/31

SALARIES/BONUSES:

Top Exec. Salary: $850,000   Bonus: $
Second Exec. Salary: $799,230   Bonus: $

OTHER THOUGHTS:

Estimated Female Officers or Directors: 3
Hot Spot for Advancement for Women/Minorities: Y

Kimpton Hotel & Restaurant Group LLC   www.kimptonhotels.com

NAIC Code: 721110

TYPES OF BUSINESS:

Hotels
Restaurants
Hotel Management Services

BRANDS/DIVISIONS/AFFILIATES:

InterContinental Hotels Group PLC
Area 31
Hotel Vintage Plaza
Hotel Burnham
Cafe Pescatore
Sazerac
Silverleaf Tavern
Scala's Bistro

CONTACTS: Note: Officers with more than one job title may be intentionally listed here more than once.

Mike DeFrino, CEO
Judy Miles, General Counsel
Joe Long, Exec. VP-Dev.
Lisa Demoney, Sr. Dir.-Digital Mktg. & Media
Stephanie Moustirats, Dir.-Hotel Public Rel.
James Alderman, Sr. VP-Acquisitions & Dev.
James Lin, Sr. VP-Restaurant Oper.
Barry Pollard, Sr. VP-Hotel Oper.
Christine Lawson, Sr. VP-Hotel Sales & Catering

GROWTH PLANS/SPECIAL FEATURES:

Kimpton Hotel & Restaurant Group LLC, based in San Francisco, owns 59 lifestyle boutique hotels in 30 U.S. cities. Its holdings also consist of more than 60 restaurants and bars next to or within its hotels. The firm specializes in renovating old, disused buildings to transform them into unique hotels as well as small, European-style restaurants. Its themed hotels include Hotel Vintage Plaza in Portland, Oregon, which has an Italian romance theme; Hotel Vintage in Seattle, highlighting local Washington wines; and Hotel Burnham in Chicago, which focuses on its significance in Chicago's history. Some notable restaurants run by Kimpton include San Francisco bistros Cafe Pescatore, Scala's Bistro and Puccini & Pinetti; Sazerac in Seattle; Atwood Cafe in Chicago; Area 31 in Miami; Firefly in Washington, D.C.; Ruby Room in Boston; and Silverleaf Tavern in New York City. The company also offers full service spas at some of its locations. Special services offered by its hotels include the Mind, Body, Spa Program, which offers in-room massage, yoga, Pilates and meditation; pet packages, which include pet-friendly amenities and services; and Hosted Evening Wine Hour. The company is also engaged in comprehensive management services for other companies, offering everything from financial management to facilities renovation. The company is owned by hotel giant InterContinental Hotels Group PLC. Kimpton plans to open eight additional hotels, located in California, Colorado, Illinois, North Carolina, Ohio, Wisconsin, as well as in the Cayman Islands. It plans to open another in Nashville, Tennessee in 2017. In January 2016, Kimpton announced the signing of its first Kimpton Hotels & Restaurants hotel outside the Americas, debuting in Amsterdam. After a complete renovation, the 270-room hotel will open in 2017.

The firm offers employees medical, dental, vision and life insurance; long- and short-term disability; paid vacation time; tuition reimbursement; and employee discounts.

FINANCIAL DATA: Note: Data for latest year may not have been available at press time.

In U.S. $	2015	2014	2013	2012	2011	2010
Revenue	1,200,000,000	1,049,880,000	1,000,000,000	945,000,000	905,000,000	780,000,000
R&D Expense						
Operating Income						
Operating Margin %						
SGA Expense						
Net Income						
Operating Cash Flow						
Capital Expenditure						
EBITDA						
Return on Assets %						
Return on Equity %						
Debt to Equity						

CONTACT INFORMATION:

Phone: 415-397-5572   Fax: 415-296-8031
Toll-Free: 800-546-7866
Address: 222 Kearny St., Ste. 200, San Francisco, CA 94108 United States

STOCK TICKER/OTHER:

Stock Ticker: Subsidiary
Employees: 7,754
Parent Company: InterContinental Hotels Group PLC

Exchange:
Fiscal Year Ends: 12/31

SALARIES/BONUSES:

Top Exec. Salary: $ Bonus: $
Second Exec. Salary: $ Bonus: $

OTHER THOUGHTS:

Estimated Female Officers or Directors: 9
Hot Spot for Advancement for Women/Minorities: Y

Sales, profits and employees may be estimates. Financial information, benefits and other data can change quickly and may vary from those stated here.

Kohler Company

www.corporate.kohler.com

NAIC Code: 327110

TYPES OF BUSINESS:

Plumbing Fixtures
Resorts
Kitchen & Bath Products
Building Materials

BRANDS/DIVISIONS/AFFILIATES:

Kohler
Sterling
Kallista
Ann Sacks
Robern
Kohler Engines
Lombardini
Somo

CONTACTS: Note: Officers with more than one job title may be intentionally listed here more than once.

David Kohler, CEO
Herbert V. Kohler, Jr., Chmn.

GROWTH PLANS/SPECIAL FEATURES:

Kohler Company founded in 1873, is an American manufacturer that operates on six continents worldwide. The firm has four divisions: kitchen & bath, power systems, interiors and hospitality. Kitchen & bath offers products under the Kohler, Sterling, Kallista, Ann Sacks and Robern brands. Products include fashionable sinks, faucets, toilets, bidets, vanities, medicine cabinets, accessories and commercial products. The products can include a number of features including luxury design, automated systems, stainless steel and touchless products. Power systems include: Kohler Engines, Lombardini, Kohler Power, Somo and Kohler Power Uninterruptible. Products include private and commercial agricultural equipment, including lawn mowers; maritime equipment, such as marine generators; home generators; and uninterruptible power solutions. Interiors include the Baker, McGuire, Ann Sacks, Kallista and Mark David brands. Products include furniture, lighting, home accessories, textiles, artesian stone design, plumbing products and high-end hospitality furnishings. Hospitality includes golf destinations and hospitality products. Hospitality includes the American Club Resort (Kohler, Wisconsin) and Old Course Hotel (St. Andrews, Scotland) offer spa services, golf, lodgings and high-end cuisine to their guests. Additional hospitality services include Kohler Original Chocolates and Kohler At Home, offering high end bedding & spa-like products for home use.

Kohler offers its employees medical, dental, life and prescription insurance; dependent and health care spending accounts; a wellness program; a pension plan; a 401(k) plan with company matching; annual bonuses; paid time off; and employee discounts.

FINANCIAL DATA: Note: Data for latest year may not have been available at press time.

In U.S. $	2015	2014	2013	2012	2011	2010
Revenue	6,000,000,000	5,210,000,000	5,000,000,000	4,800,000,000	4,500,000,000	
R&D Expense						
Operating Income						
Operating Margin %						
SGA Expense						
Net Income						
Operating Cash Flow						
Capital Expenditure						
EBITDA						
Return on Assets %						
Return on Equity %						
Debt to Equity						

CONTACT INFORMATION:

Phone: 920-457-4441   Fax:
Toll-Free: 800-456-4537
Address: 444 Highland Dr., Kohler, WI 53044 United States

STOCK TICKER/OTHER:

Stock Ticker: Private   Exchange:
Employees: 30,000   Fiscal Year Ends:
Parent Company:

SALARIES/BONUSES:

Top Exec. Salary: $ Bonus: $
Second Exec. Salary: $ Bonus: $

OTHER THOUGHTS:

Estimated Female Officers or Directors:
Hot Spot for Advancement for Women/Minorities:

Kohn Pedersen Fox Associates (KPF)   www.kpf.com

NAIC Code: 541310

TYPES OF BUSINESS:

Architectural Services

BRANDS/DIVISIONS/AFFILIATES:

GROWTH PLANS/SPECIAL FEATURES:

Kohn Pedersen Fox Associates (KPF) is an architecture firm that provides architecture, interior design, programming and master planning services for clients in both the public and private sectors. The firm operates out of six global offices in New York, London, Shanghai, Hong Kong, Seoul and Abu Dhabi. KPF has taken on projects in the field of culture & entertainment, education, corporate, government, health & science, heritage & historic, hospitality, interiors, office, residential, retail and transportation & infrastructure. Currently the firm's portfolio features over 70 projects in 35 countries that have been, or are pursing, green building certification.

CONTACTS: Note: Officers with more than one job title may be intentionally listed here more than once.

James von Klemperer, Pres.
A. Eugene Kohn, Chmn.

FINANCIAL DATA: Note: Data for latest year may not have been available at press time.

In U.S. $	2015	2014	2013	2012	2011	2010
Revenue	220,000,000					
R&D Expense						
Operating Income						
Operating Margin %						
SGA Expense						
Net Income						
Operating Cash Flow						
Capital Expenditure						
EBITDA						
Return on Assets %						
Return on Equity %						
Debt to Equity						

CONTACT INFORMATION:

Phone: 212-977-6500   Fax: 212-956-2526
Toll-Free:
Address: 11 W. 42nd St., New York, NY 10036 United States

STOCK TICKER/OTHER:

Stock Ticker: Private
Employees: 600
Parent Company:

Exchange:
Fiscal Year Ends:

SALARIES/BONUSES:

Top Exec. Salary: $ Bonus: $
Second Exec. Salary: $ Bonus: $

OTHER THOUGHTS:

Estimated Female Officers or Directors:
Hot Spot for Advancement for Women/Minorities:

Kumho Industrial Co Ltd

www.kumhoenc.com

NAIC Code: 236220

TYPES OF BUSINESS:

Commercial and Institutional Building Construction
Research & Development
Sewage Treatment

BRANDS/DIVISIONS/AFFILIATES:

Kumho Engineering & Construction
Kumho Asiana Group

CONTACTS: Note: Officers with more than one job title may be intentionally listed here more than once.

Won Ii-Woo, Co-CEO
Park Sam Koo, Co-CEO
Bak Sam Koo, CEO-Kumho Asiana Group

GROWTH PLANS/SPECIAL FEATURES:

Kumho Industrial Co. Ltd., doing business as Kumho Engineering & Construction (Kumho E&C), offers a wide range of building services. Based in Korea, it has worked on general architectural construction projects, building commercial centers, warehouses, leisure facilities and even entire cities. Kumho E&C's architectural business builds public & commercial buildings, education facilities, research & development facilities, airport/sales/logistics facilities, cultural facilities, medical facilities, as well as correctional/military facilities. Recent projects within this division include the Western Bank Tower, Sunrise Plot V and Time Square buildings in Vietnam; and the Dubai World Central Al Maktoum International Airport and Abu Dhabi International Airport Control Tower in Dubai. Kumho E&C's civil engineering business builds connections between cities and territories, such as bridges, tunnels, railways, ports and roads. The company's housing construction segment builds apartment complexes, as well as residential & commercial properties and complexes of various sizes. Kumho's plant & environment business segment develops power generation, energy storage/supply, petrochemical facilities, industrial facilities and environmental plants and facilities. It developed a high efficiency membrane filtration seawater desalination facility based on its technical skills on water treatment processes. Moreover, it is pioneering new markets for wind power generation, bio gas generation and nuclear power generation. Kumho E&C is part of the Kumho Asiana Group.

FINANCIAL DATA: Note: Data for latest year may not have been available at press time.

In U.S. $	2015	2014	2013	2012	2011	2010
Revenue	1,357,435,200	1,407,472,919	1,389,132,708	1,425,300,000	1,653,000,000	1,823,000,000
R&D Expense						
Operating Income						
Operating Margin %						
SGA Expense						
Net Income	-5,416,942	55,757,263	50,959,898	-691,700,000	-50,400,000	77,700,000
Operating Cash Flow						
Capital Expenditure						
EBITDA						
Return on Assets %						
Return on Equity %						
Debt to Equity						

CONTACT INFORMATION:

Phone: 82-2-6303-0114   Fax:
Toll-Free:
Address: Songwol-dong, Naju-si, Jeollanam-do, 1095-4 South Korea

STOCK TICKER/OTHER:

Stock Ticker: 2990   Exchange: Seoul
Employees: 3,107   Fiscal Year Ends: 12/31
Parent Company: Kumho Asiana Group

SALARIES/BONUSES:

Top Exec. Salary: $ Bonus: $
Second Exec. Salary: $ Bonus: $

OTHER THOUGHTS:

Estimated Female Officers or Directors:
Hot Spot for Advancement for Women/Minorities:

La Quinta Holdings Inc

www.lq.com

NAIC Code: 721110

TYPES OF BUSINESS:

Hotels, Motels & Suites
Hotel Management
Franchising

BRANDS/DIVISIONS/AFFILIATES:

La Quinta Properties
La Quinta Inns
La Quinta Inns and Suites
Blackstone Group LP

CONTACTS: Note: Officers with more than one job title may be intentionally listed here more than once.

James Forson, CFO
Mitesh Shah, Chairman of the Board
Julie Cary, Chief Marketing Officer
Angelo Lombardi, COO
Mark Chloupek, Executive VP
Rajiv Trivedi, Executive VP
Keith Cline, President

GROWTH PLANS/SPECIAL FEATURES:

La Quinta Holdings, Inc., owned by the Blackstone Group LP, is the operator of the La Quinta motels and suites properties. La Quinta is a leading limited-service lodging brand that provides comfortable guest rooms in convenient locations at affordable prices. The firm is one of the largest owners and operators of limited-service hotels in the U.S. It maintains 886 hotels and 87,500 rooms in 48 states, as well as Canada, Mexico and Honduras, under the brands La Quinta Inns and La Quinta Inns and Suites. The firm also licenses its brand name to franchisees for royalties and other fees. The company markets its services to both leisure guests and business travelers. All of the firm's hotels are owned through La Quinta Properties, a real estate investment trust (REIT). A typical La Quinta Inn features approximately 130 guest rooms with amenities including movies-on-demand, interactive video games, free high-speed Internet, complimentary continental breakfast, a swimming pool, fax services and 24-hour front desk message services. La Quinta Inn and Suites properties also feature deluxe two-room suites with microwaves and refrigerators as well as fitness centers, courtyards and expanded food offerings. The Blackstone Group LP holds a 28.3% interest in the company and can currently appoint 20% of the company's directors. Currently, La Quinta has a pipeline of 228 franchised hotels to be located in the U.S., Mexico, Columbia, Nicaragua, Guatemala and Chile. In May 2015, the firm expanded into the Latin American market with a franchise agreement for an LQ Hotel by La Quinta in Chile. In September of the same year, the firm sold 24 geographically dispersed hotels

The firm offers employees medical, dental and vision coverage; life insurance; long-term disability; an employee assistance program; flexible spending accounts; a 401(k) plan; tuition reimbursement; an internal referral bonus program; and room rate discounts.

FINANCIAL DATA: Note: Data for latest year may not have been available at press time.

In U.S. $	2015	2014	2013	2012	2011	2010
Revenue	1,029,974,000	976,938,000	873,893,000	818,012,000	751,541,000	705,853,000
R&D Expense						
Operating Income	128,071,000	136,669,000	156,181,000	129,595,000	101,849,000	89,084,000
Operating Margin %	12.43%	13.98%	17.87%	15.84%	13.55%	12.62%
SGA Expense	250,677,000	261,781,000	190,055,000	180,987,000	154,949,000	155,967,000
Net Income	26,365,000	-337,297,000	3,976,000	-30,954,000	63,513,000	68,828,000
Operating Cash Flow	290,495,000	286,082,000	232,858,000	248,187,000	220,429,000	200,777,000
Capital Expenditure	100,776,000	78,630,000	115,529,000	102,886,000	95,552,000	69,606,000
EBITDA	311,489,000	311,690,000	321,570,000	292,328,000	267,571,000	282,284,000
Return on Assets %	.84%	-10.48%	.12%	-.90%	1.82%	
Return on Equity %	3.40%	-59.99%	1.25%	-12.26%	33.50%	
Debt to Equity	2.27	2.31		9.00	15.17	

CONTACT INFORMATION:

Phone: 214-492-6600   Fax: 214-492-6616
Toll-Free:
Address: 909 Hidden Ridge, Ste. 600, Irving, TX 75038 United States

STOCK TICKER/OTHER:

Stock Ticker: LQ
Employees: 7,719
Parent Company: Blackstone Group LP

Exchange: NYS
Fiscal Year Ends: 12/31

SALARIES/BONUSES:

Top Exec. Salary: $675,890   Bonus: $
Second Exec. Salary: $398,288   Bonus: $18,900

OTHER THOUGHTS:

Estimated Female Officers or Directors: 2
Hot Spot for Advancement for Women/Minorities:

LafargeHolcim Ltd

NAIC Code: 327310

www.lafarge.com

TYPES OF BUSINESS:

Cement
Construction Materials

BRANDS/DIVISIONS/AFFILIATES:

Lafarge Group
Holcim Ltd

CONTACTS: *Note: Officers with more than one job title may be intentionally listed here more than once.*

Eric Olsen, CEO
Thomas Aebischer, CFO
Jean-Jacques Gauthier, Chief Integration Officer-Organization & Human Resources

GROWTH PLANS/SPECIAL FEATURES:

LafargeHolcim Ltd. (LH), a product of the 2015 merger of Lafarge Group and Holcim Ltd., is a cement and building materials firm with over 2,500 production sites in 90 countries. The firm is involved in the cement manufacturing and raw materials extracting lines of business. LH's over 2,500 production sites include over 1,600 ready mix concrete plants, over 600 aggregate plants, over 180 cement plants and 70 grinding plants. The firm undertakes projects in the Middle East & Africa, North America, Asia, Western Europe, Central & Eastern Europe and Latin America. The cement activities of LH involve extracting raw materials, grinding and milling them into 80% limestone & 20% clay and storing them into a pre-homogenization pile. This ground and milled limestone/clay mix is then fired up to 1450 degrees, which releases the carbon dioxide contained within the limestone through a process called decarbonation, and turned into hard granules called clinker. The clinker, once cooled, is then transformed into cement by the manufacturing plant. The firm's concrete and aggregate activities involve balancing the mix of cement, aggregates and water so that the cement acts as a binder and modifies the uses of concretes for purposes such as making reinforced concrete and blending cement for slabs and roads. Recently, the company returned to its gypsum activities in Algeria, Mexico, Morocco, South Africa and Turkey. The gypsum business manufactures and markets gypsum-based building products for constructing, finishing or decorating interior walls and ceilings in residential, commercial and institutional construction projects.

FINANCIAL DATA: *Note: Data for latest year may not have been available at press time.*

In U.S. $	2015	2014	2013	2012	2011	2010
Revenue	24,682,110,000	19,999,790,000	20,637,150,000	22,547,120,000	21,709,870,000	22,661,200,000
R&D Expense						
Operating Income	-773,409,000	2,424,883,000	2,466,745,000	1,900,556,000	2,023,003,000	2,740,945,000
Operating Margin %	-3.13%	12.12%	11.95%	8.42%	9.31%	12.09%
SGA Expense	8,196,669,000	6,535,777,000	6,567,173,000	7,300,813,000	6,902,073,000	6,964,867,000
Net Income	-1,537,399,000	1,346,925,000	1,331,226,000	650,961,200	713,754,900	1,696,476,000
Operating Cash Flow	2,579,774,000	2,614,310,000	2,916,767,000	2,806,878,000	2,881,184,000	3,829,369,000
Capital Expenditure	2,204,058,000	2,059,633,000	2,307,668,000	1,821,017,000	1,858,693,000	1,905,788,000
EBITDA	4,609,056,000	3,921,466,000	4,076,357,000	4,169,501,000	4,569,288,000	5,261,064,000
Return on Assets %	-2.60%	3.31%	3.20%	1.48%	.63%	2.52%
Return on Equity %	-6.02%	7.65%	7.67%	3.68%	1.57%	6.36%
Debt to Equity	0.47	0.53	0.54	0.58	0.69	0.67

CONTACT INFORMATION:

Phone: 41-588588600   Fax:
Toll-Free:
Address: Zurcherstrasse 156, Jona, 8645 Switzerland

STOCK TICKER/OTHER:

Stock Ticker: HCMLF   Exchange: PINX
Employees: 100,956   Fiscal Year Ends: 12/31
Parent Company:

SALARIES/BONUSES:

Top Exec. Salary: $ Bonus: $
Second Exec. Salary: $ Bonus: $

OTHER THOUGHTS:

Estimated Female Officers or Directors:
Hot Spot for Advancement for Women/Minorities: C

Las Vegas Sands Corp (The Venetian)

www.lasvegassands.com

NAIC Code: 721120

TYPES OF BUSINESS:

Hotel Casinos
Convention & Conference Centers
Shopping Center Development
Casino Property Development

BRANDS/DIVISIONS/AFFILIATES:

Venetian Resort Hotel Casino (The)
Sands Expo and Convention Center (The)
Sands China Ltd
Sands Macao Casino (The)
Palazzo Resort Hotel Casino (The)
Venetian Macao Resort Hotel (The)
Marina Bay Sands Pte Ltd
Cotai Strip

CONTACTS: Note: Officers with more than one job title may be intentionally listed here more than once.

George Tanasijevich, CEO, Subsidiary
Sheldon Adelson, CEO
Patrick Dumont, CFO
Randy Hyzak, Chief Accounting Officer
George Markantonis, COO, Subsidiary
Robert Goldstein, COO
Ira Raphaelson, Executive VP
Stephanie Marz, Vice President, Divisional

GROWTH PLANS/SPECIAL FEATURES:

Las Vegas Sands Corp. (LVSC) is an international hotel, resort and casino firm. Its flagship property is The Venetian Resort Hotel Casino, which is connected to The Palazzo Resort Hotel Casino. Together, The Venetian and The Palazzo offer 225,000 square feet of gaming space, with 240 table games and 2,350 slot machines, as well as 7,092 hotel suites. LVSC also runs the 1.2 million square foot convention and trade show facility, The Sands Expo and Convention Center, and a supplemental event and conference center. Additionally, the firm operates the Sands Casino Resort Bethlehem in eastern Pennsylvania, which features 145,000 square feet of gaming space, a 300-room hotel, 150,000 square feet of retail space and other amenities. Outside the U.S., LVSC has operations in Macao, through majority-owned subsidiary Sands China Ltd., and Singapore, through Marina Bay Sands Pte. Ltd. The company's largest development project, the multi-billion dollar Cotai Strip, is a collection of hotel properties, casinos and entertainment venues in Macao. Sands China runs The Sands Macao and The Venetian Macao Resort Hotel, the anchor property on the Cotai Strip. Other properties on the Cotai Strip include the Four Seasons Macao and the Plaza Casino. Its Singapore property, Marina Bay Sands features three 55-story hotel towers, gaming space, convention space, two state-of-the-art theaters and The Shoppes at Marina Bay Sands.

FINANCIAL DATA: Note: Data for latest year may not have been available at press time.

In U.S. $	2015	2014	2013	2012	2011	2010
Revenue	11,688,460,000	14,583,850,000	13,769,880,000	11,131,130,000	9,410,745,000	6,853,182,000
R&D Expense	10,372,000	14,325,000	15,809,000	19,958,000	11,309,000	1,783,000
Operating Income	2,841,475,000	4,099,226,000	3,408,243,000	2,311,382,000	2,389,887,000	1,180,586,000
Operating Margin %	24.31%	28.10%	24.75%	20.76%	25.39%	17.22%
SGA Expense	1,491,093,000	1,459,113,000	1,532,614,000	1,412,760,000	1,131,809,000	948,281,000
Net Income	1,966,236,000	2,840,629,000	2,305,997,000	1,524,093,000	1,560,123,000	599,394,000
Operating Cash Flow	3,449,971,000	4,832,844,000	4,439,412,000	3,057,757,000	2,662,496,000	1,870,151,000
Capital Expenditure	1,528,642,000	1,178,656,000	943,982,000	1,449,234,000	1,508,593,000	2,023,981,000
EBITDA	3,924,666,000	5,138,481,000	4,422,191,000	3,213,186,000	3,172,176,000	1,857,689,000
Return on Assets %	9.07%	12.60%	10.27%	6.86%	5.86%	1.95%
Return on Equity %	28.02%	38.18%	31.31%	20.44%	18.39%	6.66%
Debt to Equity	1.37	1.37	1.22	1.43	1.21	1.57

CONTACT INFORMATION:

Phone: 702 414-1000   Fax: 702 414-4884
Toll-Free:
Address: 3355 Las Vegas Blvd. S., Las Vegas, NV 89109 United States

STOCK TICKER/OTHER:

Stock Ticker: LVS   Exchange: NYS
Employees: 48,500   Fiscal Year Ends: 12/31
Parent Company:

SALARIES/BONUSES:

Top Exec. Salary: Bonus: $
$3,069,231
Second Exec. Salary: Bonus: $
$1,500,000

OTHER THOUGHTS:

Estimated Female Officers or Directors:

Hot Spot for Advancement for Women/Minorities:

LaSalle Hotel Properties

NAIC Code: 0

www.lasallehotels.com

TYPES OF BUSINESS:

Real Estate Investment Trust
Luxury Hotels
Property Investment

BRANDS/DIVISIONS/AFFILIATES:

Le Montrose Suite Hotel
Indianapolis Marriott Downtown
Westin Copley Place
Liaison Capitol Hill (The)
Hotel Solamar
Westin Michigan Avenue
Westin Market Street (The)

CONTACTS: *Note: Officers with more than one job title may be intentionally listed here more than once.*

Michael Barnello, CEO
Stuart Scott, Chairman of the Board
Alfred Young, COO
Bruce Riggins, Treasurer
William Mccalmont, Trustee
Darryl Hartley-Leonard, Trustee
Jeffrey Foland, Trustee
Denise Coll, Trustee
Donald Washburn, Trustee

GROWTH PLANS/SPECIAL FEATURES:

LaSalle Hotel Properties is a self-managed and self-administered real estate investment trust (REIT) that owns and invests in luxury hotels. The firm primarily works with hotels located in convention, resort or major urban business markets. It currently owns interests in 47 upscale and luxury full-service hotels with over 12,000 rooms/suites located in 10 states and Washington, D.C. The firm is comprised of hotel investors, asset managers and financial experts rather than hotel operators, and its hotels are operated and managed by unrelated hotel operating companies such as Westin Hotels and Resorts, Noble House Hotels & Resorts, Hilton Hotels Corporation, Hyatt Hotels Corporation, Kimpton Hotel & Restaurant Group LLC and others. The firm seeks to improve revenue growth through renovations, redevelopment and/or expansions; brand or franchise conversion; acquisitions of appropriate full-service hotels in the U.S. and abroad; and selective development of hotel properties in favorable upscale markets. It is currently focused on acquiring hotels in eight primary urban markets: Boston, Chicago, Los Angeles, New York, San Diego, San Francisco, Seattle and Washington, D.C. Properties owned by the company include the Le Montrose Suite Hotel, located in West Hollywood; Indianapolis Marriott Downtown; The Liaison Capitol Hill in Washington D.C.; Westin Copley Place in Boston; Westin Michigan Avenue in Chicago; and Hotel Solamar located in San Diego. In January 2015, it acquired The Westin Market Street in San Francisco, California and renamed it Park Central San Francisco.

FINANCIAL DATA: *Note: Data for latest year may not have been available at press time.*

In U.S. $	2015	2014	2013	2012	2011	2010
Revenue	1,164,358,000	1,052,475,000	929,456,000	818,662,000	679,351,000	563,983,000
R&D Expense						
Operating Income	134,537,000	215,803,000	90,725,000	80,639,000	49,868,000	7,696,000
Operating Margin %	11.55%	20.50%	9.76%	9.85%	7.34%	1.36%
SGA Expense	831,242,000	765,094,000	86,492,000	72,908,000	60,265,000	55,524,000
Net Income	135,552,000	212,845,000	89,935,000	71,296,000	43,617,000	1,961,000
Operating Cash Flow	337,519,000	283,236,000	245,565,000	216,364,000	165,495,000	131,572,000
Capital Expenditure	582,228,000	294,019,000	421,921,000	457,818,000	588,677,000	490,039,000
EBITDA						
Return on Assets %	3.17%	5.42%	2.07%	1.48%	.49%	-1.13%
Return on Equity %	5.12%	8.69%	3.58%	2.49%	.80%	-1.81%
Debt to Equity	0.60	0.41	0.59	0.67	0.53	0.56

CONTACT INFORMATION:

Phone: 301 941-1500   Fax: 301 941-1553
Toll-Free:
Address: 7550 Wisconsin Avenue, 10/Fl, Bethesda, MD 20814 United States

STOCK TICKER/OTHER:

Stock Ticker: LHO
Employees: 35
Parent Company:

Exchange: NYS
Fiscal Year Ends: 12/31

SALARIES/BONUSES:

Top Exec. Salary: $815,000   Bonus: $
Second Exec. Salary: $510,167   Bonus: $

OTHER THOUGHTS:

Estimated Female Officers or Directors:
Hot Spot for Advancement for Women/Minorities:

Sales, profits and employees may be estimates. Financial information, benefits and other data can change quickly and may vary from those stated here.

Layne Christensen Company

NAIC Code: 237110

TYPES OF BUSINESS:

Construction & Civil Engineering Services
Water Treatment Plant Development
Drilling Services
Oil & Gas Field Services
Unconventional Natural Gas Production

BRANDS/DIVISIONS/AFFILIATES:

Layne Inliner LLC
Inliner Technologies and Liner Products
Inliner
ALLclear

CONTACTS: Note: Officers with more than one job title may be intentionally listed here more than once.

David Brown, Chairman of the Board
Jami Phillips, Chief Accounting Officer
Michael Caliel, Director
Steven Crooke, General Counsel
Ronald Thalacker, President, Divisional
Kevin Maher, President, Divisional
Kent Wartick, President, Divisional
Mauro Chinchelli, President, Divisional
Leslie Archer, President, Subsidiary
Larry Purlee, President, Subsidiary
J. Anderson, Senior VP
Lisa Curtis, Vice President, Divisional

GROWTH PLANS/SPECIAL FEATURES:

Layne Christensen Company is a global water management, construction and drilling firm. In addition, the company is a producer of unconventional natural gas for the energy market. The company operates through six business segments: water resources, Inliner, heavy civil, geoconstruction, mineral services and energy services. Water resources includes hydrologic design and construction, source of supply exploration, well and intake construction and well and pump rehabilitation. The Inliner segment provides a range of wastewater pipeline and structure rehabilitation services, focused around the company's trademarked Inliner cured-in-place pipe (CIPP) product. Layne Inliner, LLC is the licensee of the Inliner technology and owns and operates Inliner Technologies and Liner Products, a technology company and lining tubing manufacturer. The heavy civil division serves the needs of government agencies and industrial customers by managing design and construction processes of water and wastewater treatment facilities as well as pipeline construction. Geoconstruction offers specialized foundation construction services to heavy civil, industrial and commercial construction markets, focused primarily on soil stabilization and subterranean structural support for major civil engineering and construction projects, such as dams, tunnels and highways. Mineral services segment conducts primarily above ground drilling activities, including exploratory and definitional drilling. Energy services focuses on bringing responsible water management solutions to the exploration and production industry's growing water-related challenges related to hydraulic fracturing. This segment's trademarked ALLclear family of products provide comprehensive water management solutions concerning water sources such as water well drilling, water transfer, recycling, testing and storage. In May 2015, the firm agreed to sell its geoconstruction business to a subsidiary of Keller Foundations, LLC.

FINANCIAL DATA: Note: Data for latest year may not have been available at press time.

In U.S. $	2015	2014	2013	2012	2011	2010
Revenue	797,601,000	859,283,000	1,075,624,000	1,133,147,000	1,025,659,000	866,417,000
R&D Expense						
Operating Income	-57,791,000	-59,945,000	-33,605,000	-75,878,000	55,247,000	795,000
Operating Margin %	-7.24%	-6.97%	-3.12%	-6.69%	5.38%	.09%
SGA Expense	122,240,000	146,457,000	163,962,000	167,157,000	142,808,000	124,749,000
Net Income	-110,151,000	-128,639,000	-36,651,000	-56,075,000	29,991,000	1,365,000
Operating Cash Flow	-23,102,000	-1,131,000	24,751,000	15,712,000	68,880,000	93,955,000
Capital Expenditure	16,211,000	34,409,000	77,503,000	70,826,000	67,203,000	44,825,000
EBITDA	-5,291,000	-10,916,000	50,969,000	21,967,000	109,230,000	66,871,000
Return on Assets %	-18.47%	-17.63%	-4.53%	-6.91%	3.87%	.18%
Return on Equity %	-46.80%	-36.63%	-8.50%	-11.80%	6.19%	.29%
Debt to Equity	0.72	0.37	0.24	0.11		0.01

CONTACT INFORMATION:

Phone: 281-475-2600   Fax: 281-475-2733
Toll-Free: 855-529-6301
Address: 1800 Hughes Landing Blvd. Ste. 700, The Woodlands, TX 77380 United States

STOCK TICKER/OTHER:

Stock Ticker: LAYN
Employees: 3,380
Parent Company:

Exchange: NAS
Fiscal Year Ends: 01/31

SALARIES/BONUSES:

Top Exec. Salary: $368,058   Bonus: $
Second Exec. Salary: $345,923   Bonus: $

OTHER THOUGHTS:

Estimated Female Officers or Directors:
Hot Spot for Advancement for Women/Minorities:

LB Foster Company

NAIC Code: 331110

www.lbfoster.com

TYPES OF BUSINESS:

Structural Metal Products Distribution
Railroad Materials

BRANDS/DIVISIONS/AFFILIATES:

L.B. Foster Rail Technologies, Inc
TEW Engineering Ltd
Inspection Oilfield Services
L.B. Pipe & Coupling Products LLC

CONTACTS: Note: Officers with more than one job title may be intentionally listed here more than once.

Robert Bauer, CEO
David Russo, CFO
Lee Foster, Chairman of the Board
Christopher Scanlon, Chief Accounting Officer
Patrick Guinee, General Counsel
John Kasel, Senior VP, Divisional
Konstantinos Papazoglou, Vice President, Divisional
Merry Brumbaugh, Vice President, Divisional
Samuel Fisher, Vice President, Divisional
Gregory Lippard, Vice President, Divisional
David Sauder, Vice President, Divisional
Brian Kelly, Vice President, Divisional

GROWTH PLANS/SPECIAL FEATURES:

L.B. Foster Company, founded in 1902, manufactures, fabricates and distributes products and services for the rail, construction, energy and utility industries. The company operates in three business segments: rail products and services; construction products; and tubular and energy services. The rail products and services segment provides a line of new and used rail, trackwork and related accessories to the railroad, mining and industrial markets. Subsidiary, L.B. Foster Rail Technologies, Inc. engineers, manufactures, and fabricates friction management products and application systems, railroad condition monitoring equipment, wheel impact load detection and railroad condition monitoring systems among others. The construction products division manufactures and sells steel sheet piling, H-bearing piling and pipe piling as well as providing rental sheet piling for foundation requirements. In addition, the segment supplies fabricated structural steel, bridge decking, bridge railing, expansion joints and other products for highway construction and repair. The tubular and energy services segment supplies coated pipes, precision measurement systems and upstream test and inspection services for natural gas pipelines and utilities. The firm also produces threaded pipe products for industrial water well and irrigation markets and sells micropiles for construction foundation repair and slope stabilization. L.B. Pipe & Coupling Products, LLC, a joint venture with L.B. Industries, Inc., manufactures couplings and other products for the energy, utility and construction markets. The company maintains 28 sales offices as well as 36 warehouses and plants located throughout the U.S., Canada and the U.K. In 2015, the firm acquired TEW Engineering Ltd., based in Nottingham, UK; and Inspection Oilfield Services, a Houston-based leader in non-destructive testing and inspection services.

L.B. Foster offers its employees benefits such as life, AD&D, disability, medical, dental and vision insurance; flexible spending accounts; profit sharing; a 401(k) plan; and an employee assistance plan.

FINANCIAL DATA: Note: Data for latest year may not have been available at press time.

In U.S. $	2015	2014	2013	2012	2011	2010
Revenue	624,523,000	607,192,000	597,963,000	588,541,000	590,926,000	475,050,000
R&D Expense						
Operating Income	28,760,000	37,082,000	41,571,000	23,321,000	31,795,000	31,995,000
Operating Margin %	4.60%	6.10%	6.95%	3.96%	5.38%	6.73%
SGA Expense	92,648,000	79,814,000	71,256,000	66,651,000	67,238,000	42,588,000
Net Income	-44,445,000	25,656,000	29,290,000	16,188,000	22,895,000	20,492,000
Operating Cash Flow	56,172,000	66,739,000	14,155,000	26,466,000	30,670,000	59,482,000
Capital Expenditure	14,913,000	17,138,000	9,674,000	7,160,000	11,939,000	6,160,000
EBITDA	-19,525,000	52,145,000	54,602,000	37,345,000	46,921,000	43,019,000
Return on Assets %	-8.37%	5.64%	7.14%	4.11%	6.03%	5.75%
Return on Equity %	-14.36%	7.86%	9.69%	5.80%	8.71%	8.39%
Debt to Equity	0.59	0.07		0.05		

CONTACT INFORMATION:

Phone: 412 928-3400   Fax:
Toll-Free: 800-255-4500
Address: 415 Holiday Dr., Pittsburgh, PA 15220 United States

STOCK TICKER/OTHER:

Stock Ticker: FSTR   Exchange: NAS
Employees: 1,113   Fiscal Year Ends: 12/31
Parent Company:

SALARIES/BONUSES:

Top Exec. Salary: $609,250   Bonus: $
Second Exec. Salary: Bonus: $
$297,408

OTHER THOUGHTS:

Estimated Female Officers or Directors: 3
Hot Spot for Advancement for Women/Minorities: Y

LBA Realty LLC

www.lbarealty.com

NAIC Code: 531120

TYPES OF BUSINESS:

Real Estate Operations & Development
Real Estate Management

BRANDS/DIVISIONS/AFFILIATES:

CONTACTS: *Note: Officers with more than one job title may be intentionally listed here more than once.*

David Lichtenstein, CEO
Mitchell C. Hochberg, Pres.
Donna Brandin, CFO
Jennifer D. Cline, Dir.-Human Resources & Employment
Paul Thometz, Dir.-Design & Construction
Perry Schonfeld, Principal-Oper.
Phil Belling, Principal-Strategy & Acquisitions
Tom Rutherford, Sr. VP-Finance
Alice Wilson, VP-Corp. Svcs.
Melanie Colbert, Sr. VP-Property Mgmt. Svcs.
Bill Kearns, Principal-Acquisitions & Leasing
Steve Layton, Principal-Corp. Svcs.
David Lichtenstein, Chmn.

GROWTH PLANS/SPECIAL FEATURES:

LBA Realty LLC is a leading real estate investment and management company based in California. The firm has multiple properties throughout Arizona, Nevada, California, Colorado, Oregon, Texas, Utah and Washington. LBA's services include property management, primarily focusing on acquisitions and dispositions of both office and industrial properties; construction management, focused on building and construction, office move-in and property management; and build-to-suit, a development program that facilitates corporate customer expansions, consolidations and relocations. The company's portfolio consists of office properties, such as high-rise, mid-rise, low-rise, major signage, LEED (Leadership in Energy and Environmental Design), campus, retail amenity, excess parking and single tenant buildings; and industrial, such as large, mixed-use properties, single and multi-tenant office buildings and industrial properties such as warehouse, distribution, light manufacturing and business park facilities. The firm has three primary acquisition targets: office and industrial properties, corporate-owned facilities and development. LBA has acquired more than $6 billion in real estate assets since 1995. Among the company's top 40 tenants include AT&T, Colorado State Bank, Comcast, DeVry, Ingram Micro, KPMG, Microsoft, Prudential, Sony, the State of California, Tesla, US Bank, Verizon and Zurich Insurance.

The firm offers employees medical, dental and vision coverage; a 401(k) plan; life insurance; gym membership discounts; a college 529 savings plan; a long-term care program; and a flexible spending account program.

FINANCIAL DATA: *Note: Data for latest year may not have been available at press time.*

In U.S. $	2015	2014	2013	2012	2011	2010
Revenue						
R&D Expense						
Operating Income						
Operating Margin %						
SGA Expense						
Net Income						
Operating Cash Flow						
Capital Expenditure						
EBITDA						
Return on Assets %						
Return on Equity %						
Debt to Equity						

CONTACT INFORMATION:

Phone: 212-616-9969   Fax:
Toll-Free:
Address: 460 Park Avenue, Ste. 1300, New York, NY 10022 United States

STOCK TICKER/OTHER:

Stock Ticker: Private
Employees: 195
Parent Company:

Exchange:
Fiscal Year Ends: 12/31

SALARIES/BONUSES:

Top Exec. Salary: $ Bonus: $
Second Exec. Salary: $ Bonus: $

OTHER THOUGHTS:

Estimated Female Officers or Directors: 3
Hot Spot for Advancement for Women/Minorities: Y

Lend Lease Corporation Limited

www.lendlease.com.au

NAIC Code: 237000

TYPES OF BUSINESS:

Lessors of Nonresidential Buildings (except Miniwarehouses)

BRANDS/DIVISIONS/AFFILIATES:

GROWTH PLANS/SPECIAL FEATURES:

Lend Lease Corporation Limited is a vertically integrated property group. The group operates a regional management structure focused on four major geographic regions: Australia, Asia, Europe and the Americas. The regional business units generate earnings from four lines of business, as follows. Property development, which involves the development of urban communities, inner-city mixed-use developments, apartments, retirement, retail, commercial and health care assets; construction, which involves project management, building, engineering and construction services; investment management, which involves property and infrastructure investment management, property management and asset management and includes the group's ownership interests in property and infrastructure investments; and infrastructure development, which arranges, manages and invests in Public Private Partnership (PPP) projects.

CONTACTS: *Note: Officers with more than one job title may be intentionally listed here more than once.*

Steve McCann, CEO
Dan Labbad, COO
Tony Lombardo, CFO

FINANCIAL DATA: *Note: Data for latest year may not have been available at press time.*

In U.S. $	2015	2014	2013	2012	2011	2010
Revenue	10,060,360,000	10,567,560,000	9,256,300,000	9,014,692,000	6,735,252,000	7,930,654,000
R&D Expense						
Operating Income	434,726,400	656,085,900	292,456,400	276,471,000	310,953,800	185,582,700
Operating Margin %	4.32%	6.20%	3.15%	3.06%	4.61%	2.34%
SGA Expense	1,756,185,000	1,818,604,000	1,821,573,000	1,737,535,000	1,282,941,000	1,085,408,000
Net Income	470,883,800	626,398,700	419,882,800	381,670,100	375,123,700	263,073,800
Operating Cash Flow	-126,817,400	626,018,100	72,238,720	-35,091,730	-32,123,010	173,555,600
Capital Expenditure	94,922,740	95,531,710	69,346,130	61,886,280	42,018,730	66,605,770
EBITDA	749,562,400	936,286,800	598,462,400	549,364,400	615,665,700	445,078,800
Return on Assets %	3.56%	5.49%	4.13%	4.07%	4.19%	3.51%
Return on Equity %	12.34%	17.91%	13.40%	13.36%	14.18%	12.01%
Debt to Equity	0.43	0.49	0.47	0.37	0.52	0.43

CONTACT INFORMATION:

Phone: 61 0292366111   Fax: 61 0292522192
Toll-Free:
Address: Level 4, 30 The Bond, 30 Hickson Rd, Millers Point, NSW 2000 Australia

STOCK TICKER/OTHER:

Stock Ticker: LLESY   Exchange: PINX
Employees: 12,443   Fiscal Year Ends: 06/30
Parent Company:

SALARIES/BONUSES:

Top Exec. Salary: $ Bonus: $
Second Exec. Salary: $ Bonus: $

OTHER THOUGHTS:

Estimated Female Officers or Directors:
Hot Spot for Advancement for Women/Minorities:

LendingTree LLC

NAIC Code: 522310

TYPES OF BUSINESS:

Consumer Loans & Mortgages Internet Portal
Online Financial Information & Tools
Online Realty Services
Settlement Services
Online Homeowner Resources

BRANDS/DIVISIONS/AFFILIATES:

LendingTree Inc
LendingTree.com
Tree.com Inc

CONTACTS: Note: Officers with more than one job title may be intentionally listed here more than once.

Douglas Lebda, CEO
Gabe Dalporto, CFO
Kamelia Dianati, Sr. VP-Tech.
Claudette Hampton, Sr. VP-Admin.

GROWTH PLANS/SPECIAL FEATURES:

LendingTree, LLC is an online lending and realty services exchange. The company operates LendingTree.com a leading online loan facilitator that brings consumers together with a network of lenders competing for their business. LendingTree.com provides consumers with resources to purchase and refinance home loans, home equity loans, lines of credit, auto loans, personal loans, business loans and credit cards. It also offers these services over the phone at 800-555-TREE. Customers begin by completing the firm's online loan request, which requires information concerning desired loans and personal financial information. The customer's data and credit scores are then automatically compared to the underwriting criteria of participating lenders. Customers are then connected with multiple lenders who provide customized loan offers, and are then able to discuss details with those of their choosing. LendingTree.com is a free, no-obligation service, and has financial calculators, interactive coach tools, monthly newsletters and lender ratings and reviews. Lenders pay the firm for the opportunity to compete for customer business. LendingTree LLC operates as a subsidiary of LendingTree, Inc., formerly Tree.com, Inc.

Employees of Tree.com receive medical, dental and vision insurance; flexible spending accounts; life and AD&D insurance; a 401(k); adoption assistance; and education assistance.

FINANCIAL DATA: Note: Data for latest year may not have been available at press time.

In U.S. $	2015	2014	2013	2012	2011	2010
Revenue	254,200,000	167,350,000	139,240,000	77,443,000		
R&D Expense						
Operating Income						
Operating Margin %						
SGA Expense						
Net Income	48,047,000	9,362,000	3,947,000	46,625,000		
Operating Cash Flow						
Capital Expenditure						
EBITDA						
Return on Assets %						
Return on Equity %						
Debt to Equity						

CONTACT INFORMATION:

Phone: 866-501-2397   Fax: 704-541-1824
Toll-Free: 800-813-4620
Address: 11115 Rushmore Dr., Charlotte, NC 28277 United States

STOCK TICKER/OTHER:

Stock Ticker: Subsidiary   Exchange:
Employees: 742   Fiscal Year Ends: 12/31
Parent Company: LendingTree Inc

SALARIES/BONUSES:

Top Exec. Salary: $ Bonus: $
Second Exec. Salary: $ Bonus: $

OTHER THOUGHTS:

Estimated Female Officers or Directors: 2
Hot Spot for Advancement for Women/Minorities: Y

Lennar Corporation

NAIC Code: 236117

www.lennar.com

TYPES OF BUSINESS:

Construction, Home Building and Residential
Mortgages
Title Insurance & Services

BRANDS/DIVISIONS/AFFILIATES:

Universal American Mortgage Company LLC
Eagle Home Mortgage LLC
Railto Investments
Lennar

CONTACTS: Note: Officers with more than one job title may be intentionally listed here more than once.

Stuart Miller, CEO
Bruce Gross, CFO
David Collins, Chief Accounting Officer
Jonathan Jaffe, COO
Richard Beckwitt, President
Mark Sustana, Secretary
Diane Bessette, Vice President

GROWTH PLANS/SPECIAL FEATURES:

Lennar Corporation is a U.S. homebuilder and provider of financial services operating in 20 states. The firm sells single-family attached and detached homes and, to a lesser extent, multi-level residential buildings primarily under the Lennar brand name in communities targeted to first-time, move-up and active adult homebuyers. The company also purchases, develops and sells residential land. Lennar divides its homebuilding operations into six segments: East (which includes Florida, Georgia, Maryland, New Jersey, North Carolina, South Carolina and Virginia); Central (including Arizona, Colorado and Texas, excluding Houston); West (California and Nevada); Houston, Texas; Southeast Florida; and other (including Illinois, Minnesota, Tennessee, Oregon and Washington). Lennar's homes have an average sale price of about $344,000. In 2015, it delivered 24,292 homes to buyers. Lennar generally supervises and controls the development of land and the design and building of its residential communities with a relatively small labor force, hiring subcontractors for site improvements and virtually all of the work involved in the construction of homes. Through its financial services subsidiaries, Universal American Mortgage Company, LLC and Eagle Home Mortgage, LLC, the firm provides mortgage financing, title insurance and closing services for both buyers of its homes and third parties. Lennar's subsidiaries provide loans to 82% of its homebuyers who obtain mortgage financing in areas where it offers services. The Rialto Investments division provides advisory services, ongoing asset management services and acquisition and monetization services related to distressed loans and securities portfolios. The Lennar multifamily segment, currently under development, will focus on developing a portfolio of multifamily rental properties in select U.S. markets.

The company offers employees medical, dental and vision insurance; home and auto insurance; mortgage and title benefits; short-and long-term disability; and life insurance.

FINANCIAL DATA: Note: Data for latest year may not have been available at press time.

In U.S. $	2015	2014	2013	2012	2011	2010
Revenue	9,474,008,000	7,779,812,000	5,935,095,000	4,104,706,000	3,095,385,000	3,074,022,000
R&D Expense						
Operating Income	1,086,016,000	922,038,000	685,836,000	322,176,000	103,934,000	124,367,000
Operating Margin %	11.46%	11.85%	11.55%	7.84%	3.35%	4.04%
SGA Expense	216,244,000	177,161,000	146,060,000	127,338,000	95,256,000	93,926,000
Net Income	802,894,000	638,916,000	479,674,000	679,124,000	92,199,000	95,261,000
Operating Cash Flow	-419,646,000	-788,488,000	-807,714,000	-424,648,000	-259,135,000	274,228,000
Capital Expenditure	17,623,000	14,278,000	8,126,000	2,822,000	9,936,000	5,062,000
EBITDA	1,265,736,000	1,044,877,000	806,203,000	344,548,000	210,124,000	178,670,000
Return on Assets %	5.80%	5.20%	4.43%	6.95%	1.02%	1.18%
Return on Equity %	15.16%	14.03%	12.65%	22.22%	3.47%	3.77%
Debt to Equity	0.98	0.93	0.99	1.18	1.40	1.48

CONTACT INFORMATION:

Phone: 305 559-4000   Fax: 305 227-7115
Toll-Free: 800-741-4663
Address: 700 NW 107th Ave., Ste. 400, Miami, FL 33172 United States

STOCK TICKER/OTHER:

Stock Ticker: LEN
Employees: 7,749
Parent Company:

Exchange: NYS
Fiscal Year Ends: 11/30

SALARIES/BONUSES:

Top Exec. Salary: Bonus: $
$1,000,000
Second Exec. Salary: Bonus: $
$800,000

OTHER THOUGHTS:

Estimated Female Officers or Directors: 3

Hot Spot for Advancement for Women/Minorities: Y

Sales, profits and employees may be estimates. Financial information, benefits and other data can change quickly and may vary from those stated here.

Lexington Realty Trust

www.lxp.com

NAIC Code: 0

TYPES OF BUSINESS:

Real Estate Investment Trust
Investment & Asset Management
Property Management & Leasing
Office & Industrial Properties
Retail Properties
Construction Financing

BRANDS/DIVISIONS/AFFILIATES:

Lepercq Corporate Income Fund LP
Lepercq Corporate Income Fund II LP

CONTACTS: Note: Officers with more than one job title may be intentionally listed here more than once.

T. Eglin, CEO
Patrick Carroll, CFO
Beth Boulerice, Chief Accounting Officer
Joseph Bonventre, Secretary
Claire Koeneman, Trustee
Lawrence Gray, Trustee
Kevin Lynch, Trustee
Richard Frary, Trustee
Harold First, Trustee
E. Roskind, Trustee
Richard Rouse, Trustee

GROWTH PLANS/SPECIAL FEATURES:

Lexington Realty Trust is a self-managed and self-administered real estate investment trust (REIT) that primarily acquires, owns and manages a portfolio of net leased office, industrial and retail properties. The firm also acquires and holds investments in loan assets and debt securities related to real estate. Lexington's consolidated portfolio consists of interests in 215 properties totaling 42.3 million square feet in 40 states. The majority of the company's properties are subject to triple-net or similar leases, in which the tenant bears all or substantially all of the costs and cost increases for real estate taxes, utilities, insurance and ordinary repairs. The company has diversified its portfolio by geographical location, tenant industry segment, lease term expiration and property type in an effort to insulate itself from regional recession, industry specific downturns and price fluctuations by property type. Lexington is structured as an umbrella partnership REIT, with a portion of its business conducted through its two operating subsidiaries: Lepercq Corporate Income Fund L.P. and Lepercq Corporate Income Fund II L.P. During 2015, the firm sold an office property in Orlando, Florida and sold the Transamerica Tower in Baltimore, Maryland.

Lexington offers its employees medical, dental, life and disability insurance; a 401(k) plan with company match; a stock purchase plan; flexible spending accounts; a tuition assistance program; and a 529 college savings plan.

FINANCIAL DATA: Note: Data for latest year may not have been available at press time.

In U.S. $	2015	2014	2013	2012	2011	2010
Revenue	430,839,000	424,372,000	398,440,000	344,879,000	326,914,000	342,855,000
R&D Expense						
Operating Income	178,710,000	177,607,000	133,000,000	98,834,000	80,058,000	85,106,000
Operating Margin %	41.47%	41.85%	33.38%	28.65%	24.48%	24.82%
SGA Expense	29,276,000	28,255,000	28,973,000	23,956,000	22,211,000	22,487,000
Net Income	111,703,000	93,104,000	1,630,000	180,316,000	-79,584,000	-32,960,000
Operating Cash Flow	244,930,000	214,672,000	206,304,000	163,810,000	180,137,000	164,751,000
Capital Expenditure	516,194,000	271,229,000	602,402,000	261,297,000		
EBITDA	368,950,000	312,917,000	264,372,000	429,038,000	194,972,000	271,497,000
Return on Assets %	2.76%	2.28%	-.39%	4.82%	-3.23%	-1.68%
Return on Equity %	7.67%	6.13%	-1.13%	16.38%	-11.03%	-6.08%
Debt to Equity	1.54	1.41	1.35	1.64	1.67	1.44

CONTACT INFORMATION:

Phone: 212 692-7200   Fax: 212 594-6600
Toll-Free:
Address: 1 Penn Plz., Ste. 4015, New York, NY 10119 United States

STOCK TICKER/OTHER:

Stock Ticker: LXP
Employees: 48
Parent Company:

Exchange: NYS
Fiscal Year Ends: 12/31

SALARIES/BONUSES:

Top Exec. Salary: $640,000   Bonus: $
Second Exec. Salary: $520,000   Bonus: $

OTHER THOUGHTS:

Estimated Female Officers or Directors: 2
Hot Spot for Advancement for Women/Minorities: Y

Liberty Property Trust

NAIC Code: 0

www.libertyproperty.com

TYPES OF BUSINESS:

Real Estate Investment Trust
Industrial & Office Properties
Property Management
Property Development

BRANDS/DIVISIONS/AFFILIATES:

Liberty Property LP

CONTACTS: Note: Officers with more than one job title may be intentionally listed here more than once.

William Hankowsky, CEO
George Alburger, CFO
Michael Hagan, Executive VP
Herman Fala, General Counsel
Antonio Fernandez, Independent Trustee
Thomas Deloach, Trustee
David Lingerfelt, Trustee
Daniel Garton, Trustee
Katherine Dietze, Trustee
M. Lachman, Trustee
Frederick Buchholz, Trustee
Fredric Tomczyk, Trustee

GROWTH PLANS/SPECIAL FEATURES:

Liberty Property Trust (LPT) is a self-administered and self-managed real estate investment trust (REIT). Nearly all of the firm's assets and operations are owned and conducted through its subsidiary, Liberty Property LP. LPT's portfolio consists of 482 industrial and 128 office properties, totaling 89.7 million square feet, within the U.S. and the U.K. Through joint ventures, the firm owns interests in an additional 81 properties (48 industrial and 33 office), totaling 14.0 million square feet. LPT groups its properties according to geographic segments: Carolinas, Chicago/Milwaukee, Houston, Lehigh/Central PA, Minnesota, Orlando, Philadelphia, Richmond/Hampton Roads, Southeastern PA, South Florida, Tampa, and the U.K. Industrial properties include warehouse, distribution, service, assembly, light manufacturing and research and development facilities. Office properties include single- and multi-story office buildings located principally in suburban mixed-use developments or office parks. The company provides leasing, property management, development, acquisition and other tenant-related services for its properties. LPT also owns 27 properties under development and 1,751 acres of developable land. During 2015, the company acquired a total of 38 acres in the northeast Atlanta submarket in Buford, Georgia and three buildings and 139 acres of additional land in La Porte, Texas. That same year, it sold 713,585 square foot industrial property in Orlando; and sold its entire portfolio of 41 properties in the Horsham, Pennsylvania submarket totaling 2.4 million square feet of space and 20 acres of land.

LPT offers its employees medical, dental and vision insurance; short- and long-term disability; a 401(k); flexible spending accounts; a work-life balance support program; wellness programs; an employee stock purchase plan; educational assistance; employee discounts; commuter assistance; life insurance; and dependent, supplemental and AD&D insurance.

FINANCIAL DATA: Note: Data for latest year may not have been available at press time.

In U.S. $	2015	2014	2013	2012	2011	2010
Revenue	808,773,000	792,631,000	645,930,000	685,552,000	667,594,000	746,830,000
R&D Expense						
Operating Income	257,132,000	257,423,000	203,047,000	241,705,000	245,991,000	280,145,000
Operating Margin %	31.79%	32.47%	31.43%	35.25%	36.84%	37.51%
SGA Expense	68,710,000	63,327,000	74,564,000	64,730,000	59,370,000	52,850,000
Net Income	238,039,000	217,910,000	209,738,000	137,436,000	183,999,000	127,762,000
Operating Cash Flow	385,366,000	336,484,000	315,965,000	317,166,000	317,724,000	298,957,000
Capital Expenditure	550,637,000	227,428,000	1,695,372,000	550,917,000	436,909,000	139,190,000
EBITDA	610,088,000	556,197,000	426,749,000	422,538,000	431,127,000	468,813,000
Return on Assets %	3.61%	3.25%	3.50%	2.70%	3.66%	2.48%
Return on Equity %	7.96%	7.18%	8.18%	6.55%	8.79%	6.07%
Debt to Equity	1.06	1.04	1.07	1.22	1.05	1.13

CONTACT INFORMATION:

Phone: 610 648-1700   Fax: 610 644-4129
Toll-Free:
Address: 500 Chesterfield Pkwy., Great Valley Corp. Ctr., Malvern, PA 19355 United States

STOCK TICKER/OTHER:

Stock Ticker: LPT   Exchange: NYS
Employees: 431   Fiscal Year Ends: 12/31
Parent Company:

SALARIES/BONUSES:

Top Exec. Salary: $705,000   Bonus: $
Second Exec. Salary: $455,000   Bonus: $ 500

OTHER THOUGHTS:

Estimated Female Officers or Directors: 7
Hot Spot for Advancement for Women/Minorities: Y

Lightstone Group LLC (The)

www.lightstonegroup.com

NAIC Code: 531110

TYPES OF BUSINESS:

Real Estate Company
Outlet Retail Properties
Office Building Properties
Lodging Facilities Properties
Real Estate Investment Trust

BRANDS/DIVISIONS/AFFILIATES:

Beacon Property Management

GROWTH PLANS/SPECIAL FEATURES:

The Lightstone Group LLC is one of the largest privately-owned real estate companies in the U.S. It currently owns a portfolio of diversified properties in 24 U.S. states, with holdings in excess of 11,000 residential units and 3,200 hotel keys. The company's assets also include roughly 8 million square feet of commercial space, as well as more than 12,000 land lots across the country. The firm's commercial interests generally include retail and industrial properties such as enclosed malls, outlet centers, open air centers and warehouses in excess of 150,000 square feet. The Lightstone Group prefers to acquire property around $15-$25 million. Subsidiary Beacon Property Management invests in the multi-family residential housing sector, with corporate offices in the Northeast, Southeast and Midwest regions of the U.S.

CONTACTS: Note: Officers with more than one job title may be intentionally listed here more than once.

David Lichtenstein, CEO
Mitchell C. Hochberg, Pres.
Donna Brandin, CFO
Jennifer D. Cline, Exec. VP-Human Resources
Dennis Freed, Sr. VP-Dev. & Construction
Joseph E. Teichman, General Counsel
Donna Brandin, Treas.
Akiva Elazary, VP-Acquisitions
Kasra Sanandaji, Sr. VP-Investments
Bruno de Vinck, Sr. VP-Special Projects
Christian (Gabe) Gabrielsen, Pres., Capital Markets Div.
David Lichtenstein, Chmn.

FINANCIAL DATA: Note: Data for latest year may not have been available at press time.

In U.S. $	2015	2014	2013	2012	2011	2010
Revenue						
R&D Expense						
Operating Income						
Operating Margin %						
SGA Expense						
Net Income						
Operating Cash Flow						
Capital Expenditure						
EBITDA						
Return on Assets %						
Return on Equity %						
Debt to Equity						

CONTACT INFORMATION:

Phone: 212-616-9969   Fax:
Toll-Free:
Address: 460 Park Ave., Ste. 1300, New York, NY 10022 United States

STOCK TICKER/OTHER:

Stock Ticker: Private   Exchange:
Employees: 450   Fiscal Year Ends: 12/31
Parent Company:

SALARIES/BONUSES:

Top Exec. Salary: $ Bonus: $
Second Exec. Salary: $ Bonus: $

OTHER THOUGHTS:

Estimated Female Officers or Directors: 2
Hot Spot for Advancement for Women/Minorities:

Lillibridge Healthcare Real Estate Trust   www.lillibridge.com

NAIC Code: 0

TYPES OF BUSINESS:
Real Estate Investment Trust
Property Management
Health Care Properties
Consulting Services

BRANDS/DIVISIONS/AFFILIATES:
Ventas Inc
American Realty Capital Healthcare Trust Inc

CONTACTS: Note: Officers with more than one job title may be intentionally listed here more than once.
Todd W. Lillibridge, CEO
Todd W. Lillibridge, Pres.
Carla M. Lyons, VP-Mktg.
Michael Lincoln, Exec. VP-Bus. Dev.
Carla M. Lyons, Contact-Media
Sonya Brizzolara, VP-Finance & Reporting
Vince Cozzi, Chief Investment Officer
John Montgomery, Exec. VP-Facility Dev.
Chuck Fendrich, Exec. VP-Property Mgmt. & Leasing
Kevin Geraghty, Exec. VP-Asset Mgmt.
Todd W. Lillibridge, Chmn.

GROWTH PLANS/SPECIAL FEATURES:
Lillibridge Healthcare Real Estate Trust, a wholly-owned subsidiary of Ventas Inc., is a leading national health care real estate investment trust (REIT) that focuses on nonprofit hospitals and other health care facilities. The company advises, acquires, develops and manages its properties. The firm's operations, including owned and managed property, encompasses medical office facilities across 32 states. The business serves hospital and health systems and physicians nationwide. Lillibridge's acquisition branch purchases a client's medical office buildings. After a purchase, the development branch of the company plans the campus, physician recruitment and other operations associated with the new facility or its renovation. Lillibridge also manages the property for clients and handles compliance issues that consistently come up in the health care field when new federal and state laws are passed. The firm has developed a compliance risk-assessment checklist and paperwork that aids clients in evaluating and learning about compliance issues. Lillibridge's consulting services help clients inventory strategic and non-strategic real estate assets, improve performance through operations assessment, determine the impact of a monetization on the financial position of the hospital and create an ambulatory network or leasing strategy. In early 2015, the firm acquired American Realty Capital Healthcare Trust, Inc.

FINANCIAL DATA: Note: Data for latest year may not have been available at press time.

In U.S. $	2015	2014	2013	2012	2011	2010
Revenue						
R&D Expense						
Operating Income						
Operating Margin %						
SGA Expense						
Net Income						
Operating Cash Flow						
Capital Expenditure						
EBITDA						
Return on Assets %						
Return on Equity %						
Debt to Equity						

CONTACT INFORMATION:
Phone: 312-408-1370   Fax: 312-346-2701
Toll-Free: 877-545-5430
Address: 353 N. Clark St., Ste. 3300, Chicago, IL 60654 United States

STOCK TICKER/OTHER:
Stock Ticker: Subsidiary   Exchange:
Employees: 237   Fiscal Year Ends: 12/31
Parent Company: Ventas Inc

SALARIES/BONUSES:
Top Exec. Salary: $ Bonus: $
Second Exec. Salary: $ Bonus: $

OTHER THOUGHTS:
Estimated Female Officers or Directors: 7
Hot Spot for Advancement for Women/Minorities: Y

Link Real Estate Investment Trust (The)
www.thelinkreit.com

NAIC Code: 0

TYPES OF BUSINESS:
Real Estate Investment Trust
Retail Development
Property Management

BRANDS/DIVISIONS/AFFILIATES:
Link Management Limited (The)
EC Mall

CONTACTS:
Note: Officers with more than one job title may be intentionally listed here more than once.

George Kwok Lung Hongchoy, CEO
Gordon Wu Chi Ping, COO
Andy Cheung Lee Ming, CFO
Albert Yeung Shing Wo, Dir.-Human Resources
Ricky Chan Ming Tak, Dir.-Legal
Gordon Chi Ping Wu, Head-Oper. & Property Mgmt.
Poon Kai Tik, Dir.-Corp. Dev.
Peionie Po Yan Kong, Head-Leasing
Dick Leung Yuen Dick, Dir.-Projects & Asset Dev.
Nicholas Robert Sallnow-Smith, Chmn.

GROWTH PLANS/SPECIAL FEATURES:
The Link Real Estate Investment Trust is one of Hong Kong's largest real estate investment trusts (REIT), with a portfolio that accounts for 9.5% of Hong Kong's retail market. The company's portfolio consists of retail and car park facilities that total 11 million square feet of retail space as well as 76,000 individual car spaces. The firm's facilities encompass a variety of tenants, including well known retail and restaurant brands. The company's property management activities are conducted through subsidiary The Link Management Limited. The Link's strategy is to invest in properties in Hong Kong that are primarily for car park and retail usage and to increase their value by enhancing physical structure and customer service as well as developing promotional activities. Recent asset enhancement projects include introducing more popular restaurants at Hau Tak Shopping Centre, remodeling vacant floors into shopping areas for the Tsz Wan Shan Shopping Centre, aligning the walkway with retail outlets and adding a coffee shop at Lung Cheung Mall and converting a market to a retail area and therefore adding more shopping space to the Ming Tak Shopping Centre. During 2015, the firm acquired government land in Kwun Tong; the EC Mall in Beijing; and Corporate Avenue 1 & 2 in Shanghai. In early 2016, the company won the bid to acquire a premium commercial tower with retail podium at No. 700 Nathan Road, Kowloon in core commercial district in Hong Kong.

The company offers employees retirement savings, health coverage, life insurance, annual leave time and performance-based incentives.

FINANCIAL DATA:
Note: Data for latest year may not have been available at press time.

In U.S. $	2015	2014	2013	2012	2011	2010
Revenue	995,722,100	922,490,200	838,815,000	764,809,500	690,159,400	643,358,000
R&D Expense						
Operating Income	3,601,129,000	2,375,525,000	566,387,100	504,887,700	2,098,972,000	1,678,275,000
Operating Margin %	361.66%	257.51%	67.52%	66.01%	304.12%	260.86%
SGA Expense	56,342,170	28,622,340	28,751,270	34,682,020	22,691,580	15,471,530
Net Income	3,510,749,000	2,231,124,000	2,715,512,000	1,237,981,000	1,970,042,000	1,336,096,000
Operating Cash Flow	625,694,700	600,682,300	541,374,800	486,064,000	433,847,600	386,530,500
Capital Expenditure	3,223,236	2,320,730	3,223,236	3,481,095	95,536,720	97,470,660
EBITDA	3,663,788,000	2,372,173,000	2,860,171,000	1,373,614,000	2,102,582,000	1,680,467,000
Return on Assets %	21.22%	16.25%	23.57%	12.97%	23.93%	19.26%
Return on Equity %	25.39%	19.44%	29.17%	15.30%		
Debt to Equity	0.12	0.10	0.14	0.20		

CONTACT INFORMATION:
Phone: 852 21751800   Fax: 852 21751938
Toll-Free:
Address: 100 How Ming St., AXA Tower, 33/F, Kwun Tong,   Hong Kong

STOCK TICKER/OTHER:
Stock Ticker: LKREF
Employees: 930
Parent Company:

Exchange: PINX
Fiscal Year Ends: 03/31

SALARIES/BONUSES:
Top Exec. Salary: $ Bonus: $
Second Exec. Salary: $ Bonus: $

OTHER THOUGHTS:
Estimated Female Officers or Directors: 2
Hot Spot for Advancement for Women/Minorities:

Loews Hotels Holding Corporation   www.loewshotels.com

NAIC Code: 721110

TYPES OF BUSINESS:

Hotels, Luxury
Hotel Management Services

BRANDS/DIVISIONS/AFFILIATES:

Loews Corporation
Loews Miami Beach Hotel
Loews Santa Monica Beach Hotel
Loews Royal Pacific Resort at Universal Orlando
Loews Portofino Bay Hotel at Universal Orlando
Loews Don CeSar Beach Resort & Spa
Loews Boston Hotel
Loews Chicago Hotel

CONTACTS: *Note: Officers with more than one job title may be intentionally listed here more than once.*

Kirk Kinsell, CEO
Jack Adler, Pres.
Shawn Hauver, VP-Oper.
Lark-Marie Anton, Sr. VP-Public Rel. & Mktg. Comm.

GROWTH PLANS/SPECIAL FEATURES:

Loews Hotels Holding Corporation, a subsidiary of the Loews Corporation, currently has a portfolio of 26 owned and/or operated luxury hotels and resorts. Located in 21 cities throughout the U.S. and Canada, the firm's properties include the 790-room Loews Miami Beach Hotel in Florida, the 581-room Loews Philadelphia Hotel, the 347-room Loews Santa Monica Beach Hotel in Southern California and the 142-room Loews Hotel Vogue in Montreal. Loews Hotels operates three joint venture hotels with Universal Studios in Orlando, Florida: Loews Royal Pacific Resort at Universal Orlando, its largest hotel with 1,000 rooms; the 750-room Loews Portofino Bay Hotel at Universal Orlando; the 650-room Hard Rock Hotel at Universal Orlando; and the 277-room Loews Don CeSar Beach Resort in St. Pete Beach, Florida. Loews Hotels' business amenities include high-speed Internet access, a power breakfast with notable business leaders, notarization services, private dining rooms, boardrooms and concierge services. The YouFirst Loyalty Program rewards guests based on number of stays and offers free Internet access, late checkout, guaranteed rooms and upgrades for guests who visit at least twice a year. Loews Hotels offers facilities for weddings, meetings and special events; and special programs and services designed for people traveling with pets, children and teenagers. Recently, the company completed renovations on its Boston-based hotel, located in the city's Back Bay area. Following its reopening, the hotel was renamed the Loews Boston Hotel. In early 2015, the firm agreed to purchase the 158-room Mandarin Oriental San Francisco Hotel, and the new Loews Chicago Hotel opened for business.

FINANCIAL DATA: *Note: Data for latest year may not have been available at press time.*

In U.S. $	2015	2014	2013	2012	2011	2010
Revenue	604,000,000	475,000,000	380,000,000	397,000,000	337,000,000	308,000,000
R&D Expense						
Operating Income						
Operating Margin %						
SGA Expense						
Net Income	12,000,000	11,000,000	141,000,000	61,000,000	13,000,000	1,000,000
Operating Cash Flow						
Capital Expenditure						
EBITDA						
Return on Assets %						
Return on Equity %						
Debt to Equity						

CONTACT INFORMATION:

Phone: 212-521-2000   Fax: 212-521-2525
Toll-Free: 800-235-6397
Address: 667 Madison Ave., New York, NY 10021 United States

STOCK TICKER/OTHER:

Stock Ticker: Subsidiary   Exchange:
Employees: 2,169   Fiscal Year Ends: 12/31
Parent Company: Loews Corporation

SALARIES/BONUSES:

Top Exec. Salary: $ Bonus: $
Second Exec. Salary: $ Bonus: $

OTHER THOUGHTS:

Estimated Female Officers or Directors: 2
Hot Spot for Advancement for Women/Minorities: Y

Louis Berger Group Inc (The)

www.louisberger.com

NAIC Code: 541330

TYPES OF BUSINESS:

Engineering Services
Civil Engineering
Environmental Engineering
Transportation Infrastructure
Project Management
Consulting Services
Hydrologic Engineering
Seismic & Geotechnical Services

BRANDS/DIVISIONS/AFFILIATES:

Berger Group Holdings Inc
Ammann & Whitney Consulting Engineers Inc
Berger/ABAM Engineers Inc
Klohn Crippen Berger Ltd

CONTACTS: Note: Officers with more than one job title may be intentionally listed here more than once.

James Stamatis, CEO
James G. Bach, COO
Thomas Lewis, Pres.
Meg Lassarat, CFO
Susan Knauf, Chief Learning Officer
Thomas Nicastro, Corp. Compliance & Ethics Officer
Andrew V. Bailey II, Group VP
Charles Bell, Group VP
Nichoals J. Masucci, Chmn.

GROWTH PLANS/SPECIAL FEATURES:

The Louis Berger Group, Inc. (Berger) is an infrastructure engineering, environmental science and economic development company with offices throughout the U.S. and in more than 50 countries worldwide. Berger, a subsidiary of Berger Group Holdings, Inc., operates through three primary divisions: The Louis Berger Group, international and services. Berger provides a broad range of engineering, planning, architecture, development, construction management and program management services to U.S. federal, state, local and private clients across market sectors. The international division provides the same services to national, state, local and private clients outside the U.S. The services division specializes in turn-key power projects, infrastructure operations, maintenance, logistics and ground support services globally. Specialty brands within Berger's portfolio include the following three companies. Ammann & Whitney Consulting Engineers, Inc. provides structural, civil, architectural, mechanical, electrical engineering and construction inspection services for such projects as bridges, highways, airports, transit stations, schools and government and military installations. Berger/ABAM Engineers, Inc. offers planning, environmental science, civil and structural engineering, project management and construction support consulting services. Klohn Crippen Berger Ltd. designs hydroelectric power plants, dams, tunnels and pumped storage schemes; serves mining clients internationally; and provides seismic and geotechnical services throughout Canada.

The firm offers employees medical and dental coverage, life and travel insurance, a new employee hiring referral award, tuition reimbursement, flexible spending accounts, a cafeteria plan and registration fee payment for professional licenses.

FINANCIAL DATA: Note: Data for latest year may not have been available at press time.

In U.S. $	2015	2014	2013	2012	2011	2010
Revenue	1,000,000,000	1,200,000,000	1,000,000,000	960,000,000	850,000,000	755,000,000
R&D Expense						
Operating Income						
Operating Margin %						
SGA Expense						
Net Income						
Operating Cash Flow						
Capital Expenditure						
EBITDA						
Return on Assets %						
Return on Equity %						
Debt to Equity						

CONTACT INFORMATION:

Phone: 973-407-1000   Fax: 973-267-6468
Toll-Free:
Address: 412 Mt. Kemble Ave., Morristown, NJ 07960 United States

STOCK TICKER/OTHER:

Stock Ticker: Private   Exchange:
Employees: 6,000   Fiscal Year Ends: 06/30
Parent Company: Berger Group Holdings Inc

SALARIES/BONUSES:

Top Exec. Salary: $ Bonus: $
Second Exec. Salary: $ Bonus: $

OTHER THOUGHTS:

Estimated Female Officers or Directors: 1
Hot Spot for Advancement for Women/Minorities:

Lowe's Companies Inc

NAIC Code: 444110

www.lowes.com

TYPES OF BUSINESS:

Home Centers, Retail
Home Improvement Products
Home Installation Services
Special Order Sales

BRANDS/DIVISIONS/AFFILIATES:

Orchard Supply Hardware
Iris
Kobalt
Blue Hawk
Utilitech
Aquasource
Lowes.com

CONTACTS: Note: Officers with more than one job title may be intentionally listed here more than once.

Robert Hull, CFO
Matthew Hollifield, Senior VP
Paul Ramsay, Chief Information Officer
Marshall Croom, Chief Risk Officer
Rick Damron, COO
Gaither Keener, Executive VP, Divisional
Ross Mccanless, General Counsel
N. Peace, Other Corporate Officer
Richard Maltsbarger, Other Corporate Officer
Michael Jones, Other Executive Officer
Jennifer Weber, Other Executive Officer
Robert Niblock, President

GROWTH PLANS/SPECIAL FEATURES:

Lowe's Companies, Inc. is one of the largest home improvement retailers in the world. The company owns roughly 1,840 stores in 50 states, Mexico and Canada, each carrying approximately 36,000 products and 201 million square feet of retail space. The company also operates 74 stores under the recently acquired Orchard Supply Hardware name in California and Oregon. Hundreds of thousands of items are also available through the firm's special order system. Lowe's stores chiefly serve do-it-yourself (DIY) homeowners and commercial business customers, including contractors, landscapers, electricians, painters and plumbers. Its home improvement product categories include building materials, lighting, cabinets and countertops, seasonal living, millwork, lumber, flooring, lawn and landscaping items, hardware, fashion and rough plumbing, appliances, paint, tools, plants and plant pots, outdoor power equipment, rough electrical, home environment and organization and windows and walls. Each Lowe's store carries a wide selection of national brand name merchandise such as Samsung, Whirlpool, Stainmaster, GE, Valspar, Sylvania, Dewalt, Owens Corning and Johns Manville; and exclusive brand names such as Kobalt, allen+roth, Blue Hawk, Utilitech and Aquasource. The company's Lowes.com web site facilitates customers researching, comparing and buying Lowe's products, and also allows customers to special order products not carried in its physical store locations. Lowe's entered the smarthome market with Iris, an affordable, cloud-based home management system, which allows users to interact and control their home's security cameras, thermostat, locks, lighting and appliances remotely from a smart phone or computer. In May 2015, the firm announced that it reached an agreement to lease 13 former Target Canada stores as well as an Ontario distribution center, expecting to generate 2,000 jobs in Canada.

Lowe's offers its employees life, short- and long-term disability, accident, auto, home, medical, dental and vision insurance; family assistance programs; stock purchase plan; tuition reimbursement; paid time off; 401(k); and flexible spending accounts.

FINANCIAL DATA: Note: Data for latest year may not have been available at press time.

In U.S. $	2015	2014	2013	2012	2011	2010
Revenue	56,223,000,000	53,417,000,000	50,521,000,000	50,208,000,000	48,815,000,000	
R&D Expense						
Operating Income	4,792,000,000	4,149,000,000	3,560,000,000	3,277,000,000	3,560,000,000	
Operating Margin %	8.52%	7.76%	7.04%	6.52%	7.29%	
SGA Expense	13,281,000,000	12,865,000,000	12,244,000,000	12,593,000,000	12,006,000,000	
Net Income	2,698,000,000	2,286,000,000	1,959,000,000	1,839,000,000	2,010,000,000	
Operating Cash Flow	4,929,000,000	4,111,000,000	3,762,000,000	4,349,000,000	3,852,000,000	
Capital Expenditure	880,000,000	940,000,000	1,211,000,000	1,829,000,000	1,329,000,000	
EBITDA	6,385,000,000	5,719,000,000	5,211,000,000	4,862,000,000	5,256,000,000	
Return on Assets %	8.30%	6.99%	5.91%	5.46%	6.02%	
Return on Equity %	24.58%	17.78%	12.89%	10.61%	10.81%	
Debt to Equity	1.08	0.85	0.65	0.42	0.36	

CONTACT INFORMATION:

Phone: 704 758-1000   Fax: 336 658-4766
Toll-Free: 800-445-6937
Address: 1000 Lowe's Blvd., Mooresville, NC 28117 United States

STOCK TICKER/OTHER:

Stock Ticker: LOW   Exchange: NYS
Employees: 270,000   Fiscal Year Ends: 01/31
Parent Company:

SALARIES/BONUSES:

Top Exec. Salary: Bonus: $
$1,280,000
Second Exec. Salary: Bonus: $
$780,000

OTHER THOUGHTS:

Estimated Female Officers or Directors: 3

Hot Spot for Advancement for Women/Minorities: Y

M.C. Dean Inc

www.mcdean.com

NAIC Code: 238210

TYPES OF BUSINESS:

Electrical Contractor
Design-Build
Maintenance
Cost Analysis
Communications Wiring Contractor
Security and Safety Systems Wiring

BRANDS/DIVISIONS/AFFILIATES:

CONTACTS: *Note: Officers with more than one job title may be intentionally listed here more than once.*

William H. Dean, CEO
Sharon Hawkins, Dir.-Mktg.

GROWTH PLANS/SPECIAL FEATURES:

M.C. Dean, Inc. is Virginia-based electric and wiring systems design-build and systems integration firm for complex, mission-critical organizations. The services offered by the firm are design-build, which offers design specifications, test & training plans, CADD and 3D renderings and illustrations, system schematics and block diagrams; energy, such as power systems construction, power system testing & maintenance and solar power & off-grid power systems; communications, such as fiber optics communications, voice & data networks, network management & security and information systems; security/life safety, including intrusion detection services, access control systems, closed circuit TV, sound & audio/visual systems, fire prevention and detection & alarm system installations; and automation, such as programmable logic controller (PLC) systems design & programming, supervisory control & data acquisition systems and control panel design. The markets that primarily take advantage of the services offered by M.C. Dean are those within government, corporate/special industry and commercial markets.

FINANCIAL DATA: *Note: Data for latest year may not have been available at press time.*

In U.S. $	2015	2014	2013	2012	2011	2010
Revenue	810,000,000	810,000,000	622,000,000	590,000,000	695,000,000	
R&D Expense						
Operating Income						
Operating Margin %						
SGA Expense						
Net Income						
Operating Cash Flow						
Capital Expenditure						
EBITDA						
Return on Assets %						
Return on Equity %						
Debt to Equity						

CONTACT INFORMATION:

Phone: 703-802-6231   Fax: 703-421-4670
Toll-Free:
Address: 22980 Indian Creek Dr., Dulles, VA 20166 United States

STOCK TICKER/OTHER:

Stock Ticker: Private   Exchange:
Employees: 3,800   Fiscal Year Ends:
Parent Company:

SALARIES/BONUSES:

Top Exec. Salary: $ Bonus: $
Second Exec. Salary: $ Bonus: $

OTHER THOUGHTS:

Estimated Female Officers or Directors:
Hot Spot for Advancement for Women/Minorities:

Sales, profits and employees may be estimates. Financial information, benefits and other data can change quickly and may vary from those stated here.

M.D.C. Holdings Inc

TYPES OF BUSINESS:

Construction, Home Building and Residential
Land Development
Mortgages
Title Services
Insurance

BRANDS/DIVISIONS/AFFILIATES:

Richmond American Homes
HomeAmerican Mortgage Corporation
American Home Insurance Agency
American Home Title and Escrow Company
Allegiant Insurance Company Inc
StarAmerican Insurance Ltd

CONTACTS: Note: Officers with more than one job title may be intentionally listed here more than once.

Larry Mizel, CEO
Robert Martin, CFO
David Mandarich, COO
Joseph Fretz, Other Corporate Officer
Michael Touff, Senior VP

GROWTH PLANS/SPECIAL FEATURES:

M.D.C. Holdings, Inc. is a holding company engaged in homebuilding and financial services. The company's homebuilding division, which operates through the Richmond American Homes brand, builds and sells single-family homes through the following geographic segments: West (California, Nevada, Arizona and Washington), Mountain (Colorado and Utah) and East (Virginia, Florida, Illinois and Maryland, which includes Pennsylvania, Delaware and New Jersey). Other operations of the homebuilding division include land acquisition and development, home construction, purchasing, sales and marketing and customer service. M.D.C. also acquires entitled land for development into finished lots. During 2014, the company delivered 4,366 homes, with an average selling price of $377,000. M.D.C. maintains a variety of home styles in each of its markets, generally targeting first-time and first-time move-up homebuyers. The firm's financial services operations consists principally of the operations of HomeAmerican Mortgage Corporation, which originates mortgage loans primarily for its homebuyers; American Home Title and Escrow Company, which provides title agency services to M.D.C. and its homebuyers; and American Home Insurance Agency, which offers third-party insurance products to its homebuyers. The segment also includes risk retention firm Allegiant Insurance Company, Inc. and reinsurer StarAmerican Insurance, Ltd.

M.D.C. offers its employees medical, dental, vision, life, AD&D and disability insurance; a 401(k); home purchase discounts; and flexible spending accounts.

FINANCIAL DATA: Note: Data for latest year may not have been available at press time.

In U.S. $	2015	2014	2013	2012	2011	2010
Revenue	1,860,226,000	1,650,631,000	1,629,175,000	1,156,142,000	844,168,000	958,655,000
R&D Expense						
Operating Income	71,909,000	77,438,000	76,034,000	32,617,000	-78,451,000	-59,175,000
Operating Margin %	3.86%	4.69%	4.66%	2.82%	-9.29%	-6.17%
SGA Expense	226,317,000	203,253,000	213,283,000	167,295,000	205,052,000	242,585,000
Net Income	65,791,000	63,143,000	314,385,000	62,699,000	-98,390,000	-64,770,000
Operating Cash Flow	215,000	-163,647,000	-269,549,000	-108,819,000	-80,284,000	-209,081,000
Capital Expenditure	1,491,000	3,242,000	1,785,000	1,268,000		8,149,000
EBITDA	105,493,000	105,088,000	135,415,000	66,689,000	-79,971,000	-26,219,000
Return on Assets %	2.75%	2.53%	13.84%	3.29%	-4.46%	-2.60%
Return on Equity %	5.28%	5.15%	30.02%	7.16%	-10.62%	-6.29%
Debt to Equity	0.73	0.68	0.90	0.84	0.85	1.26

CONTACT INFORMATION:

Phone: 303 773-1100   Fax: 303 793-2760
Toll-Free: 888-402-4663
Address: 4350 S. Monaco St., Ste. 500, Denver, CO 80237 United States

STOCK TICKER/OTHER:

Stock Ticker: MDC   Exchange: NYS
Employees: 1,140   Fiscal Year Ends: 12/31
Parent Company:

SALARIES/BONUSES:

Top Exec. Salary:
$1,000,000   Bonus: $
Second Exec. Salary:
$830,000   Bonus: $

OTHER THOUGHTS:

Estimated Female Officers or Directors: 2

Hot Spot for Advancement for Women/Minorities: Y

M/I Homes Inc

www.mihomes.com

NAIC Code: 236117

TYPES OF BUSINESS:

Construction, Home Building and Residential
Real Estate Development
Mortgage Services
Title Services

BRANDS/DIVISIONS/AFFILIATES:

Showcase Homes
M/I Financial
TriStone Homes
M/I Title Agency
Washington/Metro Residential Title Agency
TransOhio Residential Title Agency
Confidence Builder Program
M/I Homes

CONTACTS: Note: Officers with more than one job title may be intentionally listed here more than once.

Paul Rosen, CEO, Subsidiary
Robert Schottenstein, CEO
Phillip Creek, CFO
Ann Marie Hunker, Controller
J. Mason, Director
Fred Sikorski, President, Geographical
Thomas Jacobs, President, Geographical
Ronald Martin, President, Geographical

GROWTH PLANS/SPECIAL FEATURES:

M/I Homes, Inc. is a leading builder of single-family homes targeting first-time homebuyers, move-up buyers, empty nesters and luxury buyers under the M/I Homes, Showcase Homes and TriStone Homes trade names. The company currently offers homes for sale in 150 communities within 13 markets located in eight states, with an average 2014 sales price of $313,000. M/I sells its homes in Columbus and Cincinnati, Ohio; Tampa and Orlando, Florida; Charlotte and Raleigh, North Carolina; Indianapolis, Indiana; Chicago, Illinois; Houston, Austin, Dallas/FortWorth and San Antonio, Texas; and the Virginia and Maryland suburbs of Washington, D.C. M/I operates in four segments: Midwest homebuilding, Southern homebuilding, Mid-Atlantic homebuilding and financial services. The firm's homebuilding operations accounted for 98% of its 2014 revenues. Homebuilding operations include the acquisition and development of land, the sale and construction of single-family attached and detached homes and the occasional sale of lots and land to third parties. M/I's financial services operations generate revenue from originating and selling mortgages, collecting fees for title insurance and closing services. Subsidiary M/I Financial provides financing services in its housing markets, while title services are provided through subsidiaries TransOhio Residential Title Agency, M/I Title Agency and majority-owned Washington/Metro Residential Title Agency. The company offers roughly 630 floor plans. The firm's Confidence Builder Program includes a pre-construction conference between the client and a Personal Construction Supervisor. The supervisors manage the development and construction process, for which M/I will utilize independent subcontractors. During 2015, the firm acquired the residential homebuilding operations of Hans Hagen Homes, Inc.

The firm offers employees medical, dental and vision insurance; life insurance; disability coverage; an employee assistance program; a 401(k) plan; an employee stock purchase plan; and an employee home purchase plan.

FINANCIAL DATA: Note: Data for latest year may not have been available at press time.

In U.S. $	2015	2014	2013	2012	2011	2010
Revenue	1,418,395,000	1,215,180,000	1,036,782,000	761,905,000	566,424,000	616,377,000
R&D Expense						
Operating Income	115,432,000	82,754,000	58,693,000	28,830,000	-18,897,000	-9,611,000
Operating Margin %	8.13%	6.80%	5.66%	3.78%	-3.33%	-1.55%
SGA Expense	188,300,000	169,978,000	147,776,000	119,033,000	96,198,000	102,042,000
Net Income	51,763,000	50,789,000	151,423,000	13,347,000	-33,877,000	-26,269,000
Operating Cash Flow	-82,159,000	-132,675,000	-73,974,000	-46,995,000	-33,961,000	-37,302,000
Capital Expenditure	3,659,000	2,946,000	2,382,000	933,000	1,352,000	1,560,000
EBITDA	112,072,000	89,051,000	65,584,000	38,572,000	-11,323,000	-10,233,000
Return on Assets %	3.56%	3.95%	14.99%	1.78%	-5.10%	-3.96%
Return on Equity %	8.97%	9.76%	42.58%	6.41%	-17.63%	-12.00%
Debt to Equity	0.32	1.00	1.04	1.38	1.73	1.21

CONTACT INFORMATION:

Phone: 614 418-8000   Fax: 614 418-8080
Toll-Free: 888-644-4111
Address: 3 Easton Oval, Ste. 500, Columbus, OH 43219 United States

STOCK TICKER/OTHER:

Stock Ticker: MHO
Employees: 905
Parent Company:

Exchange: NYS
Fiscal Year Ends: 12/31

SALARIES/BONUSES:

Top Exec. Salary: $900,000   Bonus: $
Second Exec. Salary: Bonus: $
$600,000

OTHER THOUGHTS:

Estimated Female Officers or Directors: 2
Hot Spot for Advancement for Women/Minorities:

Macerich Company (The)

NAIC Code: 0

www.macerich.com

TYPES OF BUSINESS:

Real Estate Investment Trust
Regional Shopping Centers
Community Shopping Centers
Property Management
Property Redevelopment

BRANDS/DIVISIONS/AFFILIATES:

Macerich Partnership LP
Macerich Property Management Company LLC
MACW Property Management LLC
Macerich Management Company
Inland Center
443 North Wabash
Pacific Premier Retail LP
Queens JV LP)

CONTACTS: Note: Officers with more than one job title may be intentionally listed here more than once.

Arthur Coppola, CEO
Thomas OHern, CFO
Robert Perlmutter, COO
Edward Coppola, Director
Eric Salo, Executive VP
Randy Brant, Executive VP, Divisional
Thomas Leanse, Other Executive Officer

GROWTH PLANS/SPECIAL FEATURES:

The Macerich Company is a real estate investment trust (REIT) involved in the acquisition, ownership, redevelopment, management and leasing of regional and community shopping centers located in the U.S. The firm is the sole general partner in Macerich Partnership LP, which owns or maintains ownership interests in 52 regional shopping centers and eight community shopping centers, totaling approximately 55 million square feet of leasable area. Other subsidiaries include Macerich Property Management Company, MACW Property Management and Macerich Management Company. Macerich's integrated operations include in-house accounting, finance, legal, marketing, property management and redevelopment expertise. Its regional shopping centers are generally enclosed, offer a variety of small and mid-size stores anchored by several department stores or other large retailers and generally contain in excess of 400,000 square feet. The firm's community shopping centers are generally smaller open-air centers designed to attract local customers with anchors such as supermarkets and drug stores, typically containing 100,000 to 400,000 square feet. During 2014, the firm sold its 30% interest in Wilshire Boulevard and its 67.5% interest in its Camelback Colonnade joint venture (JV); formed a JV to redevelop The Gallery, a 1.4 million square foot regional shopping center in Philadelphia; formed another JV to develop a 500,000 square foot outlet center at Candlestick Point in San Francisco; acquired the remaining 49% in two JVs (Pacific Premier Retail LP and Queens JV LP) that together own five U.S. shopping centers; and acquired a 45% stake in 443 North Wabash, an undeveloped site. In 2015, it acquired the remaining 50% interest in Inland Center in California; and formed a JV that owns a 40% stake in four malls located in California (2), Texas and Oregon.

FINANCIAL DATA: Note: Data for latest year may not have been available at press time.

In U.S. $	2015	2014	2013	2012	2011	2010
Revenue	1,288,149,000	1,105,247,000	1,029,475,000	881,323,000	791,250,000	758,559,000
R&D Expense						
Operating Income	296,448,000	255,190,000	221,282,000	192,217,000	162,402,000	154,752,000
Operating Margin %	23.01%	23.08%	21.49%	21.81%	20.52%	20.40%
SGA Expense	29,870,000	29,412,000	27,772,000	20,412,000	21,113,000	20,703,000
Net Income	487,562,000	1,499,042,000	420,090,000	337,426,000	156,866,000	25,190,000
Operating Cash Flow	540,377,000	400,706,000	422,035,000	351,296,000	237,285,000	200,435,000
Capital Expenditure	351,919,000	267,363,000	726,604,000	1,249,715,000	247,011,000	185,789,000
EBITDA	1,196,104,000	2,172,067,000	757,706,000	587,695,000	681,049,000	510,902,000
Return on Assets %	3.99%	13.49%	4.56%	3.91%	2.01%	.33%
Return on Equity %	9.41%	33.28%	13.05%	11.45%	5.50%	1.06%
Debt to Equity	1.12	1.11	1.28	1.70	1.53	0.32

CONTACT INFORMATION:

Phone: 310 394-6000   Fax: 310 395-2791
Toll-Free:
Address: 401 Wilshire Blvd., Ste. 700, Santa Monica, CA 90401 United States

STOCK TICKER/OTHER:

Stock Ticker: MAC
Employees: 1,117
Parent Company:

Exchange: NYS
Fiscal Year Ends: 12/31

SALARIES/BONUSES:

Top Exec. Salary: $1,000,000   Bonus: $
Second Exec. Salary: $800,000   Bonus: $

OTHER THOUGHTS:

Estimated Female Officers or Directors: 3

Hot Spot for Advancement for Women/Minorities: Y

Mack-Cali Realty Corp

www.mack-cali.com

NAIC Code: 0

TYPES OF BUSINESS:
Real Estate Investment Trust
Office & Industrial Properties
Property Management

BRANDS/DIVISIONS/AFFILIATES:
Gale Construction Company
Roseland

CONTACTS: Note: Officers with more than one job title may be intentionally listed here more than once.
Mitchell Rudin, CEO
William Mack, Chairman of the Board
Anthony Krug, Chief Accounting Officer
Robert Marshall, COO, Subsidiary
Christopher DeLorenzo, Executive VP, Divisional
Ricardo Cardoso, Executive VP
Gary Wagner, Other Executive Officer
Marshall Tycher, President, Subsidiary
Michael DeMarco, President
Ilene Jablonski, Vice President, Divisional

GROWTH PLANS/SPECIAL FEATURES:
Mack-Cali Realty Corp. is a fully integrated, self-administered and self-managed real estate investment trust (REIT) that owns and operates a portfolio consisting predominantly of Class A office and office/flex properties located in suburban markets in the Northeast U.S. The trust performs substantially all commercial real estate leasing, management, acquisition, development and construction services on an in-house basis. The company owns or has interests in 275 properties, totaling approximately 29.9 million square feet of Class A office, office/flex properties, warehouse/industrial and standalone retail space that is leased to more than 1,9000 tenants. Mack-Cali's portfolio is approximately 86.2% leased. Many of the firm's properties have adjacent company-controlled developable land. Its major tenants include National Union Fire Insurance; Company of Pittsburgh, PA; DB Services New Jersey, Inc.; New Cingular Wireless; and Bank of Tokyo-Mitsubishi UF. Its subsidiary, Gale Construction Company, offers development management, construction advisory services, construction management services, design services and general contracting. Another subsidiary, Roseland, is a full-service residential and mixed-use developer. In 2015, the firm acquired an empty 147,241 square-foot, three-story, class A office building located in Parsipanny, New Jersey and Metroview, a class A office property in Edison, New Jersey.

FINANCIAL DATA: Note: Data for latest year may not have been available at press time.

In U.S. $	2015	2014	2013	2012	2011	2010
Revenue	594,883,000	636,799,000	667,031,000	704,743,000	724,279,000	787,480,000
R&D Expense						
Operating Income	-96,332,000	88,811,000	33,595,000	169,485,000	205,301,000	209,469,000
Operating Margin %	-16.19%	13.94%	5.03%	24.04%	28.34%	26.59%
SGA Expense	74,730,000	99,305,000	47,682,000	47,868,000	35,541,000	35,003,000
Net Income	-125,752,000	28,567,000	-14,909,000	40,922,000	71,420,000	54,900,000
Operating Cash Flow	169,455,000	159,253,000	198,693,000	244,706,000	252,065,000	223,036,000
Capital Expenditure	245,601,000	178,891,000	281,935,000	107,545,000	91,729,000	92,499,000
EBITDA	133,107,000	318,117,000	225,536,000	360,811,000	398,888,000	400,637,000
Return on Assets %	-3.04%	.65%	-.32%	.92%	1.60%	1.16%
Return on Equity %	-8.16%	1.74%	-.87%	2.23%	3.84%	2.98%
Debt to Equity	1.48	1.28	1.43	1.24	1.01	1.20

CONTACT INFORMATION:
Phone: 732 590-1000   Fax: 732 205-8237
Toll-Free:
Address: 343 Thornall St., Edison, NJ 08837-2206 United States

STOCK TICKER/OTHER:
Stock Ticker: CLI
Employees: 600
Parent Company:
Exchange: NYS
Fiscal Year Ends: 12/31

SALARIES/BONUSES:
Top Exec. Salary: $1,050,000   Bonus: $631,000
Second Exec. Salary: $325,000   Bonus: $350,000

OTHER THOUGHTS:
Estimated Female Officers or Directors: 3

Hot Spot for Advancement for Women/Minorities: Y

Mandarin Oriental International Ltd
www.mandarinoriental.com

NAIC Code: 721110

TYPES OF BUSINESS:
Hotels, Luxury
Condominiums

BRANDS/DIVISIONS/AFFILIATES:
Mandarin Oriental Hotel Group Intl Ltd
Residences at Mandarin Oriental
Spa at Mandarin Oriental

CONTACTS: Note: Officers with more than one job title may be intentionally listed here more than once.
Edouard Ettedgui, CEO
Stuart Dickie, CFO
Michael Hobson, Chief Mktg. Officer
Paul Clark, Group Dir.-Human Resources
Monika Nerger, CIO
Vincent Marot, Group Dir.-Tech. Services
Kieren Barry, Group Counsel
Terry L. Stinson, Dir.-Dev.
Jill Kluge, Group Dir.-Brand Comm.
Christoph Mares, Dir.-Oper., EMEA
Andrew Hirst, Dir.-Oper., Asia
Richard Baker, Exec. VP-Americas Oper.
David Nicholls, Group Dir.-Food & Beverage
Simon L. Keswick, Chmn.
Terry L. Stinson, Pres., The Americas

GROWTH PLANS/SPECIAL FEATURES:
Mandarin Oriental International, Ltd. (MOI) is an international hotel investment and management group. MOI operates in two business segments, hotel ownership & hotel management, and four geographical regions: Hong Kong & Macau, Other Asia, Europe and the Americas. Through subsidiary Mandarin Oriental Hotel Group International Ltd., the company either currently operates or is in the process of developing 47 luxury and first class hotels with over 11,000 rooms and a presence in 25 countries worldwide. MOI has 21 hotel properties in Asia, 10 in the Americas and 16 in Europe, the Middle East and North Africa. These include the original flagship properties of the Mandarin Oriental in Hong Kong and in Bangkok as well as locations in Singapore, Jakarta, Kuala Lumpur, Macau, Manila, London, Geneva, Tokyo, Munich, Prague and Bermuda. The company has U.S. hotels in New York City, Boston, Miami, Las Vegas, Atlanta and Washington, D.C. In addition to hotel rooms, the Mandarin Oriental New York offers the Residences at Mandarin Oriental, 65 luxury condominiums located above the hotel. Another 14 of the firm's properties also feature the Residences at Mandarin Oriental condominiums. The Spa at Mandarin Oriental can also be found in many of the firm's hotels worldwide, including in London, Miami, Boston and New York City. Each spa is unique and offers specialized treatments to clients. In 2015, the firm sold its San Francisco hotel, and acquired the Ritz Hotel in Madrid.

FINANCIAL DATA: Note: Data for latest year may not have been available at press time.

In U.S. $	2015	2014	2013	2012	2011	2010
Revenue	607,300,000	679,900,000	668,600,000	648,300,000	614,200,000	513,200,000
R&D Expense						
Operating Income	107,300,000	120,800,000	111,800,000	85,900,000	89,100,000	64,900,000
Operating Margin %	17.66%	17.76%	16.72%	13.24%	14.50%	12.64%
SGA Expense	137,900,000	149,100,000	148,400,000	147,200,000	147,800,000	121,700,000
Net Income	89,300,000	97,000,000	96,300,000	72,300,000	67,500,000	44,400,000
Operating Cash Flow	140,200,000	159,500,000	156,900,000	126,000,000	146,300,000	114,200,000
Capital Expenditure	51,500,000	32,300,000	38,800,000	55,000,000	66,200,000	74,700,000
EBITDA	169,900,000	198,800,000	192,700,000	158,500,000	151,200,000	114,700,000
Return on Assets %	4.71%	4.94%	5.08%	4.14%	3.97%	2.48%
Return on Equity %	8.18%	9.97%	9.95%	7.78%	7.45%	4.60%
Debt to Equity	0.35	0.53	0.24	0.61	0.63	0.63

CONTACT INFORMATION:
Phone: 852 28959288   Fax: 852 28373500
Toll-Free: 852 28811288
Address: 281 Gloucester Rd., 7th Fl., Hong Kong,   Hong Kong

STOCK TICKER/OTHER:
Stock Ticker: MNOIY   Exchange: PINX
Employees: Fiscal Year Ends: 12/31
Parent Company:

SALARIES/BONUSES:
Top Exec. Salary: $ Bonus: $
Second Exec. Salary: $ Bonus: $

OTHER THOUGHTS:
Estimated Female Officers or Directors: 2
Hot Spot for Advancement for Women/Minorities:

ManorCare Inc (HCR ManorCare)   www.hcr-manorcare.com

NAIC Code: 623110

TYPES OF BUSINESS:

Nursing Care Facilities
Home Health Care
Short-Term Care Facilities
Assisted Living Facilities
Rehabilitation Clinics

BRANDS/DIVISIONS/AFFILIATES:

HCR Manor Care
Heartland
ManorCare Health Services
Arden Courts
Carlyle Group (The)

CONTACTS: Note: Officers with more than one job title may be intentionally listed here more than once.

Paul A. Ormond, CEO
Paul A. Ormond, Pres.
Steven M. Cavanaugh, CFO
Spencer C. Moler, Principal Acct. Officer
Paul A. Ormond, Chmn.

GROWTH PLANS/SPECIAL FEATURES:

ManorCare, Inc., doing business as HCR ManorCare, provides a range of health care services, including skilled nursing care, assisted living, post-acute medical care, hospice care, home health care and rehabilitation therapy. ManorCare operates more than 500 properties in over 32 states, with facilities operating primarily under the Heartland, ManorCare Health Services and Arden Courts names. The firm's long-term care services consist of skilled nursing centers, assisted living services, post-hospital centers and rehabilitation care and Alzheimer's care. The skilled nursing centers use interdisciplinary teams of experienced medical professionals, including registered nurses, licensed practical nurses and certified nursing assistants, to provide services prescribed by physicians. Other services include the design of Quality of Life programs to give the highest practicable level of functional independence to patients; provide physical, speech, respiratory and occupational therapy; and provide quality nutrition services, social services, activities and housekeeping and laundry services. ManorCare's assisted living services provide personal care services and assistance with general activities of daily living such as dressing, bathing, meal preparation and medication management. The firm is owned by The Carlyle Group.

Manor Care offers its employees benefits including vision, dental, life and AD&D insurance; flexible spending accounts; legal assistance; an employee assistance program; a 401(k) savings plan; adoption assistance; and education assistance.

FINANCIAL DATA: Note: Data for latest year may not have been available at press time.

In U.S. $	2015	2014	2013	2012	2011	2010
Revenue	5,300,000,000	5,200,000,000	5,000,000,000	4,800,000,000	4,600,000,000	4,500,000,000
R&D Expense						
Operating Income						
Operating Margin %						
SGA Expense						
Net Income						
Operating Cash Flow						
Capital Expenditure						
EBITDA						
Return on Assets %						
Return on Equity %						
Debt to Equity						

CONTACT INFORMATION:

Phone: 419-252-5500   Fax: 419-252-6404
Toll-Free:
Address: 333 N. Summit St., Toledo, OH 43604 United States

STOCK TICKER/OTHER:

Stock Ticker: Private   Exchange:
Employees: 55,000   Fiscal Year Ends: 12/31
Parent Company: CARLYLE GROUP (THE)

SALARIES/BONUSES:

Top Exec. Salary: $ Bonus: $
Second Exec. Salary: $ Bonus: $

OTHER THOUGHTS:

Estimated Female Officers or Directors:
Hot Spot for Advancement for Women/Minorities:

Mapletree Investments Pte Ltd

NAIC Code: 531120

www.mapletree.com.sg

TYPES OF BUSINESS:

Lessors of Nonresidential Buildings (except Miniwarehouses)
Lessors of Residential Buildings and Dwellings

BRANDS/DIVISIONS/AFFILIATES:

GROWTH PLANS/SPECIAL FEATURES:

Mapletree Investments Pte Ltd. is Singapore-based real estate development firm with development projects across various real estate classes. Currently, the firm's assets under management of $17.78 (S$28.4) billion span retail, office, logistics, industrial, residential, corporate housing/serviced apartments and mixed-use properties. Additionally, Mapletree manages four Singapore-listed real estate investment trusts (REITs) as well as six private real estate funds. The firm's flagship developments include VivoCity, Singapore's largest retail and lifestyle destination mall, and Mapletree Business City, a business hub at Singapore's Alexandra Precinct. The firm operates in seven Asian countries, with assets in Australia, Europe and the U.S. International assets account for 52% of the firms holdings. The international assets of Mapletree include the Mapletree Business City Shanghai, VivoCity Shanghai and Nanhai Business City in China; the Saigon South Palace in Vietnam; commercial and residential properties in Australia; and various apartment complexes across the U.S.

CONTACTS: Note: Officers with more than one job title may be intentionally listed here more than once.

Hiew Yoon Khong, Group CEO
Chua Tiow Chye, Chief Investment Officer
Wong Mun Hoong, CFO
Wendy Koh Mui Ai, Regional CEO-Southeast Asia
Ho Seng Chee, Regional CEO-Vietnam
Quek Kwang Meng, Regional CEO-China & India
Edmund Cheng, Chmn.

FINANCIAL DATA: Note: Data for latest year may not have been available at press time.

In U.S. $	2015	2014	2013	2012	2011	2010
Revenue	1,633,923,000	1,521,903,000				
R&D Expense						
Operating Income						
Operating Margin %						
SGA Expense						
Net Income	1,826,956,000	1,752,577,000				
Operating Cash Flow						
Capital Expenditure						
EBITDA						
Return on Assets %						
Return on Equity %						
Debt to Equity						

CONTACT INFORMATION:

Phone: 65-6377-6111   Fax: 65-6273-2753
Toll-Free:
Address: 10 Pasir Panjang Rd. #13-01, Mapletree Business City, Singapore, 117438 Singapore

STOCK TICKER/OTHER:

Stock Ticker: Private   Exchange:
Employees: 1,804   Fiscal Year Ends:
Parent Company:

SALARIES/BONUSES:

Top Exec. Salary: $ Bonus: $
Second Exec. Salary: $ Bonus: $

OTHER THOUGHTS:

Estimated Female Officers or Directors:
Hot Spot for Advancement for Women/Minorities:

Marcus Corporation (The)

www.marcuscorp.com

NAIC Code: 721110

TYPES OF BUSINESS:

Hotels & Motels
Movie & IMAX Theaters
Hotels/Resorts

BRANDS/DIVISIONS/AFFILIATES:

Marcus Theatres Corp
MCS Capital LLC
Funset Boulevard

CONTACTS: *Note: Officers with more than one job title may be intentionally listed here more than once.*

Rolando Rodriguez, CEO, Subsidiary
Gregory Marcus, CEO
Douglas Neis, CFO
Stephen Marcus, Chairman of the Board
Thomas Kissinger, General Counsel

GROWTH PLANS/SPECIAL FEATURES:

The Marcus Corporation is an owner and operator of movie theaters, hotels and resorts. Through its Marcus Theatres Corp. subsidiary, the company owns or operates 55 movie theaters, with 681 screens in Wisconsin, Ohio, Illinois, Minnesota, North Dakota, Nebraska and Iowa. Marcus also operates a family entertainment center, Funset Boulevard, adjacent to its theater in Appleton, Wisconsin. In addition, Marcus Corporation owns or manages approximately 5,211 hotel and resort rooms, and also manages hotels and other properties for third parties. Owned hotels and resorts include the Pfister Hotel, the InterContinental Milwaukee and The Hilton Milwaukee City Center in Milwaukee, Wisconsin; the Hilton Madison at Monona Terrace in Madison, Wisconsin; The Grand Geneva Resort & Spa in Lake Geneva, Wisconsin; the Hotel Phillips in Kansas City, Missouri; the Four Points by Sheraton Chicago Downtown/Magnificent Mile in Chicago, Illinois; and the Skirvin Hilton in Oklahoma City, Oklahoma. Subsidiary MCS Capital LLC acquires and develops new hotel investments. In 2015, the firm began a partnership with Fandango, allowing its moviegoers to purchase tickets through Fandango's online web site. That same year, Marcus Corporation sold Hotel Phillips in Kansas City, Missouri.

FINANCIAL DATA: *Note: Data for latest year may not have been available at press time.*

In U.S. $	2015	2014	2013	2012	2011	2010
Revenue	488,067,000	447,939,000	412,836,000	413,898,000	377,004,000	379,069,000
R&D Expense						
Operating Income	50,194,000	48,382,000	38,204,000	46,515,000	33,497,000	36,203,000
Operating Margin %	10.28%	10.80%	9.25%	11.23%	8.88%	9.55%
SGA Expense	87,103,000	80,324,000	77,255,000	74,623,000	67,675,000	64,374,000
Net Income	23,995,000	25,001,000	17,506,000	22,734,000	13,558,000	16,115,000
Operating Cash Flow	80,452,000	66,440,000	63,202,000	69,028,000	61,502,000	52,740,000
Capital Expenditure	74,988,000	56,673,000	23,491,000	38,017,000	25,186,000	25,082,000
EBITDA	87,941,000	81,948,000	78,151,000	81,570,000	65,698,000	68,760,000
Return on Assets %	3.04%	3.29%	2.36%	3.18%	1.93%	2.27%
Return on Equity %	7.16%	7.90%	5.38%	6.65%	4.01%	4.85%
Debt to Equity	0.72	0.78	0.84	0.40	0.58	0.58

CONTACT INFORMATION:

Phone: 414 905-1000 Fax: 414 905-2879
Toll-Free:
Address: 100 E. Wisconsin Ave., Ste. 1900, Milwaukee, WI 53202-4125 United States

STOCK TICKER/OTHER:

Stock Ticker: MCS
Employees: 7,100
Parent Company:

Exchange: NYS
Fiscal Year Ends: 05/31

SALARIES/BONUSES:

Top Exec. Salary: Bonus: $
$1,002,442
Second Exec. Salary: Bonus: $
$759,327

OTHER THOUGHTS:

Estimated Female Officers or Directors: 2

Hot Spot for Advancement for Women/Minorities: Y

Sales, profits and employees may be estimates. Financial information, benefits and other data can change quickly and may vary from those stated here.

Marriott International Inc

NAIC Code: 721110

www.marriott.com

TYPES OF BUSINESS:
Hotels & Resorts
Suites Hotels
Corporate Apartments
Extended Stay Lodging
Luxury Hotels
Business Hotels

BRANDS/DIVISIONS/AFFILIATES:
Marriott Hotels
Ritz-Carlton Hotel Company LLC (The)
BVLGARI Hotels and Resorts
Renaissance Hotels
Courtyard by Marriott
Fairfield Inn & Suites by Marriott
TownePlace Suites by Marriott
JW Marriott

CONTACTS: Note: Officers with more than one job title may be intentionally listed here more than once.
Arne Sorenson, CEO
David Grissen, Pres., Divisional
Kathleen Oberg, CFO
J. Marriott, Chairman of the Board
Stephanie Linnartz, Chief Marketing Officer
Bao Giang Val Bauduin, Controller
Sterling Colton, Director Emeritus
William Shaw, Director Emeritus
Edward Ryan, Executive VP
Anthony Capuano, Executive VP
David Rodriguez, Executive VP
Amy McPherson, Managing Director, Geographical
Argiris Kyriakidis, Managing Director, Geographical
Craig Smith, Managing Director, Geographical
Bancroft Gordon, Other Corporate Officer

GROWTH PLANS/SPECIAL FEATURES:
Marriott International, Inc. operates 4,424 hotels and related lodging facilities in 79 countries and territories, totaling nearly 760,000 rooms. The company operates through three segments: North American full-service lodging, North American limited-service lodging and international. Marriott develops, operates and franchises hotels under various brand names, including Marriott Hotels, JW Marriott, The Ritz-Carlton, BuVLGARI Hotels and Resorts, Renaissance Hotels, Courtyard by Marriott, Residence Inn by Marriott, Fairfield Inn & Suites, SpringHill Suites by Marriott, EDITION, Autograph Collection Hotels, Marriott Executive Apartments, Marriott Vacation Club, Gaylor Hotels, AC Hotels by Marriott, Protea Hotels, Moxy Hotels and TownePlace Suites by Marriott. The firm also operates 41 home and condominium projects and 28 Marriott Executive Apartments located in 15 countries. Additionally, Marriott Golf manages 35 golf resorts worldwide. The company operates 15 system-wide hotel reservation centers: six in the U.S. and Canada and nine in other countries and territories. In 2015, Marriott debuted its first property in Italy, on a private island called JW Marriott Venice Resort & Spa. In February 2016, it expanded its footprint in Chhattisgarh, India with Courtyard by Marriott Raipur; signed for five hotels to open in Japan by 2018 with Mori Trust Group, which will be under the Marriott Hotels brand; and announced plans to open two new hotels in Cartagena, Colombia by 2018. As of March 2016, Marriott is in a competition with the Anbang Insurance Group for the right to acquire Starwood Hotels & Resorts Worldwide Inc.

FINANCIAL DATA: Note: Data for latest year may not have been available at press time.

In U.S. $	2015	2014	2013	2012	2011	2010
Revenue	14,486,000,000	13,796,000,000	12,784,000,000	11,814,000,000	12,317,000,000	11,691,000,000
R&D Expense						
Operating Income	1,350,000,000	1,159,000,000	988,000,000	940,000,000	526,000,000	695,000,000
Operating Margin %	9.31%	8.40%	7.72%	7.95%	4.27%	5.94%
SGA Expense	634,000,000	659,000,000	726,000,000	645,000,000	752,000,000	780,000,000
Net Income	859,000,000	753,000,000	626,000,000	571,000,000	198,000,000	458,000,000
Operating Cash Flow	1,430,000,000	1,224,000,000	1,140,000,000	989,000,000	1,089,000,000	1,151,000,000
Capital Expenditure	426,000,000	476,000,000	465,000,000	690,000,000	257,000,000	363,000,000
EBITDA	1,561,000,000	1,351,000,000	1,144,000,000	1,131,000,000	688,000,000	909,000,000
Return on Assets %	13.26%	11.02%	9.53%	9.32%	2.65%	5.41%
Return on Equity %					49.25%	33.59%
Debt to Equity						1.69

CONTACT INFORMATION:
Phone: 301 380-3000   Fax: 301 380-3967
Toll-Free: 800-721-7033
Address: 10400 Fernwood Rd., Bethesda, MD 20817 United States

STOCK TICKER/OTHER:
Stock Ticker: MAR
Employees: 123,500
Parent Company:

Exchange: NAS
Fiscal Year Ends: 12/31

SALARIES/BONUSES:
Top Exec. Salary: $3,000,000   Bonus: $
Second Exec. Salary: $1,236,000   Bonus: $

OTHER THOUGHTS:
Estimated Female Officers or Directors: 7

Hot Spot for Advancement for Women/Minorities: Y

Matrix Service Company

www.matrixservicecompany.com

NAIC Code: 237100

TYPES OF BUSINESS:

Heavy Construction for Utilities and Energy
Plant Maintenance Services
Storage Tank Services
Petrochemical Industry Services

BRANDS/DIVISIONS/AFFILIATES:

Matrix North American Construction
Calpine Garrison Energy Center

CONTACTS: Note: Officers with more than one job title may be intentionally listed here more than once.

John Hewitt, CEO
Kevin Cavanah, CFO
Michael Hall, Chairman of the Board
Rick Bennett, Chief Information Officer
Joseph Montalbano, COO
Jason Turner, President, Subsidiary
James Ryan, President, Subsidiary
Nancy Austin, Vice President, Divisional

GROWTH PLANS/SPECIAL FEATURES:

Matrix Service Company and its subsidiaries provide engineering, fabrication, construction, repair and maintenance services, primarily to the petroleum, pipeline, bulk storage terminal and industrial gas markets. Matrix operates in four segments: electrical infrastructure, oil gas & chemical, industrial and storage solutions. The electrical infrastructure segment primarily encompasses high voltage services to investor-owned utilities, such as construction of new substations, upgrades of existing substations, short-run transmission line installations, upgrades and maintenance and storm restoration services. The oil gas & chemical segment includes plant maintenance services, construction in the downstream petroleum industry and industrial cleaning services. The industrial segment encompasses work in the mining and minerals industry, thermal vacuum chambers and bulk material processing. The storage solutions segment includes new construction of crude and refined products, aboveground storage tanks as well as planned and emergency maintenance services for those tanks. Matrix targets a wide array of specialty markets, comprising the areas of liquefied natural gas (LNG)/industrial gas/liquefied petroleum gas (LPG), specialty tanks and vessels, power projects, fabrication and material handling. In February 2015, the firm announced that it consolidated its union subsidiaries, Matrix North American Construction and Matrix SME, into one company, with the new entity being Matrix North American Construction. That June, it completed the Calpine Garrison Energy Center, an acquired EPC joint venture with Parsons Brinckerhoff. The energy center is a 309 megawatt facility that employs highly efficient combined cycle gas-fired technology and advanced environmental controls and can produce enough electricity for 300,000 homes.

Matrix offers employee benefits including life, AD&D, medical, vision, dental and cancer insurance; short- and long-term disability; flexible spending accounts; retirement benefits; employee assistance and wellness programs; tuition reimbursement; and an employee stock purchase plan.

FINANCIAL DATA: Note: Data for latest year may not have been available at press time.

In U.S. $	2015	2014	2013	2012	2011	2010
Revenue	1,343,135,000	1,263,089,000	892,574,000	739,046,000	627,052,000	550,814,000
R&D Expense						
Operating Income	8,802,000	58,607,000	36,714,000	31,635,000	30,900,000	7,753,000
Operating Margin %	.65%	4.63%	4.11%	4.28%	4.92%	1.40%
SGA Expense	78,568,000	77,866,000	57,988,000	47,983,000	44,014,000	45,169,000
Net Income	17,157,000	35,810,000	24,008,000	17,188,000	18,982,000	4,876,000
Operating Cash Flow	24,438,000	76,988,000	57,084,000	2,941,000	22,749,000	4,399,000
Capital Expenditure	15,773,000	23,589,000	23,231,000	13,534,000	10,416,000	5,302,000
EBITDA	32,908,000	76,765,000	49,498,000	42,789,000	42,478,000	19,833,000
Return on Assets %	3.01%	7.31%	6.54%	5.46%	6.42%	1.71%
Return on Equity %	5.97%	13.80%	10.68%	8.36%	10.06%	2.74%
Debt to Equity	0.03	0.04				

CONTACT INFORMATION:

Phone: 918 838-8822   Fax: 918 838-8810
Toll-Free:
Address: 5100 E. Skelly Dr., Ste. 700, Tulsa, OK 74135 United States

SALARIES/BONUSES:

Top Exec. Salary: $729,904   Bonus: $
Second Exec. Salary: Bonus: $
$467,740

STOCK TICKER/OTHER:

Stock Ticker: MTRX   Exchange: NAS
Employees: 4,826   Fiscal Year Ends: 06/30
Parent Company:

OTHER THOUGHTS:

Estimated Female Officers or Directors: 1
Hot Spot for Advancement for Women/Minorities:

Maui Land & Pineapple Company Inc   www.mauiland.com
NAIC Code: 721110

TYPES OF BUSINESS:
Hotels (except Casino Hotels) and Motels
Real Estate Development
Real Estate Leasing

BRANDS/DIVISIONS/AFFILIATES:
Kapalua Resort
Kapalua Water Company Ltd
Kapalua Waste Treatment Company Ltd
Kapalua Club
Pulelehua
Hali'imaile Town

CONTACTS: *Note: Officers with more than one job title may be intentionally listed here more than once.*
Warren Haruki, CEO
Tim Esaki, CFO
Mika Miyamoto, Chief Accounting Officer

GROWTH PLANS/SPECIAL FEATURES:
Maui Land & Pineapple Company, Inc. (ML&P) has resort and real estate operations. The company owns approximately 23,000 acres of land on Maui and develops, sells and manages residential, resort, commercial and industrial real estate. The company has four reporting segments: real estate, leasing, utilities and resort amenities. Real estate operations consist of land planning and entitlement, development and sales. The firm's principal real estate development is the Kapalua Resort, a master-planned, destination resort community located in West Maui encompassing approximately 3,000 acres. Other real estate holdings include the Pulelehua, a working-class community in West Maui under development, and Hali'imaile Town, an expansion of an existing plantation town in Upcountry Maui. The real estate segment accounted for 55% of revenue in 2015. The leasing segment activities include commercial, light industrial and agricultural land leases, licensing of registered trademarks and trade names and stewardship and conservation efforts. The firm owns approximately 270,000 square feet of commercial retail and light industrial properties, and leases 1,900 acres of diversified agriculture land leases in West and Upcountry Maui. The leasing segment accounted for 24% of total revenue in 2015. The utilities segment includes the operations of two Hawaii Public Utilities Commission-regulated subsidiaries, Kapalua Water Company, Ltd. and Kapalua Waste Treatment Company, Ltd. Utilities accounted for 15% of revenue. The resort amenities segment encompasses the operations of the Kapalua Club, a private, non-equity club providing its members special programs, access and other privileges and certain amenities at the Kapalua Resort, including a 30,000 square foot full-service spa and a private pool-side dining beach club. This segment accounted for 6% of revenue in 2015.

FINANCIAL DATA: *Note: Data for latest year may not have been available at press time.*

In U.S. $	2015	2014	2013	2012	2011	2010
Revenue	22,786,000	33,007,000	15,212,000	16,164,000	14,542,000	41,954,000
R&D Expense						
Operating Income	9,332,000	19,905,000	-540,000	-2,393,000	-7,282,000	21,772,000
Operating Margin %	40.95%	60.30%	-3.54%	-14.80%	-50.07%	51.89%
SGA Expense	3,431,000	2,770,000	3,825,000	4,261,000	7,063,000	11,094,000
Net Income	6,813,000	17,635,000	-1,164,000	-4,602,000	5,078,000	24,752,000
Operating Cash Flow	10,621,000	-2,736,000	-3,586,000	-3,773,000	-10,225,000	-9,389,000
Capital Expenditure		31,000	4,000	209,000	1,025,000	4,276,000
EBITDA	11,640,000	22,419,000	2,340,000	840,000	-3,227,000	30,303,000
Return on Assets %	14.21%	34.23%	-2.02%	-7.33%	6.57%	22.66%
Return on Equity %						
Debt to Equity						

CONTACT INFORMATION:
Phone: 808 877-1608   Fax: 808 442-1116
Toll-Free:
Address: 161 S. Wakea Ave., P.O. Box 187, Maui, HI 96733 United States

STOCK TICKER/OTHER:
Stock Ticker: MLP   Exchange: NYS
Employees: 17   Fiscal Year Ends: 12/31
Parent Company:

SALARIES/BONUSES:
Top Exec. Salary: $275,000   Bonus: $
Second Exec. Salary: Bonus: $
$185,000

OTHER THOUGHTS:
Estimated Female Officers or Directors: 1
Hot Spot for Advancement for Women/Minorities:

Sales, profits and employees may be estimates. Financial information, benefits and other data can change quickly and may vary from those stated here.

McCarthy Building Companies Inc

www.mccarthy.com

NAIC Code: 236220

TYPES OF BUSINESS:

Construction & Engineering Services

BRANDS/DIVISIONS/AFFILIATES:

CONTACTS: Note: Officers with more than one job title may be intentionally listed here more than once.

Mike Bolen, CEO
Scott Wittkop, COO
Doug Audiffred, CFO
Mike Oster, CIO
Mike Bolen, Chmn.

GROWTH PLANS/SPECIAL FEATURES:

McCarthy Building Companies Inc., a wholly employee-owned firm, is one of the oldest and highly regarded commercial builders in the U.S. The firm is headquartered in St. Louis, Missouri, and has offices in Atlanta, Georgia; Houston & Dallas, Texas; Collinsville, Illinois; Kansas City, Kansas; Las Vegas, Nevada; Albuquerque, New Mexico; Newport Beach, Sacramento, San Diego, San Francisco & San Jose, California; and Phoenix, Arizona. McCarthy has undertaken projects in the markets of health care, education, higher education, green commercial construction, commercial, government, hospitality, entertainment, Native American, science & technology, parking, heavy/civil/transportation, industrial, water/wastewater and renewable energy. The firm's services include: preconstruction, construction and post-construction. The preconstruction services offered include budget development & estimating, value analysis, LEED/Green construction, constructability and virtual design & construction technology. The construction services offered are project scheduling, equipment & material procurement, self-performance, minority mentoring, quality assurance/quality planning and commissioning. The post-construction services offered are quality checks, closing out contractors, definition of an occupancy plan and following up on warranties & guarantees.

FINANCIAL DATA: Note: Data for latest year may not have been available at press time.

In U.S. $	2015	2014	2013	2012	2011	2010
Revenue	3,000,000,000	3,300,000,000	3,229,000,000			
R&D Expense						
Operating Income						
Operating Margin %						
SGA Expense						
Net Income						
Operating Cash Flow						
Capital Expenditure						
EBITDA						
Return on Assets %						
Return on Equity %						
Debt to Equity						

CONTACT INFORMATION:

Phone: 314-968-3300   Fax: 314-968-4780
Toll-Free:
Address: 1341 N. Rock Hill Rd., St. Louis, MO 63124-1498 United States

STOCK TICKER/OTHER:

Stock Ticker: Private   Exchange:
Employees: 2,800   Fiscal Year Ends:
Parent Company:

SALARIES/BONUSES:

Top Exec. Salary: $ Bonus: $
Second Exec. Salary: $ Bonus: $

OTHER THOUGHTS:

Estimated Female Officers or Directors:
Hot Spot for Advancement for Women/Minorities:

McDermott International Inc

www.mcdermott.com

NAIC Code: 541330

TYPES OF BUSINESS:

Engineering Services
Marine Construction
Procurement Services
Project Management

BRANDS/DIVISIONS/AFFILIATES:

McDermott Marine Construction Ghana Limited

CONTACTS: *Note: Officers with more than one job title may be intentionally listed here more than once.*

Stuart Spence, CFO
Gary Luquette, Chairman of the Board
Kelly Janzen, Chief Accounting Officer
David Dickson, Director
Liane Hinrichs, General Counsel
Richard Goins, Other Corporate Officer
Brian McLaughlin, Senior VP, Divisional
Jonathan Kennefick, Senior VP, Divisional
Stephen Allen, Senior VP, Divisional
Kathy Murray, Treasurer
Linh Austin, Vice President, Geographical
Hugh Cuthbertson, Vice President, Geographical
Scott Munro, Vice President, Geographical

GROWTH PLANS/SPECIAL FEATURES:

McDermott International, Inc. is a multinational engineering and construction services company specializing in offshore construction for oil and gas exploration projects. The firm operates in approximately 20 countries and divides its operations into three primary geographic segments: Asia Pacific, Americas and the Middle East. In the Asia Pacific segment, the company offers engineering, procurement, construction and installation (EPCI) services to energy companies primarily in Australia, Indonesia, Vietnam, Malaysia, India and Thailand. This division's primary fabrication facility in this region is located on Batam Island, Indonesia. The Americas segment serves customers primarily in the U.S., Brazil, Mexico and Trinidad. The Middle East segment, which includes McDermott's North Sea and Africa operations, serving customers primarily in Saudi Arabia, Qatar, the United Arab Emirates, Kuwait, India, Azerbaijan, Russia, the North Sea and Africa. All of the segments focus on the fabrication and installation of fixed and floating structures and the installation of pipelines and subsea systems. Its fabrication facilities construct a full range of offshore structures, from conventional jacket-type fixed platforms to intermediate water and deepwater platform configurations. Project installation is performed by major construction vessels stationed throughout the various regions, which McDermott either owns or operates. McDermott vessels offer a variety of complex installation services, including structural lifting/lowering and pipelay services. In April 2015, the firm formed a joint venture, McDermott Marine Construction Ghana Limited, which pursues key offshore opportunities in Ghana and other developments in the African market. Hydra Offshore Group of Accra, Ghana is the minority partner.

FINANCIAL DATA: *Note: Data for latest year may not have been available at press time.*

In U.S. $	2015	2014	2013	2012	2011	2010
Revenue	3,070,275,000	2,300,889,000	2,658,932,000	3,641,624,000	3,445,110,000	2,403,743,000
R&D Expense						
Operating Income	91,196,000	8,554,000	-464,790,000	319,327,000	250,723,000	314,905,000
Operating Margin %	2.97%	.37%	-17.48%	8.76%	7.27%	13.10%
SGA Expense	217,239,000	208,564,000	201,171,000	205,974,000	212,002,000	216,763,000
Net Income	-17,983,000	-75,994,000	-516,913,000	206,653,000	138,730,000	201,666,000
Operating Cash Flow	55,272,000	6,960,000	-256,611,000	209,784,000	97,446,000	340,174,000
Capital Expenditure	102,851,000	321,187,000	283,962,000	286,310,000	282,621,000	186,862,000
EBITDA	191,530,000	101,739,000	-361,743,000	455,115,000	334,211,000	382,830,000
Return on Assets %	-.52%	-2.43%	-16.83%	6.53%	4.96%	5.41%
Return on Equity %	-1.20%	-5.35%	-31.93%	11.65%	8.92%	12.38%
Debt to Equity	0.55	0.58	0.03	0.04	0.05	0.03

CONTACT INFORMATION:

Phone: 281 870-5000   Fax: 281 870-5095
Toll-Free:
Address: 757 N. Eldridge Pkwy., Houston, TX 77079 United States

STOCK TICKER/OTHER:

Stock Ticker: MDR   Exchange: NYS
Employees: 13,400   Fiscal Year Ends: 12/31
Parent Company:

SALARIES/BONUSES:

Top Exec. Salary: $850,000   Bonus: $
Second Exec. Salary: Bonus: $
$477,750

OTHER THOUGHTS:

Estimated Female Officers or Directors: 4
Hot Spot for Advancement for Women/Minorities: Y

McKinstry
www.mckinstry.com

NAIC Code: 238220

TYPES OF BUSINESS:
Mechanical Contractors
Engineering
Design-Build-Operate Contracts
Energy Efficient Systems Installation
Systems Monitoring

BRANDS/DIVISIONS/AFFILIATES:

CONTACTS: Note: Officers with more than one job title may be intentionally listed here more than once.
Dean C. Allen, CEO
Doug Moore, Pres.
Bill Teplicky, CFO
Ash Awad, CMO
Ron Johnson, COO

GROWTH PLANS/SPECIAL FEATURES:
McKinstry is a mechanical contractor operating as a full-service build, design, operate and maintain (BDOM) firm. The services offered by the firm include consulting, construction, energy and facility services. The consulting services offered include engineering, sustainability, energy/environmental planning, financial modeling, operational modeling and commissioning. The construction services offered include mechanical, electrical, fire protection, architectural metals, telecommunications, integrated delivery, energy retrofits and industrial. The energy services include performance-based contracting, active energy management, renewables, project funding, smart building systems, federal and Washington Energy Efficiency Grants. The facility services include mobile service, remote monitoring, operations management, facility management and 3D maintenance. The firm's current and recent projects include those related to: Enterprise School District, Montana State University, Parkrose School District, Texas State Technical College, City of Sedalia, Cameron County, City of Spokane, Harborview Medical Center, Oregon Health & Sciences University, University of Washington Medical Center and Clark Fork Valley Hospital.

FINANCIAL DATA: Note: Data for latest year may not have been available at press time.

In U.S. $	2015	2014	2013	2012	2011	2010
Revenue	500,000,000	500,000,000	523,000,000	500,000,000	459,000,000	
R&D Expense						
Operating Income						
Operating Margin %						
SGA Expense						
Net Income						
Operating Cash Flow						
Capital Expenditure						
EBITDA						
Return on Assets %						
Return on Equity %						
Debt to Equity						

CONTACT INFORMATION:
Phone: 206-762-3311   Fax: 206-762-2624
Toll-Free: 800-669-6223
Address: 5005 Third Ave. S., Seattle, WA 98134 United States

STOCK TICKER/OTHER:
Stock Ticker: Private
Employees: 1,800
Parent Company:
Exchange:
Fiscal Year Ends:

SALARIES/BONUSES:
Top Exec. Salary: $ Bonus: $
Second Exec. Salary: $ Bonus: $

OTHER THOUGHTS:
Estimated Female Officers or Directors: 2
Hot Spot for Advancement for Women/Minorities:

Meadow Valley Corporation

NAIC Code: 237310

TYPES OF BUSINESS:

Heavy Construction & Civil Engineering
Construction Materials-Concrete, Sand & Gravel
Design Services
Mining
Ready-Mix Concrete
Technical Services

BRANDS/DIVISIONS/AFFILIATES:

Meadow Valley Contractors Inc
Insight Equity Holdings LLC

CONTACTS: Note: Officers with more than one job title may be intentionally listed here more than once.

Norm Watkins, CEO
Ryan Evans, CFO
Nicole R. Smith, Controller
Norm Watkins, Dir.-Safety
Mark Krumm, Pres., Arizona Area
Robert Terril, Pres., Nevada Area
Shane Haycock, VP-Nevada Area

GROWTH PLANS/SPECIAL FEATURES:

Meadow Valley Corporation, operating as Meadow Valley Contractors, Inc., is involved in the construction industry. The construction services segment (CSS) specializes in structural concrete construction of highway bridges and overpasses and the paving of highways and airport runways. The company operates throughout Nevada, Arizona and Utah, with principal operations in the Las Vegas, Nevada and Phoenix, Arizona metropolitan areas. The firm is active in both public and private infrastructure projects, including the construction of bridges and overpasses, channels, roadways, highways and airport runways. Meadow Valley owns or leases most of the equipment used in its business, including cranes, backhoes, graders, loaders, trucks, trailers, pavers, rollers, construction materials processing plants, batch plants and related equipment. The raw materials necessary for most operations are obtained from multiple sources. The firm is owned by Dallas-based private equity firm Insight Equity Holdings LLC.

FINANCIAL DATA: Note: Data for latest year may not have been available at press time.

In U.S. $	2015	2014	2013	2012	2011	2010
Revenue						
R&D Expense						
Operating Income						
Operating Margin %						
SGA Expense						
Net Income						
Operating Cash Flow						
Capital Expenditure						
EBITDA						
Return on Assets %						
Return on Equity %						
Debt to Equity						

CONTACT INFORMATION:

Phone: 602-437-5400   Fax: 602-437-1681
Toll-Free:
Address: 3333 E. Camelback Rd., Ste. 240, Phoenix, AZ 85018 United States

STOCK TICKER/OTHER:

Stock Ticker: Private   Exchange:
Employees: 510   Fiscal Year Ends: 12/31
Parent Company: Insight Equity Holdings LLC

SALARIES/BONUSES:

Top Exec. Salary: $ Bonus: $
Second Exec. Salary: $ Bonus: $

OTHER THOUGHTS:

Estimated Female Officers or Directors: 1
Hot Spot for Advancement for Women/Minorities:

Melia Hotels International SA

www.solmelia.com

NAIC Code: 721110

TYPES OF BUSINESS:

Hotel & Resort Management
Real Estate Development

BRANDS/DIVISIONS/AFFILIATES:

Melia Hotels
TRYP by Wyndham
Sol Hotels
Paradisus Resorts
Gran Melia
ME by Melia
Innside
Club Melia

CONTACTS: *Note: Officers with more than one job title may be intentionally listed here more than once.*

Gabriel Escarrer Jaume, CEO
Luis Del Olmo Pinero, Exec. VP-Mktg.
Gabriel Canaves, Exec. VP-Human Resources
Pilar Dols, Exec. VP-Admin.
Juan Ignacio Pardo, Exec. VP-Legal & Compliance
Pilar Dols, Exec. VP-Finance
Andre Gerondeau, Exec. VP-Hotels
Mark Hoddinott, Exec. VP-Real Estate
Onofre Servera, Exec. VP-Club Melia
Gabriel Escarrer Julia, Chmn.

GROWTH PLANS/SPECIAL FEATURES:

Melia Hotels International SA is a leading Spanish hotel chain in the city and resort markets. It manages and operates more than 350 hotels and resorts under management or franchise agreements in 35 countries. The company has locations throughout Europe, the Americas, the Mediterranean, the Middle East and Asia Pacific. In the hotels segment, the firm's brands include Sol Hotels, TRYP by Wyndham, Melia Hotels, Gran Melia, Paradisus Resorts, ME by Melia and Innside. Sol Hotels are located in major tourist destinations in the Mediterranean and the Caribbean. The TRYP by Wyndham brand is designed to appeal to upscale business travelers by providing the latest in technology services and conference facilities within the setting of a scenic resort. Melia Hotels are usually in the four- to five-star range, offering luxury accommodations and amenities. Gran Melia operates in the premium luxury market of the hotel sector. Paradisus Resorts cater to destination-driven resort vacations. The ME by Melia brand of luxury hotels, located in urban and resort destinations, integrates contemporary cuisine, design and music. Innside is an upscale German hotel chain featuring modern, minimalist architecture and the latest in communications technology for business travelers. In addition, the company developed the YHI Spa brand to offer spa services at select locations. Club Melia is a chain of membership resorts offering family villas and condominiums. The firm also offers MeliaRewards, a four tier rewards program where cardholders can earn points to trade in with over 35 Melia partners. The company's Asia-Pacific division is focused on expanding the firm's brands into China and also attracting Chinese travelers to its international properties. In February 2016, it announced plans to open 25 new hotels that year, with two already opened (Sol Costa Atlantis in Spain and Melia Braco Village in Jamaica).

FINANCIAL DATA: *Note: Data for latest year may not have been available at press time.*

In U.S. $	2015	2014	2013	2012	2011	2010
Revenue		1,670,013,000	1,541,935,000	1,553,810,000	1,522,932,000	1,426,467,000
R&D Expense						
Operating Income		151,009,900	-31,507,400	181,741,800	170,050,500	161,776,200
Operating Margin %		9.04%	-2.04%	11.69%	11.16%	11.34%
SGA Expense						
Net Income		34,677,980	-83,506,120	42,571,360	47,880,380	59,301,330
Operating Cash Flow						
Capital Expenditure		46,194,730	39,733,810	45,934,700	153,738,000	187,761,300
EBITDA		260,414,400	72,430,740	298,188,900	251,303,000	271,311,900
Return on Assets %		.92%	-2.13%	1.09%	1.18%	1.53%
Return on Equity %		2.62%	-6.66%	3.47%	3.83%	5.05%
Debt to Equity		0.79	0.98	0.91	1.13	1.07

CONTACT INFORMATION:

Phone: 34 971224554   Fax: 34-971-22-44-08
Toll-Free:
Address: Calle Gremio Toneleros 24, Palma de Mallorca, 07009 Spain

STOCK TICKER/OTHER:

Stock Ticker: SMIZF
Employees: 19,514
Parent Company:

Exchange: PINX
Fiscal Year Ends: 12/31

SALARIES/BONUSES:

Top Exec. Salary: $ Bonus: $
Second Exec. Salary: $ Bonus: $

OTHER THOUGHTS:

Estimated Female Officers or Directors: 1
Hot Spot for Advancement for Women/Minorities:

Meritage Homes Corp

NAIC Code: 236117

www.meritagehomes.com

TYPES OF BUSINESS:

Construction, Home Building and Residential
Home Design Services
Home Building Services

BRANDS/DIVISIONS/AFFILIATES:

Monterey Homes
Meritage Homes
Legendary Communities

CONTACTS: Note: Officers with more than one job title may be intentionally listed here more than once.

Steven Hilton, CEO
Hilla Sferruzza, CFO
Phillippe Lord, COO
C. White, Executive VP
Javier Feliciano, Executive VP
Brent Anderson, Vice President, Divisional

GROWTH PLANS/SPECIAL FEATURES:

Meritage Homes Corp. is a designer and builder of single-family attached and detached homes in the southern and western U.S. It offers a variety of homes that are designed to appeal to a wide range of homebuyers, including first-time, move-up, luxury and active adult buyers. The company has operations in three regions: west, central and east, which are comprised of nine states: Arizona, California, Colorado, Texas, Florida, Georgia, North Carolina, South Carolina and Tennessee. Meritage's homebuilding and marketing activities are conducted under the Meritage Homes brand in each of its markets, except for certain communities in Arizona and in Texas, where it operates as Monterey Homes. The company sells homes in 229 communities, with prices ranging from roughly $130,000 to $1,230,000. Doing business primarily on an individual basis with contractors and suppliers, Meritage usually completes its homes within two to four months of construction. Main strategies of the company include purchasing land subject to complete entitlement, developing smaller parcels that can be completed within a three-year period and managing housing inventory by pre-selling and obtaining substantial customer deposits on homes prior to beginning construction. The company is also focused on offering more affordable houses in order to target first time homebuyers. In 2014, the firm acquired the homebuilding assets and operations of Legendary Communities.

FINANCIAL DATA: Note: Data for latest year may not have been available at press time.

In U.S. $	2015	2014	2013	2012	2011	2010
Revenue	2,592,556,000	2,190,633,000	1,833,879,000	1,193,674,000	861,244,000	941,656,000
R&D Expense						
Operating Income	206,713,000	207,455,000	194,146,000	55,181,000	2,012,000	31,114,000
Operating Margin %	7.97%	9.47%	10.58%	4.62%	.23%	3.30%
SGA Expense	301,267,000	261,340,000	218,226,000	163,018,000	139,096,000	136,582,000
Net Income	128,738,000	142,241,000	124,464,000	105,163,000	-21,106,000	7,150,000
Operating Cash Flow	-3,335,000	-211,248,000	-86,276,000	-220,487,000	-74,136,000	32,551,000
Capital Expenditure	16,092,000	20,788,000	15,783,000	10,863,000	7,082,000	6,389,000
EBITDA	219,670,000	225,194,000	202,698,000	61,294,000	17,201,000	44,180,000
Return on Assets %	5.14%	6.58%	6.95%	7.51%	-1.72%	.57%
Return on Equity %	10.87%	14.58%	16.21%	17.77%	-4.26%	1.45%
Debt to Equity	0.89	0.84	1.07	1.04	1.24	1.21

CONTACT INFORMATION:

Phone: 480 515-8100   Fax: 480 998-9162
Toll-Free: 877-275-6374
Address: 8800 E. Raintree Drive, Ste 300, Scottsdale, AZ 85260 United States

STOCK TICKER/OTHER:

Stock Ticker: MTH
Employees: 1,300
Parent Company:

Exchange: NYS
Fiscal Year Ends: 12/31

SALARIES/BONUSES:

Top Exec. Salary: $1,000,000   Bonus: $
Second Exec. Salary: $600,000   Bonus: $

OTHER THOUGHTS:

Estimated Female Officers or Directors:

Hot Spot for Advancement for Women/Minorities:

Meritus Hotels & Resorts Inc

www.meritushotels.com

NAIC Code: 721110

TYPES OF BUSINESS:

Hotels & Resorts

BRANDS/DIVISIONS/AFFILIATES:

Overseas Union Enterprise Ltd
Mandarin Orchard Singapore
Marina Mandarin Singapore
Meritus Pelangi Beach Resort & Spa, Langkawi

GROWTH PLANS/SPECIAL FEATURES:

Meritus Hotels & Resorts, Inc. is a hotel management firm based in Singapore. Meritus is a subsidiary of Overseas Union Enterprise Ltd., operating under the real estate developer's hospitality division. Meritus operates two hotels in Singapore, the Mandarin Orchard Singapore and the Marina Madarin Singapore; and one resort in Malaysia, the Meritus Pelangi Beach Resort & Spa, Langkawi. Mandarin Orchard is the firm's flagship five-star hotel, featuring two buildings, with 36 stories and 40 stories respectively. Together, they offer more than 1,000 rooms and suites, club lounges, free Wi-Fi, restaurants and walking distance to shopping and transportation. Marina Mandarin Singapore is a 575-room hotel that features free Wi-Fi, Halal-certified kitchens, covered walkways to city destinations and MRT stations, a swimming pool and views of Marina Bay. Meritus Pelangi is a 355-room hotel designed in the style of a traditional Malay Village. Meritus Pelangi features Wi-Fi, nearby restaurants and shops and a view of the ocean.

CONTACTS: Note: Officers with more than one job title may be intentionally listed here more than once.

Michael Ow Kum Fei, Pres.
Tan Choon Kwang, COO
Kim Seng Tan, Pres.

FINANCIAL DATA: Note: Data for latest year may not have been available at press time.

In U.S. $	2015	2014	2013	2012	2011	2010
Revenue						
R&D Expense						
Operating Income						
Operating Margin %						
SGA Expense						
Net Income						
Operating Cash Flow						
Capital Expenditure						
EBITDA						
Return on Assets %						
Return on Equity %						
Debt to Equity						

CONTACT INFORMATION:

Phone: 65-6235-7788   Fax: 65-6235 6688
Toll-Free:
Address: 333 Orchard Rd., 37th Fl., Main Tower, Singapore, 238867 Singapore

STOCK TICKER/OTHER:

Stock Ticker: Subsidiary   Exchange:
Employees: Fiscal Year Ends:
Parent Company: Overseas Union Enterprise Ltd

SALARIES/BONUSES:

Top Exec. Salary: $ Bonus: $
Second Exec. Salary: $ Bonus: $

OTHER THOUGHTS:

Estimated Female Officers or Directors: 3
Hot Spot for Advancement for Women/Minorities: Y

MFA Financial Inc

NAIC Code: 525990

www.mfafinancial.com

TYPES OF BUSINESS:

Mortgage REIT
Apartment Properties

BRANDS/DIVISIONS/AFFILIATES:

CONTACTS: Note: Officers with more than one job title may be intentionally listed here more than once.

Shira Siry, Assistant Secretary
Stephen Yarad, CFO
George Krauss, Chairman of the Board
Kathleen Hanrahan, Chief Accounting Officer
Matthew Ottinger, Controller
William Gorin, Director
Ronald Freydberg, Executive VP
Craig Knutson, President
Harold Schwartz, Secretary
Gudmundur Kristjansson, Senior Vice President
Elwin Ford, Senior VP
Bryan Wulfsohn, Senior VP
Peter Kollydas, Senior VP
Sunil Yadav, Senior VP
Terence Meyers, Senior VP, Divisional

GROWTH PLANS/SPECIAL FEATURES:

MFA Financial, Inc. is a holding company that conducts its real estate finance businesses primarily through wholly-owned subsidiaries. The subsidiaries invest, on a leveraged basis, in agency, non-agency adjustable-rate mortgage-backed securities (ARM-MBS) and residential whole loans. Mortgages collateralizing MFA's agency MBS portfolio are predominantly hybrids, 15-year fixed-rate mortgages and ARMs. Its non-agency MBS portfolio primarily consists of legacy non-agency MBS and MBS collateralized by re-performing and non-performing loans (RPL/NPL MBS). Re-performing loans are typically characterized by borrowers who have experienced payment delinquencies in the past and the amount owned on the mortgage may exceed the value of the property pledged as collateral. These loans are purchased at discount prices to the contractual loan balance in order to reflect the credit history of the borrower, the loan-to-value (LTV) of the loan and the coupon. Non-performing loans are typically characterized by borrowers who have defaulted on their obligations and/or have payment delinquencies of 60 days or more at the time MFA acquires the loan. These loans are also purchased at discounted prices. The company additionally invests in re-performing and non-performing residential whole loans through its interests in certain consolidated trusts. MFA's strategy of combining investments in agency MBS, non-agency MBS and residential whole loans is designed to generate attractive returns with less overall sensitivity to changes in the yield curve, the level of interest rates and prepayments.

FINANCIAL DATA: Note: Data for latest year may not have been available at press time.

In U.S. $	2015	2014	2013	2012	2011	2010
Revenue	365,655,000	358,794,000	340,679,000	347,908,000	347,593,000	296,086,000
R&D Expense						
Operating Income	313,226,000	313,504,000	302,709,000	306,839,000	316,414,000	269,762,000
Operating Margin %	85.66%	87.37%	88.85%	88.19%	91.03%	91.10%
SGA Expense	42,045,000	40,745,000	33,689,000	33,569,000	30,209,000	24,663,000
Net Income	313,226,000	313,504,000	302,709,000	306,839,000	316,414,000	269,762,000
Operating Cash Flow	282,174,000	255,813,000	298,088,000	310,387,000	334,409,000	245,938,000
Capital Expenditure	1,560,000	786,000	373,000	443,000	2,343,000	438,000
EBITDA						
Return on Assets %	2.33%	2.40%	2.19%	2.36%	3.01%	2.85%
Return on Equity %	9.66%	9.40%	8.83%	10.28%	12.98%	11.84%
Debt to Equity	0.21	0.22	0.27	0.37	0.47	0.09

CONTACT INFORMATION:

Phone: 212 207-6400   Fax: 212 935-8765
Toll-Free: 800-892-7547
Address: 350 Park Ave., 20th Fl., New York, NY 10022 United States

STOCK TICKER/OTHER:

Stock Ticker: MFA
Employees: 53
Parent Company:

Exchange: NYS
Fiscal Year Ends: 12/31

SALARIES/BONUSES:

Top Exec. Salary: $800,000   Bonus: $348,447
Second Exec. Salary: $700,000   Bonus: $283,795

OTHER THOUGHTS:

Estimated Female Officers or Directors: 3
Hot Spot for Advancement for Women/Minorities: Y

MGM Growth Properties LLC

www.mgmgrowthproperties.com

NAIC Code: 721120

TYPES OF BUSINESS:
Casino Hotels
Full Service Restaurants

BRANDS/DIVISIONS/AFFILIATES:
The Park in Las Vegas
Mandalay Bay
The Mirage
Monte Carlo
New York-New York
MGM Grand Detroit
Beau Rivage
Gold Strike Tunica

CONTACTS: Note: Officers with more than one job title may be intentionally listed here more than once.
James C. Stewart, CEO
Andy H. Chien, CFO

GROWTH PLANS/SPECIAL FEATURES:
MGM Growth Properties LLC is a real estate investment trust engaged in the acquisition, ownership and leasing of large-scale destination entertainment and leisure resorts, whose diverse amenities include casino gaming, hotel, convention, dining, entertainment and retail offerings. The portfolio of the firm consists of nine premier destination resorts operated by MGM Resorts International, including properties that are among the world's finest casino resorts, and The Park in Las Vegas. The Park in Las Vegas is a dining and entertainment complex located between New York-New York and Monte Carlo. Of MGM Growth's nine premier destination resorts, six are large-scale entertainment and gaming-related properties located on the Las Vegas Strip including Mandalay Bay, The Mirage, Monte Carlo, New York-New York, Luxor and Excalibur. Outside of Las Vegas, the firm owns three market leading casino resort properties including MGM Grand Detroit in Detroit, Michigan and Beau Rivage and Gold Strike Tunica, both of which are located in Mississippi. Combined, these properties comprise 24,466 hotel rooms, approximately 2.5 million convention square footage, over 100 retail outlets, over 200 food and beverage outlets and approximately 20 entertainment venues. In April 2016, MGM Growth announced both the pricing of an initial public offering (IPO) and the completion and closing of that same IPO.

FINANCIAL DATA: Note: Data for latest year may not have been available at press time.

In U.S. $	2015	2014	2013	2012	2011	2010
Revenue						
R&D Expense						
Operating Income	-261,954,000	-246,242,000				
Operating Margin %						
SGA Expense	10,351,000	11,634,000				
Net Income	-261,954,000	-246,242,000				
Operating Cash Flow	-58,473,000	-59,980,000				
Capital Expenditure	129,308,000	90,504,000				
EBITDA	-65,138,000	-59,980,000				
Return on Assets %						
Return on Equity %						
Debt to Equity						

CONTACT INFORMATION:
Phone: 702-632-7777   Fax:
Toll-Free:
Address: 3950 Las Vegas Blvd. S., Las Vegas, NV 89109 United States

STOCK TICKER/OTHER:
Stock Ticker: MGP   Exchange: NYS
Employees: Fiscal Year Ends:
Parent Company:

SALARIES/BONUSES:
Top Exec. Salary: $ Bonus: $
Second Exec. Salary: $ Bonus: $

OTHER THOUGHTS:
Estimated Female Officers or Directors:
Hot Spot for Advancement for Women/Minorities:

MGM Resorts International

www.mgmresorts.com

NAIC Code: 721120

TYPES OF BUSINESS:
Casino Hotels & Resorts
Casino & Resort Management

BRANDS/DIVISIONS/AFFILIATES:
MGM China Holdings Ltd
MGM Hospitality
MGM Growth Properties LLC
Bellagio
MGM Grand Las Vegas
Mandalay Bay
Luxor
Circus Circus Las Vegas

CONTACTS: Note: Officers with more than one job title may be intentionally listed here more than once.
Andrew Hagopian III, Assistant General Counsel
James Murren, CEO
Daniel DArrigo, CFO
Robert Selwood, Chief Accounting Officer
Corey Sanders, COO
Willie Davis, Director Emeritus
Robert Baldwin, Director
Phyllis James, Executive VP, Divisional
John McManus, Executive VP
William Hornbuckle, President

GROWTH PLANS/SPECIAL FEATURES:

MGM Resorts International (MGM) is a leading international developer and operator of integrated hotels, resorts and casinos. MGM's wholly-owned domestic resorts include 16 properties located in Las Vegas, as well as eight within Michigan and Mississippi. These resorts include the following brand lines: Bellagio, MGM Grand Las Vegas, Mandalay Bay, The Mirage, Luxor, Excalibur, New York-New York, Monte Carlo, Circus Circus Las Vegas, MGM Grand Detroit, Beau Rivage and Gold Strike. In China, MGM has a 51% stake in MGM Macau (with Macau SAR) via MGM China Holdings Ltd. Other properties include 50%-owned CityCenter, located in Las Vegas, Nevada; 50%-owned Borgata, located in Atlantic City, New Jersey; and 50%-owned Grand Victoria, located in Elgin, Illinois. These properties total more than 47,500 guest rooms and suites, 1.9 million square feet of casino gaming space, nearly 30,000 slot machines and over 1,700 gaming tables. Wholly-owned subsidiary MGM Hospitality designs, develops and manages luxury non-gaming hotels, resorts and residences in China, including the MGM Grand Sanya. MGM Growth Properties LLC is a real estate investment trust which contributes to the real estate development and management of the company's properties. Scheduled to open in late-2016, MGM National Harbor is a casino resort located in Prince George's County, Maryland, and is expected to comprise 3,600 slots, 160 gaming tables, a 300-room hotel, luxury spa, rooftop pool, 93,100 square feet of high-end retail, fine & casual dining restaurants, a 3,000-seat theater venue, as well as 50,000 square feet of meeting/event space. MGM Springfield is a casino-hotel establishment expected to open in late-2018. In 2015, the firm sold Railroad Pass, Gold Strike, Circus Circus Reno and its 50% stake in Silver Legacy.

The firm offers employees health insurance, savings plans, employee assistance & wellness programs, child development center, life & disability insurance, auto/home/renter/pet insurance and adoption assistance.

FINANCIAL DATA: Note: Data for latest year may not have been available at press time.

In U.S. $	2015	2014	2013	2012	2011	2010
Revenue	9,190,068,000	10,081,980,000	9,809,663,000	9,160,844,000	7,849,312,000	6,019,233,000
R&D Expense						
Operating Income	-156,232,000	1,323,538,000	1,111,512,000	126,908,000	-3,025,958,000	-1,080,497,000
Operating Margin %	-1.70%	13.12%	11.33%	1.38%	-38.55%	-17.95%
SGA Expense	3,158,924,000	1,637,819,000	1,633,270,000	2,182,830,000	1,535,758,000	2,708,765,000
Net Income	-447,720,000	-149,873,000	-156,606,000	-1,767,691,000	3,114,637,000	-1,437,397,000
Operating Cash Flow	1,005,079,000	1,130,670,000	1,310,448,000	909,351,000	675,126,000	504,014,000
Capital Expenditure	1,466,819,000	957,041,000	562,124,000	422,763,000	301,244,000	207,491,000
EBITDA	571,219,000	2,043,712,000	1,794,337,000	309,842,000	4,735,609,000	-469,022,000
Return on Assets %	-1.72%	-.56%	-.59%	-6.54%	13.33%	-6.93%
Return on Equity %	-9.72%	-3.60%	-3.64%	-33.82%	68.56%	-41.85%
Debt to Equity	2.41	3.15	3.17	3.15	2.24	4.08

CONTACT INFORMATION:
Phone: 702 693-7120   Fax: 702 693-8626
Toll-Free:
Address: 3600 Las Vegas Blvd. S., Las Vegas, NV 89109 United States

STOCK TICKER/OTHER:
Stock Ticker: MGM   Exchange: NYS
Employees: 62,000   Fiscal Year Ends: 12/31
Parent Company:

SALARIES/BONUSES:
Top Exec. Salary: Bonus: $
$2,000,000
Second Exec. Salary: Bonus: $
$1,650,000

OTHER THOUGHTS:
Estimated Female Officers or Directors: 5

Hot Spot for Advancement for Women/Minorities: Y

Mid-America Apartment Communities Inc

www.maac.com

NAIC Code: 0

TYPES OF BUSINESS:

Real Estate Investment Trust
Apartment Communities

BRANDS/DIVISIONS/AFFILIATES:

Mid-America Apartments LP
Mid-America Multifamily Fund II LLC

CONTACTS: Note: Officers with more than one job title may be intentionally listed here more than once.

H. Bolton, CEO
Albert Campbell, CFO
Thomas Grimes, Executive VP
Robert DelPriore, Executive VP
Leslie Wolfgang, Other Corporate Officer

GROWTH PLANS/SPECIAL FEATURES:

Mid-America Apartment Communities, Inc. (MAAC) is a self-administered and self-managed real estate investment trust (REIT) that acquires, owns and operates apartment communities in the south-east and south-central U.S. MAAC manages approximately 82,316 apartment homes in 268 communities as well as two commercial properties that total 293,242 square feet, conducting business primarily through its subsidiary and sole general partner, Mid-America Apartments LP. The firm also has a 33.33% equity stake in Mid-America Multifamily Fund II LLC, which owns eight properties located in Texas, Virginia, Florida and Georgia. MAAC's current investment focus involves acquiring relatively new properties (7-15 years old) within its present geographic area with the potential for above-average growth. It also provides onsite property management, through which the firm handles matters concerning rent, occupancy limits and regulations. A majority of MAAC property managers are Certified Apartment Managers, a designation established by the National Apartment Association. The company maintains each property's physical condition through regular landscaping and exterior improvements.

MAAC offers its employees life, disability, medical and dental insurance; a 401(k); a stock purchase plan; bonuses; apartment discounts; scholarships; and an employee assistance program. In addition, as employees stay with the firm over time, benefit packages are enhanced with higher bonuses, higher company contribution to retirement savings plans, decreased health care costs and additional vacation and leave time.

FINANCIAL DATA: Note: Data for latest year may not have been available at press time.

In U.S. $	2015	2014	2013	2012	2011	2010
Revenue	1,042,779,000	989,296,000	634,734,000	497,165,000	448,992,000	402,229,000
R&D Expense						
Operating Income	288,131,000	228,217,000	116,891,000	130,276,000	99,729,000	90,644,000
Operating Margin %	27.63%	23.06%	18.41%	26.20%	22.21%	22.53%
SGA Expense	56,706,000	53,004,000	43,754,000	35,846,000	38,823,000	30,389,000
Net Income	332,287,000	147,980,000	115,281,000	105,223,000	48,821,000	29,761,000
Operating Cash Flow	463,721,000	384,126,000	264,592,000	210,967,000	172,288,000	133,794,000
Capital Expenditure	447,636,000	420,464,000	203,646,000	373,643,000	427,562,000	323,017,000
EBITDA	769,661,000	576,815,000	306,821,000	258,511,000	219,058,000	197,336,000
Return on Assets %	4.85%	2.16%	2.40%	3.98%	2.07%	.86%
Return on Equity %	11.29%	5.07%	5.97%	12.89%	7.89%	3.80%
Debt to Equity	1.14	1.21	1.17	1.83	2.29	2.89

CONTACT INFORMATION:

Phone: 901 682-6600   Fax: 901 682-6667
Toll-Free:
Address: 6584 Poplar Ave., Memphis, TN 38138 United States

SALARIES/BONUSES:

Top Exec. Salary: $596,538   Bonus: $1,609
Second Exec. Salary: $358,385   Bonus: $1,673

STOCK TICKER/OTHER:

Stock Ticker: MAA   Exchange: NYS
Employees: 1,989   Fiscal Year Ends: 12/31
Parent Company:

OTHER THOUGHTS:

Estimated Female Officers or Directors: 7
Hot Spot for Advancement for Women/Minorities: Y

Millennium & Copthorne Hotels plc   www.mill-cop.com

NAIC Code: 721110

TYPES OF BUSINESS:

Hotels, Luxury
Restaurants
Property Management
Apartments
Conference & Event Centers
Casino
Theaters
Fitness & Spa Facilities

BRANDS/DIVISIONS/AFFILIATES:

Leng's Collection
Millennium Collection
Copthorn Collection
Hard Days Night Hotel

CONTACTS: Note: Officers with more than one job title may be intentionally listed here more than once.

Wong Hong Ren, CEO
Andrew Cherry, Interim CFO
Kwek Leng Beng, Chmn.

GROWTH PLANS/SPECIAL FEATURES:

Millennium & Copthorne Hotels plc (MCH) is a global hotel company that owns, manages and/or operates 120 hotels worldwide. MCH hotels are individually tailored to their location, yet seamlessly blend Asian hospitality with Western comfort. The company's hotels and resorts are grouped into three collections: Leng's Collection, Millennium Collection and Copthorne Collection. The Leng's Collection of hotels comprise unique historic properties as well as trendy urban properties under the following brands: The Bailey's Hotel, The Chelsea Harbour Hotel, Grand Hotel Palace Rome, M Hotels, Studio M Hotels and M Social. The Millennium Collection of hotels are created with timeless elegance, and are famous for their conference and banquet offerings. Brands include Grand Millennium Hotels and Millennium Hotels. The Copthorne Collection of hotels are reasonably-priced accommodations for both the business and leisure traveler. Brands include Copthorne Hotels and Kingsgate Hotels. Headquartered in London, England, the firm has offices in the U.S., Asia, Australia and the United Arab Emirates. In 2015, the firm acquired Beatles-inspired Hard Days Night Hotel, located in the heart of Liverpool.

FINANCIAL DATA: Note: Data for latest year may not have been available at press time.

In U.S. $	2015	2014	2013	2012	2011	2010
Revenue	1,197,495,000	1,167,805,000	1,466,825,000	1,086,228,000	1,160,029,000	1,051,448,000
R&D Expense						
Operating Income	158,346,400	275,692,400	325,741,200	234,409,200	253,778,400	168,101,700
Operating Margin %	13.22%	23.60%	22.20%	21.58%	21.87%	15.98%
SGA Expense	483,522,000	459,487,300	490,732,400	474,756,400		
Net Income	91,897,460	155,518,800	323,054,900	190,864,000	227,481,600	136,008,300
Operating Cash Flow	250,243,900	397,279,800	150,287,700	225,219,500	230,874,700	233,136,800
Capital Expenditure	120,173,600	606,523,300	113,528,700	78,890,440	152,267,000	26,720,960
EBITDA	268,623,400	370,417,500	439,976,800	308,351,300	337,758,500	240,064,400
Return on Assets %	1.59%	2.93%	6.76%	4.01%	4.94%	3.23%
Return on Equity %	2.86%	4.83%	10.23%	6.36%	8.01%	5.20%
Debt to Equity	0.29	0.22	0.10	0.07	0.15	0.16

CONTACT INFORMATION:

Phone: 44 2078722444   Fax: 44 2078722460
Toll-Free: 866-866-8086
Address: Scarsdale Place, Kensington, London, W8 5SR United Kingdom

STOCK TICKER/OTHER:

Stock Ticker: MLCTF   Exchange: GREY
Employees: 10,870   Fiscal Year Ends: 12/31
Parent Company:

SALARIES/BONUSES:

Top Exec. Salary: $ Bonus: $
Second Exec. Salary: $ Bonus: $

OTHER THOUGHTS:

Estimated Female Officers or Directors:
Hot Spot for Advancement for Women/Minorities:

MMA Capital Management LLC

www.mmacapitalmanagement.com

NAIC Code: 525990

TYPES OF BUSINESS:
Mortgage Investor
Real Estate Services

BRANDS/DIVISIONS/AFFILIATES:
MMA Energy Capital
International Housing Solutions
South Africa Workforce Housing Fund
HIS Residential Partners I
HIS Fund II

CONTACTS: Note: Officers with more than one job title may be intentionally listed here more than once.
J. Martin, Assistant Secretary
David Bjarnason, CFO
Francis Gallagher, Director
Michael Falcone, Director
Gary Mentesana, Executive VP

GROWTH PLANS/SPECIAL FEATURES:
MMA Capital Management LLC provides finance to the housing & infrastructure market in U.S. It operates through three business segments: U.S. operations, international operations and corporate operations. The U.S. operations consists of three lines of business, including leveraged bonds, low-income housing tax credits and other investments and obligations. Leveraged bonds finances affordable housing and infrastructure in the U.S. Low-income housing tax credits consists of secured loans from Morrison Grove Management LLC (MGM) and an option to purchase MGM in 2019 with its potential for long-term upside. Other investment and obligations includes legacy assets and serves as the firm's research and development unit for new business opportunities in the U.S., which has resulted in the creation of an energy capital business that is run under MMA Energy Capital. The international operations is run through International Housing Solutions (IHS), which aims to raise, invest in and manage private real estate funds. HIS currently manages three funds: the South Africa Workforce Housing Fund, a multi-investor fund; HIS Residential Partners I, a single-investor fund targeted at the emerging middle class in South Africa; and HIS Fund II, a multi-investor fund targeting investments in green affordable housing in South and Sub-Saharan Africa. The corporate operations segment include accounting, reporting, compliance and planning activities.

FINANCIAL DATA: Note: Data for latest year may not have been available at press time.

In U.S. $	2015	2014	2013	2012	2011	2010
Revenue	20,145,000	27,861,000	38,484,000	53,937,000	43,042,000	41,104,000
R&D Expense						
Operating Income	-48,032,000	-87,583,000	-49,055,000	-17,703,000	34,845,000	-34,098,000
Operating Margin %	-238.43%	-314.35%	-127.46%	-32.82%	80.95%	-82.95%
SGA Expense	18,956,000	16,155,000	17,017,000	15,460,000	16,095,000	19,645,000
Net Income	17,843,000	17,467,000	99,847,000	3,115,000	37,596,000	-28,709,000
Operating Cash Flow	-12,155,000	-1,195,000	2,866,000	4,947,000	14,400,000	20,489,000
Capital Expenditure						47,769,000
EBITDA						
Return on Assets %	2.80%	2.07%	9.07%	.17%	1.43%	-1.32%
Return on Equity %	17.18%	22.27%	231.91%	12.52%		
Debt to Equity	2.08	3.20	6.76	23.22	220.93	

CONTACT INFORMATION:
Phone: 443 263-2900   Fax: 410 727-5387
Toll-Free:
Address: 621 E. Pratt St., Ste. 600, Baltimore, MD 21202 United States

STOCK TICKER/OTHER:
Stock Ticker: MMAC   Exchange: NAS
Employees: 49   Fiscal Year Ends: 12/31
Parent Company:

SALARIES/BONUSES:
Top Exec. Salary: $555,000   Bonus: $370,000
Second Exec. Salary: $435,000   Bonus: $290,000

OTHER THOUGHTS:
Estimated Female Officers or Directors: 1
Hot Spot for Advancement for Women/Minorities:

MMR Group Inc

NAIC Code: 238210

TYPES OF BUSINESS:

Electrical Contractor
Construction
Maintenance

BRANDS/DIVISIONS/AFFILIATES:

Vector Electric and Controls Inc

GROWTH PLANS/SPECIAL FEATURES:

MMR Group, Inc., wholly owned by its management, is an electrical contractor performing services such as electrical and instrumentation construction, maintenance, management and technical services. The firm serves the alternative energy exploration & production; air separation & distribution; chemical & petrochemical; food & beverage; manufacturing; maritime/marine service; metals & minerals; oil & gas production & processing; pharmaceuticals; power generation; pulp, paper & forest production; refining; special projects; synthetic fuels; and waste & water treatment markets. Projects undertaken by the firm have included upstream, midstream and downstream oil & gas projects; communications projects; power generation projects; safety services projects; and industrial & manufacturing projects. In January 2016, the firm acquired Vector Electric and Controls, Inc.

CONTACTS: Note: Officers with more than one job title may be intentionally listed here more than once.

Allen Boudreaux, Pres.

FINANCIAL DATA: Note: Data for latest year may not have been available at press time.

In U.S. $	2015	2014	2013	2012	2011	2010
Revenue	845,000,000	800,000,000	839,000,000	712,000,000	375,000,000	
R&D Expense						
Operating Income						
Operating Margin %						
SGA Expense						
Net Income						
Operating Cash Flow						
Capital Expenditure						
EBITDA						
Return on Assets %						
Return on Equity %						
Debt to Equity						

CONTACT INFORMATION:

Phone: 225-756-5090   Fax: 225-753-7012
Toll-Free: 800-880-5090
Address: 15961 Airline Highway, Baton Rouge, LA 70817 United States

STOCK TICKER/OTHER:

Stock Ticker: Private   Exchange:
Employees: 2,000   Fiscal Year Ends:
Parent Company:

SALARIES/BONUSES:

Top Exec. Salary: $ Bonus: $
Second Exec. Salary: $ Bonus: $

OTHER THOUGHTS:

Estimated Female Officers or Directors:
Hot Spot for Advancement for Women/Minorities:

Morgans Hotel Group Co

www.morganshotelgroup.com

NAIC Code: 721110

TYPES OF BUSINESS:

Boutique Luxury Hotels & Resorts
Boutique Hotels
Property Management

BRANDS/DIVISIONS/AFFILIATES:

Mondrian
Hudson
Delano
Clift
Shore Club
SBE Entertainment Group

CONTACTS: Note: Officers with more than one job title may be intentionally listed here more than once.

Richard Szymanski, CEO
Howard Lorber, Chairman of the Board
Meredith Deutsch, Executive VP
Chadi Farhat, Former COO

GROWTH PLANS/SPECIAL FEATURES:

Morgans Hotel Group Co. (MHG) operates, owns, develops and redevelops luxury hotels. The firm, which is known for its establishment of the boutique hotel sector, primarily maintains hotels in gateway cities and select resort markets in the U.S. and Europe that feature avant-garde modern design. MHG owns or partially owns and operates 13 hotel properties in New York, Florida, California, London, Nevada and Istanbul. The firm's fully-owned and managed hotels include the Hudson, Delano South Beach and Clift. The company also owns and manages two hotels through long-term joint venture agreements. MHG has a 50% interest in Mondrian South Beach in Miami Beach and a small interest in the Shore Club in Miami Beach. Significant media attention has been devoted to MHG's hotels, which it attributes to its public spaces, modern design, celebrity guests and high-profile events for which its hotels are known. Designers of its hotels have included Phillippe Starck, Benjamin Noriega-Ortiz, Andree Putman and David Chipperfield. The firm also operates three restaurants at Mandalay Bay in Las Vegas. Upcoming hotels include the Mondrian in Doha, the Delano in Dubai, the Delano in Cartagena and the Delano in Cesme. In 2015, the firm sold its 90% interest in The Light Group and sold its 20% interest in Mondrian SoHo. In May 2016, MHG agreed to be acquired by the privately held SBE Entertainment Group.

FINANCIAL DATA: Note: Data for latest year may not have been available at press time.

In U.S. $	2015	2014	2013	2012	2011	2010
Revenue	219,982,000	234,960,992	236,486,000	189,919,008	207,332,000	236,370,000
R&D Expense						
Operating Income						
Operating Margin %						
SGA Expense						
Net Income	22,097,000	-50,724,000	-44,155,000	-55,687,000	-85,403,000	-81,409,000
Operating Cash Flow						
Capital Expenditure						
EBITDA						
Return on Assets %						
Return on Equity %						
Debt to Equity						

CONTACT INFORMATION:

Phone: 212 277-4100   Fax: 212 277-4260
Toll-Free:
Address: 475 Tenth Ave., New York, NY 10018 United States

STOCK TICKER/OTHER:

Stock Ticker: Private   Exchange:
Employees: 2,600   Fiscal Year Ends: 12/31
Parent Company: SBE Entertainment Group

SALARIES/BONUSES:

Top Exec. Salary: $ Bonus: $
Second Exec. Salary: $ Bonus: $

OTHER THOUGHTS:

Estimated Female Officers or Directors: 1
Hot Spot for Advancement for Women/Minorities:

Move Inc (Realtor.com)   www.move.com

NAIC Code: 519130

TYPES OF BUSINESS:
Online Portal-Real Estate Data
Real Estate Software
Real Estate Publishing
Real Estate Advertising

BRANDS/DIVISIONS/AFFILIATES:
Realtor.com
Move.com
News Corp
FiveStreet.com
ListHub
SeniorHousingNet.com
TigerLead.com
Moving.com

CONTACTS: *Note: Officers with more than one job title may be intentionally listed here more than once.*
Ryan O'Hara, CEO
Bryan Charap, CFO
Nate Johnson, CMO
Carol Brummer, Exec. VP-Human Resources
Raymond Picard, Executive VP, Divisional
James Caulfield, Executive VP
Errol Samuelson, President, Subsidiary

GROWTH PLANS/SPECIAL FEATURES:

Move, Inc. and its subsidiaries operate a leading network of web sites for real estate search, finance, moving and home enthusiasts and provide a resource for consumers seeking information and connections needed before, during and after a move. The firm is a subsidiary of News Corp. The company attracts approximately 30 million monthly consumers to its online network of websites. The firm's flagship consumer websites are Move.com, Realtor.com and Moving.com. The firm also offers Top Producer, a Software-as-a-Service (SaaS) program for real estate agents, complimentary to Realtor.com. Realtor.com, which is the official web site of the National Association of Realtors, offers consumers a suite of services, tools and content for all aspects of the residential real estate transaction. The company's ListHub business provides online real estate listings and performance reporting solutions for brokers, agents and real estate franchises. Doorsteps provides a step-by-step interactive guide to buying a home, and provides connects to agents. TigerLead.com delivers leads and training for realtors. FiveStreet.com is a lead follow-up software for realtors. Additionally, the company operates SeniorHousingNet.com, a searchable web site for locating housing for senior citizens.

FINANCIAL DATA: *Note: Data for latest year may not have been available at press time.*

In U.S. $	2015	2014	2013	2012	2011	2010
Revenue	260,000,000	253,000,000	227,032,992	199,232,992	191,724,000	197,503,008
R&D Expense						
Operating Income						
Operating Margin %						
SGA Expense						
Net Income		580,000	574,000	5,625,000	3,191,000	-20,855,000
Operating Cash Flow						
Capital Expenditure						
EBITDA						
Return on Assets %						
Return on Equity %						
Debt to Equity						

CONTACT INFORMATION:
Phone: 408 558-3700   Fax:
Toll-Free:
Address: 910 E. Hamilton Ave., 6th Fl., Campbell, CA 95008 United States

STOCK TICKER/OTHER:
Stock Ticker: Subsidiary   Exchange:
Employees: 943   Fiscal Year Ends: 12/31
Parent Company: News Corp

SALARIES/BONUSES:
Top Exec. Salary: $ Bonus: $
Second Exec. Salary: $ Bonus: $

OTHER THOUGHTS:
Estimated Female Officers or Directors: 5
Hot Spot for Advancement for Women/Minorities: Y

MRV Engenharia E Participacoes SA

www.mrv.com.br

NAIC Code: 531100

TYPES OF BUSINESS:
Real Estate Development

BRANDS/DIVISIONS/AFFILIATES:
Parque
Spazio
Minha Casa, Minha Vida
Village
LOG Commercial Properties
Urbamais Desenvolvimento Urbano

CONTACTS: Note: Officers with more than one job title may be intentionally listed here more than once.
Rafael Nazareth Menin Teixeira, CEO
Leonardo G. Correa, CFO
Homero Aguiar Paiva, Chief Prod. Officer
Maria F.N.M.T. de Souza Maia, Chief Legal Officer
Hudson Goncalves de Andrade, CEO-Real Estate Dev.
Monica F.G. Simao, Chief Investor Rel. Officer
Eduardo P. Barreto, Chief Commercial Officer
Jose A.T. Simao, CEO-Real Estate Financing
Junia M. de Sousa Lima Galvao, Chief Mgmt. & Shared Service Center Officer
Eduardo F.T. de Souza, Chief Regional Officer

GROWTH PLANS/SPECIAL FEATURES:
MRV Engenharia e Participacoes SA is a Brazilian developer and builder of low-income residential real estate. The firm offers two products directed to low-income customers: the Parque line, featuring units with between 430.56 and 592 square feet of usable area and a maximum sale price of approximately $55,000; and the Spazio line, which offers units with between 452 and 753.47 square feet of usable area and prices ranging between $38,500 and $77,000. Both the Parque and Spazio units are vertically designed condominiums with a maximum of five floors; the Village line of condominiums features a horizontal layout. MRV Engenharia operates in 134 cities located in the Federal District and the following states: Ceara, Maranhao, Rio Grande do Sul, Goias, Mato Grosso, Santa Catarina, Parana, Sao Paulo, Rio de Janeiro, Espirito Santo, Mato Grosso do Sul, Alagoas, Bahia, Paraiba, Minas Gerais, Pernambuco, Rio Grande do Norte and Sergipe. Approximately 84% of the firm's customers pay for their condominiums using CEF-associative credit financing; 10% utilize bank financing; 5% pay cash; and 1% use direct financing from MRV Engenharia. The company is a member of Minha Casa, Minha Vida (My Home, My Life), a program funded by the Brazilian government to promote homeownership in the country. The program aims to build 1 million new houses for families earning up to 10 times Brazil's minimum wage. Firm's subsidiaries LOG Commercial Properties develops, constructs and leases commercial properties and operates in 26 cities; and Urbamais Desenvolvimento Urbano, markets spaces in urban settlements, as well as designs single family residential areas, malls, shopping and business centers.

FINANCIAL DATA: Note: Data for latest year may not have been available at press time.

In U.S. $	2015	2014	2013	2012	2011	2010
Revenue	1,309,822,000	1,151,189,000	1,064,407,000	1,173,107,000	1,104,131,000	830,753,300
R&D Expense						
Operating Income	192,112,000	224,026,800	137,238,200	176,442,900	245,578,900	184,213,200
Operating Margin %	14.66%	19.46%	12.89%	15.04%	22.24%	22.17%
SGA Expense	202,731,800	169,301,000	141,324,700	139,292,400	117,059,700	85,242,550
Net Income	150,583,300	198,056,100	116,346,900	145,079,200	209,026,200	174,482,500
Operating Cash Flow	282,707,100	165,968,000	156,443,800	-14,935,650	-155,220,600	-103,677,000
Capital Expenditure	20,301,400	15,179,850	11,361,790	15,610,220	13,868,110	13,538,390
EBITDA	236,554,300	249,143,100	158,973,200	184,102,400	290,999,600	223,898,900
Return on Assets %	4.93%	6.85%	3.97%	5.20%	9.52%	11.38%
Return on Equity %	11.95%	16.96%	10.70%	14.60%	24.13%	24.08%
Debt to Equity	0.23	0.34	0.54	0.70	0.62	0.48

CONTACT INFORMATION:
Phone: 55-3136158153   Fax: 55 3133487155
Toll-Free:
Address: Avenida Professor Mario Werneck, 621, Estoril, Belo Horizonte, MG 30494 Brazil

STOCK TICKER/OTHER:
Stock Ticker: MRVNY
Employees:
Parent Company:

Exchange: PINX
Fiscal Year Ends: 12/31

SALARIES/BONUSES:
Top Exec. Salary: $ Bonus: $
Second Exec. Salary: $ Bonus: $

OTHER THOUGHTS:
Estimated Female Officers or Directors: 3
Hot Spot for Advancement for Women/Minorities: Y

MulvannyG2 Architecture

NAIC Code: 541310

www.mulvannyg2.com

TYPES OF BUSINESS:
Architectural Services

BRANDS/DIVISIONS/AFFILIATES:

CONTACTS: Note: Officers with more than one job title may be intentionally listed here more than once.
Mitch Smith, CEO
Celeste Lenon, COO

GROWTH PLANS/SPECIAL FEATURES:

MulvannyG2 Architecture is an architectural firm that partners with leading brands and developers around the world to craft design solutions in order to support business strategies. It creates retail, commercial, civic and mixed-used buildings. MulvannyG2's more than 50 worldwide current projects include the Shihao Center in Chengdu, China, a large-scale, mixed-use development retail-focused live/work Suning Wuhu Plaza, a one-stop shopping and entertainment destination, featuring a 10-story structure with an indoor ice skating rink, IMAX movie theater and dining options, located in Wuhu, China; Smith Dental, comprising a 4,500-square-foot (sf) office located in Nashville, Tennessee; Harborplace, a historic retail center located in Downtown Baltimore's Inner Harbor, a 1580,000 sf retail waterfront center, featuring shops and restaurants; and a Costco Wholesale headquarters campus, featuring a 529,000 sf development with a 180,000 sf office building that includes a two-story lobby, a three-story atrium in the cafeteria and a skybridge, located in Issaquah, Washington. Headquartered in Seattle, Washington, the firm has offices in California, Washington D.C. and Shanghai.

FINANCIAL DATA: Note: Data for latest year may not have been available at press time.

In U.S. $	2015	2014	2013	2012	2011	2010
Revenue	57,190,000	57,170,000	73,200,000			
R&D Expense						
Operating Income						
Operating Margin %						
SGA Expense						
Net Income						
Operating Cash Flow						
Capital Expenditure						
EBITDA						
Return on Assets %						
Return on Equity %						
Debt to Equity						

CONTACT INFORMATION:
Phone: 206-962-6500   Fax:
Toll-Free:
Address: 1101 Second Ave, Ste 100, Seattle, WA 98101 United States

STOCK TICKER/OTHER:
Stock Ticker: Private   Exchange:
Employees: 300   Fiscal Year Ends:
Parent Company:

SALARIES/BONUSES:
Top Exec. Salary: $ Bonus: $
Second Exec. Salary: $ Bonus: $

OTHER THOUGHTS:
Estimated Female Officers or Directors:
Hot Spot for Advancement for Women/Minorities:

MWH Global Inc

www.mwhglobal.com

NAIC Code: 237110

TYPES OF BUSINESS:

Engineering & Construction Services
Environmental Engineering
Water & Waste Treatment Analysis
Facilities Development
Infrastructure Asset Management
Consulting
Government Relations & Lobbying
Software & IT Services

BRANDS/DIVISIONS/AFFILIATES:

mCapitol Management
Innovyze
Hawksley Consulting

CONTACTS: *Note: Officers with more than one job title may be intentionally listed here more than once.*

Alan J. Krause, CEO
David G. Barnes, CFO
Jack Shandley, Chief Human Resources Officer
Claire Rutkowski, CIO
Jeff D'Agosta, General Counsel
Meg Vanderlaan, Chief Comm. Officer
Daniel McConville, Pres., Bus. Solutions Group
Joseph D. Adams, Pres., Energy & Industry
Paul F. Boulos, Pres., Innovyze
Bruce K. Howard, Exec. VP
Alan J. Krause, Chmn.
Wim Drossaert, Pres., Europe & Africa Gov't & Infrastructure

GROWTH PLANS/SPECIAL FEATURES:

MWH Global, Inc. is one of the world's leading providers of consulting, engineering, construction and management services with the primary aim of protecting, storing and distributing water. Through 180 offices in 35 countries, MWH specializes in proprietary software and process automation packages for the water & wastewater; energy & power; ports, waterways and coastal; industrial & commercial; transportation; natural resources & mining; and oil and gas industries. In the water &waste water market, the firm has designed or constructed almost 500 water treatment plants, 800 wastewater treatment plants, 200 reservoirs, 1,000 pumping stations and thousands of miles of pipeline on a global scale. In the energy & power market, the company has designed hydroelectric projects in 33 countries, as well as more than 300 rehabilitation projects, with a total installed capacity of 70,000 megawatts. Services for the ports, waterways and coastal industry include designing and building port infrastructure, locks, dams, levees, pump stations, canals and sea outfalls. For the industrial & commercial market, MWH provides services with the aim of limiting unnecessary resource consumption and waste generation. For the transportation market, the firm has road maintenance contracts for over 9,320 miles of highways and local roads in New Zealand and has designed provincial roads in 18 provinces in Vietnam. For the natural resources & mining and the oil and gas markets, the firm provides water/waste water storage, transmission and treatment; risk management; and energy infrastructure development. Some of its affiliates include mCapitol Management, which offers government relations, marketing and business development services; Innovyze, a wet infrastructure modeling and simulation software specialist; and Hawksley Consulting, a management consulting firm.

FINANCIAL DATA: *Note: Data for latest year may not have been available at press time.*

In U.S. $	2015	2014	2013	2012	2011	2010
Revenue	1,900,000,000	1,850,000,000	1,725,000,000	1,600,000,000	1,450,000,000	1,100,000,000
R&D Expense						
Operating Income						
Operating Margin %						
SGA Expense						
Net Income						
Operating Cash Flow						
Capital Expenditure						
EBITDA						
Return on Assets %						
Return on Equity %						
Debt to Equity						

CONTACT INFORMATION:

Phone: 303-533-1900   Fax: 303-533-1901
Toll-Free:
Address: 370 Interlocken Crescent, Ste. 200, Broomfield, CO 80021 United States

STOCK TICKER/OTHER:

Stock Ticker: Private   Exchange:
Employees: 7,000   Fiscal Year Ends: 09/30
Parent Company:

SALARIES/BONUSES:

Top Exec. Salary: $ Bonus: $
Second Exec. Salary: $ Bonus: $

OTHER THOUGHTS:

Estimated Female Officers or Directors: 2
Hot Spot for Advancement for Women/Minorities: Y

Sales, profits and employees may be estimates. Financial information, benefits and other data can change quickly and may vary from those stated here.

NAI Global Inc

NAIC Code: 531210

TYPES OF BUSINESS:

Brokerage Services
Consulting & Advisory Services
Property Management
Online Transaction Services
IT Services
Real Estate Software

BRANDS/DIVISIONS/AFFILIATES:

REALTrac Online
NAI Asia Pacific Regional Overview

CONTACTS: Note: Officers with more than one job title may be intentionally listed here more than once.

Jeffrey M. Finn, CEO
Edward J. Finn, COO
Margaret B. Smith, Exec.VP-Finance & Controller
Bobbi Jean Formosa, Exec. VP-Oper.
Mauro Keller Sarmiento, Managing Dir.-Int'l Bus.

GROWTH PLANS/SPECIAL FEATURES:

NAI Global, Inc. is an international commercial real estate network with more than 375 offices in 55 countries worldwide. The company manages over 380 million square feet of commercial space, with services providing local, regional, national and international business. The company offers leasing and brokerage agencies, disposition services, financial and investment strategies, leasing, property management, tenant representation, valuation and advice and practice groups. Some of NAI's value-added operations include auction services, build-to-suit development services, commercial real estate due diligence, cost segregation, facilities management, lease audits, relocation studies, tax appeals, roof portfolio management and title insurance services. NAI also runs several specialty segments focusing on government; life sciences; location advisory; logistics; NAFTA (North American Free Trade Agreement) advisory; process industries; supply chain services; and the office, industrial, retail, multifamily, hotel and land markets. The company also publishes market analysis reports and newsletters, including international and industry-specific reports, city-specific reports, country-specific reports and regional reports (the biggest being the NAI Asia Pacific Regional Overview). NAI runs REALTrac Online, an active platform for management of all company client transactions, with a leaning towards project coordination and transaction cycles. The company offers software for portfolio management, lease administration, online property marketing, financial analysis and demographics mapping.

FINANCIAL DATA: Note: Data for latest year may not have been available at press time.

In U.S. $	2015	2014	2013	2012	2011	2010
Revenue						
R&D Expense						
Operating Income						
Operating Margin %						
SGA Expense						
Net Income						
Operating Cash Flow						
Capital Expenditure						
EBITDA						
Return on Assets %						
Return on Equity %						
Debt to Equity						

CONTACT INFORMATION:

Phone: 212-405-2500   Fax: 609-945-4001
Toll-Free:
Address: 717 Fifth Avenue, 12th Fl., New York, NY 10022 United States

STOCK TICKER/OTHER:

Stock Ticker: Private   Exchange:
Employees: 6,700   Fiscal Year Ends: 06/30
Parent Company: C-III Capital Partners LLC

SALARIES/BONUSES:

Top Exec. Salary: $ Bonus: $
Second Exec. Salary: $ Bonus: $

OTHER THOUGHTS:

Estimated Female Officers or Directors: 1
Hot Spot for Advancement for Women/Minorities: Y

National Health Investors Inc

www.nhinvestors.com

NAIC Code: 0

TYPES OF BUSINESS:

Real Estate Investment Trust
Health Care Facilities
Long-Term Care Facilities
Retirement Centers
Mortgages

BRANDS/DIVISIONS/AFFILIATES:

CONTACTS: Note: Officers with more than one job title may be intentionally listed here more than once.

Eric Mendelsohn, CEO
Roger Hopkins, CFO
Kristin Gaines, Chief Credit Officer
W. Adams, Director
Kevin Pascoe, Executive VP, Divisional
John Spaid, Executive VP, Divisional
Susan Sidwell, Secretary

GROWTH PLANS/SPECIAL FEATURES:

National Health Investors, Inc. (NHI) is a self-managed, self-administered real estate investment trust (REIT). The firm invests in health care properties including long-term care facilities, acute care hospitals, medical office buildings, retirement centers and assisted living facilities. NHI's portfolio includes real estate and mortgage investments in 189 health care facilities located in 31 states. These facilities consist of 116 senior housing communities, 68 skilled nursing facilities, three hospitals, two medical office buildings and others. All of NHI's long-term care facilities provide some combination of skilled and intermediate nursing and rehabilitative care, including speech, physical and occupational therapy. The firm's medical office buildings are specifically configured office buildings whose tenants are primarily physicians and other medical practitioners. Each of NHI's medical office buildings is leased to one lessee, who then leases out individual office space. Its assisted living facilities are either free-standing or attached to long-term care or retirement facilities and provide basic room and board functions for the elderly. NHI's independent living centers offer specially designed residential units for the active and ambulatory elderly and provide various ancillary services for their residents, including restaurants, activity rooms and social areas. The company's hospitals offer a wide range of inpatient and outpatient services such as acute psychiatric care and rehabilitation. During 2015, the firm acquired five senior living communities in North Carolina, Indiana, Oregon, Michigan and Tennessee; and an assisted living & memory care community in Ohio.

FINANCIAL DATA: Note: Data for latest year may not have been available at press time.

In U.S. $	2015	2014	2013	2012	2011	2010
Revenue	228,988,000	177,509,000	117,828,000	96,953,000	82,702,000	78,396,000
R&D Expense						
Operating Income	126,719,000	103,123,000	75,868,000	69,741,000	61,864,000	58,498,000
Operating Margin %	55.33%	58.09%	64.38%	71.93%	74.80%	74.61%
SGA Expense	10,983,000	9,316,000	10,038,000	8,565,000	8,032,000	8,848,000
Net Income	148,862,000	101,609,000	106,183,000	90,731,000	81,132,000	69,421,000
Operating Cash Flow	164,425,000	126,143,000	104,193,000	86,266,000	76,854,000	77,544,000
Capital Expenditure	124,113,000	533,171,000	654,670,000	110,601,000	75,806,000	98,736,000
EBITDA	240,458,000	170,113,000	109,632,000	95,136,000	73,856,000	69,701,000
Return on Assets %	7.20%	5.90%	9.82%	14.11%	14.90%	14.33%
Return on Equity %	13.69%	11.24%	17.35%	20.14%	18.31%	15.82%
Debt to Equity	0.81	0.82	0.80	0.44	0.21	0.08

CONTACT INFORMATION:

Phone: 615 890-9100   Fax: 615 890-0123
Toll-Free:
Address: 222 Robert Rose Dr., Murfreesboro, TN 37129 United States

STOCK TICKER/OTHER:

Stock Ticker: NHI   Exchange: NYS
Employees: 12   Fiscal Year Ends: 12/31
Parent Company:

SALARIES/BONUSES:

Top Exec. Salary: $198,000   Bonus: $150,000
Second Exec. Salary: Bonus: $40,000
$286,841

OTHER THOUGHTS:

Estimated Female Officers or Directors: 1
Hot Spot for Advancement for Women/Minorities:

NBBJ

NAIC Code: 541310

TYPES OF BUSINESS:

Architectural Services

BRANDS/DIVISIONS/AFFILIATES:

CONTACTS: Note: Officers with more than one job title may be intentionally listed here more than once.

Steve McConnell, Managing Partner
Juli Cook, COO
Tim Leberecht, CMO

GROWTH PLANS/SPECIAL FEATURES:

NBBJ is an architectural firm that designs creative sustainable spaces. Clients of the company include Amazon, Google, Microsoft, Samsung and Tencent as well as institutional leaders such as Cambridge University, Cleveland Clinic, the Bill & Melinda Gates Foundation, Massachusetts General Hospital, NYU Medical Center and Stanford University. NBBJ specializes in architecture, interiors, planning and urban design, branding, consulting, landscape design and lighting. Its markets include civic, corporate, commercial, healthcare, higher education, science, sports and transportation. The company accepted the Architecture 2030 Challenge, a global initiative stating that all new buildings and major renovations reduce their fossil fuel GHG-emitting consumption by 50% by 2010, incrementally increasing the reduction for new buildings to carbon neutral by 2030. NBBJ was founded by four Seattle architects in 1943, and has offices in Beijing, Shanghai, London, Pune, Boston, Columbus, Los Angeles, New York, San Francisco and Seattle. Early 2016 projects occurring or about to occur include amazon.com's Denny Triangle Project in Seattle, Southeast Louisiana Veterans Health Care System Replacement Medical Center & Research Lab in New Orleans, Brigham and Women's Hospital Building for the Future (research lab and medical center) in Boston, New York University Langone Medical Center in New York (New Kimmel Pavilion) and Rupp Arena in Lexington.

FINANCIAL DATA: Note: Data for latest year may not have been available at press time.

In U.S. $	2015	2014	2013	2012	2011	2010
Revenue						
R&D Expense						
Operating Income						
Operating Margin %						
SGA Expense						
Net Income						
Operating Cash Flow						
Capital Expenditure						
EBITDA						
Return on Assets %						
Return on Equity %						
Debt to Equity						

CONTACT INFORMATION:

Phone: 206-223-5556   Fax:
Toll-Free:
Address: 223 Yale Ave. N., Seattle, WA 98109 United States

STOCK TICKER/OTHER:

Stock Ticker: Private   Exchange:
Employees: 703   Fiscal Year Ends:
Parent Company:

SALARIES/BONUSES:

Top Exec. Salary: $ Bonus: $
Second Exec. Salary: $ Bonus: $

OTHER THOUGHTS:

Estimated Female Officers or Directors:
Hot Spot for Advancement for Women/Minorities:

NCI Building Systems Inc

www.ncilp.com

NAIC Code: 332311

TYPES OF BUSINESS:

Prefabricated Metal Buildings
Metal Building Components
Engineered Building Systems

BRANDS/DIVISIONS/AFFILIATES:

Metal Building Components
American Building Components
Insulated Panel Systems
Metal Depots
Doors and Buildings Components
Ceco Building Systems
All American Systems
CENTRIA

CONTACTS: Note: Officers with more than one job title may be intentionally listed here more than once.

Norman Chambers, CEO
Mark Johnson, CFO
Bradley Little, Chief Accounting Officer
Eric Brown, Chief Information Officer
Robert Ronchetto, Other Executive Officer
Katy Theroux, Other Executive Officer
Donald Riley, President
John Kuzdal, President, Divisional
Todd Moore, Secretary

GROWTH PLANS/SPECIAL FEATURES:

NCI Building Systems, Inc. is one of North America's largest manufacturers of metal products for the nonresidential construction and building industry. Its operations are comprised of 42 manufacturing facilities located in the U.S., China and Mexico. The firm operates in three business segments: metal components, engineered building systems and metal coil coating services. The metal components segment produces metal roof and wall systems, metal partitions, metal trim, doors and other related accessories. These products are used in new construction and repair applications for industrial, commercial, institutional, agricultural and rural use. Some of this segment's brand names are Metal Building Components (MBCI), American Building Components, Insulated Panel Systems (IPS), Metal Depots and Doors and Buildings Components (DBCI). The engineered building systems segment encompasses the manufacturing of structural framing, complemented by engineering and drafting services, and includes the manufacturing of main house frames. It also offers value-added engineering and drafting services, which are typically not part of metal building component or metal coil coating products or services. This segment's brand names include A & S, All American Systems, Ceco Building Systems, Garco Building Systems, Heritage Building Systems, Mesco Building Solutions, Metallic Building Company and Mid-West Steel. The metal coil coating business is engaged in the cleaning, treating and painting of continuous steel coils before the steel is fabricated for use by construction and industrial users. In early 2015, the firm acquired CENTRIA, a provider of architectural insulated metal panel wall and roof systems and coil coating services.

The company offers employees medical, dental and vision benefits; a 401(k) plan; life and AD&D insurance; dependent life insurance; short- and long-term disability coverage; an employee assistance program; and credit union membership.

FINANCIAL DATA: Note: Data for latest year may not have been available at press time.

In U.S. $	2015	2014	2013	2012	2011	2010
Revenue	1,563,693,000	1,371,851,000	1,308,395,000	1,154,010,000	959,577,000	870,526,000
R&D Expense	4,201,000	4,998,000				
Operating Income	56,831,000	25,096,000	19,188,000	31,689,000	-1,627,000	-24,587,000
Operating Margin %	3.63%	1.82%	1.46%	2.74%	-.16%	-2.82%
SGA Expense	286,840,000	261,730,000	256,856,000	219,340,000	202,352,000	190,870,000
Net Income	17,818,000	11,185,000	-12,885,000	4,913,000	-9,950,000	-26,877,000
Operating Cash Flow	105,040,000	33,566,000	64,142,000	47,722,000	41,437,000	6,306,000
Capital Expenditure	20,683,000	18,020,000	24,426,000	28,151,000	21,040,000	14,030,000
EBITDA	106,642,000	61,006,000	35,258,000	59,664,000	32,590,000	12,215,000
Return on Assets %	1.91%	1.44%	-1.68%	-10.98%	-8.46%	-53.00%
Return on Equity %	6.80%	4.44%				-1231.58%
Debt to Equity	1.63	0.94	0.93			155.59

CONTACT INFORMATION:

Phone: 281 897-7788   Fax: 281 477-9674
Toll-Free:
Address: 10943 N. Sam Houston Pkwy. W., Houston, TX 77064 United States

STOCK TICKER/OTHER:

Stock Ticker: NCS
Employees: 4,556
Parent Company:

Exchange: NYS
Fiscal Year Ends: 10/31

SALARIES/BONUSES:

Top Exec. Salary: $433,846   Bonus: $482,272
Second Exec. Salary: $825,000   Bonus: $

OTHER THOUGHTS:

Estimated Female Officers or Directors: 1
Hot Spot for Advancement for Women/Minorities:

New World Development Company Limited   www.nwd.com.hk

NAIC Code: 531100

TYPES OF BUSINESS:
Real Estate Holdings & Development

BRANDS/DIVISIONS/AFFILIATES:
New World China Land Limited
New World Department Store China Limited
NWS Holdings Limited
New World
Rosewood Hotel Group
Rosewood Hotels & Resorts
Pentahotel (The)
Ba Li Chun Tian

CONTACTS: *Note: Officers with more than one job title may be intentionally listed here more than once.*
Adrian Chi-Kong Cheng, Co-Managing Dir.
Chen Guanzhan, Co-Managing Dir.
Cheng Chi Kong, Exec. Dir.-New World China Land Ltd.
Ki Man-Fung, Exec. Dir.
Wai-Hoi Doo, Vice Chmn.
Kar-Shun Cheng, Chmn.

GROWTH PLANS/SPECIAL FEATURES:
New World Development Company Limited is a Hong Kong-based real estate investment holding firm operating in four primary segments: property development, infrastructure & service, retail and hotels & serviced apartments. Property development comprises development projects within Hong Kong and Mainland China. These projects include shopping malls, offices, mixed-use buildings, hotels, resorts, service apartments and residential communities. Mainland China property is managed by subsidiary New World China Land Limited. This division also consists of K11 branded cinema properties that include K11 coffee restaurants/lounges. Infrastructure & services is operated by NWS Holdings Limited, which operates businesses in Hong Kong, Mainland China and Macau. Its infrastructure portfolio includes roads, energy, water and ports and logistics projects. Its services portfolio comprises facilities management, including the management of Hong Kong Convention and Exhibition Centre and Free Duty; construction and transport, including construction, bus and ferry services; and strategic investments. The retail segment operates through subsidiary New World Department Store China Limited, which operates New World department stores and Ba Li Chun Tian department stores. These department stores are located in 21 major Chinese cities, including Shanghai, Wuhan, Chengdu, Shenyang and Beijing. The hotels & serviced apartments segment operates hotels and apartment buildings throughout Hong Kong and Mainland China, as well as in the Southeast Asian countries of The Philippines and Vietnam. Subsidiary Rosewood Hotel Group manages the ultra-luxury Rosewood Hotels & Resorts brand Delux New World Hotels as well as The Pentahotel brand.

FINANCIAL DATA: *Note: Data for latest year may not have been available at press time.*

In U.S. $	2015	2014	2013	2012	2011	2010
Revenue	7,122,707,000	7,284,656,000	6,031,307,000	4,592,480,000	4,239,458,000	3,896,068,000
R&D Expense						
Operating Income	3,606,273,000	2,168,993,000	2,486,624,000	2,077,698,000	1,583,989,000	1,764,232,000
Operating Margin %	50.63%	29.77%	41.22%	45.24%	37.36%	45.28%
SGA Expense	1,264,643,000	1,123,762,000	1,005,289,000	835,798,100	731,507,000	289,459,500
Net Income	2,464,100,000	1,719,635,000	2,387,309,000	1,846,412,000	1,643,077,000	1,915,892,000
Operating Cash Flow	1,269,517,000	427,091,700	582,606,400	-2,118,027,000	50,604,810	763,081,900
Capital Expenditure		818,959,900			614,220,000	303,177,600
EBITDA	4,521,865,000	3,000,975,000	3,402,654,000	2,775,310,000	2,254,396,000	2,490,711,000
Return on Assets %	4.98%	2.77%	4.57%	3.93%	4.27%	6.17%
Return on Equity %	11.28%	6.50%	10.77%	8.92%	9.62%	14.44%
Debt to Equity	0.46	0.55	0.56	0.54	0.37	0.51

CONTACT INFORMATION:
Phone: 852-2523-1056   Fax: 852-2810-4673
Toll-Free:
Address: 18 Queen's Rd. Central, New World Tower, 30/Fl, Hong Kong, Hong Kong

STOCK TICKER/OTHER:
Stock Ticker: NDVLF   Exchange: GREY
Employees: 55,000   Fiscal Year Ends: 06/30
Parent Company:

SALARIES/BONUSES:
Top Exec. Salary: $ Bonus: $
Second Exec. Salary: $ Bonus: $

OTHER THOUGHTS:
Estimated Female Officers or Directors: 2
Hot Spot for Advancement for Women/Minorities:

Newcastle Investment Corp

www.newcastleinv.com

NAIC Code: 525990

TYPES OF BUSINESS:

Mortgage REIT
Real Estate Securities
Real Estate Loan Investments
Loans & Mortgages
Property Management

BRANDS/DIVISIONS/AFFILIATES:

Fortress Investment Group LLC
Holiday Acquisition Holdings LLC
FHC Property Management LLC
National Golf Properties
America Golf Corporation
New Media Investment Group Inc
New Senior Investment Group Inc

CONTACTS: Note: Officers with more than one job title may be intentionally listed here more than once.

Kenneth Riis, CEO
Justine Cheng, CFO
Wesley Edens, Chairman of the Board
Eun Nam, Chief Accounting Officer
Randal Nardone, Secretary

GROWTH PLANS/SPECIAL FEATURES:

Newcastle Investment Corp. is a real estate investment and finance firm that invests in real estate securities and other real estate-related assets such as loans and excess mortgage servicing rights (MSR). An affiliate of Fortress Investment Group LLC provides day-to-day management of the firm's operations. Newcastle primarily invests in senior housing assets, real estate debt, golf and other investments. The firm's objective is to leverage its longstanding investment expertise to drive attractive risk adjusted returns that will help deliver strong and growing dividends to its shareholders. The company targets stable long-term cash flows and seeks to employ conservative capital structures to generate returns throughout different interest rate environments. Currently, the firm's managed properties are managed by affiliates or subsidiaries of either Holiday Acquisition Holdings LLC or FHC Property Management LLC. In 2015, Newcastle leased 49 public and private properties, owned 27 public and private properties and managed 11 properties. The company invests in real estate debt and other investments, managing its collateralized debt obligations (CDOs) and investments which are primarily secured by or related to commercial real estate. The golf business segment, through the acquisition of National Golf Properties and its affiliate, America Golf Corporation, is one of the largest owners and operators of gold courses in the U.S. The golf business either operates or owns a portfolio of 27 golf courses across eight states. American Golf also leases an additional 49 golf courses and manages 11 courses, all owned by third parties. During 2014, the company spun off New Media Investment Group, Inc. into a publicly-traded company (Nasdaq: NEWM); sold its manufactured housing portfolio; and spun off New Senior Investment Group, Inc. (Nasdaq: SNR).

FINANCIAL DATA: Note: Data for latest year may not have been available at press time.

In U.S. $	2015	2014	2013	2012	2011	2010
Revenue	349,718,000	416,023,000	308,846,000	481,404,000	290,051,000	892,056,000
R&D Expense						
Operating Income	21,253,000	67,791,000	121,109,000	434,178,000	259,141,000	621,670,000
Operating Margin %	6.07%	16.29%	39.21%	90.18%	89.34%	69.68%
SGA Expense	299,540,000	284,375,000	176,533,000	64,838,000	30,233,000	29,528,000
Net Income	21,847,000	33,246,000	151,413,000	434,110,000	259,447,000	621,662,000
Operating Cash Flow	-2,641,000	40,380,000	106,186,000	97,334,000	57,031,000	48,846,000
Capital Expenditure					2,268,000	
EBITDA						
Return on Assets %	1.00%	.83%	3.31%	11.28%	6.91%	18.25%
Return on Equity %	9.41%	4.25%	13.79%	75.05%		
Debt to Equity	3.71	4.44	2.39	1.83	18.81	

CONTACT INFORMATION:

Phone: 212 798-6100   Fax: 212 798-6133
Toll-Free:
Address: 1345 Ave. of the Americas, New York, NY 10105 United States

SALARIES/BONUSES:

Top Exec. Salary: $ Bonus: $
Second Exec. Salary: $ Bonus: $

STOCK TICKER/OTHER:

Stock Ticker: NCT
Employees:
Parent Company:

Exchange: NYS
Fiscal Year Ends: 12/31

OTHER THOUGHTS:

Estimated Female Officers or Directors:
Hot Spot for Advancement for Women/Minorities:

Newhall Land & Farming Company

www.valencia.com

NAIC Code: 237210

TYPES OF BUSINESS:

Real Estate Investment & Development
Master-Planned Communities
Commercial Property Management
Agricultural Land Management
Residential Development
Land Sales

BRANDS/DIVISIONS/AFFILIATES:

Villa Metro
Newhall Land Development LLC
West Hills
RiverVillage
Valencia Water Company
West Hills
West Creek

CONTACTS: Note: Officers with more than one job title may be intentionally listed here more than once.

Marlee Lauffer, Pres.
Genji Nakata, VP-Finance

GROWTH PLANS/SPECIAL FEATURES:

Newhall Land & Farming Company, a subsidiary of Newhall Land Development LLC, develops new towns and master-planned, mixed-use communities. The land, initially used for ranching, agriculture and oil production, today includes the towns of Valencia, 30 miles north of downtown Los Angeles, and Newhall Ranch, which is in the planning stages. Since 1965, Newhall has been developing the town of Valencia on a portion of the company's landholdings, which now provides over 60,000 jobs. Valencia is notable for its environmentally sensitive planning, with natural amenities such as an extensive network of pedestrian walkways, 10 square miles of open space, long range water planning and 30 miles of the free-flowing Santa Clara River. Within its communities, Newhall sells residential lots to merchant builders, operates a portfolio of commercial properties, provides building-ready sites for sale to industrial and commercial developers and users and owns a public water utility, the Valencia Water Company. Valencia is comprised of villages West Creek, West Hills, RiverVillage and Villa Metro.

FINANCIAL DATA: Note: Data for latest year may not have been available at press time.

In U.S. $	2015	2014	2013	2012	2011	2010
Revenue						
R&D Expense						
Operating Income						
Operating Margin %						
SGA Expense						
Net Income						
Operating Cash Flow						
Capital Expenditure						
EBITDA						
Return on Assets %						
Return on Equity %						
Debt to Equity						

CONTACT INFORMATION:

Phone: 661-255-4000   Fax: 661-255-3960
Toll-Free: 888-897-6815
Address: 25124 Springfield Ct., Fl. 3, Valencia, CA 91355 United States

STOCK TICKER/OTHER:

Stock Ticker: Subsidiary   Exchange:
Employees: 39   Fiscal Year Ends: 12/31
Parent Company: Newhall Land Development LLC

SALARIES/BONUSES:

Top Exec. Salary: $ Bonus: $
Second Exec. Salary: $ Bonus: $

OTHER THOUGHTS:

Estimated Female Officers or Directors:
Hot Spot for Advancement for Women/Minorities:

NH Hotel Group SA

www.nh-hotels.com

NAIC Code: 721110

TYPES OF BUSINESS:

Hotels
Golf Courses
Restaurants

GROWTH PLANS/SPECIAL FEATURES:

NH Hotel Group SA, based in Spain, is a hotel operator with nearly 400 hotels with 60,000 rooms in 29 countries across Europe, the Americas and Africa, ranging from budget hotels to four-star resorts. In Europe, NH has hotels in Spain, Andorra, Germany, Italy, Luxembourg, Switzerland, Belgium, Austria, the U.K., France, Portugal, Poland, the Czech Republic, Romania and Hungary. In the Americas, the firm has hotels in Mexico, the U.S., Argentina, Colombia, Venezuela, Chile, Uruguay and the Dominican Republic. In South Africa, the company has two hotels, one in Plettenberg Bay and the other in Cape Town. The firm's hotel brands include NH Hotels, a three to four star urban chain; Hesperia, a chain of resorts located in privileged surroundings; nhow, unconventional and cosmopolitan hotels, each with a unique personality in major international cities; and NH Collection, premium hotels located in major capitals across Europe and America. In November 2015, the firm signed with AXA Investment Mangers-Real Assets operate a new hotel development under its nhow brand, to be located in the Shoreditch district of London and operational in 2019.

BRANDS/DIVISIONS/AFFILIATES:

NH Hotels
Hesperia
nhow
NH Collection

CONTACTS: Note: Officers with more than one job title may be intentionally listed here more than once.

Frederico Gonzalez Tejera, CEO
Ramon Aragones Marin, COO
Roberto Chollet Ibarra, CFO
Isidoro Martinez de la Escalera, CMO
Fernando Cordova, Head-Human Resources
Inigo Capell Arrieta, Chief Resources Officer

FINANCIAL DATA: Note: Data for latest year may not have been available at press time.

In U.S. $	2015	2014	2013	2012	2011	2010
Revenue	1,570,048,000	1,422,148,000	1,437,589,000	1,466,718,000	1,517,987,000	1,462,050,000
R&D Expense						
Operating Income	11,769,940	-30,038,430	-37,450,530	-446,614,400	27,608,030	10,441,260
Operating Margin %	.74%	-2.11%	-2.60%	-30.44%	1.81%	.71%
SGA Expense						
Net Income	1,069,787	-10,891,760	-45,412,350	-333,150,900	11,976,370	-54,109,780
Operating Cash Flow	102,984,700	36,624,810	52,784,520	109,780,900	151,246,000	114,129,600
Capital Expenditure	200,822,300	125,331,600	45,164,860	60,490,870	77,047,480	80,586,440
EBITDA	194,347,700	143,321,800	83,851,690	-319,618,800	228,331,100	151,722,700
Return on Assets %	.03%	-.35%	-1.44%	-9.90%	.19%	-1.20%
Return on Equity %	.08%	-.90%	-4.31%	-29.50%	.54%	-3.51%
Debt to Equity	0.74	0.66	0.77	0.21	0.19	0.57

CONTACT INFORMATION:

Phone: 34 914519718   Fax:
Toll-Free: 800-232-9860
Address: Santa Engracia 120, Edifico Central, Madrid, 28003 Spain

STOCK TICKER/OTHER:

Stock Ticker: NHHEY
Employees: 13,795
Parent Company:

Exchange: PINX
Fiscal Year Ends: 12/31

SALARIES/BONUSES:

Top Exec. Salary: $ Bonus: $
Second Exec. Salary: $ Bonus: $

OTHER THOUGHTS:

Estimated Female Officers or Directors:
Hot Spot for Advancement for Women/Minorities: Y

Nomura Real Estate Master Fund Inc

www.nre-mf.co.jp/en/

NAIC Code: 531120

TYPES OF BUSINESS:

Equity Real Estate Investment Trusts (REITs), Primarily Leasing Nonresidential Buildings

BRANDS/DIVISIONS/AFFILIATES:

Nomura Real Estate Asset Management Co Ltd
Nomura Real Estate Development Co Ltd
Nomura Real Estate Office Fund Inc
Nomura Real Estate Residential Fund Inc
Sumitomo Mitsui Trust Bank Ltd

GROWTH PLANS/SPECIAL FEATURES:

Nomura Real Estate Master Fund, Inc. (NMF) is a real estate investment trust (REIT). By the acquisition and merging of former Nomura Real Estate Office Fund, Inc. and Nomura Real Estate Residential Fund, Inc. into itself in early 2016, the company became one of Japan's largest REIT, with $7.3 billion (900 billion yen) of assets. NMF's portfolio consists of 264 properties that offer office, retail, logistics and residential spaces. Properties are located in Tokyo, Kanagawa, Hokkaido, Sendai, Osaka, Hiroshima, Chiba, Saitama, Miyagi, Gunma, Fukuoka, and Aichi. Subsidiaries of the firm include Sumitomo Mitsui Trust Bank Ltd., Nomura Real Estate Asset Management Co. Ltd. and Nomura Real Estate Development Co. Ltd.

CONTACTS: Note: Officers with more than one job title may be intentionally listed here more than once.

Satoshi Yanagita, Exec. Dir.

FINANCIAL DATA: Note: Data for latest year may not have been available at press time.

In U.S. $	2015	2014	2013	2012	2011	2010
Revenue		211,224,848	208,609,840	209,371,296	222,793,216	225,255,200
R&D Expense						
Operating Income						
Operating Margin %						
SGA Expense						
Net Income		63,047,164	63,805,344	65,079,688	75,404,160	79,224,096
Operating Cash Flow						
Capital Expenditure						
EBITDA						
Return on Assets %						
Return on Equity %						
Debt to Equity						

CONTACT INFORMATION:

Phone: 81 333650507   Fax:
Toll-Free:
Address: 8-5-1 Nishi Shinjuku, Tokyo, 160-0023 Japan

STOCK TICKER/OTHER:

Stock Ticker: NREOF   Exchange: GREY
Employees: Fiscal Year Ends:
Parent Company:

SALARIES/BONUSES:

Top Exec. Salary: $ Bonus: $
Second Exec. Salary: $ Bonus: $

OTHER THOUGHTS:

Estimated Female Officers or Directors:
Hot Spot for Advancement for Women/Minorities:

NRT LLC

NAIC Code: 531210

www.nrtinc.com

TYPES OF BUSINESS:

Real Estate Brokerage
Mortgage Lending
Real Estate Services

BRANDS/DIVISIONS/AFFILIATES:

Realogy Holdings Corporation
Coldwell Banker Real Estate Corp
Sotheby's International Realty
Corcoran Group (The)
Citi Habitats
ZipRealty Inc
Coldwell Banker United Realtors
Aptsandlofts.com

CONTACTS: Note: Officers with more than one job title may be intentionally listed here more than once.

Bruce G. Zipf, CEO
Bruce G. Zipf, Pres.
Kevin Greene, CFO
Dan Barnett, Sr. VP-Mktg.
Dina Di Maria, Sr. VP-IT
Ken Hoffert, General Counsel
Ryan Gorman, Sr. VP-Strategic Oper.
Kathy Borruso, Contact-Media Rel.
Jeff Culbertson, Exec. VP-Southwest Region
Kate Rossi, Exec. VP-Southeast Region
Greg Macres, Exec. VP-Western Region
Monty D. Smith, Sr. VP-Strategic Initiatives
Robert M. Becker, Chmn.

GROWTH PLANS/SPECIAL FEATURES:

NRT LLC, a subsidiary of Realogy Holdings Corporation, is a leading residential real estate brokerage company. NRT, through subsidiaries Coldwell Banker Real Estate Corp., Sotheby's International Realty, The Corcoran Group, Citi Habitats and ZipRealty Inc., has operations in 50 of the largest markets in the U.S., with about 790 offices in its extended family. Coldwell Banker operates through a handful of various divisions, which offer commercial and residential real estate brokerage services and mortgage lending. Coldwell Banker Commercial NRT is the firm's commercial real estate brokerage division. Additionally, Coldwell Banker commercial affiliates operate throughout North America, South America, Europe, Africa, Asia and Australia. Sotheby's International Realty, the owners of Sotheby Auction House, specializes in luxury commercial and residential real estate worldwide. The Corcoran Group operates in New York, the Hamptons and Palm Beach. Citi Habitats is a residential real estate brokerage firm in New York City that specializes in the rental market. ZipRealty is a real estate brokerage firm with licensed agents in offices nationwide who sell real estate on ZipRealty.com NRT's growth strategy involves an aggressive attitude towards expansion, focusing on companies that have established themselves locally. Acquired companies are then affiliated with one of NRT's brands and allowed to continue a relatively independent management, simply building upon pre-established success and local identity. In 2015, NRT acquired Countrywood Reality Inc., Coldwell Banker United Realtors (Texas), Fenwick Keats Real Estate, Aptsandlofts.com, Chicago Apartment Finders, Curry Realty Inc., B.T. Edgar & Son Realtors and Island Estates Realty.

FINANCIAL DATA: Note: Data for latest year may not have been available at press time.

In U.S. $	2015	2014	2013	2012	2011	2010
Revenue	7,515,000,000					
R&D Expense						
Operating Income						
Operating Margin %						
SGA Expense						
Net Income						
Operating Cash Flow						
Capital Expenditure						
EBITDA						
Return on Assets %						
Return on Equity %						
Debt to Equity						

CONTACT INFORMATION:

Phone: 973-407-5296   Fax: 973-407-7999
Toll-Free:
Address: 175 Park Ave., Madison, NJ 07940 United States

STOCK TICKER/OTHER:

Stock Ticker: Subsidiary   Exchange:
Employees: 5,100   Fiscal Year Ends: 12/31
Parent Company: Realogy Holdings Corporation

SALARIES/BONUSES:

Top Exec. Salary: $ Bonus: $
Second Exec. Salary: $ Bonus: $

OTHER THOUGHTS:

Estimated Female Officers or Directors: 5
Hot Spot for Advancement for Women/Minorities: Y

NVR Inc

NAIC Code: 236117

www.nvrinc.com/

TYPES OF BUSINESS:

Construction, Home Building and Residential
Mortgages
Townhouse Construction
Condominium Construction

BRANDS/DIVISIONS/AFFILIATES:

Ryan Homes
NVHomes
Fox Ridge Homes
NVR Mortgage Finance Inc
VR Mortgage Finance Inc

CONTACTS: *Note: Officers with more than one job title may be intentionally listed here more than once.*

Paul Saville, CEO
Daniel Malzahn, CFO
Dwight Schar, Chairman of the Board
Eugene Bredow, Chief Accounting Officer
Jeffrey Martchek, President, Divisional
Robert Henley, President, Subsidiary
James Sack, Secretary

GROWTH PLANS/SPECIAL FEATURES:

NVR, Inc. is primarily engaged in the construction and sale of single-family detached homes, townhomes and condominium buildings. Additionally, NVR offers mortgage banking services through its subsidiary NVR Mortgage Finance, Inc. (NVRM). NVRM originates mortgage loans for NVR's homebuilding customers and sells all mortgage loans it closes to investors in the secondary markets on a servicing released basis. The company operates in 14 states, with concentration in the Washington, D.C. and Baltimore, Maryland metropolitan areas, which accounted for 31% and 13% of its 2015 homebuilding revenues. NVR's homebuilding operations include the sale and construction of single-family detached homes, townhomes and condominium buildings under four brand names: Ryan Homes, Fox Ridge Homes, NVHomes and Heartland Homes. The Ryan Homes and Fox Ridge products are moderately priced and marketed primarily to first-time homeowners and first-time move-up buyers. Ryan Homes are currently sold in 28 metropolitan areas located primarily in the eastern U.S. Fox Ridge Homes are sold solely in the Nashville, Tennessee metropolitan area. NVHomes are marketed primarily to move-up and upscale buyers and are sold in Delaware, Washington, D.C., Baltimore and Philadelphia metropolitan areas. Heartland Homes are sold in Pittsburgh. The firm's houses range from approximately 1,000 to 7,000 square feet, typically including two to four bedrooms, and are priced between $136,000 and $1.9 million. NVR also provides mortgage-related services through its mortgage banking operations, which include subsidiaries that broker title insurance and perform title searches.

FINANCIAL DATA: *Note: Data for latest year may not have been available at press time.*

In U.S. $	2015	2014	2013	2012	2011	2010
Revenue	5,169,562,000	4,449,508,000	4,220,908,000	3,193,204,000	2,663,906,000	3,051,958,000
R&D Expense						
Operating Income	626,771,000	473,055,000	441,081,000	282,606,000	185,605,000	328,422,000
Operating Margin %	12.12%	10.63%	10.44%	8.85%	6.96%	10.76%
SGA Expense	424,009,000	407,867,000	355,623,000	334,959,000	294,515,000	290,655,000
Net Income	382,927,000	281,630,000	266,477,000	180,588,000	129,420,000	206,005,000
Operating Cash Flow	203,391,000	184,549,000	270,222,000	264,384,000	1,463,000	55,388,000
Capital Expenditure	18,277,000	31,672,000	19,016,000	12,365,000	11,444,000	6,943,000
EBITDA	648,305,000	494,300,000	454,472,000	290,706,000	216,140,000	335,685,000
Return on Assets %	15.73%	11.64%	10.46%	8.23%	6.40%	8.84%
Return on Equity %	32.40%	23.61%	19.43%	12.64%	8.30%	11.77%
Debt to Equity	0.48	0.53	0.47	0.40		

CONTACT INFORMATION:

Phone: 703 956-4000   Fax: 703 956-4750
Toll-Free:
Address: 11700 Plz. America Dr., Ste. 500, Reston, VA 20190 United States

STOCK TICKER/OTHER:

Stock Ticker: NVR
Employees: 3,942
Parent Company:

Exchange: NYS
Fiscal Year Ends: 12/31

SALARIES/BONUSES:

Top Exec. Salary: $1,528,125   Bonus: $
Second Exec. Salary: $462,500   Bonus: $

OTHER THOUGHTS:

Estimated Female Officers or Directors:

Hot Spot for Advancement for Women/Minorities:

Oakwood Worldwide

www.oakwood.com

NAIC Code: 721110

TYPES OF BUSINESS:

Rental Housing
Temporary Housing
Corporate Apartments

BRANDS/DIVISIONS/AFFILIATES:

Oakwood Corporate Housing
Oakwood Premier
Oakwood Residence
Oakwood Apartment
ExecuStay
Oakwood Asia Pacific Pte Ltd
Personal Property Protection
Insurance Housing Solutions

CONTACTS: Note: Officers with more than one job title may be intentionally listed here more than once.

Howard Ruby, CEO
Ric Villarreal, Pres.
Chris Brenk, CFO
Patricia Hintze, VP-Global Sales
Marina Lubinsky, CIO
Howard Ruby, Chmn.

GROWTH PLANS/SPECIAL FEATURES:

Oakwood Worldwide offers temporary housing, corporate housing and multifamily property management. The firm provides approximately 25,000 furnished and unfurnished accommodations throughout the U.S., the U.K. and the Asia Pacific region through its Oakwood Corporate Housing, Oakwood Premier, Oakwood Residence and Oakwood Apartment brands. Each apartment is equipped with fine furnishings, housewares, telephone service and cable television service. Oakwood services for temporary housing include weekly housekeeping, an on-call concierge and maintenance services. These services are available in most areas nationwide on a day-by-day pay schedule, and are often used by production crews for the footage of television shows or movies. The company also manages traditional long-term apartment residences through its apartment communities in the U.S., with both furnished and unfurnished apartments. Oakwood has apartments of all styles, residing in differing natural environments, and will custom fit its models to its clients. The company's ExecuStay subsidiary offer temporary housing for a stay of at least a month in over 300 cities across the U.S., including both urban and suburban locations. Its Insurance Housing Solutions subsidiary is a resource for insurance professionals and displaced policyholders seeking emergency housing assistance. In 2014, the firm launched Personal Property Protection, a new travel insurance service tailored to meet the needs of extended stay business travelers by protecting their personal belongings and reducing their financial responsibility for accidental damage to their accommodations; it also formed a corporate and serviced apartment joint venture with Mapletree Group, with Mapletree acquiring a 40% stake in Oakwood Asia Pacific Pte. Ltd. Oakwood Asia will focus on acquiring and developing within Asia, Europe and North America.

Oakwood offers its employees medical, dental, vision, prescription, life, disability and AD&D insurance; a 401(k) plan; educational reimbursement; an employee assistance program; corporate discounts and bonus opportunities; flexible spending accounts; and referral bonuses.

FINANCIAL DATA: Note: Data for latest year may not have been available at press time.

In U.S. $	2015	2014	2013	2012	2011	2010
Revenue	860,000,000	840,000,000	821,000,000	750,000,000		
R&D Expense						
Operating Income						
Operating Margin %						
SGA Expense						
Net Income						
Operating Cash Flow						
Capital Expenditure						
EBITDA						
Return on Assets %						
Return on Equity %						
Debt to Equity						

CONTACT INFORMATION:

Phone: 310-478-1021   Fax: 310-444-2210
Toll-Free: 877-902-0832
Address: 2222 Corinth Ave., Los Angeles, CA 90064 United States

STOCK TICKER/OTHER:

Stock Ticker: Private
Employees: 1,064
Parent Company:

Exchange:
Fiscal Year Ends:

SALARIES/BONUSES:

Top Exec. Salary: $ Bonus: $
Second Exec. Salary: $ Bonus: $

OTHER THOUGHTS:

Estimated Female Officers or Directors: 3
Hot Spot for Advancement for Women/Minorities: Y

Sales, profits and employees may be estimates. Financial information, benefits and other data can change quickly and may vary from those stated here.

Oberoi Group (EIH Ltd)

NAIC Code: 721110

www.oberoihotels.com

TYPES OF BUSINESS:

Luxury Hotels
Commercial Hotels
Cruise Ships
Travel Agency
Charter Aircraft
Tour Services
In-flight Catering
Car Rental

BRANDS/DIVISIONS/AFFILIATES:

EIH Ltd
Oberoi
Trident
Oberoi Philae (The)
Oberoi Zahra (The)
Maidens Hotel
Clarke's Hotel
Connections

CONTACTS: *Note: Officers with more than one job title may be intentionally listed here more than once.*

Kapil Chopra, Pres.
Rai Bahadur Mohan Singh Oberoi, Chmn.

GROWTH PLANS/SPECIAL FEATURES:

Oberoi Group, also known as EIH, Ltd., owns or manages 30 Oberoi and Trident brand luxury hotels and two small cruise ships in six countries. Additionally, the group operates airport restaurants, travel and tour services, in-flight catering, car rentals, project management and corporate air charters. Oberoi has eleven hotels in India, two hotels in Indonesia, two hotels and a Nile river boat in Egypt (The Oberoi Philae and The Oberoi Zahra), one hotel in Mauritius, one hotel in Saudi Arabia and one hotel in the UAE. The company also operates two heritage hotels, the Maidens Hotel and the Clarke's Hotel, which are located in historic colonial areas. The firm offers a member program called Connections, which rewards points for staying at the hotels. These points can be redeemed for a variety of things, including additional stay at the group's hotels. Many of the hotels feature spas, which can also be used by non-guests. Upcoming hotels include The Oberoi Sukhvilas in Chandigarh, India; The Oberoi in Marrakech, Morocco; and The Oberoi in Al Zorah, United Arab Emirates. The Oberoi aviation segment of the group offers an aircraft for charter, a Hawker 850 XP. The planes can be booked with as little as 48 hours notice and fly between locations within India along with a limited international service. The company has a long term agreement with Lufthansa Airlines to integrate its hotels with Lufthansa's frequent flyer program.

Oberoi maintains The Oberoi Centre of Learning and Development (OCLD), which offers a specialized two year course in hospitality management.

FINANCIAL DATA: *Note: Data for latest year may not have been available at press time.*

In U.S. $	2015	2014	2013	2012	2011	2010
Revenue	250,608,522	241,972,062	201,797,000	215,129,000	202,036,000	258,585,000
R&D Expense						
Operating Income						
Operating Margin %						
SGA Expense						
Net Income	9,487,804	16,838,324	6,571,853	23,412,000	12,482,000	11,929,000
Operating Cash Flow						
Capital Expenditure						
EBITDA						
Return on Assets %						
Return on Equity %						
Debt to Equity						

CONTACT INFORMATION:

Phone: 91-11-2389-0505   Fax: 91-11-2389-0582
Toll-Free: 800-562-3764
Address: 7 Sham Nath Marg, Delhi, 110 054 India

STOCK TICKER/OTHER:

Stock Ticker: EIHOTEL   Exchange: India
Employees: 9,851   Fiscal Year Ends: 03/31
Parent Company:

SALARIES/BONUSES:

Top Exec. Salary: $ Bonus: $
Second Exec. Salary: $ Bonus: $

OTHER THOUGHTS:

Estimated Female Officers or Directors:
Hot Spot for Advancement for Women/Minorities: Y

Odebrecht SA

www.odebrecht.com.br

NAIC Code: 237990

TYPES OF BUSINESS:

Engineering & Heavy Construction
Environmental Engineering
Real Estate Development
Insurance and Warranties Services
Ethanol and Sugar
Energy Operations and Infrastructure
Transportation Infrastructure

BRANDS/DIVISIONS/AFFILIATES:

Odebrecht Transport
Odebrecht Energia
Odebrecht Engenharia Industrial
Odebrecht Oil & Gas
Odebrecht Realizacoes Imobiliarias
Braskem
Odebrecht Ambiental
Odebrecht Defesa e Tecnologia

CONTACTS: Note: Officers with more than one job title may be intentionally listed here more than once.

Newton Sergio De Souza, Interim-CEO
Paulo Lacerda De Melo, COO
Marcelo Bahia Odebrecht, Pres.
Marcela Drehmer, Sr. Dir.-Finance
Marco Rabello, VP-Eng. & Construction Oper.
Newton de Souza, Head-Legal
Paulo Lacerda de Melo, Dir.-Oper.
Euzenando Azevedo, Sr. Dir.-Bus. Dev.
Marcio Polidoro, Head-Comm.
Benedicto Barbosa da Silva Junior, CEO-Odebrecht Infrastructure-Brazil
Marcio Faria, CEO-Odebrecht Engenharia Industrial
Roberto Ramos, CEO-Odebrecht Oil & Gas
Paulo Cesena, CEO-Odebrecht Transport
Emilio Odebrecht, Chmn.
Luiz Mameri, CEO-Odebrecht Infrastructure-Latin America

GROWTH PLANS/SPECIAL FEATURES:

Odebrecht SA, a Brazilian conglomerate, has operations across a diversified portfolio of industries. These include engineering & construction, oil & gas, real estate development, ethanol & sugar, chemicals & petrochemicals, environmental engineering and support services. Odebrecht Transport, the company's transport and logistics segment, focuses on investing in highways, urban transport, railroads, ports and airports. Other businesses in this sector include Odebrecht Energia, which designs, constructs and manages power generation projects; and Odebrecht Engenharia Industrial, which specializes in developing oil and gas, mining, steel mills, petrochemicals, fertilizer, metalworking and pulp and paper complexes. Subsidiary Odebrecht Oil & Gas is responsible for the company's oil and gas operations. Through Odebrecht Realizacoes Imobiliarias, the firm conducts real estate development activities that include residential, corporate, commercial, tourism and mixed use developments. Odebrecht Agroindustrial manages the ethanol and sugar businesses, including sugarcane cultivation, harvesting, ethanol production and energy cogeneration. Subsidiary Braskem operates in the chemicals and petrochemicals sector, focusing primarily on thermoplastic resins. Odebrecht Ambiental operates in the environmental engineering sector, providing waste and industrial effluent treatment, sanitation services and urban solid waste disposal. Odebrecht Infraestrutura invests in infrastructure development projects, including energy, transportation and logistics. Odebrecht Administradora e Corretora de Seguros Ltda provides insurance and warranties services that focus on protecting assets and managing risks. The company offers pension fund services through Odebrecht Previdencia. Odebrecht's newer defense and security company is Odebrecht Defesa e Tecnologia. In March 2016, the firm announced it was seeking to sell its energy-generating assets.

FINANCIAL DATA: Note: Data for latest year may not have been available at press time.

In U.S. $	2015	2014	2013	2012	2011	2010
Revenue	45,800,000,000	46,564,417,000	41,396,493,000			
R&D Expense						
Operating Income						
Operating Margin %						
SGA Expense						
Net Income		10,362,887,000	10,362,887,000			
Operating Cash Flow						
Capital Expenditure						
EBITDA						
Return on Assets %						
Return on Equity %						
Debt to Equity						

CONTACT INFORMATION:

Phone: 55-71-2105-1111   Fax: 55-71-2105-1112
Toll-Free:
Address: Ave. Luis Viana, 2841, Edificio Odebrecht, Parelela, Salvador, Bahia 41730-900 Brazil

STOCK TICKER/OTHER:

Stock Ticker: Private
Employees: 276,224
Parent Company:

Exchange:
Fiscal Year Ends:

SALARIES/BONUSES:

Top Exec. Salary: $ Bonus: $
Second Exec. Salary: $ Bonus: $

OTHER THOUGHTS:

Estimated Female Officers or Directors: 1
Hot Spot for Advancement for Women/Minorities:

Oldcastle Inc

NAIC Code: 327320   **www.oldcastle.com**

TYPES OF BUSINESS:

Ready-Mix Concrete Manufacturing
Aggregates
Asphalt
Paving Services
Construction Services
Building Materials Distribution
Concrete Technologies
Envelope and Curtain Wall Technologies

BRANDS/DIVISIONS/AFFILIATES:

CRH plc
Oldcastle BuildingEnvelope

CONTACTS: Note: Officers with more than one job title may be intentionally listed here more than once.

Mark Towe, CEO

GROWTH PLANS/SPECIAL FEATURES:

Oldcastle Inc. is a subsidiary of CRH plc and engages in the production and supply of building materials in the U.S. Its operations include more than 1,800 locations in 50 U.S. states and six Canadian provinces. The firm supplies aggregates, asphalt, ready mix concrete, and construction and paving services in the U.S. The company is divided into three segments: materials, building products and distribution. The materials segment supplies aggregates, asphalt, ready mix concrete and construction and paving services. The building products segment is a broad group comprised of the leading North American manufacturers of concrete masonry, precast, paving and lawn and garden products as well as a regional leader in clay brick. This segment also includes Oldcastle BuildingEnvelope, a supplier of building envelope solutions, including curtain wall, architectural windows, storefronts, skylights, entrance doors and architectural glass. The distribution segment is a national distributor of roofing, siding, window and interior products for specialty contractors in the construction and remodeling industries.

The company offers its employees medical, dental and vision insurance; a flexible spending account; life insurance & short- and long-term disability; and a 401(k) plan.

FINANCIAL DATA: Note: Data for latest year may not have been available at press time.

In U.S. $	2015	2014	2013	2012	2011	2010
Revenue	11,000,000,000	13,250,000,000	12,250,000,000	12,000,000,000	11,500,000,000	
R&D Expense						
Operating Income						
Operating Margin %						
SGA Expense						
Net Income						
Operating Cash Flow						
Capital Expenditure						
EBITDA						
Return on Assets %						
Return on Equity %						
Debt to Equity						

CONTACT INFORMATION:

Phone: 770-804-3363   Fax:
Toll-Free: 800-899-8455
Address: 900 Ashwood Pkwy., Ste. 600, Atlanta, GA 30338 United States

STOCK TICKER/OTHER:

Stock Ticker: Subsidiary   Exchange:
Employees: 39,000   Fiscal Year Ends:
Parent Company: CRH plc

SALARIES/BONUSES:

Top Exec. Salary: $ Bonus: $
Second Exec. Salary: $ Bonus: $

OTHER THOUGHTS:

Estimated Female Officers or Directors:
Hot Spot for Advancement for Women/Minorities:

Omega Healthcare Investors Inc

www.omegahealthcare.com

NAIC Code: 0

TYPES OF BUSINESS:

Real Estate Investment Trust
Health Care Facilities
Long-Term Care Facilities
Lease & Mortgage Financing

BRANDS/DIVISIONS/AFFILIATES:

Autumn Corporation

CONTACTS: Note: Officers with more than one job title may be intentionally listed here more than once.

C. Pickett, CEO
Robert Stephenson, CFO
Bernard Korman, Chairman of the Board
Michael Ritz, Chief Accounting Officer
Daniel Booth, COO
Steven Insoft, Other Executive Officer

GROWTH PLANS/SPECIAL FEATURES:

Omega Healthcare Investors, Inc. is a self-administered real estate investment trust (REIT) investing in health care facilities, particularly long-term care facilities located in the U.S. and the U.K. The firm provides lease or mortgage financing to qualified operators of skilled nursing facilities and, to a lesser extent, assisted living facilities, independent living facilities and rehabilitation and acute care facilities. Omega's portfolio of investments is comprised of 949 health care facilities located in 42 states and run by 83 third-party operators. The portfolio includes 782 skilled nursing facilities (SNFs), 85 assisted living facilities and 16 specialty hospitals and other facilities. The company seeks to maintain a diversified investment portfolio in terms of geography and operators, and generally focuses on established, creditworthy, middle-market health care operators. Factors considered by Omega in evaluating potential investments include the quality and experience of management; the facility's historical and forecasted cash flow; the construction quality, condition and design of the facility; the geographic area of the facility; the occupancy and demand for similar health care facilities in the same or nearby communities; and the payer mix of private, Medicare and Medicaid patients. One of Omega's fundamental investment strategies is to obtain contractual rent escalations under long-term, non-cancelable, triple-net leases and fixed-rate mortgage loans and to obtain substantial liquidity deposits. The firm's highest concentrations of investments are located in Ohio, Florida and Indiana. In 2016, the firm acquired long-term nursing home care provider, Autumn Corporation.

FINANCIAL DATA: Note: Data for latest year may not have been available at press time.

In U.S. $	2015	2014	2013	2012	2011	2010
Revenue	743,617,000	504,787,000	418,714,000	350,460,000	292,204,000	258,321,000
R&D Expense						
Operating Income	411,269,000	345,311,000	265,679,000	214,966,000	137,795,000	148,937,000
Operating Margin %	55.30%	68.40%	63.45%	61.33%	47.15%	57.65%
SGA Expense	38,568,000	25,888,000	21,588,000	21,330,000	19,432,000	15,054,000
Net Income	224,524,000	221,349,000	172,521,000	120,698,000	52,606,000	58,436,000
Operating Cash Flow	463,885,000	337,540,000	279,949,000	208,271,000	169,771,000	157,563,000
Capital Expenditure	484,805,000	149,606,000	63,862,000	426,059,000	106,301,000	379,205,000
EBITDA	628,437,000	471,475,000	404,327,000	339,777,000	239,842,000	233,661,000
Return on Assets %	3.76%	5.99%	5.35%	4.35%	1.95%	2.49%
Return on Equity %	8.73%	16.38%	14.92%	12.77%	5.35%	5.97%
Debt to Equity	0.95	1.69	1.55	1.80	1.76	1.31

CONTACT INFORMATION:

Phone: 410 427-1700   Fax: 410 887-0201
Toll-Free:
Address: 200 International Cir., Ste. 3500, Hunt Valley, MD 21030 United States

STOCK TICKER/OTHER:

Stock Ticker: OHI
Employees: 27
Parent Company:

Exchange: NYS
Fiscal Year Ends: 12/31

SALARIES/BONUSES:

Top Exec. Salary: $717,500   Bonus: $322,875
Second Exec. Salary: $451,000   Bonus: $135,300

OTHER THOUGHTS:

Estimated Female Officers or Directors: 1
Hot Spot for Advancement for Women/Minorities:

Opendoor

NAIC Code: 519130

www.opendoor.com

TYPES OF BUSINESS:

Online Real Estate Portal

BRANDS/DIVISIONS/AFFILIATES:

GROWTH PLANS/SPECIAL FEATURES:

Opendoor is a San Francisco, California-based online real estate portal that purchases homes from users who answer basic questions about the property they wish to sell. The company quickly makes offers via email based on comparative market analysis and then schedules free inspections to confirm a property's sellable condition. Closings are scheduled at the user's convenience, usually between three and 60 days after an offer is agreed upon. Opendoor then makes improvements or renovates the purchased properties and lists them for sale. As of early 2015, Opendoor offered its services in the Phoenix, Arizona area, and planned to launch in Portland, Oregon and Denver, Colorado.

CONTACTS:
Note: Officers with more than one job title may be intentionally listed here more than once.

Eric Wu, CEO
Keith Rabois, Exec. Chmn.

FINANCIAL DATA:
Note: Data for latest year may not have been available at press time.

In U.S. $	2015	2014	2013	2012	2011	2010
Revenue	18,685,720	16,795,434				
R&D Expense						
Operating Income						
Operating Margin %						
SGA Expense						
Net Income						
Operating Cash Flow						
Capital Expenditure						
EBITDA						
Return on Assets %						
Return on Equity %						
Debt to Equity						

CONTACT INFORMATION:

Phone: 602-396-5668   Fax:
Toll-Free:
Address: 116 New Montgomery St., Ste. 820, San Francisco, CA 94105 United States

STOCK TICKER/OTHER:

Stock Ticker: 3926   Exchange: Tokyo
Employees: 137   Fiscal Year Ends:
Parent Company:

SALARIES/BONUSES:

Top Exec. Salary: $ Bonus: $
Second Exec. Salary: $ Bonus: $

OTHER THOUGHTS:

Estimated Female Officers or Directors:
Hot Spot for Advancement for Women/Minorities:

Orient Overseas (International) Ltd   www.ooilgroup.com

NAIC Code: 483111

TYPES OF BUSINESS:

Deep Sea Shipping
Support Services
Port & Terminal Operations
Logistics Software & Services
Real Estate Investment
Office Buildings

BRANDS/DIVISIONS/AFFILIATES:

Orient Overseas Container Line Limited
OOCL China Domestic Ltd
CargoSmart Ltd
OOCL Logistics Ltd
Kaohsiung Terminal
Long Beach Container Terminal

CONTACTS: Note: Officers with more than one job title may be intentionally listed here more than once.

Chee Chen Tung, CEO
Chee Chen Tung, Pres.
Lieh Sing Alan Tung, CFO
Lammy Chee Fun Lee, Corp. Sec.
Chee Chen Tung, Chmn.

GROWTH PLANS/SPECIAL FEATURES:

Orient Overseas (International) Ltd. (OOIL) is engaged in international container transport and logistics services, port and terminal operations and property investment. It has more than 270 offices in 60 countries. The firm has three main business units: container transport and logistics services; ports and terminals; and property investment. The container transport and logistics services consists of firm's subsidiaries, Orient Overseas Container Line Limited (OOCL), which is an integrated international logistics company offering intermodal transport services worldwide; OOCL China Domestic, Ltd., which provides a full range of logistics and transportation services throughout the country; CargoSmart, Ltd., which uses information technology and e-commerce to manage the entire cargo process; and OOCL Logistics, Ltd. , which provides international freight consolidation, IT solutions and logistics services. The ports and terminals business unit operates two private terminals: the Kaosiung Terminal in Taiwan and the Long Beach Container Terminal in California. The firm's property investment division holds a 7.9% interest in the Beijing Oriental Plaza and a long-standing ownership of the Wall Street Plaza. The company has a grand alliance with Hapag-Lloyd, MISC Berhad, P&O Nedlloyd and Nippon Yusen Kaisha, in which they share the same routes for cargos. In addition, the company has strategic alliances with St. Lawrence Coordinated Services members; Hapag-Lloyd; Tokyo Senpaku Kaisha Ltd.; and COSCO Container Lines Co., Ltd.

FINANCIAL DATA: Note: Data for latest year may not have been available at press time.

In U.S. $	2015	2014	2013	2012	2011	2010
Revenue	5,953,444,000	6,521,589,000	6,231,583,000	6,459,059,000	6,011,836,000	6,033,402,000
R&D Expense						
Operating Income	353,068,000	329,147,000	90,314,000	328,621,000	174,598,000	918,807,000
Operating Margin %	5.93%	5.04%	1.44%	5.08%	2.90%	15.22%
SGA Expense	441,917,000	432,448,000	414,913,000	440,765,000	401,964,000	485,322,000
Net Income	283,851,000	270,538,000	47,036,000	296,359,000	181,645,000	1,866,780,000
Operating Cash Flow	458,281,000	478,531,000	413,178,000	424,850,000	245,754,000	1,174,734,000
Capital Expenditure	366,189,000	374,676,000	574,357,000	737,423,000	744,190,000	218,356,000
EBITDA	690,463,000	680,375,000	419,329,000	589,933,000	444,585,000	1,193,849,000
Return on Assets %	2.93%	2.90%	.54%	3.70%	2.16%	22.76%
Return on Equity %	6.01%	5.94%	1.04%	6.74%	3.68%	39.22%
Debt to Equity	0.76	0.77	0.73	0.51	0.52	0.43

CONTACT INFORMATION:

Phone: 852 28333838   Fax:
Toll-Free:
Address: 25 Harbour Rd., Harbour Center, 33rd Fl., Wanchai, Wanchai, Hong Kong

STOCK TICKER/OTHER:

Stock Ticker: OROVF
Employees: 9,889
Parent Company:

Exchange: PINX
Fiscal Year Ends: 12/31

SALARIES/BONUSES:

Top Exec. Salary: $ Bonus: $
Second Exec. Salary: $ Bonus: $

OTHER THOUGHTS:

Estimated Female Officers or Directors: 1
Hot Spot for Advancement for Women/Minorities:

Orion Marine Group Inc

NAIC Code: 237990

www.orionmarinegroup.com

TYPES OF BUSINESS:

Marine Construction & Specialty Services
Dredging Services

BRANDS/DIVISIONS/AFFILIATES:

CONTACTS: *Note: Officers with more than one job title may be intentionally listed here more than once.*

Peter Buchler, Chief Administrative Officer
Richard Daerr, Director
L. Breaux, Executive VP
Mark Stauffer, Pres.
Christopher DeAlmeida, Vice President

GROWTH PLANS/SPECIAL FEATURES:

Orion Marine Group, Inc. is a marine specialty contractor that serves the heavy civil marine infrastructure industry. The firm offers an array of marine construction services on, over and under the water along the Atlantic Seaboard, Canada, Alaska, the Gulf Coast and in the Caribbean Basin. The company's customers, for which it acts as a single-source turnkey solution, include private commercial and industrial enterprises; and federal, state and municipal governments. The firm offers three types of services: heavy civil marine construction, dredging and specialty services. Its heavy civil marine construction services include construction of marine transportation facilities, bridges, causeways, marine pipelines and marine environmental structures. Marine transportation facility construction projects include public port facilities, private and special-use Navy terminals, cruise ship port facilities and recreational use marinas and docks. The firm's bridge and causeway projects include the construction, repair and maintenance of bridges and causeways and the development of fendering systems in marine environments. Its marine pipeline service projects generally include the installation and removal of underwater buried pipeline transmission lines; installation of pipeline intakes and outfalls for industrial facilities; construction of pipeline outfalls for wastewater and industrial discharges; river crossing and directional drilling; creation of hot taps and tie-ins; and inspection, maintenance and repair services. Orion Marine's dredging services involve removing mud and silt from the channel floor by means of pipeline systems and a mechanical backhoe, crane/bucket or cutter suction dredge. The company's specialty services include demolition, surveying, diving, salvage, towing and underwater inspection, excavation and repair.

Orion Marine employees receive medical, dental, vision, life, AD&D and disability insurance; an employee assistance plan; and a 401(k).

FINANCIAL DATA: *Note: Data for latest year may not have been available at press time.*

In U.S. $	2015	2014	2013	2012	2011	2010
Revenue	466,498,000	385,818,000	354,544,000	292,042,000	259,852,000	353,135,000
R&D Expense						
Operating Income	-7,533,000	9,903,000	-106,000	-14,203,000	-19,341,000	32,587,000
Operating Margin %	-1.61%	2.56%	-.02%	-4.86%	-7.44%	9.22%
SGA Expense	47,715,000	34,691,000	32,110,000	28,573,000	29,519,000	32,646,000
Net Income	-8,060,000	6,877,000	331,000	-11,866,000	-13,114,000	21,882,000
Operating Cash Flow	25,179,000	11,945,000	13,033,000	24,438,000	32,676,000	13,839,000
Capital Expenditure	20,802,000	43,792,000	12,760,000	24,647,000	14,894,000	29,050,000
EBITDA	20,652,000	34,197,000	21,457,000	5,807,000	2,980,000	53,733,000
Return on Assets %	-1.97%	2.08%	.10%	-3.95%	-4.44%	7.53%
Return on Equity %	-3.47%	2.96%	.14%	-5.17%	-5.45%	9.34%
Debt to Equity	0.43	0.01				

CONTACT INFORMATION:

Phone: 713 852-6500   Fax: 713 852-6530
Toll-Free:
Address: 12000 Aerospace, Ste. 300, Houston, TX 77034 United States

STOCK TICKER/OTHER:

Stock Ticker: ORN   Exchange: NYS
Employees: 1,200   Fiscal Year Ends: 12/31
Parent Company:

SALARIES/BONUSES:

Top Exec. Salary: $528,615   Bonus: $
Second Exec. Salary: $341,135   Bonus: $

OTHER THOUGHTS:

Estimated Female Officers or Directors:
Hot Spot for Advancement for Women/Minorities:

Pacific Coast Building Products Inc

www.paccoast.com

NAIC Code: 444190

TYPES OF BUSINESS:

Building Materials-Manufacturing & Distribution
Contracting Services
Transportation Services
Charter Aviation Services

BRANDS/DIVISIONS/AFFILIATES:

Basalite Concrete Products LLC
Epic Plastics Inc
PABCO Building Products LLC
Pacific Coast Supply LLC
Transportation Services Inc
Pacific Coast Jet Charter Inc
Alcal/Arcade Contracting Inc
Pacific Coast Companies Inc

CONTACTS: Note: Officers with more than one job title may be intentionally listed here more than once.

David J. Lucchetti, CEO
David J. Lucchetti, Pres.
Josh Kimerer, CFO
Elaine Keane, Dir.-Mktg.
Hilda Watson, Dir.-Human Resources
Megan Vincent, Dir.-Community Rel.
James B. Thompson, Chmn.

GROWTH PLANS/SPECIAL FEATURES:

Pacific Coast Building Products, Inc. is a manufacturer and distributor of building materials for residential, commercial and industrial construction on the west coast. The firm operates through its family of companies: Basalite Concrete Products, LLC; PABCO Building Products, LLC; PABCO Clay Products, LLC; Pacific Coast Building Services, Inc.; Pacific Coast Supply, LLC; Transportation Services, Inc.; Pacific Coast Jet Charter, Inc.; Pacific Coast Companies, Inc.; and PCBP Properties, Inc. Basalite Concrete Products specializes in wall, roof and block products and oversees companies such as Blocklite; Columbia Roof Tile; Epic Plastics, Inc.; and Patterson Whittaker Architectural Profiles. PABCO Building Products manufactures wallboard and asphalt roof shingle products and controls the companies PABCO Gypsum, PABCO Paper and PABCO Roofing products. PABCO Clay Products provides clay veneers and related products and operates companies such as Gladding McBean, H.C. Muddox and Interstate Brick. Pacific Coast Building Services offers windows, insulation and related products. It directs companies such as Alcal/Arcade Contracting, Inc. and Pacific Coast Contracting Specialties, Inc. Lumber provider Pacific Coast Supply operates through divisions Anderson Lumber, Diamond Pacific, Pacific Supply, P.C. Wholesale and Weyrick Pacific. The firm's transportation unit operates through Material Transport, which moves products and raw materials to and from the company's subsidiaries; and Pacific Coast Jet Charter, which operates a Cessna Citation XLS jet that is available for charter to destinations throughout the U.S., Canada and Mexico. The Pacific Coast Companies team is in charge of much of the group's administrative work. Finally, PCBP Properties invests in and manages a number of properties in the western U.S.

FINANCIAL DATA: Note: Data for latest year may not have been available at press time.

In U.S. $	2015	2014	2013	2012	2011	2010
Revenue						
R&D Expense						
Operating Income						
Operating Margin %						
SGA Expense						
Net Income						
Operating Cash Flow						
Capital Expenditure						
EBITDA						
Return on Assets %						
Return on Equity %						
Debt to Equity						

CONTACT INFORMATION:

Phone: 916-631-6500   Fax:
Toll-Free:
Address: 10600 White Rock Rd., Bldg. B, Ste. 100, Rancho Cordova, CA 95670-6032 United States

STOCK TICKER/OTHER:

Stock Ticker: Private
Employees:
Parent Company:

Exchange:
Fiscal Year Ends: 03/31

SALARIES/BONUSES:

Top Exec. Salary: $ Bonus: $
Second Exec. Salary: $ Bonus: $

OTHER THOUGHTS:

Estimated Female Officers or Directors: 3
Hot Spot for Advancement for Women/Minorities: Y

Palm Harbor Homes Inc

www.palmharbor.com

NAIC Code: 321992

TYPES OF BUSINESS:

Manufactured & Modular Housing
Mortgages
Property & Casualty Insurance

BRANDS/DIVISIONS/AFFILIATES:

Cavco Industries Inc
CountryPlace Mortgage Ltd
Standard Casualty Company
Nationwide Custom Homes Inc
Fleetwood Homes
Palm Harbor
SmartPlus Construction
Keystone

CONTACTS: Note: Officers with more than one job title may be intentionally listed here more than once.

Larry Keener, CEO
Jennifer Jones, COO
Larry Keener, Pres.
Kelly Tacke, CFO
Maury Kennedy, VP-Mktg. & Sales
Larry Keener, Chmn.

GROWTH PLANS/SPECIAL FEATURES:

Palm Harbor Homes, Inc. is a U.S. manufacturer and marketer of factory-built and modular homes. The firm markets nationwide through vertically-integrated operations encompassing manufactured housing, modular housing, retailing, chattel and mortgage bank financing and insurance. Subsidiary CountryPlace Mortgage, Ltd. offers chattel and non-conforming land and home mortgages to purchasers of manufactured homes sold by Palm Harbor. In addition, the firm offers property and casualty insurance through subsidiary Standard Casualty Company. Palm Harbor's products include single- and multi-section manufactured homes sold under brand names including Palm Harbor, Masterpiece, Keystone, CountryPlace, River Bend and Windsor Homes. The firm also helps customers find an appropriate lot or look for a pre-owned home. Palm Harbor's building system, SmartPlus Construction, uses a process called EnerGmiser, an energy management system that includes added insulation and other efficiency-increasing products. Modular homes are manufactured through subsidiary Nationwide Custom Homes, Inc., principally under the brand name Discovery Custom Homes. Modular offerings include single-story ranch homes, split-levels and two- and three-story homes. The firm is a subsidiary of Fleetwood Homes, which is itself a subsidiary of Cavco Industries, Inc. In 2015, the firm's Plant City, Florida manufacturing plant earned FORTIFIED Home-Hurricane certification to construct and deliver the first-ever FORTIFIED Home labeled, disaster resistant modular homes in hurricane-prone regions.

The company offers its employees benefits including medical, dental, life, disability and vision coverage; a 401(k) plan; and paid time off.

FINANCIAL DATA: Note: Data for latest year may not have been available at press time.

In U.S. $	2015	2014	2013	2012	2011	2010
Revenue						
R&D Expense						
Operating Income						
Operating Margin %						
SGA Expense						
Net Income						
Operating Cash Flow						
Capital Expenditure						
EBITDA						
Return on Assets %						
Return on Equity %						
Debt to Equity						

CONTACT INFORMATION:

Phone: 972-991-2422   Fax: 972-991-5949
Toll-Free: 800-456-8744
Address: 15301 Spectrum Dr., Ste. 500, Addison, TX 75001 United States

STOCK TICKER/OTHER:

Stock Ticker: Subsidiary   Exchange:
Employees: 2,900   Fiscal Year Ends: 03/31
Parent Company: Cavco Industries Inc

SALARIES/BONUSES:

Top Exec. Salary: $ Bonus: $
Second Exec. Salary: $ Bonus: $

OTHER THOUGHTS:

Estimated Female Officers or Directors: 1
Hot Spot for Advancement for Women/Minorities:

Paramount Group Inc

www.paramount-group.com

NAIC Code: 531120

TYPES OF BUSINESS:
Lessors of Nonresidential Buildings (except Miniwarehouses)
Lessors of Residential Buildings and Dwellings

GROWTH PLANS/SPECIAL FEATURES:
Paramount Group Inc. is a fully-integrated real estate investment trust (REIT) focused on owning, operating and managing high-quality office properties in central business district submarkets of New York City, Washington, D.C. and San Francisco. It conducts its business through majority-owned (80.4%) Paramount Group Operating Partnership LP. Paramount's portfolio consists of 12 Class A office properties aggregating approximately 10.4 million square feet that was 93.9% leased. In 2015, the firm acquired the remaining 35.8% that it did not own in 31 West 52nd Street, New York, from its joint venture for $230 million.

BRANDS/DIVISIONS/AFFILIATES:
Paramount Group Operating Partnership LP

CONTACTS: Note: Officers with more than one job title may be intentionally listed here more than once.
Albert Behler, CEO
Wilbur Paes, CFO
Jolanta Bott, Executive VP, Divisional
Theodore Koltis, Executive VP, Divisional
Daniel Lauer, Executive VP
Gage Johnson, General Counsel
Ralph DiRuggiero, Senior VP, Divisional
Vito Messina, Senior VP, Divisional

FINANCIAL DATA: Note: Data for latest year may not have been available at press time.

In U.S. $	2015	2014	2013	2012	2011	2010
Revenue	662,408,000	227,389,000	419,890,000	246,815,000	605,940,000	
R&D Expense						
Operating Income	70,619,000	150,397,000	331,591,000	168,812,000	471,361,000	
Operating Margin %	10.66%	66.14%	78.97%	68.39%	77.78%	
SGA Expense	42,056,000	30,912,000	33,504,000	28,374,000	25,556,000	
Net Income	-4,419,000	21,510,000	16,514,000	2,295,000	53,878,000	
Operating Cash Flow	-16,969,000	-84,495,000	33,485,000	-83,464,000	-328,503,000	
Capital Expenditure	107,859,000	65,916,000	147,000	188,000	280,000	
EBITDA	486,699,000	166,647,000	354,257,000	194,168,000	489,124,000	
Return on Assets %	-.04%		.59%	.08%		
Return on Equity %	-.11%		4.44%	.54%		
Debt to Equity	0.79		1.55	1.23		

CONTACT INFORMATION:
Phone: 212-237-3100 Fax:
Toll-Free:
Address: 1633 Broadway, Ste. 1801, New York, NY 10019 United States

STOCK TICKER/OTHER:
Stock Ticker: PGRE Exchange: NYS
Employees: 219 Fiscal Year Ends:
Parent Company:

SALARIES/BONUSES:
Top Exec. Salary: $114,521 Bonus: $426,293
Second Exec. Salary: $62,466 Bonus: $117,069

OTHER THOUGHTS:
Estimated Female Officers or Directors:
Hot Spot for Advancement for Women/Minorities:

Parkway Properties Inc

NAIC Code: 0

www.pky.com

TYPES OF BUSINESS:

Real Estate Investment Trust
Office Buildings
Property Management

BRANDS/DIVISIONS/AFFILIATES:

Parkway Properties Office Fund II LP
Courvoisier Centre

CONTACTS: Note: Officers with more than one job title may be intentionally listed here more than once.

James Heistand, CEO
David OReilly, CFO
Scott Francis, Chief Accounting Officer
M. Lipsey, COO
James Thomas, Director
Jeremy Dorsett, Executive VP
Jason Bates, Executive VP

GROWTH PLANS/SPECIAL FEATURES:

Parkway Properties, Inc. is a self-administered real estate investment trust (REIT) specializing in the operation, acquisition, ownership and leasing of office properties in six states (Arizona, Texas, North Carolina, Georgia, Florida and Pennsylvania). Parkway owns or has interests in 36 office properties totaling 14.3 million square feet. Additionally, the company manages and leases properties it owns as well as for third-party real estate owners. The company maintains one joint venture partnership: Parkway Properties Office Fund II, L.P. (Fund II), for which Parkway provides asset management, property management, construction management and leasing services and is paid market-based fees. Parkway leases its office properties to customers in the banking, insurance, professional services, legal, accounting, consulting, energy, financial services and telecommunications industries. The company provides investment, administrative, management and maintenance services internally. In November 2015, the firm agreed to sell an 80% stake in Courvoisier Centre, and agreed to form a joint venture with a foreign capital source, with Parkway retaining its 20% interest in Courvoisier. Parkway will continue to manage the joint venture.

Parkway offers its employees health, life and disability insurance; a 401(k) plan; wellness programs and health club benefits; and paid holidays.

FINANCIAL DATA: Note: Data for latest year may not have been available at press time.

In U.S. $	2015	2014	2013	2012	2011	2010
Revenue	473,983,000	456,701,000	291,578,000	226,517,000	165,896,000	256,263,000
R&D Expense						
Operating Income	50,632,000	19,623,000	6,488,000	-23,599,000	-4,759,000	30,675,000
Operating Margin %	10.68%	4.29%	2.22%	-10.41%	-2.86%	11.97%
SGA Expense	31,194,000	32,660,000	25,653,000	16,420,000	18,805,000	7,382,000
Net Income	67,335,000	42,943,000	-19,650,000	-39,395,000	-126,903,000	-2,618,000
Operating Cash Flow	82,648,000	53,004,000	65,086,000	61,573,000	35,527,000	68,059,000
Capital Expenditure	89,001,000	52,569,000	35,373,000	718,532,000	533,090,000	36,335,000
EBITDA	358,494,000	283,441,000	131,493,000	59,691,000	108,129,000	124,839,000
Return on Assets %	1.83%	1.29%	-1.22%	-2.89%	-8.45%	-.55%
Return on Equity %	4.84%	3.56%	-3.75%	-12.68%	-43.72%	-2.29%
Debt to Equity	1.28	1.30	1.37	1.53	2.59	2.30

CONTACT INFORMATION:

Phone: 407 650-0593   Fax: 601 949-4077
Toll-Free: 800-748-1667
Address: 390 North Orange Avenue, Ste 2400, Orlando, FL 32901 United States

STOCK TICKER/OTHER:

Stock Ticker: PKY
Employees: 322
Parent Company:

Exchange: NYS
Fiscal Year Ends: 12/31

SALARIES/BONUSES:

Top Exec. Salary: $750,000   Bonus: $525,000
Second Exec. Salary: $430,000   Bonus: $250,000

OTHER THOUGHTS:

Estimated Female Officers or Directors:
Hot Spot for Advancement for Women/Minorities: Y

Parsons Corp

www.parsons.com

NAIC Code: 541330

TYPES OF BUSINESS:

Civil Engineering
Construction Management
Facility Operations and Maintenance
Environmental Services
Analytical, Technical and Training Services
Transportation Infrastructure Project Design and Construction

BRANDS/DIVISIONS/AFFILIATES:

Delcan
Secure Mission Solutions

CONTACTS: Note: Officers with more than one job title may be intentionally listed here more than once.

Charles L. Harrington, CEO
Charles L. Harrington, Chmn.

GROWTH PLANS/SPECIAL FEATURES:

Parsons Corp. is an employee-owned engineering, construction, technical, and management services firm. The firm is headquartered in Pasadena, California but has additional domestic offices in 32 states in the U.S., as well as in Washington D.C. and Guam. Internationally, Parsons has offices in 14 countries throughout South America, Europe, the Middle East and Asia and is currently undertaking over 3,000 projects in 29 countries. The firm's projects are primarily focused upon the diversified markets of transportation, environmental, defense/security and resources. Within the transportation market, the firm has designed and built airports; bridges & tunnels; rail systems; and various roads, highways and toll ways. In the environmental market, Parsons has completed decontamination & demolition, environmental studies, sediment management and site investigation & remediation projects. Within the market of defense/security, the company has constructed readiness training centers and offers a wide range of systems engineering and advisory/assistance services to the U.S. Missile Defense Agency. The resource market work undertaken by Parsons encompasses energy; hydroelectric, nuclear, petroleum, delivery and renewable; and water/wastewater; industrial wastewater and groundwater treatment; and projects. The services that the firm offers to any project include asset management; commissioning; qualifications and validation; condition assessments; construction; design; disaster response; information technology; intelligence/security; operations & services; planning; and program/construction management.

Parsons offers its employees medical and life insurance, tuition reimbursement, an ESOP for eligible employees, a 401(k) and membership to a Federal Credit Union.

FINANCIAL DATA: Note: Data for latest year may not have been available at press time.

In U.S. $	2015	2014	2013	2012	2011	2010
Revenue	3,200,000,000	3,098,000,000	2,992,000,000	3,001,000,000	2,876,000,000	2,709,000,000
R&D Expense						
Operating Income						
Operating Margin %						
SGA Expense						
Net Income		159,000,000	144,000,000	143,000,000	125,000,000	165,000,000
Operating Cash Flow						
Capital Expenditure						
EBITDA						
Return on Assets %						
Return on Equity %						
Debt to Equity						

CONTACT INFORMATION:

Phone: 626-440-2000   Fax: 626-440-2630
Toll-Free:
Address: 100 W. Walnut St., Pasadena, CA 91124 United States

SALARIES/BONUSES:

Top Exec. Salary: $ Bonus: $
Second Exec. Salary: $ Bonus: $

STOCK TICKER/OTHER:

Stock Ticker: Private
Employees: 14,000
Parent Company:

Exchange:
Fiscal Year Ends:

OTHER THOUGHTS:

Estimated Female Officers or Directors:
Hot Spot for Advancement for Women/Minorities:

PCL Construction Group Inc

NAIC Code: 237000

TYPES OF BUSINESS:

Heavy Construction
Financial and Accounting Reporting
Development, Support and Project Management
Engineering Services

BRANDS/DIVISIONS/AFFILIATES:

Melloy Industrial Services Inc
PCL Energy Inc
PCL Civil Constructors Inc
PCL Industrial Services Inc
PCL Constructors Bahamas Ltd

CONTACTS: Note: Officers with more than one job title may be intentionally listed here more than once.

Paul Douglas, CEO
Paul Douglas, Pres.
Gordon Panas, CFO
Mark Bryant, CIO
Steve Richards, General Counsel
Lee Clayton, VP-Global Strategic Initiatives
Gordon Stephenson, VP-Corp. Finance
Luis Ventoza, COO-Civil Infrastructure
Ian Johnson, COO-Heavy Industrial
Rob Hoimberg, COO-Buildings
Ross Grieve, Chmn.

GROWTH PLANS/SPECIAL FEATURES:

PCL Construction Group, Inc., founded in 1906, is an employee-owned group of construction companies throughout Canada, Australia, the U.S. and the Caribbean. The firm focuses on three main areas of construction: buildings construction, civil infrastructure and heavy industrial. The buildings construction segment conducts projects throughout North America and is able to work on an array of projects including commercial; institutional; educational; residential; adaptive reuse, which entails upgrading and converting an existing facility; cultural consideration, including onsite and on-the-job employment and training opportunities; green building; high-tech projects, for meeting cleanliness protocols in the medical, biotech and research working environments; and historical preservation, including repair, exterior masonry and renovations combining typical construction methods with scenic construction technology. PCL's building operations include larger projects, such as airports, sports facilities and office towers, and smaller projects, such as renovations, restorations and repairs. The civil infrastructure segment undertakes various civil structure projects including bridges, overpasses, tunnels, interchanges, water treatment facilities and light rail transportation projects. The heavy industrial division offers construction assistance to the petrochemical, oil and gas, pulp and paper, mining and power and generation industries. Subsidiaries include Melloy Industrial Services, Inc.; PCL Energy, Inc.; PCL Civil Constructors, Inc.; PCL Industrial Services, Inc.; and PCL Constructors Bahamas Ltd.

The firm offers employees medical, vision and dental insurance; flexible spending accounts; a prescription drug plan; a 401(k); a profit sharing bonus; and an employee assistance program.

FINANCIAL DATA: Note: Data for latest year may not have been available at press time.

In U.S. $	2015	2014	2013	2012	2011	2010
Revenue	7,232,900,000					
R&D Expense						
Operating Income						
Operating Margin %						
SGA Expense						
Net Income						
Operating Cash Flow						
Capital Expenditure						
EBITDA						
Return on Assets %						
Return on Equity %						
Debt to Equity						

CONTACT INFORMATION:

Phone: 780-733-5000   Fax: 780-733-5075
Toll-Free:
Address: 5410 99 St., Edmonton, AB T6E 3P4 Canada

STOCK TICKER/OTHER:

Stock Ticker: Private   Exchange:
Employees: 5,499   Fiscal Year Ends:
Parent Company:

SALARIES/BONUSES:

Top Exec. Salary: $ Bonus: $
Second Exec. Salary: $ Bonus: $

OTHER THOUGHTS:

Estimated Female Officers or Directors:
Hot Spot for Advancement for Women/Minorities:

PDG Realty SA Empreendimentos e Participacoes

www.pdgrealty.com.br

NAIC Code: 531100

TYPES OF BUSINESS:

Real Estate Investments

BRANDS/DIVISIONS/AFFILIATES:

TGLT
REP Desenvolvimento Imobiliario SA
Goldfarb Incorporacoes e Construcoes SA
CHL Desenvolvimento Imobiliaro SA
AGRE
PDG Sau Paulo

CONTACTS: Note: Officers with more than one job title may be intentionally listed here more than once.

Marcio Tabatchnik Trigueiro, CEO
Mauricio Fernandes Teixeira, CFO
Marcus Vinicius Medeiros Cardoso de Sa, Chief Admin. Officer
Caue Castello Veiga Innocencio Cardoso, General Counsel
Antonio Guedes, Chief Dev. Officer
Michel Wurman, Investor Rel. Officer
Michel Wurman, VP-Finance
Frederico Marinho Carneiro da Cunha, Dir.-Real Estate Dev.
Gilberto Sayao da Silva, Chmn.

GROWTH PLANS/SPECIAL FEATURES:

PDG Realty SA Empreemdimentos e Participacoes is a Brazilian firm that operates in five segments of the real estate industry: the development of residential projects targeted at income classes ranging from the low middle-income class to the high-income class, the development of residential lots, investments in commercial developments for the generation of rental income, the purchase of commercial & residential units for subsequent resale and the offering of rendering services as a real estate brokerage & consulting firm. PDG Realty develops real estate projects through investments in its portfolio companies and is involved in co-development real estate projects with several other Brazilian real estate developers. Its portfolio firms include wholly-owned Goldfarb Incorporacoes e Construcoes SA, which develops residential housing for middle and low middle-income classes in the states of Rio de Janeiro and Sao Paulo; commercial developer REP Real Estate Partners Desenvolvimento Imobiliario SA (58.10% owned); wholly-owned CHL Desenvolvimento Imobiliaro SA, which develops commercial real estate and housing for middle, high-middle and high-income classes in the state of Rio de Janeiro; wholly-owned homebuilder AGRE; wholly-owned apartment realty PDG Sau Paulo; and Argentinian residential building developer TGLT (27.18% owned). In 2015, the firm began restructuring its debt. In January 2016, it executed a non-binding memorandum of understanding with Banco Votorantim SA and BV Empreendimentos e Participacoes SA for a possible sale.

FINANCIAL DATA: Note: Data for latest year may not have been available at press time.

In U.S. $	2015	2014	2013	2012	2011	2010
Revenue	501,677,800	1,170,554,000	1,462,141,000	1,198,575,000	1,891,261,000	1,438,185,000
R&D Expense						
Operating Income	-770,042,100	-88,397,320	-7,589,925	-507,669,700	251,850,800	263,466,100
Operating Margin %	-153.49%	-7.55%	-.51%	-42.35%	13.31%	18.31%
SGA Expense	114,250,400	154,158,000	118,665,200	195,497,800	220,184,300	161,913,700
Net Income	-760,197,500	-145,540,400	-74,520,690	-598,698,200	193,718,800	220,231,000
Operating Cash Flow	211,382,100	20,906,120	-200,293,700	-363,705,900	-150,270,300	-197,877,600
Capital Expenditure	2,683,973	8,056,870	1,679,408	2,551,700	49,927,950	55,097,620
EBITDA	-411,258,100	-71,245,460	18,926,960	-459,505,800	291,904,900	270,328,600
Return on Assets %		-3.23%	-1.65%	-13.08%	4.33%	7.34%
Return on Equity %		-11.91%	-5.57%	-38.07%	11.50%	17.88%
Debt to Equity		0.77	0.78		0.54	0.73

CONTACT INFORMATION:

Phone: 55-2135043800   Fax: 55 2132545600
Toll-Free:
Address: 501 Praia de Botafogo, Botafogo, RJ 22250-040 Brazil

STOCK TICKER/OTHER:

Stock Ticker: PDGRY
Employees:
Parent Company:

Exchange: PINX
Fiscal Year Ends: 12/31

SALARIES/BONUSES:

Top Exec. Salary: $ Bonus: $
Second Exec. Salary: $ Bonus: $

OTHER THOUGHTS:

Estimated Female Officers or Directors:
Hot Spot for Advancement for Women/Minorities:

Pearlmark Real Estate Partners
NAIC Code: 531120

www.pearlmarkrealestate.com

TYPES OF BUSINESS:
Real Estate Investment Trust, Commercial
Industrial/Flex Properties
Multi-Family Residential Properties
Retail Properties
Mezzanine Loans

BRANDS/DIVISIONS/AFFILIATES:
Aslan Realty Partners IV LLC
Pearlmark Multifamily Partners LLC
Mezzanine Realty Partners III LLC
Mezzanine Realty Partners IV LLC
Pearlmark Investment Advisers LLC

CONTACTS: Note: Officers with more than one job title may be intentionally listed here more than once.
Stephen R. Quazzo, CEO
Jodi A. Stuart, Sr. VP-Finance & Acct.
Linda Bursic, Sr. VP-Capital Markets
Timothy E. McChesney, Managing Dir.-Asset Mgmt.
Ronald I. Miles, Managing Dir.-Due Diligence & Asset Mgmt.
Douglas Crocker, II, Chmn.-Multifamily Investments

GROWTH PLANS/SPECIAL FEATURES:

Pearlmark Real Estate Partners is a Chicago-based real estate principal investment firm specializing in commercial real estate. Since its founding in 1996, the company has invested in nearly 500 properties totaling over $12 billion throughout the U.S. through its 15 sponsored investment funds. The firm is currently offering two private equity funds: Aslan Realty Partners IV, LLC (Aslan IV) and Pearlmark Multifamily Partners, LLC (MP I). Aslan IV focuses on investing in office, retail and industrial properties in major metropolitan markets in the U.S., and has a typical transaction size of $20 to $150 million. Aslan IV's current portfolio is comprised of assets with a gross investment cost of $330 million. MP I's investment focus is on Class A or B multifamily assets with a minimum of 200 units and located in either Atlanta, Florida, the greater Washington D.C. area, Charlotte, Denver, Phoenix or Raleigh/Durham. The typical transaction size ranges between $15 and $50 million. In addition to Aslan IV and MP I, Pearlmark also invests in high-yield debt funds through Mezzanine Realty Partners III & IV, LLC (Mezz Fund III/Mezz IV). The Mezz Fund III invests in various properties, such as medical offices, student housing and hotels, through transaction types ranging from partner buy-out, development acquisitions, discounted loan payoffs and recapitalization. The typical Mezz Fund III loan size is $5 to $40 million. Mezz IV serves to fulfill borrower demand for mezzanine debt solutions for refinancings, recapitalizations, acquisitions and development projects. Pearlmark's high-yield and fixed income real estate debt investment management funds have made 117 loans totaling approximately $1.5 billion in commitments. Subsidiary Pearlmark Investment Advisers LLC, is an SEC-registered investment adviser.

FINANCIAL DATA: Note: Data for latest year may not have been available at press time.

In U.S. $	2015	2014	2013	2012	2011	2010
Revenue						
R&D Expense						
Operating Income						
Operating Margin %						
SGA Expense						
Net Income						
Operating Cash Flow						
Capital Expenditure						
EBITDA						
Return on Assets %						
Return on Equity %						
Debt to Equity						

CONTACT INFORMATION:
Phone: 312-499-1900   Fax: 312-499-1901
Toll-Free:
Address: 200 W. Madison, Ste. 3200, Chicago, IL 60606 United States

STOCK TICKER/OTHER:
Stock Ticker: Private   Exchange:
Employees: 31   Fiscal Year Ends: 12/31
Parent Company:

SALARIES/BONUSES:
Top Exec. Salary: $ Bonus: $
Second Exec. Salary: $ Bonus: $

OTHER THOUGHTS:
Estimated Female Officers or Directors: 2
Hot Spot for Advancement for Women/Minorities: Y

Pennsylvania REIT

www.preit.com

NAIC Code: 0

TYPES OF BUSINESS:

Real Estate Investment Trust
Retail Properties
Industrial Properties
Multi-Family Residential Properties
Property Development & Redevelopment

BRANDS/DIVISIONS/AFFILIATES:

PREIT Associates LP
PREIT Services LLC
PREIT-RUBIN Inc

CONTACTS: Note: Officers with more than one job title may be intentionally listed here more than once.

Joseph Coradino, CEO
Ronald Rubin, Chairman of the Board
Jonathen Bell, Chief Accounting Officer
Andrew Ioannou, Executive VP, Divisional
Mario Ventresca, Executive VP, Subsidiary
Robert Mccadden, Executive VP
Bruce Goldman, Executive VP
Rosemarie Greco, Trustee
M. DAlessio, Trustee
Michael DeMarco, Trustee
John Roberts, Trustee
Leonard Korman, Trustee
Mark Pasquerilla, Trustee

GROWTH PLANS/SPECIAL FEATURES:

Pennsylvania REIT (PREIT) is a fully integrated, self-administered and self-managed real estate investment trust (REIT) whose primary investment focus is on retail shopping malls, strip centers and power centers in the Eastern U.S. The firm's portfolio currently consists of 33 retail properties in 10 states, of which 29 are operating properties and four are developed properties. The 29 include 25 shopping malls and four retail properties, having a total of 24.3 million square feet. PREIT holds interests in its portfolio through its 89.2% controlling interest in PREIT Associates LP. The company provides management, leasing and real estate development services through two of its subsidiaries: PREIT Services, LLC and PREIT-RUBIN, Inc. PREIT Services generally develops and manages properties that the firm intends to consolidate for financial reporting purposes. PREIT-RUBIN develops and manages properties that the firm does not consolidate for financial reporting purposes, including properties in which it owns interests through partnerships with third parties and properties that are owned by third parties in which the firm does not own an interest. In 2015, the firm acquired Springfield Town Center in Springfield, Virginia, a newly-redeveloped 1.35 million square foot premier regional mall. In March 2016, it sold four non-core malls.

PREIT employees receive benefits including a 401(k) plan; an employee share purchase plan; flexible spending accounts; corporate flex time hours; tuition reimbursement; disability, life and AD&D insurance; and an employee assistance program.

FINANCIAL DATA: Note: Data for latest year may not have been available at press time.

In U.S. $	2015	2014	2013	2012	2011	2010
Revenue	425,411,000	432,703,000	438,678,000	427,182,000	456,560,000	455,641,000
R&D Expense						
Operating Income	69,686,000	62,556,000	68,504,000	70,888,000	-102,160,000	-84,064,000
Operating Margin %	16.38%	14.45%	15.61%	16.59%	-22.37%	-18.44%
SGA Expense	34,836,000	35,518,000	40,711,000	39,474,000	39,865,000	38,973,000
Net Income	-116,683,000	-13,830,000	35,859,000	-40,837,000	-90,161,000	-51,927,000
Operating Cash Flow	135,661,000	145,075,000	136,219,000	120,324,000	105,262,000	116,791,000
Capital Expenditure	90,215,000	119,960,000	83,303,000	82,528,000	61,807,000	55,964,000
EBITDA	214,940,000	207,156,000	214,569,000	215,043,000	44,238,000	90,565,000
Return on Assets %	-4.95%	-1.12%	.71%	-1.68%	-3.01%	-1.61%
Return on Equity %	-18.34%	-3.47%	2.55%	-8.01%	-15.04%	-8.42%
Debt to Equity	2.84	1.88	1.82	2.81	3.97	3.40

CONTACT INFORMATION:

Phone: 215 875-0700   Fax: 215 546-7311
Toll-Free: 866-875-0700
Address: 200 S. Broad St., Philadelphia, PA 19102 United States

SALARIES/BONUSES:

Top Exec. Salary: $650,000   Bonus: $162,000
Second Exec. Salary: $300,000   Bonus: $300,000

STOCK TICKER/OTHER:

Stock Ticker: PEI   Exchange: NYS
Employees: 429   Fiscal Year Ends: 12/31
Parent Company:

OTHER THOUGHTS:

Estimated Female Officers or Directors: 8
Hot Spot for Advancement for Women/Minorities: Y

PennyMac Mortgage Investment Trust   www.pennymac-reit.com

NAIC Code: 525990

TYPES OF BUSINESS:

Mortgage REIT

BRANDS/DIVISIONS/AFFILIATES:

PNMAC Capital Management LLC
Private National Mortgage Acceptance Company LLC
PennyMac Loan Services LLC

CONTACTS: Note: Officers with more than one job title may be intentionally listed here more than once.

Stanford Kurland, CEO
Anne McCallion, CFO
Gregory Hendry, Chief Accounting Officer
Jeffrey Grogin, Chief Administrative Officer
David Walker, Chief Credit Officer
David Spector, COO
Doug Jones, Other Executive Officer
Steven Bailey, Other Executive Officer
Daniel Perotti, Other Executive Officer
Vandad Fartaj, Other Executive Officer
Andrew Chang, Other Executive Officer
Stacey Stewart, Trustee
Scott Carnahan, Trustee
Randall Hadley, Trustee
Clay Halvorsen, Trustee
Frank Willey, Trustee
Preston DuFauchard, Trustee
Nancy McAllister, Trustee

GROWTH PLANS/SPECIAL FEATURES:

PennyMac Mortgage Investment Trust is a newly-formed finance company that invests mainly in residential mortgage loans and mortgage-related assets. The firm's objective is to provide attractive risk-adjusted returns to its investors over the long-term, mainly through dividends and secondarily through capital appreciation. The company is managed by PNMAC Capital Management, LLC (PCM), a wholly-owned subsidiary of Private National Mortgage Acceptance Company, LLC (PennyMac) and an SEC-registered investment adviser that specializes in, and focuses on, residential mortgage loans. The company's main focus is on investing in distressed residential mortgage loans. The firm seeks to maximize the value of the mortgage loans that it acquires through proprietary and federally sponsored loan modification programs, special servicing and other initiatives focused on keeping borrowers in their homes. It intends to achieve its investment objective by investing in mortgage loans, a substantial portion of which might be distressed and acquired at discounts to their unpaid principal balances. All of the loans the firm acquires in its correspondent lending operations are acquired, pooled for sale, sold and/or securitized on its behalf by wholly-owned subsidiary PennyMac Loan Services, LLC, which also services most of the loans PennyMac holds.

FINANCIAL DATA: Note: Data for latest year may not have been available at press time.

In U.S. $	2015	2014	2013	2012	2011	2010
Revenue	373,473,000	442,330,000	470,740,000	335,168,000	128,614,000	44,054,000
R&D Expense						
Operating Income	73,304,000	179,464,000	214,635,000	186,822,000	72,495,000	27,026,000
Operating Margin %	19.62%	40.57%	45.59%	55.73%	56.36%	61.34%
SGA Expense	143,896,000	152,984,000	135,412,000	108,719,000	30,521,000	13,849,000
Net Income	90,100,000	194,544,000	200,190,000	138,249,000	64,439,000	24,483,000
Operating Cash Flow	-863,188,000	-366,036,000	-242,832,000	-820,400,000	-275,131,000	-22,789,000
Capital Expenditure	2,335,000		1,419,000	23,000		
EBITDA	198,012,000	265,053,000	302,499,000	218,464,000	89,441,000	27,852,000
Return on Assets %	1.67%	4.22%	5.82%	7.00%	6.52%	5.35%
Return on Equity %	5.86%	12.77%	15.00%	15.82%	14.88%	7.72%
Debt to Equity	2.24	2.00	1.43		0.99	0.77

CONTACT INFORMATION:

Phone: 818 224-7442   Fax:
Toll-Free:
Address: 6101 Condor Drive, Moorpark, CA 93021 United States

STOCK TICKER/OTHER:

Stock Ticker: PMT   Exchange: NYS
Employees: 1,816   Fiscal Year Ends: 12/31
Parent Company:

SALARIES/BONUSES:

Top Exec. Salary: $ Bonus: $
Second Exec. Salary: $ Bonus: $

OTHER THOUGHTS:

Estimated Female Officers or Directors:
Hot Spot for Advancement for Women/Minorities:

Perkins Eastman

www.perkinseastman.com

NAIC Code: 541310

TYPES OF BUSINESS:

Architectural Services
Planning
Design
Architecture
Consulting

BRANDS/DIVISIONS/AFFILIATES:

GROWTH PLANS/SPECIAL FEATURES:

Perkins Eastman is an international planning, design and consulting firm founded in 1981. It has nine offices across the U.S. and five overseas. Its staff consists of architects, interior designers, planners, urban designers, landscape architects, graphic designers, construction specification writers, construction administrators, economists, environmental analysts as well as traffic and transportation engineers. Perkins Eastman has completed projects in 46 states and in more than 40 countries; projects range from small buildings and interiors for non-profit organizations to large new health care and educational campuses, major mixed-use commercial developments and entire cities. In March 2016, the firm received approval to design the Turkevi Center, a new 200,000 square-foot, 32-story mission, consulate and residential building for the Republic of Turkey in the Manhattan First Avenue/New York City region. The building is scheduled for completion in 2018.

CONTACTS: Note: Officers with more than one job title may be intentionally listed here more than once.

Bradford Perkins, Co-CEO
Mary-Jean Eastman, Co-CEO
Candace Carroll, CFO

FINANCIAL DATA: Note: Data for latest year may not have been available at press time.

In U.S. $	2015	2014	2013	2012	2011	2010
Revenue	182,900,000					
R&D Expense						
Operating Income						
Operating Margin %						
SGA Expense						
Net Income						
Operating Cash Flow						
Capital Expenditure						
EBITDA						
Return on Assets %						
Return on Equity %						
Debt to Equity						

CONTACT INFORMATION:

Phone: 212-353-7200   Fax: 212-353-7676
Toll-Free:
Address: 115 Fifth Ave., New York, NY 10003 United States

SALARIES/BONUSES:

Top Exec. Salary: $ Bonus: $
Second Exec. Salary: $ Bonus: $

STOCK TICKER/OTHER:

Stock Ticker: Private   Exchange:
Employees: 1,000   Fiscal Year Ends:
Parent Company:

OTHER THOUGHTS:

Estimated Female Officers or Directors:
Hot Spot for Advancement for Women/Minorities:

Populous Holdings Inc

NAIC Code: 541310

TYPES OF BUSINESS:

Architectural Services
Architecture
Design
Planning

BRANDS/DIVISIONS/AFFILIATES:

CONTACTS: Note: Officers with more than one job title may be intentionally listed here more than once.

Jim Walters, Founder

GROWTH PLANS/SPECIAL FEATURES:

Populous Holdings, Inc. provides design services for sports facilities. These services include sports architecture, conference and exhibition center architecture, special events overlay and bids, environmental branding and graphics, events planning and overlay, master planning and sustainable design consulting as well as facility operations, evaluation and analysis. Populous also provides planning services such as site selection, site design, district plans, landscape plans and design guidelines. Its interior design services include master planning, strategic planning, brand development and space needs analysis and planning. The company's project portfolio includes aquatic, athletics/track and field, football, soccer, rugby, tennis, baseball, basketball, collegiate, convention and exhibition centers, fairgrounds, equestrian centers, horse racing, ice hockey, speed skating and motor sports. It has designed more than 2,000 projects worth $30 billion, including 50 arenas, which include places like the O2 Arena, Yankee Stadium, Wembley Stadium, Philippine Arena and the London Olympic Games stadium. In early 2016, Populous obtained approval by the mayor of London to complete the final planning stages of The Northumberland Development Project, which comprises the Populous-designed Tottenham Hotspur Stadium.

FINANCIAL DATA: Note: Data for latest year may not have been available at press time.

In U.S. $	2015	2014	2013	2012	2011	2010
Revenue						
R&D Expense						
Operating Income						
Operating Margin %						
SGA Expense						
Net Income						
Operating Cash Flow						
Capital Expenditure						
EBITDA						
Return on Assets %						
Return on Equity %						
Debt to Equity						

CONTACT INFORMATION:

Phone: 816-221-1500   Fax: 816-221-1578
Toll-Free:
Address: 4800 Main Street, Ste 300, Kansas City, MO 64112 United States

STOCK TICKER/OTHER:

Stock Ticker: Private   Exchange:
Employees: 541   Fiscal Year Ends:
Parent Company:

SALARIES/BONUSES:

Top Exec. Salary: $ Bonus: $
Second Exec. Salary: $ Bonus: $

OTHER THOUGHTS:

Estimated Female Officers or Directors:
Hot Spot for Advancement for Women/Minorities:

Post Properties Inc

www.postproperties.com

NAIC Code: 0

TYPES OF BUSINESS:

Real Estate Investment Trust
Apartment Communities
Landscaping & Design Services
Apartment Management & Development
Corporate Apartments
Condominiums
Retail Space

BRANDS/DIVISIONS/AFFILIATES:

CONTACTS: Note: Officers with more than one job title may be intentionally listed here more than once.

David Stockert, CEO
Christopher Papa, CFO
Robert Goddard, Chairman of the Board
Arthur Quirk, Chief Accounting Officer
Charles Konas, Executive VP, Divisional
S. Teabo, Executive VP, Divisional
David Ward, Executive VP
Sherry Cohen, Executive VP

GROWTH PLANS/SPECIAL FEATURES:

Post Properties, Inc. is a self-administered and self-managed equity real estate investment trust (REIT). The firm owns, develops and manages upscale multifamily apartment communities primarily in Atlanta, Georgia; Dallas, Texas; and the greater Washington, D.C. metropolitan areas. Other communities are located in Austin and Houston, Texas; Charlotte and Raleigh, North Carolina; and Orlando and Tampa, Florida. The company owns approximately 22,000 apartment units in 58 apartment communities, including units in communities held in unconsolidated entities as well as communities currently under construction. Post's major operating divisions include Post Apartment Management; Post Investment Group; Post Corporate Services; and Post Construction & Property Services. Post Apartment Management is responsible for the day-to-day operations of all Post communities and also conducts short-term corporate apartment leasing activities. Post Investment Group handles the company's acquisition, development, rehabilitation, disposition, condominium sales and asset management operations. Post Corporate Services provides all compliance, information systems, human resources, accounting, management reporting, legal, security, personnel recruiting, training/development, risk management and insurance services for Post. The Post Construction & Property Services division oversees all construction and physical asset maintenance activities at each Post community.

The firm offers employees benefits including medical, dental and vision coverage; education assistance; a 401(k); a stock purchase program; an employee assistance program; and an apartment discount program.

FINANCIAL DATA: Note: Data for latest year may not have been available at press time.

In U.S. $	2015	2014	2013	2012	2011	2010
Revenue	384,006,000	377,812,000	362,737,000	334,911,000	305,316,000	285,138,000
R&D Expense	616,000	2,366,000	1,755,000	1,317,000	1,161,000	2,415,000
Operating Income	111,499,000	106,796,000	99,127,000	92,028,000	78,947,000	22,897,000
Operating Margin %	29.03%	28.26%	27.32%	27.47%	25.85%	8.03%
SGA Expense	19,077,000	18,666,000	17,245,000	16,342,000	16,100,000	16,443,000
Net Income	80,623,000	215,120,000	110,534,000	83,939,000	25,466,000	-6,960,000
Operating Cash Flow	173,205,000	163,339,000	150,374,000	134,189,000	102,384,000	77,111,000
Capital Expenditure	148,137,000	105,449,000	163,620,000	161,394,000	151,010,000	91,857,000
EBITDA	199,838,000	363,228,000	211,961,000	210,855,000	157,649,000	97,394,000
Return on Assets %	3.35%	9.00%	4.50%	3.56%	.90%	-.67%
Return on Equity %	6.08%	17.33%	9.40%	7.40%	1.91%	-1.46%
Debt to Equity	0.71	0.69	0.95	0.98	0.92	1.06

CONTACT INFORMATION:

Phone: 404 846-5000   Fax: 404 846-6282
Toll-Free:
Address: 4401 Northside Pkwy., Ste. 800, Atlanta, GA 30327 United States

STOCK TICKER/OTHER:

Stock Ticker: PPS
Employees: 625
Parent Company:

Exchange: NYS
Fiscal Year Ends: 12/31

SALARIES/BONUSES:

Top Exec. Salary: $450,000   Bonus: $
Second Exec. Salary: $342,000   Bonus: $

OTHER THOUGHTS:

Estimated Female Officers or Directors: 6
Hot Spot for Advancement for Women/Minorities: Y

Sales, profits and employees may be estimates. Financial information, benefits and other data can change quickly and may vary from those stated here.

ProLogis Inc

NAIC Code: 531120

www.prologis.com

TYPES OF BUSINESS:

Real Estate Operations- Industrial
Port Facilities
Distribution Centers

BRANDS/DIVISIONS/AFFILIATES:

ProLogis LP

CONTACTS: Note: Officers with more than one job title may be intentionally listed here more than once.

Michael Blair, Assistant Secretary
Eugene Reilly, CEO, Geographical
Gary Anderson, CEO, Geographical
Hamid Moghadam, CEO
Thomas Olinger, CFO
Lori Palazzolo, Chief Accounting Officer
Michael Curless, Other Executive Officer
Edward Nekritz, Other Executive Officer

GROWTH PLANS/SPECIAL FEATURES:

ProLogis, Inc. is a real estate investment trust (REIT) that owns, acquires, develops and operates industrial distribution properties globally. Its investment portfolio includes various types of properties, such as logistics and distribution facilities and warehouses, throughout the Americas, Europe and Asia. The company owns and manages a portfolio of industrial properties and development projects totaling 671 million square feet in 21 countries. Through its primary operating subsidiary, ProLogis, L.P., the firm invests in high-throughput distribution centers that have a variety of characteristics allowing for the rapid point-to-point transport of goods, including numerous dock doors, shallower building depths, fewer columns, large truck courts and more space for trailer parking. These facilities function best when located in convenient proximity to transportation infrastructure such as major airports and seaports. The company leases its facilities to more than 5,200 customers, including manufacturers, retailers, transportation companies, third-party logistics providers and other enterprises. The firm has approximately $57 billion in total assets under management. These funds are focused exclusively on industrial real estate in North America, Europe and Asia.

The company offers employees medical, dental and vision insurance; flexible spending accounts; life insurance; disability coverage; a retirement plan; education reimbursement; and an employee assistance program.

FINANCIAL DATA: Note: Data for latest year may not have been available at press time.

In U.S. $	2015	2014	2013	2012	2011	2010
Revenue	2,197,074,000	1,760,787,000	1,750,486,000	2,005,961,000	1,533,291,000	633,500,000
R&D Expense						
Operating Income	380,172,000	319,808,000	304,412,000	108,447,000	127,430,000	114,928,000
Operating Margin %	17.30%	18.16%	17.39%	5.40%	8.31%	18.14%
SGA Expense	238,199,000	247,768,000	229,207,000	228,068,000	195,161,000	125,155,000
Net Income	869,439,000	636,183,000	342,921,000	-39,720,000	-153,414,000	27,119,000
Operating Cash Flow	963,410,000	704,531,000	484,989,000	463,492,000	207,064,000	252,760,000
Capital Expenditure	2,313,438,000	1,742,121,000	1,442,455,000	1,128,375,000	1,081,496,000	272,919,000
EBITDA	2,130,341,000	1,664,974,000	1,379,596,000	1,185,321,000	829,939,000	311,564,000
Return on Assets %	3.01%	2.46%	1.21%	-.29%	-1.07%	.14%
Return on Equity %	6.05%	4.52%	2.41%	-.63%	-2.32%	.34%
Debt to Equity	0.79	0.67	0.66	0.94	0.87	1.07

CONTACT INFORMATION:

Phone: 415-394-9000   Fax: 415 394-9001
Toll-Free:
Address: Pier 1, Bay 1, San Francisco, CA 94111 United States

STOCK TICKER/OTHER:

Stock Ticker: PLD   Exchange: NYS
Employees: 1,555   Fiscal Year Ends: 12/31
Parent Company:

SALARIES/BONUSES:

Top Exec. Salary: $950,000   Bonus: $2,315,625
Second Exec. Salary: Bonus: $1,240,000
$575,000

OTHER THOUGHTS:

Estimated Female Officers or Directors: 3
Hot Spot for Advancement for Women/Minorities: Y

PS Business Parks Inc

www.psbusinessparks.com

NAIC Code: 0

TYPES OF BUSINESS:

Real Estate Investment Trust
Industrial Properties
Office Properties
Retail Properties
Business Parks
Property Management

BRANDS/DIVISIONS/AFFILIATES:

PS Business Parks LP
Public Storage Inc

CONTACTS: Note: Officers with more than one job title may be intentionally listed here more than once.

Joseph Russell, CEO
Edward Stokx, CFO
Ronald Havner, Chairman of the Board
John Petersen, Executive VP
Maria Hawthorne, President

GROWTH PLANS/SPECIAL FEATURES:

PS Business Parks, Inc. (PSB) is a self-advised and self-managed REIT. It acquires, owns, operates and develops commercial properties, primarily multi-tenant flex, office and industrial space. The company does business through an operating partnership, PS Business Parks LP, in which PSB holds a 77.8% share, with the remainder being held by Public Storage, Inc. As the sole general partner of PS Business Parks LP, PSB maintains complete responsibility and discretion in managing and controlling the operating partnership. The company wholly owns and operates approximately 28 million rentable square feet of commercial space, comprising 99 business parks located in six states: California, Florida, Maryland, Texas, Virginia and Washington. Approximately 14.6 million square feet of the firm's owned property is classified as flex space, which includes buildings that are configured with a combination of office, assembly, showroom, laboratory, light manufacturing and warehouse space suitable to a variety of business needs. Approximately 8.8 million square feet of the firm's owned property is classified as industrial space, also suitable for a variety of uses. The remaining 4.6 million square feet consists of low-rise office space. The company also manages rentable square feet on behalf of Public Storage and its affiliated entities. PSB targets properties in select sub-markets located in high-growth areas across the U.S. with above average education and personal income levels, access to critical infrastructure and with universities and major transportation arteries in close proximity. In 2015, the firm sold its Northgate business park in Sacramento, California.

FINANCIAL DATA: Note: Data for latest year may not have been available at press time.

In U.S. $	2015	2014	2013	2012	2011	2010
Revenue	373,675,000	376,915,000	359,885,000	347,197,000	298,503,000	279,089,000
R&D Expense						
Operating Income	133,475,000	125,548,000	130,825,000	114,772,000	104,777,000	100,036,000
Operating Margin %	35.71%	33.30%	36.35%	33.05%	35.10%	35.84%
SGA Expense	13,582,000	13,639,000	5,312,000	8,919,000	9,036,000	9,651,000
Net Income	130,475,000	173,971,000	103,192,000	89,079,000	94,088,000	85,325,000
Operating Cash Flow	238,072,000	227,771,000	222,294,000	209,127,000	181,876,000	177,941,000
Capital Expenditure	46,463,000	98,996,000	172,872,000	107,794,000	347,361,000	336,629,000
EBITDA	267,694,000	328,650,000	241,227,000	224,507,000	189,459,000	178,904,000
Return on Assets %	3.09%	5.06%	1.99%	.92%	2.77%	2.44%
Return on Equity %	9.36%	15.70%	6.83%	3.47%	8.87%	6.57%
Debt to Equity	0.33	0.34	0.34	0.83	1.23	0.24

CONTACT INFORMATION:

Phone: 818 244-8080   Fax: 818 242-0566
Toll-Free:
Address: 701 Western Ave., Glendale, CA 91201-2397 United States

STOCK TICKER/OTHER:

Stock Ticker: PSB
Employees: 142
Parent Company:

Exchange: NYS
Fiscal Year Ends: 12/31

SALARIES/BONUSES:

Top Exec. Salary: $568,301   Bonus: $
Second Exec. Salary: $355,801   Bonus: $

OTHER THOUGHTS:

Estimated Female Officers or Directors: 4
Hot Spot for Advancement for Women/Minorities: Y

Public Storage Inc

NAIC Code: 531130

www.publicstorage.com

TYPES OF BUSINESS:

Real Estate Investment Trust
Self-Storage Facilities
Commercial Properties
Online Storage Reservations
Transportation Equipment

BRANDS/DIVISIONS/AFFILIATES:

Public Storage
Shurgard Self Storage Europe Limited
PS Business Parks Inc

CONTACTS: Note: Officers with more than one job title may be intentionally listed here more than once.

Ronald Havner, CEO
John Reyes, CFO
Candace Krol, Other Executive Officer
Lily Hughes, Other Executive Officer
David Doll, President, Divisional
Ronald Spogli, Trustee
Avedick Poladian, Trustee
B. Hughes, Trustee
Daniel Staton, Trustee
Tamara Gustavson, Trustee
Uri Harkham, Trustee
Gary Pruitt, Trustee

GROWTH PLANS/SPECIAL FEATURES:

Public Storage, Inc. is a fully integrated, self-administered and self-managed equity real estate investment trust (REIT) that acquires, develops, owns and operates self-storage facilities primarily used for month-to-month personal and business use. It is one of the largest owners and operators of self-storage space in the U.S. The company operates in three segments: domestic self-storage, Europe self-storage and commercial. The domestic self-storage segment, accounting for 93% of revenues in 2014, consists of the firm's direct and indirect equity interests in 2,277 self-storage facilities, containing approximately 148 million square feet of net rentable space, in 38 states under the Public Storage brand name. The self-storage facilities consist of 3-7 buildings containing 350-750 storage spaces, most of which have 25-400 square feet and an interior height of 8-12 feet. The Europe self-storage segment comprises the firm's 49% interest in Shurgard Self Storage Europe Limited, which owns 216 storage facilities in seven countries, with 12 million square feet of net rentable space. This segment also manages one facility in the U.K. that is wholly owned by Public Storage. The commercial segment includes the company's direct and indirect equity interests in 28 million net rentable square feet of commercial space, partially held through its 42% interest in PS Business Parks, Inc. (PSB), a publicly-traded REIT.

The company offers employees medical, dental and vision insurance; short- and long-term disability coverage; life insurance; medical care and dependent care spending plans; an employee assistance program; a 401(k) plan; and employee discounts.

FINANCIAL DATA: Note: Data for latest year may not have been available at press time.

In U.S. $	2015	2014	2013	2012	2011	2010
Revenue	2,381,696,000	2,195,404,000	1,981,746,000	1,826,729,000	1,752,102,000	1,646,722,000
R&D Expense	10,006,000					
Operating Income	1,232,009,000	1,068,111,000	962,504,000	871,982,000	798,232,000	724,238,000
Operating Margin %	51.72%	48.65%	48.56%	47.73%	45.55%	43.98%
SGA Expense	72,799,000	71,459,000	66,679,000	56,837,000	52,410,000	38,487,000
Net Income	1,311,244,000	1,144,204,000	1,052,453,000	939,258,000	823,842,000	672,038,000
Operating Cash Flow	1,732,601,000	1,606,758,000	1,430,339,000	1,285,659,000	1,203,452,000	1,093,221,000
Capital Expenditure	242,670,000	640,393,000	71,270,000	67,737,000	69,777,000	77,500,000
EBITDA	1,744,307,000	1,593,850,000	1,451,377,000	1,308,077,000	1,216,789,000	1,078,624,000
Return on Assets %	10.74%	9.22%	9.04%	7.55%	6.09%	4.13%
Return on Equity %	20.50%	17.49%	16.11%	12.83%	10.74%	7.38%
Debt to Equity	0.06	0.01	0.16	0.08	0.07	0.10

CONTACT INFORMATION:

Phone: 818 244-8080   Fax: 818 244-9530
Toll-Free: 800-688-8057
Address: 701 Western Ave., Glendale, CA 91201 United States

STOCK TICKER/OTHER:

Stock Ticker: PSA
Employees: 5,300
Parent Company:

Exchange: NYS
Fiscal Year Ends: 12/31

SALARIES/BONUSES:

Top Exec. Salary: $1,000,000   Bonus: $
Second Exec. Salary: $600,000   Bonus: $

OTHER THOUGHTS:

Estimated Female Officers or Directors: 2

Hot Spot for Advancement for Women/Minorities:

PulteGroup Inc

pultegroupinc.com

NAIC Code: 236117

TYPES OF BUSINESS:

Construction, Home Building and Residential
Financial Services
Mortgages
Land Development

BRANDS/DIVISIONS/AFFILIATES:

Pulte Home Corp
Del Webb Corp
Centex Corp
Pulte Mortgage LLC

CONTACTS: Note: Officers with more than one job title may be intentionally listed here more than once.

Robert OShaughnessy, CFO
James Ossowski, Chief Accounting Officer
Richard Dugas, Director
James Ellinghausen, Executive VP, Divisional
Harmon Smith, Executive VP
Steven Cook, Executive VP
Ryan Marshall, President

GROWTH PLANS/SPECIAL FEATURES:

PulteGroup, Inc. is a holding company with subsidiaries in the homebuilding and financial services industries. These subsidiaries include Del Webb Corp.; Pulte Home Corp.; Centex Corp.; and Pulte Mortgage LLC. PulteGroup's core homebuilding business is engaged in the acquisition and development of land, primarily for residential purposes within the U.S. The firm builds a wide variety of homes targeted for first-time, first and second move-up and active adult home buyers, including detached units, townhouses, condominium apartments and duplexes, with varying prices, models, options and lot sizes. Its homebuilding business operates in 50 markets spanning 26 states and offers homes in about 620 communities. During 2015, the firm closed 17,127 homes, with the average unit selling price of $338,000. Sales prices range from less than $100,000 to more than $1,500,000, 85% of which fall between $150,000 and $500,000. PulteGroup's homebuilding operations consist of six geographic segments: Northeast (including Connecticut, Maryland, Massachusetts, New Jersey, New York, Pennsylvania, Rhode Island and Virginia), Florida, Texas, Midwest (including Illinois, Indiana, Kentucky, Michigan, Minnesota, Missouri and Ohio), Southeast (consisting of Georgia, North Carolina, South Carolina and Tennessee) and West (including Arizona, California, Nevada, New Mexico and Washington). The firm's strategy is based on extensive market research that reveals well-defined buying profiles, job demographics and lifestyle choices. PulteGroup's financial services segment consists principally of mortgage operations conducted through Pulte Mortgage LLC and its subsidiaries. In 2015, the firm acquired 507 acres in North Naples, Florida that includes an existing 18-hole golf course and 548 new single family home sites; and agreed to acquire assets of John Wieland Homes & Neighborhood that December.

The company offers employees medical, dental and vision insurance; a 401(k); life and AD&D insurance; business travel accident insurance; short- and long-term disability; tuition reimbursement; an employee assistance program; and time off for volunteering.

FINANCIAL DATA: Note: Data for latest year may not have been available at press time.

In U.S. $	2015	2014	2013	2012	2011	2010
Revenue	5,981,964,000	5,822,363,000	5,679,595,000	4,819,998,000	4,136,690,000	4,569,290,000
R&D Expense						
Operating Income	820,486,000	703,279,000	587,540,000	241,699,000	-24,236,000	-1,163,766,000
Operating Margin %	13.71%	12.07%	10.34%	5.01%	-.58%	-25.46%
SGA Expense	589,780,000	667,815,000	568,500,000	514,457,000	519,583,000	895,102,000
Net Income	494,090,000	474,338,000	2,620,116,000	206,145,000	-210,388,000	-1,096,729,000
Operating Cash Flow	-348,129,000	309,249,000	881,136,000	760,140,000	17,222,000	580,256,000
Capital Expenditure	45,440,000	48,790,000	28,899,000	13,942,000	21,238,000	15,179,000
EBITDA	863,033,000	730,471,000	560,121,000	214,400,000	-276,889,000	-1,186,157,000
Return on Assets %	5.59%	5.44%	33.87%	3.02%	-2.88%	-12.35%
Return on Equity %	10.26%	9.96%	76.62%	9.98%	-10.32%	-41.15%
Debt to Equity	0.49	0.40	0.46	1.14	1.59	1.58

CONTACT INFORMATION:

Phone: 404-978-6400   Fax:
Toll-Free: 866-785-8325
Address: 3350 Peachtree Road NE, Ste 150, Atlanta, GA 30326 United States

STOCK TICKER/OTHER:

Stock Ticker: PHM
Employees: 4,542
Parent Company:

Exchange: NYS
Fiscal Year Ends: 12/31

SALARIES/BONUSES:

Top Exec. Salary:
$1,200,000
Second Exec. Salary:
$700,000

Bonus: $

Bonus: $

OTHER THOUGHTS:

Estimated Female Officers or Directors: 4

Hot Spot for Advancement for Women/Minorities: Y

RailWorks Corp

www.railworks.com

NAIC Code: 237990

TYPES OF BUSINESS:

Railroad Construction
Rail Technologies
Electrical & Mechanical Installations
Communications Technologies

BRANDS/DIVISIONS/AFFILIATES:

Wind Point Partners
Railworks Track Systems Inc
RailWorks Signals & Communications Inc
HSQ Technology
LK Comstock & Company Inc
PNR Railworks Inc
NY Transit
LK Comstock National Transit Inc

CONTACTS: Note: Officers with more than one job title may be intentionally listed here more than once.

Greg Muldoon, Interim-CEO
Jeffrey M. Levy, Pres.
Veronica Lubatkin, CFO
Kirk Johnson, VP-Human Resources
Robert Cummings, VP-Information Systems
Benjamin D. Levy, General Counsel
James R. Hansen, VP-Bus. Dev.
Kathleen Simpson, Dir.-Corp. Comm.
Gene Cellini, Sr. VP-Tax
Tamara Mathews, Dir.-Corp. Safety, Health & Environmental
Michael P. Holt, Pres., RailWorks Systems, Inc.
Ben D'Alessandro, Pres., L.K. Comstock & Company, Inc.
Scott Brace, Exec. VP-RailWorks Track Systems, Inc.

GROWTH PLANS/SPECIAL FEATURES:

RailWorks Corp. and its network of affiliated companies provide track and transit systems construction and maintenance services for the rail and rail-transit industries in the U.S. and Canada. The firm is owned by Wind Point Partners, a private equity investment firm. The company's business serves a range of railroads, public transit authorities and commuter railroads as well as private industries. The firm divides its operations into two business units: transit & systems and tracks. The transit & systems segment handles automatic train controls and systems; traction power systems, including overhead catenary; transit facilities, including general and mechanical contracting; communication systems, including systems that incorporate fiber/Sonet, radio/microwave, cellular and WiFi technology; Supervisory Control and Data Acquisition System (SCADA) and system integration; and rain construction, maintenance and rehabilitation. In addition to these services, the transit & systems unit provides design support, construction engineering and installation, testing, start-up and maintenance services for heavy rail, light rail and automated people mover facilities. The tracks segment designs, builds, manages and supplies railways and railway construction, maintenance and rehabilitation projects. This business involves, in addition to rail and tie installation techniques, the development of signals and crossings; production involving steel and tie gangs; maintenance-of-way services; communication systems; scheduled maintenance work; and emergency repairs, which include derailment response. In total, RailWorks has 11 subsidiaries: HSQ Technology; L.K. Comstock & Company, Inc.; PNR RailWorks, Inc.; L.K. Comstock National Transit, Inc.; PNR RailWorks, Inc.; RailWorks Signals & Communications, Inc.; RailWorks Track Systems, Inc.; RailWorks Track Services, Inc.; NY Transit; RailWorks Maintenance of Way, Inc.; and RailWorks Transit, Inc.

RailWorks offers its employees medical and dental insurance and a 401(k) plan with company match.

FINANCIAL DATA: Note: Data for latest year may not have been available at press time.

In U.S. $	2015	2014	2013	2012	2011	2010
Revenue	703,000,000	750,000,000	730,000,000	684,000,000	550,000,000	450,000,000
R&D Expense						
Operating Income						
Operating Margin %						
SGA Expense						
Net Income						
Operating Cash Flow						
Capital Expenditure						
EBITDA						
Return on Assets %						
Return on Equity %						
Debt to Equity						

CONTACT INFORMATION:

Phone: 212-502-7900   Fax: 212-502-1865
Toll-Free:
Address: 5 Penn Plz., 15/Fl, New York, NY 10001 United States

STOCK TICKER/OTHER:

Stock Ticker: Private   Exchange:
Employees: 3,850   Fiscal Year Ends: 12/31
Parent Company: Wind Point Partners

SALARIES/BONUSES:

Top Exec. Salary: $ Bonus: $
Second Exec. Salary: $ Bonus: $

OTHER THOUGHTS:

Estimated Female Officers or Directors: 3
Hot Spot for Advancement for Women/Minorities: Y

Ramada Worldwide Inc

www.ramada.com

NAIC Code: 721110

TYPES OF BUSINESS:

Hotels & Motels

BRANDS/DIVISIONS/AFFILIATES:

Wyndham Worldwide
Ramada
Ramada Limited Hotel
Ramada Hotel
Ramada Plaza Hotel
Ramada Resort
Ramada Hotel & Resorts
Ramada Resort Akiriki

CONTACTS: Note: Officers with more than one job title may be intentionally listed here more than once.

Keith J. Pierce, Pres.
Stephen P. Holmes, CEO

GROWTH PLANS/SPECIAL FEATURES:

Ramada Worldwide, Inc., a subsidiary of Wyndham Worldwide, is a leading franchisor of hotel and motel properties. The company owns nearly 900 properties throughout the U.S. and in more than 50 other countries, and offers discounts for seniors, groups, government employees and AAA and AARP members. The franchise also participates in the Wyndham Rewards program, in which guests can earn rewards points, gift cards or resort vacations by staying in participating hotels. The company's core brand, Ramada, offers room service, bell service, onsite restaurants and cocktail lounges and full-service meeting facilities. Ramada Limited Hotel properties offer less expensive rates and limited amenities. These properties provide accommodations for the mid-market traveler and include swimming pools, on-site restaurants or a-la-carte food service, plus business services, meeting rooms and fitness facilities. Ramada Plaza Hotel properties are full-service hotels, offering on-site restaurants and lounges, conference and banquet facilities, as well as business center and fitness facilities. Ramada Resort properties are designed for leisure travelers on extended stays and offer playgrounds, pools, spas, fitness centers, dining room services and car rental facilities. The franchise's resort properties, principally located outside the U.S. and Canada, include Ramada Hotel & Resorts, which feature oversized rooms, pools, saunas, golf and tennis facilities, restaurants and lounges; and Ramada Hotel & Suites, offering luxury suites, restaurants and meeting and banquet facilities. In 2015, Ramada expanded its brand in the South Pacific with waterfront, 80-room Ramada Resort Akiriki, located in Vanautu's capital Port Vila.

Wyndham Worldwide and its subsidiaries offer employees a choice of medical, dental and vision plans; flexible spending accounts; adoption reimbursement; business travel accident insurance; an educational assistance program; and a 401(k) plan.

FINANCIAL DATA: Note: Data for latest year may not have been available at press time.

In U.S. $	2015	2014	2013	2012	2011	2010
Revenue						
R&D Expense						
Operating Income						
Operating Margin %						
SGA Expense						
Net Income						
Operating Cash Flow						
Capital Expenditure						
EBITDA						
Return on Assets %						
Return on Equity %						
Debt to Equity						

CONTACT INFORMATION:

Phone: 973-428-9700   Fax: 605-496-7658
Toll-Free: 800-854-9517
Address: 1 Sylvan Way, Parsippany, NJ 07054 United States

STOCK TICKER/OTHER:

Stock Ticker: Subsidiary
Employees: 30,000
Parent Company: Wyndham Worldwide

Exchange:
Fiscal Year Ends: 12/31

SALARIES/BONUSES:

Top Exec. Salary: $ Bonus: $
Second Exec. Salary: $ Bonus: $

OTHER THOUGHTS:

Estimated Female Officers or Directors:
Hot Spot for Advancement for Women/Minorities:

RE/MAX Holdings Inc

NAIC Code: 531210

www.remax.com

TYPES OF BUSINESS:

Real Estate Brokerage
Moving & Relocation Services
Online Database

BRANDS/DIVISIONS/AFFILIATES:

RE/MAX Commercial Network

CONTACTS: *Note: Officers with more than one job title may be intentionally listed here more than once.*

David Liniger, Chairman of the Board
Karri Callahan, Chief Accounting Officer
David Metzger, Co-CFO
Gail Liniger, Co-Founder
Adam Contos, COO
Adam Scoville, General Counsel
Geoffrey Lewis, President

GROWTH PLANS/SPECIAL FEATURES:

RE/MAX Holdings, Inc. is a leading franchisor of real estate brokerage services. Its focus is to aid clients in buying, selling, financing and moving into properties. Residential sales make up the majority of the firm's revenue, but it also operates a commercial brokerage division. The firm's franchise network consists of more than 100,000 sales agents in approximately 6,986 offices operating in 98 countries, including Canada, Australia, the U.K. and Mexico as well as countries in Central and South America, the Caribbean, Europe, the Middle East, China and southern Africa. The company's commercial properties division gives customers access to the RE/MAX Commercial Network database of commercial properties, which allows clients to find commercial office space as well as industrial locations, shopping centers, vacant land, hospitality and farm/ranch properties. RE/MAX's commercial agents aid clients in the selling, buying or leasing of their property interests. The company uses a hot air balloon as its corporate logo, one of the most recognized real estate logos in the world. In 2015, global real estate franchisor RE/MAX LLC sold its RE/MAX 100 brokerage firm. In early 2016, RE/MAX LLC agreed to re-acquire master franchise RE/MAX of New York. It also sold its last two remaining corporate-owned brokerages, RE/MAX Northwest in Seattle, Washington and RE/MAX Equity Group in Portland, Oregon. As a result of these three brokerage sales, RE/MAX franchise offices are now completely independently owned.

FINANCIAL DATA: *Note: Data for latest year may not have been available at press time.*

In U.S. $	2015	2014	2013	2012	2011	2010
Revenue	176,868,000	170,984,000	158,862,000	143,677,000	138,302,000	140,217,000
R&D Expense						
Operating Income	74,155,000	63,835,000	47,080,000	45,546,000	38,471,000	38,410,000
Operating Margin %	41.92%	37.33%	29.63%	31.70%	27.81%	27.39%
SGA Expense	90,986,000	91,847,000	96,243,000	84,337,000	85,291,000	81,353,000
Net Income	16,655,000	13,436,000	1,506,000	33,324,000		7,312,000
Operating Cash Flow	74,588,000	63,709,000	50,069,000	51,259,000	43,589,000	36,422,000
Capital Expenditure	3,628,000	2,148,000	1,340,000	1,816,000	1,404,000	1,378,000
EBITDA	88,917,000	78,538,000	60,909,000	59,238,000	53,097,000	38,332,000
Return on Assets %	4.47%	3.77%	.49%	8.23%		
Return on Equity %	4.72%	5.40%	2.07%			
Debt to Equity	0.41	0.79	0.87			

CONTACT INFORMATION:

Phone: 303-796-3800   Fax: 303-796-3599
Toll-Free: 800-525-7452
Address: 5075 S. Syracuse St., Denver, CO 80237 United States

STOCK TICKER/OTHER:

Stock Ticker: RMAX   Exchange: NYS
Employees: 399   Fiscal Year Ends: 12/31
Parent Company:

SALARIES/BONUSES:

Top Exec. Salary: $575,000   Bonus: $60,000
Second Exec. Salary: Bonus: $
$489,083

OTHER THOUGHTS:

Estimated Female Officers or Directors: 3
Hot Spot for Advancement for Women/Minorities: Y

Realogy Holdings Corp

www.realogy.com

NAIC Code: 531210

TYPES OF BUSINESS:
Real Estate Services
Real Estate Brokerages
Settlement Services
Property Financing Services

BRANDS/DIVISIONS/AFFILIATES:
Century 21
Coldwell Banker
ERA Sotheby's International Realty
Better Homes and Gardens Real Estate
Corcoran
ZipRealty
Citi Habitats
PHH Home Loans LLC

CONTACTS: Note: Officers with more than one job title may be intentionally listed here more than once.
Alexander Perriello, CEO, Divisional
Bruce Zipf, CEO, Subsidiary
Donald Casey, CEO, Subsidiary
Kevin Kelleher, CEO, Subsidiary
Richard Smith, CEO
Anthony Hull, CFO
Timothy Gustavson, Chief Accounting Officer
Stephen Fraser, Chief Information Officer
Marilyn Wasser, Executive VP
Sunita Holzer, Executive VP
David Weaving, Other Corporate Officer
Dea Benson, Senior VP, Divisional

GROWTH PLANS/SPECIAL FEATURES:
Realogy Holdings Corp. is an integrated provider of residential real estate services in the U.S. The company's operating platform is supported by industry-leading brokerage brands such as Century 21, Coldwell Banker, Coldwell Banker Commercial, ERA, Sotheby's International Realty and Better Homes and Gardens Real Estate. Realogy also owns and operates the Corcoran, Citi Habitats and ZipRealty brands. The company operates via four segments, each of which receives fees based upon services performed for its customers: real estate franchise services (REF), company-owned real estate brokerage services (CREB), relocation services and title and settlement services (TSS). REF has 13,600 offices worldwide in 110 countries, including 6,000 brokerage offices in the U.S. and 256,800 independent sales associates worldwide. CREB owns and operates its residential real estate brokerage business under the Coldwell Banker, Corcoran, Sotheby's, ZipRealty and Citi Habitats brands, offering full-service residential brokerage services through 790 company-owned offices and 47,000 independent sales associates. This segment is also engaged in the mortgage process through 49.9%-owned PHH Home Loans LLC, which serves real estate brokerage and relocation customers. The relocation services segment is a global provider of outsources employee relocation services, and also services affinity organizations such as insurance companies and credit unions that provide Realogy services to their members. Title and settlement services assist with the closing of real estate transactions by providing title and escrow services to customers and real estate companies.

The firm offers employees medical, dental and vision coverage; a 401(k) plan; business travel accident insurance; and flexible spending accounts for dependent care and health care.

FINANCIAL DATA: Note: Data for latest year may not have been available at press time.

In U.S. $	2015	2014	2013	2012	2011	2010
Revenue	5,706,000,000	5,328,000,000	5,289,000,000	4,672,000,000	4,093,000,000	4,090,000,000
R&D Expense						
Operating Income	282,000,000	225,000,000	175,000,000	-563,000,000	-433,000,000	6,000,000
Operating Margin %	4.94%	4.22%	3.30%	-12.05%	-10.57%	.14%
SGA Expense	2,021,000,000	1,857,000,000	1,897,000,000	1,830,000,000	1,709,000,000	1,658,000,000
Net Income	184,000,000	143,000,000	438,000,000	-543,000,000	-441,000,000	-99,000,000
Operating Cash Flow	544,000,000	423,000,000	492,000,000	-103,000,000	-192,000,000	-118,000,000
Capital Expenditure	84,000,000	71,000,000	62,000,000	54,000,000	49,000,000	49,000,000
EBITDA	483,000,000	415,000,000	351,000,000	138,000,000	-247,000,000	203,000,000
Return on Assets %	2.44%	1.92%	5.93%	-7.34%	-5.73%	-1.23%
Return on Equity %	8.00%	6.82%	24.84%	-7240.00%		
Debt to Equity	1.22	1.78	1.93	2.80		

CONTACT INFORMATION:
Phone: 973 407-2000   Fax:
Toll-Free:
Address: 175 Park Ave., Madison, NJ 07940 United States

STOCK TICKER/OTHER:
Stock Ticker: RLGY
Employees: 10,700
Parent Company:

Exchange: NYS
Fiscal Year Ends: 12/31

SALARIES/BONUSES:
Top Exec. Salary: $1,000,000   Bonus: $103,544
Second Exec. Salary: $650,000   Bonus: $

OTHER THOUGHTS:
Estimated Female Officers or Directors: 2

Hot Spot for Advancement for Women/Minorities: Y

RealPage Inc

NAIC Code: 0

www.realpage.com

TYPES OF BUSINESS:

Online Real Estate Management Services
Customer Service Centers

BRANDS/DIVISIONS/AFFILIATES:

OneSite
LeaseStar
YieldStar
MPF Research
Velocity
LeasingDesk
Notivus Multi-Family LLC
Kigo

CONTACTS: Note: Officers with more than one job title may be intentionally listed here more than once.

Stephen Winn, CEO
William Chaney, Executive VP, Divisional
Daryl Rolley, Executive VP
David Monk, Other Corporate Officer
Brian Shelton, Senior VP, Divisional
Rhett Butler, Vice President, Divisional
Janine Jovanovic, Vice President, Divisional

GROWTH PLANS/SPECIAL FEATURES:

RealPage, Inc. is a provider of on-demand software services for the rental housing industry. The company's products enable owners and managers of rental property to manage their marketing, pricing, screening, leasing, accounting, purchasing and other property operations. Its systems manage conventional, student, military, senior, commercial, urban and rural housing as well as tax credit-compliant housing. The firm's primary products include OneSite, LeaseStar, MPF Research, YieldStar, Velocity, LeasingDesk and OpsTechnology. OneSite is a centralized product suite that enables users to track leasing, accounting, purchasing and facilities management data. LeaseStar offers multichannel managed marketing which enables owners to efficiently originate, syndicate, manage and capture leads. MPF Research specializes in apartment market research, statistics and analysis of the most vital industry trends. YieldStar uses current pricing, revenue and occupancy trends to help a rental manager determine the best price for an individual unit. Velocity is a utility and billing service that consolidates payments, invoicing and submetering to improve records and reduce potential labor costs. LeasingDesk provides automated applicant screening, document management and other similar renting service necessities in a single online setting. OpsTechnology tracks order, procurement and other budget functions in order to provide a complete and transparent spending management system. RealPage also provides professional services, such as customer support, technical support, consulting and training services to ease the implementation of its products. In 2014, the firm acquired Notivus Multi-Family, LLC, a risk mitigation and compliance service platform; and Kigo, a software-as-a-service vacation rental booking system. In June 2015, RealPage acquired select assets of ICIM Corporation, including Indatus, the cloud-based, smart answer automation and routing solutions that handle maintenance calls for over 11,000 apartment communities.

FINANCIAL DATA: Note: Data for latest year may not have been available at press time.

In U.S. $	2015	2014	2013	2012	2011	2010
Revenue	468,520,000	404,551,000	377,022,000	322,172,000	257,979,000	188,274,000
R&D Expense	68,799,000	64,418,000	50,638,000	48,177,000	44,561,000	36,922,000
Operating Income	-11,615,000	-15,503,000	21,559,000	11,448,000	1,810,000	6,287,000
Operating Margin %	-2.47%	-3.83%	5.71%	3.55%	.70%	3.33%
SGA Expense	191,922,000	180,765,000	156,504,000	133,985,000	105,891,000	66,021,000
Net Income	-9,218,000	-10,274,000	20,692,000	5,183,000	-1,231,000	67,000
Operating Cash Flow	96,012,000	69,972,000	69,209,000	58,412,000	49,226,000	27,690,000
Capital Expenditure	33,384,000	37,322,000	33,879,000	22,149,000	17,997,000	12,178,000
EBITDA	34,276,000	25,803,000	53,304,000	43,917,000	30,957,000	27,243,000
Return on Assets %	-1.54%	-1.91%	4.57%	1.29%	-.33%	-1.18%
Return on Equity %	-2.81%	-3.19%	7.26%	2.16%	-.62%	-5.42%
Debt to Equity	0.12	0.06		0.03	0.22	0.32

CONTACT INFORMATION:

Phone: 972 820-3000   Fax:
Toll-Free: 877-325-7243
Address: 4000 International Pkwy., Carrollton, TX 75007 United States

STOCK TICKER/OTHER:

Stock Ticker: RP
Employees: 3,875
Parent Company:

Exchange: NAS
Fiscal Year Ends: 12/31

SALARIES/BONUSES:

Top Exec. Salary: $500,000   Bonus: $
Second Exec. Salary: $365,000   Bonus: $

OTHER THOUGHTS:

Estimated Female Officers or Directors: 6
Hot Spot for Advancement for Women/Minorities: Y

Realty Income Corp

www.realtyincome.com

NAIC Code: 0

TYPES OF BUSINESS:

Real Estate Investment Trust
Retail Property
Property Management

BRANDS/DIVISIONS/AFFILIATES:

Crest Net Lease Inc

CONTACTS: Note: Officers with more than one job title may be intentionally listed here more than once.

John Case, CEO
Michael Mckee, Chairman of the Board
Sean Nugent, Chief Accounting Officer
Paul Meurer, Executive VP
Neil Abraham, Executive VP
Michael Pfeiffer, Executive VP
Sumit Roy, President
Dawn Nguyen, Senior VP, Divisional
Benjamin Fox, Senior VP, Divisional
Joel Tomlinson, Senior VP, Divisional
Cary Wenthur, Senior VP, Divisional
Robert Israel, Senior VP, Divisional

GROWTH PLANS/SPECIAL FEATURES:

Realty Income Corp. (RIC) is a real estate investment trust (REIT). The firm's primary objective is to generate stable monthly cash distributions based on its portfolio of retail properties leased to regional and national chains. The company's portfolio management focus includes contractual rent increases on existing leases; rent increases at the termination of existing leases, when market conditions permit; the re-leasing of vacant properties; and the selective sale of other properties. The firm owns a portfolio of 4,538 properties located in 49 states, totaling over 76 million square feet of leasable space and leased to 240 different commercial enterprises in 47 separate industries. Of the 4,538 retail properties owned by the company, 4,467 (approximately 98.4%) are single-tenant retail locations; the remaining are multi-tenant, distribution and office properties. RIC leases its properties under long-term, net-lease agreements (usually 15-20 years) that generally require the tenant to pay for maintenance, minimum monthly rents and property operating expenses such as taxes and insurance. The company's acquisition strategy includes a focus on freestanding, commercially-zoned, single-tenant properties in important retail locations that can be purchased with the simultaneous execution or assumption of long-term, net-lease agreements. Additionally, the company's wholly-owned taxable REIT subsidiary, Crest Net Lease, Inc., owns properties currently classified as held for investment. Crest Net Lease primarily engages in the buying and selling of properties, mainly to individual investors involved in tax-deferred exchanges. During 2015, RIC invested $1.26 billion in 506 new properties and properties under development or expansion. In 2015, RIC was added to the Standard & Poor's 500 list.

FINANCIAL DATA: Note: Data for latest year may not have been available at press time.

In U.S. $	2015	2014	2013	2012	2011	2010
Revenue	1,023,285,000	933,505,000	778,375,000	475,510,000	421,059,000	345,009,000
R&D Expense						
Operating Income	265,781,000	233,396,000	181,914,000	147,401,000	152,607,000	122,809,000
Operating Margin %	25.97%	25.00%	23.37%	30.99%	36.24%	35.59%
SGA Expense	49,298,000	51,085,000	56,827,000	37,998,000	30,954,000	25,311,000
Net Income	283,766,000	270,635,000	245,564,000	159,152,000	157,032,000	130,784,000
Operating Cash Flow	692,303,000	627,692,000	518,906,000	326,469,000	298,952,000	243,368,000
Capital Expenditure	1,278,426,000	1,234,275,000	8,507,000	6,554,000		15,385,000
EBITDA	930,318,000	863,628,000	669,407,000	419,540,000	382,659,000	311,559,000
Return on Assets %	2.24%	2.17%	2.65%	2.32%	3.33%	3.30%
Return on Equity %	4.52%	4.55%	6.18%	6.15%	7.75%	8.01%
Debt to Equity	0.78	0.94	0.87	1.59	1.07	1.06

CONTACT INFORMATION:

Phone: 858 284-5000   Fax:
Toll-Free: 877- 924-6266
Address: 11995 El Camino Real, San Diego, CA 92130 United States

STOCK TICKER/OTHER:

Stock Ticker: O
Employees: 132
Parent Company:

Exchange: NYS
Fiscal Year Ends: 12/31

SALARIES/BONUSES:

Top Exec. Salary: $825,000   Bonus: $
Second Exec. Salary: Bonus: $
$475,000

OTHER THOUGHTS:

Estimated Female Officers or Directors: 7
Hot Spot for Advancement for Women/Minorities: Y

Rebel Design+Group

NAIC Code: 541310

www.rebeldesign.com

TYPES OF BUSINESS:
Architectural Services

BRANDS/DIVISIONS/AFFILIATES:

CONTACTS: *Note: Officers with more than one job title may be intentionally listed here more than once.*
Douglas DeBoer, CEO

GROWTH PLANS/SPECIAL FEATURES:

Rebel Design+Group is a boutique firm that offers the hospitality market elite design and consulting services. Services offered by the firm range throughout the entire lifespan of a given project. Rebel operates through four divisions: architecture & interior design, development & trend analysis, branding & marketing and food & beverage consulting. Architecture & interior design services include design for hospitality, restaurant and architectural projects, as well as providing construction documents, renderings, 3D visualization, contract bidding and administration services. Development & trend analysis services include advisory services, value engineering, feasibility studies, programming and scheduling, market research, analysis and design construction management. Branding & marketing services include logo design, naming services, branding services, web site development, packaging, advertising and environmental graphics. Food & beverage consulting services include interior design, kitchen and bar design, menu conceptualization, operational troubleshooting, key talent and team recruiting and turn-key development services. The firm's boutique size allows the flexibility required to custom tailor its services and provide a one-on-one relationship to a given client. Clients include An Restaurants, JAX Hotels and Pronto EuroCafes. While headquartered in California, Rebel Design+Group also has offices in Florida, London and Kuala Lumpur.

FINANCIAL DATA: *Note: Data for latest year may not have been available at press time.*

In U.S. $	2015	2014	2013	2012	2011	2010
Revenue	18,792,889					
R&D Expense						
Operating Income						
Operating Margin %						
SGA Expense						
Net Income						
Operating Cash Flow						
Capital Expenditure						
EBITDA						
Return on Assets %						
Return on Equity %						
Debt to Equity						

CONTACT INFORMATION:
Phone: Fax:
Toll-Free: 800-927-3235
Address: 2554 Lincoln Blvd., Marina Del Rey, CA 90292 United States

STOCK TICKER/OTHER:
Stock Ticker: Private   Exchange:
Employees: 90   Fiscal Year Ends:
Parent Company:

SALARIES/BONUSES:
Top Exec. Salary: $ Bonus: $
Second Exec. Salary: $ Bonus: $

OTHER THOUGHTS:
Estimated Female Officers or Directors:
Hot Spot for Advancement for Women/Minorities:

Red Lion Hotels Corporation

www.redlion.com

NAIC Code: 721110

TYPES OF BUSINESS:

Hotels
Event Ticketing Services
Property Management Services
Entertainment Productions
Guest Loyalty Programs

BRANDS/DIVISIONS/AFFILIATES:

Red Lion Hotels
WestCoast Entertainment
Leo Hotel Collection
TicketsWest
GuestHouse International
Settle Inn
Red Lion Inn & Suites
Hotel RL

CONTACTS: Note: Officers with more than one job title may be intentionally listed here more than once.

Gregory Mount, CEO
James Bell, CFO
Robert Wolfe, Chairman of the Board
William Linehan, Chief Marketing Officer
Harry Sladich, Executive VP, Divisional
Thomas Mckeirnan, Executive VP
Jack Lucas, President, Divisional
Pam Scott, Vice President, Divisional

GROWTH PLANS/SPECIAL FEATURES:

Red Lion Hotels Corporation owns, operates and franchises midscale, full, select and limited service hotels in the western U.S. Its hotel brands include Red Lion Hotels, Red Lion Inn & Suites, Hotel RL, GuestHouse and Settle Inn. The firm currently holds interests in approximately 55 hotels across nine states and one Canadian province, offering a total of 11,820 rooms and over 662,322 square feet of meeting space. Red Lion operates in three segments: hotels, franchise and entertainment. The hotels segment, deriving 81.5% revenue, consists of the operations of its 19 company-operated hotels, of which 14 are wholly-owned and 5 are leased. The franchise segment (6.6%) licenses the firm's 36 franchised hotels. To support its hotels, Red Lion provides services in marketing, sales, advertising, guest loyalty programs, revenue management, reservation systems, quality assurance and brand standards. This segment's Leo Hotel Collection is a unique, historic and boutique style of hotels. The entertainment segment (11.8%), operating through TicketsWest and WestCoast Entertainment, offers ticketing services, ticketing inventory management systems, call center services, promotion services and outlet/electronic channel distribution for event locations. TicketsWest offers tickets for live music, sporting events, family events (such as circuses, expos and fairs) and theater events. The company has also developed an electronic ticketing platform that integrates with its electronic hotel distribution system. In 2015, the firm sold its Bellevue and Wenatchee hotels in Washington; formed a joint venture to accelerate its national growth strategy, which includes selling a 45% stake in 12 wholly-owned hotels to the joint venture and concurrently refinancing all of the company's secured debt; and acquired GuestHouse International, as well as Settle Inn.

The firm offers employees life, medical, dental and vision insurance; a flexible spending plan; an associate travel program; and an employee assistance program.

FINANCIAL DATA: Note: Data for latest year may not have been available at press time.

In U.S. $	2015	2014	2013	2012	2011	2010
Revenue	142,920,000	145,426,000	120,055,000	150,707,000	156,080,000	163,494,000
R&D Expense						
Operating Income	13,880,000	6,759,000	-4,428,000	-9,870,000	7,480,000	-4,302,000
Operating Margin %	9.71%	4.64%	-3.68%	-6.54%	4.79%	-2.63%
SGA Expense	16,388,000	13,563,000	12,409,000	10,881,000	12,755,000	12,144,000
Net Income	2,719,000	2,303,000	-17,047,000	-14,674,000	-7,148,000	-8,609,000
Operating Cash Flow	9,680,000	10,578,000	7,087,000	13,470,000	1,881,000	19,491,000
Capital Expenditure	16,542,000	24,891,000	13,193,000	8,442,000	46,278,000	10,615,000
EBITDA	24,395,000	19,984,000	10,193,000	6,174,000	26,937,000	17,026,000
Return on Assets %	1.06%	1.00%	-6.87%	-5.18%	-2.24%	-2.52%
Return on Equity %	1.97%	1.65%	-11.75%	-9.20%	-4.24%	-4.96%
Debt to Equity	0.65	0.43	0.51	0.20	0.58	0.48

CONTACT INFORMATION:

Phone: 509 459-6100   Fax: 509 325-7324
Toll-Free: 800-733-5466
Address: 201 W. North River Dr., Ste. 100, Spokane, WA 99201 United States

STOCK TICKER/OTHER:

Stock Ticker: RLH
Employees: 1,783
Parent Company:

Exchange: NYS
Fiscal Year Ends: 12/31

SALARIES/BONUSES:

Top Exec. Salary: $298,045   Bonus: $162,500
Second Exec. Salary: $253,071   Bonus: $125,000

OTHER THOUGHTS:

Estimated Female Officers or Directors: 2
Hot Spot for Advancement for Women/Minorities: Y

Red Rock Resorts Inc

redrock.sclv.com/Home/Corporate/Corporate

NAIC Code: 721120

TYPES OF BUSINESS:

Casino Hotels

BRANDS/DIVISIONS/AFFILIATES:

Red Rock Casino, Resort & Spa
Green Valley Ranch, Resort, Casino & Spa
Boulder Station
Sunset Station
Texas Station
Wildfire Casino & Lanes
Wild Wild West
Gun Lake Casino

CONTACTS: *Note: Officers with more than one job title may be intentionally listed here more than once.*

Frank J. Fertitta III, CEO
Richard J. Haskins, Pres.
Marc J. Falcone, CFO
Daniel J. Roy, COO
Frank J. Fertitta III, Chmn.

GROWTH PLANS/SPECIAL FEATURES:

Red Rock Resorts, Inc., formerly Station Casinos Corp., is a leading gaming, development and management company operating 21 strategically-located casino and entertainment properties. In total, the firm has developed over $5 billion of regional gaming and entertainment destinations in multiple jurisdictions. Additionally, Red Rock is an established leader in the Native American gaming, managing facilities in northern California and western Michigan. The Las Vegas portfolio of the firm consists of nine major gaming and entertainment facilities as well as 10 smaller casinos. Combined, the Las Vegas portfolio of casinos offers approximately 19,300 slot machines, 300 table games and 4,000 hotel rooms. Properties of the firm include: Red Rock Casino, Resort & Spa; Green Valley Ranch, Resort, Casino & Spa; Boulder Station; Palace Station; Santa Fe Station; Sunset Station; Texas Station; Fiesta Henderson; Fiesta Ranch; Barley's; Wildfire Sunset; Wildfire Boulder; Wildfire Casino & Lanes; Wild Wild West; Gun Lake Casino; and Graton Resort & Casino. Moreover, Red Rock also controls seven gaming-entitled development sites consisting of approximately 398 acres in Las Vegas and Reno, Nevada.

FINANCIAL DATA: *Note: Data for latest year may not have been available at press time.*

In U.S. $	2015	2014	2013	2012	2011	2010
Revenue						
R&D Expense						
Operating Income						
Operating Margin %						
SGA Expense						
Net Income						
Operating Cash Flow						
Capital Expenditure						
EBITDA						
Return on Assets %						
Return on Equity %						
Debt to Equity						

CONTACT INFORMATION:

Phone: 702-495-3000   Fax:
Toll-Free: 866-767-7773
Address: 1505 S. Pavilion Ctr. Dr., Las Vegas, NV 89135 United States

STOCK TICKER/OTHER:

Stock Ticker: RRR   Exchange: NAS
Employees: 11,800   Fiscal Year Ends:
Parent Company:

SALARIES/BONUSES:

Top Exec. Salary: $ Bonus: $
Second Exec. Salary: $ Bonus: $

OTHER THOUGHTS:

Estimated Female Officers or Directors:
Hot Spot for Advancement for Women/Minorities:

Redfin

www.redfin.com

NAIC Code: 531210

TYPES OF BUSINESS:

Real Estate Brokerage
Online Real Estate Listings

BRANDS/DIVISIONS/AFFILIATES:

Redfin Home Value Tool
Redfin Collections
Book It Now

CONTACTS: *Note: Officers with more than one job title may be intentionally listed here more than once.*

Glenn Kelman, CEO
Chris Nielsen, CFO
Bridget Frey, CTO

GROWTH PLANS/SPECIAL FEATURES:

Redfin is a technology-powered real estate brokerage firm which provides information on real estate listings, home appraisal and past record sales while allowing buyers, sellers and renters to connect with real estate and mortgage professionals. In contrast to traditional firms, Redfin agents do not receive a full commission from transactions completed. Instead, a portion of the commission is refunded to the customer and usually applied to closing costs. On the firm's web page, consumers can also browse, share and create albums of houses for sale through the free service Redfin Collections. The Redfin Home Value Tool gives home owners real-time data on the estimated value of their home. Recently-released Book It Now is a fast and easy way to book a home tour. Book It Now turns every listing into an open house that can be seen at one's own schedule. Each website page showing a home for sale on Redfin.com displays the option to Go Tour This Home. Customers select a preferred date and time, how long they are available to view the property and click to request a tour. Confirmation via email and a phone call from a Redfin tour coordinator are returned shortly. The company offers services in 80 markets across 42 states and Washington, D.C. and partners with many other real estate agents across the country to service areas not covered.

Redfin offers its employees medical, dental and vision coverage; life insurance; short- and long-term disability; financial support for books, classes and career development; and a 401(k) retirement plan.

FINANCIAL DATA: *Note: Data for latest year may not have been available at press time.*

In U.S. $	2015	2014	2013	2012	2011	2010
Revenue	125,000,000	100,000,000	63,500,000	52,500,000	31,000,000	
R&D Expense						
Operating Income						
Operating Margin %						
SGA Expense						
Net Income						
Operating Cash Flow						
Capital Expenditure						
EBITDA						
Return on Assets %						
Return on Equity %						
Debt to Equity						

CONTACT INFORMATION:

Phone: 206-588-6863   Fax:
Toll-Free: 877-973-3346
Address: 2025 1st Ave., Ste. 500, Seattle, WA 98121 United States

STOCK TICKER/OTHER:

Stock Ticker: Private   Exchange:
Employees: 1,778   Fiscal Year Ends:
Parent Company:

SALARIES/BONUSES:

Top Exec. Salary: $ Bonus: $
Second Exec. Salary: $ Bonus: $

OTHER THOUGHTS:

Estimated Female Officers or Directors: 2
Hot Spot for Advancement for Women/Minorities:

Redwood Trust Inc

www.redwoodtrust.com

NAIC Code: 525990

TYPES OF BUSINESS:

Mortgage REIT
Investments-Real Estate Loans & Securities

BRANDS/DIVISIONS/AFFILIATES:

CONTACTS: *Note: Officers with more than one job title may be intentionally listed here more than once.*

Martin Hughes, CEO
Christopher Abate, CFO
Richard Baum, Chairman of the Board
Collin Cochrane, Chief Accounting Officer
Douglas Hansen, Director
Fred Matera, Executive VP, Divisional
Andrew Stone, General Counsel
Brett Nicholas, President

GROWTH PLANS/SPECIAL FEATURES:

Redwood Trust, Inc. is a mortgage real estate investment trust (REIT). The firm primarily invests in high-quality commercial and residential real estate loans. It also invests in securities backed by those loans. Redwood Trust purchases the loans from originators that do not have the necessary liquidity to continue making loans of that size without first selling the old loan to a capital market company. It does not deal directly with consumers. The firm acquires most of its real estate loans from banks and other large mortgage origination companies throughout the U.S. Its revenues are derived from monthly loan payments made by the property owners. The firm's primary real estate investments include investments in senior and subordinate non-agency residential mortgage-backed securities (RMBS), real estate loans, commercial mortgage-backed securities (CMBS), investments in a private fund and self-sponsored securitization entities and other structured credit risk transfer investments. Redwood Trust funds the majority of its real estate loan investments through securitizations, whereby it issues non-recourse long-term debt in the form of securities, using the real estate loan investments as collateral.

The company offers employees benefits including term life, dental, vision, medical and short- and long-term disability insurance; flexible spending accounts; a 401(k) plan; and an employee stock purchase plan.

FINANCIAL DATA: *Note: Data for latest year may not have been available at press time.*

In U.S. $	2015	2014	2013	2012	2011	2010
Revenue	188,803,000	190,616,000	170,444,000	165,511,000	129,088,000	208,886,000
R&D Expense						
Operating Income	91,742,000	99,532,000	75,209,000	86,430,000	25,238,000	111,482,000
Operating Margin %	48.59%	52.21%	44.12%	52.22%	19.55%	53.36%
SGA Expense	81,988,000					
Net Income	102,088,000	100,569,000	173,246,000	131,769,000	26,343,000	110,052,000
Operating Cash Flow	-1,250,210,000	-1,791,133,000	-221,829,000	-450,227,000	17,810,000	29,893,000
Capital Expenditure	32,388,000	46,113,000	3,106,000			
EBITDA						
Return on Assets %	1.68%	1.91%	3.82%	2.58%	.48%	2.11%
Return on Equity %	8.49%	8.03%	14.52%	12.96%	2.69%	10.80%
Debt to Equity	2.69	2.18	1.94	2.34	4.79	3.66

CONTACT INFORMATION:

Phone: 415 389-7373   Fax: 415 381-1773
Toll-Free: 866-269-4976
Address: 1 Belvedere Pl., Ste. 300, Mill Valley, CA 94941 United States

STOCK TICKER/OTHER:

Stock Ticker: RWT   Exchange: NYS
Employees: 221   Fiscal Year Ends: 12/31
Parent Company:

SALARIES/BONUSES:

Top Exec. Salary: $750,000   Bonus: $
Second Exec. Salary: Bonus: $
$600,000

OTHER THOUGHTS:

Estimated Female Officers or Directors: 1
Hot Spot for Advancement for Women/Minorities: Y

Regency Centers Corp

www.regencycenters.com

NAIC Code: 0

TYPES OF BUSINESS:
Real Estate Investment Trust
Shopping Centers
Property Management

BRANDS/DIVISIONS/AFFILIATES:
Regency Centers LP

CONTACTS: *Note: Officers with more than one job title may be intentionally listed here more than once.*
Martin Stein, CEO
Lisa Palmer, CFO
J. Leavitt, Chief Accounting Officer
Dan Chandler, Executive VP, Divisional
James Thompson, Executive VP, Divisional
Barbara Johnston, General Counsel

GROWTH PLANS/SPECIAL FEATURES:
Regency Centers Corp. (REG) is a real estate investment trust (REIT) engaged in the ownership, operation and development of grocery-anchored and community shopping centers located in areas with above-average household incomes and population densities. The firm's operations are conducted primarily through its majority-owned operating partnership Regency Centers, L.P., of which it owns 99.8%. REG directly owns 318 shopping centers in 27 states, representing 38 million square feet of rentable space. Through co-investment partnerships, it owns partial interests in 200 shopping centers in 23 states and Washington, D.C., representing 28.4 million square feet. REG leases space in its shopping centers to grocers, specialty side-shop retailers, restaurants and major retail anchors. The company's largest tenants include Kroger, which represents 4.7% of its directly owned leasable area; Publix, 3.7%; and Albertsons/Safeway, 2.9%.

The firm offers employees medical, dental, vision, life and disability insurance; a 401(k) plan; profit sharing; stock grant awards; wellness programs; and educational assistance.

FINANCIAL DATA: *Note: Data for latest year may not have been available at press time.*

In U.S. $	2015	2014	2013	2012	2011	2010
Revenue	569,763,000	537,898,000	489,007,000	496,920,000	500,417,000	486,806,000
R&D Expense						
Operating Income	204,665,000	184,550,000	164,320,000	175,662,000	174,279,000	175,184,000
Operating Margin %	35.92%	34.30%	33.60%	35.35%	34.82%	35.98%
SGA Expense	65,600,000	60,242,000	61,234,000	61,700,000	56,117,000	56,324,000
Net Income	150,056,000	187,390,000	149,804,000	25,867,000	51,370,000	12,014,000
Operating Cash Flow	275,637,000	277,742,000	250,731,000	257,215,000	217,633,000	141,208,000
Capital Expenditure	248,086,000	350,357,000	321,072,000	320,614,000	152,698,000	90,458,000
EBITDA	381,076,000	391,266,000	329,360,000	231,565,000	298,849,000	273,079,000
Return on Assets %	3.07%	4.09%	3.31%	-.17%	.79%	-.19%
Return on Equity %	7.79%	10.70%	8.80%	-.45%	2.14%	-.50%
Debt to Equity	1.08	1.27	1.23	1.38	1.29	1.46

CONTACT INFORMATION:
Phone: 904 598-7000   Fax: 904 634-3428
Toll-Free: 800-950-6333
Address: 1 Independent Dr., Ste. 114, Jacksonville, FL 32202 United States

STOCK TICKER/OTHER:
Stock Ticker: REG
Employees: 371
Parent Company:

Exchange: NYS
Fiscal Year Ends: 12/31

SALARIES/BONUSES:
Top Exec. Salary: $790,000   Bonus: $
Second Exec. Salary: $570,000   Bonus: $

OTHER THOUGHTS:
Estimated Female Officers or Directors: 1
Hot Spot for Advancement for Women/Minorities: Y

Reis Inc

NAIC Code: 0

www.reis.com

TYPES OF BUSINESS:

Real Estate Information & Analytics

GROWTH PLANS/SPECIAL FEATURES:

Reis, Inc. is a provider of commercial real estate market information and analytical tools. The company offers these services through its subsidiary, Reis Services. Reis Services maintains a proprietary database that contains detailed information on commercial properties in metropolitan markets and neighborhoods. This database offers information on retail, apartment, office and industrial properties. The database is utilized by real estate investors, lenders and other professionals that wish to make informed selling, buying and financing decisions. The firm's data is also used by debt and equity investors to assess, quantify and manage the risks of default and loss associated with individual mortgages, properties, portfolios and real estate backed securities. Reis Services, through both its flagship product Reis SE and its new ReisReports product, provides web-based access to commercial real estate information and analytical tools designed to facilitate debt/equity transactions and ongoing evaluations. Reis SE offers trend and forecast analysis at metropolitan and neighborhood levels, and also provides detailed building-specific information, such as vacancy rates, rents, property sales, lease terms, new construction listings and property valuation estimates. The product is designed to meet the needs of banks, developers/builders, property owners, non-bank lenders and equity investors. ReisReports, dependent on the package the subscriber chooses, provides users with a set number of monthly market reports. Reis currently serves 1,000 companies, most of which have multiple users entitled to access Reis SE.

BRANDS/DIVISIONS/AFFILIATES:

Reis Services LLC
Reis SE
ReisReports

CONTACTS: Note: Officers with more than one job title may be intentionally listed here more than once.

Lloyd Lynford, CEO
Mark Cantaluppi, CFO
William Sander, COO
M. Mitchell, Director
Jonathan Garfield, Director

FINANCIAL DATA: Note: Data for latest year may not have been available at press time.

In U.S. $	2015	2014	2013	2012	2011	2010
Revenue	50,890,440	41,335,160	34,721,090	31,228,640	27,180,480	27,576,200
R&D Expense	3,711,054	3,472,875	3,121,729	2,485,168	2,093,303	1,810,845
Operating Income	12,129,890	7,549,569	4,366,581	2,689,801	983,048	731,605
Operating Margin %	23.83%	18.26%	12.57%	8.61%	3.61%	2.65%
SGA Expense	25,967,870	22,275,690	20,259,010	19,436,740	17,799,530	15,841,790
Net Income	10,304,980	4,047,122	17,596,940	-4,283,582	1,886,427	667,853
Operating Cash Flow	24,235,650	14,788,860	11,441,800	-6,554,862	11,960,940	9,665,189
Capital Expenditure	6,187,149	4,203,063	4,498,923	4,036,929	3,623,162	2,657,067
EBITDA	17,745,790	12,782,870	9,406,077	7,726,120	6,200,332	5,651,303
Return on Assets %	8.01%	3.34%	16.30%	-4.09%	1.73%	.61%
Return on Equity %	10.42%	4.28%	21.02%	-5.63%	2.48%	.90%
Debt to Equity						0.07

CONTACT INFORMATION:

Phone: 212 921-1122   Fax:
Toll-Free: 800-366-7347
Address: 530 5th Ave., 5th Fl., New York, NY 10036 United States

STOCK TICKER/OTHER:

Stock Ticker: REIS   Exchange: NAS
Employees: 205   Fiscal Year Ends: 12/31
Parent Company:

SALARIES/BONUSES:

Top Exec. Salary: $450,000   Bonus: $
Second Exec. Salary: Bonus: $
$415,000

OTHER THOUGHTS:

Estimated Female Officers or Directors:
Hot Spot for Advancement for Women/Minorities:

Related Group (The)

www.relatedgroup.com

NAIC Code: 531110

TYPES OF BUSINESS:

Condominium Construction & Management
Multi-Family Residence Development
Mixed-Use Urban Environments
Financing Services
Property Management
Construction

BRANDS/DIVISIONS/AFFILIATES:

CityPlace
Plaza on Brickell (The)
Trump Hollywood
St. Regis Resort & Residences
Related Financial
Related Cervera Realty Services
TRG Management
Realty Asset Advisors

CONTACTS: *Note: Officers with more than one job title may be intentionally listed here more than once.*

Jorge M. Perez, CEO
Matthew J. Allen, COO
Ben Gerber, Dir.-Finance
James M. Werbelow, Sr. VP-Construction
Betsy McCoy, General Counsel
Sonia Figueroa, VP-Dev.
Jeffery Hoyos, Chief Acct. Officer
Steve Patterson, CEO
Carlos Rosso, Pres., Condominium Dev. Div.
Larry Lennon, Pres., TRG Mgmt.
Robert (Bob) Dorfman, Sr. VP-Acquisitions
Jorge M. Perez, Chmn.

GROWTH PLANS/SPECIAL FEATURES:

The Related Group is a multi-family real estate development firm and one of the nation's leading builders of luxury condominiums. Since its inception in 1979, it has built and managed over 80,000 residential units throughout its home state of Florida. The Related Group's buildings have signature characteristics, such as very artistic, often decadent designs that are often located in unknown or under-developed neighborhoods, although it has set up some run-of-the-mill country club villas and other smaller scale projects as well. Some of the company's buildings are rental properties. Also in Related Group's portfolio are so-called urban mixed-use environments, which are planned neighborhoods that integrate housing, shopping and entertainment. The company's flagship community, considered to have changed the cityscape forever, is CityPlace in West Palm Beach, Florida. Other major Florida projects include The Plaza on Brickell, Trump Hollywood and St. Regis Resort & Residences. The group operates various subsidiaries to supplement its chief real estate business, including Related Financial, which offers financing services; Related Cervera Realty Services and TRG Management, which cover sales and property management; Realty Asset Advisors, which offers property management, sales, construction management, leasing and loan workout services; Fortune Construction Company, which serves as the firm's own private contractor; and Related International, which was created to drive the company's $1 billion urbanization project in Mexico. Related International has a property in Puerto Vallarta, Icon Vallarta, a beachfront luxury condominium featuring three towers, and plans to open condominiums and hotels in other locations such as Panama, Argentina, Uruguay, India and the Caribbean.

FINANCIAL DATA: *Note: Data for latest year may not have been available at press time.*

In U.S. $	2015	2014	2013	2012	2011	2010
Revenue						
R&D Expense						
Operating Income						
Operating Margin %						
SGA Expense						
Net Income						
Operating Cash Flow						
Capital Expenditure						
EBITDA						
Return on Assets %						
Return on Equity %						
Debt to Equity						

CONTACT INFORMATION:

Phone: 305-460-9900   Fax: 305-460-9911
Toll-Free:
Address: 315 S. Biscayne Blvd., Miami, FL 33131 United States

STOCK TICKER/OTHER:

Stock Ticker: Private   Exchange:
Employees: Fiscal Year Ends: 12/31
Parent Company:

SALARIES/BONUSES:

Top Exec. Salary: $ Bonus: $
Second Exec. Salary: $ Bonus: $

OTHER THOUGHTS:

Estimated Female Officers or Directors: 3
Hot Spot for Advancement for Women/Minorities: Y

ResortQuest International Inc

www.wyndhamvacationrentals.com/resortquest/index.htm
NAIC Code: 561599

TYPES OF BUSINESS:
Timeshare Rentals
Property Management Services
Real Estate Sales

BRANDS/DIVISIONS/AFFILIATES:
Wyndham Worldwide
Wyndham Exchange & Rentals
Vacation Palm Springs

CONTACTS: *Note: Officers with more than one job title may be intentionally listed here more than once.*
Park Brady, COO
Robert Adams, Sr. VP-Mktg.
Steve Caron, CIO
John W McConmy, Sr. VP
Geoff Ballotti, CEO-Pres., Wyndham Exchange & Rentals

GROWTH PLANS/SPECIAL FEATURES:
ResortQuest International, Inc. provides vacation condominium and home rental property management services. The firm is a subsidiary of Wyndham Exchange & Rentals, itself a division of travel services giant Wyndham Worldwide. The company markets and provides management services for 7,000 premier beach, golf, ski and tennis destination resort locations. In conjunction with Partner Affiliates in North America and Europe, ResortQuest provides management services to approximately 50,000 vacation rental properties. ResortQuest conducts its business through two divisions: vacation rentals and real estate sales. Vacation rental properties are generally second homes or investment properties owned by individuals who assign ResortQuest the responsibility of managing, marketing and renting their properties. Vacation properties include hotels, lodges, condominiums, town homes, cottages, villas and vacation homes. Properties are located in over 100 locations across the U.S., Canada, Mexico, the Caribbean and Europe. The company offers real estate brokerage services throughout its U.S. resort locations, primarily in Delaware and Florida. This division allows customers to list a home as a ResortQuest property or to buy a vacation home. In June 2015, Wyndham acquired Vacation Palm Springs, which manages more than 450 upscale vacation properties and marks the firm's first entry into California.

The company offers its employees benefits that include health coverage, paid time off and a 401(k) plan.

FINANCIAL DATA: *Note: Data for latest year may not have been available at press time.*

In U.S. $	2015	2014	2013	2012	2011	2010
Revenue	505,000,000	500,000,000	475,000,000	450,000,000	435,000,000	
R&D Expense						
Operating Income						
Operating Margin %						
SGA Expense						
Net Income						
Operating Cash Flow						
Capital Expenditure						
EBITDA						
Return on Assets %						
Return on Equity %						
Debt to Equity						

CONTACT INFORMATION:
Phone: 850-837-4774   Fax: 850-837-5390
Toll-Free: 800-467-3529
Address: 546 Mary Esther Cut-Off NW, Ste. 3, Fort Walton Beach, FL 32548 United States

STOCK TICKER/OTHER:
Stock Ticker: Subsidiary   Exchange:
Employees: 5,000   Fiscal Year Ends: 12/31
Parent Company: WYNDHAM WORLDWIDE

SALARIES/BONUSES:
Top Exec. Salary: $ Bonus: $
Second Exec. Salary: $ Bonus: $

OTHER THOUGHTS:
Estimated Female Officers or Directors:
Hot Spot for Advancement for Women/Minorities:

Rezidor Hotel Group AB

www.rezidor.com

NAIC Code: 721110

TYPES OF BUSINESS:

Hotel Management

BRANDS/DIVISIONS/AFFILIATES:

Radisson Blu
Park Inn by Radisson
Quorvus Collection
Filini
RBG Bar & Grill
Verres en Vers
Sure Bar

CONTACTS: Note: Officers with more than one job title may be intentionally listed here more than once.

Wolfgang Neumann, CEO
Olivier Harnisch, COO
Wolfgang Neumann, Pres.
Knut Kleiven, CFO
Eric De Neef, Sr. VP
Michael Farrell, Sr. VP-Human Resources
Eugene Staal, Sr. VP-Tech. Dev.
Marianne Ruhngard, General Counsel
Elie Younes, Head-Group Dev.
Trudy Rautio, Chmn.

GROWTH PLANS/SPECIAL FEATURES:

Rezidor Hotel Group AB is a hospitality management company that franchises, leases and manages hotel properties in 73 countries across the Europe, Middle East and Africa (EMEA) region. The company currently has over 432 available hotels, with 95,609 rooms in operation or under development. The company's brands include Radisson Blu and Park Inn by Radisson, which are operated under a franchise agreement with Carlson Hotels Worldwide, and Quorvus Collection. Radisson Blu is a first-class full-service hotel brand that currently operates over 228 hotels in EMEA. Park Inn by Radisson is an up and coming mid-market hotel brand primarily located in EMEA. Its operations currently include more than 107 locations. The Quorvus Collection is a new generation of luxury hotels for contemporary global travelers. The Quorvus Collection brand expects a portfolio growth to 20 hotels in operation and development by 2020. Rezidor's hotels are located countries such as Azerbaijan, Bulgaria, Croatia, Czech Republic, Estonia, Georgia, Hungary, Kazakhstan, Latvia, Lithuania, Poland, Romania, Russia, Slovakia, Turkey, Ukraine and Uzbekistan. The firm also operates four bar and restaurant brands: Filini, an Italian restaurant chain; RBG Bar & Grill, a trendy casual restaurant; Verres En Vers, a brasserie-style French restaurant; and Sure Bar, a style-focused casual bar.

FINANCIAL DATA: Note: Data for latest year may not have been available at press time.

In U.S. $	2015	2014	2013	2012	2011	2010
Revenue		1,068,990,000	1,048,663,000	1,053,441,000	985,633,200	896,106,300
R&D Expense						
Operating Income		35,011,000	50,363,250	-1,056,101	-8,744,198	4,460,488
Operating Margin %		3.27%	4.80%	-.10%	-.88%	.49%
SGA Expense		16,351,320	17,031,060	17,740,450		
Net Income		16,156,290	26,424,200	-19,215,110	-13,577,630	-3,026,882
Operating Cash Flow		46,998,780	62,250,660	18,826,200	16,067,340	54,313,930
Capital Expenditure		61,369,050	35,913,140	31,210,870	15,899,680	17,397,160
EBITDA		81,338,030	89,842,720	47,257,670	42,122,010	41,343,050
Return on Assets %		3.50%	6.11%	-4.49%	-3.16%	-.67%
Return on Equity %		7.56%	15.20%	-10.71%	-7.00%	-1.45%
Debt to Equity		0.02	0.11	0.11	0.06	0.04

CONTACT INFORMATION:

Phone: 33-2-702-9200   Fax: 32-2-702-9300
Toll-Free:
Address: Ave. du Bourget 44, Brussels, B-1130 Belgium

STOCK TICKER/OTHER:

Stock Ticker: REZIF   Exchange: GREY
Employees: 5,518   Fiscal Year Ends: 12/31
Parent Company:

SALARIES/BONUSES:

Top Exec. Salary: $ Bonus: $
Second Exec. Salary: $ Bonus: $

OTHER THOUGHTS:

Estimated Female Officers or Directors: 3
Hot Spot for Advancement for Women/Minorities: Y

Rio Properties Inc
NAIC Code: 721120

www.riolasvegas.com

TYPES OF BUSINESS:
Hotels & Casinos
Wine Shop
Golf Course

BRANDS/DIVISIONS/AFFILIATES:
Caesars Entertainment Corporation
Rio All-Suite Hotel & Casino
Rio Spa & Salon
Rio Secco Golf Club

CONTACTS: Note: Officers with more than one job title may be intentionally listed here more than once.
Gary W. Loveman, CEO
Gary W. Loveman, Pres.
Gary W. Loveman, Chmn.

GROWTH PLANS/SPECIAL FEATURES:
Rio Properties LLC, a subsidiary of Caesars Entertainment Corporation, operates the Rio All-Suite Hotel & Casino in Las Vegas. Caesars Entertainment is a diversified U.S. gaming company that operates numerous gaming and resort establishments in the U.S. and abroad under the brand names Caesars, Harrah's, Showboat, Horseshoe, Flamingo and Bally's, among others. The Rio All-Suite Hotel & Casino features more than 600 square feet of space for every room as part of its all-suite concept. The suites are equipped with such amenities as a separate dressing area, a couch, a 32-inch TV, a table with chairs, a hairdryer, a refrigerator, an iron and ironing board and a safe. Rio's hotel features 120,000 square feet of gaming space, including slot machines and table games; and other amenities such as the All-American Bar & Grille, Martorano's and Carnival World Buffet restaurants; the Voodoo Rooftop Nightclub; four pools; the Rio Spa & Salon; the Masquerade Village Shops, with over 60,000 square feet of shops; and access to nearby Rio Secco Golf Club. Rio's Wine Cellar & Tasting Room, a wine bar and retail shop, showcases approximately 100 wines. Additionally, Rio hosts the annual World Series of Poker main event, which is owned by Caesars Entertainment. The company actively markets its services and facilities to both local residents and Las Vegas visitors. Rio believes that its all-suite concept, diverse high-quality dining, easy access and ample parking provide an attractive alternative to the Las Vegas Strip, which is 15 minutes away. In January 2015, parent Caesars filed for Chapter 11 bankruptcy.

FINANCIAL DATA: Note: Data for latest year may not have been available at press time.

In U.S. $	2015	2014	2013	2012	2011	2010
Revenue						
R&D Expense						
Operating Income						
Operating Margin %						
SGA Expense						
Net Income						
Operating Cash Flow						
Capital Expenditure						
EBITDA						
Return on Assets %						
Return on Equity %						
Debt to Equity						

CONTACT INFORMATION:
Phone: 702-252-7777   Fax:
Toll-Free: 888-746-7671
Address: 3700 W. Flamingo Rd., Las Vegas, NV 89103 United States

STOCK TICKER/OTHER:
Stock Ticker: Subsidiary   Exchange:
Employees: Fiscal Year Ends: 12/31
Parent Company: CAESARS ENTERTAINMENT CORPORATION

SALARIES/BONUSES:
Top Exec. Salary: $ Bonus: $
Second Exec. Salary: $ Bonus: $

OTHER THOUGHTS:
Estimated Female Officers or Directors:
Hot Spot for Advancement for Women/Minorities:

Ritz-Carlton Hotel Company LLC (The)   www.ritzcarlton.com

NAIC Code: 721110

TYPES OF BUSINESS:

Hotels, Luxury
Condominiums
Golf Courses
Spas
Time Share Units

BRANDS/DIVISIONS/AFFILIATES:

Marriott International Inc
Six Senses
La Prairie
ESPA
Ritz-Carlton Destination Club (The)
Residencies at the Ritz-Carlton (The)
Ritz-Carlton Koh Samui
Ritz-Carlton Reserve

CONTACTS: Note: Officers with more than one job title may be intentionally listed here more than once.

Herve Humler, COO
Herve Humler, Pres.

GROWTH PLANS/SPECIAL FEATURES:

The Ritz-Carlton Hotel Company, LLC, a subsidiary of Marriott International, Inc., is one of the world's best-known luxury hotel chains, operating 91 hotels in 30 countries and territories. The firm maintains international sales offices in locations such as Chicago, New York, Los Angeles, Dubai, Shanghai, Tokyo and London. In an attempt to cater to an upscale client base, full-service luxury spas are offered at most of the company's resorts. Some spas at Ritz-Carlton hotels operate under the brand names Six Senses, La Prairie and ESPA. Besides its hotels, the firm provides vacation properties and residential suites under The Ritz-Carlton Destination Club and The Residencies at the Ritz-Carlton. The Ritz-Carlton Destination Club is the firm's time-share ownership unit, offering a flexible alternative to a second home. Membership is currently available in locations such as Aspen Highlands, Bachelor Gulch and Vail, Colorado; St. Thomas, U.S. Virgin Islands; San Francisco and North Lake Tahoe, California; Jupiter, Florida; Abaco, Bahamas; and Kauai Lagoons and Maui, Hawaii. The Residencies at the Ritz-Carlton offer luxury condominiums and estate homes throughout the U.S. and in Canada, Thailand, Israel, the Bahamas and Malaysia. Ritz-Carlton also markets its 12 luxury golf courses (many designed by leading names in the golf world such as Greg Norman and Jack Nicklaus) and fitness facilities to both local residents and visitors and hosts many PGA and Senior PGA tournaments. In November 2015, the firm signed with YTL Hotels for the development of two new hotels in Asia Pacific, the Ritz-Carlton, Koh Samui, in Thailand, and Ritz-Carlton Reserve in Niseko Village, Japan.

The firm offers employees health care & dependent care spending accounts; tuition reimbursement; dental & vision insurance; short- and long-term disability coverage; credit union membership; and employee assistance programs.

FINANCIAL DATA: Note: Data for latest year may not have been available at press time.

In U.S. $	2015	2014	2013	2012	2011	2010
Revenue	2,400,000,000	2,355,320,000	2,222,213,400	2,126,520,000	2,217,060,000	2,104,380,000
R&D Expense						
Operating Income						
Operating Margin %						
SGA Expense						
Net Income						
Operating Cash Flow						
Capital Expenditure						
EBITDA						
Return on Assets %						
Return on Equity %						
Debt to Equity						

CONTACT INFORMATION:

Phone: 301-547-4700   Fax:
Toll-Free: 888-241-3333
Address: 4445 Willard Ave., Ste. 800, Chevy Chase, MD 20815 United States

STOCK TICKER/OTHER:

Stock Ticker: Subsidiary   Exchange:
Employees: 35,000   Fiscal Year Ends: 12/31
Parent Company: Marriott International Company

SALARIES/BONUSES:

Top Exec. Salary: $ Bonus: $
Second Exec. Salary: $ Bonus: $

OTHER THOUGHTS:

Estimated Female Officers or Directors: 1
Hot Spot for Advancement for Women/Minorities:

Riverview Realty Partners

www.rrpchicago.com

NAIC Code: 0

TYPES OF BUSINESS:

Real Estate Investment Trust
Office & Industrial Properties
Land Development
Property Management

BRANDS/DIVISIONS/AFFILIATES:

Five Mile Capital Partners LLC

CONTACTS: Note: Officers with more than one job title may be intentionally listed here more than once.

Jeffrey A. Patterson, CEO
James F. Hoffman, Sr. Exec. VP
Jeffrey A. Patterson, Pres.
Victoria A. Cory, Sr. VP-Loan Admin., Real Estate Tax & Due Diligence
Paul G. Del Vecchio, Exec. VP-Capital Markets
Steven R. Baron, Exec. VP-Office Leasing
James F. Hoffman, General Counsel
Steven R. Baron, Exec. VP-Office Leasing
Paul G. Del Vecchio, Exec. VP-Capital Markets
Victoria A. Cory, Sr VP-Loan Admin., Real Estate Tax & Due Diligence

GROWTH PLANS/SPECIAL FEATURES:

Riverview Realty Partners (RRP), wholly-owned by Five Mile Capital Partners LLC, is a real estate investment trust (REIT). RRP owns, manages, leases and redevelops office real estate, including Class A properties such as the AMA Plaza in Chicago, and the Olympian Office Center in Lisle, Illinois; the Park Center Plaza I, II & III and the Corporate Plaza I & II in Independence, Ohio; 411 East Wisconsin in Milwaukee, Wisconsin; 3000 Post Oak Blvd. in Houston, Texas; and 100 Peachtree Street in Atlanta, Georgia. The firm's offices range in size from 775 to 2,350 square feet; its hotels offers state-of-the-art space specifically designed for meetings and social gatherings (business, conference or wedding), a ballroom, restaurants, lounges, fitness center and indoor pool and spa. Tenants include the American Medical Association, BDO, Canon Business Solutions, CBIZ, Inc., Continental Casualty Corporation, Farmers Insurance, American Appraisal Associates, Charles Schwab & Co., Inc., DeVry, Inc., Bechtel Corporation and Koch Properties Company LP.

FINANCIAL DATA: Note: Data for latest year may not have been available at press time.

In U.S. $	2015	2014	2013	2012	2011	2010
Revenue						
R&D Expense						
Operating Income						
Operating Margin %						
SGA Expense						
Net Income						
Operating Cash Flow						
Capital Expenditure						
EBITDA						
Return on Assets %						
Return on Equity %						
Debt to Equity						

CONTACT INFORMATION:

Phone: 312-917-1300   Fax: 312-917-1310
Toll-Free:
Address: 330 N. Wabash Ave., Ste. 2800, Chicago, IL 60611 United States

STOCK TICKER/OTHER:

Stock Ticker: Private   Exchange:
Employees: Fiscal Year Ends: 12/31
Parent Company: Five Mile Capital Partners LLC

SALARIES/BONUSES:

Top Exec. Salary: $ Bonus: $
Second Exec. Salary: $ Bonus: $

OTHER THOUGHTS:

Estimated Female Officers or Directors: 1
Hot Spot for Advancement for Women/Minorities:

Robert A.M. Stern Architects LLP

www.ramsa.com

NAIC Code: 541310

TYPES OF BUSINESS:
Architectural Services
Architecture

BRANDS/DIVISIONS/AFFILIATES:

GROWTH PLANS/SPECIAL FEATURES:
Robert A.M. Stern Architects LLP (RSA) is a firm of architects, interior designers and support staff. It has an international reputation as a leading designer with experience in residential, commercial and institutional work. Its geographical scope includes projects in Europe, Asia, South America and the U.S. Among the company's extensive list of completed projects include the George W. Bush Presidential Center at Southwestern Methodist University in Dallas, Texas; the Normal Rockwell Museum in Stockbridge, Massachusetts; and the Disney Feature Animation Building in Burbank, California. International projects include Schwarzman College at Tsinghua University; One Horizon Center, a 650,000 square foot office tower in Gurgaon, India; and Heart of Lake, a high-rise garden suburb in Xiamen, China. Currently, RSA has projects underway in 32 U.S. states as well as in India and Canada.

CONTACTS: Note: Officers with more than one job title may be intentionally listed here more than once.
Robert A. M. Stern, Founder

FINANCIAL DATA: Note: Data for latest year may not have been available at press time.

In U.S. $	2015	2014	2013	2012	2011	2010
Revenue	62,860,000					
R&D Expense						
Operating Income						
Operating Margin %						
SGA Expense						
Net Income						
Operating Cash Flow						
Capital Expenditure						
EBITDA						
Return on Assets %						
Return on Equity %						
Debt to Equity						

CONTACT INFORMATION:
Phone: 212-967-5100   Fax: 212-967-5588
Toll-Free:
Address: 460 W. 34th St., Ste. 18, New York, NY 10001 United States

STOCK TICKER/OTHER:
Stock Ticker: Private   Exchange:
Employees: 298   Fiscal Year Ends:
Parent Company:

SALARIES/BONUSES:
Top Exec. Salary: $ Bonus: $
Second Exec. Salary: $ Bonus: $

OTHER THOUGHTS:
Estimated Female Officers or Directors:
Hot Spot for Advancement for Women/Minorities:

Rosendin Electric
NAIC Code: 238210

www.rosendin.com

TYPES OF BUSINESS:
Electrical Contractor

BRANDS/DIVISIONS/AFFILIATES:
KST Electric

GROWTH PLANS/SPECIAL FEATURES:
Rosendin Electric is a San Jose, California-based electrical contractor that has been operating since the early 1900s. The services provided by the firm include preconstruction services that assist general contractors and owners during the development phase of a project. Other services include engineering, prefabrication, 24-hour service response, network services, electric utility work along highways and the construction of solar & wind farms. Its project portfolio includes services for the following industries: biotechnology, pharmaceuticals, data centers, education, health care, high tech, institutions, multi-family residences, solar power, transportation, commercial, design-build, entertainment, heavy industrial, hotels, power, telecom and wind energy. Rosendin is also experienced in meeting LEED regulations. Subsidiary KST Electric is a Texas-based electrical, data and communication contracting company.

Rosendin is an employee-owned firm.

CONTACTS: Note: Officers with more than one job title may be intentionally listed here more than once.
Tom Sorley, CEO
Larry Beltramo, Pres.

FINANCIAL DATA: Note: Data for latest year may not have been available at press time.

In U.S. $	2015	2014	2013	2012	2011	2010
Revenue	1,300,000,000	1,100,000,000	935,000,000	887,000,000	820,000,000	
R&D Expense						
Operating Income						
Operating Margin %						
SGA Expense						
Net Income						
Operating Cash Flow						
Capital Expenditure						
EBITDA						
Return on Assets %						
Return on Equity %						
Debt to Equity						

CONTACT INFORMATION:
Phone: 480-286-2800　Fax:
Toll-Free:
Address: 880 Mabury Rd., San Jose, CA 95133 United States

STOCK TICKER/OTHER:
Stock Ticker: Private　Exchange:
Employees: 5,300　Fiscal Year Ends:
Parent Company:

SALARIES/BONUSES:
Top Exec. Salary: $　Bonus: $
Second Exec. Salary: $　Bonus: $

OTHER THOUGHTS:
Estimated Female Officers or Directors:
Hot Spot for Advancement for Women/Minorities:

Rosewood Hotels & Resorts LLC

www.rosewoodhotels.com

NAIC Code: 721110

TYPES OF BUSINESS:

Hotel & Resort Management
Spas
Private Residences

BRANDS/DIVISIONS/AFFILIATES:

New World Hospitality
Rosewood Inn of the Anasazi
Rosewood Siem Reap
Rosewood Phnom Penh
Rosewood Luang Prabang

CONTACTS: Note: Officers with more than one job title may be intentionally listed here more than once.

Sonia Cheng, CEO
Radha Arora, Pres.
Sheri Line, Corp. Dir.-Human Resources
George Fong, Sr. VP-Architecture & Design
Susan Aldridge, General Counsel
Elias Assaly, VP-Oper.
Stephen Miano, Sr. VP-Finance
Katherine Blaisdell, VP-Construction Dev.
Erin Green, VP-Dev., Americas
James Simmonds, VP-Dev., Asia Pacific
Paul Arnold, VP-Dev., EMEA

GROWTH PLANS/SPECIAL FEATURES:

Rosewood Hotels & Resorts, LLC, a subsidiary of New World Hospitality, operates ultra-luxury boutique hotels and resorts worldwide. It has 18 hotels and resorts in the U.S., Canada, Mexico, the Caribbean, France, the UAE, Saudi Arabia, Thailand, Cambodia, Indonesia and China. Several facilities are currently under construction including Rosewood Beijing, Rosewood Phuket and Rosewood Dubai. Besides constructing its own facilities, Rosewood has acquired existing properties and management contracts, including the Rosewood Inn of the Anasazi in Santa Fe, New Mexico. Its facilities are generally small, featuring less than 200 accommodations. The company uses architecture and decor to attempt to capture the unique history, geography and culture of each hotel or resort location. Amenities offered at Rosewood's facilities might include tennis courts, a courtesy car with 5-mile radius, unpacking and packing services, babysitting services, twice-daily housekeeping with nightly turndown service, pools and fitness centers as well as shops and various dining facilities. Some locations also offer business centers stocked with computers, printers, faxes and copiers. In addition, several hotels and resorts also feature private residences, which offer owners the same services and amenities as resort guests. The private residences often come fully furnished and feature floor plan inclusions such as full kitchens, fireplaces, private pools and terraces. Rosewood plans to expand to 50 properties by 2020. In October 2015, the firm was appointed to manage Rosewood Siem Reap, scheduled to open in 2019, which will be the company's second property in Cambodia, following Rosewood Phnom Penh, which opened in 2016. Also, Rosewood Luang Prabang in Laos is scheduled to open in 2017.

FINANCIAL DATA: Note: Data for latest year may not have been available at press time.

In U.S. $	2015	2014	2013	2012	2011	2010
Revenue	575,000,000	563,000,000	546,000,000	500,000,000	471,000,000	
R&D Expense						
Operating Income						
Operating Margin %						
SGA Expense						
Net Income						
Operating Cash Flow						
Capital Expenditure						
EBITDA						
Return on Assets %						
Return on Equity %						
Debt to Equity						

CONTACT INFORMATION:

Phone: 214-880-4231   Fax:
Toll-Free: 888-767-3966
Address: 500 Crescent Court, Ste. 300, Dallas, TX 75201 United States

STOCK TICKER/OTHER:

Stock Ticker: Subsidiary   Exchange:
Employees: 6,000   Fiscal Year Ends:
Parent Company: New World Hospitality

SALARIES/BONUSES:

Top Exec. Salary: $ Bonus: $
Second Exec. Salary: $ Bonus: $

OTHER THOUGHTS:

Estimated Female Officers or Directors: 4
Hot Spot for Advancement for Women/Minorities: Y

Rossi Residencial SA

NAIC Code: 531100

www.rossiresidencial.com.br

TYPES OF BUSINESS:
Real Estate Development

BRANDS/DIVISIONS/AFFILIATES:
Grupo Rossi
Rossi Commercial Properties
Rossi Urbanizadora
Norcon Rossi

GROWTH PLANS/SPECIAL FEATURES:
Rossi Residencial SA is a Brazilian holding company involved in real estate and construction. The firm is part of construction, engineering and development conglomerate Grupo Rossi. Rossi Residencial is primarily involved in the construction, development, sale and commercialization of residential and commercial real estate properties, and has delivered more than 100,000 units. Rossi Commercial Properties manages convenience centers, shopping malls and retail in medium markets, while Rossi Urbanizadora focuses on the development of single family and multifamily housing. The company also established Norcon Rossi, a joint venture with Construction Norcon, to build in Alagoas, Bahiam Pernambuco and Sergipe. Between Rossi and Rossi Norcon, the firm has a combined land bank of more than 26 million square feet dedicated to residential planning. Rossi currently has 53 projects under construction.

CONTACTS:
Note: Officers with more than one job title may be intentionally listed here more than once.

Leonardo Nogueira Diniz, CEO
Rodrigo Medeiros Ferreira da Silva, CFO
Rodrigo Moraes Martins, Sales Officer
Renato Gamba Rock Diniz, Engineering Officer
Rodrigo Medeiros Ferreira da Silva, Investor Rel. Officer
John Rossi Cuppoloni, Chmn.

FINANCIAL DATA:
Note: Data for latest year may not have been available at press time.

In U.S. $	2015	2014	2013	2012	2011	2010
Revenue	337,185,400	444,681,900	586,508,700	746,504,300	844,835,600	686,346,400
R&D Expense						
Operating Income	-146,372,200	-155,170,000	30,218,350	1,855,957	103,480,900	97,672,700
Operating Margin %	-43.41%	-34.89%	5.15%	.24%	12.24%	14.23%
SGA Expense	70,023,930	47,371,580	97,135,360	139,423,100	144,619,100	111,412,900
Net Income	-153,466,600	-170,341,300	11,290,840	-56,571,890	93,423,440	96,185,790
Operating Cash Flow	56,521,010	91,712,960	-92,424,660	-177,590,500	-88,012,870	-172,942,800
Capital Expenditure	2,109,504	5,791,168	8,474,041	8,925,036	19,693,100	12,550,880
EBITDA	-141,732,200	-152,404,600	35,554,670	19,374,930	150,275,000	133,133,900
Return on Assets %	-9.26%	-8.61%	.51%	-2.62%	5.16%	6.92%
Return on Equity %	-37.21%	-29.73%	1.75%	-8.09%	12.80%	14.55%
Debt to Equity	0.41	0.61	0.57	0.09	0.92	0.59

CONTACT INFORMATION:
Phone: 55-1140582685   Fax: 55 1140582100
Toll-Free:
Address: 5200 Ave. Major Sylvio de Magalhaes Padilha, Fl. 3, Sao Paulo, SP 05693-000 Brazil

STOCK TICKER/OTHER:
Stock Ticker: RSRZD   Exchange: PINX
Employees: Fiscal Year Ends: 12/31
Parent Company:

SALARIES/BONUSES:
Top Exec. Salary: $ Bonus: $
Second Exec. Salary: $ Bonus: $

OTHER THOUGHTS:
Estimated Female Officers or Directors:
Hot Spot for Advancement for Women/Minorities:

Rouse Properties Inc

www.rouseproperties.com

NAIC Code: 531120

TYPES OF BUSINESS:

Lessors of Nonresidential Buildings
Real Estate
Properties

BRANDS/DIVISIONS/AFFILIATES:

Mt. Shasta Mall

CONTACTS: Note: Officers with more than one job title may be intentionally listed here more than once.

Andrew Silberfein, CEO
John Wain, CFO
Richard Clark, Chairman of the Board
Michael Grant, Chief Accounting Officer
Timothy Salvemini, Chief Administrative Officer
Brian Harper, COO
Susan Elman, General Counsel

GROWTH PLANS/SPECIAL FEATURES:

Rouse Properties, Inc. owns and manages dominant Class B regional malls in secondary and tertiary markets and repositions Class B regional malls in primary markets. Its portfolio consists of 36 regional malls in 23 states totaling over 25.5 million square feet of retail and ancillary space. The company actively manages all of its properties, performing the day-to-day functions, operations, leasing, maintenance, marketing and promotional services. The firm's malls are anchored by operators across the retail spectrum, including departments stores such as Macy's, JC Penney, Belk, Sears, Dillard's and Target; mall shop tenants like Victoria's Secret, Bath & Body Works, H&M, francesca's, Chico's, The Children's Place, Gap/Old Navy, Foot Locker, Maurices, Justice, Ulta and Forever 21; restaurants such as food court leaders like Sarku Japan, Panda Express and Chick-Fil-A; best in class fast-casual chains like Chipotle, Panera Bread and Starbucks; and proven sit-down restaurants including BJ's Restaurants, Olive Garden, Red Lobster and Buffalo Wild Wings. The majority of the income from properties is derived from rents received through long-term leases with retail tenants. These long-term leases generally require the tenants to pay base rent which is a fixed amount specified in the lease. The company's direct competitors include other publicly-traded retail mall development and operating companies, retail real estate companies, commercial property developers, internet retail sales and other owners of retail real estate that engage in similar businesses. In 2015, the firm acquired Mt. Shasta Mall in Redding, California, and sold The Shoppes at Knollwood in Minneapolis, Minnesota. In February 2016, Rouse agreed to be acquired by Canada's Brookfield Asset Management, Inc.

FINANCIAL DATA: Note: Data for latest year may not have been available at press time.

In U.S. $	2015	2014	2013	2012	2011	2010
Revenue	305,384,000	292,127,000	243,542,000	233,974,000	234,816,000	255,281,000
R&D Expense						
Operating Income	52,275,000	31,438,000	37,248,000	27,920,000	44,505,000	85,198,000
Operating Margin %	17.11%	10.76%	15.29%	11.93%	18.95%	33.37%
SGA Expense	27,963,000	29,586,000	25,705,000	24,439,000	15,391,000	12,111,000
Net Income	41,699,000	-51,756,000	-54,745,000	-68,659,000	-26,976,000	-23,896,000
Operating Cash Flow	77,183,000	82,301,000	62,602,000	38,277,000	80,723,000	48,468,000
Capital Expenditure	334,095,000	109,330,000		64,343,000	25,167,000	23,475,000
EBITDA	221,588,000	145,164,000	122,028,000	123,918,000	122,757,000	153,726,000
Return on Assets %	1.69%	-2.41%	-2.78%	-3.93%	-1.67%	-1.41%
Return on Equity %	7.86%	-10.65%	-11.08%	-14.31%	-7.13%	-6.96%
Debt to Equity						

CONTACT INFORMATION:

Phone: 212 608-5108   Fax:
Toll-Free:
Address: 1114 Ave. of the Americas, Suite 2800, New York, NY 10036 United States

STOCK TICKER/OTHER:

Stock Ticker: RSE
Employees: 331
Parent Company:

Exchange: NYS
Fiscal Year Ends: 12/31

SALARIES/BONUSES:

Top Exec. Salary: $750,000   Bonus: $1,375,000
Second Exec. Salary: $500,000   Bonus: $775,000

OTHER THOUGHTS:

Estimated Female Officers or Directors: 1
Hot Spot for Advancement for Women/Minorities:

Royal Group Inc

www.royalbuildingproducts.com

NAIC Code: 326122

TYPES OF BUSINESS:
Plastics Pipe and Pipe Fitting Manufacturing

BRANDS/DIVISIONS/AFFILIATES:
Axiall Corporation
Royal Mouldings Limited
Celect Cellular Exteriors
Zuri Premium Decking
Royal Kor Flo
Overture
Genesis Cellular Window System

CONTACTS: Note: Officers with more than one job title may be intentionally listed here more than once.
Guy Prentice, VP-Finance
Stephen Ryan, Dir.-Human Resources, Royal Building Prod.
Michel Boulanger, Dir.-Innovation, Royal Building Prod.
Michel Boulanger, Dir.-Tech., Royal Building Prod.
Mark J. Orcutt, Exec. VP-Royal Building Prod.
Paul Czachor, VP
Shane Short, VP
Simon Bates, VP

GROWTH PLANS/SPECIAL FEATURES:

Royal Group, Inc. manufactures and sells polyvinyl chloride (PVC) and vinyl construction, improvement and building products. The firm is a subsidiary of Axiall Corporation, a North American manufacturer and international marketer of commodity chemicals, polymers and durable, custom and other vinyl-based building and home improvement products. Royal operates primarily in Canada and the U.S., and markets its products under the Royal Building Products, Celect Cellular Exteriors, Zuri Premium Decking, Royal Kor Flo, Overture, Genesis Cellular Window System, Royal S4S Trimboard and Exterior Portfolio brand names. The company's building product offerings include soffits, shutters, premium vinyl siding, mounts, vents, vinyl/aluminum columns and accessories. Royal Mouldings Limited manufactures decorative polymer and cellular vinyl molding extrusion components and systems. Royal's outdoor products include railings, decking systems, columns and fencing. The firm's pipe systems offerings include pipes for municipal potable water, sewer systems and storm drains; plumbing pipes; ducts, conduit and other pipes and fittings for electrical construction; and fittings for backwater valves, inspection chambers and controlled settlement joints. Royal's window and door products utilize polymer fenestration technologies to produce patio door systems, custom window profiles and window systems.

FINANCIAL DATA: Note: Data for latest year may not have been available at press time.

In U.S. $	2015	2014	2013	2012	2011	2010
Revenue						
R&D Expense						
Operating Income						
Operating Margin %						
SGA Expense						
Net Income						
Operating Cash Flow						
Capital Expenditure						
EBITDA						
Return on Assets %						
Return on Equity %						
Debt to Equity						

CONTACT INFORMATION:
Phone: 905-850-9700   Fax: 905-850-9184
Toll-Free: 800-461-0849
Address: 91 Royal Group Crescent, Woodbridge, ON L4H 1X9 Canada

STOCK TICKER/OTHER:
Stock Ticker: Subsidiary   Exchange:
Employees: 628   Fiscal Year Ends: 12/31
Parent Company: Axiall Corporation

SALARIES/BONUSES:
Top Exec. Salary: $ Bonus: $
Second Exec. Salary: $ Bonus: $

OTHER THOUGHTS:
Estimated Female Officers or Directors: 1
Hot Spot for Advancement for Women/Minorities:

RXR Realty

www.rxrrealty.com

NAIC Code: 0

TYPES OF BUSINESS:

Real Estate Investment Trust
Office & Industrial Properties
Commercial Construction & Development
Property Management

BRANDS/DIVISIONS/AFFILIATES:

Tri-State Prime Property
Core Plus Value Enhanced Strategy
Opportunity Fund I
Opportunity Fund II
75 Rockefeller Plaza

CONTACTS: Note: Officers with more than one job title may be intentionally listed here more than once.

Scott Rechler, CEO
Richard Conniff, COO
Michael Maturo, Pres.
Michael Maturo, CFO
F.D. Rich, III, Chief Admin. Officer
Jason Barnett, General Counsel
Glenn Wasserman, Exec. VP-Strategic Planning & Capital Markets
Tom Carey, Controller
Kenneth W. Bauer, Exec. VP-Leasing
Todd Rechler, Co-COO
William Elder, Exec. VP
Frank Adipirtro, Exec. VP
Scott Rechler, Chmn.

GROWTH PLANS/SPECIAL FEATURES:

RXR Realty is a self-administered and self-managed real estate investment trust (REIT) that acquires, owns, develops, constructs, manages and leases office and industrial properties in the New York Tri-State area. RXR's primary investment funds include the Tri-State Prime Property joint venture, which owns office properties located in Long Island and New Jersey; and Core Plus Value Enhanced Strategy, which includes the Opportunity Fund I, Opportunity Fund II and Value Enhanced Portfolio. The Core Plus portfolios include developed properties in New Jersey, Long Island, New York City, Westchester and Connecticut as well as several active development projects. The Opportunity Fund I and Opportunity Fund II portfolios invest in properties and other interests. The firm offers its tenants cross-leasing opportunities to provide enhanced flexibility and the convenience of being able to work with a single owner/manager over multiple spaces. RXR's 99-year lease of 75 Rockefeller Plaza has involved completely renovating the building, it reopened in 2015. In all, RXR manages 88 commercial real estate properties, and its investments total 23.1 million square feet. In February 2015, the firm agreed to sell its 50% stake in a portfolio of six New York office buildings to Blackstone Group LP. RXR continues to manage the properties and uses the proceeds of the sale to return about all of its investors' equity. In March 2016, the firm announced that it seeks to sell its assets over the next 12-24 months.

FINANCIAL DATA: Note: Data for latest year may not have been available at press time.

In U.S. $	2015	2014	2013	2012	2011	2010
Revenue						
R&D Expense						
Operating Income						
Operating Margin %						
SGA Expense						
Net Income						
Operating Cash Flow						
Capital Expenditure						
EBITDA						
Return on Assets %						
Return on Equity %						
Debt to Equity						

CONTACT INFORMATION:

Phone: 516-506-6000   Fax: 516-506-6800
Toll-Free:
Address: 625 RXR Plz., Uniondale, NY 11556 United States

SALARIES/BONUSES:

Top Exec. Salary: $ Bonus: $
Second Exec. Salary: $ Bonus: $

STOCK TICKER/OTHER:

Stock Ticker: Private   Exchange:
Employees: 263   Fiscal Year Ends: 12/31
Parent Company:

OTHER THOUGHTS:

Estimated Female Officers or Directors:
Hot Spot for Advancement for Women/Minorities: Y

Ryman Hospitality Properties Inc

rymanhp.com

NAIC Code: 721110

TYPES OF BUSINESS:
Hotels & Convention Centers
Vacation Property Management
Live Entertainment Venues
Golf Courses
Radio Station Operation

BRANDS/DIVISIONS/AFFILIATES:
Gaylord Opryland Resort & Convention Center
Gaylord Palms Resort & Convention Center
Gaylord Texan Resort & Convention Center
Gaylord National Resort & Convention Center
Grand Ole Opry (The)
General Jackson Showboat
Gaylord Springs Golf Links
AC Hotel

CONTACTS: *Note: Officers with more than one job title may be intentionally listed here more than once.*
Colin Reed, CEO
Mark Fioravanti, CFO
Jennifer Hutcheson, Controller
Scott Lynn, General Counsel
Patrick Chaffin, Senior VP, Divisional
Bennett Westbrook, Senior VP, Divisional
Todd Siefert, Treasurer

GROWTH PLANS/SPECIAL FEATURES:
Ryman Hospitality Properties, Inc. is a real estate investment trust (REIT) that owns hospitality assets focused on the large group meetings and conventions sector of the lodging market. The firm's hospitality assets include four upscale, meetings-focused resorts totaling 7,795 rooms and 1.9 million square feet of meeting and exhibit space that are managed by Marriott International, Inc. under the Gaylord Hotels brand. Its hotels consist of the Gaylord Opryland Resort & Convention Center in Nashville, Tennessee; the Gaylord Palms Resort & Convention Center near Orlando, Florida; the Gaylord Texan Resort & Convention Center near Dallas, Texas; and the Gaylord National Resort & Convention Center near Washington, D.C. Other assets managed by Marriott include the General Jackson Showboat; the Gaylord Springs Golf Links, a championship golf course; the Wildhorse Saloon; and the Inn at Opryland, a 303-room overflow hotel adjacent to Gaylord Opryland. Additionally, the firm owns The Grand Ole Opry, a live country music variety show and one of the longest-running radio shows in the U.S.; the Ryman Auditorium; and the country radio station WSM-AM radio. In December 2014, the firm acquired a 192-room hotel that it rebranded AC Hotel at National Harbor, Washington D.C.

FINANCIAL DATA: *Note: Data for latest year may not have been available at press time.*

In U.S. $	2015	2014	2013	2012	2011	2010
Revenue	1,092,124,000	1,040,991,000	954,562,000	986,594,000	952,144,000	769,961,000
R&D Expense						
Operating Income	162,062,000	153,105,000	76,188,000	-4,754,000	79,531,000	-65,986,000
Operating Margin %	14.83%	14.70%	7.98%	-.48%	8.35%	-8.56%
SGA Expense	96,277,000	87,388,000	82,820,000	186,590,000	179,301,000	200,490,000
Net Income	111,511,000	126,452,000	118,352,000	-26,644,000	10,177,000	-89,128,000
Operating Cash Flow	234,362,000	247,004,000	137,699,000	176,470,000	153,919,000	139,484,000
Capital Expenditure	79,815,000	79,583,000	36,959,000	95,233,000	132,592,000	194,647,000
EBITDA	277,940,000	298,725,000	203,259,000	160,604,000	217,450,000	54,071,000
Return on Assets %	4.70%	5.00%	4.56%	-1.04%	.39%	-3.37%
Return on Equity %	28.55%	20.88%	14.08%	-2.80%	.98%	-8.45%
Debt to Equity	3.77	3.34	1.52	1.05	1.02	1.06

CONTACT INFORMATION:
Phone: 615-316-6000   Fax: 615-316-6555
Toll-Free:
Address: 1 Gaylord Dr., Nashville, TN 37214 United States

STOCK TICKER/OTHER:
Stock Ticker: RHP   Exchange: NYS
Employees: 682   Fiscal Year Ends: 12/31
Parent Company:

SALARIES/BONUSES:
Top Exec. Salary: $782,830   Bonus: $156,544
Second Exec. Salary: $469,407   Bonus: $63,817

OTHER THOUGHTS:
Estimated Female Officers or Directors: 1
Hot Spot for Advancement for Women/Minorities:

Saha Pathana Inter-Holding PCL

www.spi.co.th

NAIC Code: 531100

TYPES OF BUSINESS:
Industrial Real Estate Holdings

BRANDS/DIVISIONS/AFFILIATES:
Brahma Park
Health Park

CONTACTS: *Note: Officers with more than one job title may be intentionally listed here more than once.*
Chantra Pumarikaha, Pres.
Boonsithi Chokwatana, Chmn.

GROWTH PLANS/SPECIAL FEATURES:

Saha Pathana Inter-Holding PCL is a Thailand-based real estate holding and development company investing primarily in industrial parks. The company is currently invested in five industrial parks in Sriracha, Kabinburi, Lamphun, Ratchaburi and Measod. The Sriracha park features a 174 megawatt (MW) power plant, a centralized water treatment plant with a daily capacity of 706,293 cubic feet, a private air field, a trade exhibition center and a 30 million cubic foot water reservoir. The park located in Kabinburi contains two 50 megavolt ampere (MVA) electrical substations, a centralized waste water treatment plant with a daily capacity of 565,035 cubic feet, a private air field, a 35.3 million cubic foot water reservoir and an industrial incinerator with a 220 pound per hour capacity. The Lamphun park features two 50 megavolt ampere (MVA) electrical substations; a centralized water treatment plant with a daily capacity of 300,175 cubic feet; a private water supply of 127,133 cubic feet per day; a private air field; a 17.6 million cubic foot water reservoir; a golf driving range; a meeting room; Brahma Park, a peace park for relaxation and meditative purposes; and Health Park, for exercise and recreation. The Ratchaburi and Measod industrial parks are still under construction. The company, to a lesser extent, is invested in consumer products, distribution, manufacturing, services and investment firms.

FINANCIAL DATA: *Note: Data for latest year may not have been available at press time.*

In U.S. $	2015	2014	2013	2012	2011	2010
Revenue	115,940,467	119,087,719	117,151,333			
R&D Expense						
Operating Income						
Operating Margin %						
SGA Expense						
Net Income	35,823,135	32,728,574	37,014,393			
Operating Cash Flow						
Capital Expenditure						
EBITDA						
Return on Assets %						
Return on Equity %						
Debt to Equity						

CONTACT INFORMATION:
Phone: 0-2293-0030   Fax: 0-2293-0040
Toll-Free:
Address: 530 Soi Sathupradit 58, Bangpongpang, Bangkok, 10120 Thailand

STOCK TICKER/OTHER:
Stock Ticker: SPI.F
Employees: 115
Parent Company:

Exchange: SET
Fiscal Year Ends: 12/31

SALARIES/BONUSES:
Top Exec. Salary: $ Bonus: $
Second Exec. Salary: $ Bonus: $

OTHER THOUGHTS:
Estimated Female Officers or Directors: 1
Hot Spot for Advancement for Women/Minorities: Y

Sales, profits and employees may be estimates. Financial information, benefits and other data can change quickly and may vary from those stated here.

Salini Impregilo SpA

NAIC Code: 237000

www.salini-impregilo.com/en/#

TYPES OF BUSINESS:

Heavy Construction
Civil Engineering
Environmental Engineering
Infrastructure Management
Airport Operations

BRANDS/DIVISIONS/AFFILIATES:

CONTACTS: *Note: Officers with more than one job title may be intentionally listed here more than once.*

Pietro Salini, CEO
Rosario Fiumara, CFO
Claudio Costamagna, Chmn.

GROWTH PLANS/SPECIAL FEATURES:

Salini Impregilo SpA is a leading Italian civil engineering and construction company, active in over 50 countries. Salini Impregilo provides design and construction services to all sectors requiring complex, large-scale infrastructures. Infrastructure operations include dams, hydroelectric plants, motorways, roads, bridges, railways, metro systems, civil buildings, industrial buildings and airports. Salini has built 230 dams and hydroelectric plants in five different continents, with an installed capacity of more than 37,200 megawatts (MW) of clean energy. It has constructed over 23,000 miles (37,200 km) of roads and motorways, as well as over 205 miles (330 km) of bridges and viaducts worldwide. Salini has built more than 4,100 miles (6,730 km) of railways, 233 miles (375 km) of metro systems and 838 miles (1,350 km) of underground works. Salini covers an important role in the construction of innovative and iconic buildings worldwide, including the Great Mosque in Abu Dhabi and the Kingdom Centre in Riyadh. The group has also built hospital complexes, university campuses, government buildings as well as buildings which play an important cultural role such as the Nigeria Cultura Centre and Millennium Tower. The firm currently has construction projects in Georgia, Poland, Qatar, Turkey, the U.S. and Nigeria. It is currently working on the Nenskra Hydroelectric Project, which will have an installed power equal to 280MW in the country of Georgia; and the A1 Motorway project, comprising a motorway section south of Warsaw, and includes three junctions, four bridges, a railway bridge and 21 viaducts. Salini's expansion project of the Panama Canal is 95% complete as of late 2015, and expects structural stress tests and navigational testing at the start of 2016.

FINANCIAL DATA: *Note: Data for latest year may not have been available at press time.*

In U.S. $	2015	2014	2013	2012	2011	2010
Revenue	5,404,678,000	4,783,375,000	2,649,757,000	2,601,465,000	2,404,097,000	2,202,815,000
R&D Expense						
Operating Income	310,963,600	294,685,300	180,059,500	-29,057,610	257,629,400	255,264,000
Operating Margin %	5.75%	6.16%	6.79%	-1.11%	10.71%	11.58%
SGA Expense					2,051,163,000	
Net Income	69,105,050	106,947,900	214,126,200	687,331,300	205,049,000	149,128,100
Operating Cash Flow	485,718,700	163,496,100	4,191,330	-18,738,380	-11,861,180	-26,189,250
Capital Expenditure	317,287,600	349,891,100	43,598,960	83,428,560	185,661,700	209,916,600
EBITDA	572,685,100	499,714,900	297,807,900	96,117,740	357,299,700	392,708,800
Return on Assets %	.87%	1.77%	4.35%	13.11%	3.93%	2.96%
Return on Equity %	5.49%	7.56%	11.78%	39.43%	14.93%	12.74%
Debt to Equity	1.09	0.85	0.20	0.18	0.42	0.60

CONTACT INFORMATION:

Phone: 39 244422111   Fax: 39 244422293
Toll-Free:
Address: Via de Missaglia, 97, Milan, 20142 Italy

STOCK TICKER/OTHER:

Stock Ticker: IMPJY   Exchange: PINX
Employees: 17,794   Fiscal Year Ends: 12/31
Parent Company:

SALARIES/BONUSES:

Top Exec. Salary: $ Bonus: $
Second Exec. Salary: $ Bonus: $

OTHER THOUGHTS:

Estimated Female Officers or Directors: 1
Hot Spot for Advancement for Women/Minorities:

Sales, profits and employees may be estimates. Financial information, benefits and other data can change quickly and may vary from those stated here.

Samsung C&T Corporation

www.samsungcnt.com

NAIC Code: 237000

TYPES OF BUSINESS:

Heavy Construction and Engineering
Trading and Investments

BRANDS/DIVISIONS/AFFILIATES:

Cheil Industries Inc
Galaxy
Beanpole
KUHO
LeBeige
8seconds

CONTACTS: *Note: Officers with more than one job title may be intentionally listed here more than once.*

Jung Yeon-Joo, CEO
Jung Yeon-Joo, Pres.
Dong-hee Lee, Sr. Exec. VP
Dong-hee Lee, Exec. VP-Treasury & Financial Mgmt.
Shin Kim, CEO
Shin Kim, Sr. Exec. VP-Energy & Minerals Bus. Unit

GROWTH PLANS/SPECIAL FEATURES:

Samsung C&T Corporation engages in construction and trading businesses worldwide. The company operates through four segments: engineering & construction, trading & investment, fashion and resort. Engineering & construction is experienced in architecture, civil engineering, plant and housing development. This division has erected two of the world's most famous skyscapers: Petronas Twin Towers in Malaysia and Burj Khalifa in Dubai. Trading & investment focuses on trading and organizing the following: trading industrial commodities such as chemicals, steel and natural resources, and organizing projects such as power plant, renewable energy and infrastructure development. The fashion segment provides accessories, children's and outdoor clothing lines, as well as women's clothing. Brands in this segment include Galaxy, Beanpole, KUHO, LeBeige and 8seconds. The resort segment develops resorts, golf courses, amusement parks, as well as food and beverage establishments. In 2015, the firm was acquired by Cheil Industries, Inc.

FINANCIAL DATA: *Note: Data for latest year may not have been available at press time.*

In U.S. $	2015	2014	2013	2012	2011	2010
Revenue		16,051,993,600	15,795,988,480	14,451,881,984	13,157,328,896	10,933,843,968
R&D Expense						
Operating Income						
Operating Margin %						
SGA Expense						
Net Income		302,446,368	117,511,376	348,805,760	249,557,552	392,838,592
Operating Cash Flow						
Capital Expenditure						
EBITDA						
Return on Assets %						
Return on Equity %						
Debt to Equity						

CONTACT INFORMATION:

Phone: 82 221452114   Fax: 82 221453114
Toll-Free:
Address: 14 Seocho-daero 72-gil, Seocho-Gu, Seoul, 137-857 South Korea

STOCK TICKER/OTHER:

Stock Ticker: SSGQF
Employees:
Parent Company: Cheil Industries Inc

Exchange: GREY
Fiscal Year Ends: 12/31

SALARIES/BONUSES:

Top Exec. Salary: $ Bonus: $
Second Exec. Salary: $ Bonus: $

OTHER THOUGHTS:

Estimated Female Officers or Directors:
Hot Spot for Advancement for Women/Minorities:

Samwhan Corporation

NAIC Code: 237990

www.samwhan.co.kr/sw/english

TYPES OF BUSINESS:

Subway Construction
General Importing & Exporting
Precast Concrete

BRANDS/DIVISIONS/AFFILIATES:

Samwhan Camus Co
Samwhan Machinery Co Ltd
Sinmin Mutual Savings Bank
Whoihyun Co Ltd
Woosung Development Co Ltd

CONTACTS: Note: Officers with more than one job title may be intentionally listed here more than once.

Yong-Kwon Choi, Chmn.

GROWTH PLANS/SPECIAL FEATURES:

Samwhan Corporation, established in 1946, is a Korean construction company. It has experience building housing units; commercial buildings; offices; ports and airports; industrial plants; fabrication facilities; power distribution and transmission systems; pipelines and pumping systems; and highways and bridges. More specifically, its projects have included resorts and recreational complexes; hospital and medical healthcare facilities; retail complexes, department stores and shopping malls; remodeling and rehabilitation; real estate development; thermal, nuclear and hydraulic power plants; municipal solid waste treatment plants; natural gas pipelines; oil refineries; subways and underground works; dams; and land reclamation. The company has branches in the U.S., Afghanistan, Vietnam, Bangladesh, Japan, Saudi Arabia, Libya, Turkey, Russia and Yemen. Besides these locations, Samwhan has completed major projects in Jordan, Papua New Guinea, Guam, Malaysia, Indonesia and Singapore. It was one of the first Korean contractors to work in Mongolia, and has worked on numerous U.S. military bases in Korea. Samwhan's affiliates include Samwhan Camus Co., which focuses on pre-cast concrete production and construction; Samwhan Machinery Co., Ltd., a manufacturer of materials for metal structures; Sinmin Mutual Savings Bank, a banking and financing company; Whoihyun Co., Ltd., a real estate leasing service provider; and Woosung Development Co., Ltd., a company engaged in land development.

FINANCIAL DATA: Note: Data for latest year may not have been available at press time.

In U.S. $	2015	2014	2013	2012	2011	2010
Revenue		471,849,329	525,324,000	527,315,000	756,000,000	1,080,000,000
R&D Expense						
Operating Income						
Operating Margin %						
SGA Expense						
Net Income		-46,593,889	-135,989,000	-249,023,000	-115,700,000	-4,200,000
Operating Cash Flow						
Capital Expenditure						
EBITDA						
Return on Assets %						
Return on Equity %						
Debt to Equity						

CONTACT INFORMATION:

Phone: 82-2-740-2114   Fax: 82-2-742-1849
Toll-Free:
Address: 98-20 Wooni-dong, Jongno-Gu, Seoul, 110-742 South Korea

STOCK TICKER/OTHER:

Stock Ticker: Private   Exchange:
Employees: 370   Fiscal Year Ends: 12/31
Parent Company:

SALARIES/BONUSES:

Top Exec. Salary: $ Bonus: $
Second Exec. Salary: $ Bonus: $

OTHER THOUGHTS:

Estimated Female Officers or Directors:
Hot Spot for Advancement for Women/Minorities:

Sawyer Realty Holdings LLC

www.sawyerapts.com

NAIC Code: 0

TYPES OF BUSINESS:

Real Estate Investment Trust
Apartment Communities
Property Management

BRANDS/DIVISIONS/AFFILIATES:

Forewinds Hospitality LLC

CONTACTS: Note: Officers with more than one job title may be intentionally listed here more than once.

David M. Rosenberg, CEO
David Rosenberg, Pres.
Margie Nichols, Dir.-Admin.

GROWTH PLANS/SPECIAL FEATURES:

Sawyer Realty Holdings LLC is a privately-held real estate investment and management firm that acquires, maintains and operates undervalued multifamily properties. The company works to add value to its acquisitions through teams of mostly autonomous property specialists across the east coast. Sawyer's multifamily portfolio consists of apartment communities located in Massachusetts and Maryland, primarily within the Boston, Baltimore, Frederick and Washington, D.C. metropolitan areas. In addition to its multi-family property focus, the company invests in ground up development projects for hotel and commercial projects, which includes the purchase or joint venture of permitted or raw land and acquisition of existing properties for significant redevelopment. The company provides property management services for its own properties as well as for third-party real estate owners. Sawyer has also invested in numerous golf course assets through its wholly-owned subsidiary Forewinds Hospitality, LLC.

The firm offers employees medical, dental, vision and prescription drug plans; flexible spending accounts for health care and dependent care; a 401(k) retirement savings plan; long-term and short-term disability; an employee assistance program; life insurance; and a discount program.

FINANCIAL DATA: Note: Data for latest year may not have been available at press time.

In U.S. $	2015	2014	2013	2012	2011	2010
Revenue						
R&D Expense						
Operating Income						
Operating Margin %						
SGA Expense						
Net Income						
Operating Cash Flow						
Capital Expenditure						
EBITDA						
Return on Assets %						
Return on Equity %						
Debt to Equity						

CONTACT INFORMATION:

Phone: 781-449-6650   Fax: 781-449-6573
Toll-Free:
Address: 1215 Chestnut St., Newton, MA 02464 United States

STOCK TICKER/OTHER:

Stock Ticker: Private   Exchange:
Employees: 70   Fiscal Year Ends: 12/31
Parent Company:

SALARIES/BONUSES:

Top Exec. Salary: $ Bonus: $
Second Exec. Salary: $ Bonus: $

OTHER THOUGHTS:

Estimated Female Officers or Directors: 1
Hot Spot for Advancement for Women/Minorities:

Sales, profits and employees may be estimates. Financial information, benefits and other data can change quickly and may vary from those stated here.

Scandic Hotels AB

NAIC Code: 721110

www.scandichotels.com

TYPES OF BUSINESS:

Hotels

BRANDS/DIVISIONS/AFFILIATES:

EQT Partners AB
Scandic Antwerpen
Hotel Scandic The Reef
Scandic Palace Hotel Copenhagen
Scandic Hamburg Emporio

CONTACTS: Note: Officers with more than one job title may be intentionally listed here more than once.

Frank Fiskers, CEO
Martin Creydt, COO
Anders Ehrling, Pres.
Gunilla Rudebjer, CFO
Johan Michelson, VP-Mktg. & Brand
Pelle Ekman, Sr. VP-Commercial Oper.
Martin Creydt, Chief Dev. Officer
Margareta Thorgren, VP-Group Comm.
Gunilla Rudebjer, Sr. VP-Finance
Joakim Nilsson, VP-Sweden
Ulrika Garbrant, VP-Food & Beverage
Aki Kayhko, VP-Finland
Svein Arild Steen-Mevold, VP-Norway
Jens Mathiesen, VP-Denmark

GROWTH PLANS/SPECIAL FEATURES:

Scandic Hotels AB, which is owned by private equity firm EQT Partners AB, is one of Scandinavia's largest hotel operators. The firm has 226 hotels, featuring 41,735 rooms, located in seven countries: Sweden, Norway, Finland, Denmark, Germany, Belgium and Poland. Scandic Hotels also has partnerships with numerous hotel, travel and entertainment companies to provide more options to its customers. Its hotels include Scandic Antwerpen, Scandic Palace Hotel Copenhagen, Hotel Scandic The Reef and Scandic Hamburg Emporio. The firm has a long history of being environmentally conscious, and all of its Swedish hotels carry the Swan eco-label, a certification earned for effort and results toward sustainable practices. In an effort to simplify services for customers, Scandic Hotels' pricing plan offers two price levels: early and flex. Early means a reduced price reservation without the option to cancel or change, while flex is a higher, flexible reservation price.

FINANCIAL DATA: Note: Data for latest year may not have been available at press time.

In U.S. $	2015	2014	2013	2012	2011	2010
Revenue	1,494,083,703	1,082,590,000	1,385,330,000	1,405,090,000		
R&D Expense						
Operating Income						
Operating Margin %						
SGA Expense						
Net Income	25,843,517	38,261,000	31,866,200	48,709,700		
Operating Cash Flow						
Capital Expenditure						
EBITDA						
Return on Assets %						
Return on Equity %						
Debt to Equity						

CONTACT INFORMATION:

Phone: 46-85-17-350-00   Fax: 46-85-17-350-11
Toll-Free:
Address: Sveavagen 167, Stockholm, SE-102 33 Sweden

STOCK TICKER/OTHER:

Stock Ticker: Private   Exchange:
Employees: 9,887   Fiscal Year Ends: 12/31
Parent Company: EQT PARTNERS AB

SALARIES/BONUSES:

Top Exec. Salary: $ Bonus: $
Second Exec. Salary: $ Bonus: $

OTHER THOUGHTS:

Estimated Female Officers or Directors: 3
Hot Spot for Advancement for Women/Minorities: Y

Schuff International Inc

www.schuff.com

NAIC Code: 238120

TYPES OF BUSINESS:

Steel Contractor
Structural Steel Fabrication & Erection
Project Management Services
Design & Engineering Consulting Services
Subcontract Management

BRANDS/DIVISIONS/AFFILIATES:

CONTACTS: Note: Officers with more than one job title may be intentionally listed here more than once.

Scott A. Schuff, CEO
Scott A. Schuff, Pres.
Michael R. Hill, CFO
Ryan S. Schuff, CEO
Robert N. Waldrep, VP
Rustin Roach, Pres., Western Region, Schuff Steel Company
David A. Schuff, Chmn.

GROWTH PLANS/SPECIAL FEATURES:

Schuff International, Inc. is a fully integrated steel fabrication and erection company. The firm fabricates and erects structural steel for construction projects such as arenas and stadiums, bridges, casinos and hotels, convention centers, hospitals, power plants, office buildings, parking structures, retail centers, warehouses and distribution facilities. Schuff International maintains 16 offices and facilities in California, Arizona, Georgia, Kansas and Texas. The company's services include design-assist/design-build, fabrication, safety, preconstruction design and budgeting, erection, steel management and building information modeling (BIM). Design-assist/design-build utilizes the latest technology and BIM systems to design, construct and deliver cost effective steel designs. Fabrication consists of ten structural steel fabrication plants with an annual capacity of producing over 300,000 tons of steel. This division's production systems allow Schuff to track each piece of steel through the entire process of raw material procurement, 3D modeling/detailing, fabrication and erection. Safety aims to meet the goal of total elimination of accidents from all its fabrication facilities and job sites via planning, organization and effective management. Preconstruction design and budgeting provides clients a detailed cost breakdown as well as a thorough analysis of the structure process before construction begins. Erection includes creative options for erecting projects, and emphasizing clear, detailed planning and coordination between the firm, facilities and the customer in order to allow the erection process flow smoothly and efficiently. Steel management offers its products at competitive prices and through reliable, state-of-the-art systems. BIM tools for 3D modeling and detailing are used for every project, and embeds vital data that allows the firm to manage jobs efficiently, both upstream and downstream.

FINANCIAL DATA: Note: Data for latest year may not have been available at press time.

In U.S. $	2015	2014	2013	2012	2011	2010
Revenue						
R&D Expense						
Operating Income						
Operating Margin %						
SGA Expense						
Net Income						
Operating Cash Flow						
Capital Expenditure						
EBITDA						
Return on Assets %						
Return on Equity %						
Debt to Equity						

CONTACT INFORMATION:

Phone: 602 252-7787   Fax: 602 251-0335
Toll-Free: 800-435-8528
Address: 420 S. 19th Ave., Phoenix, AZ 85007 United States

SALARIES/BONUSES:

Top Exec. Salary: $ Bonus: $
Second Exec. Salary: $ Bonus: $

STOCK TICKER/OTHER:

Stock Ticker: SHFK   Exchange: PINX
Employees: 1,800   Fiscal Year Ends: 12/31
Parent Company: HC2 Holdings Inc

OTHER THOUGHTS:

Estimated Female Officers or Directors:
Hot Spot for Advancement for Women/Minorities:

Sales, profits and employees may be estimates. Financial information, benefits and other data can change quickly and may vary from those stated here.

Sekisui House Ltd

NAIC Code: 236110

www.sekisuihouse.co.jp

TYPES OF BUSINESS:

Residential Construction
Real Estate Sales
Real Estate Leasing
Urban Construction
Garden Design
Pre-fabricated Housing Construction

BRANDS/DIVISIONS/AFFILIATES:

Sekiwa Real Estate Ltd
Sekisui House Umeda Operation Co Ltd
Sekiwa Woods Ltd
Sekisui House Remodeling LTd
Sekiwa Real Estate Tohoku Ltd
North America Sekisui House LLC

CONTACTS: Note: Officers with more than one job title may be intentionally listed here more than once.

Isami Wada, CEO
Toshinori Abe, COO
Toshinori Abe, Pres.
Shiro Inagaki, CFO
Isami Wada, Chmn.

GROWTH PLANS/SPECIAL FEATURES:

Sekisui House, Ltd., based in Japan, and its subsidiaries engage primarily in the contract design, construction and leasing of pre-fabricated houses. These residences are built-to-order and include both single- and multi-family houses, condominiums and low-rise apartments, with Sekisui building over 2.25 million homes since 1960. The company is also involved in the construction of urban development projects and the management of recreation facilities, training facilities, restaurants and hotels. Sekisui House manufactures and sells construction materials; purchases and sells materials for reforestation, landscaping and other civil engineering works; and undertakes research, planning, design and consulting activities related to regional development, land preparation and environmental improvement. The firm also contracts, executes and supervises civil engineering works; carpentry works; plasterwork; earth, stone and concrete work; roofing; electrical and piping work; tile, brick and blockwork; steel structure and reinforcing; paving; sheet metal and glasswork; painting; waterproofing; interior finishing; furnishing; machinery and tool installation; heating insulation work; telecommunications installation; waterworks; and firefighting equipment construction. In real estate, the company buys and sells properties, manages and appraises real estate and acts as a broker for property purchase and sale. Sekisui House also engages in the cultivation, purchase and sale of plants and the design and cultivation of gardens. The company has more than 200 subsidiaries and affiliate companies. Subsidiaries include Sekiwa Real Estate, Ltd.; Sekisui House Umeda Operation Co., Ltd.; Sekiwa Woods, Ltd.; Sekisui House Remodeling, Ltd.; Sekiwa Real Estate Tohoku, Ltd.; and North America Sekisui House, LLC.

FINANCIAL DATA: Note: Data for latest year may not have been available at press time.

In U.S. $	2015	2014	2013	2012	2011	2010
Revenue	17,436,080,000	16,455,050,000	14,711,310,000	13,952,520,000	13,567,750,000	12,335,440,000
R&D Expense						
Operating Income	1,336,339,000	1,202,654,000	785,750,100	646,286,700	513,714,800	-353,275,800
Operating Margin %	7.66%	7.30%	5.34%	4.63%	3.78%	-2.86%
SGA Expense	2,022,361,000	2,065,370,000	1,944,457,000	1,863,508,000	1,830,937,000	1,778,421,000
Net Income	822,468,700	757,810,000	442,912,000	264,204,800	277,523,000	-266,647,800
Operating Cash Flow	1,069,818,000	711,702,100	752,805,400	239,801,600	775,403,600	505,036,500
Capital Expenditure	1,093,875,000	692,166,800	498,755,700	278,124,700	199,965,400	197,175,900
EBITDA	1,631,792,000	1,422,948,000	976,271,400	748,584,800	663,953,200	-245,006,800
Return on Assets %	4.87%	4.82%	3.11%	2.07%	2.25%	-2.13%
Return on Equity %	9.02%	9.18%	5.99%	3.90%	4.18%	-3.98%
Debt to Equity	0.16	0.21	0.24	0.24	0.25	0.12

CONTACT INFORMATION:

Phone: 81 664403111   Fax: 81 664403369
Toll-Free:
Address: 1-88 Oyodonaka, 1-Chome, Kita-ku, Osaka, 531-0076 Japan

STOCK TICKER/OTHER:

Stock Ticker: SKHSY   Exchange: PINX
Employees: 22,913   Fiscal Year Ends: 01/31
Parent Company:

SALARIES/BONUSES:

Top Exec. Salary: $ Bonus: $
Second Exec. Salary: $ Bonus: $

OTHER THOUGHTS:

Estimated Female Officers or Directors:
Hot Spot for Advancement for Women/Minorities:

Sembcorp Industries Ltd

www.sembcorp.com.sg

NAIC Code: 221112

TYPES OF BUSINESS:

Electric Utility
Marine Construction & Shipbuilding
Heavy Construction
Environmental Engineering & Waste Management
Industrial Parks and Real Estate Development
Internet Service Provider
Floating Oil Production Platforms
Pipelines

BRANDS/DIVISIONS/AFFILIATES:

Smoe Pte Ltd
NCC Power Projects
Sembmarine SLP
GraviFloat AS
Sembcorp Utilities Pte ltd
Sembcorp Marine Ltd
Sembcorp Development Ltd

CONTACTS: Note: Officers with more than one job title may be intentionally listed here more than once.

Tang Kin Fei, CEO
Tang Kin Fei, Pres.
Koh Chiap Khiong, CFO
Tan Cheng Guan, Exec. VP-Bus. Dev., Commercial & Corp. Planning
Wong Weng Sun, Pres.
Ng Meng Poh, Head-Utilities
Ang Kong Hua, Chmn.

GROWTH PLANS/SPECIAL FEATURES:

Sembcorp Industries, Ltd. is one of Singapore's leading utilities and marine groups. The firm's primary business segments include utilities, marine and urban development. Utilities provides integrated utilities, energy and water solutions to the chemical and petrochemical industries primarily in Singapore, but also in the U.K., Middle East and Africa, the Americas, Asia and Australia and China. In all, the segment operates in 14 countries through wholly-owned subsidiary, Sembcorp Utilities Pte. Ltd. The segment also offers process steam production and supply; natural gas import, supply and retail; wastewater treatment; and solid waste management in select countries. The marine segment, operated through 61%-owned Sembcorp Marine Ltd., has one of the largest ship repair, shipbuilding and ship conversion operations in East Asia. Other services include rig building and offshore engineering and construction. In Singapore, its marine hub consists of Jurong Shipyard, Sembawang Shipyard, Smoe Pte. Ltd. and PPL Shipyard. Overseas, the firm operates four hubs and yards, including the U.K based Sembmarine SLP. The urban development segment owns, develops, markets and manages integrated urban developments. These developments are comprised of industrial parks as well as business, commercial and residential space in Vietnam, Indonesia and China through wholly-owned subsidiary Sembcorp Development Ltd. Sembcorp owns a 12% stake in GraviFloat AS, a technology provider focused on developing & delivering unique near-shore liquid natural gas terminal solutions. In September 2015, the firm sold its 40% stake in SembSita Pacific Pte. Ltd. to 60% joint venture partner, Suez Environment Asia for $485 million.

Sembcorp Industries offers employees health and welfare benefits, term life insurance, personal accident insurance and a loan and interest subsidy.

FINANCIAL DATA: Note: Data for latest year may not have been available at press time.

In U.S. $	2015	2014	2013	2012	2011	2010
Revenue	7,090,784,000	8,093,740,000	8,021,649,000	7,569,542,000	6,721,145,000	6,510,567,000
R&D Expense						
Operating Income	464,525,300	846,082,600	704,441,800	787,395,800	791,795,300	876,316,800
Operating Margin %	6.55%	10.45%	8.78%	10.40%	11.78%	13.45%
SGA Expense	389,561,400	261,865,000	251,877,300	238,013,200	214,434,000	253,380,200
Net Income	407,749,200	595,141,400	609,518,100	559,620,700	601,222,800	589,030,900
Operating Cash Flow	-522,883,800	-42,654,860	1,092,267,000	460,925,900	724,343,600	1,264,702,000
Capital Expenditure	1,063,964,000	993,894,800	889,972,900	839,394,200	809,922,300	469,753,900
EBITDA	790,030,800	1,210,217,000	1,215,068,000	1,155,669,000	1,167,157,000	1,241,101,000
Return on Assets %	2.95%	5.11%	6.15%	6.11%	7.14%	7.84%
Return on Equity %	9.11%	14.59%	16.85%	17.48%	20.41%	22.22%
Debt to Equity	0.78	0.64	0.28	0.48	0.45	0.40

CONTACT INFORMATION:

Phone: 65 6723-3113   Fax: 65 6822-3254
Toll-Free:
Address: 30 Hill St., #05-04, Singapore, 179360 Singapore

SALARIES/BONUSES:

Top Exec. Salary: $ Bonus: $
Second Exec. Salary: $ Bonus: $

STOCK TICKER/OTHER:

Stock Ticker: SCRPF   Exchange: PINX
Employees: 8,000   Fiscal Year Ends: 12/31
Parent Company:

OTHER THOUGHTS:

Estimated Female Officers or Directors: 1
Hot Spot for Advancement for Women/Minorities:

Senior Housing Properties Trust

www.snhreit.com

NAIC Code: 0

TYPES OF BUSINESS:

Real Estate Investment Trust
Senior Housing Properties

BRANDS/DIVISIONS/AFFILIATES:

Reit Management and Research LLC

CONTACTS: Note: Officers with more than one job title may be intentionally listed here more than once.

David Hegarty, COO
Jennifer Clark, Secretary
Richard Siedel, Treasurer
Barry Portnoy, Trustee
Lisa Jones, Trustee
John Harrington, Trustee
Jeffrey Somers, Trustee
Adam Portnoy, Trustee

GROWTH PLANS/SPECIAL FEATURES:

Senior Housing Properties Trust (SHPT) is a real estate investment trust (REIT) that invests in properties such as apartments for aged residents, assisted living facilities and nursing homes. SHPT owns 370 properties with a total of 397 buildings in 38 states and Washington, D.C. The firm's portfolio consists of 261 senior living communities, with 31,414 units; 99 properties (126 buildings) leased to medical providers, medical-related businesses, clinics and biotech laboratory tenants with 9.5 million square feet of space; and 10 wellness centers with approximately 812,000 square feet of interior space plus outdoor developed facilities. SHPT's senior apartments are properties in which residents care for themselves; independent living properties include services such as meals and maid service; assisted living properties typically offer private units with 24-hour availability of various types of assistance to residents; nursing homes generally provide extensive nursing and health care services; and wellness centers feature amenities such as gymnasiums, strength and cardiovascular equipment areas, tennis and racquet sports facilities, pools and spas. All SHPT properties are triple-net leased, meaning that each tenant pays rent and is also responsible for expenses such as taxes, insurance and maintenance costs. The company's day-to-day operations are conducted by Reit Management & Research LLC (RMR), a real estate management company that oversees a large portfolio of publicly owned real estate in the U.S., Puerto Rico and Canada.

FINANCIAL DATA: Note: Data for latest year may not have been available at press time.

In U.S. $	2015	2014	2013	2012	2011	2010
Revenue	998,773,000	844,887,000	761,438,000	644,842,000	450,017,000	339,009,000
R&D Expense						
Operating Income	313,534,000	291,389,000	264,510,000	258,141,000	227,515,000	203,930,000
Operating Margin %	31.39%	34.48%	34.73%	40.03%	50.55%	60.15%
SGA Expense	42,830,000	38,946,000	32,657,000	31,517,000	26,041,000	21,865,000
Net Income	123,968,000	158,637,000	151,164,000	135,884,000	151,419,000	116,485,000
Operating Cash Flow	405,521,000	350,901,000	306,703,000	283,302,000	255,559,000	215,305,000
Capital Expenditure	1,205,162,000	1,290,524,000	245,573,000	366,900,000		467,441,000
EBITDA	534,138,000	477,280,000	421,905,000	396,552,000	363,258,000	288,102,000
Return on Assets %	1.88%	2.95%	3.17%	2.97%	3.89%	3.65%
Return on Equity %	3.92%	5.53%	5.57%	5.30%	6.58%	5.78%
Debt to Equity	1.04	0.94	0.68	0.75	0.73	0.56

CONTACT INFORMATION:

Phone: 617 796-8350   Fax: 617 796-8349
Toll-Free:
Address: 2 Newton Place, 255 Washington St., Ste 300, Newton, MA 02458 United States

STOCK TICKER/OTHER:

Stock Ticker: SNH   Exchange: NYS
Employees: 850   Fiscal Year Ends: 12/31
Parent Company:

SALARIES/BONUSES:

Top Exec. Salary: $ Bonus: $
Second Exec. Salary: $ Bonus: $

OTHER THOUGHTS:

Estimated Female Officers or Directors: 1
Hot Spot for Advancement for Women/Minorities:

Shangri-La Asia Ltd

www.shangri-la.com

NAIC Code: 721110

TYPES OF BUSINESS:

Hotels, Luxury
Property Management
Health Spas
Golf Courses
Commercial Real Estate Leasing
Serviced Apartments

BRANDS/DIVISIONS/AFFILIATES:

Shangri-La
Traders Hotels
Kerry Hotels
Aberdeen Marina Club
Xili Golf and Country Club
Chi, The Spa at the Shangri-La
Shangri-La Hotel (Malaysia) Berhad
Shangri-La Hotel PCL

CONTACTS: Note: Officers with more than one job title may be intentionally listed here more than once.

Greg Dogan, CEO
Lothar Nessmann, COO
Steven Taylor, CMO
Ching Leun Teo, Sec.
Madhu Rao, Chmn.
Judy Reeves, Dir.-Public Rel., North America

GROWTH PLANS/SPECIAL FEATURES:

Shangri-La Asia, Ltd. is an investment holding company that owns and manages hotels and resorts, primarily under the Shangri-La, Traders Hotels, Kerry Hotels and Hotel Jen brands. Its portfolio consists of 92 luxury hotels and resorts (amounting to a room inventory of over 38,000) located in Australia, China, Hong Kong, Japan, Malaysia, the Philippines, Singapore, Thailand and France, among other countries. The company also manages the Xili Golf and Country Club in Shenzhen and the Aberdeen Marina Club in Hong Kong. Additionally, the firm leases office, commercial and serviced apartments in various locations, such as mainland China, Singapore, Malaysia, Thailand and Mongolia. Subsidiaries Shangri-La Hotels (Malaysia) Berhad and Shangri-La Hotel PCL are engaged in the operation of hotels, beach resorts, golf courses, clubhouses and related facilities in Malaysia and Thailand respectively. Shangri-La Asia has whole or partial interests in various projects in development, including new hotels in China, Cambodia, Hong Kong, India, Myanmar, Philippines, Qatar and Sri Lanka.

FINANCIAL DATA: Note: Data for latest year may not have been available at press time.

In U.S. $	2015	2014	2013	2012	2011	2010
Revenue		2,111,584,000	2,081,081,000	2,057,249,000	1,912,089,000	1,575,095,000
R&D Expense						
Operating Income		224,574,000	217,197,000	302,008,000	239,465,000	220,710,000
Operating Margin %		10.63%	10.43%	14.68%	12.52%	14.01%
SGA Expense		288,817,000	279,516,000	263,649,000	252,586,000	199,895,000
Net Income		180,889,000	392,298,000	358,895,000	252,979,000	287,076,000
Operating Cash Flow		348,108,000	383,038,000	292,740,000	454,694,000	326,134,000
Capital Expenditure		198,825,000	689,784,000	998,368,000	289,109,000	586,581,000
EBITDA		749,205,000	983,531,000	868,253,000	700,602,000	644,691,000
Return on Assets %		1.35%	3.16%	3.27%	2.73%	3.51%
Return on Equity %		2.73%	6.35%	6.17%	4.93%	6.47%
Debt to Equity		0.63	0.70	0.63	0.42	0.30

CONTACT INFORMATION:

Phone: 852 25993000   Fax: 852 25993131
Toll-Free:
Address: 683 Kings Rd., Kerry Ctr., 28F, Quarry Bay, Quarry Bay, Hong Kong

STOCK TICKER/OTHER:

Stock Ticker: SHALY
Employees: 28,100
Parent Company:

Exchange: PINX
Fiscal Year Ends: 12/31

SALARIES/BONUSES:

Top Exec. Salary: $ Bonus: $
Second Exec. Salary: $ Bonus: $

OTHER THOUGHTS:

Estimated Female Officers or Directors: 1
Hot Spot for Advancement for Women/Minorities: Y

Shimao Property Holdings Ltd

www.shimaoproperty.com

NAIC Code: 531100

TYPES OF BUSINESS:
Real Estate Holdings & Development

BRANDS/DIVISIONS/AFFILIATES:
Shanghai Shimao Real Estate Co Ltd
Shanghai Shimao International Plaza Co Ltd
Shanghai Shimao Manor Real Estate Co Ltd
Shanghai Shimao Jianse Co Ltd
Beijing Changyang
Chengdu Chenghua
Changzhou Shimao Champagne Lake
Hangszhou Shimao Imperial Landscape

CONTACTS:
Note: Officers with more than one job title may be intentionally listed here more than once.

Liao Lujiang, COO
Sai Tan Hui, Group Sales Controller
Tammy Tam, Investor Rel.
James Yu, Corp. Finance
Sai Tan Hui, Vice Chmn.
Liu Sai Fei, VP
Xu You Nong, VP
Wing Mau Hui, Chmn.

GROWTH PLANS/SPECIAL FEATURES:
Shimao Property Holdings Ltd., along with its subsidiaries, is engaged in investment holding and develops residential, hotel, office and commercial properties in large- to medium-sized cities in China. The company's current projects under development include: Beijing Changyang, Beijing La Villa, Beijing Shimao Alhambra Palace, Beijing Tongzhou, Dalian Shimao Glory City, Chengdu Chenghua, Chengdu Shimao Royal Bay, Changzhou Shimao Champagne Lake, Hangzhou Born with Legend, Hangzhou Above the Lake,Hangzhou East No. 1, Hangzhou Gate of Wisdom, Hangzhou Shimao Imperial Landscape, Fuzhou Pingtan, Guangzhou Asian Games City, Guangzhou FCC, Nanjing Straits City and more. The firm utilizes its Riviera Model, which combines landscape, waterside, gardening and architectural design. Subsidiaries of the company that are involved in property development include Shanghai Shimao Real Estate Co., Ltd.; Shanghai Shimao International Plaza Co., Ltd.; and Shanghai Shimao Manor Real Estate Co., Ltd. Subsidiary Shanghai Shimao Jianse Co., Ltd, is focused on investment holding operations.

FINANCIAL DATA:
Note: Data for latest year may not have been available at press time.

In U.S. $	2015	2014	2013	2012	2011	2010
Revenue		8,666,576,000	6,413,738,000	4,427,860,000	4,022,845,000	3,367,296,000
R&D Expense						
Operating Income		2,421,836,000	1,906,378,000	1,550,845,000	1,701,471,000	1,435,652,000
Operating Margin %		27.94%	29.72%	35.02%	42.29%	42.63%
SGA Expense		665,066,700	554,642,800	442,397,800	327,490,900	254,527,500
Net Income		1,252,350,000	1,142,014,000	890,854,700	884,386,200	721,929,900
Operating Cash Flow		-1,813,528,000	-1,670,723,000	1,297,240,000	-627,704,400	-573,010,200
Capital Expenditure		663,985,900	739,272,400	660,639,600	400,354,500	193,808,600
EBITDA		2,512,073,000	2,130,826,000	1,657,112,000	1,788,014,000	1,485,543,000
Return on Assets %		4.08%	4.82%	4.64%	5.37%	5.76%
Return on Equity %		18.29%	19.03%	17.33%	19.97%	18.66%
Debt to Equity		0.92	0.89	0.79	0.90	0.92

CONTACT INFORMATION:
Phone: 852-2511-9968   Fax: 852-2511-0287
Toll-Free:
Address: Fl. 38, Tower 1, Lippo Ctr., 89 Queensway, Hong Kong,   Hong Kong

STOCK TICKER/OTHER:
Stock Ticker: SHMAY
Employees: 7,191
Parent Company:

Exchange: PINX
Fiscal Year Ends: 12/31

SALARIES/BONUSES:
Top Exec. Salary: $ Bonus: $
Second Exec. Salary: $ Bonus: $

OTHER THOUGHTS:
Estimated Female Officers or Directors: 3
Hot Spot for Advancement for Women/Minorities: Y

Shun Tak Holdings Limited

NAIC Code: 721110

TYPES OF BUSINESS:

Investment Holding Company
High-Speed Ferry Services
Real Estate Investment
Hotel Management
Casino Management

BRANDS/DIVISIONS/AFFILIATES:

Shun Tak & CITS Coach (Macao) Limited
Shun Tak-China Travel Ship Management Limited
TuboJET
Shun Tak Real Estate Ltd
Shun Tak Property Management Ltd
Shun Tak Macau Services Ltd
Clean Living (Macau) Ltd
Macau Tower Convention & Entertainment Center

CONTACTS: Note: Officers with more than one job title may be intentionally listed here more than once.

Pansy Ho, Managing Dir.
Daisy Ho, Deputy Managing Dir.
Maisy Ho, Exec. Dir.
David Shum, Exec. Dir.
Rogier Verhoeven, Exec. Dir.
Stanley Ho, Chmn.

GROWTH PLANS/SPECIAL FEATURES:

Shun Tak Holdings Limited is a publicly traded Hong Kong-based conglomerate with core businesses in the transportation, property, hospitality and investment sectors. The transportation segment maintains ferry and air services through the joint venture Shun Tak-China Travel Ship Management Limited, which is known under the brand name TurboJET. The company shares ownership of TurboJET with China Travel International Investment Hong Kong Limited. TurboJET operates one of the largest fleets of high-speed passenger ferries in Asia. It also maintains airport routes that connect Hong Kong International airport with Shenzhen and Macau. The segment additionally maintains bus services in Macau and the Guangdong province through joint venture Shun Tak & CITS Coach (Macao) Limited, operating 144 vehicles. Shun Tak's property division develops and invests in property in Hong Kong and Macau. Its major operations include property development and sales, developing residential, retail and commercial properties; property leasing and asset management through Shun Tak Real Estate Ltd., which markets and leases residential, retail and commercial properties; property management services through Shun Tak Property Management Ltd., which maintains the company's owned properties; cleaning services through Shun Tak Macau Services, Ltd.; and laundry services through Clean Living (Macau), Ltd. Shun Tak's hospitality segment is engaged in hotel and casino management, managing the Macau Tower Convention & Entertainment Center and owning a 70% interest in Hong Kong SkyCity Marriot Hotel. The firm's investments segment holds interests in several hotels in Macau as well as Macau International Airport and Macau Matters Co. Ltd., which specializes in retail facility operations.

FINANCIAL DATA: Note: Data for latest year may not have been available at press time.

In U.S. $	2015	2014	2013	2012	2011	2010
Revenue		1,229,801,000	461,016,400	708,426,000	382,703,600	412,085,600
R&D Expense						
Operating Income		537,921,400	134,290,100	249,703,000	90,223,660	75,367,510
Operating Margin %		43.74%	29.12%	35.24%	23.57%	18.28%
SGA Expense						
Net Income		574,111,100	181,332,400	330,419,600	100,641,200	110,021,200
Operating Cash Flow		335,477,500	18,129,540	96,709,340	-38,901,110	34,493,140
Capital Expenditure		109,145,900	3,514,230	4,683,234	12,699,810	3,016,691
EBITDA		768,371,400	278,636,300	469,127,900	171,875,700	170,845,200
Return on Assets %		10.37%	3.87%	8.06%	2.76%	3.18%
Return on Equity %		18.72%	6.83%	14.27%	5.01%	5.82%
Debt to Equity		0.35	0.34	0.18	0.35	0.28

CONTACT INFORMATION:

Phone: 852-2859-3111   Fax: 852-2857-7181
Toll-Free:
Address: 200 Connaught Rd., Penthouse, 39 Fl. W. Tower, Hong Kong, Hong Kong

STOCK TICKER/OTHER:

Stock Ticker: SHTGF   Exchange: GREY
Employees: 3,370   Fiscal Year Ends: 12/31
Parent Company:

SALARIES/BONUSES:

Top Exec. Salary: $ Bonus: $
Second Exec. Salary: $ Bonus: $

OTHER THOUGHTS:

Estimated Female Officers or Directors: 4
Hot Spot for Advancement for Women/Minorities: Y

Siemens Building Technologies

www.buildingtechnologies.siemens.com/bt/global/en/pages/home.aspx

NAIC Code: 334512

TYPES OF BUSINESS:

Building Control Systems
Environmental Control Systems

BRANDS/DIVISIONS/AFFILIATES:

Energy Saving Performance Contracting.
Intelligence Response
Siemens AG

CONTACTS: Note: Officers with more than one job title may be intentionally listed here more than once.

Matthias Rebellius, CEO
Rene Jungbluth, Head-Solutions and Service Portfolio

GROWTH PLANS/SPECIAL FEATURES:

Siemens Building Technologies (SBT) provides products, services and solutions for public, industrial, commercial and residential buildings. Based in Switzerland, SBT is a division of Siemens AG's infrastructure & cities segment. The division's building specialties include services related to comfort, automation, safety, security and operations. SBT also offers energy solutions aimed at improving buildings' energy costs, reliability and performance while minimizing their impact on the environment. The division offers products for building automation, HVAC (heating, ventilation and air conditioning) systems and fire safety and security. The building automation unit offers computerized, distributed control systems that monitor and control a building's mechanical and lighting systems. The HVAC systems and products unit includes controls for heating, ventilations, air conditioning and cooling applications. The fire safety and security products unit offers fire detectors, protection alarm systems and non-water based fire extinguishers. The company's proprietary Intelligence Response system evaluates building security and suggest what measures can be put into place for emergency situations. The firm also provides consultation for all levels of the energy management cycle and finances modernization through Energy Saving Performance Contracting.

FINANCIAL DATA: Note: Data for latest year may not have been available at press time.

In U.S. $	2015	2014	2013	2012	2011	2010
Revenue	6,891,651,200	6,396,525,555				
R&D Expense						
Operating Income						
Operating Margin %						
SGA Expense						
Net Income	635,173,035	587,036,800				
Operating Cash Flow						
Capital Expenditure						
EBITDA						
Return on Assets %						
Return on Equity %						
Debt to Equity						

CONTACT INFORMATION:

Phone: 41-41-724-2424   Fax: 41-41-724-3522
Toll-Free:
Address: Gubelstrasse 22, Zug, CH-6301 Switzerland

STOCK TICKER/OTHER:

Stock Ticker: Subsidiary   Exchange:
Employees: Fiscal Year Ends: 09/30
Parent Company: Siemens AG

SALARIES/BONUSES:

Top Exec. Salary: $ Bonus: $
Second Exec. Salary: $ Bonus: $

OTHER THOUGHTS:

Estimated Female Officers or Directors:
Hot Spot for Advancement for Women/Minorities:

Simon Property Group Inc

www.simon.com

NAIC Code: 0

TYPES OF BUSINESS:

Real Estate Investment Trust
Malls, Shopping & Outlet Centers
Office & Mixed-Use Properties
Real Estate Development
Real Estate Consulting
Marketing Services
Property Management

BRANDS/DIVISIONS/AFFILIATES:

Simon Property Group LP
Simon Brand Ventures
Klepierre SA

CONTACTS: Note: Officers with more than one job title may be intentionally listed here more than once.

Steven Fivel, Assistant General Counsel
David Simon, CEO
Andrew Juster, CFO
Herbert Simon, Chairman Emeritus
Steven Broadwater, Chief Accounting Officer
John Rulli, Chief Administrative Officer
Richard Sokolov, COO
James Barkley, General Counsel
David Contis, President, Subsidiary
Brian McDade, Senior VP

GROWTH PLANS/SPECIAL FEATURES:

Simon Property Group, Inc. (SPG) is a self-managed real estate investment trust (REIT) engaged in owning, managing and developing retail real estate, primarily regional enclosed malls, outlet centers and community/lifestyle centers. The company owns all of its real estate properties through Simon Property Group LP, a majority-owned subsidiary. The company owns or holds interest in 209 properties in 37 states and Puerto Rico. Its portfolio consists of 108 malls, four lifestyle centers, 71 premium outlet centers, 14 Mills and 12 other shopping centers or outlet centers. Additionally, the firm has ownership interests in nine premium outlets in Japan, three in South Korea, two in Canada, one in Mexico and one in Malaysia. SPG also holds a 20.3% equity stake in Klepierre SA, a publicly-traded Paris-based real estate firm that owns or has an interest in shopping centers located in 16 European countries. SPG's strategy includes aggressively marketing available space and renewing existing leases at higher rents, pursuing acquisitions, developing new properties in major metropolitan areas and expanding and renovating existing properties. Through Simon Brand Ventures, the company develops national, regional and local marketing and media plans for clients.

SPG employees receive medical, dental, vision and prescription drug coverage; basic life and AD&D insurance; short- and long-term disability; flexible spending accounts; and a 401(k).

FINANCIAL DATA: Note: Data for latest year may not have been available at press time.

In U.S. $	2015	2014	2013	2012	2011	2010
Revenue	5,266,103,000	4,870,818,000	5,170,138,000	4,880,084,000	4,306,432,000	3,957,630,000
R&D Expense						
Operating Income	2,668,873,000	2,385,342,000	2,415,689,000	2,220,600,000	1,935,142,000	1,744,239,000
Operating Margin %	50.68%	48.97%	46.72%	45.50%	44.93%	44.07%
SGA Expense	349,999,000	355,190,000	326,944,000	416,028,000	395,435,000	296,747,000
Net Income	1,827,720,000	1,408,588,000	1,319,641,000	1,434,496,000	1,024,799,000	617,038,000
Operating Cash Flow	3,024,685,000	2,730,420,000	2,700,996,000	2,513,072,000	2,005,887,000	1,755,210,000
Capital Expenditure	1,020,924,000	796,736,000	841,209,000	802,427,000	445,495,000	256,312,000
EBITDA	4,071,940,000	3,770,327,000	4,061,413,000	3,521,904,000	3,345,447,000	2,477,330,000
Return on Assets %	6.06%	4.47%	3.99%	4.86%	3.99%	2.40%
Return on Equity %	38.50%	25.89%	22.55%	27.33%	21.75%	13.27%
Debt to Equity	5.08	4.13	4.06	4.06	4.15	3.65

CONTACT INFORMATION:

Phone: 317 636-1600   Fax:
Toll-Free:
Address: 225 W. Washington St., Indianapolis, IN 46204 United States

STOCK TICKER/OTHER:

Stock Ticker: SPG   Exchange: NYS
Employees: 5,250   Fiscal Year Ends: 12/31
Parent Company:

SALARIES/BONUSES:

Top Exec. Salary: Bonus: $
$1,250,000
Second Exec. Salary: Bonus: $
$800,000

OTHER THOUGHTS:

Estimated Female Officers or Directors: 5

Hot Spot for Advancement for Women/Minorities: Y

Sino Land Company Limited

NAIC Code: 531100

www.sino.com

TYPES OF BUSINESS:

Real Estate Holdings & Development
Commercial & Residential Construction
Property Management Services

BRANDS/DIVISIONS/AFFILIATES:

Fullerton Hotel
Fullerton Bay Hotel
Conrad Hotel
Sino Group

CONTACTS: *Note: Officers with more than one job title may be intentionally listed here more than once.*

Valencia Lee, CFO
Alice Ip Mo Lin, Exec. Dir.-Human Resources
Alice Ip Mo Lin, Exec. Dir.-Admin. Function
Gordan Lee Ching Keung, Exec. Dir.-Dev. Div.
Ringo Chan Wing Kwong, Exec. Dir.-Treasury
Daryl Ng Win Kong, Exec. Dir.
Robert Ng Chee Siong, Chmn.
Alice Ip Mo Lin, Exec. Dir.-Procurement

GROWTH PLANS/SPECIAL FEATURES:

Sino Land Company Limited, a member of the Sino Group, is a Hong Kong-based real estate firm whose core business involves the development of property for both sale and investment purposes. The company holds a property portfolio totaling approximately 76.5 million square feet, consisting of five primary property types: residential, accounting for 61.9% of the firm's portfolio; commercial, accounting for 24.7%; industrial, accounting for 6%; car parks, accounting for 4%; and hotels, accounting for 3.4%. Commercial and industrial properties, as well as hotels and car parks, are typically held by Sino Land as long-term investments to generate recurrent income, while the firm generally acquires residential properties with the intention of developing them for resale. The firm aims to acquire residential properties in popular urban areas, with convenient access to various forms of transportation, such as subways and rail lines. Of the company's total holdings, approximately 24 million square feet consist of properties currently under development, approximately 11.5 million square feet represent properties held either as investments or for the firm's own use, and approximately 2.2 million square feet are completed properties held for sale. Sino Land's hotels include the Fullterton Hotel and Fullterton Bay Hotel in Singapore as well as the Conrad Hotel in Hong Kong.

FINANCIAL DATA: *Note: Data for latest year may not have been available at press time.*

In U.S. $	2015	2014	2013	2012	2011	2010
Revenue	2,815,628,000	960,642,000	1,008,046,000	1,082,469,000	766,376,100	992,481,900
R&D Expense						
Operating Income	649,016,800	514,843,000	343,738,800	485,823,000	337,480,700	812,486,800
Operating Margin %	23.05%	53.59%	34.09%	44.88%	44.03%	81.86%
SGA Expense	88,209,230	85,801,300	82,977,550	81,241,410	94,198,410	78,087,280
Net Income	1,208,325,000	1,150,242,000	1,506,815,000	1,277,774,000	1,359,477,000	785,656,100
Operating Cash Flow	1,007,997,000	199,194,200	207,324,700	441,967,000	122,057,700	-14,830,550
Capital Expenditure	9,995,792	7,899,280	6,259,164	4,139,313	6,400,890	42,131,180
EBITDA	1,407,483,000	1,371,167,000	1,651,353,000	1,483,019,000	1,649,622,000	951,992,800
Return on Assets %	6.72%	6.70%	9.64%	8.92%	10.30%	6.59%
Return on Equity %	8.14%	8.24%	12.01%	11.63%	14.44%	9.72%
Debt to Equity	0.04	0.07	0.05	0.08	0.15	0.21

CONTACT INFORMATION:

Phone: 852 27218388   Fax: 852 27235901
Toll-Free:
Address: Salisbury Rd., 12th Fl., Tsim Sha Tsui Centre, Kowloon, Hong Kong

STOCK TICKER/OTHER:

Stock Ticker: SNLAY   Exchange: PINX
Employees: 9,300   Fiscal Year Ends: 06/30
Parent Company:

SALARIES/BONUSES:

Top Exec. Salary: $ Bonus: $
Second Exec. Salary: $ Bonus: $

OTHER THOUGHTS:

Estimated Female Officers or Directors: 2
Hot Spot for Advancement for Women/Minorities:

Skanska USA Inc

www.usa.skanska.com

NAIC Code: 237000

TYPES OF BUSINESS:
Heavy & Civil Engineering Construction

BRANDS/DIVISIONS/AFFILIATES:
Skanska AB
Skanska USA Building Inc
Skanska USA Civil Inc
Bayshore Concrete
Underpinning & Foundation Skanska
Skanska Koch
MachineryTrader.com
Skanska Financial Services

CONTACTS: Note: Officers with more than one job title may be intentionally listed here more than once.
Richard Cavallaro, CEO
Scott MacLeod, COO
Mats Johansson, Pres., Skanska USA Commercial Dev.

GROWTH PLANS/SPECIAL FEATURES:
Skanska USA, Inc., a subsidiary of Skanska AB, operates as world-class construction, civil and development firm. The company operates in four business units: USA building, USA civil, infrastructure development and commercial development. Services offered by Skanska USA Building Inc. include construction management, design-build, pre-construction, pharmaceutical validation, privately developed public infrastructure projects, commercial development and Skanska Financial Services. Skanska USA Civil Inc. is a leading contractor in the U.S. market for civil engineering construction and infrastructure projects. The business unit provides construction services to public and private clients in the civil, mechanical, industrial, marine, foundation and environmental sectors. Within this unit are the contractors Bayshore Concrete, Underpinning & Foundation Skanska and Skanska Koch as well as MachineryTrader.com, a used construction equipment website. The infrastructure development business encompasses the firm's Public Private Partnership (PPP) business. PPP develops public infrastructure projects, for example hospitals, schools and roads, to meet a public demand through private solutions. Skanska Commercial Development acquires, develops, leases and divests commercial properties located in select core urban markets in the U.S. and Europe. The firm invests on average $500 million of Skanska capital annually.

FINANCIAL DATA: Note: Data for latest year may not have been available at press time.

In U.S. $	2015	2014	2013	2012	2011	2010
Revenue	7,100,000,000	7,300,000,000	6,718,100,000	6,400,000,000	6,250,000,000	
R&D Expense						
Operating Income						
Operating Margin %						
SGA Expense						
Net Income						
Operating Cash Flow						
Capital Expenditure						
EBITDA						
Return on Assets %						
Return on Equity %						
Debt to Equity						

CONTACT INFORMATION:
Phone: 917-438-4500   Fax:
Toll-Free:
Address: 350 Fifth Ave., New York, NY 10118 United States

STOCK TICKER/OTHER:
Stock Ticker: Subsidiary   Exchange:
Employees: 11,000   Fiscal Year Ends:
Parent Company: Skanska AB

SALARIES/BONUSES:
Top Exec. Salary: $ Bonus: $
Second Exec. Salary: $ Bonus: $

OTHER THOUGHTS:
Estimated Female Officers or Directors:
Hot Spot for Advancement for Women/Minorities:

Skidmore Owings & Merrill LLP

www.som.com

NAIC Code: 541330

TYPES OF BUSINESS:

Architectural & Engineering Services
Urban Design Services
Transportation Planning
Seismic Analysis & Consulting
Environmental Engineering
Digital Design
Graphics
Interior Design

BRANDS/DIVISIONS/AFFILIATES:

CONTACTS: *Note: Officers with more than one job title may be intentionally listed here more than once.*

Kenneth A. Lewis, Managing Partner
William F. Baker, Partner-Civil & Structural Eng.
Gene Schnair, Managing Partner-San Francisco
Thomas Behr, Managing Dir-New York
Rod Garrett, Managing Dir.-Washington, D.C.
Xuan Fu, Managing Partner-Chicago
Silas Chiow, Dir.-China

GROWTH PLANS/SPECIAL FEATURES:

Skidmore, Owings & Merrill, LLP (SOM) is a leading global architecture, urban design, engineering and interior architecture firm. Its services span architectural design and engineering of individual buildings to the master planning and design of entire communities. The firm's projects include corporate offices, banking and financial institutions, government buildings, public and private institutions, health care facilities, religious buildings, airports, recreational and sports facilities, university buildings and residential developments. SOM's graphics group provides branding, corporate identity and graphic design services, with expansive projects for clients including the Bay Area Discovery Museum and Dublin Airport. Additionally, its digital design service provides virtual modeling for potential projects. The SOM Interiors group provides interior design solutions for corporate facilities, educational facilities, retail spaces, performing arts spaces, hotels and residential complexes, among others. The company's sustainable designs include extensive uses of natural light and low-impact environmental building systems designed to reduce waste and energy use. SOM's Mechanical and Electrical Engineering Group (MEP Engineering) provides design engineering services for new and existing buildings, such as developing solar heating and cooling for new and existing buildings, modeling energy consumption patterns, forecasting probable operating costs and developing energy recovery systems. Its urban planning services assist city officials in providing plans for efficient commuting, potable water, adequate housing and related problems. Signature projects include Bronzeville Metra Station, Centennial Tower, 7 World Trade Center, Willis Tower (formerly known as the Sears Tower) and John Hancock Tower. Recent projects include Denver Union Station in Colorado, Manhattan Loft Gardens in the U.K., The University Center-The New School in New York, China Merchants Tower & Woods Park in Shenzhen, China and Poly International Plaza in Beijing, China. In December 2015, the firm was hired to develop the downtown Bakersfield High-Speed Rail Station in California.

FINANCIAL DATA: *Note: Data for latest year may not have been available at press time.*

In U.S. $	2015	2014	2013	2012	2011	2010
Revenue	356,070,000	283,000,000	277,000,000	255,000,000	236,000,000	337,000,000
R&D Expense						
Operating Income						
Operating Margin %						
SGA Expense						
Net Income						
Operating Cash Flow						
Capital Expenditure						
EBITDA						
Return on Assets %						
Return on Equity %						
Debt to Equity						

CONTACT INFORMATION:

Phone: 312-554-9090   Fax: 312-360-4545
Toll-Free: 866-269-2688
Address: 224 S. Michigan Ave., Ste. 1000, Chicago, IL 60604 United States

STOCK TICKER/OTHER:

Stock Ticker: Private
Employees: 1,421
Parent Company:

Exchange:
Fiscal Year Ends: 09/30

SALARIES/BONUSES:

Top Exec. Salary: $ Bonus: $
Second Exec. Salary: $ Bonus: $

OTHER THOUGHTS:

Estimated Female Officers or Directors: 8
Hot Spot for Advancement for Women/Minorities: Y

Skyline Corporation

www.skylinecorp.com

NAIC Code: 321992

TYPES OF BUSINESS:

Prefabricated Wood Building Manufacturing (Manufactured Housing)
Recreational Vehicles
Prefabricated Metal Buildings & Components

BRANDS/DIVISIONS/AFFILIATES:

Kensington
Shore Park
Stone Harbor
Vacation Villa

CONTACTS: Note: Officers with more than one job title may be intentionally listed here more than once.

Jon Pilarski, CFO
Martin Fransted, Controller
John Firth, Director
Richard Florea, Director
Jeffrey Newport, Senior VP, Divisional
Robert Davis, Vice President, Divisional
Terrence Decio, Vice President, Divisional

GROWTH PLANS/SPECIAL FEATURES:

Skyline Corporation, along with its consolidated subsidiaries, designs, produces and distributes manufactured housing, modular housing and park models to independent dealers and communities throughout the U.S. and Canada. The firm sold 2,691 manufactured homes, 344 modular homes and 391 park models in 2015. The manufactured and modular homes are marketed under a number of trademarks, and are available in a variety of dimensions. Manufactured housing models are built according to standards established by the U.S. Department of Housing and Urban Development. Modular homes are build according to state, provincial or local buildings codes. Each of these homes typically includes two to four bedrooms, kitchen, dining area, living room, one or two bathrooms, kitchen appliances, central heating and cooling. Custom options may include exterior dormers & windows, interior or exterior accent columns, fireplaces and whirlpool tubs. Skyline's park housing models are marketed under the following trademarks: Kensington, Shore Park, Stone Harbor and Vacation Villa. Park models are intended to provide temporary living accommodations for individuals seeking leisure travel and outdoor recreation. The company sells homes that are Energy-Star compliant. Skyline has nine manufacturing facilities and three corporate facilities, all of which are owned, located in California, Florida, Indiana, Kansas, Ohio, Oregon, Pennsylvania, Wisconsin and Texas. The firm's manufactured homes are distributed by approximately 270 independent dealers at 510 locations. In 2015, the firm sold its recreational business to Evergreen Recreational Vehicles, LLC.

FINANCIAL DATA: Note: Data for latest year may not have been available at press time.

In U.S. $	2015	2014	2013	2012	2011	2010
Revenue	186,985,000	191,730,000	177,574,000	182,846,000	162,327,000	136,230,000
R&D Expense						
Operating Income	-3,857,000	-11,893,000	-10,577,000	-19,382,000	-26,688,000	-19,414,000
Operating Margin %	-2.06%	-6.20%	-5.95%	-10.60%	-16.44%	-14.25%
SGA Expense	21,194,000	22,886,000	23,681,000	27,450,000	28,490,000	26,200,000
Net Income	-10,414,000	-11,864,000	-10,513,000	-19,365,000	-26,627,000	-28,993,000
Operating Cash Flow	-4,335,000	-17,596,000	-14,039,000	-18,597,000	-25,575,000	-15,251,000
Capital Expenditure	473,000	753,000	75,000	614,000	816,000	891,000
EBITDA	-2,487,000	-10,077,000	-8,575,000	-17,023,000	-24,005,000	-17,225,000
Return on Assets %	-17.92%	-17.74%	-14.32%	-21.76%	-23.17%	-19.40%
Return on Equity %	-36.43%	-29.86%	-20.65%	-29.07%	-28.51%	-22.78%
Debt to Equity	0.18	0.18				

CONTACT INFORMATION:

Phone: 574-294-6521   Fax: 574 293-0693
Toll-Free: 800-755-6521
Address: 2520 By-Pass Rd., Elkhart, IN 46515 United States

STOCK TICKER/OTHER:

Stock Ticker: SKY   Exchange: ASE
Employees: 1,200   Fiscal Year Ends: 05/31
Parent Company:

SALARIES/BONUSES:

Top Exec. Salary: $300,000   Bonus: $
Second Exec. Salary: Bonus: $
$240,000

OTHER THOUGHTS:

Estimated Female Officers or Directors:
Hot Spot for Advancement for Women/Minorities:

SL Green Realty Corp

NAIC Code: 0

www.slgreen.com

TYPES OF BUSINESS:

Real Estate Investment Trust
Commercial Properties
Property Management

BRANDS/DIVISIONS/AFFILIATES:

CONTACTS: Note: Officers with more than one job title may be intentionally listed here more than once.

Marc Holliday, CEO
Matthew DiLiberto, CFO
Stephen Green, Chairman of the Board
Andrew Levine, Executive VP
Heidi Gillette, Other Corporate Officer
Andrew Mathias, President

GROWTH PLANS/SPECIAL FEATURES:

SL Green Realty Corp. is a self-managed REIT (real estate investment trust) investing primarily in commercial office properties in the New York Metropolitan area. The firm's core business is the ownership of office buildings that are strategically located in close proximity to midtown Manhattan's primary commuter stations. Its property portfolio consists of fee- and lease-based ownership in 117 properties in Manhattan, totaling over 43.6 million square feet; and 37 properties in the New York Metro area, including Brooklyn, Long Island, Westchester County, Connecticut and Northern New Jersey, totaling roughly 5.9 million square feet. Additionally, the firm owns interests in 29 million square feet of commercial buildings, as well as debt and preferred equity investments secured by 14.6 million square feet of buildings. In 2015, the firm acquired 29-story, 2.3 million square foot Class A office property Eleven Madison Avenue in NYC; and a 90% interest in The SoHo Building; two mixed-use properties on Broadway and 5-7 Dey Street in downtown Manhattan. That same year, it sold Tower 45; sold two Fifth Avenue retail development sites; sold its interest in 315 West 36th Street, NYC; sold its Pace University dormitory tower on Beekman Street, NYC; and sold its 90% stake in residential condo 248-252 Bedford Ave, Brooklyn. Additionally, SL Green formed a joint venture with Invesco Real Estate in order to acquire 131-137 Sprint Street in SoHo.

FINANCIAL DATA: Note: Data for latest year may not have been available at press time.

In U.S. $	2015	2014	2013	2012	2011	2010
Revenue	1,662,829,000	1,519,978,000	1,469,077,000	1,400,255,000	1,263,428,000	1,101,246,000
R&D Expense						
Operating Income	77,261,000	174,963,000	142,024,000	82,524,000	122,580,000	111,132,000
Operating Margin %	4.64%	11.51%	9.66%	5.89%	9.70%	10.09%
SGA Expense	139,137,000	133,502,000	126,118,000	120,706,000	118,583,000	119,012,000
Net Income	284,084,000	536,523,000	135,371,000	196,405,000	647,410,000	300,575,000
Operating Cash Flow	526,484,000	490,381,000	386,203,000	353,743,000	312,860,000	321,058,000
Capital Expenditure	3,059,753,000	1,409,417,000	791,166,000	692,716,000	605,856,000	378,759,000
EBITDA	1,217,966,000	1,083,628,000	824,533,000	877,684,000	1,205,489,000	754,489,000
Return on Assets %	1.45%	3.25%	.69%	1.11%	4.98%	2.48%
Return on Equity %	3.90%	8.01%	1.64%	2.68%	12.20%	6.26%
Debt to Equity	1.47	1.16	1.07	1.07	1.01	1.02

CONTACT INFORMATION:

Phone: 212-594-2700   Fax: 212 216-1785
Toll-Free:
Address: 420 Lexington Ave., New York, NY 10170 United States

STOCK TICKER/OTHER:

Stock Ticker: SLG   Exchange: NYS
Employees: 1,177   Fiscal Year Ends: 12/31
Parent Company:

SALARIES/BONUSES:

Top Exec. Salary: $525,000   Bonus: $1,100,000
Second Exec. Salary: Bonus: $
$1,050,000

OTHER THOUGHTS:

Estimated Female Officers or Directors:
Hot Spot for Advancement for Women/Minorities:

Smallwood Reynolds Stewart Stewart & Associates Inc

www.srssa.com
NAIC Code: 541310

TYPES OF BUSINESS:

Architectural Services
Architecture
Interior Design
Master Planning
Graphic Design
Landscape

BRANDS/DIVISIONS/AFFILIATES:

CONTACTS: Note: Officers with more than one job title may be intentionally listed here more than once.

Howard Stewart, Principal
Toni Bagala, CFO

GROWTH PLANS/SPECIAL FEATURES:

Smallwood Reynolds Stewart Stewart & Associates, Inc. (SRSSA) is an international architecture and design firm. It provides post-modern designs and skyscrapers for the corporate, commercial, hospitality, residential, industrial, government and educational sectors. SRSSA is headquartered in Atlanta, Georgia, with offices in Dubai and Singapore. The company's design services consist of architecture, interior design, master planning, graphic design and landscape architecture. Its architecture services include schematic design, development, contract documents, specifications, bidding review and analysis, contract administration, construction contract negotiation, due diligence inspections, renovation assessment studies, thematic design, animations, roof consulting, building rating systems, building materials and planning. Interior design includes programming, interior architectural design, furniture/fixture/finishes design, artwork and accessories, construction documents, contract administration, benchmarking, building evaluation, merger and acquisition studies, engineer and specialty consultant coordination, interior renderings and interior animations. Master planning includes site analysis, land use analysis, master planning, urban planning and design, recreation planning and design, thematic planning, long-range master planning, feasibility studies, functional programming and energy conservation studies. Graphics design provides design solutions such as brand identity, signage, environmental and print, and is interactive by listening to client needs and responds with innovative design and implementation packages. Landscape architecture includes site inventory and analysis, site planning, design guidelines, planting design, hardscape design, streetscape design, water feature design, irrigation design, landscape lighting, interior landscape design, lake design, plant material selection and tagging, specifications and contract administration.

FINANCIAL DATA: Note: Data for latest year may not have been available at press time.

In U.S. $	2015	2014	2013	2012	2011	2010
Revenue	22,740,000					
R&D Expense						
Operating Income						
Operating Margin %						
SGA Expense						
Net Income						
Operating Cash Flow						
Capital Expenditure						
EBITDA						
Return on Assets %						
Return on Equity %						
Debt to Equity						

CONTACT INFORMATION:

Phone: 404-233-5453   Fax: 404-264-0929
Toll-Free:
Address: 3565 Piedmont Rd., Ste. 303, Atlanta, GA 30305 United States

STOCK TICKER/OTHER:

Stock Ticker: Private   Exchange:
Employees: 109   Fiscal Year Ends:
Parent Company:

SALARIES/BONUSES:

Top Exec. Salary: $ Bonus: $
Second Exec. Salary: $ Bonus: $

OTHER THOUGHTS:

Estimated Female Officers or Directors:
Hot Spot for Advancement for Women/Minorities:

Societe Anonyme des Bains de Mer et du Cercle des Etrangers a Monaco

www.montecarlosbm.com

NAIC Code: 721120

TYPES OF BUSINESS:

Hotels
Casinos
Restaurants
Resorts

BRANDS/DIVISIONS/AFFILIATES:

Monte-Carlo SBM
Hotel Hermitage
Hotel de Paris
Monte-Carlo Beach Hotel
Monte-Carlo Bay Hotel & Resort
Le Louis XV-Alain Ducasse
Casino de Monte Carlo
Cinq Mondes

CONTACTS: *Note: Officers with more than one job title may be intentionally listed here more than once.*

Jean-Louis Masurel, CEO
Yves de Toytot, CFO
Agness Puons, Corp. Sec.
Veronique Burki-Despont, Dir.-Comm.
Luca Allegri, General Dir.
Pascal Camia, Dir.-Hotel Hermitage
Patrick Nayrolles, Gen. Mgr.-Monte-Carlo Beach Club, Saadiyat
Isabelle Simon, Deputy CEO
Jean-Luc Biamonti, Chmn.
Jean-Pierre Siri, Dir.-Int'l Dev. & Hotel Svcs.

GROWTH PLANS/SPECIAL FEATURES:

Societe Anonyme des Bains de Mer et du Cercle des Etrangers a Monaco (SBM) is a hotel and resort company that manages luxury properties in the Principality of Monaco, located next to France on the Mediterranean coast. It generally refers to all its properties as a single resort: Monte-Carlo SBM. The firm owns and manages four hotels: Hotel Hermitage, with 280 rooms and suites; Hotel de Paris, 182 rooms and suites; Monte-Carlo Beach Hotel, 40 rooms; and Monte-Carlo Bay Hotel & Resort, 334 rooms. Each hotel features luxury accommodations. The Hotel Hermitage and Hotel de Paris directly link to the firm's Les Thermes Marins de Monte-Carlo, one of the most prestigious spas in Europe; the Beach Hotel features an Olympic-sized seawater swimming pool; and the newer, high-tech Bay Hotel & Resort, located on 10 acres of land, features a sand-bottomed lagoon and the Cinq Mondes spa. SBM's five casinos include the legendary Casino de Monte-Carlo as well as Cafe de Paris Casino, Sun Casino, Casino Monte-Carlo Bay and Casino la Rascasse. The casinos, some of which require guests to follow a strict dress code, feature a wide variety of European and American table games as well as slot machines. SBM also offers more than 30 bars and restaurants, including Le Louis XV-Alain Ducasse, the oldest restaurant in Monaco; numerous conference rooms; an 18-hole golf course; 28 tennis courts; retail outlets; cocktail rooms for up to 1,000 people; an opera company; a live performance arena with regular entertainment; and multiple night clubs. The firm also hosts the annual Grand Prix de Monaco, a Formula One auto race. Shares in the company are 69.1% owned by the royal family of Monte Carlo.

FINANCIAL DATA: *Note: Data for latest year may not have been available at press time.*

In U.S. $	2015	2014	2013	2012	2011	2010
Revenue	514,150,000	538,250,350	482,490,800			
R&D Expense						
Operating Income						
Operating Margin %						
SGA Expense						
Net Income	11,373,400	19,342,600	-58,027,800			
Operating Cash Flow						
Capital Expenditure						
EBITDA						
Return on Assets %						
Return on Equity %						
Debt to Equity						

CONTACT INFORMATION:

Phone: 377-98-06-4151   Fax: 377-98-06-2626
Toll-Free:
Address: Place du Casino, Monte Carlo, 98000 Monaco

STOCK TICKER/OTHER:

Stock Ticker: BAIN   Exchange: Paris
Employees: 4,400   Fiscal Year Ends: 03/31
Parent Company:

SALARIES/BONUSES:

Top Exec. Salary: $ Bonus: $
Second Exec. Salary: $ Bonus: $

OTHER THOUGHTS:

Estimated Female Officers or Directors: 3
Hot Spot for Advancement for Women/Minorities: Y

Sonesta International Hotels Corp

www.sonesta.com

NAIC Code: 721110

TYPES OF BUSINESS:

Hotels
River Cruise Ships
Hotels
Resorts

BRANDS/DIVISIONS/AFFILIATES:

Sonesta Art Collection
Just Us Kids
Nile
Sonesta Collection

CONTACTS: Note: Officers with more than one job title may be intentionally listed here more than once.

Carlos Flores, CEO
Scott Weiler, VP-Mktg. & Communications
Lorie Juliano, Dir.-Comm.

GROWTH PLANS/SPECIAL FEATURES:

Sonesta International Hotels Corp., doing business as Sonesta Collection, specializes in providing upscale accommodation. There are currently more than 65 Sonesta hotels and resorts as well as Nile cruise ships, totaling more than 12,000 rooms. The properties are located in the U.S, the Caribbean, Chile, Colombia, Ecuador, Panama, Peru and Egypt. Locations feature the Sonesta Art Collection, a collection of art that consists of more than 7,000 contemporary paintings, sculptures, original prints and tapestries by world-renowned artists, which are placed in public places and guestrooms inside its hotels. Some of the company's resorts also offer Just Us Kids, a complimentary supervised children's program for ages 5-13. Family packages, children's menus, baby-sitting and discounts on a second room add to the appeal for families. Sonesta's strategy is to consolidate assets and position the company for opportunities to expand with more hotels and resorts. The firm intends for its growth to be continual, but at a pace that will preserve the character that distinguishes each Sonesta property.

FINANCIAL DATA: Note: Data for latest year may not have been available at press time.

In U.S. $	2015	2014	2013	2012	2011	2010
Revenue						
R&D Expense						
Operating Income						
Operating Margin %						
SGA Expense						
Net Income						
Operating Cash Flow						
Capital Expenditure						
EBITDA						
Return on Assets %						
Return on Equity %						
Debt to Equity						

CONTACT INFORMATION:

Phone: 617-421-5400   Fax: 617-421-5402
Toll-Free: 800-766-3378
Address: 225 Washington Street, Ste 270, Newton, MA 02458 United States

STOCK TICKER/OTHER:

Stock Ticker: Private   Exchange:
Employees: 766   Fiscal Year Ends: 12/31
Parent Company: HOSPITALITY PROPERTIES TRUST

SALARIES/BONUSES:

Top Exec. Salary: $ Bonus: $
Second Exec. Salary: $ Bonus: $

OTHER THOUGHTS:

Estimated Female Officers or Directors: 2
Hot Spot for Advancement for Women/Minorities: Y

Southern Energy Homes Inc

www.sehomes.com

NAIC Code: 321992

TYPES OF BUSINESS:

Manufactured Housing
Home Financing

BRANDS/DIVISIONS/AFFILIATES:

Southern Homes
Clayton Homes Inc
Setexas
Southern Estates
Southern Energy Homes Renew Center
Energy Homes

GROWTH PLANS/SPECIAL FEATURES:

Southern Energy Homes, Inc. (SE Homes), a subsidiary of Clayton Homes, Inc., builds and sells pre-manufactured homes and operates five home manufacturing facilities: four in Alabama and one in Texas. SE Homes custom-designs and manufactures homes intended for primary residences in 20 states, primarily in the Southeastern and South Central U.S., and sold through over 300 independent dealers. The company offers a plethora of different floor plans for clients to choose from, ranging from under 1,000 to over 2,000 square feet. The firm's homes are marketed under four brand names: Energy Homes, Southern Estates, Southern Homes and SEtexas. Additionally, SE Homes sells pre-owned homes through its Southern Energy Homes Renew Center division. Since its inception in 1982, the firm has built more than 100,000 homes.

CONTACTS: *Note: Officers with more than one job title may be intentionally listed here more than once.*

Keith O. Holdbrooks, CEO
Keith O. Holdbrooks, Pres.
James Stariha, CFO
Dan Batchelor, Exec. VP

FINANCIAL DATA: *Note: Data for latest year may not have been available at press time.*

In U.S. $	2015	2014	2013	2012	2011	2010
Revenue						
R&D Expense						
Operating Income						
Operating Margin %						
SGA Expense						
Net Income						
Operating Cash Flow						
Capital Expenditure						
EBITDA						
Return on Assets %						
Return on Equity %						
Debt to Equity						

CONTACT INFORMATION:

Phone: 256-747-8589　　　Fax: 256-747-7586
Toll-Free: 866-896-2737
Address: 144 Corporate Way, Addison, AL 35540 United States

STOCK TICKER/OTHER:

Stock Ticker: Subsidiary　　　　　　　Exchange:
Employees: 3　　　　　　　　　　　Fiscal Year Ends: 12/31
Parent Company: Clayton Homes Inc

SALARIES/BONUSES:

Top Exec. Salary: $　　　Bonus: $
Second Exec. Salary: $　　　Bonus: $

OTHER THOUGHTS:

Estimated Female Officers or Directors:
Hot Spot for Advancement for Women/Minorities:

Sovran Self Storage Inc

www.unclebobs.com

NAIC Code: 531130

TYPES OF BUSINESS:

Real Estate Investment Trust
Self-Storage Properties
Truck Rental
Moving & Storage Products
Property Management

BRANDS/DIVISIONS/AFFILIATES:

Uncle Bob's Self Storage
Dri-guard
Uncle Bob's Rental Trucks
AllBoxesny.com

CONTACTS: *Note: Officers with more than one job title may be intentionally listed here more than once.*

David Rogers, CEO
Andrew Gregoire, CFO
Robert Attea, Chairman of the Board
Edward Killeen, COO
Kenneth Myszka, Director
Paul Powell, Other Executive Officer

GROWTH PLANS/SPECIAL FEATURES:

Sovran Self Storage, Inc. is a self-administered and self-managed REIT (real estate investment trust) that acquires, owns and manages self-storage properties. It owns and manages 518 self-storage facilities, comprising over 35.5 million square feet, making it one of the largest self-storage companies in the U.S. Sovran operates its stores under the trade name Uncle Bob's Self Storage, serving over 250,000 customers in 25 states. The states with the largest number of Uncle Bob's properties include Texas, with 133 stores; Florida, 72; Georgia, 30; New York, 35; Ohio, 23; Alabama, 22; New Jersey, 29; Virginia, 19; and North Carolina, 20. Stores range in size from 23,000-181,000 rentable square feet, with an average size of 68,000 rentable square feet. Sovran's stores typically offer various value-added products and services. Dri-guard, its state-of-the art centrally controlled dehumidification system, is available either singly or paired with the firm's climate controlled storage system. Dri-guard has been used by the Smithsonian, Boeing, Volvo and the U.S. Military, including for the storage of Air Force One and Two. Uncle Bob's Rental Trucks offers moving trucks with 14 x 8 feet of storage space on a two-way, local basis only. The firm also sells locks, boxes, tarps and other ancillary items used by its customers, offering these products both at its stores and through its affiliate web site, AllBoxesny.com.

FINANCIAL DATA: *Note: Data for latest year may not have been available at press time.*

In U.S. $	2015	2014	2013	2012	2011	2010
Revenue	366,602,000	326,080,000	273,507,000	236,007,000	211,156,000	192,072,000
R&D Expense						
Operating Income	147,285,000	116,333,000	101,063,000	80,487,000	68,950,000	66,366,000
Operating Margin %	40.17%	35.67%	36.95%	34.10%	32.65%	34.55%
SGA Expense	38,659,000	40,792,000	34,939,000	32,313,000	25,986,000	21,857,000
Net Income	112,524,000	88,531,000	74,126,000	55,128,000	30,592,000	40,642,000
Operating Cash Flow	186,210,000	146,068,000	120,646,000	99,004,000	80,310,000	73,671,000
Capital Expenditure	321,749,000	316,828,000	128,648,000	223,715,000	178,508,000	56,233,000
EBITDA	208,707,000	175,384,000	149,018,000	123,793,000	106,656,000	100,876,000
Return on Assets %	5.65%	5.18%	4.86%	3.89%	2.41%	3.42%
Return on Equity %	10.33%	9.58%	9.26%	7.96%	4.75%	6.40%
Debt to Equity	0.69	0.82	0.71	0.93	0.94	0.64

CONTACT INFORMATION:

Phone: 716 633-1850   Fax: 716 633-1860
Toll-Free: 800-242-1715
Address: 6467 Main St., Williamsville, NY 14221 United States

STOCK TICKER/OTHER:

Stock Ticker: SSS
Employees: 1,378
Parent Company:

Exchange: NYS
Fiscal Year Ends: 12/31

SALARIES/BONUSES:

Top Exec. Salary: $548,000   Bonus: $
Second Exec. Salary: $548,000   Bonus: $

OTHER THOUGHTS:

Estimated Female Officers or Directors: 4
Hot Spot for Advancement for Women/Minorities: Y

Sales, profits and employees may be estimates. Financial information, benefits and other data can change quickly and may vary from those stated here.

St Joe Company (The)

NAIC Code: 237210

www.joe.com

TYPES OF BUSINESS:

Real Estate Development, Investment & Operations
Residential Communities
Commercial Real Estate Development
Retail, Office & Industrial Properties
Forestry Operations

BRANDS/DIVISIONS/AFFILIATES:

WaterColor Inn
WaterSound Beach
Breakfast Point (The)
Watersound Origins
Pier Park North
VentureCrossings Enterprise Centre
Southwood
St. Joe Club & Resorts

CONTACTS: Note: Officers with more than one job title may be intentionally listed here more than once.

Marek Bakun, CFO
Bruce Berkowitz, Chairman of the Board
Kenneth Borick, General Counsel
Jorge Gonzalez, President
Patrick Murphy, Senior VP, Divisional
David Harrelson, Senior VP, Divisional

GROWTH PLANS/SPECIAL FEATURES:

The St. Joe Company, founded in 1936, is a real estate development company, which is engaged in town and resort development; commercial and industrial development, with significant interests in timber. The firm operates through four segments: residential real estate; commercial real estate; forestry; and resorts, leisure and leasing operations. The residential real estate segment develops large-scale, mixed-use resort, seasonal and primary residential communities, including large tracts of land in Northwest Florida and significant Gulf of Mexico beach frontage and waterfront properties. The residential resort communities are aimed to attract second-home buyers and include WaterColor and WaterSound Beach communities, while properties, such as Watersound Origins, Breakfast Point and SouthWood communities are aimed at primary home buyers. The commercial real estate segment develops and sells real estate for commercial purposes primarily in Northwest Florida. The operations of this segment include offering land for commercial and light industrial uses, develop commercial parcels, industrial and commerce parks. The forestry segment manages and harvests timber holdings. The firm grows and sells sawtimber, wood fiber and forest products and provides land management services for conservation of properties. The resorts, leisure and leasing segment consists of the WaterColor Inn and vacation rentals and the management of the firm's four golf courses as well as the firm's retail and commercial leasing operations. In addition, this segment operates St. Joe Club & Resorts, a private membership club, as well as oversees retail and commercial leasing operations. Properties include VentureCrossings Enterprise Centre, a 1,000 acre commercial and industrial development and Pier Park North, a joint venture retail center.

FINANCIAL DATA: Note: Data for latest year may not have been available at press time.

In U.S. $	2015	2014	2013	2012	2011	2010
Revenue	103,871,000	701,873,000	131,256,000	139,396,000	145,285,000	99,540,000
R&D Expense						
Operating Income	-6,135,000	513,250,000	777,000	2,134,000	-386,807,000	-51,554,000
Operating Margin %	-5.90%	73.12%	.59%	1.53%	-266.24%	-51.79%
SGA Expense	33,426,000	43,403,000	29,355,000	33,325,000	55,908,000	65,099,000
Net Income	-1,731,000	406,453,000	4,990,000	6,012,000	-330,279,000	-35,864,000
Operating Cash Flow	22,418,000	331,035,000	16,333,000	23,041,000	-9,839,000	16,312,000
Capital Expenditure	3,304,000	2,483,000	3,594,000	475,000	2,426,000	1,282,000
EBITDA	19,752,000	538,851,000	15,616,000	19,353,000	-366,112,000	-33,177,000
Return on Assets %	-.15%	41.20%	.75%	.92%	-38.56%	-3.33%
Return on Equity %	-.21%	53.14%	.89%	1.09%	-46.66%	-4.05%
Debt to Equity	0.34	0.24	0.07	0.06	0.09	0.06

CONTACT INFORMATION:

Phone: 904 301-4200   Fax:
Toll-Free: 866-417-7133
Address: 133 S. WaterSound Pkwy., WaterSound, FL 32413 United States

STOCK TICKER/OTHER:

Stock Ticker: JOE   Exchange: NYS
Employees: 61   Fiscal Year Ends: 12/31
Parent Company:

SALARIES/BONUSES:

Top Exec. Salary: $250,000   Bonus: $1,000,000
Second Exec. Salary: $350,000   Bonus: $250,000

OTHER THOUGHTS:

Estimated Female Officers or Directors: 1
Hot Spot for Advancement for Women/Minorities: Y

Starwood Capital Group Global LLC

www.starwoodcapital.com

NAIC Code: 721110

TYPES OF BUSINESS:

Real Estate Investments
Energy Investments
Commercial and Residential Development
Industrial Properties
Recreational Properties
Retail Properties
Office Buildings
Golf Courses

BRANDS/DIVISIONS/AFFILIATES:

Starwood Real Estate Securities
Starwood Energy Group
SH Group
1 Hotels & Resorts
Starwood Property Trust Inc
iStar Financial
Starwood Hotels & Resorts Worldwide
Colony Starwood Homes

CONTACTS: *Note: Officers with more than one job title may be intentionally listed here more than once.*

Barry S. Sternlicht, CEO
Steven M. Hankin, COO
Jerome C. Silvey, CFO
Ellis F. Rinaldi, Co-General Counsel
Jeffrey G. Dishner, Head-Real Estate Acquisitions & Debt Investments
Madison F. Grose, Co-General Counsel
Christopher D. Graham, Sr. Managing Dir.-Acquisitions
Barry S. Sternlicht, Chmn.

GROWTH PLANS/SPECIAL FEATURES:

Starwood Capital Group Global LLC is a private equity real estate investment firm with approximately $45 billion in assets under management. The company specializes in real estate-related investments on behalf of select private and institutional investor partners. It has invested in a wide range of property types, including multifamily, office, retail, hotel, industrial, residential and commercial land, senior housing, mixed-use and golf property, through equity, preferred equity, mezzanine debt and senior debt capital structures. Starwood operates through a number of subsidiaries and affiliated companies. Subsidiary Starwood Real Estate Securities is a hedge fund designed for global public real estate securities investment. Starwood Energy Group is an energy fund focused on investment in transmission, natural gas, wind and solar power generation facilities in North America. SH Group manages the eco-focused brand 1 Hotels & Resorts and also holds the license for Baccarat-brand hotels. Starwood Property Trust, Inc. is a real estate investment trust (REIT) which originates and invests in commercial mortgage loans and other commercial-related debt investments. iStar Financial is the firm's mezzanine debt centered portfolio. Starwood Hotels & Resorts Worldwide, Inc. encompasses the firm's Starwood and Westin hotel chain operations. Starwood is headquartered in Greenwich, Connecticut, with additional offices in Georgia, California, Washington, D.C., the U.K., Luxembourg, France and Brazil. In January 2016, the firm merged Starwood Waypoint Residential Trust with Colony American Homes, forming Colony Starwood Homes, a publicly-traded company under ticker SFR.

FINANCIAL DATA: *Note: Data for latest year may not have been available at press time.*

In U.S. $	2015	2014	2013	2012	2011	2010
Revenue						
R&D Expense						
Operating Income						
Operating Margin %						
SGA Expense						
Net Income						
Operating Cash Flow						
Capital Expenditure						
EBITDA						
Return on Assets %						
Return on Equity %						
Debt to Equity						

CONTACT INFORMATION:

Phone: 203-422-7700   Fax: 203-422-7784
Toll-Free:
Address: 591 W. Putnam Ave., Greenwich, CT 06830 United States

STOCK TICKER/OTHER:

Stock Ticker: Private   Exchange:
Employees: 2,100   Fiscal Year Ends:
Parent Company:

SALARIES/BONUSES:

Top Exec. Salary: $ Bonus: $
Second Exec. Salary: $ Bonus: $

OTHER THOUGHTS:

Estimated Female Officers or Directors: 1
Hot Spot for Advancement for Women/Minorities:

Starwood Hotels & Resorts Worldwide Inc www.starwoodhotels.com

NAIC Code: 721110

TYPES OF BUSINESS:
Hotels & Resorts
Financial Services
Hotel Management & Franchising
Spa Services

BRANDS/DIVISIONS/AFFILIATES:
Sheraton
W
Four Points
Westin
Le Meridien
St. Regis
Luxury Collection (The)
Aloft

CONTACTS: Note: Officers with more than one job title may be intentionally listed here more than once.
Thomas Mangas, CEO
Bruce Duncan, Chairman of the Board
Alan Schnaid, Chief Accounting Officer
Kenneth Siegel, Chief Administrative Officer
Martha Poulter, Chief Information Officer
Robyn Arnell, Controller
Jeffrey Cava, Executive VP
Kristen Prohl, Other Executive Officer
Simon Turner, President, Divisional
Sergio Rivera, President, Geographical

GROWTH PLANS/SPECIAL FEATURES:
Starwood Hotels & Resorts Worldwide, Inc. is involved in the global operation of hotels and resorts, primarily in the luxury and upscale segments of the industry. It owns, leases, manages or franchises approximately 1,222 properties with approximately 354,200 rooms in nearly 100 countries. The company's hotel brand names include St. Regis, The Luxury Collection, Phoenician, Tremont, Hotel Alfonso, Hotel Imperial, Hotel Maria Cristina, The Gritti Palace, Hotel Goldener Hirsch, W, Westin, Sheraton, Four Points, Le Meridien, Aloft and Element. The firm's earnings are derived mainly from its hotel and leisure operations; the receipt of franchise fees; and the development, ownership and operation of vacation ownership resorts. Additionally, Starwood provides financing to customers who purchase interests in resorts. The firm's frequent guest loyalty program, Starwood Preferred Guest, is unique in the hotel industry for its lack of capacity controls and blackout dates. In November 2015, the firm received an acquisition offer from Marriott. In March 2016, Starwood received a competing acquisition offer from Anbang Insurance Group, a China-based firm that has a history of acquiring hotels for its investment portfolio. The following week, Mariott countered with another offer valuing Starwood at $13.6 billion.

The firm offers employees dental, vision and health insurance; life insurance; wellness programs; dependent care flexible spending accounts; an employee assistance program; and adoption assistance.

FINANCIAL DATA: Note: Data for latest year may not have been available at press time.

In U.S. $	2015	2014	2013	2012	2011	2010
Revenue	5,763,000,000	5,983,000,000	6,115,000,000	6,321,000,000	5,624,000,000	5,071,000,000
R&D Expense						
Operating Income	740,000,000	883,000,000	925,000,000	912,000,000	630,000,000	600,000,000
Operating Margin %	12.84%	14.75%	15.12%	14.42%	11.20%	11.83%
SGA Expense	388,000,000	402,000,000	384,000,000	370,000,000	352,000,000	344,000,000
Net Income	489,000,000	633,000,000	635,000,000	562,000,000	489,000,000	477,000,000
Operating Cash Flow	890,000,000	994,000,000	1,151,000,000	1,184,000,000	641,000,000	764,000,000
Capital Expenditure	261,000,000	327,000,000	364,000,000	362,000,000	385,000,000	227,000,000
EBITDA	1,065,000,000	1,162,000,000	1,198,000,000	1,039,000,000	906,000,000	858,000,000
Return on Assets %	5.77%	7.26%	7.20%	6.10%	5.05%	23.53%
Return on Equity %	34.66%	25.91%	19.54%	18.45%	18.02%	101.55%
Debt to Equity	1.75	1.68	0.45	0.52	0.87	1.30

CONTACT INFORMATION:
Phone: 914 640-8100   Fax: 914 640-8310
Toll-Free:
Address: 1 StarPoint, White Plains, NY 06902 United States

STOCK TICKER/OTHER:
Stock Ticker: HOT   Exchange: NYS
Employees: 180,400   Fiscal Year Ends: 12/31
Parent Company:

SALARIES/BONUSES:
Top Exec. Salary: Bonus: $
$1,250,000
Second Exec. Salary: Bonus: $500,000
$365,909

OTHER THOUGHTS:
Estimated Female Officers or Directors: 4

Hot Spot for Advancement for Women/Minorities: Y

Starwood Property Trust Inc

www.starwoodpropertytrust.com

NAIC Code: 525990

TYPES OF BUSINESS:
Mortgage REIT

BRANDS/DIVISIONS/AFFILIATES:
SPT Management LLC
LNR Property LLC
Hatfield Philips International
Starwood Waypoint Residential Trust

CONTACTS: *Note: Officers with more than one job title may be intentionally listed here more than once.*
Barry Sternlicht, CEO
Rina Paniry, CFO
Andrew Sossen, COO
Zach Tanenbaum, Other Corporate Officer
Jeffrey DiModica, President

GROWTH PLANS/SPECIAL FEATURES:
Starwood Property Trust, Inc. is a major commercial mortgage real estate investment trust in the U.S. It is externally managed and advised by SPT Management, LLC. Starwood's core business focuses on originating, acquiring, financing and managing commercial mortgage loans and other commercial real estate debt and equity investments. Through its subsidiaries LNR Property LLC and Hatfield Philips International, it also operates as a commercial mortgage special servicer in the U.S. and across Europe. Since inception, the firm has deployed nearly $20 billion of total capital. Starwood's target assets may also include residential mortgage-backed securities, as well as commercial mortgage-backed securities, certain residential mortgage loans, distressed or non-performing commercial loans, commercial properties subject to net leases and equity interests in commercial real estate. The firm has two reportable segments: real estate lending, which includes all business activities of the company, excluding real estate investing and servicing; and real estate investing & servicing, which includes all business activities of LNR excluding the consolidation of securitization variable interest entities. In 2014, the firm spun off its former single family residential segment, creating Starwood Waypoint Residential Trust., which trades on NYSE under the SWAY symbol.

FINANCIAL DATA: *Note: Data for latest year may not have been available at press time.*

In U.S. $	2015	2014	2013	2012	2011	2010
Revenue	803,118,000	848,034,000	638,881,000	280,098,000	172,318,000	89,320,000
R&D Expense						
Operating Income	469,389,000	525,129,000	334,383,000	204,705,000	121,398,000	59,268,000
Operating Margin %	58.44%	61.92%	52.33%	73.08%	70.45%	66.35%
SGA Expense	279,361,000	287,393,000	261,111,000	69,022,000	48,349,000	30,052,000
Net Income	450,697,000	495,021,000	305,030,000	201,195,000	119,377,000	57,046,000
Operating Cash Flow	612,506,000	220,709,000	326,314,000	265,582,000	79,404,000	-99,671,000
Capital Expenditure						
EBITDA						
Return on Assets %	.44%	.43%	.52%	5.49%	4.68%	3.55%
Return on Equity %	11.26%	12.15%	8.71%	8.98%	7.73%	5.14%
Debt to Equity	0.32	0.36	0.23	0.48	0.62	0.43

CONTACT INFORMATION:
Phone: 203-422-7700   Fax: 203-422-8159
Toll-Free:
Address: 591 West Putnam Ave., Greenwich, CT 06830 United States

STOCK TICKER/OTHER:
Stock Ticker: STWD
Employees: 450
Parent Company:

Exchange: NYS
Fiscal Year Ends: 12/31

SALARIES/BONUSES:
Top Exec. Salary: $400,000   Bonus: $1,800,000
Second Exec. Salary: $350,000   Bonus: $1,200,000

OTHER THOUGHTS:
Estimated Female Officers or Directors:
Hot Spot for Advancement for Women/Minorities:

Station Casinos LLC

www.stationcasinos.com

NAIC Code: 721120

TYPES OF BUSINESS:

Casino Hotel
Casino Management
Restaurants
Movie Theaters & Entertainment Venues

BRANDS/DIVISIONS/AFFILIATES:

Palace Station Hotel & Casino
Texas Station Gambling Hall & Hotel
Boulder Station Hotel & Casino
Santa Fe Station Hotel & Casino
Barley's Casino & Brewing Company
Sunset Station Hotel & Casino
Fiesta Rancho Casino Hotel
Green Valley Ranch Station Resort

CONTACTS: Note: Officers with more than one job title may be intentionally listed here more than once.

Frank J. Fertitta, III, CEO
Daniel J. Roy, COO
Stephen L. Cavallaro, Pres.
Marc J. Falcone, Exec. VP
Richard J. Haskins, General Counsel
Scott M. Nielson, Chief Dev. Officer
Wes D. Allison, Chief Acct. Officer
Thomas M. Friel, Exec. VP
Frank J. Fertitta, III, Chmn.

GROWTH PLANS/SPECIAL FEATURES:

Station Casinos LLC is a gaming and entertainment company concentrated in the Las Vegas area, mainly targeting locals and repeat customers. Its properties include nine major casinos and hotels and 10 smaller casinos, featuring more than 4,000 hotel rooms, 2,909 slot machines and 131 gaming tables. Other offerings include movie screens, bowling lanes, live entertainment venues, retail outlets, sports betting and convention banquet space. Station Casinos' owned and operated properties in Las Vegas include Palace Station Hotel & Casino; Boulder Station Hotel & Casino; Santa Fe Station Hotel & Casino; Red Rock Casino, Resort & Spa; Wild Wild West Gambling Hall & Hotel; Wildfire Rancho; Texas Station Gambling Hall & Hotel; and Fiesta Rancho Casino Hotel. In Henderson, Nevada, it owns Sunset Station Hotel & Casino; Fiesta Henderson Casino Hotel; Wildfire Boulder; and Green Valley Ranch Station Resort, Spa & Casino. The firm also owns 50% of three properties in Henderson: Barley's Casino & Brewing Company, The Greens Gaming & Dining and Wildfire Casino & Lanes. Additionally, the company manages Graton Resort and Casino in Sonoma County, California and Gun Lake Casino in Allegan County, Michigan. In October 2015, the firm filed a registration statement with the Securities and Exchange Commission relating to the proposed initial public offering of its common stock.

The firm offers employees benefits including onsite childcare, tuition reimbursement, a 401(k) and employee discounts.

FINANCIAL DATA: Note: Data for latest year may not have been available at press time.

In U.S. $	2015	2014	2013	2012	2011	2010
Revenue	1,352,135,000	1,291,616,000	1,261,478,000	1,229,476,000	1,178,148,000	944,955,000
R&D Expense						
Operating Income						
Operating Margin %						
SGA Expense						
Net Income	132,504,000	71,326,000	-113,493,000	13,318,000	-40,816,000	-563,769,000
Operating Cash Flow						
Capital Expenditure						
EBITDA						
Return on Assets %						
Return on Equity %						
Debt to Equity						

CONTACT INFORMATION:

Phone: 702-495-3000   Fax: 702-495-3530
Toll-Free: 800-634-3101
Address: 1505 S. Pavilion Ctr. Dr., Las Vegas, NV 89135 United States

STOCK TICKER/OTHER:

Stock Ticker: Private   Exchange:
Employees: 12,000   Fiscal Year Ends: 12/31
Parent Company:

SALARIES/BONUSES:

Top Exec. Salary: $ Bonus: $
Second Exec. Salary: $ Bonus: $

OTHER THOUGHTS:

Estimated Female Officers or Directors:
Hot Spot for Advancement for Women/Minorities:

Steelman Partners LLP

www.steelmanpartners.com

NAIC Code: 541310

TYPES OF BUSINESS:

Architectural Services

GROWTH PLANS/SPECIAL FEATURES:

Steelman Partners LLP is a Las Vegas, Nevada-based international design firm specializing in the multi-disciplinary facets of entertainment architecture, interior design, graphic design, planning, theater design, 3D design and lighting. The services offered by the firm are architectural, interior design, lighting, branding and 3D visualization. Steelman's clients have included a number of casino companies, including Ameristar, Cantor Gaming, Boyd Gaming (Treasure Chest), Harrah's Entertainment and Caesar's Entertainment; hotel companies, including Four Seasons, Hyatt and Global Resorts; and entertainment companies, including Playboy, MGM Resorts International and Disney. Select completed projects of the firm include the JW Marriott in Las Vegas, the Sands in Macau and Harrah's Atlantic City in New Jersey. Steelman has designed well over 4,000 casino and integrated resort designs around the world. The affiliated companies of the firm are DSAA, SHOP12, Inviro Studios, Marqi Branding Studio and Steelman Architecture Asia.

BRANDS/DIVISIONS/AFFILIATES:

DSAA
SHOP12
Inviro Studios
Marqi Branding Studio
Steelman Architecture Asia

CONTACTS: Note: Officers with more than one job title may be intentionally listed here more than once.

Paul Steelman, CEO
Ethan Nelson, Pres.

FINANCIAL DATA: Note: Data for latest year may not have been available at press time.

In U.S. $	2015	2014	2013	2012	2011	2010
Revenue	42,000,000					
R&D Expense						
Operating Income						
Operating Margin %						
SGA Expense						
Net Income						
Operating Cash Flow						
Capital Expenditure						
EBITDA						
Return on Assets %						
Return on Equity %						
Debt to Equity						

CONTACT INFORMATION:

Phone: 702-873-0221   Fax: 702-367-3565
Toll-Free:
Address: 3330 W. Desert Inn Rd., Las Vegas, NV 89102 United States

STOCK TICKER/OTHER:

Stock Ticker: Private   Exchange:
Employees: 200   Fiscal Year Ends:
Parent Company:

SALARIES/BONUSES:

Top Exec. Salary: $ Bonus: $
Second Exec. Salary: $ Bonus: $

OTHER THOUGHTS:

Estimated Female Officers or Directors:
Hot Spot for Advancement for Women/Minorities:

Sterling Construction Company Inc   www.sterlingconstructionco.com
NAIC Code: 237110

TYPES OF BUSINESS:
Heavy Construction & Reconstruction, Civil
Transportation & Water Infrastructure
Municipal Construction

BRANDS/DIVISIONS/AFFILIATES:
Texas Sterling Construction Co
Road and Highway Builders LLC
Ralph L. Wadsworth Construction Co LLC
Road and Highway Builders of California Inc
J. Banicki Construction Inc
Myers & Sons Construction LP

CONTACTS: *Note: Officers with more than one job title may be intentionally listed here more than once.*
Paul Varello, CEO
Ronald Ballschmiede, CFO
Milton Scott, Director
Con Wadsworth, Executive VP
Jennifer Maxwell, Other Corporate Officer
Roger Barzun, Secretary
Kevan Blair, Senior VP, Divisional

GROWTH PLANS/SPECIAL FEATURES:
Sterling Construction Company, Inc. is a leading heavy civil construction firm that specializes in the building and reconstruction of transportation and water infrastructure projects. The company is active in Texas, Utah, Nevada, Arizona, California and other states where there are construction opportunities. The firm operates through six primary subsidiaries: Texas Sterling Construction Co.; Road and Highway Builders LLC; Road and Highway Builders of California, Inc.; J. Banicki Construction, Inc.; Myers & Sons Construction LP; and Ralph L. Wadsworth Construction Company, LLC. Sterling's activities include concrete and asphalt paving, concrete slip forming, concrete crushing, concrete batch plant operations, installation of large-diameter water and wastewater distribution systems, bridge construction and the building of light rail infrastructure. The company provides general contracting services to both public sector and municipal clients, utilizing its own employees and equipment for activities including excavating, paving, pipe installation and concrete placement. The firm normally purchases its own materials, using subcontractors only for ancillary services.

FINANCIAL DATA: *Note: Data for latest year may not have been available at press time.*

In U.S. $	2015	2014	2013	2012	2011	2010
Revenue	623,595,000	672,230,000	556,236,000	630,507,000	501,156,000	459,893,000
R&D Expense						
Operating Income	-14,387,000	-4,224,000	-69,589,000	14,979,000	-52,234,000	35,910,000
Operating Margin %	-2.30%	-.62%	-12.51%	2.37%	-10.42%	7.80%
SGA Expense	41,880,000	36,897,000	40,951,000	35,496,000	25,005,000	24,895,000
Net Income	-20,402,000	-9,781,000	-73,929,000	-297,000	-35,900,000	19,087,000
Operating Cash Flow	8,969,000	-10,513,000	-21,562,000	24,789,000	20,988,000	47,073,000
Capital Expenditure	8,086,000	13,509,000	14,900,000	37,359,000	23,989,000	13,409,000
EBITDA	2,362,000	14,878,000	-49,538,000	37,074,000	-33,163,000	53,451,000
Return on Assets %	-13.65%	-3.37%	-24.45%	-.09%	-10.70%	5.07%
Return on Equity %	-34.13%	-7.44%	-43.61%	-.14%	-15.48%	7.93%
Debt to Equity	0.16	0.27	0.06	0.11		

CONTACT INFORMATION:
Phone: 281-214-0800   Fax:
Toll-Free:
Address: 1800 Hughes Landing Blvd, Ste 250, The Woodlands, TX 77380 United States

STOCK TICKER/OTHER:
Stock Ticker: STRL   Exchange: NAS
Employees: 1,799   Fiscal Year Ends: 12/31
Parent Company:

SALARIES/BONUSES:
Top Exec. Salary: $304,504   Bonus: $75,000
Second Exec. Salary: Bonus: $
$250,000

OTHER THOUGHTS:
Estimated Female Officers or Directors: 1
Hot Spot for Advancement for Women/Minorities:

Stewart Information Services Corp

www.stewart.com

NAIC Code: 524127

TYPES OF BUSINESS:

Title Insurance
Real Estate Information Services & Software

BRANDS/DIVISIONS/AFFILIATES:

Stewart Title Guaranty Company
Stewart Title Company
Stewart Lenders Services
DataQuick Lending Solutions
Wetzel Trott Inc
LandSafe Title

CONTACTS: *Note: Officers with more than one job title may be intentionally listed here more than once.*

Matthew Morris, CEO
Joseph Berryman, CFO
Thomas Apel, Chairman of the Board
Brian Glaze, Chief Accounting Officer
Murshid Khan, Chief Information Officer
Brad Rable, Chief Information Officer
John Arcidiacono, Chief Marketing Officer
Malcolm Morris, Director
Stewart Morris, Director
Nat Otis, Other Corporate Officer
Susan McLauchlan, Other Executive Officer
Jay Milligan, Other Executive Officer
John Killea, Other Executive Officer
Steven Lessack, President, Divisional
Glenn Clements, President, Divisional
Patrick Beall, President, Divisional
David Fauth, President, Divisional

GROWTH PLANS/SPECIAL FEATURES:

Stewart Information Services Corp. (Stewart) is a holding company that specializes in title insurance and real estate services globally. Stewart provides these services to homebuyers and sellers, residential and commercial real estate professionals, mortgage lenders and servicers, title agencies and real estate attorneys, home builders and the U.S. government. The firm also provides loan origination and servicing support, loan review services, loss mitigation, REO (real estate owned) asset management, home and personal insurance services, performance management services and technology to streamline the real estate process. Stewart closes transactions and issue title policies in the U.S. and globally. The company operates in three main segments: title insurance and related services, which includes searching, examining, closing and insuring the condition of the title to real property; mortgage services, offering mortgage origination support, component servicing, default and REO services (including post-closing outsourcing), portfolio due diligence, mortgage compliance solutions and servicer oversight to residential mortgage lenders; and corporate, consisting of the expenses of the parent holding company, certain other corporate overhead expenses and the costs of its centralized support operations not otherwise allocated to the lines of business. The firm's international division delivers products and services such as title insurance, various types of policies and escrow services involving residential and commercial real estate. With more than 100 subsidiaries, its principal U.S. subsidiaries include Stewart Title Company, Stewart Title Guaranty Company and Stewart Lenders Services. In 2014, Stewart acquired DataQuick Lending Solutions, Wetzel Trott, Inc. and LandSafe Title.

The firm offers its employees medical, dental and vision insurance; life insurance; short- and long-term disability; flexible spending accounts; a 401(k) plan; a stock purchase plan; and employee as well as educational assistance programs.

FINANCIAL DATA: *Note: Data for latest year may not have been available at press time.*

In U.S. $	2015	2014	2013	2012	2011	2010
Revenue	2,033,885,000	1,870,830,000	1,927,980,000	1,910,412,000	1,634,906,000	1,672,390,000
R&D Expense						
Operating Income	9,693,000	51,812,000	101,065,000	89,338,000	18,019,000	2,925,000
Operating Margin %	.47%	2.76%	5.24%	4.67%	1.10%	.17%
SGA Expense	658,266,000	624,326,000	571,026,000	542,461,000	726,033,000	467,491,000
Net Income	-6,204,000	29,753,000	63,026,000	109,182,000	2,348,000	-12,582,000
Operating Cash Flow	80,514,000	63,989,000	87,187,000	120,522,000	23,409,000	41,194,000
Capital Expenditure	15,444,000	18,122,000	17,282,000	16,752,000	17,704,000	16,339,000
EBITDA						
Return on Assets %	-.45%	2.18%	4.81%	8.92%	.20%	-1.00%
Return on Equity %	-.93%	4.41%	10.29%	21.42%	.53%	-2.84%
Debt to Equity	0.16	0.09	0.04	0.12	0.14	0.16

CONTACT INFORMATION:

Phone: 713 625-8100   Fax: 713 629-2244
Toll-Free: 800-729-1900
Address: 1980 Post Oak Blvd., Houston, TX 77056 United States

SALARIES/BONUSES:

Top Exec. Salary: $500,000   Bonus: $
Second Exec. Salary: $420,000   Bonus: $

STOCK TICKER/OTHER:

Stock Ticker: STC   Exchange: NYS
Employees: 7,400   Fiscal Year Ends: 12/31
Parent Company:

OTHER THOUGHTS:

Estimated Female Officers or Directors: 3
Hot Spot for Advancement for Women/Minorities: Y

Stock Building Supply Inc

www.stockbuildingsupply.com

NAIC Code: 444190

TYPES OF BUSINESS:
Building Materials/Hardware Stores
Design & Installation Services
Lending & Insurance

BRANDS/DIVISIONS/AFFILIATES:
Gores Group, LLC (The)
Building Materials Holding Corporation

CONTACTS: Note: Officers with more than one job title may be intentionally listed here more than once.
Peter Alexander, CEO
James Major, CFO
William Gay, Chief Accounting Officer
Paul Street, Chief Administrative Officer
David Bullock, Director
Lisa Hamblet, Executive VP, Divisional
Mark Necaise, Other Corporate Officer
Walter Randolph, President, Divisional
Duff Wakefield, President, Divisional
Steven Wilson, President, Divisional
C. Ball, Senior VP
Michael Farmer, Vice President, Divisional

GROWTH PLANS/SPECIAL FEATURES:
Stock Building Supply, Inc. (SBS) is a supplier of building materials to professional homebuilders and contractors in the U.S. SBS is a wholly-owned subsidiary of Gores Group, LLC. The company serves the single- and multi-family residential, repair, remodeling and light commercial construction industries through over 69 stores in 13 U.S. states. The firm divides its products into five categories: structural components; millwork and other interior products; lumber and lumber sheet goods; windows and other exterior products; and other building products and services. In 2013, the company's combined sales of structural components, millwork and other interior products, and windows and other exterior products represented 53% of net sales. Structural components are factory-built substitutes for job-site framing and include floor trusses, roof trusses, wall panels and engineered wood that are designed and cut for each home. Millwork and other interior products includes interior doors, interior trim, custom millwork, moldings, stairs, stair parts, flooring, cabinets, gypsum and other products. Lumber and lumber sheet goods include dimensional lumber, plywood and oriented strand board products. Windows and other exterior products includes exterior door units, as well as roofing and siding products. Other building products and services consist of hardware, boards, insulation and other products. The firm also provides professional estimating, product advisory and product display services. In June 2015, the firm reached a definitive agreement to merge with Building Materials Holding Corporation.

SBS offers its employees life, disability, medical, dental and vision insurance; a health reimbursement account; a health care flexible spending account; a 401(k) plan; employee discounts from 10%-15%; health and wellness programs; a 24/7 nurse line; and tobacco cessation programs.

FINANCIAL DATA: Note: Data for latest year may not have been available at press time.

In U.S. $	2015	2014	2013	2012	2011	2010
Revenue	1,576,746,000	1,295,716,000	1,197,037,000	942,398,000	759,982,000	751,706,000
R&D Expense						
Operating Income	12,248,000	18,324,000	761,000	-18,907,000	-59,301,000	-122,834,000
Operating Margin %	.77%	1.41%	.06%	-2.00%	-7.80%	-16.34%
SGA Expense	306,843,000	279,717,000	254,935,000	221,192,000	213,036,000	246,338,000
Net Income	-4,831,000	10,419,000	-4,635,000	-14,533,000	-42,133,000	-69,994,000
Operating Cash Flow	743,000	16,941,000	-40,264,000	-12,243,000	-7,001,000	-57,999,000
Capital Expenditure	31,319,000	43,306,000	7,448,000	2,741,000	1,339,000	2,506,000
EBITDA	37,621,000	32,454,000	13,694,000	-6,860,000	-44,707,000	-71,587,000
Return on Assets %	- .55%	3.02%	-2.14%	-8.88%	-18.19%	
Return on Equity %	-1.25%	7.75%	-7.99%	-56.11%	-90.07%	
Debt to Equity	0.67	0.68	0.50	0.16	0.01	

CONTACT INFORMATION:
Phone: 919-431-1000   Fax:
Toll-Free:
Address: 8020 Arco Corporate Dr., Raleigh, NC 27617 United States

STOCK TICKER/OTHER:
Stock Ticker: STCK   Exchange: NAS
Employees: 9,600   Fiscal Year Ends: 07/31
Parent Company:

SALARIES/BONUSES:
Top Exec. Salary: $600,000   Bonus: $
Second Exec. Salary: Bonus: $200,000
$310,000

OTHER THOUGHTS:
Estimated Female Officers or Directors:
Hot Spot for Advancement for Women/Minorities:

Strategic Hotels & Resorts Inc

www.strategichotels.com

NAIC Code: 531120

TYPES OF BUSINESS:
Real Estate Investment Trust
Hotels

BRANDS/DIVISIONS/AFFILIATES:
Strategic Hotel Capital LLC
Blackstone Real Estate Partners VIII LP
Blackstone Group L P

CONTACTS: Note: Officers with more than one job title may be intentionally listed here more than once.
Stephen A. Schwarzman, CEO-Blackstone Real Estate
Diane Morefield, CFO
Jon Stanner, CFO
George Stowers, Sr. VP-Design & Construction
Bryce White, VP-IT
Paula Maggio, Executive VP

GROWTH PLANS/SPECIAL FEATURES:
Strategic Hotels & Resorts, Inc. (SHR) is a portfolio company of The Blackstone Group, which owns and manages upper upscale and luxury hotels in North America and Europe through its investment in the operating partnership Strategic Hotel Capital LLC and its subsidiaries. The company owns or leases 16 hotels; owns 53.5% and 51.0% interests in affiliates that each own one hotel where it asset manages such hotels; owns land held for development including 50.7 acres of oceanfront land in Nayarit, Mexico, 13.8 acres of land in Scottsdale, Arizona and a 20,000 square-foot oceanfront land parcel in Santa Monica, California. In total, the 16 hotels and resorts consists of 7,532 rooms, 807,000 square feet of multi-purpose meeting and banqueting space, world class restaurants, wine and cocktail bars, high-end spas and retail offerings. The firm does not operate any of its hotels directly; instead, it employs internationally known hotel management companies to operate them under management contracts or operating leases. The company's existing hotels are operated under the Fairmont, Four Seasons, Hyatt, InterContinental, Loews, JW Marriot, Marriott, Montage, Renaissance, Ritz-Carlton and Westin brands. In December 2015, the firm was acquired by Blackstone Real Estate Partners VIII L.P., part of The Blackstone Group L.P.

FINANCIAL DATA: Note: Data for latest year may not have been available at press time.

In U.S. $	2015	2014	2013	2012	2011	2010
Revenue	1,381,600,000	1,089,081,984	900,012,992	808,316,992	763,838,016	686,292,992
R&D Expense						
Operating Income						
Operating Margin %						
SGA Expense						
Net Income		344,483,008	10,975,000	-55,306,000	-5,206,000	-231,051,008
Operating Cash Flow						
Capital Expenditure						
EBITDA						
Return on Assets %						
Return on Equity %						
Debt to Equity						

CONTACT INFORMATION:
Phone: 312 658-5000   Fax: 312 658-5799
Toll-Free:
Address: 200 W. Madison St., Ste. 1700, Chicago, IL 60606 United States

STOCK TICKER/OTHER:
Stock Ticker: Private   Exchange:
Employees: 35   Fiscal Year Ends: 12/31
Parent Company: Blackstone Real Estate Partners VIII LP

SALARIES/BONUSES:
Top Exec. Salary: $ Bonus: $
Second Exec. Salary: $ Bonus: $

OTHER THOUGHTS:
Estimated Female Officers or Directors: 3
Hot Spot for Advancement for Women/Minorities: Y

Structure Tone Inc

www.structuretone.com

NAIC Code: 236220

TYPES OF BUSINESS:

Construction & Engineering Services

BRANDS/DIVISIONS/AFFILIATES:

GROWTH PLANS/SPECIAL FEATURES:

Structure Tone, Inc. provides comprehensive construction services for commercial, education, mission critical, health care, hospitality, retail, life sciences, and government clients. For its clients, the firm can act as construction management, general contracting, the design/build team, a special projects team, technology management and as a global services project delivery program. The services offered by Structure Tone include building evaluation, preconstruction, building information modeling (BIM), value management, procurement, construction, commissioning, sustainability/LEED and quality assurance/quality control. The Structure Tone portfolio encompasses projects in a variety of areas of market sector expertise. These sectors include broadcast/media, commercial, cultural, education, government, health care, hospitality, law, life sciences, mission critical, nonprofit, residential, retail and sustainable.

CONTACTS: Note: Officers with more than one job title may be intentionally listed here more than once.

Robert W. Mullen, CEO
Anthony M. Carvette, Pres.
Brett Phillips, CFO
Rebecca Leonardis, VP-Mktg.
Jim Donaghy, Chmn.

FINANCIAL DATA: Note: Data for latest year may not have been available at press time.

In U.S. $	2015	2014	2013	2012	2011	2010
Revenue	3,500,000,000	3,200,000,000	3,152,100,000			
R&D Expense						
Operating Income						
Operating Margin %						
SGA Expense						
Net Income						
Operating Cash Flow						
Capital Expenditure						
EBITDA						
Return on Assets %						
Return on Equity %						
Debt to Equity						

CONTACT INFORMATION:

Phone: 212-481-6100   Fax:
Toll-Free:
Address: 770 Broadway, New York, NY 10003 United States

STOCK TICKER/OTHER:

Stock Ticker: Private   Exchange:
Employees: 1,625   Fiscal Year Ends:
Parent Company:

SALARIES/BONUSES:

Top Exec. Salary: $ Bonus: $
Second Exec. Salary: $ Bonus: $

OTHER THOUGHTS:

Estimated Female Officers or Directors:
Hot Spot for Advancement for Women/Minorities:

STUDIOS Architecture

NAIC Code: 541310

TYPES OF BUSINESS:
Architectural Services

BRANDS/DIVISIONS/AFFILIATES:

GROWTH PLANS/SPECIAL FEATURES:
STUDIOS Architecture is an international architecture design firm that provides architectural, master planning, interiors, workplace strategy and renovation services for commercial, mixed-use, civic and institutional projects around the world. The project types completed include cultural, educational, governmental, commercial office, fitness + spa, hospitality, residential and retail facilities and buildings. Clients of STUDIOS have included the likes of the California Department of Public Health, the International Monetary Fund, Lowe Enterprises, the American Institution of Architects, the Shanghai Theatre Company, the University of Texas, UC Berkley, the Waldorf Astoria, Coca Cola, Nike; American Express, Barclays Capital, Bloomberg, Pfizer and The Walt Disney Company. The firm has offices in Los Angeles, New York, Paris, San Francisco, Washington, D.C. and Mumbai.

CONTACTS: Note: Officers with more than one job title may be intentionally listed here more than once.
Todd DeGarmo, CEO

FINANCIAL DATA: Note: Data for latest year may not have been available at press time.

In U.S. $	2015	2014	2013	2012	2011	2010
Revenue						
R&D Expense						
Operating Income						
Operating Margin %						
SGA Expense						
Net Income						
Operating Cash Flow						
Capital Expenditure						
EBITDA						
Return on Assets %						
Return on Equity %						
Debt to Equity						

CONTACT INFORMATION:
Phone: 202-736-5900 Fax: 202-736-5959
Toll-Free:
Address: 1625 M St., N.W., Washington, DC 20036 United States

STOCK TICKER/OTHER:
Stock Ticker: Private Exchange:
Employees: 334 Fiscal Year Ends:
Parent Company:

SALARIES/BONUSES:
Top Exec. Salary: $ Bonus: $
Second Exec. Salary: $ Bonus: $

OTHER THOUGHTS:
Estimated Female Officers or Directors:
Hot Spot for Advancement for Women/Minorities:

STV Group Inc

NAIC Code: 541330

www.stvinc.com

TYPES OF BUSINESS:

Architectural & Engineering Services
Construction Management
Infrastructure Design
Defense Systems Engineering
Industrial Process Engineering

BRANDS/DIVISIONS/AFFILIATES:

STV Security Solutions Group Inc

CONTACTS: Note: Officers with more than one job title may be intentionally listed here more than once.

Milo E. Riverso, CEO
Milo E. Riverso, Pres.
Thomas Butcher, CFO
Debra B. Trace, Dir.-Corp. Comm.
Michael D. Garz, Sr. VP-Buildings & Facilities Div.
William F. Matts, Exec. VP
Steve Pressler, Exec. VP
Gerald Donnelly, Exec. VP-STV Energy Svcs. Div.
Dominick M. Servedio, Chmn.

GROWTH PLANS/SPECIAL FEATURES:

STV Group, Inc. is an architectural, engineering, planning, environmental and construction management firm. The firm specializes in constructing airports, highways, bridges, ports, railroad systems and schools, almost all of which are in the U.S. It operates through four divisions: construction management, energy services, buildings/facilities and transportation/infrastructure. Construction management undertakes design and building contracts in nearly every field of industry; the firm's personnel oversee construction programs through administrative, inspection and surveillance. The firm's energy services division offers services related to engineering design, permitting and the environment, construction support and industry specialty support. The buildings/facilities division works directly with architects, to address the safety, practicality, cost and efficiency of its buildings. Transportation/infrastructure focuses on the management, planning and design of transportation systems and facilities. Specialties of the firm include aviation/transportation architecture, defense systems and sustainable design. Subsidiary STV Security Solutions Group, Inc. offers security as well as risk, crisis and emergency management to its clients. Some of the STV's representative projects include the MetroLink in St. Louis, Missouri; the Villanova University Center for Engineering Education and Research; Sprint PCS Environmental Site Assessments in seven states; the Metra Inner Circumferential Rail Study in Chicago, Illinois; Shea Stadium in Queens, New York; and engineering and technical services for the U.S. Naval Air Warfare Center in Patuxent River, Maryland.

The company offers employees health and life insurance, short- and long-term disability benefits, an employee assistance program, paid time off and a credit union membership.

FINANCIAL DATA: Note: Data for latest year may not have been available at press time.

In U.S. $	2015	2014	2013	2012	2011	2010
Revenue	450,000,000	430,000,000	410,000,000	373,538,000	320,000,000	250,000,000
R&D Expense						
Operating Income						
Operating Margin %						
SGA Expense						
Net Income						
Operating Cash Flow						
Capital Expenditure						
EBITDA						
Return on Assets %						
Return on Equity %						
Debt to Equity						

CONTACT INFORMATION:

Phone: 610-385-8200   Fax: 610-385-8500
Toll-Free:
Address: 205 W. Welsh Dr., Douglassville, PA 19518 United States

STOCK TICKER/OTHER:

Stock Ticker: Private   Exchange:
Employees: 1,770   Fiscal Year Ends: 09/30
Parent Company:

SALARIES/BONUSES:

Top Exec. Salary: $ Bonus: $
Second Exec. Salary: $ Bonus: $

OTHER THOUGHTS:

Estimated Female Officers or Directors: 2
Hot Spot for Advancement for Women/Minorities: Y

Sun Communities Inc

www.suncommunities.com

NAIC Code: 531110

TYPES OF BUSINESS:
Real Estate Investment Trust
Manufactured Housing Communities
Recreational Vehicle Communities
Manufactured Home Sales
Property Management

BRANDS/DIVISIONS/AFFILIATES:
Sun Home Services Inc

CONTACTS: Note: Officers with more than one job title may be intentionally listed here more than once.
Gary Shiffman, CEO
Karen Dearing, CFO
John McLaren, COO
Jonathan Colman, Executive VP

GROWTH PLANS/SPECIAL FEATURES:
Sun Communities, Inc. is a self-administered and self-managed real estate investment trust (REIT) that owns, operates and develops manufactured housing communities concentrated in the midwestern, southern and southeastern U.S. Sun, together with its affiliates and predecessors, has been in the business of acquiring, operating and expanding manufactured housing communities since 1975. The company owns and operates a portfolio of 217 properties in 29 states, including 183 manufactured housing communities, 25 recreational vehicle (RV) communities and nine multi-use properties containing both manufactured housing and RV sites. The properties contain a total of 79,554 developed sites, including 61,231 developed manufactured home sites, 9,297 permanent RV sites, 9,026 transient RV sites and an additional 7,000 manufactured home sites suitable for development. The firm leases individual sites and utility access to customers for the placement of manufactured homes and RVs. Most of its communities include amenities such as a clubhouse, a swimming pool and laundry facilities. Some offer additional amenities, including sauna/whirlpool spas, tennis and basketball courts and exercise rooms. Subsidiary Sun Home Services, Inc. markets, sells and leases new and pre-owned manufactured homes for placement in its properties. The firm expands primarily through the acquisition of manufactured housing and RV communities. Headquartered in Michigan, the company also maintains management offices in Texas, Ohio, Michigan, North Carolina, Colorado, Indiana and Florida. During 2015, the firm acquired 26 properties of an American land lease manufactured housing portfolio from Green Courte Partners; six properties in the Orlando, Florida area; three recreational vehicle resorts (one in Florida and two in Maryland); and sold three manufactured home communities (two in Ohio and one in Michigan).

Sun employees receive medical and prescription coverage; dental and vision plans; life, AD&D and identity theft insurance; a 401(k); an employee assistance program; tuition reimbursement; and health and dependent care reimbursement accounts.

FINANCIAL DATA: Note: Data for latest year may not have been available at press time.

In U.S. $	2015	2014	2013	2012	2011	2010
Revenue	674,731,000	471,675,000	415,222,000	339,616,000	289,185,000	263,140,000
R&D Expense						
Operating Income	38,755,000	10,109,000	18,132,000	4,344,000	-2,485,000	-1,855,000
Operating Margin %	5.74%	2.14%	4.36%	1.27%	-.85%	-.70%
SGA Expense	90,253,000	60,881,000	35,854,000	29,017,000	27,860,000	24,810,000
Net Income	155,446,000	31,444,000	16,666,000	4,958,000	-1,086,000	-2,883,000
Operating Cash Flow	182,263,000	133,320,000	114,683,000	87,251,000	63,521,000	59,121,000
Capital Expenditure	517,701,000	604,457,000	301,589,000	374,392,000	164,891,000	50,863,000
EBITDA	452,646,000	242,922,000	204,787,000	165,198,000	139,647,000	131,928,000
Return on Assets %	3.85%	.90%	.56%	.31%	-.08%	-.24%
Return on Equity %	11.69%	3.51%	3.70%	15.05%		
Debt to Equity	1.56	2.04	3.44	7.39		

CONTACT INFORMATION:
Phone: 248 208-2500   Fax: 248 932-4070
Toll-Free:
Address: 27777 Franklin Rd., Ste. 200, Southfield, MI 48034 United States

STOCK TICKER/OTHER:
Stock Ticker: SUI
Employees: 1,525
Parent Company:

Exchange: NYS
Fiscal Year Ends: 12/31

SALARIES/BONUSES:
Top Exec. Salary: $691,418   Bonus: $
Second Exec. Salary: $488,892   Bonus: $

OTHER THOUGHTS:
Estimated Female Officers or Directors: 2
Hot Spot for Advancement for Women/Minorities:

Sun Hung Kai Properties Ltd

NAIC Code: 531100

www.shkp.com.hk

TYPES OF BUSINESS:

Real Estate Operations & Development
Land Development
Engineering Services
Hotels & Shopping Malls
Insurance & Financial Services
Logistics & Transportation
Ocean Port Terminals
Infrastructure

BRANDS/DIVISIONS/AFFILIATES:

Royal Garden
Royal Park Hotel
Royal Plaza Hotel
Royal View Hotel
Crowne Plaza Hong Koong Kowloon East
Holiday Inn Express Hong Kong Kowloon
Sun Hung Kai Properties Insurance Limited

CONTACTS: Note: Officers with more than one job title may be intentionally listed here more than once.

AdamThomas, Co-Managing Dir.
Chan Kai-ming, Exec. Dir.-Architecture & Eng.
Ping-luen Kwok, Co-Chmn.
Shau-kee Lee, Vice Chmn.
Chik-wing Wong, Deputy Managing Dir.
Ting Lui, Deputy Managing Dir.
Raymond Kwok, Chmn

GROWTH PLANS/SPECIAL FEATURES:

Sun Hung Kai Properties Ltd. (SHKP) is one of the largest property developers in Hong Kong. The company's core areas of business include construction, land acquisition and property management; non-core businesses include hotels, insurance and financial services. SHKP's land bank is one of the largest private landholders in Hong Kong, with 28.7 million square feet (sf) of completed investment property, 22.1 million sf of space currently under development and more than 30 million sf of farmland in various states of land use conversion. The firm's mainland China land bank has nearly 80 million sf in terms of attributable gross floor area, including 68 million sf of properties for development and 11.6 million sf of completed properties for rent. The firm also owns the following hotels in Hong Kong: Royal Garden, Royal View Hotel, Royal Park Hotel, Royal Plaza Hotel, Four Seasons Hotel at IFC, The Ritz-Carlton, Crowne Plaza Hong Kong Kowloon East and Holiday Inn Express Hong Kong Kowloon East. Subsidiary Sun Hung Kai Properties Insurance Limited offers a comprehensive range of insurance services such as contractor's risks, employee compensation, third-party liability, property risks, commercial, medical, householders comprehensive, fire, travel, personal accident and motor vehicles. SHKP is active in a number of additional business areas, including waste management, port operations, department stores, broadband telecommunications infrastructure development and IT venture capital investments. Moreover, the firm is involved in air transportation and logistics, with subsidiaries focused on international air cargo consolidation and freight forwarding; airport services such as storage, X-ray scanning, loading and unloading, collection and delivery, palletization and containerization; and aviation services such as aircraft parking, marshalling, towing, fueling, potable water supply and other turn-around operations.

FINANCIAL DATA: Note: Data for latest year may not have been available at press time.

In U.S. $	2015	2014	2013	2012	2011	2010
Revenue	8,610,296,000	9,682,601,000	6,935,502,000	8,818,775,000		
R&D Expense						
Operating Income	2,936,755,000	3,220,915,000	4,962,108,000	5,733,493,000	5,986,968,000	3,907,981,000
Operating Margin %	34.10%	33.26%	71.54%	65.01%		
SGA Expense	723,294,200	791,369,000	718,265,900	624,792,100	580,440,400	372,606,100
Net Income	4,007,385,000	4,321,715,000	5,199,596,000	5,554,280,000	6,201,120,000	3,615,569,000
Operating Cash Flow	1,707,413,000	-412,187,500	2,343,680,000	1,050,646,000	-560,327,400	1,025,505,000
Capital Expenditure	243,289,900	436,941,900	789,692,900	808,000,800	1,254,484,000	371,187,900
EBITDA	5,348,638,000	5,765,209,000	6,050,143,000	6,741,076,000	7,558,489,000	4,703,991,000
Return on Assets %	5.35%	6.26%	8.22%	9.78%	12.57%	8.47%
Return on Equity %	7.17%	8.37%	11.01%	13.18%	17.42%	12.00%
Debt to Equity	0.16	0.17	0.14	0.17	0.16	0.13

CONTACT INFORMATION:

Phone: 85 228278111   Fax: 85 228272862
Toll-Free:
Address: Sun Hung Kai Ctr., 30 Harbour Rd., 45/Fl, Hong Kong,   Hong Kong

STOCK TICKER/OTHER:

Stock Ticker: SUHJF
Employees: 37,000
Parent Company:

Exchange: PINX
Fiscal Year Ends: 06/30

SALARIES/BONUSES:

Top Exec. Salary: $ Bonus: $
Second Exec. Salary: $ Bonus: $

OTHER THOUGHTS:

Estimated Female Officers or Directors:
Hot Spot for Advancement for Women/Minorities:

Sunburst Hospitality Corporation

www.snbhotels.com

NAIC Code: 721110

TYPES OF BUSINESS:

Hotels
Hotel Management

GROWTH PLANS/SPECIAL FEATURES:

Sunburst Hospitality Corporation owns, acquires and manages hotels, a championship golf course and townhomes. The company maintains operations in seven states: Arkansas, Maryland, California, New York, Florida, Ohio and Virginia. The firm's hotels are extended-stay, full-service and limited service. On average, Sunburst Hospitality produces annual revenues of $100 million. Sunburst Hospitality owns and operates hotels under several brand names including Comfort Inns & Suites, Holiday Inn Express, Crowne Plaza, Quality Inn, Clarion, Sleep Inn, Super 8, Clarion, Arlington Court and Best Western. Sunburst constantly seeks to expand its hotel business through acquisitions. The firm seeks to acquire existing 70-300 room extended-stay, limited-service and full-service hotels, with or without food and beverage service. Additionally, the company develops residential real estate and operates The Vista on Courthouse, a 252 unit residential complex in Arlington, Virginia.

BRANDS/DIVISIONS/AFFILIATES:

Comfort Inns & Suites
Holiday Inn Express
Crowne Plaza
Quality Inn
The Vista on Courthouse
Sleep Inn
Clarion
Best Western

CONTACTS: Note: Officers with more than one job title may be intentionally listed here more than once.

Pamela Williams, CEO
Ned Heiss, VP-Oper.
Pam Williams, Pres.
Joe Smith, CFO
Tonio Noonan, VP-Mktg. & Sales
Mark Elbaum, VP-Information Systems
Leon Vainikos, Regional General Counsel
Ned Heiss, VP-Oper.
Chris Milke, VP-Acquisitions & Dev.
Joe Smith, Treas.
Chris Milke, VP-Acquisitions & Dev.

FINANCIAL DATA: Note: Data for latest year may not have been available at press time.

In U.S. $	2015	2014	2013	2012	2011	2010
Revenue						
R&D Expense						
Operating Income						
Operating Margin %						
SGA Expense						
Net Income						
Operating Cash Flow						
Capital Expenditure						
EBITDA						
Return on Assets %						
Return on Equity %						
Debt to Equity						

CONTACT INFORMATION:

Phone: 301-592-3800   Fax: 301-592-3830
Toll-Free:
Address: 10770 Columbia Pike, Ste. 300, Silver Spring, MD 20901 United States

STOCK TICKER/OTHER:

Stock Ticker: Private
Employees: 3,500
Parent Company:

Exchange:
Fiscal Year Ends: 12/31

SALARIES/BONUSES:

Top Exec. Salary: $ Bonus: $
Second Exec. Salary: $ Bonus: $

OTHER THOUGHTS:

Estimated Female Officers or Directors: 2
Hot Spot for Advancement for Women/Minorities:

Sales, profits and employees may be estimates. Financial information, benefits and other data can change quickly and may vary from those stated here.

Sunrise Senior Living

NAIC Code: 623310

www.sunriseseniorliving.com

TYPES OF BUSINESS:

Assisted Living Facilities
Assisted Living Centers
Independent Living Centers
Nursing Homes

BRANDS/DIVISIONS/AFFILIATES:

Welltower Inc
Health Care REIT Inc

CONTACTS: Note: Officers with more than one job title may be intentionally listed here more than once.

Chris Winkle, CEO
Edward Burnett, CFO
Felipe Mestre, Chief Administrative Officer
Vanessa Forsythe, General Counsel
Jeff Fischer, Sr. VP-Oper.
Farinaz Tehrani, General Counsel
Paul J. Klaassen, Chmn.

GROWTH PLANS/SPECIAL FEATURES:

Sunrise Senior Living is an international provider of senior living services. The firm operates 300 communities, including 258 communities in the U.S., 15 communities in Canada and 27 communities in the U.K., with a total unit capacity of approximately 30,000. Sunrise offers services tailored to the unique needs of each of its residents, typically in apartment-like assisted living environments. Upon move-in, the company assists the resident in developing an individualized service plan, including selection of resident accommodations and the appropriate level of care. Services provided range from basic care, consisting of assistance with activities of daily living, to reminiscence care, which consists of programs and services to help cognitively impaired residents such as those with Alzheimer's. The firm targets sites for development located in major metropolitan areas and their surrounding suburban communities, considering factors such as population, age demographics and estimated level of market demand. Sunrise is a subsidiary of Welltower, Inc., (formerly Health Care REIT, Inc.) a real estate investment trust that primarily invests in assisted living facilities and other forms of housing and medical facilities for senior citizens. In September 2015, parent Welltower's name change reflects its commitment to environments that promote wellness.

The firm offers employees medical, prescription drug, dental and vision plans; health care and dependent care flexible spending accounts; wellness programs; tuition assistance; and short- and long-term disability coverage.

FINANCIAL DATA: Note: Data for latest year may not have been available at press time.

In U.S. $	2015	2014	2013	2012	2011	2010
Revenue	1,600,000,000	1,456,200,000	1,342,000,000	1,315,882,667	1,312,212,992	1,406,701,056
R&D Expense						
Operating Income						
Operating Margin %						
SGA Expense						
Net Income						
Operating Cash Flow						
Capital Expenditure						
EBITDA						
Return on Assets %						
Return on Equity %						
Debt to Equity						

CONTACT INFORMATION:

Phone: 703 273-7500   Fax: 703 744-1628
Toll-Free:
Address: 7902 Westpark Dr., Ste. T-900, McLean, VA 22102 United States

STOCK TICKER/OTHER:

Stock Ticker: Subsidiary   Exchange:
Employees: 10,000   Fiscal Year Ends: 12/31
Parent Company: Welltower Inc

SALARIES/BONUSES:

Top Exec. Salary: $ Bonus: $
Second Exec. Salary: $ Bonus: $

OTHER THOUGHTS:

Estimated Female Officers or Directors: 1
Hot Spot for Advancement for Women/Minorities: Y

Sunstone Hotel Investors Inc

www.sunstonehotels.com

NAIC Code: 0

TYPES OF BUSINESS:

Real Estate Investment Trust
Hotel Ownership
Online Purchasing Systems

BRANDS/DIVISIONS/AFFILIATES:

Sunstone Hotel TRS Lessee Inc
BuyEfficient LLC
Wailea Beach Marriott Resort & Spa
Sunstone Hotel Partnership LLC

CONTACTS: Note: Officers with more than one job title may be intentionally listed here more than once.

Bryan Giglia, CFO
Marc Hoffman, COO
Douglas Pasquale, Director
Robert Springer, Executive VP
John Arabia, President

GROWTH PLANS/SPECIAL FEATURES:

Sunstone Hotel Investors, Inc. is a real estate investment trust (REIT) that acquires, owns, manages and renovates luxury, upper upscale and upscale full service hotels. Sunstone maintains interests in 30 hotels, with a total of 14,303 rooms, located in 13 states and Washington, D.C. The company's hotels are marketed under the brand names Hilton, Fairmont, Sheraton, Hyatt and Marriott. Third-party managers operate the company's hotels through agreements with wholly-owned subsidiary Sunstone Hotel TRS Lessee, Inc. During 2014, the firm's 30 hotels had an average of 477 rooms and generated a comparable RevPAR (revenue per available room) of $160.11. Sunstone specializes in acquiring under-performing properties and developing them into profitable upscale and luxury properties. The firm prefers to acquire major brand properties situated in one of the top 25 U.S. markets, with upscale potential and a location with high barriers to entry. In addition, the company owns BuyEfficient LLC, an electronic purchasing management platform that helps to simplify procurement and accounting activities of operating supplies, furniture, equipment, fixtures and food; and Sunstone Hotel Partnership LLC, the entity that directly or indirectly owns Sunstone's hotel properties. In 2014, Sunstone acquired the 544-room Wailea Beach Marriott Resort & Spa in Maui, Hawaii.

FINANCIAL DATA: Note: Data for latest year may not have been available at press time.

In U.S. $	2015	2014	2013	2012	2011	2010
Revenue	1,249,180,000	1,141,998,000	923,824,000	829,084,000	834,729,000	643,090,000
R&D Expense						
Operating Income	180,436,000	156,743,000	99,198,000	79,010,000	58,718,000	32,001,000
Operating Margin %	14.44%	13.72%	10.73%	9.52%	7.03%	4.97%
SGA Expense	312,202,000	282,880,000	237,369,000	249,137,000	260,235,000	213,841,000
Net Income	347,355,000	81,231,000	65,956,000	47,765,000	80,957,000	38,542,000
Operating Cash Flow	300,061,000	278,595,000	171,119,000	171,496,000	156,390,000	45,402,000
Capital Expenditure	164,232,000	402,533,000	568,238,000	229,324,000	218,688,000	199,394,000
EBITDA	571,619,000	315,708,000	241,450,000	227,520,000	192,621,000	133,214,000
Return on Assets %	8.68%	1.93%	1.39%	.57%	1.91%	.71%
Return on Equity %	15.77%	3.70%	3.14%	1.65%	5.58%	2.17%
Debt to Equity	0.46	0.62	0.78	1.11	1.55	1.22

CONTACT INFORMATION:

Phone: 949 330-4000   Fax: 949 369-3134
Toll-Free:
Address: 120 Vantis, Ste. 350, Aliso Viejo, CA 92656 United States

STOCK TICKER/OTHER:

Stock Ticker: SHO
Employees: 50
Parent Company:

Exchange: NYS
Fiscal Year Ends: 12/31

SALARIES/BONUSES:

Top Exec. Salary: $691,486   Bonus: $
Second Exec. Salary: Bonus: $
$507,018

OTHER THOUGHTS:

Estimated Female Officers or Directors: 3
Hot Spot for Advancement for Women/Minorities: Y

Super 8 Worldwide Inc

NAIC Code: 721110

www.super8.com

TYPES OF BUSINESS:

Hotel Franchising
Economy Motels
Online Reservations & Services

BRANDS/DIVISIONS/AFFILIATES:

Wyndham Worldwide
Wyndham Rewards
Super 8 Motels Inc

CONTACTS: Note: Officers with more than one job title may be intentionally listed here more than once.

Stephen P. Holmes, CEO-Wyndham Worldwide
John Valletta, Pres.
Jim Darby, VP-Oper.
Stephen P. Holmes, Chmn.-Wyndham Worldwide

GROWTH PLANS/SPECIAL FEATURES:

Super 8 Worldwide, Inc., formerly Super 8 Motels, Inc., is a subsidiary of Wyndham Worldwide and is one of the world's largest franchised economy lodging chains, with a portfolio of nearly 1,800 motels. The firm offers lodging for the budget traveler, with many rooms below $50 per night. Super 8 motels are found in every domestic state besides Hawaii and in every Canadian province. Super 8 also operates 400 properties in China, two in Brazil and one in Saudi Arabia. The company offers numerous promotions, such as guaranteed best rates, group and corporate rates and AAA and AARP discounts. Super 8's web site offers customers a wide array of integrated travel resources, including flight tracking, driving directions, street maps, airport maps, destination guides and weather information, in addition to online reservations, location information and special promotional programs. A search feature can automatically find the Super 8 motel and room with the best available rate for a given date in the travel area. The company offers several in-room products and services, including in-room coffee makers, bath amenities and expanded breakfast choices. In addition, the firm is a member of Wyndham Rewards, one of the largest hotel reward programs in the world based on the 7,400 participating hotels. The program allows customers to earn points toward hotel stays as well as airline miles, gifts, meals and other incentives.

FINANCIAL DATA: Note: Data for latest year may not have been available at press time.

In U.S. $	2015	2014	2013	2012	2011	2010
Revenue						
R&D Expense						
Operating Income						
Operating Margin %						
SGA Expense						
Net Income						
Operating Cash Flow						
Capital Expenditure						
EBITDA						
Return on Assets %						
Return on Equity %						
Debt to Equity						

CONTACT INFORMATION:

Phone: 973-428-9700   Fax: 973-496-7307
Toll-Free: 800-800-8000
Address: 1 Sylvan Way, Parsippany, NJ 07054 United States

STOCK TICKER/OTHER:

Stock Ticker: Subsidiary   Exchange:
Employees: 1,190   Fiscal Year Ends: 12/31
Parent Company: WYNDHAM WORLDWIDE

SALARIES/BONUSES:

Top Exec. Salary: $ Bonus: $
Second Exec. Salary: $ Bonus: $

OTHER THOUGHTS:

Estimated Female Officers or Directors:
Hot Spot for Advancement for Women/Minorities:

Swire Pacific Ltd

www.swirepacific.com

NAIC Code: 483111

TYPES OF BUSINESS:

Deep Sea Shipping
Airlines and Air Freight
International & Regional Airlines
Apparel Retail
Real Estate, Hotels, Commercial Properties
Aircraft Maintenance
Airline Catering Service
Beverage Manufacturing & Distribution

BRANDS/DIVISIONS/AFFILIATES:

Swire Group
Swire Properties
Swire Properties Inc
Swire Hotels
Cathay Pacific Airways Limited
Hong Kong Aircraft Engineering Company Limited
HUD Group (The)
Swire Beverages Ltd

CONTACTS: Note: Officers with more than one job title may be intentionally listed here more than once.

Martin Cubbon, Dir.-Finance
Martin Cubbon, CEO-Swire Properties Limited
Kenny Tang, CEO-Dragon Airlines
Kin Wing Tang, CEO-Hong Kong Aircraft Engineering Company Limited
John Robert Slosar, Exec. Chmn.

GROWTH PLANS/SPECIAL FEATURES:

Swire Pacific Ltd. is part of the Swire Group and one of Hong Kong's leading listed companies. It operates several core businesses, organized into five divisions: property, aviation, beverages, marine services and trading and industrial. Swire Properties develops and manages retail, office, hotel and residential properties primarily throughout mainland China and Hong Kong. Swire Properties, Inc., its U.S. subsidiary, develops and trades properties in Florida, specifically downtown Miami, including office space, retail space and hotels. The Swire Hotels subsidiary develops and manages urban hotels throughout Mainland China, Hong Kong and the U.K., including The Upper House and EAST in Hong Kong and The Opposite House in Beijing. The company's aviation holdings include international passenger and freight airline Cathay Pacific Airways Limited, with nearly 700 worldwide destinations; Hong Kong Dragon Airlines, with more than 29 destinations and cargo operations in Asia, Europe, the U.S. and the Middle East; and Hong Kong Aircraft Engineering Company Limited, a provider of base and line maintenance at Hong Kong International Airport. Swire Beverages, Ltd., a joint venture between Swire Pacific (owning 87.5%) and The Coca-Cola Company, manufactures and distribute Coca-Cola products in Mainland China and Hong Kong. The firm's marine services holdings include Swire Pacific Offshore Operations (Pte) Ltd.(SPO), one of the largest offshore energy support fleets in the world, with a fleet of 71 vessels; and the HUD Group, which provides ship repair, towage and salvage, mechanical and electrical engineering and steelwork services. Trading and industrial subsidiary Swire Resources Limited acts as the holding company for various retail and wholesale interests in sports and active footwear and apparel, operating more than 180 retail locations in China. In October 2013, the company agreed to acquire TIMCO Aviation Services., Inc., and that December also agreed to acquire DCH Commercial Centre.

FINANCIAL DATA: Note: Data for latest year may not have been available at press time.

In U.S. $	2015	2014	2013	2012	2011	2010
Revenue	7,849,870,000	7,903,504,000	6,631,744,000	5,654,717,000	4,678,334,000	3,764,869,000
R&D Expense						
Operating Income	2,122,308,000	1,765,947,000	2,151,317,000	3,002,380,000	4,051,479,000	4,379,863,000
Operating Margin %	27.03%	22.34%	32.43%	53.09%	86.60%	116.33%
SGA Expense	1,672,473,000	1,408,554,000	1,274,210,000	985,278,800	908,050,100	766,098,800
Net Income	1,731,394,000	1,427,120,000	1,713,601,000	2,254,202,000	4,152,818,000	4,931,809,000
Operating Cash Flow	1,542,383,000	1,866,125,000	1,528,717,000	1,067,665,000	1,231,018,000	928,678,800
Capital Expenditure	554,267,700	802,199,000	828,887,500	928,549,900	1,257,836,000	899,154,000
EBITDA	3,327,540,000	2,562,344,000	2,838,769,000	3,556,390,000	4,839,883,000	5,566,142,000
Return on Assets %	3.73%	3.13%	3.96%	5.69%	11.43%	15.66%
Return on Equity %	6.14%	5.04%	6.18%	7.99%	14.91%	21.43%
Debt to Equity	0.27	0.27	0.23	0.19	0.11	0.14

CONTACT INFORMATION:

Phone: 852 2840808   Fax: 852 25269365
Toll-Free:
Address: 1 Pacific Place, 88 Queensway, 33rd Fl., Hong Kong, Hong Kong

STOCK TICKER/OTHER:

Stock Ticker: SWRAY
Employees: 82,000
Parent Company:

Exchange: PINX
Fiscal Year Ends: 12/31

SALARIES/BONUSES:

Top Exec. Salary: $ Bonus: $
Second Exec. Salary: $ Bonus: $

OTHER THOUGHTS:

Estimated Female Officers or Directors:
Hot Spot for Advancement for Women/Minorities:

Tanger Factory Outlet Centers Inc

www.tangeroutlet.com

NAIC Code: 0

TYPES OF BUSINESS:

Real Estate Investment Trust-Nonresidential
Factory Outlet Centers
Property Management

BRANDS/DIVISIONS/AFFILIATES:

Tanger Outlet Centers
Tanger LP Trust
Tanger GP Trust
Family Limited Partnership

CONTACTS: *Note: Officers with more than one job title may be intentionally listed here more than once.*

Steven Tanger, CEO
Ricky Farrar, VP, Divisional
Frank Marchisello, CFO
William Benton, Chairman of the Board
James Williams, Chief Accounting Officer
Thomas McDonough, Executive VP
Chad Perry, Executive VP
Brian Auger, General Counsel
Charles Worsham, Senior VP, Divisional
Lisa Morrison, Senior VP, Divisional
Manuel Jessup, Senior VP, Divisional
Virginia Summerell, Senior VP, Divisional
Carrie Geldner, Senior VP
Leigh Boyer, Vice Chairman, Divisional
Joshua Cox, Vice President, Divisional
Cyndi Holt, Vice President, Divisional
Quentin Pell, Vice President, Divisional
Mary Shifflette, Vice President, Divisional
Beth Lippincott, Vice President, Divisional
Gary Block, Vice President, Divisional
Laura Atwell, Vice President, Divisional

GROWTH PLANS/SPECIAL FEATURES:

Tanger Factory Outlet Centers, Inc. (Tanger) is a self-administered and self-managed real estate investment trust (REIT) that owns and operates outlet centers. Through its two wholly-owned subsidiaries, Tanger LP Trust and Tanger GP Trust, the firm maintains the majority stake in the operating partnership known as the Family Limited Partnership. It is through Family Limited Partnership that the company develops, acquires, owns, operates and manages factory outlet shopping centers under the brand name Tanger Outlet Center. The firm owns 36 centers with a total gross leasable area of 11.3 million square feet of retail space containing over 2,400 stores (98% occupied). Tanger also operates and owns interest in nine outlet centers totaling approximately 2.6 million square feet, including four outlet centers in Canada. Tanger Outlet Centers typically incorporate a mix of leading designer and brand name manufacturers, allowing customers to buy discounted products directly from the manufacturer. The roughly 380 brand names sold at Tanger Outlet Centers include Polo Ralph Lauren, Ann Taylor, GAP and Nike. In addition, the company owns The Walk outlets in Atlantic City, New Jersey; Ocean City Front Factory Outlets in Ocean City, Maryland; The Outlets at Hershey in Hershey, Pennsylvania; and Prime Outlets at Jeffersonville in Jeffersonville, Ohio. In 2015, the firm sold five non-core outlet centers for $150.7 million, located in Barstow, California; Kittery, Maine (2); Tuscola, Illinois; and West Branch, Michigan.

FINANCIAL DATA: *Note: Data for latest year may not have been available at press time.*

In U.S. $	2015	2014	2013	2012	2011	2010
Revenue	439,369,000	418,558,000	385,009,000	356,997,000	315,223,000	276,303,000
R&D Expense		2,365,000			158,000	
Operating Income	144,461,000	131,863,000	127,895,000	109,585,000	97,936,000	79,631,000
Operating Margin %	32.87%	31.50%	33.21%	30.69%	31.06%	28.82%
SGA Expense	44,469,000	44,469,000	39,119,000	37,452,000	30,132,000	24,553,000
Net Income	211,200,000	74,011,000	107,557,000	53,228,000	44,641,000	34,249,000
Operating Cash Flow	220,755,000	188,771,000	187,486,000	165,765,000	135,994,000	118,500,000
Capital Expenditure	238,706,000	147,976,000	47,436,000	41,283,000	326,525,000	77,487,000
EBITDA	368,808,000	229,462,000	249,643,000	208,268,000	181,951,000	151,052,000
Return on Assets %	9.43%	3.51%	5.84%	3.23%	3.14%	2.85%
Return on Equity %	38.95%	14.15%	21.40%	11.28%	10.78%	9.07%
Debt to Equity	2.71	2.90	2.54	2.26	2.22	1.94

CONTACT INFORMATION:

Phone: 336 292-3010   Fax: 336 852-2096
Toll-Free:
Address: 3200 Northline Ave., Ste. 360, Greensboro, NC 27408 United States

STOCK TICKER/OTHER:

Stock Ticker: SKT   Exchange: NYS
Employees: 289   Fiscal Year Ends: 12/31
Parent Company:

SALARIES/BONUSES:

Top Exec. Salary: $824,000   Bonus: $
Second Exec. Salary: $417,665   Bonus: $

OTHER THOUGHTS:

Estimated Female Officers or Directors: 4
Hot Spot for Advancement for Women/Minorities: Y

Taubman Centers Inc

www.taubman.com

NAIC Code: 0

TYPES OF BUSINESS:

Real Estate Investment Trust
Malls & Shopping Centers
Property Management

BRANDS/DIVISIONS/AFFILIATES:

Taubman Realty Group LP (The)
Taubman Company LLC (The)
Taubman Properties Asia LLC
Country Club Plaza

CONTACTS: *Note: Officers with more than one job title may be intentionally listed here more than once.*

Robert Taubman, CEO
Simon Leopold, CFO
David Wolff, Chief Accounting Officer
William Taubman, COO
Lisa Payne, Director
Maria Mainville, Other Corporate Officer
Ryan Hurren, Other Corporate Officer
Rene Tremblay, President, Subsidiary
Chris Heaphy, Senior Vice President
David Joseph, Senior VP, Subsidiary
Stephen Kieras, Senior VP, Subsidiary

GROWTH PLANS/SPECIAL FEATURES:

Taubman Centers, Inc. is a real estate investment trust (REIT) that owns, develops, acquires and operates regional luxury shopping centers. Headquartered in Michigan, the company's portfolio includes 24 urban and suburban shopping centers in major metropolitan areas often featuring highly affluent communities such as Los Angeles, Miami, Denver, Detroit, Nashville, New York City, Orlando, San Francisco, St. Louis, Tampa and Washington, D.C. Taubman owns a majority interest in The Taubman Realty Group LP, through which all operations are conducted. The Taubman Realty Group also owns certain regional retail shopping center development projects and majority interest in The Taubman Company LLC, which manages the shopping centers and provides other services to the operating partnership. 90% of the firm's revenues come from leasing to mall tenants, 96% of which consist of national retail chains such as Forever 21, Limited Brands and The Gap. Taubman Properties Asia LLC, headquartered in Hong Kong, is the platform for Taubman's expansion into China and South Korea, where it is actively engaged in projects that leverage the company's retail planning, design and operational capabilities. In March 2016, the firm acquired the Country Club Plaza (Kansas City, Missouri) from Highwoods Properties, Inc.

The firm offers employees tuition assistance; medical, prescription, dental and vision coverage; life and long-term disability insurance; health care and dependent care flexible spending accounts; and an employee assistance program.

FINANCIAL DATA: *Note: Data for latest year may not have been available at press time.*

In U.S. $	2015	2014	2013	2012	2011	2010
Revenue	557,172,000	679,129,000	767,154,000	747,974,000	644,918,000	654,558,000
R&D Expense						
Operating Income	132,886,000	154,640,000	138,964,000	114,010,000	94,693,000	54,847,000
Operating Margin %	23.85%	22.77%	18.11%	15.24%	14.68%	8.37%
SGA Expense	196,759,000	244,631,000	271,160,000	268,628,000	222,645,000	217,726,000
Net Income	134,127,000	893,013,000	132,590,000	106,174,000	192,871,000	63,868,000
Operating Cash Flow	307,685,000	363,686,000	371,372,000	324,349,000	270,166,000	264,608,000
Capital Expenditure	440,678,000	442,991,000	283,864,000			72,152,000
EBITDA						
Return on Assets %	3.21%	25.70%	3.24%	2.52%	6.00%	1.84%
Return on Equity %	50.57%	871.76%				
Debt to Equity	23.44	6.36				

CONTACT INFORMATION:

Phone: 248 258-6800   Fax: 248 258-7596
Toll-Free:
Address: 200 E. Long Lake Rd., Ste. 300, Bloomfield Hills, MI 48304
United States

STOCK TICKER/OTHER:

Stock Ticker: TCO
Employees: 615
Parent Company:

Exchange: NYS
Fiscal Year Ends: 12/31

SALARIES/BONUSES:

Top Exec. Salary: Bonus: $
$1,000,000
Second Exec. Salary: Bonus: $
$901,765

OTHER THOUGHTS:

Estimated Female Officers or Directors: 4

Hot Spot for Advancement for Women/Minorities: Y

TDIndustries

NAIC Code: 236220

www.tdindustries.com

TYPES OF BUSINESS:

Commercial and Institutional Building Construction
Facilities Services
HVAC

BRANDS/DIVISIONS/AFFILIATES:

CONTACTS: *Note: Officers with more than one job title may be intentionally listed here more than once.*

Harold F. MacDowell, CEO
Evelyn Henry Miller, CFO
Steve Canter, IT
Phil Claybrooke, VP-Special Projects, Dallas
Jason Cinek, Sr. VP-Tech.
Dave Youden, Sr. VP-Major Projects Dallas
Nikki Morgan, VP-Houston
Robert G. Wilken, Exec. VP-Dallas Service
Bill Parten, Exec. VP-Facilities Mgmt. Services
Ed White, Sr. VP-Phoenix
Tim McNew, Sr. VP-Fort Worth
Ben Simmons, Exec. VP-Multifamily
Randee Herrin, VP-New Construction, Houston
Rod Johannsen, Exec. VP
Graham T. Moore, Pres., Houston
Bob Richards, Pres., Central Texas
Jack Lowe, Chmn.

GROWTH PLANS/SPECIAL FEATURES:

TDIndustries is a leading facilities service and specialty construction company. The Dallas-based firm offers construction, installation and operations services for commercial, industrial and institutional buildings in the following industries: education, food service, government, health care, historic properties, hotels, industrial, multifamily, sports and assembly, pharmaceutical, semiconductor manufacturing and telecom. Its services fall into a number of key areas, including automation and control, energy solutions, facility services, truck-based services and mechanical construction. The company's automation and control services consist of design and development of fully integrated building control systems, which covers building security, lighting, metering and property irrigation systems. Energy solutions comprise energy retrofitting, energy audits and energy efficiency modeling, central plant optimization, new building design, energy efficient upgrades and systems monitoring and measurement. It is also involved in the design and construction of LEED (Leadership in Energy and Environmental Design) buildings. TDIndustries' facility services include comprehensive operations and maintenance for airports, schools, government and private sector clients. Its truck-based services cover planned maintenance services for a facility's heating, ventilation and air conditioning (HVAC), plumbing and electrical systems; large scale capital projects; scheduled repair and replacement of machinery and systems; and emergency on-demand services. Mechanical construction services include conceptual design, design-build and design-assist; estimating; constructability; planning; and manufacturing for large scale buildings and their mechanical systems, such as stadiums, high-rises, hospitals, schools and corporate offices.

FINANCIAL DATA: *Note: Data for latest year may not have been available at press time.*

In U.S. $	2015	2014	2013	2012	2011	2010
Revenue	433,000,000					
R&D Expense						
Operating Income						
Operating Margin %						
SGA Expense						
Net Income						
Operating Cash Flow						
Capital Expenditure						
EBITDA						
Return on Assets %						
Return on Equity %						
Debt to Equity						

CONTACT INFORMATION:

Phone: 972-888-9500　　　Fax: 972-888-9507
Toll-Free:
Address: 13850 Diplomat Dr., Dallas, TX 75234 United States

STOCK TICKER/OTHER:

Stock Ticker: Private　　　　　　Exchange:
Employees: 2,030　　　　　　　　Fiscal Year Ends: 12/31
Parent Company:

SALARIES/BONUSES:

Top Exec. Salary: $　　　Bonus: $
Second Exec. Salary: $　　Bonus: $

OTHER THOUGHTS:

Estimated Female Officers or Directors: 4
Hot Spot for Advancement for Women/Minorities: Y

Sales, profits and employees may be estimates. Financial information, benefits and other data can change quickly and may vary from those stated here.

TMI Hospitality Inc

www.tmihospitality.com

NAIC Code: 721110

TYPES OF BUSINESS:

Hotels
Construction Services
Communications Services
Property Management

BRANDS/DIVISIONS/AFFILIATES:

Starwood Capital Group
TMI Communications Inc
Room In The Inn Program
Hampton Inn
Staybridge Suites
Sleep Inn
TownePlace Suites
Comfort Inn and Suites

CONTACTS: Note: Officers with more than one job title may be intentionally listed here more than once.

Lauris Molbert, CEO
Rick Larson, Pres.
Lisa Helbling, CFO
Tracy Koenig, CIO
Robert McConn, Jr., General Counsel
Douglas Dobmeier, Sr. VP-Oper.
Charles A. Krumwiede, Sr. VP-Property Support & Quality

GROWTH PLANS/SPECIAL FEATURES:

TMI Hospitality, Inc. builds and operates select-service and extended-service hotels across the U.S. It is a leading hotel property-management company, operating 180 hotels in 25 states. TMI Hospitality operates properties under brand names such as Courtyard, Residence Inn, Fairfield Inn & Suites, Springhill Suites and TownePlace Suites by Marriott; Homewood Suites, Home2 Suites and Hampton Inn by Hilton; Country Inn and Suites by Carlson; Holiday Inn Express and Staybridge Suites by InterContinental Hotels Group (IHG); Hyatt Place by Hyatt; and Comfort Inn and Suites, Sleep Inn and Cambria Suites by Choice Hotels International. The company's properties offer many amenities, including cable television, indoor and outdoor pools, complimentary breakfast, free local phone calls and 24-hour coffee service. In addition, children under 18 stay free. The firm also provides its Room In The Inn Program, which offers free rooms to guests visiting friends or family members in a hospital, nursing home or treatment facility during the Christmas and Thanksgiving seasons. Through subsidiary TMI Communications, Inc., the company provides nationwide telecommunications services to the hospitality industry. In January 2015, Starwood Capital Group, a leading global private investment firm acquired all of the outstanding stock of TMI Hospitality from the TMI Hospitality Employee Stock Ownership Plan. This acquisition includes all the hotels, the management company and the development platform.

TMI Hospitality offers its employees medical, dental and vision coverage; an employee stock ownership program; lodging discounts; a flexible spending account; training; and an education assistance program.

FINANCIAL DATA: Note: Data for latest year may not have been available at press time.

In U.S. $	2015	2014	2013	2012	2011	2010
Revenue						
R&D Expense						
Operating Income						
Operating Margin %						
SGA Expense						
Net Income						
Operating Cash Flow						
Capital Expenditure						
EBITDA						
Return on Assets %						
Return on Equity %						
Debt to Equity						

CONTACT INFORMATION:

Phone: 701-235-1060   Fax: 701-235-0948
Toll-Free:
Address: 4850 32nd Avenue South, Fargo, ND 58104 United States

SALARIES/BONUSES:

Top Exec. Salary: $ Bonus: $
Second Exec. Salary: $ Bonus: $

STOCK TICKER/OTHER:

Stock Ticker: Private   Exchange:
Employees: 4,000   Fiscal Year Ends: 12/31
Parent Company:

OTHER THOUGHTS:

Estimated Female Officers or Directors: 2
Hot Spot for Advancement for Women/Minorities:

Toll Brothers Inc

NAIC Code: 236117

www.tollbrothers.com

TYPES OF BUSINESS:

Construction, Home Building and Residential
Mortgages & Insurance
Property Management
Landscaping
Country Club Communities
Golf Courses
Security Monitoring
Lumber Distribution

BRANDS/DIVISIONS/AFFILIATES:

Gramercy Park

CONTACTS: Note: Officers with more than one job title may be intentionally listed here more than once.

Douglas Yearley, CEO
Martin Connor, CFO
Robert Toll, Chairman of the Board
Richard Hartman, COO
Michael Snyder, Secretary
Joseph Sicree, Senior VP

GROWTH PLANS/SPECIAL FEATURES:

Toll Brothers, Inc. designs, builds, markets and arranges financing for single-family detached and attached homes in luxury residential communities. The firm is also involved, both directly and through joint ventures, in building or converting existing rental apartment buildings into high-, mid- and low-rise luxury homes. Toll Brothers markets its services to move-up, empty-nester, active-adult, age-qualified and second-home buyers through its operations in 19 U.S. states. The company is present in major suburban and urban residential areas including the Philadelphia, Pennsylvania metropolitan area; Virginia and Maryland suburbs of Washington, D.C.; central and northern New Jersey; Boston, Massachusetts; Westchester, Dutchess and Ulster Counties, New York; San Diego and Palm Springs, California; the San Francisco Bay area; Phoenix, Arizona; Las Vegas and Reno, Nevada; and Chicago, Illinois. The average base sales price of the company's homes is roughly $646,000. Toll Brothers operates its own land development, architectural, engineering, mortgage, title, landscaping, lumber distribution, house component assembly and manufacturing operations. In addition, the company owns and operates golf courses in conjunction with several of its master planned communities. The company operates a portfolio of 539 communities with 44,253 homes sites. In 2015, the firm acquired Gramercy Park in order to expand its presence in New York City.

The firm offers employees discounts on homes, mortgages, titles, home appliances and kitchen cabinets; life, medical and dental insurance; long- and short-term disability coverage; a 401(k); and educational reimbursement. After two years of service, employees are able to use the company's furnished resort luxury guesthouses for their personal vacations.

FINANCIAL DATA: Note: Data for latest year may not have been available at press time.

In U.S. $	2015	2014	2013	2012	2011	2010
Revenue	4,171,248,000	3,911,602,000	2,674,299,000	1,882,781,000	1,475,881,000	1,494,771,000
R&D Expense						
Operating Income	446,870,000	397,249,000	201,067,000	63,429,000	-47,748,000	-174,279,000
Operating Margin %	10.71%	10.15%	7.51%	3.36%	-3.23%	-11.65%
SGA Expense	455,108,000	432,516,000	339,932,000	287,257,000	261,355,000	263,224,000
Net Income	363,167,000	340,032,000	170,606,000	487,146,000	39,795,000	-3,374,000
Operating Cash Flow	60,182,000	313,200,000	-568,963,000	-168,962,000	52,850,000	-146,284,000
Capital Expenditure	9,447,000	15,074,000	26,567,000	14,495,000	9,553,000	4,830,000
EBITDA	559,119,000	528,237,000	292,907,000	135,528,000	-4,720,000	-74,392,000
Return on Assets %	4.12%	4.46%	2.62%	8.67%	.77%	-.06%
Return on Equity %	8.99%	9.46%	5.28%	17.06%	1.54%	-.13%
Debt to Equity	0.89	0.88	0.75	0.72	0.63	0.66

CONTACT INFORMATION:

Phone: 215 938-8000   Fax: 215 938-8023
Toll-Free:
Address: 250 Gibraltar Rd., Horsham, PA 19044 United States

STOCK TICKER/OTHER:

Stock Ticker: TOL
Employees: 3,500
Parent Company:

Exchange: NYS
Fiscal Year Ends: 10/31

SALARIES/BONUSES:

Top Exec. Salary:
$1,000,000
Second Exec. Salary:
$1,000,000

Bonus: $

Bonus: $

OTHER THOUGHTS:

Estimated Female Officers or Directors:

Hot Spot for Advancement for Women/Minorities:

Trammell Crow Company

www.trammellcrow.com

NAIC Code: 531120

TYPES OF BUSINESS:

Real Estate Development & Investment Services
Building Management
Brokerage Services
Project Management Services
Development Services
Property & Facility Management

BRANDS/DIVISIONS/AFFILIATES:

CB Richard Ellis Group Inc

CONTACTS: *Note: Officers with more than one job title may be intentionally listed here more than once.*

Matt Khourie, CEO
Michael Duffy, COO
Craig Cheney, CIO
John Stirek, Pres., Western Oper.
Chris Roth, Pres., Eastern Oper.
Craig Cheney, Chief Investment Officer
Adam Saphier, Pres., Central Oper.

GROWTH PLANS/SPECIAL FEATURES:

Trammell Crow Company, a subsidiary of CB Richard Ellis Group, Inc., is one of the largest diversified commercial real estate service companies in the world, providing real estate services primarily in North America. The company has 16 offices throughout the U.S. Trammel Crow serves the office, industrial, retail, health care, residential, mixed use and land development industries. As the manager of development projects, the firm offers services including the evaluation of project feasibility, budgeting and scheduling; site identification, due diligence and acquisition; cash flow analysis; the procurement of approvals and permits, including zoning and other entitlements; project finance advisory services; coordination of project design and engineering; construction bidding and management; tenant finish coordination; and project close-out and tenant move coordination. Trammell Crow's outsourcing is done on a contractual basis, aiming at long-term and full-service agreements with its clients. The company has developed or acquired nearly 2,600 buildings valued at nearly $60 billion and over 565 million square feet. In early 2016, the firm acquired a 90-acre site in Pennsylvania, for the development of Lehigh Valley Trade Center, comprising two Class A warehouse/distribution buildings totaling more than 1.2 million square feet.

Employees receive medical, vision and dental insurance for themselves and their families/domestic partners; life and disability insurance; educational assistance; and several employee discount programs.

FINANCIAL DATA: *Note: Data for latest year may not have been available at press time.*

In U.S. $	2015	2014	2013	2012	2011	2010
Revenue	1,600,000,000	1,500,000,000				
R&D Expense						
Operating Income						
Operating Margin %						
SGA Expense						
Net Income						
Operating Cash Flow						
Capital Expenditure						
EBITDA						
Return on Assets %						
Return on Equity %						
Debt to Equity						

CONTACT INFORMATION:

Phone: 214-863-4101   Fax: 214-863-4493
Toll-Free:
Address: 2100 McKinney Ave., Ste. 800, Dallas, TX 75201 United States

STOCK TICKER/OTHER:

Stock Ticker: Subsidiary   Exchange:
Employees: 1,308   Fiscal Year Ends: 12/31
Parent Company: CB Richard Ellis Group Inc (CBRE)

SALARIES/BONUSES:

Top Exec. Salary: $ Bonus: $
Second Exec. Salary: $ Bonus: $

OTHER THOUGHTS:

Estimated Female Officers or Directors:
Hot Spot for Advancement for Women/Minorities:

Transcontinental Realty Investors Inc www.transconrealty-invest.com
NAIC Code: 531110

TYPES OF BUSINESS:
Real Estate Investment Trust
Warehouses
Apartments
Office Properties
Retail Properties

BRANDS/DIVISIONS/AFFILIATES:
American Realty Investment Inc
Prime Asset Management
Income Opportunity Realty Investors Inc

CONTACTS: *Note: Officers with more than one job title may be intentionally listed here more than once.*
Daniel Moos, CEO
Gene Bertcher, CFO
Henry Butler, Director
Alfred Crozier, Executive VP, Divisional
Louis Corna, Executive VP

GROWTH PLANS/SPECIAL FEATURES:
Transcontinental Realty Investors, Inc. (TCI) is an externally advised real estate investment company. The firm specializes in the acquisition, development and ownership of income-producing residential and commercial real estate properties. TCI also acquires land for future development in in-fill or high-growth suburban markets. The company's portfolio consists of 46 residential apartment communities, totaling roughly 6,078 units, and nine commercial properties, consisting of six office buildings, two retail properties and one industrial warehouse, totaling approximately 2.3 million square feet. In addition, TCI owns 4,100 acres of undeveloped and partially developed land located in Texas, Louisiana and Florida. The company produces revenue by managing previously undervalued or underperforming office buildings, apartments, retail centers and warehouses. Day-to-day management of TCI's business is contracted to Prime Asset Management, whose employees serve as company officers. Subsidiaries of American Realty Investment, Inc. own approximately 83.8% of the company's stock. TCI owns roughly 81.1% interest in Income Opportunity Realty Investors, Inc. In November 2015, the firm began construction on a high-end residential development in Rowlett, Texas.

FINANCIAL DATA: *Note: Data for latest year may not have been available at press time.*

In U.S. $	2015	2014	2013	2012	2011	2010
Revenue	102,220,000	75,858,000	86,237,000	116,051,000	114,087,000	129,862,000
R&D Expense						
Operating Income	9,301,000	771,000	-2,111,000	18,181,000	-30,951,000	-15,783,000
Operating Margin %	9.09%	1.01%	-2.44%	15.66%	-27.12%	-12.15%
SGA Expense	13,876,000	14,536,000	14,817,000	14,341,000	19,171,000	20,400,000
Net Income	-7,636,000	41,578,000	58,530,000	-8,324,000	-46,321,000	-67,196,000
Operating Cash Flow	-50,919,000	-29,382,000	-66,695,000	-21,119,000	6,538,000	-8,271,000
Capital Expenditure	239,140,000	91,877,000	9,315,000	8,068,000	83,441,000	50,394,000
EBITDA	55,002,000	30,237,000	11,297,000	65,636,000	16,692,000	9,999,000
Return on Assets %	-.83%	4.43%	5.91%	-.85%	-3.72%	-4.56%
Return on Equity %	-4.04%	20.85%	39.59%	-7.79%	-33.28%	-35.30%
Debt to Equity	3.73	2.73	3.23	6.28	6.59	5.22

CONTACT INFORMATION:
Phone: 469 522-4200   Fax: 469 522-4299
Toll-Free: 800-400-6407
Address: 1603 Lyndon B Johnson Fwy., Ste. 800, Dallas, TX 75234
United States

STOCK TICKER/OTHER:
Stock Ticker: TCI
Employees:
Parent Company:

Exchange: NYS
Fiscal Year Ends: 12/31

SALARIES/BONUSES:
Top Exec. Salary: $ Bonus: $
Second Exec. Salary: $ Bonus: $

OTHER THOUGHTS:
Estimated Female Officers or Directors: 2
Hot Spot for Advancement for Women/Minorities:

Transswestern Commercial Services   www.transwestern.net

NAIC Code: 531120

TYPES OF BUSINESS:

Real Estate Operations
Property Management
Advisory Services
Development & Research Services
Tenant Representation
Investment & Finance Services

BRANDS/DIVISIONS/AFFILIATES:

Transwestern Investment Service Group
Delta Associates

CONTACTS: *Note: Officers with more than one job title may be intentionally listed here more than once.*

Larry P. Heard, CEO
Mark Doran, COO
Larry P. Heard, Pres.
Steve Harding, CFO
Kim Croley, National Dir.-Mktg.
Rob Bagguley, Chief Innovation Officer
Tom McNearney, CIO
Kim Croley, National Dir.-Comm.
Mike McLain, Chief Acct. Officer
Tom McNearney, Chief Investment Officer
Al Skodowski, Sr. VP
Robert Duncan, Chmn.
Chip Clarke, Pres., Americas

GROWTH PLANS/SPECIAL FEATURES:

Transwestern Commercial Services is one of the largest privately held full-service real estate firms in the U.S. The company operates in 34 offices across the U.S. and in 180 offices in 37 countries worldwide. Transwestern's global presence is due in part to a strategic alliance with Paris-based BNP Paribas Real Estate. The firm offers a range of services, including agency leasing, development, property and facility management, investment services, tenant advisory and research. Transwestern's agency leasing services utilize local market knowledge to provide customized marketing plans for each asset being considered by its clients. The development offerings include site analysis and acquisition, construction management and project marketing. Property and facility management services aim at maximizing property values through tenant relations programs, strategic planning, building enhancement and accurate accounting. The company offers its investment services through its Investment Services Group, which provides clients with the firm's experience with real estate capital markets and investment banking to improve the financial performance of their assets. The firm's tenant advisory services include strategic planning, needs analysis, market research/analysis, disposition services, document management/negotiation, lease administration and project management. Transwestern's research services, provided by wholly-owned subsidiary Delta Associates, include national economic updates and regional market reports designed for the commercial real estate industry.

The company offers its employees health, dental and vision insurance; a 401(k) plan; flexible spending accounts; life and AD&D insurance; long-term disability insurance; and an employee assistance program.

FINANCIAL DATA: *Note: Data for latest year may not have been available at press time.*

In U.S. $	2015	2014	2013	2012	2011	2010
Revenue						
R&D Expense						
Operating Income						
Operating Margin %						
SGA Expense						
Net Income						
Operating Cash Flow						
Capital Expenditure						
EBITDA						
Return on Assets %						
Return on Equity %						
Debt to Equity						

CONTACT INFORMATION:

Phone: 713-270-7700   Fax: 713-270-6285
Toll-Free:
Address: 1900 W. Loop S., Ste. 1300, Houston, TX 77027 United States

SALARIES/BONUSES:

Top Exec. Salary: $ Bonus: $
Second Exec. Salary: $ Bonus: $

STOCK TICKER/OTHER:

Stock Ticker: Private   Exchange:
Employees: 2,123   Fiscal Year Ends: 12/31
Parent Company:

OTHER THOUGHTS:

Estimated Female Officers or Directors: 2
Hot Spot for Advancement for Women/Minorities: Y

Trevi-Finanziaria Industriale SpA (Trevi Group) www.trevifin.com

NAIC Code: 541330

TYPES OF BUSINESS:
Engineering & Construction Services
Underground Construction Services
Foundation & Drilling Machinery & Services
Wind Farms
Geothermal Energy Projects

BRANDS/DIVISIONS/AFFILIATES:
Petreven CA
Trevi SpA
Soilmec SpA
Drillmec SpA
Trevi Energy SpA
Petreven SpA
Petreven Peru SA
Petreven Chile SpA

CONTACTS: Note: Officers with more than one job title may be intentionally listed here more than once.
Davide Trevisani, Managing Dir.
Gianluigi Trevisani, VP
Daniele Forti, CFO
Franco Cicognani, Head-Corp. Comm. Dept.
Stefano Trevisani, Managing Dir.
Cesare Trevisani, Managing Dir.
Davide Trevisani, Chmn.

GROWTH PLANS/SPECIAL FEATURES:
Trevi-Finanziaria Industriale SpA (Trevi Group) is a global leader in foundation engineering and drilling and provides project management services supporting related projects around the world. The group operates through five main subsidiaries: Trevi SpA, Petreven SpA, Soilmec SpA, Drillmec SpA and Trevi Energy SpA. Trevi provides construction services for underground engineering projects, such as special foundations for bridges, railways, dams, industrial systems and tunnels. Petreven is the company's segment in charge of oil drilling, with operations concentrated in South America, including Argentina (Petreven SA), Chile (Petreven Chile SpA), Peru (Petreven Peru SA) and Venezuela (Petreven CA). Soilmec manufactures plants and rigs used for foundation engineering. Drillmec specializes in drilling projects and manufactures hydraulic rigs for oil, geothermal and water drilling. It also has operations in well drilling for water research as well as having completed numerous automated car park projects. Trevi Energy is engaged in the design, development and engineering of offshore wind farms and the development of geothermal energy projects. Trevi Group's strong worldwide presence is illustrated by such past projects as the foundation works for the Third Mainland Bridge of Lagos in Nigeria, the Alicura in Argentina, the Port of Bandar Abbas in Iran and the Khao dam in Thailand.

FINANCIAL DATA: Note: Data for latest year may not have been available at press time.

In U.S. $	2015	2014	2013	2012	2011	2010
Revenue		1,381,138,000	1,418,728,000	1,236,984,000	1,174,811,000	1,040,480,000
R&D Expense						
Operating Income		71,370,080	91,593,390	54,130,310	79,021,680	96,157,660
Operating Margin %		5.16%	6.45%	4.37%	6.72%	9.24%
SGA Expense						
Net Income		27,845,260	15,696,670	12,320,800	29,310,800	52,095,660
Operating Cash Flow		-32,732,290	99,286,050	75,969,700	19,472,860	113,116,900
Capital Expenditure						
EBITDA		136,419,500	154,510,100	117,173,600	133,341,300	149,201,100
Return on Assets %		1.42%	.90%	.71%	1.83%	3.74%
Return on Equity %		4.63%	3.33%	2.55%	6.58%	14.12%
Debt to Equity		0.40	0.52	0.57	0.67	0.61

CONTACT INFORMATION:
Phone: 39 547319111   Fax: 39 547319313
Toll-Free:
Address: 201 Via Larga, Cesena, FO 47522 Italy

STOCK TICKER/OTHER:
Stock Ticker: TVFZF   Exchange: GREY
Employees: 7,493   Fiscal Year Ends: 12/31
Parent Company:

SALARIES/BONUSES:
Top Exec. Salary: $ Bonus: $
Second Exec. Salary: $ Bonus: $

OTHER THOUGHTS:
Estimated Female Officers or Directors: 1
Hot Spot for Advancement for Women/Minorities:

Triple Five Group

www.triplefive.com

NAIC Code: 531100

TYPES OF BUSINESS:

Shopping Malls
Hotels
Office Buildings
Mixed-Use Developments
Entertainment Venues & Amusement Parks
Banking
Venture Capital
Oil & Gas Exploration

BRANDS/DIVISIONS/AFFILIATES:

Mall of America
West Edmonton Mall
First Nuclear Corporation
American Dream Meadowlands
T5 Equity Partners LLC
MEI Triplefive Engineering
People's Trust
American Dream

CONTACTS: *Note: Officers with more than one job title may be intentionally listed here more than once.*

Don Ghermezian, Pres.
Nader Ghermezian, Chmn.

GROWTH PLANS/SPECIAL FEATURES:

Triple Five Group is a diversified corporation engaged in real estate, resource development and mining, venture capital, hotels, banking, private equity, engineering, manufacturing and retail operations. Together with its group of affiliates, the firm owns major shopping centers and other commercial real estate. It also focuses on planning and development, tourism projects, planned communities and technology. Triple Five's primary projects include the 5.3-million-square-foot West Edmonton Mall in Canada, the 4.2-million-square-foot Mall of America in Minnesota and the 1 million square foot American Dream in Metropolitan New York, which are three of the largest mixed-use tourism, retail and entertainment complexes in the world. Annually, Triple Five's three supermalls attract roughly 112 million visitors. Subsidiary People's Trust operates throughout Canada and provides investment and lending services, estate and trust asset management and mortgage servicing. Affiliate T5 Equity Partners LLC provides venture and equity capital investments. Venture Capital provides funds for development in the wireless, Internet and high technology industries; and subsidiary First Nuclear Corporation is active in resource development and exploration within the oil, gas and metallurgical industries. Other subsidiaries include MEI Triplefive Engineering, Celonex Pharmaceuticals and DiamondGear. Projects under development include Phase II of the Mall of America, which will expand the mall to over 10 million square feet of space; and the American Dream Meadowlands Mall in New Jersey, with a projected square footage of over 2.9 square feet of retail, entertainment, dining and attractions. In 2015, the firm announced plans to build the largest shopping mall in the U.S., the American Dream Miami. It would be built on 200 acres of grazing land, and include amusement and gaming areas consisting of a ski slope, skating rink, roller coaster, Ferris wheel, an indoor lake with submarine trips, hotels, condominiums, a mini golf course, water park and sea lion exhibit.

FINANCIAL DATA: *Note: Data for latest year may not have been available at press time.*

In U.S. $	2015	2014	2013	2012	2011	2010
Revenue						
R&D Expense						
Operating Income						
Operating Margin %						
SGA Expense						
Net Income						
Operating Cash Flow						
Capital Expenditure						
EBITDA						
Return on Assets %						
Return on Equity %						
Debt to Equity						

CONTACT INFORMATION:

Phone: 780-444-8100   Fax: 780-444-5232
Toll-Free:
Address: 8882-170 St., Ste. 3000, Edmonton, AB T5T 4M2 Canada

SALARIES/BONUSES:

Top Exec. Salary: $ Bonus: $
Second Exec. Salary: $ Bonus: $

STOCK TICKER/OTHER:

Stock Ticker: Private   Exchange:
Employees: 5,000   Fiscal Year Ends: 07/31
Parent Company:

OTHER THOUGHTS:

Estimated Female Officers or Directors: 1
Hot Spot for Advancement for Women/Minorities:

TRT Holdings Inc

www.omnihotels.com

NAIC Code: 721110

TYPES OF BUSINESS:
Hotels
Gymnasiums
Oil & Gas Exploration

BRANDS/DIVISIONS/AFFILIATES:
Omni Hotels & Resorts
Mokara Hotel and Spa
Gold's Gym International
Tana Exploration Company LLC
Waldo's Dollar Mart

CONTACTS: *Note: Officers with more than one job title may be intentionally listed here more than once.*
Robert B. Rowling, CEO
James D. Caldwell, Pres.
Robert B. Bean, CFO
James D. Caldwell, CEO-Omni Hotels & Resorts
Michael J. Deitemeyer, Pres., Omni Hotels & Resorts
Kevin D. Talley, Pres., Tana Exploration Company LLC
Jim Snow, Pres., Gold's Gym
Robert B. Rowling, Chmn.

GROWTH PLANS/SPECIAL FEATURES:
TRT Holdings, Inc. is a diversified holding company with interests in the lodging and hospitality, fitness and oil industries. The firm owns and franchises the Omni Hotels & Resorts chain, Gold's Gym International, an oil and gas exploration firm Tana Exploration Company, Waldo's Dollar Mart in Mexico and many investments in other companies. Omni hotels offer luxury and first-class accommodations at approximately 60 locations across North America, totaling approximately 21,000 rooms. A typical Omni hotel has over 300 rooms and offers amenities such as marble bathrooms, high-speed Internet access, gourmet dining, fitness centers, 24-hour room service and valet ordering, among other services. The hotel chain caters primarily to corporate business and upscale leisure travelers. Apart from the Omni brand, the firm has introduced the Mokara Hotel and Spa brand, which currently has one location in San Antonio, Texas. Gold's Gym International is a franchiser of more than 260 fitness centers throughout the U.S. and worldwide. Tana Exploration Company LLC is an independent oil and gas exploration firm headquartered in Houston, Texas. The firm has working interests in a number of wells in the offshore Gulf of Mexico region, as well as in the Fort Worth basin. Waldo's Dollar Mart is a line of convenience stores with headquarters in Tijuana and Baja California, Mexico.

FINANCIAL DATA: *Note: Data for latest year may not have been available at press time.*

In U.S. $	2015	2014	2013	2012	2011	2010
Revenue						
R&D Expense						
Operating Income						
Operating Margin %						
SGA Expense						
Net Income						
Operating Cash Flow						
Capital Expenditure						
EBITDA						
Return on Assets %						
Return on Equity %						
Debt to Equity						

CONTACT INFORMATION:
Phone: 972-730-6664   Fax: 972-871-5665
Toll-Free: 800-843-6664
Address: 4001 Maple Ave., Dallas, TX 75219 United States

STOCK TICKER/OTHER:
Stock Ticker: Private   Exchange:
Employees: 47   Fiscal Year Ends: 12/31
Parent Company:

SALARIES/BONUSES:
Top Exec. Salary: $ Bonus: $
Second Exec. Salary: $ Bonus: $

OTHER THOUGHTS:
Estimated Female Officers or Directors:
Hot Spot for Advancement for Women/Minorities:

True Home Value Inc

www.truehomevalue.com

NAIC Code: 236118

TYPES OF BUSINESS:

Window Installation
Door & Window Assemblies
Retail & Installation
Roofing & Siding Materials
Coatings
Remodeling

BRANDS/DIVISIONS/AFFILIATES:

Soft-Lite LLC

GROWTH PLANS/SPECIAL FEATURES:

True Home Value, Inc. (THV) is one of the largest full-service national home improvement companies in the U.S. The company sells and installs thermal replacement windows, doors, textured coatings, vinyl siding, patio decks, patio enclosures, cabinet re-facings, bathroom and kitchen remodeling products and residential roofing. The firm's primary business is in the replacement window market. Its windows use high-technology glass that allows sunlight in while keeping damaging ultraviolet rays out. THV offers a 50-year non-prorated glass and window warranty as well as a 25% fuel savings pledge in writing on all its window products. The firm offers custom-made insulated steel doors with a wood grain finish in different styles and color combinations as well as several variations of sliding glass doors. In November 2015, the firm was acquired out of bankruptcy by Soft-Lite LLC, a manufacturer of high quality and energy efficient vinyl replacement windows and doors.

CONTACTS: *Note: Officers with more than one job title may be intentionally listed here more than once.*

Kyle Pozek, CFO-Soft-Lite

FINANCIAL DATA: *Note: Data for latest year may not have been available at press time.*

In U.S. $	2015	2014	2013	2012	2011	2010
Revenue						
R&D Expense						
Operating Income						
Operating Margin %						
SGA Expense						
Net Income						
Operating Cash Flow						
Capital Expenditure						
EBITDA						
Return on Assets %						
Return on Equity %						
Debt to Equity						

CONTACT INFORMATION:

Phone: 502 968-2020   Fax: 502 968-7798
Toll-Free: 800-669-2020
Address: 5611 Fern Valley Rd., Louisville, KY 40228 United States

STOCK TICKER/OTHER:

Stock Ticker: Subsidiary   Exchange:
Employees: 500   Fiscal Year Ends: 12/31
Parent Company: Soft-Lite LLC

SALARIES/BONUSES:

Top Exec. Salary: $ Bonus: $
Second Exec. Salary: $ Bonus: $

OTHER THOUGHTS:

Estimated Female Officers or Directors:
Hot Spot for Advancement for Women/Minorities:

Trump Entertainment Resorts Inc

www.trumpcasinos.com/

NAIC Code: 531120

TYPES OF BUSINESS:
Real Estate Investment, Development & Operations
Property Management
Residential & Commercial Brokerage
Hotel, Casino & Resort Management
Golf Courses

BRANDS/DIVISIONS/AFFILIATES:
Trump Taj Mahal
Taj Mahal's Trump One

GROWTH PLANS/SPECIAL FEATURES:
Trump Entertainment Resorts, Inc. is a gaming and hospitality company that owns and operates the Trump Taj Mahal hotel and casino in Atlantic City, New Jersey. The firm was initially founded by Donald J. Trump, who is no longer involved in the company, but is now a subsidiary of Icahn Enterprises. The Trump Taj Mahal operates approximately 5,700 slot machines, 280 gaming tables and 2,700 hotel rooms. Taj Mahal's Trump One card is a free, with a valid photo ID, which provides special offers and benefits such as free show tickets, the display of account balances online, free parking, as well as opportunities to earn slot/table dollars. The firm is currently undergoing operational and debt restructuring.

CONTACTS: Note: Officers with more than one job title may be intentionally listed here more than once.
Robert F. Griffin, CEO
Michael P. Mellon, VP-Oper.
Donald J. Trump, Pres.
Daniel M. McFadden, CFO
Donald Trump, Jr., Exec. VP-Dev. & Acquisitions
Ivanka Trump, Exec. VP-Dev. & Acquisitions
Eric Trump, Exec. VP-Dev. & Acquisitions
Cathy Hoffman Glosser, Exec. VP-Global Licensing
Robert F. Griffin, Chmn.

FINANCIAL DATA: Note: Data for latest year may not have been available at press time.

In U.S. $	2015	2014	2013	2012	2011	2010
Revenue						
R&D Expense						
Operating Income						
Operating Margin %						
SGA Expense						
Net Income						
Operating Cash Flow						
Capital Expenditure						
EBITDA						
Return on Assets %						
Return on Equity %						
Debt to Equity						

CONTACT INFORMATION:
Phone: 609-449-5534   Fax:
Toll-Free:
Address: 1000 Boardwalk at Virginia Avenue, Atlantic City, NY 08401 United States

STOCK TICKER/OTHER:
Stock Ticker: Subsidiary   Exchange:
Employees: 5,500   Fiscal Year Ends: 12/31
Parent Company: Icahn Enterprises LP

SALARIES/BONUSES:
Top Exec. Salary: $ Bonus: $
Second Exec. Salary: $ Bonus: $

OTHER THOUGHTS:
Estimated Female Officers or Directors: 2
Hot Spot for Advancement for Women/Minorities:

Turner Corp (The)

www.turnerconstruction.com

NAIC Code: 236220

TYPES OF BUSINESS:

Commercial & Institutional Building Construction

BRANDS/DIVISIONS/AFFILIATES:

Hochtief AG
Turner Construction Company
Turner International LLC

GROWTH PLANS/SPECIAL FEATURES:

The Turner Corp., a subsidiary of Hochtief AG, operates as a construction management company. Turner operates its construction business through Turner Construction Company, which is one of the largest construction companies in the U.S. The services offered by the firm include design & build, procurement services, medical planning & procurement, building information modeling, facilities management solutions and lean construction. Turner has completed projects within the categories of aviation/transportation, commercial, culture & entertainment, data center, education, government, health care, infrastructure, interiors, industrial/manufacturing, pharmaceutical, public assembly, religious, research & development, residential/hotel, retail & restaurant and sports. In North America, the firm has offices throughout the U.S. and in Toronto, Canada. Turner's international presence, via Turner International LLC, has offices in East Asia, Europe, Latin America, the Caribbean, India and the Middle East.

CONTACTS:
Note: Officers with more than one job title may be intentionally listed here more than once.

Peter J. Davoren, CEO
Karen O. Gould, CFO

FINANCIAL DATA:
Note: Data for latest year may not have been available at press time.

In U.S. $	2015	2014	2013	2012	2011	2010
Revenue	10,797,500,000	10,000,000,000	9,979,400,000	9,650,000,000	9,275,000,000	
R&D Expense						
Operating Income						
Operating Margin %						
SGA Expense						
Net Income						
Operating Cash Flow						
Capital Expenditure						
EBITDA						
Return on Assets %						
Return on Equity %						
Debt to Equity						

CONTACT INFORMATION:

Phone: 212-229-6000   Fax:
Toll-Free:
Address: 375 Hudson St., New York, NY 10014 United States

STOCK TICKER/OTHER:

Stock Ticker: Subsidiary   Exchange:
Employees: 5,200   Fiscal Year Ends:
Parent Company: Hochtief AG

SALARIES/BONUSES:

Top Exec. Salary: $ Bonus: $
Second Exec. Salary: $ Bonus: $

OTHER THOUGHTS:

Estimated Female Officers or Directors:
Hot Spot for Advancement for Women/Minorities:

Tutor Perini Corporation

NAIC Code: 237310

www.tutorperini.com

TYPES OF BUSINESS:
Construction Services
Hospitality & Casino Construction
Construction Management Services
Civic & Infrastructure Construction
Design Services

BRANDS/DIVISIONS/AFFILIATES:

CONTACTS: Note: Officers with more than one job title may be intentionally listed here more than once.
Craig Shaw, CEO, Divisional
Ronald Tutor, CEO
Gary Smalley, CFO
Ronald Marano, Chief Accounting Officer
James Frost, COO
Michael Klein, Director
Robert Band, Executive VP, Subsidiary
William Sparks, Secretary
Jorge Casado, Vice President, Divisional

GROWTH PLANS/SPECIAL FEATURES:

Tutor Perini Corporation and its subsidiaries provide general contracting, construction management and design-build services worldwide. It operates in three segments: building, civil and specialty contractors. The building segment focuses on large, complex projects in the hospitality and gaming, transportation, health care, municipal offices, sports and entertainment, education, correctional facilities, biotech, pharmaceutical, industrial and high-tech markets. The civil segment focuses on public works construction, including the new construction, repair, replacement and reconstruction of public infrastructure such as highways, bridges, mass transit systems and wastewater treatment facilities. The company's customers primarily award contracts through the public competitive bid, in which price is the major determining factor; or through a request for proposals, where contracts are awarded based on a combination of technical capability and price. The specialty contractors segment engages in electrical, mechanical, HVAC (heating, ventilation and air conditioning), plumbing and pneumatically paced concrete for construction projects in the commercial, industrial, hospitality, transportation and gaming markets.

The firm offers employees medical, dental, vision and life insurance; a flexible spending account; an employee assistance program; educational assistance; and reimbursement on health club memberships.

FINANCIAL DATA: Note: Data for latest year may not have been available at press time.

In U.S. $	2015	2014	2013	2012	2011	2010
Revenue	4,920,472,000	4,492,309,000	4,175,672,000	4,111,471,000	3,716,317,000	3,199,210,000
R&D Expense						
Operating Income	105,413,000	241,690,000	203,822,000	-221,811,000	168,376,000	172,312,000
Operating Margin %	2.14%	5.38%	4.88%	-5.39%	4.53%	5.38%
SGA Expense	250,840,000	263,752,000	263,082,000	260,369,000	226,965,000	165,536,000
Net Income	45,292,000	107,936,000	87,296,000	-265,400,000	86,148,000	103,500,000
Operating Cash Flow	14,072,000	-56,678,000	50,728,000	-67,863,000	-30,524,000	26,272,000
Capital Expenditure	35,912,000	75,013,000	42,360,000	41,352,000	66,747,000	25,200,000
EBITDA	161,595,000	288,126,000	244,657,000	-162,211,000	220,428,000	201,366,000
Return on Assets %	1.15%	3.01%	2.60%	-7.68%	2.69%	3.69%
Return on Equity %	3.25%	8.26%	7.30%	-20.86%	6.35%	7.95%
Debt to Equity	0.51	0.57	0.49	0.58	0.43	0.28

CONTACT INFORMATION:
Phone: 818 362-8391   Fax:
Toll-Free:
Address: 15901 Olden St., Sylmar, CA 91342 United States

STOCK TICKER/OTHER:
Stock Ticker: TPC
Employees: 10,939
Parent Company:

Exchange: NYS
Fiscal Year Ends: 12/31

SALARIES/BONUSES:
Top Exec. Salary: Bonus: $
$1,500,058
Second Exec. Salary: Bonus: $
$796,875

OTHER THOUGHTS:
Estimated Female Officers or Directors:

Hot Spot for Advancement for Women/Minorities:

UDR Inc

NAIC Code: 0

www.udr.com

TYPES OF BUSINESS:

Real Estate Investment Trust
Apartment Communities

BRANDS/DIVISIONS/AFFILIATES:

Heritage Communities LP
United Dominion Realty LP

CONTACTS: *Note: Officers with more than one job title may be intentionally listed here more than once.*

Thomas Toomey, CEO
Thomas Herzog, CFO
James Klingbeil, Chairman of the Board
Shawn Johnston, Chief Accounting Officer
Jerry Davis, COO
Lynne Sagalyn, Director
Warren Troupe, Senior Executive VP
Mark Schumacher, Senior VP
Harry Alcock, Senior VP, Divisional

GROWTH PLANS/SPECIAL FEATURES:

UDR, Inc. is a self-administered real estate investment trust (REIT) that owns, acquires, renovates, develops and manages multifamily apartment communities nationwide. UDR's property portfolio includes 139 communities located in 20 markets in the U.S. and Washington, D.C. The firm currently owns 39,851 completed apartment homes and maintains an ownership interest in 10,055 apartment units (36 communities) through joint ventures. UDR is also developing one wholly-owned community with 369 apartment homes and three unconsolidated joint venture community with 1,018 apartment homes, none of which have been completed. Its subsidiaries include two operating partnerships: Heritage Communities LP and United Dominion Realty LP. UDR's upgrade and rehabilitation programs enable it to raise rents and attract residents with higher levels of disposable income who are more likely to accept the transfer of expenses, such as water and sewer costs, from the landlord to the resident. In October 2015, the firm acquired six Washington, D.C. communities for $901 million from Home Properties LP.

UDR offers its employees benefits including life, disability, medical, business travel accident, AD&D, dental and vision coverage; a 401(k); flexible spending accounts; apartment discounts; and educational assistance.

FINANCIAL DATA: *Note: Data for latest year may not have been available at press time.*

In U.S. $	2015	2014	2013	2012	2011	2010
Revenue	894,638,000	818,046,000	758,926,000	729,363,000	708,685,000	646,596,000
R&D Expense						
Operating Income	159,591,000	126,764,000	116,920,000	-64,739,000	-110,931,000	-107,109,000
Operating Margin %	17.83%	15.49%	15.40%	-8.87%	-15.65%	-16.56%
SGA Expense	59,690,000	47,800,000	42,238,000	43,059,000	45,915,000	42,710,000
Net Income	340,383,000	154,334,000	44,812,000	212,177,000	20,023,000	-102,899,000
Operating Cash Flow	431,615,000	392,360,000	339,902,000	317,341,000	244,236,000	214,180,000
Capital Expenditure	465,423,000	582,479,000	441,918,000	506,420,000	1,192,455,000	518,043,000
EBITDA	604,748,000	495,545,000	469,355,000	419,979,000	413,814,000	201,180,000
Return on Assets %	4.64%	2.20%	.60%	2.98%	.17%	-2.10%
Return on Equity %	12.14%	5.52%	1.43%	7.92%	.57%	-7.95%
Debt to Equity	1.25	1.33	1.27	1.15	1.79	2.41

CONTACT INFORMATION:

Phone: 720 283-6120   Fax: 720 283-2451
Toll-Free:
Address: 1745 Shea Center Dr., Ste. 200, Highlands Ranch, CO 80129
United States

STOCK TICKER/OTHER:

Stock Ticker: UDR
Employees: 1,582
Parent Company:

Exchange: NYS
Fiscal Year Ends: 12/31

SALARIES/BONUSES:

Top Exec. Salary: $800,000   Bonus: $
Second Exec. Salary: Bonus: $
$500,000

OTHER THOUGHTS:

Estimated Female Officers or Directors: 9
Hot Spot for Advancement for Women/Minorities: Y

Unibail-Rodamco SE

www.unibail-rodamco.com

NAIC Code: 531120

TYPES OF BUSINESS:

Commercial Property Investment Company
Office Buildings
Shopping Center Real Estate
Convention Center Real Estate

BRANDS/DIVISIONS/AFFILIATES:

Rodamco Europe NV
Unibail Holdings SA
Viparis
So Ouest

GROWTH PLANS/SPECIAL FEATURES:

Unibail-Rodamco SE, formed by the merger of Rodamco Europe NV and Unibail Holdings SA, is a Paris-based commercial property investment company that spans across Europe. It operates within three business sectors: offices, shopping centers and exhibition & convention complexes. The firm's office portfolio is comprised of office buildings, mainly located in the Paris central business district and its western outskirts. This segment has a portfolio of 15 office buildings and is focused on large buildings which have been refurbished and equipped with advanced technical installations. Unibail's portfolio of 72 shopping centers is almost exclusively comprised of very large shopping centers. The exhibition & convention centers segment operates through subsidiary Viparis, owning and operating 11 world-class convention & exhibition centers in and around Paris. This division manages over 100 events yearly, with over 3.5 million visitors. In February 2016, the firm agreed to sell its So Ouest office building located in Levallois-Perret, Paris.

CONTACTS: Note: Officers with more than one job title may be intentionally listed here more than once.

Christophe Cuvillier, CEO
Jean-Marie Tritant, COO
Jaap L. Tonckens, CFO
Catherine Pourre, Chief Resources Officer
Olivier Bossard, Chief Dev. Officer
Fabrice Mouchel, Deputy CFO
Robert Ter Haar, Chmn.

FINANCIAL DATA: Note: Data for latest year may not have been available at press time.

In U.S. $	2015	2014	2013	2012	2011	2010
Revenue	1,921,739,000	1,941,127,000	1,806,891,000	1,765,833,000	1,668,663,000	1,693,069,000
R&D Expense	5,132,241	4,676,042	4,561,992	5,132,241	5,930,589	8,211,585
Operating Income	3,704,565,000	2,490,619,000	1,800,276,000	2,846,455,000	2,411,355,000	3,415,335,000
Operating Margin %	192.77%	128.30%	99.63%	161.19%	144.50%	201.72%
SGA Expense	118,611,800	99,337,370	91,810,080	94,547,280	92,494,380	108,575,400
Net Income	2,661,922,000	1,905,202,000	1,471,927,000	1,663,530,000	1,514,353,000	2,834,707,000
Operating Cash Flow	1,614,489,000	1,811,567,000	1,471,470,000	1,518,345,000	1,390,267,000	1,393,802,000
Capital Expenditure			1,733,899,000	1,553,472,000	1,544,462,000	940,796,700
EBITDA	3,495,512,000	2,526,887,000	1,821,717,000	2,575,929,000	2,330,265,000	3,425,599,000
Return on Assets %	6.33%	4.91%	4.16%	5.21%	5.16%	9.18%
Return on Equity %	15.27%	11.83%	9.70%	11.88%	11.71%	19.58%
Debt to Equity	0.81	0.94	0.79	0.68	0.72	0.66

CONTACT INFORMATION:

Phone: 33 153437437   Fax: 33 153437438
Toll-Free:
Address: 7 Place du Chancelier Adenauer, Paris, CS 31622-75722 France

STOCK TICKER/OTHER:

Stock Ticker: UNRDY   Exchange: PINX
Employees: 2,089   Fiscal Year Ends: 12/31
Parent Company:

SALARIES/BONUSES:

Top Exec. Salary: $ Bonus: $
Second Exec. Salary: $ Bonus: $

OTHER THOUGHTS:

Estimated Female Officers or Directors: 3
Hot Spot for Advancement for Women/Minorities: Y

Sales, profits and employees may be estimates. Financial information, benefits and other data can change quickly and may vary from those stated here.

Unitech Limited

www.unitechgroup.com

NAIC Code: 531100

TYPES OF BUSINESS:

Real Estate Development
Construction and Civil Engineering Services
Telecommunications

BRANDS/DIVISIONS/AFFILIATES:

CONTACTS: Note: Officers with more than one job title may be intentionally listed here more than once.

Sanjay Chandra, Joint Managing Dir.
Ajay Chandra, Joint Managing Dir.
Ramesh Chandra, Chmn.

GROWTH PLANS/SPECIAL FEATURES:

Unitech Limited is an India-based real estate development and construction firm. The company's operations are organized into four segments: residential, commercial, retail and hospitality. The residential segment includes sold out housing units in Chennai, Greater Noida, Gurgaon, Kolkata, Lucknow, Mohali and Noida; completed housing projects in Greater Noida, Gurgaon, Kolkata and Lucknow; and current housing projects in Ambala, Bangalore, Bhopal, Chennai, Dehradun, Greater Noida, Gurgaon, Kolkata, Mohali, Noida and Rewari (up to one bedroom) â€" Gurgaon, Kolkata and Noida (one to three bedrooms) â€" and Gurgaon and Noida (more than three bedrooms). The commercial segment leases office space buildings including Unitech Business Park, Unitech Trade Centre, Signature Towers and Unitech Cyber Park. The retail segment includes amusement park, hotel and shopping centers such as Adventure Island in Rohini and Worlds of Wonder in Noida. The hospitality segment builds hotels, clubs and serviced apartments, with completed projects located in Gurgaon and Kolkata, and current projects located in Bangalore and Noida.

FINANCIAL DATA: Note: Data for latest year may not have been available at press time.

In U.S. $	2015	2014	2013	2012	2011	2010
Revenue		459,400,000	382,200,000	378,700,000	636,712,000	673,260,000
R&D Expense						
Operating Income						
Operating Margin %						
SGA Expense						
Net Income		10,900,000	32,800,000	37,200,000	112,402,000	150,730,000
Operating Cash Flow						
Capital Expenditure						
EBITDA						
Return on Assets %						
Return on Equity %						
Debt to Equity						

CONTACT INFORMATION:

Phone: 91-124-4125200   Fax: 91-124-2383332
Toll-Free:
Address: Unitech House L Block, South City-1, Gurgaon, 122007 India

STOCK TICKER/OTHER:

Stock Ticker: UNTHY   Exchange: GREY
Employees: 1,577   Fiscal Year Ends: 03/31
Parent Company:

SALARIES/BONUSES:

Top Exec. Salary: $ Bonus: $
Second Exec. Salary: $ Bonus: $

OTHER THOUGHTS:

Estimated Female Officers or Directors: 1
Hot Spot for Advancement for Women/Minorities:

Ventas Inc

NAIC Code: 0

TYPES OF BUSINESS:

Real Estate Investment Trust
Hospitals
Senior Housing Properties
Medical Office Buildings
Skilled Nursing Facilities

BRANDS/DIVISIONS/AFFILIATES:

Lillibridge Healthcare Services Inc
PMB Real Estate Services LLC
American Realty Capital Healthcare Trust Inc
Ardent Medical Services Inc
Care Capital Properties Inc

CONTACTS: Note: Officers with more than one job title may be intentionally listed here more than once.

Todd Lillibridge, CEO, Subsidiary
Robert Probst, CFO
Debra Cafaro, Chairman of the Board
Gregory Liebbe, Chief Accounting Officer
T. Riney, Chief Administrative Officer
John Cobb, Executive VP

GROWTH PLANS/SPECIAL FEATURES:

Ventas, Inc. is a health care real estate investment trust (REIT) with health care and senior living facilities in operation throughout the U.S. and Canada. The company primarily engages in acquiring, owning and leasing health care and senior living properties. Additionally, through its subsidiaries Lillibridge Healthcare Services, Inc. and its interest in PMB Real Estate Services LLC, Ventas engages in medical office building (MOB) management, leasing, marketing, facility development and advisory services. The company also makes mortgage loans to health care and senior housing operators. The firm maintains a portfolio of more than 1,190 properties, including seniors housing communities, skilled nursing and other facilities, medical office buildings and hospitals throughout the U.S., Canada and the U.K. The company's facilities in California accounted for 15.4% of its 2015 total revenues. The company leases its facilities either under triple-net or absolute-net leases, which require the tenants to pay all property-related expenses, such as maintenance, taxes and insurance, or operates its properties through third-party managers such as Sunrise Senior Living LLC, Atria Senior Living, Inc., Brookdale Senior Living and Kindred Healthcare, Inc. In 2015, the firm acquired American Realty Capital Healthcare Trust, Inc.; acquired Ardent Medical Services, Inc.; and began to spin off its Care Capital Properties, Inc. (CCP) business by selling or disposing seven of nine CCP-related properties by years' end.

FINANCIAL DATA: Note: Data for latest year may not have been available at press time.

In U.S. $	2015	2014	2013	2012	2011	2010
Revenue	3,286,398,000	3,075,746,000	2,751,845,000	2,485,299,000	1,764,991,000	1,016,867,000
R&D Expense						
Operating Income	486,987,000	538,057,000	462,349,000	280,112,000	330,493,000	227,797,000
Operating Margin %	14.81%	17.49%	16.80%	11.27%	18.72%	22.40%
SGA Expense	128,035,000	121,746,000	115,106,000	98,801,000	74,537,000	49,830,000
Net Income	417,843,000	475,767,000	453,509,000	362,800,000	364,493,000	246,167,000
Operating Cash Flow	1,391,767,000	1,254,845,000	1,194,755,000	992,816,000	773,197,000	447,622,000
Capital Expenditure	2,758,275,000	1,437,494,000	1,518,616,000	1,636,497,000	629,669,000	294,296,000
EBITDA	1,696,480,000	1,653,375,000	1,582,847,000	1,356,442,000	1,026,952,000	620,703,000
Return on Assets %	1.92%	2.32%	2.34%	2.00%	3.16%	4.32%
Return on Equity %	4.58%	5.43%	5.08%	3.96%	6.25%	10.14%
Debt to Equity	1.17	1.25	1.06	0.93	0.69	1.21

CONTACT INFORMATION:

Phone: 877 483-6827   Fax:
Toll-Free: 877-483-6827
Address: 353 N. Clark St., Ste. 3300, Chicago, IL 60654 United States

STOCK TICKER/OTHER:

Stock Ticker: VTR   Exchange: NYS
Employees: 466   Fiscal Year Ends: 12/31
Parent Company:

SALARIES/BONUSES:

Top Exec. Salary: Bonus: $
$1,075,000
Second Exec. Salary: Bonus: $
$575,000

OTHER THOUGHTS:

Estimated Female Officers or Directors: 3

Hot Spot for Advancement for Women/Minorities: Y

VINCI SA

www.vinci.com

NAIC Code: 237310

TYPES OF BUSINESS:

Highway, Street, and Bridge Construction
Infrastructure Management
Information & Energy Technologies
Commercial Construction
Engineering Services
Highway Construction
Airport Management & Support Services
Power Transmission Services

BRANDS/DIVISIONS/AFFILIATES:

VINCI Concessions SA
VINCI Contracting LLC
VINCI Autoroutes
VINCI Airports
VINCI Energies
Eurovia
VINCI Construction

CONTACTS: Note: Officers with more than one job title may be intentionally listed here more than once.

Xavier Huillard, CEO
Pierre Coppey, COO
Christian Labeyrie, CFO
Franck Mougin, VP-Human Resources and Corporate Social Responsibility
Pierre Anjolras, Chmn.

GROWTH PLANS/SPECIAL FEATURES:

VINCI SA is one of the largest companies operating in construction and related services worldwide. The company designs, finances, builds and operates infrastructures and facilities that help improve daily life and mobility for all. It consists of two major subsidiary divisions: VINCI Concessions SA, comprising VINCI Autoroutes, VINCI Airports and other concessions; and VINCI Contracting LLC, comprising VINCI Energies, Eurovia and VINCI Construction. The concessions division is Europe's leading transport infrastructure concession operator, operating in the motorway, airport, bridge & tunnel, rail, stadium and parking facility sectors. VINCI Autoroutes operates more than 2,700 miles of motorways in France, carrying over 2 million customers each day; VINCI Airports operates 23 airports in Portugal, France and Cambodia, transporting nearly 47 million passengers each year; and other concessions include motorways, bridges, tunnels, rail and parking facilities. The contracting division is engaged in building contracting, value engineering, interior design, insulation works, metal & steel structures, electrical works, plumbing works and air conditioning (HVAC), having worked on more than 260,000 projects in some 110 countries. VINCI Energies serves public authorities and business clients by helping them to deploy electrical power & heating, HVAC, mechanical engineering and communication equipment, as well as operating their energy, transport and communication infrastructures; Eurovia is a world leader in transport infrastructure and urban development, building and upgrading roads, motorways, rail systems, airport hard surfaces and industrial facilities; and VINCI Construction brings together more than 790 consolidated companies in some 100 countries by delivering capabilities in building, civil engineering, hydraulic engineering and contracting-related specialties. In 2015, VINCI and Orix won a 45-year contract to operate Itami Airport and Kansai International Airport in Japan, at a price of approximately $18 billion.

FINANCIAL DATA: Note: Data for latest year may not have been available at press time.

In U.S. $	2015	2014	2013	2012	2011	2010
Revenue	44,663,040,000	44,528,460,000	46,463,890,000	44,688,360,000	42,935,300,000	38,780,130,000
R&D Expense						
Operating Income	4,236,950,000	4,839,133,000	4,296,256,000	4,163,958,000	4,106,933,000	3,910,882,000
Operating Margin %	9.48%	10.86%	9.24%	9.31%	9.56%	10.08%
SGA Expense					1,150,990,000	12,892,190,000
Net Income	2,333,459,000	2,835,278,000	2,237,657,000	2,185,992,000	2,276,434,000	2,167,630,000
Operating Cash Flow	5,157,331,000	4,143,429,000	4,160,536,000	4,407,682,000	4,490,825,000	3,860,927,000
Capital Expenditure	1,864,714,000	1,718,730,000	886,166,900	992,917,500	2,126,002,000	1,745,988,000
EBITDA	6,577,252,000	7,223,914,000	6,673,053,000	6,384,850,000	6,393,745,000	5,918,956,000
Return on Assets %	3.26%	3.94%	3.14%	3.13%	3.25%	3.26%
Return on Equity %	13.70%	17.21%	14.28%	14.61%	15.11%	16.06%
Debt to Equity	0.99	1.16	1.24	0.52	1.35	1.43

CONTACT INFORMATION:

Phone: 33 147163500   Fax: 33 147519102
Toll-Free:
Address: 1 Cours Ferdinand-de-Lesseps, Cedex, 92851 France

SALARIES/BONUSES:

Top Exec. Salary: $ Bonus: $
Second Exec. Salary: $ Bonus: $

STOCK TICKER/OTHER:

Stock Ticker: VCISY   Exchange: PINX
Employees: 214   Fiscal Year Ends: 12/31
Parent Company:

OTHER THOUGHTS:

Estimated Female Officers or Directors: 2
Hot Spot for Advancement for Women/Minorities:

VOA Associates Inc

www.voa.com

NAIC Code: 541310

TYPES OF BUSINESS:
Architectural Services

GROWTH PLANS/SPECIAL FEATURES:

VOA Associates, Inc. is an international architecture, planning and design firm with offices in Chicago, Illinois; Highland, Indiana; New York, New York; Orlando, Florida; Los Angeles, California; Washington D.C.; Sao Paulo, Brazil; and Beijing, China. The firm's services include architectural, master planning, interior design, landscape architecture and renovation for the commercial and mixed-use, corporate interiors, nonprofit, education, governmental, entertainment and cultural, health care, hospitality, retail, multi-family housing and residential markets. Select completed projects of VOA include the Harley Davidson-Financial Services Center, a 103,000 square foot facility with an open road concept; Hanley Hall at Purdue University, housing the Human Development Institute and the College of Consumer and Family Sciences; and the Air Force Officers Club in Seoul, a 90,000 square foot, 130-room hotel for the Korean Air Force.

BRANDS/DIVISIONS/AFFILIATES:

CONTACTS: Note: Officers with more than one job title may be intentionally listed here more than once.

Michael Toolis, CEO
Percy (Rebel) Roberts III, COO
Nicholas Ciotola, CFO
Theresa Danko, Sr. VP-Mktg.
Michael Toolis, Chmn.

FINANCIAL DATA: Note: Data for latest year may not have been available at press time.

In U.S. $	2015	2014	2013	2012	2011	2010
Revenue	70,000,000	55,630,000				
R&D Expense						
Operating Income						
Operating Margin %						
SGA Expense						
Net Income						
Operating Cash Flow						
Capital Expenditure						
EBITDA						
Return on Assets %						
Return on Equity %						
Debt to Equity						

CONTACT INFORMATION:
Phone: 312-554-1400 Fax: 312-554-1412
Toll-Free:
Address: 224 S. Michigan Ave., Ste. 1400, Chicago, IL 60604 United States

STOCK TICKER/OTHER:
Stock Ticker: Private
Employees: 366
Parent Company:

Exchange:
Fiscal Year Ends:

SALARIES/BONUSES:
Top Exec. Salary: $ Bonus: $
Second Exec. Salary: $ Bonus: $

OTHER THOUGHTS:
Estimated Female Officers or Directors:
Hot Spot for Advancement for Women/Minorities:

Vornado Realty Trust

www.vno.com

NAIC Code: 0

TYPES OF BUSINESS:

Real Estate Investment Trust
Office Properties
Retail Properties
Toys
Merchandise

BRANDS/DIVISIONS/AFFILIATES:

Vornado Realty LP
Hotel Pennsylvania
Alexander's Inc
Merchandise Mart of Chicago
Toys R Us Inc
555 California Street
Vornado Capital Partners
Crown Plaza Times Square Hotel

CONTACTS: *Note: Officers with more than one job title may be intentionally listed here more than once.*

Stephen Theriot, CFO
Steven Roth, Chairman of the Board
Joseph Macnow, Chief Administrative Officer
Michael Franco, Executive VP
David Greenbaum, President, Divisional
Mitchell Schear, President, Divisional
Alan Rice, Secretary
Michael Fascitelli, Trustee
Daniel Tisch, Trustee
Candace Beinecke, Trustee
David Mandelbaum, Trustee
Michael Fascitelli, Trustee
Russell Wight, Trustee
Robert Kogod, Trustee
Michael Lynne, Trustee
Richard West, Trustee

GROWTH PLANS/SPECIAL FEATURES:

Vornado Realty Trust, which conducts its business through 94%-owned Vornado Realty L.P., is a fully integrated real estate investment trust (REIT). It is one of the largest property owners in New York City. The firm has three platforms: New York; Washington, D.C.; and other real estate and related investments. In New York, the firm manages 21.3 million square feet of Manhattan office space in 35 properties and four residential properties, 2.6 million square feet of Manhattan street retail space in 65 properties, Hotel Pennsylvania and owns a 32.4% interest in Alexander's, Inc. In Washington D.C., the company owns 15.8 million square feet of office space in 57 properties and 2,414 units in seven residential properties. The other real estate and related investments division maintains the Merchandise Mart of Chicago (3.6 million square feet). This division also owns a 32.5% interest in Toys R Us, Inc.; a 70% controlling interest in 555 California Street, a three-building office complex in San Francisco's financial district; and a 25.0% interest in Vornado Capital Partners, its real estate fund. In 2015, the firm spun off all of its retail segment to Urban Edge Properties; and sold several properties. That same year it acquired 150 West 34th Street, The Center Building in Long Island City, 260 Eleventh Avenue, 265 West 34th Street, increased its Crown Plaza Times Square Hotel ownership to 33% and entered a joint venture in which it has a 55% ownership at 512 West 22nd Street to develop a Class-A office building.

The firm offers employees medical, prescription and dental coverage; life insurance; flexible spending accounts for medical and dependent care; an employee assistance program; and tuition reimbursement.

FINANCIAL DATA: *Note: Data for latest year may not have been available at press time.*

In U.S. $	2015	2014	2013	2012	2011	2010
Revenue	2,502,267,000	2,325,483,000	2,020,540,000	2,426,765,000	2,678,806,000	2,724,752,000
R&D Expense						
Operating Income	760,248,000	505,185,000	133,821,000	416,731,000	619,294,000	750,887,000
Operating Margin %	30.38%	21.72%	6.62%	17.17%	23.11%	27.55%
SGA Expense	1,186,556,000	1,250,677,000	1,265,997,000	1,223,613,000	1,301,578,000	1,313,703,000
Net Income	760,434,000	864,852,000	475,971,000	617,260,000	662,302,000	647,883,000
Operating Cash Flow	672,150,000	1,135,310,000	1,040,789,000	825,049,000	702,499,000	771,086,000
Capital Expenditure	792,232,000	823,393,000	469,417,000	362,525,000	349,604,000	474,982,000
EBITDA						
Return on Assets %	3.20%	3.78%	1.86%	2.59%	2.93%	2.93%
Return on Equity %	12.48%	14.30%	7.06%	9.62%	10.61%	10.89%
Debt to Equity	2.04	1.99	1.81	0.45	1.81	1.96

CONTACT INFORMATION:

Phone: 212 894-7000   Fax:
Toll-Free:
Address: 888 7th Ave., New York, NY 10019 United States

STOCK TICKER/OTHER:

Stock Ticker: VNO
Employees: 4,503
Parent Company:

Exchange: NYS
Fiscal Year Ends: 12/31

SALARIES/BONUSES:

Top Exec. Salary:
$1,000,000
Second Exec. Salary:
$1,000,000

Bonus: $1,000,000

Bonus: $811,043

OTHER THOUGHTS:

Estimated Female Officers or Directors: 2

Hot Spot for Advancement for Women/Minorities:

Sales, profits and employees may be estimates. Financial information, benefits and other data can change quickly and may vary from those stated here.

Walsh Group (The)

NAIC Code: 236220

TYPES OF BUSINESS:
Commercial and Institutional Building Construction

BRANDS/DIVISIONS/AFFILIATES:
Walsh Construction Company
Archer Western

CONTACTS: Note: Officers with more than one job title may be intentionally listed here more than once.
Matthew M Walsh, CEO
Daniel Walsh, Pres.

GROWTH PLANS/SPECIAL FEATURES:
The Walsh Group is a family-held general building construction firm operating in the U.S. The Walsh Group is composed of two companies, the original Walsh Construction Company and its subsidiary Archer Western. The Walsh Construction Company, one of the largest construction firms in Chicago, is a general contracting, construction management and design-build firm. The firm has experience with a wide variety of building, civil, and transportation sector projects. The Walsh Construction Company has offices throughout North America. Archer Western, headquartered in Atlanta, Georgia, performs as an open-shop contractor. The firm has experience with a wide variety of building, civil, and transportation sector projects. Archer Western has seven regional offices nationwide. The Walsh Group is consistently recognized as one of the top construction firms in the U.S. Projects include buildings such as correctional, data centers, education, entertainment, federal, health care, high-rise residential, hospitality, laboratories, multi-family residences, office, parking, retail, senior living and warehouse distribution; transportation, such as bridges, highways and transit; aviation, such as airports and terminals; water, such as pump stations, waste water treatment plants and water treatment plants; industrial, such as mining; and power, such as hydroplants.

FINANCIAL DATA: Note: Data for latest year may not have been available at press time.

In U.S. $	2015	2014	2013	2012	2011	2010
Revenue	4,608,100,000	3,600,000,000				
R&D Expense						
Operating Income						
Operating Margin %						
SGA Expense						
Net Income						
Operating Cash Flow						
Capital Expenditure						
EBITDA						
Return on Assets %						
Return on Equity %						
Debt to Equity						

CONTACT INFORMATION:
Phone: 312-563-5400   Fax: 312-563-5466
Toll-Free:
Address: 929 W. Adams, Chicago, IL 60607 United States

STOCK TICKER/OTHER:
Stock Ticker: Private   Exchange:
Employees: 6,500   Fiscal Year Ends:
Parent Company:

SALARIES/BONUSES:
Top Exec. Salary: $ Bonus: $
Second Exec. Salary: $ Bonus: $

OTHER THOUGHTS:
Estimated Female Officers or Directors:
Hot Spot for Advancement for Women/Minorities:

Washington Real Estate Investment Trust

www.writ.com

NAIC Code: 531120

TYPES OF BUSINESS:

Real Estate Investment Trust, Commercial
Office Buildings
Medical Office Buildings
Retail Properties
Apartments

BRANDS/DIVISIONS/AFFILIATES:

Glimcher Realty Trust

CONTACTS: Note: Officers with more than one job title may be intentionally listed here more than once.

Paul McDermott, CEO
Laura Franklin, Executive VP, Divisional
Stephen Riffee, Executive VP
Thomas Bakke, Executive VP
Thomas Morey, General Counsel
Tejal Engman, Other Corporate Officer
Benjamin Butcher, Trustee
Anthony Winns, Trustee
Edward Civera, Trustee
Wendelin White, Trustee
Thomas Nolan, Trustee
William Byrnes, Trustee
Charles Nason, Trustee
W. Hammond, Vice President

GROWTH PLANS/SPECIAL FEATURES:

Washington Real Estate Investment Trust (WREIT) is a self-administered, self-managed, equity real estate investment trust successor to a trust organized in 1960. Its business consists of the ownership and operation of income-producing real property in the greater Washington metro region. The company owns a diversified portfolio of office buildings, multifamily buildings and retail centers. In 2015, it owned a diversified portfolio of 54 properties, totaling approximately 7.2 million square feet of commercial space and 3,258 residential units, and land held for development. These 54 properties consist of 25 office properties, 16 retail centers and 13 multifamily properties. WREIT's principal objective is to invest in high quality properties in prime locations, then proactively manage, lease and direct ongoing capital improvement programs to improve their economic performance. In 2015, the firm acquired Glimcher Realty Trust for $4.3 billion, which will continue to trade under its WPG ticker.

The company offers employees medical, vision and dental coverage; tuition reimbursement; flexible spending accounts for medical expenses and dependent care; short- and long-term disability insurance; annual incentive bonuses; an employee assistance program; and annual company picnics.

FINANCIAL DATA: Note: Data for latest year may not have been available at press time.

In U.S. $	2015	2014	2013	2012	2011	2010
Revenue	306,427,000	288,637,000	263,024,000	304,983,000	289,527,000	297,977,000
R&D Expense						
Operating Income	148,143,000	64,030,000	65,191,000	80,821,000	83,310,000	90,657,000
Operating Margin %	48.34%	22.18%	24.78%	26.50%	28.77%	30.42%
SGA Expense	20,257,000	19,761,000	17,535,000	15,488,000	15,728,000	14,406,000
Net Income	89,740,000	111,639,000	37,346,000	23,708,000	104,884,000	37,426,000
Operating Cash Flow	107,355,000	80,701,000	113,318,000	131,103,000	117,855,000	111,933,000
Capital Expenditure	72,710,000	297,329,000	119,855,000	110,371,000	315,137,000	182,655,000
EBITDA	257,668,000	160,866,000	161,281,000	185,730,000	183,838,000	186,403,000
Return on Assets %	4.16%	5.46%	1.82%	1.11%	4.89%	1.77%
Return on Equity %	10.84%	14.18%	4.83%	2.87%	12.22%	4.67%
Debt to Equity	1.51	1.42	1.51	1.57	1.37	1.43

CONTACT INFORMATION:

Phone: 202-774-3200   Fax:
Toll-Free: 800-565-9748
Address: 1775 Eye Street NW, Ste 1000, Washington D.C., MD 20006
United States

STOCK TICKER/OTHER:

Stock Ticker: WRE
Employees: 181
Parent Company:

Exchange: NYS
Fiscal Year Ends: 12/31

SALARIES/BONUSES:

Top Exec. Salary: $500,000   Bonus: $
Second Exec. Salary: $386,045   Bonus: $

OTHER THOUGHTS:

Estimated Female Officers or Directors: 2
Hot Spot for Advancement for Women/Minorities:

Watson Land Company

www.watsonlandcompany.com

NAIC Code: 531120

TYPES OF BUSINESS:

Office & Industrial Properties
Land Development
Commercial Construction
Property Management
Building Maintenance & Operations

BRANDS/DIVISIONS/AFFILIATES:

Heritage Customer Service Program

CONTACTS: *Note: Officers with more than one job title may be intentionally listed here more than once.*

Bruce A. Choate, CEO
Jeffrey Jennison, Pres.
Roger von Ting, CFO
Lance P. Ryan, VP-Mktg. & Leasing
Kirk R. Johnson, CIO
Bradley D. Frazier, General Counsel
Kirk R. Johnson, Exec. VP-Real Estate Oper.
Pilar M. Hoyos, VP-Public Affairs
Jeffrey R. Jennison, VP-Real Estate Asset Mgmt.
Christopher L. Trujillo, VP-Construction
Craig B. Halverson, VP-Acquisitions
Robert W. Huston, Chmn.

GROWTH PLANS/SPECIAL FEATURES:

Watson Land Company is a private REIT that develops, owns and manages industrial properties in Southern California. It is one of the largest developers of industrial centers in Los Angeles County, having planned acres of industrial and commercial property in addition to owning and managing over 17 million square feet of industrial, office and technology buildings and business centers. Watson specializes in building master-planned facilities for warehousing, distribution, assembly and manufacturing. It also builds office buildings and multi-tenant business centers that serve as corporate headquarters; administrative and general offices; and computer operations, data processing, technology and service-related centers. Watson's Heritage Customer Service Program offers clients a unique lease structure through which Watson assumes and manages the maintenance and operating functions typically performed by the customer. The Heritage program is tailored to the client and can include roofing, yard, lighting, painting and HVAC (heating, ventilation and air conditioning) maintenance, repair and replacement services. Watson has also completed the incorporation of several master-planned business centers into a General Purpose Zone Site as part of Foreign Trade Zone (FTZ) 202, sponsored by the Port of Los Angeles. FTZ status is intended to provide approved customers enhanced cost savings and operating benefits to maximize efficiency. In addition, the company has land holdings available for development in the Inland Empire region of California.

FINANCIAL DATA: *Note: Data for latest year may not have been available at press time.*

In U.S. $	2015	2014	2013	2012	2011	2010
Revenue						
R&D Expense						
Operating Income						
Operating Margin %						
SGA Expense						
Net Income						
Operating Cash Flow						
Capital Expenditure						
EBITDA						
Return on Assets %						
Return on Equity %						
Debt to Equity						

CONTACT INFORMATION:

Phone: 310-952-6400   Fax: 310-522-8788
Toll-Free:
Address: 22010 S. Wilmington Ave., Carson, CA 90745 United States

STOCK TICKER/OTHER:

Stock Ticker: Private   Exchange:
Employees: 30   Fiscal Year Ends: 12/31
Parent Company:

SALARIES/BONUSES:

Top Exec. Salary: $ Bonus: $
Second Exec. Salary: $ Bonus: $

OTHER THOUGHTS:

Estimated Female Officers or Directors: 1
Hot Spot for Advancement for Women/Minorities:

WCI Communities Inc

www.wcicommunities.com

NAIC Code: 236117

TYPES OF BUSINESS:

Construction, Home Building and Residential
Country Club Management
Title Insurance
Design & Engineering Services
Residential Brokerage
Land Development
Mortgages

BRANDS/DIVISIONS/AFFILIATES:

Berkshire Hathaway HomeServices

CONTACTS: *Note: Officers with more than one job title may be intentionally listed here more than once.*

Reinaldo Mesa, CEO, Subsidiary
Keith Bass, CEO
Russell Devendorf, CFO
Stephen Plavin, Chairman of the Board
John Ferry, Chief Accounting Officer
Vivien Hastings, General Counsel
Paul Erhardt, President, Divisional
David Ivin, President, Divisional
Jonathan Rapaport, President, Divisional
John Mcgoldrick, Senior VP, Divisional

GROWTH PLANS/SPECIAL FEATURES:

WCI Communities, Inc. designs, builds and sells traditional and luxury high-rise homes, which are typically part of master planned communities in Florida. WCI owns or controls approximately 13,300 home sites. Its businesses are divided into homebuilding, real estate services and amenities. Homebuilding, accounting for 77.7% of net revenue, designs, sells and builds homes ranging in price from $170,000 to $1.1 million and in size from approximately 1,100 square feet to 5,100 square feet. This division acquires and creates master-planned communities, and is actively selling in 47 neighborhoods situated in 18 master-planned communities. During 2015, homebuilding delivered 938 homes with an average price of $466,000. Real estate services operates a full-service real estate brokerage business under the Berkshire Hathaway HomeServices brand as well as title services that complement the homebuilding operations. During 2015, real estate services represented either the buyer or seller in approximately 9,600 home sale transactions. Amenities consists of amenities either owned and/or operated by WCI within the communities that are built. Amenities include resort-style club and fitness facilities, championship golf courses, country clubs and marinas. Current active communities of the firm include Heron Bay, Estates at Tuscany, Woodlands at Ibis Golf & Country Club, Livingston Lakes, Artesia Naples, Raffia Preserve, LaMorada, Talis Park, The Colony Golf & Bay Club, Pelican Preserve, Timberwood Preserve, Shadow Wood Preserve, Arborwood Preserve, Venetian Golf & River Club, Sarasota National, Country Club East at Lakewood Ranch, The Links at Rosedale, Tidewater Preserve, Westshore Yacht Club and Lost Key Golf & Beach Club.

WCI offers its employees medical and dental coverage; a vision reimbursement plan; life and AD&D insurance; flexible spending accounts; short- and long-term disability; and discounts for Dell computers, Sprint Wireless, dry cleaners, hotel accommodations and more.

FINANCIAL DATA: *Note: Data for latest year may not have been available at press time.*

In U.S. $	2015	2014	2013	2012	2011	2010
Revenue	562,768,000	405,863,000	314,812,000	234,030,000	127,318,000	
R&D Expense						
Operating Income	55,654,000	36,036,000	20,776,000	-4,305,000	-54,453,000	
Operating Margin %	9.88%	8.87%	6.59%	-1.83%	-42.76%	
SGA Expense	507,924,000	371,431,000	116,520,000	103,804,000	99,120,000	
Net Income	35,400,000	21,597,000	146,648,000	50,823,000	-47,125,000	
Operating Cash Flow	-36,082,000	-105,925,000	-22,581,000	22,014,000	-19,721,000	
Capital Expenditure	3,007,000	2,977,000	2,554,000	1,077,000	1,323,000	
EBITDA						
Return on Assets %	4.24%	2.89%	24.58%	15.58%	-15.45%	
Return on Equity %	7.82%	5.14%	44.26%	43.79%	-71.48%	
Debt to Equity	0.52	0.58	0.49	0.73	2.11	

CONTACT INFORMATION:

Phone: 239-498-8200   Fax: 239-498-8338
Toll-Free: 800-924-4005
Address: 24301 Walden Ctr. Dr., Bonita Springs, FL 34134 United States

STOCK TICKER/OTHER:

Stock Ticker: WCIC   Exchange: NYS
Employees: 689   Fiscal Year Ends: 12/31
Parent Company:

SALARIES/BONUSES:

Top Exec. Salary: $675,000   Bonus: $432,720
Second Exec. Salary: $357,500   Bonus: $108,750

OTHER THOUGHTS:

Estimated Female Officers or Directors: 1
Hot Spot for Advancement for Women/Minorities:

Weingarten Realty Investors

NAIC Code: 0

www.weingarten.com

TYPES OF BUSINESS:

Real Estate Investment Trust
Industrial & Office Properties
Shopping Centers
Commercial Construction
Property Management

BRANDS/DIVISIONS/AFFILIATES:

CONTACTS: *Note: Officers with more than one job title may be intentionally listed here more than once.*

Andrew Alexander, CEO
Stanford Alexander, Chairman of the Board
Joe Shafer, Chief Accounting Officer
Johnny Hendrix, COO
Stephen Richter, Executive VP
Thomas Ryan, Trustee
C. Shaper, Trustee
Douglas Schnitzer, Trustee
James Crownover, Trustee
Marc Shapiro, Trustee
Melvin Dow, Trustee
Robert Cruikshank, Trustee
Shelaghmichael Brown, Trustee
Stephen Lasher, Trustee

GROWTH PLANS/SPECIAL FEATURES:

Weingarten Realty Investors, a real estate investment trust (REIT), primarily leases available space to tenants within the neighborhood and community shopping centers it owns or leases. It owns or operates under long-term leases, directly and through interests in joint ventures or partnerships, a total of 232 developed income-producing properties and three properties under development. The firm's portfolio totals roughly 45.6 million square feet, including neighborhood and community shopping centers and other operating properties located in 18 states spanning the country from coast to coast. In Addition, Weingarten owns 30 parcels of land totaling approximately 24.1 million square feet. 17.9% of Weingarten's income is derived from properties located in the Houston metropolitan area and other parts of Texas although the company is attempting to branch out and expand in other states. The company's operating strategy consists of intensive hands-on management with a focus on long-term ownership. When acquiring properties, it attempts to accumulate enough to establish a regional office, enabling the company to obtain in-depth knowledge of the market from a leasing perspective and to have easy access to the property and tenants from a management standpoint. Weingarten's criteria for retail acquisition includes an anchored community, neighborhood or power shopping center; over 100,000 square feet; multiple tenants, including an existing anchor store, preferably a supermarket, or accommodations for one; properties that can be leased-up, re-merchandised or renovated; a trade area with a minimum population of 50,000; and a location at a major intersection of two thoroughfares.

The company offers employees medical, prescription, dental and vision coverage; flexible spending accounts for both medical expenses and dependents; a 529 college savings program; and an educational reimbursement program.

FINANCIAL DATA: *Note: Data for latest year may not have been available at press time.*

In U.S. $	2015	2014	2013	2012	2011	2010
Revenue	512,844,000	514,406,000	497,725,000	503,538,000	541,561,000	554,667,000
R&D Expense						
Operating Income	184,694,000	182,038,000	163,400,000	167,516,000	138,416,000	174,583,000
Operating Margin %	36.01%	35.38%	32.82%	33.26%	25.55%	31.47%
SGA Expense	27,524,000	24,902,000	25,371,000	28,554,000	25,528,000	25,000,000
Net Income	174,352,000	288,008,000	220,262,000	146,640,000	15,621,000	46,206,000
Operating Cash Flow	244,416,000	240,769,000	233,992,000	227,330,000	214,731,000	214,625,000
Capital Expenditure	305,481,000	144,513,000	182,757,000	293,914,000	143,988,000	142,972,000
EBITDA	346,726,000	360,445,000	407,924,000	336,491,000	309,429,000	348,269,000
Return on Assets %	4.16%	6.89%	4.37%	2.48%	-.42%	.22%
Return on Equity %	11.18%	19.35%	12.46%	6.75%	-1.14%	.57%
Debt to Equity	1.52	1.30	1.66	1.39	1.52	1.43

CONTACT INFORMATION:

Phone: 713 866-6000   Fax: 713 866-6049
Toll-Free: 800-688-8865
Address: 2600 Citadel Plaza Dr., Ste. 125, Houston, TX 77008 United States

STOCK TICKER/OTHER:

Stock Ticker: WRI
Employees: 315
Parent Company:

Exchange: NYS
Fiscal Year Ends: 12/31

SALARIES/BONUSES:

Top Exec. Salary: $700,000   Bonus: $
Second Exec. Salary: $525,000   Bonus: $

OTHER THOUGHTS:

Estimated Female Officers or Directors: 3
Hot Spot for Advancement for Women/Minorities: Y

Wells Real Estate Funds Inc

www.wellsref.com

NAIC Code: 0

TYPES OF BUSINESS:

Real Estate Investment Trust
Office & Industrial Properties
Property Management
Construction Management

BRANDS/DIVISIONS/AFFILIATES:

Wells Limited Partnership
Wells Capital

CONTACTS: Note: Officers with more than one job title may be intentionally listed here more than once.

Leo F. Wells III, Pres.

GROWTH PLANS/SPECIAL FEATURES:

Wells Real Estate Funds, Inc. is a national real estate investment management firm that purchases corporate and industrial property. It leases its holdings on a long-term basis to tenants with a net worth of at least $100 million. In total, Wells manages more than $12 billion in assets for almost 300,000 investors nationwide. The firm generally maintains a conservative investment strategy that focuses on acquiring and managing high-quality properties. Part of its strategy is leasing only to credit-worthy tenants; securing leases with major clients before acquiring a property; and using net leases in which the tenant pays all cost of taxes, insurance and maintenance. The firm maintains a diversified portfolio consisting of various geographic regions, tenants, industries and staggered lease terms. Complementing the investment business, property acquisitions and leasing, Wells also oversees property management, construction and tenant relations. Wells Limited Partnership is a wholly-owned subsidiary of the company and is involved in acquiring, developing, owning, operating, leasing, and managing income-producing commercial properties for investment purposes. Wells Capital manages investment assets for its clients.

FINANCIAL DATA: Note: Data for latest year may not have been available at press time.

In U.S. $	2015	2014	2013	2012	2011	2010
Revenue						
R&D Expense						
Operating Income						
Operating Margin %						
SGA Expense						
Net Income						
Operating Cash Flow						
Capital Expenditure						
EBITDA						
Return on Assets %						
Return on Equity %						
Debt to Equity						

CONTACT INFORMATION:

Phone: 770 449-7800   Fax: 770 243-8199
Toll-Free: 800-448-1010
Address: 5445 Triangle Pkwy, Ste 320, Peachtree Corners, GA 30092
United States

STOCK TICKER/OTHER:

Stock Ticker: Private
Employees: 117
Parent Company:

Exchange:
Fiscal Year Ends: 12/31

SALARIES/BONUSES:

Top Exec. Salary: $ Bonus: $
Second Exec. Salary: $ Bonus: $

OTHER THOUGHTS:

Estimated Female Officers or Directors:
Hot Spot for Advancement for Women/Minorities:

Welltower Inc

www.hcreit.com

NAIC Code: 531110

TYPES OF BUSINESS:

Real Estate Investment Trust, Senior Living
Health Care Properties

GROWTH PLANS/SPECIAL FEATURES:

Welltower, Inc. formerly Health Care REIT, Inc., is a self-managed and self-administered real estate investment trust (REIT) that owns, acquires, manages and develops real estate properties in the health care industry. HCR's investment portfolio consists of 1,400 properties across the U.S., Canada and the U.K. The company divides its portfolio into three segments: triple-net, seniors housing operating and outpatient medical. The firm's triple-net segment consists of independent living, independent supportive living, continuing care retirement, assisted living, Alzheimer's/dementia care and skilled nursing and long-term/post-acute care facilities. The seniors housing operating segment includes the same facilities as above, but held in consolidated joint venture entities with operating partners. Its medical facilities segment includes medical office buildings, hospitals and life science facilities. Outpatient medical office buildings house ambulatory care and surgery centers, physician's clinics, diagnostic facilities, specialty outpatient and inpatient facilities and other specialty health care operations. The firm's hospital investments include acute care hospitals, inpatient rehabilitation hospitals and long-term acute care hospitals. In September 2015, the company changed its name from Health Care REIT, Inc. to Welltower, Inc. in order to better position itself in healthcare infrastructure business. During 2015, the firm sold its life science portfolio and acquired Regal Lifestyle Communities, Inc. through joint venture with Revera, Inc.

BRANDS/DIVISIONS/AFFILIATES:

Health Care REIT Inc
Regal Lifestyle Communities Inc

CONTACTS: Note: Officers with more than one job title may be intentionally listed here more than once.

Thomas Derosa, CEO
Scott Estes, CFO
Jeffrey Donahue, Chairman of the Board
Paul Nungester, Chief Accounting Officer
Jeffrey Miller, Executive VP
Erin Ibele, Executive VP
Scott Brinker, Executive VP

FINANCIAL DATA: Note: Data for latest year may not have been available at press time.

In U.S. $	2015	2014	2013	2012	2011	2010
Revenue	3,859,826,000	3,343,546,000	2,880,608,000	1,822,099,000	1,421,162,000	680,530,000
R&D Expense						
Operating Income	660,818,000	402,538,000	105,806,000	183,053,000	155,450,000	112,214,000
Operating Margin %	17.12%	12.03%	3.67%	10.04%	10.93%	16.48%
SGA Expense	258,342,000	212,481,000	241,719,000	97,341,000	147,425,000	54,626,000
Net Income	883,750,000	512,153,000	145,050,000	297,255,000	217,610,000	128,527,000
Operating Cash Flow	1,373,468,000	1,138,670,000	988,497,000	818,133,000	588,224,000	364,741,000
Capital Expenditure	440,983,000	205,031,000	390,092,000	3,354,888,000	4,918,286,000	2,094,968,000
EBITDA	1,959,517,000	1,716,353,000	1,442,662,000	1,101,506,000	915,280,000	454,863,000
Return on Assets %	3.02%	1.85%	.36%	1.28%	1.28%	1.35%
Return on Equity %	6.35%	3.95%	.80%	2.88%	3.01%	2.73%
Debt to Equity	0.95	0.88	1.02	0.92	1.18	1.03

CONTACT INFORMATION:

Phone: 419 247-2800   Fax: 419 247-2826
Toll-Free:
Address: 4500 Dorr St., Toledo, OH 43615 United States

STOCK TICKER/OTHER:

Stock Ticker: HCN   Exchange: NYS
Employees: 476   Fiscal Year Ends: 12/31
Parent Company:

SALARIES/BONUSES:

Top Exec. Salary: $950,000   Bonus: $
Second Exec. Salary: Bonus: $
$510,000

OTHER THOUGHTS:

Estimated Female Officers or Directors: 3
Hot Spot for Advancement for Women/Minorities: Y

Westfield Corporation

www.westfieldcorp.com

NAIC Code: 531120

TYPES OF BUSINESS:

Real Estate Operations
Shopping Centers
Design & Construction
Asset Management
Property Management
Leasing
Marketing

BRANDS/DIVISIONS/AFFILIATES:

Scentre Group
Westfield Group

CONTACTS: Note: Officers with more than one job title may be intentionally listed here more than once.

Steven M. Lowy, Co-CEO
Michael Gutman, COO
Elliot Rusanow, CFO
Bett Ann Kaminkow, CMO
Peter S. Lowry, Co-CEO
Simon Tuxen, General Counsel
Kevin McKenzie, Chief Digital Officer
Mark Ryan, Dir.-Corp. Affairs
Elliott Rusanow, Head-Corp. Finance
Peter S. Lowy, Co-CEO
Robert Jordan, Managing Dir.-Australia, U.S. & New Zealand
Eamonn Cunningham, Chief Risk Officer
David Temby, Group Tax Counsel
Frank P. Lowy, Chmn.
Michael Gutman, Managing Dir.-U.K., Europe & New Markets

GROWTH PLANS/SPECIAL FEATURES:

Westfield Corporation is an internally-managed retail property developer with offices and investments throughout the U.K. and the U.S. It has investment interests in 34 shopping centers in iconic retail destinations such as London, New York, San Francisco and Los Angeles. Within the 34 malls, the corporation has a combined total of 6,480 retail outlets, 400 million annual shopping visits and a portfolio value of $29 billion. Westfield's strategy is to develop and own superior retail destinations in major cities by integrating food, fashion, leisure and entertainment, as well as by using technology to better connect retailers with consumers. The company works with the world's leading retail and luxury brands in order to create a unique shopping and leisure experience. Westfield manages every aspect of its portfolio from design, construction & development to leasing, management and marketing. Sister company Scentre Group owns and manages Westfield Group's shopping centers located in Australia and New Zealand. In 2015, the firm sold six of its U.S. malls for $1.3 billion to help fund an $11.4 billion property development program as part of its strategy to divest from non-core assets and reduce debt.

FINANCIAL DATA: Note: Data for latest year may not have been available at press time.

In U.S. $	2015	2014	2013	2012	2011	2010
Revenue	1,296,431,000	759,723,900	1,815,559,000	1,733,881,000	3,049,403,000	2,594,733,000
R&D Expense						
Operating Income	1,380,511,000	283,991,600	2,033,722,000	2,085,636,000	1,813,428,000	1,708,914,000
Operating Margin %	106.48%	37.38%	112.01%	120.28%	59.46%	65.86%
SGA Expense						31,818,530
Net Income	2,420,844,000	-199,536,600	1,219,989,000	1,307,833,000	1,166,705,000	847,986,600
Operating Cash Flow	907,907,700	515,732,400	1,256,984,000	1,555,835,000	1,777,651,000	1,869,986,000
Capital Expenditure		7,053,386	33,645,430	17,964,530	21,466,090	31,590,170
EBITDA	1,389,917,000	308,121,600	2,086,854,000	2,131,842,000	1,859,253,000	1,758,849,000
Return on Assets %	14.01%	-.93%	4.64%	4.84%	4.21%	2.68%
Return on Equity %	28.70%	-2.13%	10.52%	10.71%	9.21%	5.48%
Debt to Equity	0.56	0.68	0.81	0.70	0.71	0.85

CONTACT INFORMATION:

Phone: 310-478-4456   Fax: 310-478-1267
Toll-Free:
Address: 2049 Century Park E., 4/Fl, Century City, CA 90067 United States

STOCK TICKER/OTHER:

Stock Ticker: WEFIF   Exchange: PINX
Employees: 4,974   Fiscal Year Ends: 12/31
Parent Company:

SALARIES/BONUSES:

Top Exec. Salary: $ Bonus: $
Second Exec. Salary: $ Bonus: $

OTHER THOUGHTS:

Estimated Female Officers or Directors: 10
Hot Spot for Advancement for Women/Minorities: Y

Westgate Resorts

NAIC Code: 561599

www.westgateresorts.com

TYPES OF BUSINESS:

Condominium Time Share Exchange Services

BRANDS/DIVISIONS/AFFILIATES:

Central Florida Investments Inc
Wild Bear Inn

CONTACTS: *Note: Officers with more than one job title may be intentionally listed here more than once.*

David Siegel, CEO
Mark Waltrip, COO
Tom Dugan, CFO

GROWTH PLANS/SPECIAL FEATURES:

Westgate Resorts is a Florida-based timeshare resort. The company operates as a subsidiary of Central Florida Investments, Inc. Westgate comprises more than 13,500 villas at 28 resorts in premiere travel destinations throughout the U.S. Its resorts are located in Florida, Tennessee, South Carolina, Missouri, Utah, Nevada and Arizona, providing a diverse range of vacation experiences. Westgate owners can enjoy world-class theme parks in Orlando, relax on sandy beaches in Myrtle Beach, hike the Smoky Mountains in Gatlinburg, snow ski the slops of Park City, horseback ride on trails at a dude ranch or gamble in Las Vegas. There are vacation and budgets for every family. Each villa is furnished, features fully-equipped kitchens or kitchenettes and most offer separate living areas with sleeper sofas, leather furnishings, jetted tubs and private balconies or patios. Partners of the firm include Interval International, providing vacation owners exchange services; Westgate Cruise & Travel Collection, a premium vacation loyalty program that offers members vacation savings; Tickets2You.com, which provides discounted prices on tickets to Walt Disney World, Universal Orlando and Busch Gardens; and Enterprise, Alamo and National, each providing car rental services. In February 2016, the firm acquired Wild Bear Inn, a resort in Pigeon Forge, Tennessee, which is very close to Great Smoky Mountains National Park.

FINANCIAL DATA: *Note: Data for latest year may not have been available at press time.*

In U.S. $	2015	2014	2013	2012	2011	2010
Revenue						
R&D Expense						
Operating Income						
Operating Margin %						
SGA Expense						
Net Income						
Operating Cash Flow						
Capital Expenditure						
EBITDA						
Return on Assets %						
Return on Equity %						
Debt to Equity						

CONTACT INFORMATION:

Phone: 407-351-3351   Fax:
Toll-Free:
Address: 5601 Windhover Dr., Orlando, FL 32819 United States

SALARIES/BONUSES:

Top Exec. Salary: $ Bonus: $
Second Exec. Salary: $ Bonus: $

STOCK TICKER/OTHER:

Stock Ticker: Private   Exchange:
Employees: 6,500   Fiscal Year Ends:
Parent Company: Central Florida Investments Inc

OTHER THOUGHTS:

Estimated Female Officers or Directors:
Hot Spot for Advancement for Women/Minorities:

WeWork Companies Inc

www.wework.com

NAIC Code: 531120

TYPES OF BUSINESS:

Shared Office Space Services

BRANDS/DIVISIONS/AFFILIATES:

GROWTH PLANS/SPECIAL FEATURES:

WeWork Companies, Inc. is a real estate firm that operates as a provider of shared office space. The company divides rented office space that it then sublets to other companies, primarily startups. WeWork operates by having its clients pay a monthly membership fee following one of two tiers. The first tier of membership offers on-demand access to workspaces and conference rooms, when and where they are needed. The second tier of membership provides workspaces, offices and desks for teams of all sizes. With both tiers of membership, members gain access to all of the firm's locations around the world. Currently, WeWork has locations in the U.S., including Atlanta, Austin, Berkeley, Boston, Chicago, Denver, Los Angeles, Miami, New York City, Philadelphia, Portland, San Francisco, Seattle and Washington, D.C.; in Israel, including Be-er Sheva, Herzliya and Tel Aviv; London, U.K.; Amsterdam, Netherlands; Montreal, Canada; and Berlin, Germany. Locations coming soon include Mexico City and India.

CONTACTS: Note: Officers with more than one job title may be intentionally listed here more than once.

Adam Neumann, CEO
Arthur T. Minson, Pres.

FINANCIAL DATA: Note: Data for latest year may not have been available at press time.

In U.S. $	2015	2014	2013	2012	2011	2010
Revenue	251,000,000	74,600,000				
R&D Expense						
Operating Income						
Operating Margin %						
SGA Expense						
Net Income		4,200,000				
Operating Cash Flow						
Capital Expenditure						
EBITDA						
Return on Assets %						
Return on Equity %						
Debt to Equity						

CONTACT INFORMATION:

Phone: 877-713-1804   Fax:
Toll-Free:
Address: 115 W. 18th St., Fl. 4, New York, NY 10011 United States

STOCK TICKER/OTHER:

Stock Ticker: Private   Exchange:
Employees: 1,094   Fiscal Year Ends:
Parent Company:

SALARIES/BONUSES:

Top Exec. Salary: $ Bonus: $
Second Exec. Salary: $ Bonus: $

OTHER THOUGHTS:

Estimated Female Officers or Directors:
Hot Spot for Advancement for Women/Minorities:

Wharf (Holdings) Limited The

NAIC Code: 531100

www.wharfholdings.com

TYPES OF BUSINESS:
Real Estate Holdings & Development

BRANDS/DIVISIONS/AFFILIATES:
Wheelock and Company Limited
i-Cable Communications Limited
Harbour Centre Development Limited
Modern Terminals Limited
Harbour City
Times Square
Plaza Hollywood
Marco Polo Hotels

CONTACTS: *Note: Officers with more than one job title may be intentionally listed here more than once.*
Stephen T H Ng, Managing Dir.
Peter K C Woo, CFO
T Y Ng, Exec. Dir.
Stephen Tin Hoi Ng, Deputy Chmn.
Doreen Yuk Fong Lee, Exec. Dir.
Peter Kwong Ching Woo, Chmn.

GROWTH PLANS/SPECIAL FEATURES:
The Wharf (Holdings) Limited is a Hong Kong-based real estate investment holding firm. The company's two core properties, both located in Hong Kong, are Harbour City and Times Square. Harbour City, located in the commercial center of Tsim Sha Tsui, is Wharf Holding's flagship property, typically generating 60% of the company's annual rental income. With roughly 700 shops, the center encompasses approximately 8.4 million total square feet of commercial space consisting of offices, retail stores, serviced apartments, clubs, hotels and parking facilities. The Times Square property, located in the Causeway Bay commercial district of northern Hong Kong, encompasses approximately 900,000 square feet of retail space and over 1 million square feet of office space. Other Hong Kong properties include Plaza Hollywood, a retail center totaling over 562,000 square feet, as well as a number of upscale residential properties. In addition to its Hong Kong investments, Wharf Holding has been expanding its activities in Mainland China in recent years, having acquired a land bank totaling approximately 119 million square feet, primarily in or near metropolitan areas such as Beijing, Shanghai, Chongqing, Wuhan, Dalian, Chengdu, Suzhou, Wuxi, Hangzhou, Nanjing and Changzhou. The company also holds a 67.6% share in Modern Terminals Limited, a provider of container terminal services in the South China region since 1969. The firm also owns Marco Polo Hotels, which operates 13 owned or managed hotels in the Asia Pacific region. Additional subsidiaries include i-Cable Communications Limited, a telecommunications firm, and Harbour Centre Development Limited, a Hong Kong and Mainland China property development and investment business. The company is 50%-owned by Wheelock and Company Limited. In December 2015, the firm sold its minority stake in Sino-Ocean Land Holdings Ltd.

FINANCIAL DATA: *Note: Data for latest year may not have been available at press time.*

In U.S. $	2015	2014	2013	2012	2011	2010
Revenue	5,269,991,000	4,916,854,000	4,111,173,000	3,978,247,000	3,094,822,000	2,498,653,000
R&D Expense						
Operating Income	1,914,989,000	1,841,499,000	1,712,183,000	1,826,930,000	1,468,249,000	1,208,327,000
Operating Margin %	36.33%	37.45%	41.64%	45.92%	47.44%	48.35%
SGA Expense	411,542,800	368,609,300	389,495,900	341,920,900	274,104,000	207,318,600
Net Income	2,065,966,000	4,632,435,000	3,787,947,000	6,093,593,000	3,941,115,000	4,609,228,000
Operating Cash Flow	3,101,140,000	2,353,349,000	2,037,730,000	1,719,790,000	-5,157,178	189,139,500
Capital Expenditure	167,221,500	218,664,400	728,193,500	1,896,423,000	1,540,707,000	474,589,300
EBITDA	3,024,814,000	5,551,573,000	4,828,150,000	7,135,343,000	4,977,064,000	5,305,060,000
Return on Assets %		8.35%	7.49%	13.75%	10.91%	16.52%
Return on Equity %		12.36%	11.21%	20.92%	16.68%	25.69%
Debt to Equity		0.22	0.26	0.27	0.33	0.25

CONTACT INFORMATION:
Phone: 852 21188118   Fax: 852 21188018
Toll-Free:
Address: Canton Rd., Ocean Centre, 16/Fl, Kowloon, Hong Kong

STOCK TICKER/OTHER:
Stock Ticker: WARFY   Exchange: PINX
Employees: 15,800   Fiscal Year Ends: 12/31
Parent Company:

SALARIES/BONUSES:
Top Exec. Salary: $ Bonus: $
Second Exec. Salary: $ Bonus: $

OTHER THOUGHTS:
Estimated Female Officers or Directors: 1
Hot Spot for Advancement for Women/Minorities:

Wheelock and Company Limited

www.wheelockcompany.com

NAIC Code: 531100

TYPES OF BUSINESS:

Real Estate Holdings & Development

BRANDS/DIVISIONS/AFFILIATES:

Wharf (Holdings) Limited (The)
Wheelock Properties Limited
Landmark Harbour City
Times Square
Harbour Centre Development Ltd
i-CABLE Communications
Wheelock Properties (Singapore) Limited

CONTACTS: Note: Officers with more than one job title may be intentionally listed here more than once.

Douglas C.K. Woo, Managing Dir.
Paul Y. C. Tsui, CFO
Peter K.C. Woo, Sr. Dir.
Douglas C.K. Woo, Chmn.

GROWTH PLANS/SPECIAL FEATURES:

Wheelock and Company Limited (Wheelock & Co), founded in 1857, is an investment holding company based in Hong Kong, with services offered in China and Singapore. The firm operates through two main subsidiaries: 58%-owned The Wharf (Holdings) Limited (Wharf) and 100%-owned Wheelock Properties Limited. Wharf focuses on developing property and infrastructure in Hong Kong and China. Its properties include Landmark Harbour City and Times Square in Hong Kong, which offer retail and commercial real estate space. Wharf's group of companies includes Harbour Centre Development, Ltd., a real estate property investor and developer, and i-CABLE Communications, a telecommunications company. Wharf also has over 119 million square feet of investment properties and developable land in China. It is currently developing container terminals in the Pearl River Delta and the Yangtze River Delta and other projects along China's coast. Wheelock Properties Limited's primary business is property with a land bank of 7.4 million square feet in Hong Kong. It also has a presence in Singapore through 75.8%-owned Wheelock Properties (Singapore) Limited.

FINANCIAL DATA: Note: Data for latest year may not have been available at press time.

In U.S. $	2015	2014	2013	2012	2011	2010
Revenue	7,404,547,000	5,280,048,000	4,521,685,000	4,270,659,000	4,455,544,000	3,118,288,000
R&D Expense						
Operating Income	2,585,422,000	2,027,931,000	1,925,948,000	2,007,432,000	2,285,919,000	1,467,733,000
Operating Margin %	34.91%	38.40%	42.59%	47.00%	51.30%	47.06%
SGA Expense	508,497,800	404,967,400	416,571,100	363,581,100	301,308,100	227,560,500
Net Income	1,834,924,000	2,837,608,000	2,185,870,000	3,472,715,000	2,948,101,000	2,603,601,000
Operating Cash Flow	4,212,899,000	1,796,374,000	-20,499,780	1,493,519,000	-435,136,900	437,199,800
Capital Expenditure	168,381,900	220,211,500	729,225,000	1,935,231,000	1,560,046,000	520,746,000
EBITDA	3,779,696,000	5,916,186,000	5,117,855,000	7,555,266,000	5,992,641,000	5,770,366,000
Return on Assets %		4.38%	3.69%	6.78%	7.04%	7.92%
Return on Equity %		12.30%	10.64%	19.61%	20.51%	23.74%
Debt to Equity		0.56	0.67	0.63	0.70	0.49

CONTACT INFORMATION:

Phone: 852 21182118   Fax: 852 21182018
Toll-Free:
Address: 20 Pedder St., Wheelock House, 23/Fl, Hong Kong,   Hong Kong

STOCK TICKER/OTHER:

Stock Ticker: WHLKF
Employees: 2,600
Parent Company:

Exchange: PINX
Fiscal Year Ends: 12/31

SALARIES/BONUSES:

Top Exec. Salary: $ Bonus: $
Second Exec. Salary: $ Bonus: $

OTHER THOUGHTS:

Estimated Female Officers or Directors:
Hot Spot for Advancement for Women/Minorities:

Whiting-Turner Contracting Co (The)   www.whiting-turner.com
NAIC Code: 236220

TYPES OF BUSINESS:
Commercial and Institutional Building Construction

BRANDS/DIVISIONS/AFFILIATES:

GROWTH PLANS/SPECIAL FEATURES:

The Whiting-Turner Contracting Company (WT) is a large, nationwide general contracting firm specializing in the construction of commercial and institutional buildings. The company offers a myriad of services and expertise to its clients. Through its expertise in sustainability, the firm has become one of the largest builders of sustainable buildings in the U.S., with completed projects across every region. These projects have included office buildings, stadiums, libraries, laboratories, dining halls and retail centers and stores, to name a few. In addition its headquarters in Maryland, the firm has an additional 30 offices in California, Colorado, Connecticut, Delaware, Washington, D.C., Florida, Georgia, Maryland, Massachusetts, Missouri, New Jersey, New York, North Carolina, Ohio, Pennsylvania, Texas and Virginia. Recently-completed projects include Stanford-Central Energy, Holy Cross Hospital, KIPP College Prep High School, LEGOLAND Florida Hotel, Gordon Food Services and Johns Hopkins Bayview.

CONTACTS: *Note: Officers with more than one job title may be intentionally listed here more than once.*
Timothy Regan, CEO
Timothy Regan, Pres.

FINANCIAL DATA: *Note: Data for latest year may not have been available at press time.*

In U.S. $	2015	2014	2013	2012	2011	2010
Revenue	6,347,100,000	5,100,000,000	5,062,700,000			
R&D Expense						
Operating Income						
Operating Margin %						
SGA Expense						
Net Income						
Operating Cash Flow						
Capital Expenditure						
EBITDA						
Return on Assets %						
Return on Equity %						
Debt to Equity						

CONTACT INFORMATION:
Phone: 410-821-1100   Fax: 410-337-5770
Toll-Free: 800-638-4279
Address: 300 E. Joppa Rd., Baltimore, MD 21286 United States

STOCK TICKER/OTHER:
Stock Ticker: Private   Exchange:
Employees: 2,100   Fiscal Year Ends:
Parent Company:

SALARIES/BONUSES:
Top Exec. Salary: $ Bonus: $
Second Exec. Salary: $ Bonus: $

OTHER THOUGHTS:
Estimated Female Officers or Directors:
Hot Spot for Advancement for Women/Minorities:

William Lyon Homes Inc

www.lyonhomes.com

NAIC Code: 236117

TYPES OF BUSINESS:

Residential Construction
Land Development
Home Sales
Home Financing
Home Design
Title Reinsurance

BRANDS/DIVISIONS/AFFILIATES:

William Lyon Mortgage LLC
Duxford Escrow Inc
Duxford Title Reinsurance Company

CONTACTS: *Note: Officers with more than one job title may be intentionally listed here more than once.*

William H. Lyon, Jr., CEO
Matthew R. Zaist, COO
Matthew R. Zaist, Pres.
Colin T. Severn, CFO
Colin T. Severn, Corp. Sec.
William Lyon, Sr., Chmn.

GROWTH PLANS/SPECIAL FEATURES:

William Lyon Homes, Inc. and its subsidiaries design, construct and sell single-family detached and attached homes in California, Arizona, Nevada, Colorado, Washington and Oregon. The company has five homebuilding operations, which are geographically divided into Southern California, Northern California, Arizona, Nevada and Colorado, and include both wholly-owned projects and projects being developed in unconsolidated joint ventures. The majority of the firm's home closings occur in California. William Lyon markets its homes through about 73 sales locations; its homes range in price from $166,000 to $2.7 million. The company offers a wide range of homes, primarily emphasizing sales to the entry-level and move-up homebuyer markets. Since its founding in 1956 as a company that provided homes for military families, William Lyon has sold over 93,000 homes. Annually, the company delivers approximately 1,700 homes, with an average selling price of more than $400,000. Subsidiaries William Lyon Mortgage LLC, Duxford Escrow, Inc. and Duxford Title Reinsurance Company provide home financing loans, title reinsurance and financial assistance to William Lyon customers.

FINANCIAL DATA: *Note: Data for latest year may not have been available at press time.*

In U.S. $	2015	2014	2013	2012	2011	2010
Revenue	1,106,552,000	896,679,000	572,535,000	372,760,000	226,823,000	294,698,000
R&D Expense						
Operating Income	81,018,000	76,425,000	55,857,000	4,666,000	-148,015,000	-117,843,000
Operating Margin %	7.32%	8.52%	9.75%	1.25%	-65.25%	-39.98%
SGA Expense	120,700,000	106,361,000	66,872,000	40,023,000	39,259,000	52,680,000
Net Income	57,336,000	44,625,000	129,132,000	-8,859,000	-193,330,000	-136,786,000
Operating Cash Flow	-172,908,000	-160,160,000	-174,534,000	49,993,000	-38,651,000	24,119,000
Capital Expenditure	4,800,000	2,078,000	3,754,000	312,000	128,000	64,000
EBITDA	89,730,000	81,197,000	59,698,000	8,908,000	-164,484,000	-111,829,000
Return on Assets %	3.18%	3.32%	16.03%	-2.15%	-33.74%	-18.12%
Return on Equity %	9.53%	8.94%	51.98%			-166.39%
Debt to Equity	1.47	1.58	1.09	5.39		37.62

CONTACT INFORMATION:

Phone: 949-833-3600   Fax: 949-476-2178
Toll-Free:
Address: 4695 Mac Arthur Court, 8th Fl, Newport Beach, CA 92660 United States

STOCK TICKER/OTHER:

Stock Ticker: WLH
Employees:
Parent Company:

Exchange: NYS
Fiscal Year Ends: 12/31

SALARIES/BONUSES:

Top Exec. Salary: $1,000,000   Bonus: $500,000
Second Exec. Salary: $600,000   Bonus: $

OTHER THOUGHTS:

Estimated Female Officers or Directors:

Hot Spot for Advancement for Women/Minorities: Y

Wolseley PLC

NAIC Code: 444190   **www.wolseley.com**

TYPES OF BUSINESS:

Building Materials Wholesale Distribution
Plumbing & Heating Products Distribution

BRANDS/DIVISIONS/AFFILIATES:

Ferguson Enterprises Inc
Wolseley Canada
Tobler
Wasco
BathEmpire.com

CONTACTS: Note: Officers with more than one job title may be intentionally listed here more than once.

Ian K. Meakins, CEO
John Martin, CFO
Bob Morrison, Dir.-Group Human Resources
Tony England, CIO
Richard Shoylekov, General Counsel
Steve Ashmore, Managing Dir.-Wolseley U.K.
Ole Mikael Jensen, CEO
Gareth Davis, Chmn.
Frank W. Roach, CEO-North America

GROWTH PLANS/SPECIAL FEATURES:

Wolseley plc is one of the world's largest distributors of residential construction materials, including plumbing, building and heating equipment. It operates over 40 branches across Europe and North America, dividing its business into five geographic divisions: the U.S., Canada, U.K., Nordic and Central Europe. Its U.S. subsidiary, Ferguson Enterprises, Inc., primarily sells equipment such as heating, ventilation and air conditioning (HVAC) supplies; pipes, valves, hydrants and other plumbing supplies; and fire protection systems. Wolseley's U.S. building materials distribution segment is one of the largest wholesalers and retailers of lumber and building materials to professional contractors in the U.S. This segment supplies value-added pre-assembled components such as roof and floor trusses and wall panels and floor systems, doors, windows and staircases to house builders and professional contractors. The company's Canadian operations function similarly to its U.S. businesses under the Wolseley Canada brand. The firm's primary U.K. operations include plumbing and heating parts; construction materials; pipe and climate products; and other businesses involved in interior walls, electrical materials and bathroom products. Wolseley's Nordic operations primarily provide building materials to the Danish, Swedish, Norwegian and Finnish markets. Its Central Europe division operates Tobler and Wasco, which are Swiss, Austrian and Dutch providers of plumbing and HVAC materials respectively. In 2015, Wolseley acquired a substantial shareholding in BathEmpire.com, a U.K. bathroom retailer; and sold its France-based business to OpenGate Capital.

FINANCIAL DATA: Note: Data for latest year may not have been available at press time.

In U.S. $	2015	2014	2013	2012	2011	2010
Revenue			21,884,127,127	22,328,794,195	20,993,200,000	22,003,900,000
R&D Expense						
Operating Income						
Operating Margin %						
SGA Expense						
Net Income			507,425,619	94,832,074	419,600,000	-566,640,000
Operating Cash Flow						
Capital Expenditure						
EBITDA						
Return on Assets %						
Return on Equity %						
Debt to Equity						

CONTACT INFORMATION:

Phone: 41-41-723-22-30   Fax: 41-41-723-22-31
Toll-Free:
Address: Grafenauweg 10, Zug, CH-6301 Switzerland

STOCK TICKER/OTHER:

Stock Ticker: WOSYY   Exchange: OTC
Employees: 44,000   Fiscal Year Ends: 07/31
Parent Company:

SALARIES/BONUSES:

Top Exec. Salary: $ Bonus: $
Second Exec. Salary: $ Bonus: $

OTHER THOUGHTS:

Estimated Female Officers or Directors: 2
Hot Spot for Advancement for Women/Minorities:

WorldMark by Wyndham Inc

www.worldmarkbywyndham.com/

NAIC Code: 561599

TYPES OF BUSINESS:

Timeshare Resorts
Property Management

BRANDS/DIVISIONS/AFFILIATES:

Wyndham Worldwide
WorldMark The Club
Wyndham Vacation Ownership
www.worldmarktheclub.com

CONTACTS: Note: Officers with more than one job title may be intentionally listed here more than once.

Stephen P. Holmes, CEO-Wyndham Worldwide
Franz S. Hanning, Pres.

GROWTH PLANS/SPECIAL FEATURES:

WorldMark by Wyndham, Inc., a wholly-owned subsidiary of Wyndham Worldwide, develops and markets the vacation ownership program WorldMark, The Club. It is one of the three brands operated by Wyndham Vacation Ownership, another subsidiary of Wyndham Worldwide. Unlike traditional timeshare owners, who have access to a specific time at a specific place, WorldMark member-owners purchase annually-renewed credits, which can be spent like currency to reserve vacations in WorldMark resorts. Among other benefits, this gives member-owners flexibility over vacation dates and locations. The firm operates 60 WorldMark resorts, mostly in the U.S., with additional locations in Canada and Mexico. Most of the resorts are designed to be a short drive away from major metropolitan areas, and the remaining resorts are in exotic locations, such as those in Hawaii and Fiji. Resorts are often located near golf courses, beaches, sightseeing locations, hiking trails, romantic getaways, shopping centers, skiing slopes or family-friendly entertainment. A typical resort room includes a stocked kitchen, laundry facilities, television and video players, stereos, a fireplace, a private deck with access to swimming pools, indoor and outdoor spas, exercise rooms, arcade games and sports facilities. Its WorldMark, The Club program has over 234,000 member-owners. The Club is based in Redmond, Washington, and its website is www.worldmarktheclub.com. The firm also operates a vacation exchange partnership with Resort Condominiums International (RCI), one of the largest vacation ownership companies in the world. This partnership grants WorldMark, The Club member-owners access to thousands of resorts worldwide for an exchange fee.

Employees receive medical, dental and vision plans; a 401(k) plan; long- and short-term disability benefits; an educational assistance plan; adoption assistance; domestic partner benefits; paid time off; an employee assistance program; and discounts on various products and services, including hotels.

FINANCIAL DATA: Note: Data for latest year may not have been available at press time.

In U.S. $	2015	2014	2013	2012	2011	2010
Revenue						
R&D Expense						
Operating Income						
Operating Margin %						
SGA Expense						
Net Income						
Operating Cash Flow						
Capital Expenditure						
EBITDA						
Return on Assets %						
Return on Equity %						
Debt to Equity						

CONTACT INFORMATION:

Phone: 425-498-2500   Fax: 425-498-3675
Toll-Free:
Address: 6277 Sea Harbor Dr, Orlando, FL 32821 United States

STOCK TICKER/OTHER:

Stock Ticker: Subsidiary   Exchange:
Employees: 27,000   Fiscal Year Ends: 12/31
Parent Company: Wyndham Worldwide

SALARIES/BONUSES:

Top Exec. Salary: $ Bonus: $
Second Exec. Salary: $ Bonus: $

OTHER THOUGHTS:

Estimated Female Officers or Directors:
Hot Spot for Advancement for Women/Minorities:

WP Carey & Co LLC

NAIC Code: 531120

www.wpcarey.com

TYPES OF BUSINESS:

Real Estate Operations
Commercial Properties
Property Management & Leasing
Real Estate Financing
Commercial Construction
Commercial Brokerage

BRANDS/DIVISIONS/AFFILIATES:

Carey Asset Management Corp
CPA 17 Inc
CPA 16 Inc

CONTACTS: Note: Officers with more than one job title may be intentionally listed here more than once.

Mark Goldberg, Chairman of the Board, Subsidiary
ToniAnn Sanzone, Chief Accounting Officer
Hisham Kader, Controller
Thomas Zacharias, COO
Mark Decesaris, Director
Benjamin Griswold, Director
Susan Hyde, Managing Director
John Miller, Other Executive Officer
Jason Fox, President

GROWTH PLANS/SPECIAL FEATURES:

W.P. Carey & Co. LLC is a real estate advisory and investment company that invests primarily in commercial properties that are each triple-net leased to single corporate tenants, domestically and internationally. The company has nearly $10.8 billion in assets, owning commercial and industrial facilities worldwide. In additional to its owned portfolio, W.P. Carey manages a series of non-traded publicly registered investment programs with assets under management of approximately $11.6 billion. The company operates in two segments: investment management and real estate ownership. The investment management segment, operating through Carey Asset Management Corp., acts as an advisor to non-traded REITs (real estate investment trusts) that it sponsors, including Corporate Property Associates 16 Incorporated (CPA 16, Inc.) and CPA 17, Inc., which it collectively refers to as the CPA REITs. The firm provides various services to the CPA REITs, including day-to-day management, transaction related services and asset management. The real estate ownership segment typically acquires a property and leases it back to the tenant company (referred to as sale-leaseback) on a triple-net long-term basis, meaning the tenant company bears the responsibility for maintaining the premises, insuring the building and paying real estate taxes, among other operating costs. As a result, the tenant company preserves operational control of the property and benefits from the immediate access to capital, which can be used to improve its balance sheet, fund future growth or reduce debt. Most of the firm's real estate acquisitions in recent years have been done on behalf of the CPA REITs. In April 2016, the firm acquired a 4 million square foot, 40-property industrial portfolio located in the U.S. and Canada for $217 million.

The firm offers employees benefits including medical, dental, life and long-term disability insurance; flexible spending accounts; an employee share purchase plan; and a firm-sponsored profit sharing plan.

FINANCIAL DATA: Note: Data for latest year may not have been available at press time.

In U.S. $	2015	2014	2013	2012	2011	2010
Revenue	938,383,000	906,193,000	489,851,000	373,995,000	336,409,000	273,910,000
R&D Expense						
Operating Income	357,554,000	268,815,000	137,145,000	48,793,000	114,898,000	87,968,000
Operating Margin %	38.10%	29.66%	27.99%	13.04%	34.15%	32.11%
SGA Expense	226,173,000	128,164,000	121,392,000	243,054,000	158,536,000	133,452,000
Net Income	172,258,000	239,826,000	98,876,000	62,132,000	139,079,000	73,972,000
Operating Cash Flow	477,277,000	399,092,000	207,908,000	80,643,000	80,116,000	86,417,000
Capital Expenditure	711,988,000	942,828,000	279,422,000	53,155,000	13,239,000	107,812,000
EBITDA	714,593,000	700,293,000	339,281,000	191,841,000	230,152,000	146,078,000
Return on Assets %	1.98%	3.58%	2.12%	2.04%	10.55%	6.52%
Return on Equity %	4.79%	8.44%	5.08%	4.65%	21.27%	11.82%
Debt to Equity	1.31	1.09	0.78	0.86	0.52	0.40

CONTACT INFORMATION:

Phone: 212-492-1100   Fax: 212-492-8922
Toll-Free: 800-972-2739
Address: 50 Rockefeller Plz., New York, NY 10020 United States

STOCK TICKER/OTHER:

Stock Ticker: WPC
Employees: 272
Parent Company:

Exchange: NYS
Fiscal Year Ends: 12/31

SALARIES/BONUSES:

Top Exec. Salary: $700,000   Bonus: $1,909,000
Second Exec. Salary: $350,000   Bonus: $1,336,000

OTHER THOUGHTS:

Estimated Female Officers or Directors: 9
Hot Spot for Advancement for Women/Minorities: Y

Sales, profits and employees may be estimates. Financial information, benefits and other data can change quickly and may vary from those stated here.

WP GLIMCHER Inc

www.glimcher.com

NAIC Code: 0

TYPES OF BUSINESS:

Real Estate Investment Trust
Malls & Shopping Centers

BRANDS/DIVISIONS/AFFILIATES:

Washington Prime Group Inc

GROWTH PLANS/SPECIAL FEATURES:

WP GLIMCHER, Inc. is a fully integrated, self-administered and self-managed real estate investment trust (REIT) that owns, leases, acquires, develops and operates a portfolio across 30 U.S. states. The company's portfolio consists of 121 enclosed regional malls and open-air lifestyle community centers. Properties are located in Arizona, California, Colorado, Connecticut, Florida, Georgia, Hawaii, Illinois, Indiana, Iowa, Kansas, Kentucky, Maryland, Michigan, Minnesota, New Jersey, New Mexico, New York, North Carolina, Ohio, Oklahoma, Pennsylvania, South Carolina, South Dakota, Tennessee, Texas, Virginia, Washington, West Virginia and Wisconsin. WP GLIMCHER is headquartered in Columbus, Ohio and has additional corporate offices in Bethesda, Maryland and Indianapolis, Indiana. In January 2015, the firm merged with Washington Prime Group, Inc. and subsequently changed its name to WP GLIMCHER.

CONTACTS: Note: Officers with more than one job title may be intentionally listed here more than once.

Robert Demchak, Assistant General Counsel
Michael Glimcher, CEO
Mark Yale, CFO
Mark Ordan, Chairman of the Board
Melissa Indest, Chief Accounting Officer
Farinaz Tehrani, Executive VP, Divisional
Keric Knerr, Executive VP
Gregory Gorospe, Executive VP
Thomas Drought, Executive VP
Victor Pildes, Senior VP, Divisional
Paul Ajdaharian, Senior VP, Divisional
Grace Schmitt, Senior VP, Divisional
Armand Mastropietro, Senior VP, Divisional

FINANCIAL DATA: Note: Data for latest year may not have been available at press time.

In U.S. $	2015	2014	2013	2012	2011	2010
Revenue	921,656,000	661,126,000	626,289,000	623,927,000		
R&D Expense						
Operating Income	33,741,000	177,161,000	227,020,000	214,371,000		
Operating Margin %	3.66%	26.79%	36.24%	34.35%		
SGA Expense	99,750,000	64,171,000	12,980,000	13,655,000		
Net Income	-85,297,000	170,029,000	155,481,000	129,731,000		
Operating Cash Flow	310,763,000	277,640,000	336,434,000	350,703,000		
Capital Expenditure	160,512,000	89,453,000	93,292,000	67,841,000		
EBITDA	366,210,000	488,056,000	411,487,000	405,114,000		
Return on Assets %	-2.24%	5.20%	5.17%			
Return on Equity %	-11.23%	14.44%	9.93%			
Debt to Equity	3.61	2.97	0.58			

CONTACT INFORMATION:

Phone: 614 621-9000   Fax: 614 621-9311
Toll-Free:
Address: 180 E. Broad St., Columbus, OH 43215 United States

STOCK TICKER/OTHER:

Stock Ticker: WPG   Exchange: NYS
Employees: 81   Fiscal Year Ends: 12/31
Parent Company:

SALARIES/BONUSES:

Top Exec. Salary: $597,945   Bonus: $1,793,836
Second Exec. Salary: Bonus: $393,750
$283,562

OTHER THOUGHTS:

Estimated Female Officers or Directors: 3
Hot Spot for Advancement for Women/Minorities: Y

Wyndham Vacation Ownership www.wyndhamworldwide.com/about-wyndham-worldwide/wyndham-vacation-ownership

NAIC Code: 561599

TYPES OF BUSINESS:

Resorts
Timeshare Resorts
Property Management

BRANDS/DIVISIONS/AFFILIATES:

Wyndham Worldwide
Wyndham Vacation Resorts Inc
Club Wyndham
WorldMark by Wyndham
Club Wyndham Asia
WorldMark South Pacific Club by Wyndham
Wyndham Club Brasil
Margaritaville Vacation Club

CONTACTS: Note: Officers with more than one job title may be intentionally listed here more than once.

Franz S. Hanning, CEO
Franz S. Hanning, Pres.

GROWTH PLANS/SPECIAL FEATURES:

Wyndham Vacation Ownership (WVO), a subsidiary of Wyndham Worldwide, is one of the world's leading time share companies. WVO primarily operates through subsidiary Wyndham Vacation Resorts, Inc. (WVR). WVR markets point-based vacation ownership units through the brands Club Wyndham, WorldMark by Wyndham, Club Wyndham Asia, WorldMark South Pacific Club by Wyndham, Wyndham Club Brasil, Shell Vacations Club and Margaritaville Vacation Club. The firm maintains 213 resort properties as well as over 24,000 individual vacation ownership units and 897,000 owners of vacation ownership interests across the U.S., Canada, Mexico, the Caribbean and the South Pacific. Domestic locations can be found in every region of the U.S., including Texas, California, Hawaii, Florida, South Carolina and Massachusetts. WVR uses a points-based internal reservation system called Club Wyndham Plus to provide owners with flexibility as to resort location, length of stay, unit type and time of year. Through the Wyndham Asset Affiliation Model (WAAM), WVO also offers a fee-for-service timeshare sales model. Vacation units are fully furnished, and many include washers/dryers, whirlpool spas, fireplaces and home entertainment systems. WVR manages the majority of property owners' associations at resorts in which it develops, markets and sells vacation ownership interests as well as at resorts developed by third parties. Additionally, the company offers travel discounts and other member specials to all vacation ownership members.

Through Wyndham Worldwide Corp., the company offers employees a variety of benefits, including medical and life insurance, short- and long-term disability, a 401(k), an employee stock purchase program, adoption assistance, educational assistance and discounts.

FINANCIAL DATA: Note: Data for latest year may not have been available at press time.

In U.S. $	2015	2014	2013	2012	2011	2010
Revenue	2,660,000,000	2,600,000,000				
R&D Expense						
Operating Income						
Operating Margin %						
SGA Expense						
Net Income						
Operating Cash Flow						
Capital Expenditure						
EBITDA						
Return on Assets %						
Return on Equity %						
Debt to Equity						

CONTACT INFORMATION:

Phone: 407-370-5200   Fax: 407-370-5143
Toll-Free: 800-251-8736
Address: 6277 Sea Harbor Drive, Orlando, FL 32821 United States

STOCK TICKER/OTHER:

Stock Ticker: Subsidiary   Exchange:
Employees: 16,000   Fiscal Year Ends: 12/31
Parent Company: Wyndham Worldwide

SALARIES/BONUSES:

Top Exec. Salary: $ Bonus: $
Second Exec. Salary: $ Bonus: $

OTHER THOUGHTS:

Estimated Female Officers or Directors:
Hot Spot for Advancement for Women/Minorities:

Wyndham Worldwide Corporation

www.wyndhamworldwide.com

NAIC Code: 721110

TYPES OF BUSINESS:

Hotels, Motels & Resorts
Property Management
Hotel Development
Vacation Property Exchange and Rental
Timeshare Resorts
Franchising
Vacation Ownership

BRANDS/DIVISIONS/AFFILIATES:

Wyndham Hotel Group
Wyndham Destination Network
Wyndham Vacation Ownership
Margaritaville Vacation Club
Baymont Inn & Suites
Wyndham Vacation Rentals
Club Wyndham
WorldMark

CONTACTS: Note: Officers with more than one job title may be intentionally listed here more than once.

Franz Hanning, CEO, Divisional
Geoffrey Ballotti, CEO, Divisional
Gail Mandel, CEO, Divisional
Stephen Holmes, CEO
Thomas Conforti, CFO
Scott McLester, Executive VP
Mary Falvey, Executive VP
Thomas Anderson, Executive VP
Nicola Rossi, Senior VP

GROWTH PLANS/SPECIAL FEATURES:

Wyndham Worldwide Corporation (WW) is a hospitality company offering individual consumers and business customers an array of hospitality products and services as well as various accommodation alternatives and price ranges through its portfolio of world-renowned brands. WW's Wyndham Hotel Group is a world renowned hotel company with 7,812 hotels and more than 678,000 hotel rooms. The group franchises in the upscale, upper midscale, midscale, economy and extended stay segments with a concentration on economy brands. It also provides property management services for full-service and select limited-service hotels, which is predominantly a fee-for-service business. Group brands include Dolce Hotels & Resorts, Wyndham Grand Hotels & Resorts, Wyndham Hotels & Resorts, Wyndham Garden Hotels, TRYP, Wingate, Hawthorn, Microtel, Ramada Worldwide, Baymont Inn & Suites, DaysInn, Super 8, Howard Johnson, Travelodge and Knights Inn. Wyndham Destination Network (formerly Wyndham Exchange & Rentals) operates more than 110,000 vacation properties worldwide under the following brands: cottages4you, Hoseasons, James Villa Hollidays, Landal GreenParks, Novasol, RCI, The Registry Collection and Wyndham Vacation Rentals. Wyndham Vacation Ownership (WVO) is a timeshare/vacation ownership business with 213 resorts and approximately 897,000 owners. WVO develops and markets vacation ownership interests (VOI) to individual consumers, provides consumer financing, as well as property management services at the resorts. WVO brands include Club Wyndham, WorldMark, Shell Vacations Club, Margaritaville Vacation Club, Club Wyndham Asia, WorldMark South Pacific Club and Wyndham Club Brasil. In January 2016, Wyndham Exchange & Rentals changed its name to Wyndham Destination Network.

The firm offers employees medical, dental, vision and life insurance; domestic partner benefits; flexible spending accounts; an educational assistance program; an employee assistance program; adoption reimbursement; and travel discounts on Wyndham properties and the firm's affiliated car rental partners.

FINANCIAL DATA: Note: Data for latest year may not have been available at press time.

In U.S. $	2015	2014	2013	2012	2011	2010
Revenue	5,536,000,000	5,281,000,000	5,009,000,000	4,534,000,000	4,254,000,000	3,851,000,000
R&D Expense						
Operating Income	1,015,000,000	941,000,000	910,000,000	852,000,000	767,000,000	718,000,000
Operating Margin %	18.33%	17.81%	18.16%	18.79%	18.03%	18.64%
SGA Expense	1,574,000,000	1,557,000,000	1,471,000,000	1,389,000,000	1,221,000,000	1,071,000,000
Net Income	612,000,000	529,000,000	432,000,000	400,000,000	417,000,000	379,000,000
Operating Cash Flow	991,000,000	984,000,000	1,008,000,000	1,004,000,000	1,003,000,000	635,000,000
Capital Expenditure	222,000,000	235,000,000	238,000,000	208,000,000	239,000,000	167,000,000
EBITDA	1,275,000,000	1,191,000,000	1,030,000,000	945,000,000	980,000,000	903,000,000
Return on Assets %	6.31%	5.44%	4.49%	4.32%	4.52%	4.03%
Return on Equity %	55.51%	36.76%	24.31%	19.22%	16.19%	13.52%
Debt to Equity	5.21	3.81	2.83	2.08	1.69	1.20

CONTACT INFORMATION:

Phone: 973 753-6000   Fax: 973 496-7658
Toll-Free:
Address: 22 Sylvan Way, Parsippany, NJ 07054 United States

STOCK TICKER/OTHER:

Stock Ticker: WYN   Exchange: NYS
Employees: 37,700   Fiscal Year Ends: 12/31
Parent Company:

SALARIES/BONUSES:

Top Exec. Salary: Bonus: $
$1,595,784
Second Exec. Salary: Bonus: $
$794,824

OTHER THOUGHTS:

Estimated Female Officers or Directors: 2

Hot Spot for Advancement for Women/Minorities: Y

Sales, profits and employees may be estimates. Financial information, benefits and other data can change quickly and may vary from those stated here.

Wynn Resorts Limited
NAIC Code: 721120   www.wynnresorts.com

TYPES OF BUSINESS:
Hotel Casinos
Online Poker

BRANDS/DIVISIONS/AFFILIATES:
Wynn Las Vegas
Encore at Wynn Macau
Wynn Macau
Encore Theater
Palo Real Estate Company Limited
Wynn Resorts
Wynn Palace

CONTACTS: *Note: Officers with more than one job title may be intentionally listed here more than once.*
Stephen Wynn, CEO
Stephen Cootey, CFO
John Strzemp, Chief Administrative Officer
Linda Chen, COO, Subsidiary
Kim Sinatra, Executive VP
Matt Maddox, President

GROWTH PLANS/SPECIAL FEATURES:
Wynn Resorts Limited is a developer, owner and operator of destination casino resorts. It owns and operates two destination casino resorts: The Wynn Las Vegas on the Strip in Las Vegas, Nevada, which includes Encore at Wynn Las Vegas; and the Wynn Macau in the Macau Special Administrative Region of China. The firm's Las Vegas operations offer 4,748 rooms and suites. The 186,000 square foot casino features 232 table games, a poker room, 1,849 slot machines and a race and sports book. The resort also features 34 food and beverage outlets; three nightclubs; two spas and salons; a Ferrari and Maserati automobile dealership; wedding chapels; an 18-hole golf course; 290,000 square feet of meeting space; and a 99,000 square foot retail promenade featuring boutiques from Alexander McQueen, Cartier, Chanel and Louis Vuitton. At the Encore Theater, the company offers headlining entertainment acts from personalities such as Beyonce. The company's Wynn Macau resort operations, including Encore at Wynn Macau, features 1,008 rooms and suites, approximately 284,000 square feet of casino gaming space with 625 slot machines and 498 table games, eight restaurants, two luxury spas and 57,000 square feet of retail space. Wynn Palace is a resort development scheduled to open by mid-2016, featuring 1,700 rooms and suites, an 8-acre performance lake, air-conditioned SkyCabs, floral sculptures, gaming space, meeting facilities, spa, salon, retail spaces and fine dining.

FINANCIAL DATA: *Note: Data for latest year may not have been available at press time.*

In U.S. $	2015	2014	2013	2012	2011	2010
Revenue	4,075,883,000	5,433,661,000	5,620,936,000	5,154,284,000	5,269,792,000	4,184,698,000
R&D Expense						
Operating Income	658,814,000	1,266,278,000	1,290,091,000	1,029,276,000	1,008,240,000	625,252,000
Operating Margin %	16.16%	23.30%	22.95%	19.96%	19.13%	14.94%
SGA Expense	552,951,000	533,047,000	469,095,000	482,143,000	519,702,000	425,969,000
Net Income	195,290,000	731,554,000	728,652,000	502,036,000	613,371,000	160,127,000
Operating Cash Flow	572,813,000	1,098,317,000	1,676,642,000	1,185,718,000	1,515,835,000	1,057,312,000
Capital Expenditure	1,925,152,000	1,345,940,000	506,786,000	240,985,000	184,146,000	283,828,000
EBITDA	912,782,000	1,588,043,000	1,656,596,000	1,394,956,000	1,433,524,000	965,464,000
Return on Assets %	1.99%	8.38%	9.30%	7.08%	9.03%	2.24%
Return on Equity %				54.86%	28.35%	6.07%
Debt to Equity					1.34	1.45

CONTACT INFORMATION:
Phone: 702 770-7555   Fax: 702 733-4681
Toll-Free:
Address: 3131 Las Vegas Blvd. S., Las Vegas, NV 89109 United States

STOCK TICKER/OTHER:
Stock Ticker: WYNN   Exchange: NAS
Employees: 16,800   Fiscal Year Ends: 12/31
Parent Company:

SALARIES/BONUSES:
Top Exec. Salary: Bonus: $
$2,500,000
Second Exec. Salary: Bonus: $
$1,500,000

OTHER THOUGHTS:
Estimated Female Officers or Directors: 3

Hot Spot for Advancement for Women/Minorities: Y

Sales, profits and employees may be estimates. Financial information, benefits and other data can change quickly and may vary from those stated here.

Xanterra Parks & Resorts Inc   www.xanterra.com
NAIC Code: 721110

TYPES OF BUSINESS:
Hotel & Restaurant Management
Conference Center Management

BRANDS/DIVISIONS/AFFILIATES:
Anshutz Company (The)
Windstar Cruises
VBT Bicycling and Walking Vacations
Austin Adventures
Country Walkers

CONTACTS: *Note: Officers with more than one job title may be intentionally listed here more than once.*
Andrew N. Todd, CEO
Andrew N. Todd, Pres.
Michael F. Welch, CFO
Betsy O'Rourke, VP-Mktg. & Sales
Shannon Dierenbach, VP-Human Resources
Kirk H. Anderson, General Counsel
Robert T. Tow, VP-Bus. Dev.
Michael F. Welch, VP-Finance
James W. McCaleb, VP-Parks, North
Gordon R. Taylor, VP-Parks, South
Tim Schoonover, VP-Retail
Catherine Greener, VP-Sustainability

GROWTH PLANS/SPECIAL FEATURES:
Xanterra Parks & Resorts, Inc. is a parks concessions management company operating in national parks and state parks across the U.S. The company operates hotels, restaurants, cruises and adventures. Xanterra hotels offer a wide range of experiences, from conference facilities with premium golf, tennis and spa amenities, to historic park lodges & cabins. Some of the firm's hotels include Furnace Creek Resort, Grand Canyon Railway, Yellowstone's Old Faithful Inn and El Tovar Hotel. Restaurants include dozens of restaurants across the U.S., ranging from full-service to quick-service outlets and cafes. Restaurants can be found in the Grand Canyon Lodge, the Old Faithful Lodge and the Zion Lodge. Through wholly-owned subsidiary Windstar Cruises, Xanterra operates a fleet of luxury yachts and ships that cruise to 50 nations and 100 ports-of-call throughout Europe, the Caribbean and the Americas. The firm's fleet of small ships carry between 148 and 310 guests. Adventures include watching wild buffalo at Yellowstone, viewing Zion, experiencing Death Valley and exploring Crater Lake. Subsidiary VBT Bicycling and Walking Vacations offers deluxe, small group bicycling, walking and cross-country skiing tours worldwide, including destinations in Europe, Costa Rica, New Zealand, Vietnam, Thailand, Peru, Argentina, Canada and the U.S. Other experiences at select destinations worldwide include Austin Adventures tour operator, as well as Country Walkers, each providing cultural and artistic adventures. National parks include Crater Lake, Furnace Creek-Death Valley, Glacier, Grand Canyon, Mount Rushmore, Yellowstone, Rocky Mountain and Zion. Ohio state parks include Ohio State Park Lodges, Deer Creek Lodge, Geneva State Park, Maumee Bay Lodge, Mohican Lodge, Punderson Manor Lodge and Salt Fork Lodge. The firm is owned by The Anschutz Company.

Xanterra offers employee benefits including dental, vision, life, disability, AD&D and medical coverage; an employee assistance program; education assistance; and a prescription drug plan.

FINANCIAL DATA: *Note: Data for latest year may not have been available at press time.*

In U.S. $	2015	2014	2013	2012	2011	2010
Revenue	106,000,000	103,000,000	98,000,000	90,000,000	89,000,000	86,000,000
R&D Expense						
Operating Income						
Operating Margin %						
SGA Expense						
Net Income						
Operating Cash Flow						
Capital Expenditure						
EBITDA						
Return on Assets %						
Return on Equity %						
Debt to Equity						

CONTACT INFORMATION:
Phone: 303-600-3400   Fax: 303-600-3600
Toll-Free:
Address: 6312 South Fiddlers Green Cir., Ste. 600N, Greenwood Village, CO 80111 United States

STOCK TICKER/OTHER:
Stock Ticker: Subsidiary   Exchange:
Employees: 7,500   Fiscal Year Ends:
Parent Company: Anschutz Company (The)

SALARIES/BONUSES:
Top Exec. Salary: $ Bonus: $
Second Exec. Salary: $ Bonus: $

OTHER THOUGHTS:
Estimated Female Officers or Directors: 3
Hot Spot for Advancement for Women/Minorities: Y

YIT Corporation

NAIC Code: 541330

TYPES OF BUSINESS:

Engineering and Maintenance Services
Building Systems
Construction Services
Industrial Services
Network Services

BRANDS/DIVISIONS/AFFILIATES:

CONTACTS: *Note: Officers with more than one job title may be intentionally listed here more than once.*

Kari Kauniskangas, CEO
Kari Kauniskangas, Pres.
Timo Lehtinen, CFO
Pii Raulo, Sr. VP-Human Resources
Juhani Nummi, Sr. VP-Bus. Dev.
Tero Kiviniemi, Exec. VP
Jouni Forsman, Head-Infrastructure Bus. Div.
Harri Isoviita, Head-Residential Construction Bus. Div.
Matti Koskela, Head-Bldg. Construction Bus. Div.
Reino Hanhinen, Chmn.
Teemu Helppolainen, Head-Russia

GROWTH PLANS/SPECIAL FEATURES:

YIT Corporation, which operates as a group of companies, offers technical infrastructure investment and upkeep services for the property, construction, industry and telecommunications sectors. In all sectors of operations, YIT's services cover the entire lifecycle of its projects. The firm's primary market areas are the Nordic countries, Central Europe, the Baltic countries and Russia. YIT's operations are divided into four business segments: building services, Northern Europe; building services, Central Europe; construction services, Finland; and international construction services. The building services, Northern Europe segment offers technical building systems, infrastructure upgrading, building maintenance, industrial piping, tanks and pulp towers, access-control systems, industrial maintenance and energy-efficiency services. It operates in Finland, Sweden, Norway, Denmark, Russia, Estonia, Latvia and Lithuania. Building services, Central Europe offers similar services in Germany, Austria, Poland, the Czech Republic and Romania. Both of the construction services segments are primarily focused on the construction of residential, commercial and industrial properties as well as the provision of civil engineering, water and environmental services.

FINANCIAL DATA: *Note: Data for latest year may not have been available at press time.*

In U.S. $	2015	2014	2013	2012	2011	2010
Revenue	1,975,570,000	2,028,489,000	1,987,888,000	5,596,537,000	4,997,776,000	4,319,779,000
R&D Expense						
Operating Income	93,064,630	108,119,200	118,611,800	295,617,100	228,099,600	251,070,400
Operating Margin %	4.71%	5.33%	5.96%	5.28%	4.56%	5.81%
SGA Expense	883,771,800	976,608,400	1,058,496,000		111,768,800	
Net Income	53,831,500	63,753,830	407,956,100	203,807,000	142,562,200	160,378,000
Operating Cash Flow	221,712,800	176,549,100	-140,509,300	138,570,500	19,844,660	7,755,386
Capital Expenditure	13,115,730	13,457,880	17,221,520	40,031,480	44,479,420	32,196,260
EBITDA	90,441,490	122,489,500	138,456,400	346,825,400	273,377,300	287,917,500
Return on Assets %	2.24%	2.33%	11.47%	4.97%	3.75%	4.88%
Return on Equity %	8.91%	9.06%	41.35%	18.32%	13.83%	16.72%
Debt to Equity	0.50	0.51	0.43	0.50	0.56	0.57

CONTACT INFORMATION:

Phone: 358 20433111 Fax: 358 204333700
Toll-Free:
Address: P.O. Box 36, Panuntie 11, Helsinki, 00621 Finland

STOCK TICKER/OTHER:

Stock Ticker: YITYF   Exchange: GREY
Employees: 5,881   Fiscal Year Ends: 12/31
Parent Company:

SALARIES/BONUSES:

Top Exec. Salary: $ Bonus: $
Second Exec. Salary: $ Bonus: $

OTHER THOUGHTS:

Estimated Female Officers or Directors: 2
Hot Spot for Advancement for Women/Minorities: Y

Zillow Inc

NAIC Code: 519130

www.zillow.com

TYPES OF BUSINESS:
Online Real Estate Information

BRANDS/DIVISIONS/AFFILIATES:
Zillow.com
Zestimates
Rent Zestimates
Zillow Mortgage Marketplace
Zillow Mobile
Trulia
Zillow Digs
Dotloop

CONTACTS: *Note: Officers with more than one job title may be intentionally listed here more than once.*
Spencer Rascoff, CEO
Kathleen Philips, CFO
Richard Barton, Chairman of the Board
David Beitel, Chief Technology Officer
Amy Bohutinsky, COO
Lloyd Frink, Director
Errol Samuelson, Other Executive Officer
Greg Schwartz, Other Executive Officer
Paul Levine, President, Subsidiary

GROWTH PLANS/SPECIAL FEATURES:
Zillow, Inc. operates a real estate information marketplace dedicated to providing information about homes, real estate listings and mortgages and enabling homeowners, buyers, sellers and renters to connect with real estate and mortgage professionals. The company maintains a database of over 110 million homes in the U.S. that are either for sale, for rent or not currently on the market. Individuals and businesses that use Zillow have updated information on more than 59 million homes and added more than 343 million home photos, creating exclusive home profiles that are available nowhere else. These profiles include detailed information about homes, such as property facts, listing information and purchase and sale. In conjuncture with the database, the firm offers its users its proprietary automated valuation models, Zestimates and Rent Zestimates, on more than 100 million homes. In addition to its primary website, Zillow.com, the company also operates Zillow Mortgage Marketplace, connecting borrowers with lenders in order to find loans and good mortgage rates; Zillow Mobile, a real estate mobile platform; Zillow Rentals, a marketplace and suite of tools for rental professionals; Postlets, which allows postings and listings to appear on sites such as Zillow, Trulia, Yahoo! Homes and HotPads; and Diverse Solutions, a provider of easy-to-implement technology and websites that help real estate professionals manage their brands and businesses. Additionally, the company operates Zillow Digs, a home improvement marketplace where consumers can find visual inspiration and local cost estimates. In 2015, the firm acquired competitor Trulia, as well as digital transaction management firm Dotloop.

FINANCIAL DATA: *Note: Data for latest year may not have been available at press time.*

In U.S. $	2015	2014	2013	2012	2011	2010
Revenue	644,677,000	325,893,000	197,545,000	116,850,000	66,053,000	30,467,000
R&D Expense	198,565,000	86,406,000	48,498,000	26,614,000	14,143,000	10,651,000
Operating Income	-149,531,000	-44,695,000	-16,949,000	5,797,000	997,000	-6,837,000
Operating Margin %	-23.19%	-13.71%	-8.57%	4.96%	1.50%	-22.44%
SGA Expense	477,534,000	233,228,000	147,186,000	70,396,000	40,338,000	21,680,000
Net Income	-148,874,000	-43,610,000	-12,453,000	5,939,000	1,102,000	-6,774,000
Operating Cash Flow	22,659,000	45,519,000	31,298,000	32,298,000	14,826,000	2,258,000
Capital Expenditure	68,108,000	44,242,000	25,972,000	16,750,000	8,821,000	5,526,000
EBITDA	-72,644,000	-9,071,000	6,305,000	18,570,000	8,187,000	-1,575,000
Return on Assets %	-7.86%	-6.93%	-2.73%	2.82%	1.56%	-27.86%
Return on Equity %	-9.11%	-7.54%	-2.93%	3.11%	1.85%	-35.12%
Debt to Equity	0.08					

CONTACT INFORMATION:
Phone: 206 470-7000   Fax:
Toll-Free:
Address: 1301 Second Ave., Fl. 31, Seattle, WA 98101 United States

STOCK TICKER/OTHER:
Stock Ticker: Z   Exchange: NAS
Employees: 2,204   Fiscal Year Ends: 12/31
Parent Company:

SALARIES/BONUSES:
Top Exec. Salary: $285,445   Bonus: $395,000
Second Exec. Salary: $512,553   Bonus: $

OTHER THOUGHTS:
Estimated Female Officers or Directors: 2
Hot Spot for Advancement for Women/Minorities: Y

ZipRealty Inc

NAIC Code: 531210

www.ziprealty.com

TYPES OF BUSINESS:

Real Estate Brokerage

BRANDS/DIVISIONS/AFFILIATES:

Powered by Zip
ZipAgent Platform
Realogy Holdings Corporation

CONTACTS: Note: Officers with more than one job title may be intentionally listed here more than once.

Lanny Baker, CEO
Lanny Baker, Pres.
Jamie Wilson, VP-IT
Samantha Harnett, General Counsel
Stefan Peterson, VP-Oper.
Genni Combes, Sr. VP-Corp. Dev.
Van Davis, Pres., Brokerage Oper.

GROWTH PLANS/SPECIAL FEATURES:

ZipRealty, Inc. is a full-service residential real estate brokerage firm that utilizes the Internet, proprietary technology, local agents and efficient business processes to provide home buyers and sellers with the appropriate tools for the home-buying process. The company employs a client-centric approach, a sophisticated web site and a proprietary business management technology platform. It also serves markets through its Powered by Zip network of third-party local brokerages. ZipRealty's web site provides users with direct access to comprehensive local Multiple Listing Services (MLS) home listings data, including photos, asking prices and home layouts, and allows the user the option of receiving automatic email notifications whenever a property that meets their search criteria becomes available on an MLS. The firm attracts users to its web site through a number of marketing channels, including online advertising, word-of-mouth and traditional media advertisements. The proprietary ZipAgent Platform, or ZAP, automatically matches registered users with local agents who market and provide the company's comprehensive real estate brokerage services, including showing properties to buyers and listing and marketing properties on behalf of its sellers as well as negotiating, advisory, transaction processing and closing services. The company also provides further information such as neighborhood attributes, school district information, comparable home sales data, maps and driving directions. ZipRealty is owned by Realogy Holdings Corporation.

FINANCIAL DATA: Note: Data for latest year may not have been available at press time.

In U.S. $	2015	2014	2013	2012	2011	2010
Revenue	85,000,000	80,000,000	75,853,000	73,820,000	85,149,000	118,696,000
R&D Expense						
Operating Income						
Operating Margin %						
SGA Expense						
Net Income			-5,436,000	-9,678,000	-9,731,000	-15,550,000
Operating Cash Flow						
Capital Expenditure						
EBITDA						
Return on Assets %						
Return on Equity %						
Debt to Equity						

CONTACT INFORMATION:

Phone: 510 735-2600   Fax: 510 735-2850
Toll-Free: 800-225-5947
Address: 2000 Powell St., Ste. 300, Emeryville, CA 94608 United States

STOCK TICKER/OTHER:

Stock Ticker: Subsidiary   Exchange:
Employees: 1,800   Fiscal Year Ends: 12/31
Parent Company: Realogy Holdings Corporation

SALARIES/BONUSES:

Top Exec. Salary: $ Bonus: $
Second Exec. Salary: $ Bonus: $

OTHER THOUGHTS:

Estimated Female Officers or Directors: 3
Hot Spot for Advancement for Women/Minorities: Y

ADDITIONAL INDEXES

Contents:

INDEX OF FIRMS NOTED AS HOT SPOTS FOR ADVANCEMENT FOR WOMEN & MINORITIES

ABM Industries Inc
Accor SA
AECOM Technology Corporation
Aegion Corp
Alexandria Real Estate Equities Inc
Amec Foster Wheeler plc
American Campus Communities Inc
American Casino & Entertainment Properties LLC
Ameristar Casinos Inc
AMLI Residential Properties Trust
Apartment Investment and Management Co
Arcadis NV
AV Homes Inc
AvalonBay Communities Inc
Bank of America Home Loans
Banyan Tree Holdings Limited
Barcelo Crestline Corporation
Beazer Homes Usa Inc
Bechtel Group Inc
Belmond Ltd
Best Western International Inc
Black & Veatch Holding Company
Boardwalk Real Estate Investment Trust
Boston Properties Inc
Bouygues SA
Boyd Gaming Corp
Brandywine Realty Trust
Brixmor Property Group Inc
Brookfield Asset Management Inc
Caesars Entertainment Corporation
CalAtlantic Group Inc
Camden Property Trust
CAPREIT Inc
Carlson Companies Inc
Carlson Rezidor Hotel Group
CBL & Associates Properties Inc
CH2M HILL Companies Ltd
Chicago Bridge & Iron Company NV (CB&I)
China Lodging Group Ltd
China Overseas Land & Investment Limited
Chinese Estates Holdings Ltd
Choice Hotels International Inc
Christies International Real Estate
Club Mediterranee SA (Club Med)
CNL Lifestyle Properties Inc
Colliers International Group Inc
Comfort Systems USA Inc
Commune Hotels & Resorts
Community Development Trust
Condor Hospitality Trust Inc
Corporate Office Properties Trust
CoStar Group Inc
Crescent Real Estate Equities LP

CRH plc
DCT Industrial Trust Inc
DDR Corp
Duke Realty Corp
EastGroup Properties Inc
EMCOR Group Inc
Empresas ICA SAB de CV
ENGlobal Corp
Equity Lifestyle Properties Inc
Equity One Inc
Essex Property Trust Inc
Evergrande Real Estate Group
Extendicare Inc
Fannie Mae (Federal National Mortgage Association)
Federal Realty Investment Trust
Ferguson Enterprises Inc
Fluor Corp
Fomento de Construcciones y Contratas SA (FCC)
Forest Realty Trust Inc
Four Seasons Hotels Inc
Freddie Mac (Federal Home Loan Mortgage Corp, FHLMC)
Gables Residential Trust
GE Capital
General Growth Properties Inc
Hang Lung Group Ltd
Hang Lung Properties Limited
HCP Inc
Healthcare Realty Trust Inc
Henderson Land Development Co Ltd
Highwoods Properties Inc
Hines Interests LP
Home Depot Inc
Home Properties Inc
HomeServices of America Inc
Hongkong and Shanghai Hotels Ltd
Hongkong Land Holdings Ltd
Host Hotels & Resorts LP
Hyatt Hotels Corporation
Hysan Development Co Ltd
IDI Gazeley
Inland Real Estate Corporation
InterContinental Hotels Group plc
InvenTrust Properties Corp
Investors Real Estate Trust
Investors Title Company
iStar Inc
Jacobs Engineering Group Inc
Janus Hotels and Resorts Inc
Jones Lang LaSalle Inc
KB Home
KBR Inc
Kimco Realty Corp
Kimpton Hotel & Restaurant Group LLC
LB Foster Company
LBA Realty LLC
LendingTree LLC
Lennar Corporation

Lexington Realty Trust
Liberty Property Trust
Lillibridge Healthcare Real Estate Trust
Loews Hotels Holding Corporation
Lowe's Companies Inc
M.D.C. Holdings Inc
Macerich Company (The)
Mack-Cali Realty Corp
Marcus Corporation (The)
Marriott International Inc
McDermott International Inc
Meritus Hotels & Resorts Inc
MFA Financial Inc
MGM Resorts International
Mid-America Apartment Communities Inc
Move Inc (Realtor.com)
MRV Engenharia E Participacoes SA
MWH Global Inc
NAI Global Inc
NH Hotel Group SA
NRT LLC
Oakwood Worldwide
Oberoi Group (EIH Ltd)
Pacific Coast Building Products Inc
Parkway Properties Inc
Pearlmark Real Estate Partners
Pennsylvania REIT
Post Properties Inc
ProLogis Inc
PS Business Parks Inc
PulteGroup Inc
RailWorks Corp
RE/MAX Holdings Inc
Realogy Holdings Corp
RealPage Inc
Realty Income Corp
Red Lion Hotels Corporation
Redwood Trust Inc
Regency Centers Corp
Related Group (The)
Rezidor Hotel Group AB
Rosewood Hotels & Resorts LLC
RXR Realty
Saha Pathana Inter-Holding PCL
Scandic Hotels AB
Shangri-La Asia Ltd
Shimao Property Holdings Ltd
Shun Tak Holdings Limited
Simon Property Group Inc
Skidmore Owings & Merrill LLP
Societe Anonyme des Bains de Mer et du Cercle des
Etrangers a Monaco
Sonesta International Hotels Corp
Sovran Self Storage Inc
St Joe Company (The)
Starwood Hotels & Resorts Worldwide Inc
Stewart Information Services Corp
Strategic Hotels & Resorts Inc

STV Group Inc
Sunrise Senior Living
Sunstone Hotel Investors Inc
Tanger Factory Outlet Centers Inc
Taubman Centers Inc
TDIndustries
Transwestern Commercial Services
UDR Inc
Unibail-Rodamco SE
Ventas Inc
Weingarten Realty Investors
Welltower Inc
Westfield Corporation
William Lyon Homes Inc
WP Carey & Co LLC
WP GLIMCHER Inc
Wyndham Worldwide Corporation
Wynn Resorts Limited
Xanterra Parks & Resorts Inc
YIT Corporation
Zillow Inc
ZipRealty Inc

INDEX OF SUBSIDIARIES, BRAND NAMES AND AFFILIATIONS

Brand or subsidiary, followed by the name of the related corporation

INDEX OF SUBSIDIARIES, BRAND NAMES AND AFFILIATIONS, CONT.

INDEX OF SUBSIDIARIES, BRAND NAMES AND AFFILIATIONS, CONT.

INDEX OF SUBSIDIARIES, BRAND NAMES AND AFFILIATIONS, CONT.

INDEX OF SUBSIDIARIES, BRAND NAMES AND AFFILIATIONS, CONT.

INDEX OF SUBSIDIARIES, BRAND NAMES AND AFFILIATIONS, CONT.

Country Walkers; **Xanterra Parks & Resorts Inc**
CountryPlace Mortgage Ltd; **Palm Harbor Homes Inc**
Courtyard by Marriott; **Hospitality Properties Trust**
Courtyard by Marriott; **Marriott International Inc**
Courvoisier Centre; **Parkway Properties Inc**
Cove Atlantis (The); **Kerzner International Holdings Limited**
CP Masters Inc; **Brand Energy & Infrastructure Services Inc**
CPA 16 Inc; **WP Carey & Co LLC**
CPA 17 Inc; **WP Carey & Co LLC**
CRAVE; **HDR Inc**
Crest Homes; **Clayton Homes Inc**
Crest Net Lease Inc; **Realty Income Corp**
Crestline Hotels & Resorts LLC; **Barcelo Crestline Corporation**
CRH plc; **Oldcastle Inc**
Crillon; **Groupe du Louvre**
Croke Park Hotel (The); **Doyle Collection (The)**
Crossland Economy Studios; **Extended Stay America Inc**
Crown Plaza Times Square Hotel; **Vornado Realty Trust**
Crowne Plaza; **Sunburst Hospitality Corporation**
Crowne Plaza Hong Koong Kowloon East; **Sun Hung Kai Properties Ltd**
Crowne Plaza Hotels & Resorts; **Hospitality Properties Trust**
CSCEC Property Management Co; **China State Construction Engineering Corp**
CuautiPark II; **Fibra Uno Administracion SA de CV**
Cury Construtora e Incorporadora; **Cyrela Brazil Realty SA**
Cushman & Wakefield Sonnenblick Goldman; **Cushman & Wakefield Inc**
Cyrela; **Cyrela Brazil Realty SA**
Daeryun Power; **Hanjin Heavy Industries and Construction Co Ltd**
DataQuick Lending Solutions; **Stewart Information Services Corp**
Days Hotel; **Days Inn Worldwide Inc**
Days Inn; **Days Inn Worldwide Inc**
Days Inn & Suites; **Days Inn Worldwide Inc**
Days Inn Business Place; **Days Inn Worldwide Inc**
DCT Industrial Trust Inc; **Dividend Capital Group LLC**
Del Webb Corp; **PulteGroup Inc**
Delano; **Morgans Hotel Group Co**
Delcan; **Parsons Corp**
Delta Associates; **Transwestern Commercial Services**
Delta Downs Racetrack and Casino; **Boyd Gaming Corp**
DEPFA Bank Plc; **Hypo Real Estate Holding AG (Hypo Bank)**
Deutsche Pfandbriefbank AG; **Hypo Real Estate Holding AG (Hypo Bank)**
DHI Mortgage; **DR Horton Inc**
Diamond Resorts Corporation; **Diamond Resorts Holdings LLC**

DiamondResorts.com; **Diamond Resorts Holdings LLC**
DIM Vastgoed NV; **Equity One Inc**
Direct Win Development Limited; **Chinese Estates Holdings Ltd**
Dividend Capital Diversified Property Fund; **Dividend Capital Group LLC**
DLF Utilities Limited; **DLF Limited**
Doherty & Associates; **HDR Inc**
Doors and Buildings Components; **NCI Building Systems Inc**
Doosan Babcock; **Doosan Heavy Industry & Construction Co**
Doosan IMGB; **Doosan Heavy Industry & Construction Co**
Doosan Power Systems; **Doosan Heavy Industry & Construction Co**
Doosan Power Systems Europe; **Doosan Heavy Industry & Construction Co**
Doosan Skoda Power; **Doosan Heavy Industry & Construction Co**
Dotloop; **Zillow Inc**
Double ZeroHouse 3.0; **KB Home**
DoubleTree; **Hilton Worldwide Inc**
Dri-guard; **Sovran Self Storage Inc**
Drillmec SpA; **Trevi-Finanziaria Industriale SpA (Trevi Group)**
DSAA; **Steelman Partners LLP**
DTZ; **Cushman & Wakefield Inc**
Duke Construction LP; **Duke Realty Corp**
Duke Realty LP; **Duke Realty Corp**
Duke Realty Services LLC; **Duke Realty Corp**
Duke Realty Services LP; **Duke Realty Corp**
Dupont Circle Hotel (The); **Doyle Collection (The)**
Duxford Escrow Inc; **William Lyon Homes Inc**
Duxford Title Reinsurance Company; **William Lyon Homes Inc**
E&P Financing Limited Partnership; **Condor Hospitality Trust Inc**
E&P REIT Trust; **Condor Hospitality Trust Inc**
E.T. Archer Corporation; **HDR Inc**
Eagle Home Mortgage LLC; **Lennar Corporation**
Eastern & Oriental Express Railway; **Belmond Ltd**
EC Mall; **Link Real Estate Investment Trust (The)**
Eclat; **GS Engineering & Construction Corp**
Econo Lodge; **Choice Hotels International Inc**
Edgemoor Infrastructure & Real Estate; **Clark Construction Group LLC**
Edina Realty; **HomeServices of America Inc**
EIH Ltd; **Oberoi Group (EIH Ltd)**
Elan Hotel; **China Lodging Group Ltd**
Eldorado Casino; **Boyd Gaming Corp**
Emaar Hospitality Group LLC; **Emaar Properties PJSC**
Emaar Hotels and Resorts LLC; **Emaar Properties PJSC**
Emaar International; **Emaar Properties PJSC**
Emaar Malls Group LLC; **Emaar Properties PJSC**

INDEX OF SUBSIDIARIES, BRAND NAMES AND AFFILIATIONS, CONT.

INDEX OF SUBSIDIARIES, BRAND NAMES AND AFFILIATIONS, CONT.

INDEX OF SUBSIDIARIES, BRAND NAMES AND AFFILIATIONS, CONT.

Gwynt y Mor Offshore Transmission Project; **Balfour Beatty plc**
Haband; **Bluestem Group Inc**
Hali'imaile Town; **Maui Land & Pineapple Company Inc**
Hampton Bay; **Home Depot Inc**
Hampton Inn; **TMI Hospitality Inc**
Hampton Inn; **Hilton Worldwide Inc**
Hang Lung Group Ltd; **Hang Lung Properties Limited**
Hang Lung Properties Limited; **Hang Lung Group Ltd**
Hang Lung-Hakuyosha Dry Cleaning; **Hang Lung Group Ltd**
Hangszhou Shimao Imperial Landscape; **Shimao Property Holdings Ltd**
HanTing Hi Inn; **China Lodging Group Ltd**
HanTing Hotels; **China Lodging Group Ltd**
Harbour Centre Development Limited; **Wharf (Holdings) Limited The**
Harbour Centre Development Ltd; **Wheelock and Company Limited**
Harbour City; **Wharf (Holdings) Limited The**
Harbour Plaza Hotels & Resorts; **Hutchison Whampoa Properties Ltd**
HarbourSide; **Hang Lung Properties Limited**
Hard Days Night Hotel; **Millennium & Copthorne Hotels plc**
Harrah's; **Caesars Entertainment Corporation**
Hatfield Philips International; **Starwood Property Trust Inc**
Hawksley Consulting; **MWH Global Inc**
Haystack Coal Company; **Kiewit Corp**
Haystack Coal Company; **Kiewit Corp**
HC Green Development Fund; **Hines Interests LP**
HC Trading; **HeidelbergCement AG**
HCM Holdings LP; **Hines Interests LP**
HCR Manor Care; **ManorCare Inc (HCR ManorCare)**
Health Care REIT Inc; **Welltower Inc**
Health Care REIT Inc; **Sunrise Senior Living**
Health Park; **Saha Pathana Inter-Holding PCL**
Heartland; **ManorCare Inc (HCR ManorCare)**
HeidelbergCement AG; **HeidelbergCement North America**
Henderson Group plc; **John Laing plc**
Henderson Infrastructure Holdco Ltd; **John Laing plc**
Henderson Investment Limited; **Henderson Land Development Co Ltd**
Heritage Communities LP; **UDR Inc**
Heritage Customer Service Program; **Watson Land Company**
Hesperia; **NH Hotel Group SA**
Hhonors; **Hilton Worldwide Inc**
Highwoods Realty LP; **Highwoods Properties Inc**
Hijos De Antonio Barcelo SA; **Acciona SA**
Hillstate; **Hyundai Engineering & Construction Company Ltd**

Hilton; **Hilton Worldwide Inc**
Hilton Grand Vacations Co LLC; **Hilton Worldwide Inc**
Hines European Development Fund LP; **Hines Interests LP**
HIS Fund II; **MMA Capital Management LLC**
HIS Residential Partners I; **MMA Capital Management LLC**
HNA Group Co Ltd; **Carlson Rezidor Hotel Group**
HNA Tourism Holding (Group) Co Ltd; **Carlson Rezidor Hotel Group**
Hochtief AG; **Turner Corp (The)**
HOCHTIEF AirPort; **HOCHTIEF AG**
HOCHTIEF Solutions AG; **HOCHTIEF AG**
HoJo; **Howard Johnson International Inc**
HoJo.com Booking Advantage; **Howard Johnson International Inc**
Holcim Ltd; **LafargeHolcim Ltd**
Holiday Acquisition Holdings LLC; **Newcastle Investment Corp**
Holiday Inn; **InterContinental Hotels Group plc**
Holiday Inn Express; **Sunburst Hospitality Corporation**
Holiday Inn Express Hong Kong Kowloon; **Sun Hung Kai Properties Ltd**
Hollister Commons; **KB Home**
Home Affordable Modification Program; **CitiMortgage Inc**
Home Affordable Unemployment Program; **CitiMortgage Inc**
Home Inns; **Homeinns Hotel Group**
Home Properties LP; **Home Properties Inc**
Home Properties Resident Services Inc; **Home Properties Inc**
HomeAmerican Mortgage Corporation; **M.D.C. Holdings Inc**
Hometown America Age-Qualified Communities; **Hometown America LLC**
Hometown America Family Communities; **Hometown America LLC**
Hometown America Foundation; **Hometown America LLC**
Hometown Inn; **Extended Stay America Inc**
Homewood Suites; **Hilton Worldwide Inc**
Hong Kong & China Gas Company Limited (The); **Henderson Land Development Co Ltd**
Hong Kong Aircraft Engineering Company Limited; **Swire Pacific Ltd**
Hong Kong Ferry (Holdings) Company Limited; **Henderson Land Development Co Ltd**
Hongkong & Whampoa Dock Company Limited; **Hutchison Whampoa Properties Ltd**
Horseshoe; **Caesars Entertainment Corporation**
Horseshu Hotel & Casino; **Ameristar Casinos Inc**
Hotel Burnham; **Kimpton Hotel & Restaurant Group LLC**

INDEX OF SUBSIDIARIES, BRAND NAMES AND AFFILIATIONS, CONT.

INDEX OF SUBSIDIARIES, BRAND NAMES AND AFFILIATIONS, CONT.

International Hardwood Resource; **Anderson-Tully Lumber Company**
International Housing Solutions; **MMA Capital Management LLC**
Intown Suites; **InTown Suites Management Inc**
Investment Corporation of Dubai; **Kerzner International Holdings Limited**
Investors Capital Management Company; **Investors Title Company**
Investors Title Accommodation Corporation (ITAC); **Investors Title Company**
Investors Title Exchange Corporation (ITEC); **Investors Title Company**
Investors Title Insurance Company (ITIC); **Investors Title Company**
Investors Title Management Services Inc; **Investors Title Company**
Investors Trust Company; **Investors Title Company**
Inviro Studios; **Steelman Partners LLP**
IP Casino Resort Spa; **Boyd Gaming Corp**
IRET Properties; **Investors Real Estate Trust**
Iris; **Lowe's Companies Inc**
Island Hospitality Management; **Innkeepers USA Trust**
iStar Financial; **Starwood Capital Group Global LLC**
iStar Financial Inc; **iStar Inc**
iSWIM; **Anthony & Sylvan Pools**
I-XEL; **Hyundai Elevator Co Ltd**
J L Patterson & Associates; **Jacobs Engineering Group Inc**
J. Alexander's Corporation; **Fidelity National Financial Inc**
J. Banicki Construction Inc; **Sterling Construction Company Inc**
J. Koski Company; **APi Group Inc**
Jameson Inn; **Jameson Inn Inc**
Jardine House; **Hongkong Land Holdings Ltd**
Jardine Matheson Ltd; **Hongkong Land Holdings Ltd**
JI Hotel; **China Lodging Group Ltd**
Jiva Spas; **Indian Hotels Company Limited (The)**
John Holland; **China Communications Construction Company Ltd**
John Laing Capital Management; **John Laing plc**
John Laing Environmental Assets Group; **John Laing plc**
John Laing Infrastructure Fund; **John Laing plc**
John P Picone Inc; **ACS Actividades de Construccion y Servicios SA**
Joie de Vivre Hotels; **Commune Hotels & Resorts**
Joya Hotel; **China Lodging Group Ltd**
Just Us Kids; **Sonesta International Hotels Corp**
JW Marriott; **Marriott International Inc**
Kabuki Springs & Spa; **Commune Hotels & Resorts**
Kallista; **Kohler Company**
Kaohsiung Terminal; **Orient Overseas (International) Ltd**
Kapalua Club; **Maui Land & Pineapple Company Inc**

Kapalua Resort; **Maui Land & Pineapple Company Inc**
Kapalua Waste Treatment Company Ltd; **Maui Land & Pineapple Company Inc**
Kapalua Water Company Ltd; **Maui Land & Pineapple Company Inc**
KB Home Studios; **KB Home**
KBnxt; **KB Home**
KBR Training Solutions; **KBR Inc**
Kenny Construction Company; **Granite Construction Inc**
Kensington; **Skyline Corporation**
Kensington Hotel (The); **Doyle Collection (The)**
Kerry Hotels; **Shangri-La Asia Ltd**
Keystone; **Palm Harbor Homes Inc**
Kigo; **RealPage Inc**
Kilroy Realty Finance Partnership LP; **Kilroy Realty Corporation**
Kilroy Services LLC; **Kilroy Realty Corporation**
Kimpton Hotels & Restaurants; **InterContinental Hotels Group plc**
Klepierre SA; **Simon Property Group Inc**
Klohn Crippen Berger Ltd; **Louis Berger Group Inc (The)**
Kobalt; **Lowe's Companies Inc**
Kohler; **Kohler Company**
Kohler Engines; **Kohler Company**
KST Electric; **Rosendin Electric**
KUHO; **Samsung C&T Corporation**
Kumho Asiana Group; **Kumho Industrial Co Ltd**
Kumho Engineering & Construction; **Kumho Industrial Co Ltd**
Kwong Sang Hong International Limited (The); **Chinese Estates Holdings Ltd**
Kyriad; **Golden Tulip Hospitality Group**
Kyriad Prestige; **Groupe du Louvre**
L.B. Foster Rail Technologies, Inc; **LB Foster Company**
L.B. Pipe & Coupling Products LLC; **LB Foster Company**
La Prairie; **Ritz-Carlton Hotel Company LLC (The)**
La Quinta Inns; **La Quinta Holdings Inc**
La Quinta Inns and Suites; **La Quinta Holdings Inc**
La Quinta Properties; **La Quinta Holdings Inc**
Lafarge Group; **LafargeHolcim Ltd**
Landmark (The); **Hongkong and Shanghai Hotels Ltd**
Landmark Harbour City; **Wheelock and Company Limited**
Landmark Mandarin Oriental; **Hongkong Land Holdings Ltd**
LandSafe Title; **Stewart Information Services Corp**
LaSalle Investment Management; **Jones Lang LaSalle Inc**
Layne Inliner LLC; **Layne Christensen Company**
Le Louis XV-Alain Ducasse; **Societe Anonyme des Bains de Mer et du Cercle des Etrangers a Monaco**
Le Melezin; **Amanresorts International Pte Ltd (Aman Resorts)**

INDEX OF SUBSIDIARIES, BRAND NAMES AND AFFILIATIONS, CONT.

INDEX OF SUBSIDIARIES, BRAND NAMES AND AFFILIATIONS, CONT.

Marina Bay Suites; **Hongkong Land Holdings Ltd**
Marina Mandarin Singapore; **Meritus Hotels & Resorts Inc**
Marlette; **Clayton Homes Inc**
Marqi Branding Studio; **Steelman Partners LLP**
Marque; **Far East Organization**
Marriott Hotels; **Marriott International Inc**
Marriott International Inc; **Ritz-Carlton Hotel Company LLC (The)**
Marylebone Hotel (The); **Doyle Collection (The)**
MATCOR Inc; **Brand Energy & Infrastructure Services Inc**
Matrix North American Construction; **Matrix Service Company**
Mayfaire Town Center and Community Center; **CBL & Associates Properties Inc**
Mazagan Beach Resort; **Kerzner International Holdings Limited**
mCapitol Management; **MWH Global Inc**
McDermott Marine Construction Ghana Limited; **McDermott International Inc**
MCL Land; **Hongkong Land Holdings Ltd**
MCS Capital LLC; **Marcus Corporation (The)**
ME by Melia; **Melia Hotels International SA**
Meadow Valley Contractors Inc; **Meadow Valley Corporation**
MEI Triplefive Engineering; **Triple Five Group**
Melia Hotels; **Melia Hotels International SA**
Melloy Industrial Services Inc; **PCL Construction Group Inc**
Merchandise Mart of Chicago; **Vornado Realty Trust**
Meritage Homes; **Meritage Homes Corp**
Meritus Pelangi Beach Resort & Spa, Langkawi; **Meritus Hotels & Resorts Inc**
Metal Building Components; **NCI Building Systems Inc**
Metal Depots; **NCI Building Systems Inc**
Mezzanine Realty Partners III LLC; **Pearlmark Real Estate Partners**
Mezzanine Realty Partners IV LLC; **Pearlmark Real Estate Partners**
MGallery; **Accor SA**
MGM China Holdings Ltd; **MGM Resorts International**
MGM Grand Detroit; **MGM Growth Properties LLC**
MGM Grand Las Vegas; **MGM Resorts International**
MGM Growth Properties LLC; **MGM Resorts International**
MGM Hospitality; **MGM Resorts International**
MHC Operating Limited Partnership; **Equity Lifestyle Properties Inc**
Mid-America Apartments LP; **Mid-America Apartment Communities Inc**
Mid-America Multifamily Fund II LLC; **Mid-America Apartment Communities Inc**
Millenium; **Hyundai Elevator Co Ltd**

Millennium Collection; **Millennium & Copthorne Hotels plc**
Millennium Restaurant; **Commune Hotels & Resorts**
Minha Casa, Minha Vida; **MRV Engenharia E Participacoes SA**
Miramar Hotel & Investment Company, Limited; **Henderson Land Development Co Ltd**
MMA Energy Capital; **MMA Capital Management LLC**
Modern Terminals Limited; **Wharf (Holdings) Limited The**
Modular; **Hyundai Elevator Co Ltd**
Moduline Industries; **Champion Enterprises Holdings LLC**
Mokara Hotel and Spa; **TRT Holdings Inc**
Molokai Properties Ltd; **Guocoleisure Ltd**
Mondrian; **Morgans Hotel Group Co**
Monte Carlo; **MGM Growth Properties LLC**
Monte-Carlo Bay Hotel & Resort; **Societe Anonyme des Bains de Mer et du Cercle des Etrangers a Monaco**
Monte-Carlo Beach Hotel; **Societe Anonyme des Bains de Mer et du Cercle des Etrangers a Monaco**
Monte-Carlo SBM; **Societe Anonyme des Bains de Mer et du Cercle des Etrangers a Monaco**
Monterey Homes; **Meritage Homes Corp**
Morgan Stanley; **AMLI Residential Properties Trust**
Morgan Stanley Real Estate; **Glenborough LLC**
Morgan Stanley Smith Barney; **CitiMortgage Inc**
Motel 168; **Homeinns Hotel Group**
Move.com; **Move Inc (Realtor.com)**
Moving.com; **Move Inc (Realtor.com)**
MPF Research; **RealPage Inc**
Mt. Shasta Mall; **Rouse Properties Inc**
My Home Industries Ltd; **CRH plc**
Myers & Sons Construction LP; **Sterling Construction Company Inc**
NAI Asia Pacific Regional Overview; **NAI Global Inc**
National Golf Properties; **Newcastle Investment Corp**
National Investors Title Insurance (NITIC); **Investors Title Company**
Nationwide Custom Homes Inc; **Palm Harbor Homes Inc**
NCC Power Projects; **Sembcorp Industries Ltd**
New Media Investment Group Inc; **Newcastle Investment Corp**
New Senior Investment Group Inc; **Newcastle Investment Corp**
New World; **New World Development Company Limited**
New World China Land Limited; **New World Development Company Limited**
New World Department Store China Limited; **New World Development Company Limited**
New World Hospitality; **Rosewood Hotels & Resorts LLC**
New York-New York; **MGM Growth Properties LLC**

INDEX OF SUBSIDIARIES, BRAND NAMES AND AFFILIATIONS, CONT.

Newhall Land Development LLC; **Newhall Land & Farming Company**
News Corp; **Move Inc (Realtor.com)**
NH Collection; **NH Hotel Group SA**
NH Hotels; **NH Hotel Group SA**
nhow; **NH Hotel Group SA**
Nile; **Sonesta International Hotels Corp**
Nomura Real Estate Asset Management Co Ltd; **Nomura Real Estate Master Fund Inc**
Nomura Real Estate Development Co Ltd; **Nomura Real Estate Master Fund Inc**
Nomura Real Estate Office Fund Inc; **Nomura Real Estate Master Fund Inc**
Nomura Real Estate Residential Fund Inc; **Nomura Real Estate Master Fund Inc**
Norcon Rossi; **Rossi Residencial SA**
Norm Thompson; **Bluestem Group Inc**
North America Sekisui House LLC; **Sekisui House Ltd**
North Park; **KB Home**
Notivus Multi-Family LLC; **RealPage Inc**
Novolac; **Five Star Products Inc**
Novotel; **Accor North America**
Novotel; **Accor SA**
Nutrional Support System; **Extendicare Inc**
NVHomes; **NVR Inc**
NVR Mortgage Finance Inc; **NVR Inc**
NWS Holdings Limited; **New World Development Company Limited**
NY Transit; **RailWorks Corp**
O2O e-commerce; **Dalian Wanda Group Co Ltd**
Oak Grove Capital; **Jones Lang LaSalle Inc**
Oaks Collection; **84 Lumber Company**
Oakwood Apartment; **Oakwood Worldwide**
Oakwood Asia Pacific Pte Ltd; **Oakwood Worldwide**
Oakwood Corporate Housing; **Oakwood Worldwide**
Oakwood Premier; **Oakwood Worldwide**
Oakwood Residence; **Oakwood Worldwide**
Oasia; **Far East Organization**
Oberoi; **Oberoi Group (EIH Ltd)**
Oberoi Philae (The); **Oberoi Group (EIH Ltd)**
Oberoi Zahra (The); **Oberoi Group (EIH Ltd)**
Odebrecht Ambiental; **Odebrecht SA**
Odebrecht Defesa e Tecnologia; **Odebrecht SA**
Odebrecht Energia; **Odebrecht SA**
Odebrecht Engenharia Industrial; **Odebrecht SA**
Odebrecht Oil & Gas; **Odebrecht SA**
Odebrecht Realizacoes Imobiliarias; **Odebrecht SA**
Odebrecht Transport; **Odebrecht SA**
OfficeMax; **Boise Cascade Corp**
Oldcastle BuildingEnvelope; **Oldcastle Inc**
Olympia 66; **Hang Lung Properties Limited**
Omni Hotels & Resorts; **TRT Holdings Inc**
One Hysan Avenue; **Hysan Development Co Ltd**
One&Only; **Kerzner International Holdings Limited**
OneSite; **RealPage Inc**

OOCL China Domestic Ltd; **Orient Overseas (International) Ltd**
OOCL Logistics Ltd; **Orient Overseas (International) Ltd**
Opportunity Fund I; **RXR Realty**
Opportunity Fund II; **RXR Realty**
Orchard Brands Corporation; **Bluestem Group Inc**
Orchard Supply Hardware; **Lowe's Companies Inc**
Orient Overseas Container Line Limited; **Orient Overseas (International) Ltd**
Overseas Union Enterprise Ltd; **Meritus Hotels & Resorts Inc**
Overture; **Royal Group Inc**
Oxford Life Insurance Company; **AMERCO (U-Haul)**
PABCO Building Products LLC; **Pacific Coast Building Products Inc**
Pacific Coast Companies Inc; **Pacific Coast Building Products Inc**
Pacific Coast Jet Charter Inc; **Pacific Coast Building Products Inc**
Pacific Coast Supply LLC; **Pacific Coast Building Products Inc**
Pacific Premier Retail LP; **Macerich Company (The)**
Pacific Property Net; **Hutchison Whampoa Properties Ltd**
Palace 66; **Hang Lung Properties Limited**
Palace Station Hotel & Casino; **Station Casinos LLC**
Palazzo Resort Hotel Casino (The); **Las Vegas Sands Corp (The Venetian)**
Palm Harbor; **Palm Harbor Homes Inc**
Palo Real Estate Company Limited; **Wynn Resorts Limited**
Paradisus Resorts; **Melia Hotels International SA**
ParaMed; **Extendicare Inc**
Paramount Group Operating Partnership LP; **Paramount Group Inc**
Park Hyatt; **Hyatt Hotels Corporation**
Park Inn by Radisson; **Rezidor Hotel Group AB**
Park Inn by Radisson; **Carlson Companies Inc**
Park Inn Hotels; **Carlson Rezidor Hotel Group**
Park Plaza; **Carlson Companies Inc**
Park Plaza Hotels & Resorts; **Carlson Rezidor Hotel Group**
Parkway Properties Office Fund II LP; **Parkway Properties Inc**
Parque; **MRV Engenharia E Participacoes SA**
Patton Tully Transportation LLC; **Anderson-Tully Lumber Company**
Paul Davis Restoration; **FirstService Corporation**
PCIC Insurance; **Centex Corp**
PCL Civil Constructors Inc; **PCL Construction Group Inc**
PCL Constructors Bahamas Ltd; **PCL Construction Group Inc**
PCL Energy Inc; **PCL Construction Group Inc**

INDEX OF SUBSIDIARIES, BRAND NAMES AND AFFILIATIONS, CONT.

INDEX OF SUBSIDIARIES, BRAND NAMES AND AFFILIATIONS, CONT.

RCC Consultants; **Black & Veatch Holding Company**
RE/MAX Commercial Network; **RE/MAX Holdings Inc**
Ready Mix USA LLC; **CEMEX Inc**
Realogy Corporation; **Century 21 Real Estate LLC**
Realogy Corporation; **Coldwell Banker Real Estate LLC**
Realogy Holdings Corporation; **NRT LLC**
Realogy Holdings Corporation; **ZipRealty Inc**
Realtor.com; **Move Inc (Realtor.com)**
REALTrac Online; **NAI Global Inc**
Realty Asset Advisors; **Related Group (The)**
Realty South; **HomeServices of America Inc**
Red Lion Hotels; **Red Lion Hotels Corporation**
Red Lion Inn & Suites; **Red Lion Hotels Corporation**
Red Rock Casino, Resort & Spa; **Red Rock Resorts Inc**
Redfin Collections; **Redfin**
Redfin Home Value Tool; **Redfin**
Regal Lifestyle Communities Inc; **Welltower Inc**
Regency Centers LP; **Regency Centers Corp**
Regent Hotels; **Four Seasons Hotels Inc**
Reis SE; **Reis Inc**
Reis Services LLC; **Reis Inc**
ReisReports; **Reis Inc**
Reit Management and Research LLC; **Senior Housing Properties Trust**
Related Cervera Realty Services; **Related Group (The)**
Related Financial; **Related Group (The)**
Reliance Fire Protection Inc; **APi Group Inc**
Remy International Inc; **Fidelity National Financial Inc**
Renaissance Hotels; **Marriott International Inc**
Rendezvous Hotels; **Far East Organization**
Rent Zestimates; **Zillow Inc**
Renton Howard Hood Levin; **Aedas**
REP Desenvolvimento Imobiliario SA; **PDG Realty SA Empreendimentos e Participacoes**
Repwest Insurance Company; **AMERCO (U-Haul)**
Residence Inn by Marriott; **Hospitality Properties Trust**
Residences at Mandarin Oriental; **Mandarin Oriental International Ltd**
Residencies at the Ritz-Carlton (The); **Ritz-Carlton Hotel Company LLC (The)**
Resolve Technology Inc; **CoStar Group Inc**
Retevision; **Abertis Infraestructuras SA**
Revera Home Health; **Extendicare Inc**
Rezidor Hotel Group; **Carlson Companies Inc**
Richmond American Homes; **M.D.C. Holdings Inc**
Ridge at Bandera (The); **KB Home**
RIDGID; **Home Depot Inc**
Rio All-Suite Hotel & Casino; **Rio Properties Inc**
Rio Secco Golf Club; **Rio Properties Inc**
Rio Spa & Salon; **Rio Properties Inc**
Rise & Dine; **Howard Johnson International Inc**
Ritz-Carlton Destination Club (The); **Ritz-Carlton Hotel Company LLC (The)**
Ritz-Carlton Hotel Company LLC (The); **Marriott International Inc**

Ritz-Carlton Koh Samui; **Ritz-Carlton Hotel Company LLC (The)**
Ritz-Carlton Reserve; **Ritz-Carlton Hotel Company LLC (The)**
River Lee Hotel (The); **Doyle Collection (The)**
RiverVillage; **Newhall Land & Farming Company**
Road and Highway Builders LLC; **Sterling Construction Company Inc**
Road and Highway Builders of California Inc; **Sterling Construction Company Inc**
Robern; **Kohler Company**
Rodamco Europe NV; **Unibail-Rodamco SE**
Rodeway Inn; **Choice Hotels International Inc**
Room In The Inn Program; **TMI Hospitality Inc**
Roots Corporation Ltd; **Indian Hotels Company Limited (The)**
Roseland; **Mack-Cali Realty Corp**
Rosewood Hotel Group; **New World Development Company Limited**
Rosewood Hotels & Resorts; **New World Development Company Limited**
Rosewood Inn of the Anasazi; **Rosewood Hotels & Resorts LLC**
Rosewood Luang Prabang; **Rosewood Hotels & Resorts LLC**
Rosewood Phnom Penh; **Rosewood Hotels & Resorts LLC**
Rosewood Siem Reap; **Rosewood Hotels & Resorts LLC**
Rossi Commercial Properties; **Rossi Residencial SA**
Rossi Urbanizadora; **Rossi Residencial SA**
Royal Garden; **Sun Hung Kai Properties Ltd**
Royal Kor Flo; **Royal Group Inc**
Royal Mouldings Limited; **Royal Group Inc**
Royal Park Hotel; **Sun Hung Kai Properties Ltd**
Royal Plaza Hotel; **Sun Hung Kai Properties Ltd**
Royal Tulip; **Golden Tulip Hospitality Group**
Royal View Hotel; **Sun Hung Kai Properties Ltd**
Royco Hotels Inc; **Condor Hospitality Trust Inc**
Ryan Homes; **NVR Inc**
Ryland Group (The); **CalAtlantic Group Inc**
Ryobi; **Home Depot Inc**
S2N Technology Group; **Clark Construction Group LLC**
Samwhan Camus Co; **Samwhan Corporation**
Samwhan Machinery Co Ltd; **Samwhan Corporation**
Sands China Ltd; **Las Vegas Sands Corp (The Venetian)**
Sands Expo and Convention Center (The); **Las Vegas Sands Corp (The Venetian)**
Sands Macao Casino (The); **Las Vegas Sands Corp (The Venetian)**
Sanef; **Abertis Infraestructuras SA**
Sankaty Advisors LLC; **Champion Enterprises Holdings LLC**
Santa Fe Station Hotel & Casino; **Station Casinos LLC**
Sazerac; **Kimpton Hotel & Restaurant Group LLC**

INDEX OF SUBSIDIARIES, BRAND NAMES AND AFFILIATIONS, CONT.

INDEX OF SUBSIDIARIES, BRAND NAMES AND AFFILIATIONS, CONT.

INDEX OF SUBSIDIARIES, BRAND NAMES AND AFFILIATIONS, CONT.

INDEX OF SUBSIDIARIES, BRAND NAMES AND AFFILIATIONS, CONT.

INDEX OF SUBSIDIARIES, BRAND NAMES AND AFFILIATIONS, CONT.

A Short Real Estate & Construction Industry Glossary

Abatement (Property Tax): A decrease in property taxes, due to a problem with the property or to an agreement by taxing authorities to delay taxing the property at full value.

Acceleration Clause: A clause in a contract of debt, such as a mortgage, which causes the entire amount to become due upon the borrower's default.

Accessibility: The degree to which customers can easily get into and out of a shopping center, store, home or office's various rooms and facilities. Accessibility is an issue in providing proper accommodation to the elderly and the physically challenged.

Acre: A measure of land equaling 43,560 square feet.

Adjustable-Rate Mortgage (ARM): A mortgage that changes interest rates periodically to match a specific index.

Adjusted Funds From Operations (AFFO): A measure of cash flow from operations, typically used to gauge REITs. AFFO is the result of subtracting from funds from operations (FFO) certain costs that otherwise would have been capitalized for depreciation, but are actually needed to maintain a property on a day-to-day basis, such as the replacement of carpeting, or leasing expenses such as remodeling allowances.

Air Rights: A sellable right to the air space above a property.

Alt-A: See "Alternative Documentation Mortgage (Alt-A)."

Alternative Documentation Mortgage (Alt-A): An alternative method of documenting a loan file, often referred to as "alt doc" or "Alt-A," that relies on unsubstantiated information provided by the borrower regarding personal income and assets, rather than positive proof such as income tax returns, W2 forms from employers, bank statements and investment account statements.

Amortization Schedule: A printed table showing the principal and interest payments and due dates of the payments owed on a mortgage or loan.

Amortization through Equal Monthly Payments: A repayment method in which the amount borrowed is repaid gradually through regular monthly payments of principal and interest. During the first few years, most of each payment is applied toward the interest owed. During the final years of the loan, payment amounts are applied almost exclusively to the remaining principal.

Anticompetitive Leasing Arrangement: A lease that limits the type and amount of competition a particular retailer faces within a trading area (e.g., a lease that will not allow two supermarkets in one shopping center).

Arbitration: A legal process wherein an impartial third party decides a dispute. This practice is used instead of a standard trial in court.

Assumable Mortgage (Assumable Loan): An existing mortgage that can be assumed by a future purchaser of a piece of real estate.

Atmosphere: Architecture, layout, signs and displays, color, lighting, music and scents which together create an image of a store in the customer's mind.

Average Daily Rate (ADR): In hotels, room revenue divided by rooms sold.

Baby Boomer: Generally refers to people born from 1946 to 1964. In the U.S., the initial number of Baby Boomers totaled about 78 million. The term evolved to describe the children of soldiers and war industry workers who were involved in World War II and who began forming families after the war's end. In 2011, the oldest Baby Boomers began reaching the traditional retirement age of 65.

Back-End Ratio: See "Debt-to-Income Ratio (Mortgages)."

Balloon Loan: A loan in which a large payment for the remaining balance is due at the end of the loan.

Balloon Payment: The final principal payment due at the end of a balloon mortgage. Typically, lower monthly payments are provided for under a balloon mortgage, with a substantial final payment due later.

Basis Point: One 100th of 1%.

Below-Market Interest Rate (BMIR): An interest rate that is lower than rates generally charged on similar loans in the marketplace. In real estate, this is typically a subsidized interest rate provided by the Federal Government for low-income housing.

Biennial Payments: Loan payments that are due every other year.

Blanket Mortgage: A mortgage covering at least two pieces of real estate as security for the same mortgage.

Boiler Plate: A standard clause that appears in all similar contracts.

BPO: See "Business Process Outsourcing (BPO)."

BRAC: Base realignment and closure. The process whereby military bases are closed and repurposed into commercial, industrial or other real estate uses.

Brand: A marketing strategy that places a focus on the brand name of a product, service or firm in order to increase the brand's market share, increase sales, establish credibility, improve satisfaction, raise the profile of the firm and increase profits. Also, see "Brand."

Branding: A marketing strategy that places a focus on the brand name of a product, service or firm in order to increase the brand's market share, increase sales, establish credibility, improve satisfaction, raise the profile of the firm and increase profits. Also, see "Brand."

B-to-B, or B2B: See "Business-to-Business."

B-to-C, or B2C: See "Business-to-Consumer."

Build-to-Suit: A building created by a landlord specifically for one tenant, designed to suit the tenant's particular needs.

Business Process Outsourcing (BPO): The process of hiring another company to handle business activities. BPO is one of the fastest-growing segments in the offshoring sector. Services include

human resources management, billing and purchasing and call centers, as well as many types of customer service or marketing activities, depending on the industry involved. Also, see "Knowledge Process Outsourcing (KPO)" and "Business Transformation Outsourcing (BTO)."

Business Transformation Outsourcing (BTO): A segment within outsourcing in which the client company revamps its business processes with the goal of transforming its business by following a collaborative approach with its outsourced services provider.

Business-to-Business: An organization focused on selling products, services or data to commercial customers rather than individual consumers. Also known as B2B.

Business-to-Consumer: An organization focused on selling products, services or data to individual consumers rather than commercial customers. Also known as B2C.

Buy-down (Buydown): With a "temporary" buydown, a lender or homebuilder subsidizes a mortgage by lowering the interest rate during the first few years of the loan. While the payments are initially low, they increase when the subsidy expires. A "permanent" buydown, however, reduces the interest rate over the entire life of the loan.

Buying Power Index (BPI): An index indicating the percentage of total U.S. retail sales occurring in a specific geographic area. Used to forecast demand for new stores and to evaluate the performance of existing stores.

Capital Gain on Real Estate: The amount of increase in value of a property, other than a primary residence, that is taxable.

Capitalization Rate (Real Estate): A measure of a property's sale price, determined by dividing the annual net income of a property by the purchase price.

Cash Available for Distribution (CAD): A measure of a REIT's ability to generate cash and to distribute dividends. Also known as FAD (Funds Available for Distribution.)

Caveat Emptor: Latin for "buyer beware." In real estate it is generally incumbent upon the buyer to discover defects with the property unless those defects are known to the seller.

CDO: See "Collateralized Debt Obligation (CDO)."

Central Business District (CBD): The traditional downtown business area of a city or town.

Certificate of Eligibility: The document given to qualified veterans that entitles them to obtain VA-guaranteed loans for homes, businesses and mobile homes. Certificates of eligibility may be obtained by sending Form DD-214 (Separation Paper) to the local VA office with VA form 1880 (request for Certificate of Eligibility).

Certificate of Insurance: A document detailing the insurance coverage provided for a property, asset or business.

Certificate of Occupancy: A document issued by the local government stating that a property is ready for occupancy.

Chattel: An item not attached to the land or buildings in a purchase agreement unless otherwise specified.

Class A Building: The most visible, best-located, most prestigious buildings that rent to tenants at rates above typical buildings in the area.

Class B Building: Typical buildings within a given area, which rent to tenants seeking to pay average rates for the neighborhood. These buildings are well-maintained and well-located but are below the standards of Class A buildings.

Class C Building: Buildings that are maintained and located in a fashion that is satisfactory for tenants seeking to pay rents that are below average for the neighborhood. These buildings have a presence and level of finish that is well below those of Class A or B buildings.

Close-Out Store: A retailer offering low-priced merchandise obtained through liquidations.

Closing: The meeting between the buyer, seller and lender or their agents where the property or asset and funds legally change hands. Also called "settlement."

Closing Costs: Costs associated with the closing of a mortgage, including legal fees, taxes, mortgage application fees and other items.

CMO: See "Collateralized Mortgage Obligation (CMO)."

Collateralized Debt Obligation (CDO): A method of taking a pool of debts, such as mortgages, and selling pieces of that pool to multiple investors. CDOs can also be created for pools of bonds, loans, leases and other types of financial assets.

Collateralized Mortgage Obligation (CMO): A bond that is a debt instrument backed by a pool of underlying mortgages that pass through payments received to the holder of the CMO. A CMO is a popular way to trade large amounts of mortgages at once. The underlying pool may represent thousands of individual mortgages.

Combination Store: A store that combines grocery and drug products in the same building. See "Food/Drug Combo."

Commitment: 1) A promise by a lender to make a loan on specific terms or conditions to a borrower. 2) A promise by an investor to purchase mortgages from a lender with specific terms or conditions. 3) An agreement, often in writing, between a lender and a borrower to loan money at a future date subject to the completion of paperwork or compliance with stated conditions.

Common Area Maintenance (CAM) Assessments: Fees paid to a landlord or property owners association for the maintenance of common areas such as hallways, elevators, exterior lighting, sidewalks or recreational facilities.

Common Areas: Areas shared by tenants or property owners. For example, in a shopping center, the parking lots, sidewalks and hallways are common areas. In a housing development, recreational facilities, greenbelts and hike or bike trails may be common areas.

Community Center: A large shopping center that includes a discount store, specialty department store, super drugstore, home-improvement center and other convenience and shopping goods stores.

Compact Fluorescent Lamp (CFL): A type of light bulb that provides considerable energy savings over traditional incandescent light bulbs.

Condemnation: The claiming of private land by a government for public use or because of code violations. Also see "Eminent Domain."

Condominium: 1) In the travel industry, lodging similar to furnished, private apartments that are available to rent for days or weeks. 2) In real estate, a kind of property ownership in which the owner holds title to an individual unit in a multi-unit dwelling and shares ownership of common areas such as hallways or swimming pools.

Conforming Loan: See "Non-Conforming Loan."

Construction Loan: A short-term interim loan to pay for the construction a building or home. Such loans are usually designed to provide periodic disbursements to the builder as progress with construction is made.

Construction/Permanent Loan: A mortgage loan combining short-term financing of construction with long-term financing of the completed property.

Consumerism: The activities of government, business and independent organizations designed to protect individuals from practices that infringe upon their rights as consumers.

Contract for Deed: A property transaction wherein the seller retains the title to the property until the buyer has paid a certain amount, usually the entire balance owned. This is generally a risky way for a buyer to enter into a purchase.

Convenience Center: A shopping center that typically includes such stores as a convenience store and a dry cleaner.

Convenience Stores: Stores between 3,000 and 8,000 square feet in size providing a limited assortment of merchandise at a convenient location and time (e.g., 7-Eleven). Many also sell gasoline.

Conventional Mortgage: A loan other than a government-financed loan (e.g., a loan that is not a VA mortgage).

Conventional Supermarket: A market that offers a complete line of groceries, meat and produce with a minimum of $2 million in annual revenue, at least 9% of which comes from GM/HBC. Stores typically carry approximately 15,000 items. Many stores offer bakery, deli, banking and other services.

Cooperative (Co-Op): In real estate, a type of multiple ownership in which members of the co-operative own shares in the corporation owning the property. Co-op owners typically have the right to occupy one unit in the co-op building. It is a distinctly different type of ownership from a condominium, wherein each unit's owner takes title to the condominium and then shares ownership of the common areas such as hallways.

Cost Plus Contract: A contract that sets the contractor's compensation as a percentage of the total cost of labor and materials.

Credit Risk: The risk assumed by the lender that the borrower may default on a loan or mortgage. The apparent credit risk is considered when setting an interest rate to be charged.

CRM: See "Customer Relationship Management (CRM)."

Customer Relationship Management (CRM): Refers to the automation, via sophisticated software, of business processes involving existing and prospective customers. CRM may cover aspects such as sales (contact management and contact history), marketing (campaign management and telemarketing) and customer service (call center history and field service history). Well known providers of CRM software include Salesforce, which delivers via a Software as a Service model (see "Software as a Service (Saas)"), Microsoft and Oracle.

Debt-to-Income Ratio (Mortgages): Also called the "back-end" ratio, this ratio, expressed as a percentage, is calculated by dividing a borrower's monthly payment obligations on long-term debt (including housing expense) by his or her gross monthly income.

Deed of Trust: In conjunction with a, this instrument is used in many western states to pledge the home or other real estate as security for a loan.

Deferred-Interest Mortgage: A mortgage that defers some of the interest to a later date.

Demographics: The breakdown of the population into statistical categories such as age, income, education and sex.

Department Stores: Very large stores carrying a wide variety and deep assortment while offering considerable customer services (e.g., Dillard's or Saks Fifth Avenue). Stores are organized into separate departments for displaying merchandise. However, the variety of departments has lessened in recent years, and these stores now tend to focus on apparel.

Depreciation: A method of amortizing the cost of an asset in equal dollar amounts over the useful life of an asset. For example, an item with a five-year life would be charged with a certain amount of its cost every year for five years.

Discount Broker: A broker or brokerage firm that executes buy and sell transactions at commission rates lower than a full-service broker or brokerage.

Discount Points (Mortgages): The amount paid either to maintain or to lower the interest rate charged on a mortgage. Each point is equal to 1% of the loan amount. (For example, two points on a $100,000 mortgage would equal $2,000.)

Discount Store: A general merchandise retailer offering a wide variety of merchandise, limited service and low prices (e.g., Target or Kmart).

Discount-Anchored Center: A shopping center that contains one or more discount stores plus smaller retail tenants.

Down Payment: The difference between the purchase price and that portion of the purchase price being financed.

Down REIT: A structure that enables an REIT to acquire properties in exchange for partnership units.

Earnest Money: Good faith money provided to the seller by the potential buyer to show that he or she is serious about purchasing a piece of real estate. This amount can be applied to the purchase at closing, but if the deal does not go through it may be forfeited, or in some cases returned, depending on the terms of the purchase contract.

Easement: A right of way giving persons other than the owner access to a property. A common use is to provide a driveway through one property owner's land for access to an adjacent property.

Echo Boomers: See "Generation Y."

E-Commerce: The use of online, Internet-based sales methods. The phrase is used to describe both business-to-consumer and business-to-business sales.

Eminent Domain: The right of a governmental unit to force an owner to sell his or her property for fair market value for public use. This is the legal basis for the condemnation of private property. For example, if a state needs property for the construction or widening of a highway, it may condemn the needed property under this feature of the law.

Encroachment: An intrusion onto a property. For example, if one property owner erects a fence along his property border and fails to properly stay inside of his property line, a portion of the fence lies on the adjacent property, forming an encroachment.

Encumbrance: Any claim on a title or property that might prevent the owner from passing good title at a sale.

End Caps: Display fixtures located at the end of an aisle. Also, in real estate, the end unit or corner unit in a shopping center.

Energy Service Company (ESCO): A company that provides energy-efficiency-related and other value-added services and for which performance contracting is a core part of its energy-efficiency services business. In a performance contract, the ESCO guarantees energy and/or dollar savings for the project and the ESCO's compensation is therefore linked in some fashion to the performance of the project.

Enterprise Resource Planning (ERP): An integrated information system that helps manage all aspects of a business, including accounting, ordering and human resources, typically across all locations of a major corporation or organization. ERP is considered to be a critical tool for management of large organizations. Suppliers of ERP tools include SAP and Oracle.

Equal Credit Opportunity Act (ECOA): A federal law that requires lenders to make credit equally available without discrimination based on race, religion, national origin, age, sex, marital status or receipt of income from public assistance programs.

Equity: 1) The net value of the common stockholders' interest in a company as listed on a company's balance sheet. 2) The difference between a company's assets and liabilities (it is possible for a company to have negative equity). 3) The difference between the value of a property and the amount owed on that property. 4) A share of stock.

Equity REIT: A REIT that owns real estate, as opposed to an REIT that specializes in making mortgage loans.

ERP: See "Enterprise Resource Planning (ERP)."

ESCO: See "Energy Service Company (ESCO)."

Escrow Account: A mortgage lender's method of accounting for escrow monies received from the borrower. See "Escrow(s)."

Escrow Waiver Fee: A fee paid by a home mortgage borrower who elects to pay insurance and property taxes directly, rather than paying them into an escrow account at the mortgage company. There is a one-time charge by the mortgage company of 0.25% to 0.375% of the loan amount.

Escrow(s): That portion of a borrower's monthly payments held by the lender or servicer to pay for taxes, hazard insurance, mortgage insurance and other items as they become due. Also known as impound(s).

EU: See "European Union (EU)."

EU Competence: The jurisdiction in which the European Union (EU) can take legal action.

European Union (EU): A consolidation of European countries (member states) functioning as one body to facilitate trade. Previously known as the European Community (EC). The EU has a unified currency, the Euro. See europa.eu.int.

Expense Ratio: The comparison of the cost of operating a property, business or organization to its gross income.

Facilities Management: The management of a company's physical buildings and/or information systems on an outsourced basis.

Factory Outlet Stores: Off-price retail stores owned by manufacturers.

Fair Credit Reporting Act: A consumer protection law that regulates consumer credit report providers. For example, the act sets up procedures for correcting mistakes on an individual's credit record, provides certain restrictions on publishing credit reports and generally governs a consumer's rights.

Fair Market Value (FMV): A term used to indicate the value of a property on the open market.

Fannie Mae: See "Federal National Mortgage Association (FNMA)."

Farmers Home Administration (FmHA): A part of the U.S. Department of Agriculture that provides financing for farmers and residents of rural areas.

FASB: See "Financial Accounting Standards Board (FASB)."

Federal Home Loan Bank Board (FHLBB): The agency of the federal government that supervises all federal savings and loan associations and federally insured state-chartered savings and loan associations. The FHLBB also operates the Federal Savings and Loan Insurance Corporation, which insures accounts at federal savings and loan associations and those state-chartered associations that apply and are accepted. In addition, the FHLBB directs the Federal Home Loan Bank System, which provides a flexible credit facility for member savings institutions to promote the availability of home financing. The FHL Banks also own the Federal Home Loan Mortgage Corporation, established in 1970 to promote secondary markets for mortgages.

Federal Home Loan Mortgage Corporation (FHLMC): A U.S. government-sponsored agency that purchases conventional mortgages from lending institutions, thus adding liquidity to the market. Also known as "Freddie Mac."

Federal Housing Administration (FHA): An agency of the U.S. Department of Housing and Urban Development (HUD). The FHA primarily insures housing loans.

Federal National Mortgage Association (FNMA): A major, government-sponsored investor that purchases mortgage loans from mortgage bankers. It is similar to FHLMC. Also known as "Fannie Mae."

Fee Simple: Absolute or sole ownership of a property.

FHA: See "Federal Housing Administration (FHA)."

FHLMC: See "Federal Home Loan Mortgage Corporation (FHLMC)."

Financial Accounting Standards Board (FASB): An independent organization that establishes the Generally Accepted Accounting Principles (GAAP).

First Mortgage: The mortgage that has legal precedence over other mortgage claims, usually, but not always, chronologically first in the history of the property. Also see "Second Mortgage."

Fixed-Rate Mortgage or Loan: A mortgage (or loan) that has an interest rate that does not change over the life of the loan.

FmHA: See "Farmers Home Administration (FmHA)."

FNMA: See "Federal National Mortgage Association (FNMA)."

Food/Drug Combo: A superstore and drug store that share checkout lanes. GM/HBC (General Merchandise/Health and Beauty Care) takes up at least one-third of retail space in the store, and accounts for 15% or more of revenue. Food/drug combos offer pharmacy services.

Forbearance (Mortgages): A temporary reprieve from paying a mortgage, usually granted because of some kind of hardship.

Foreclosure: The process whereby a borrower in default under a mortgage is deprived of his or her interest in a property, which is taken away by the lender via a legal procedure.

Franchise: 1) A contractual agreement between a franchisor (for example, a company or organization owning all rights to a brand, type of business, retail operation, restaurant concept or sports league) and a franchisee (person or organization desiring to license the use of those rights for a specific purpose within a specific region) that allows the franchisee to operate a retail outlet or other type of business using a brand, trade secrets, formulas and format developed and supported by the franchisor. Typically, a franchisee pays an upfront fee and then continuing fees to the franchisor. 2) A generic term used to describe a very well established business or brand.

Franchisee: See "Franchise."

Franchisor: See "Franchise."

Freddie Mac: See "Federal Home Loan Mortgage Corporation (FHLMC)."

Free-Standing Retailer: A location for a retailer that is a building by itself, frequently on a pad site in front of a shopping center.

Front-End Ratio: See "Housing-Expenses-to-Income Ratio."

Full-Service Leasing: A program under which a vehicle or building is leased and the operation and maintenance are included in the lease fee.

GAAP: See "Generally Accepted Accounting Principles (GAAP)."

GDP: See "Gross Domestic Product (GDP)."

General Merchandise, Apparel, Furniture and Other (GAFO): Usually used in reference to the retail sector, excluding automotive and food stores, and includes general merchandise, department, discount apparel, furniture and miscellaneous specialty stores.

Generally Accepted Accounting Principles (GAAP): A set of accounting standards administered by the Financial Accounting Standards Board (FASB) and enforced by the U.S. Security and Exchange Commission (SEC). GAAP is primarily used in the U.S.

Generation M: A very loosely defined term that is sometimes used to refer to young people who have grown up in the digital age. "M" may refer to any or all of media-saturated, mobile or multi-tasking. The term was most notably used in a Kaiser Family Foundation report published in 2005, "Generation M: Media in the Lives of 8-18 year olds." Also, see "Generation Y" and "Generation Z."

Generation X: A loosely-defined and variously-used term that describes people born between approximately 1965 and 1980, but other time frames are recited. Generation X is often referred to as a group influential in defining tastes in consumer goods, entertainment and/or political and social matters.

Generation Y: Refers to people born between approximately 1982 and 2002. In the U.S., they number more than 90 million, making them the largest generation segment in the nation's history. They are also known as Echo Boomers, Millennials or the Millennial Generation. These are children of the Baby Boom generation who will be filling the work force as Baby Boomers retire.

Generation Z: Some people refer to Generation Z as people born after 1991. Others use the beginning date of 2001, or refer to the era of 1994 to 2004. Members of Generation Z are considered to be natural and rapid adopters of the latest technologies.

Geological Information System (GIS): A computer software system which captures, stores, updates, manipulates, analyzes, and displays all forms of geographically referenced information.

Geotextile: A synthetic material used for such construction purposes as strengthening a roadway base or embankment, erosion control and drainage liners. Sometimes referred to as geosynthetics, they are typically made of polymers. These textiles may reduce costs, enhance the useful life of a project, save weight or increase strength.

Ginnie Mae: See "Government National Mortgage Association (GNMA)."

GM/HBC: General Merchandise/Health and Beauty Care.

GNMA: See "Government National Mortgage Association (GNMA)."

Good Faith Estimate: A written estimate of closing costs that a lender must provide for the borrower within three days of submitting a new loan application.

Government National Mortgage Association (GNMA): A government-owned corporation within the U.S. Department of Housing and Urban Development (HUD) that specializes in the purchase of FHA and VA loans. Also known as "Ginnie Mae."

Graduated Payment Mortgage (GPM): A type of flexible-payment mortgage where the payments increase for a specified period of time and then level off. This type of mortgage has negative amortization built into it. For example, a homebuyer who believes that his or her income may increase a few years later may enter into a mortgage that has lower payments in the first five years but substantially higher monthly payments later in the loan.

Gross Domestic Product (GDP): The total value of a nation's output, income and expenditures produced with a nation's physical borders.

Gross Leasable Area (GLA): The total space that is leasable in a property. GLA may include common areas. Also see "Net Leasable Area (NLA)."

Gross National Product (GNP): A country's total output of goods and services from all forms of economic activity measured at market prices for one calendar year. It differs from Gross Domestic Product (GDP) in that GNP includes income from investments made in foreign nations.

Hard-Lines: Durable, non-apparel items, such as furniture, appliances and housewares.

Hectare (ha): A measurement of area equal to 10,000 square meters or 2.471 acres.

Historic District: A zoning classification for neighborhoods of historic value. This classification may limit what changes an owner can make to the property. For example, the colors of exterior paint or types of doors, windows and fences that may be used may be regulated.

Home Equity Loan: A fixed- or adjustable-rate loan obtained for a variety of purposes, secured by the equity in a home. Interest paid is usually tax-deductible. Often used for home improvement or the freeing of equity for investment. Home equity loans are tax-advantaged alternatives to consumer loans whose interest is not tax-deductible, such as auto or boat loans, credit card debt, medical debt and education loans.

Home Improvement Center: A category specialist combining the traditional hardware store and lumberyard (e.g., Home Depot).

Home Inspection: A thorough inspection by a professional that ensures that a property is mechanically and structurally sound. Home inspectors generally are licensed by the states in which they work.

Homeowners' Association: A nonprofit association that manages the common areas of a housing development or condominium project. In addition to owning and maintaining common areas, it may enforce deed restrictions and covenants.

Housing-Expenses-to-Income Ratio: Used in evaluating the income of a potential mortgage borrower. It is calculated by dividing the anticipated housing expense by the gross monthly income of the borrower. Also known as the "front-end" ratio.

HUD: The Department of Housing and Urban Development, a U.S. Government agency. HUD's mission is to increase homeownership, support community development and increase access to affordable housing free from discrimination.

HUD-I Settlement Statement: A form utilized at mortgage closing to itemize the costs associated with purchasing a home. Used universally by mandate of HUD (the U.S. Department of Housing and Urban Development).

Hybrid Mortgage: A type of mortgage that includes some compensation to the lender, such as a portion of income, in addition to the principal and interest on the loan.

Hybrid REIT: A REIT that combines the strategies of both an equity REIT and a mortgage REIT.

Hypermarket: A very large retail store that offers low prices and combines a discount store and a superstore food retailer in one warehouse-like building (e.g., a Wal-Mart Supercenter). These stores may be as large as 200,000 square feet.

IFRS: See "International Financials Reporting Standards (IFRS)."

Implied Equity Market Cap (REITs): The market value of all outstanding common stock of a company plus the value of all UPREIT partnership units as if they had been converted into stock.

Independent Retailer: A retailer that owns only one or a few retail stores and is not part of a large chain.

Indexed Lease: A rental agreement whereby the rent changes in accordance with a certain index, such as the consumer price index.

Industry Code: A descriptive code assigned to any company in order to group it with firms that operate in similar businesses. Common industry codes include the NAICS (North American Industrial Classification System) and the SIC (Standard Industrial Classification), both of which are standards widely used in America, as well as the International Standard Industrial Classification of all Economic Activities (ISIC), the Standard International Trade Classification established by the United Nations (SITC) and the General Industrial Classification of Economic Activities within the European Communities (NACE).

Infrastructure: 1) The equipment that comprises a system. 2) Public-use assets such as roads, bridges, water systems, sewers and other assets necessary for public accommodation and utilities. 3) The underlying base of a system or network. 4) Transportation and shipping support systems such as ports, airports and railways.

Initial Public Offering (IPO): A company's first effort to sell its stock to investors (the public). Investors in an up-trending market eagerly seek stocks offered in many IPOs because the stocks of newly public companies that seem to have great promise may appreciate very rapidly in price, reaping great profits for those who were able to get the stock at the first offering. In the United States, IPOs are regulated by the SEC (U.S. Securities Exchange Commission) and by the state-level regulatory agencies of the states in which the IPO shares are offered.

Intellectual Property (IP): The exclusive ownership of original concepts, ideas, designs, engineering plans or other assets that are protected by law. Examples include items covered by trademarks, copyrights and patents. Items such as software, engineering plans, fashion designs and architectural designs, as well as games, books, songs and other entertainment items are among the many things that may be considered to be intellectual property. (Also, see "Patent.")

Interim Loan: A loan that provides proceeds for the construction costs of a project. Often paid out in installments as the work progresses.

International Financials Reporting Standards (IFRS): A set of accounting standards established by the International Accounting Standards Board (IASB) for the preparation of public financial statements. IFRS has been adopted by much of the world, including the European Union, Russia and Singapore.

Interstate Sales Full Disclosure Act: A law that requires a residential development of 100 or more lots to file a disclosure statement with HUD.

IP: See "Intellectual Property (IP)."

Jumbo Loan: Also known as "non-conforming" loans, mortgage loans over the maximum "conforming" amount as set by FNMA are considered jumbo and are subject to different underwriting criteria. The benchmark loan amount is evaluated on a yearly basis by FNMA and adjusted accordingly. Interest rates on jumbo loans are generally 0.25% higher than their conforming counterparts. Also see "Non-Conforming Loan."

Knowledge Process Outsourcing (KPO): The use of outsourced and/or offshore workers to perform business tasks that require judgment and analysis. Examples include such professional tasks as patent research, legal research, architecture, design, engineering, market research, scientific research, accounting and tax return preparation. Also, see "Business Process Outsourcing (BPO)."

LAC: Latin America and the Caribbean.

LDCs: See "Least Developed Countries (LDCs)."

Leased Department: A department in a retail store operated by an outside party. The outside party either pays fixed rent or a percentage of sales to the retailer for the space.

Least Developed Countries (LDCs): Nations determined by the U.N. Economic and Social Council to be the poorest and weakest members of the international community. There are currently 50 LDCs, of which 34 are in Africa, 15 are in Asia Pacific and the remaining one (Haiti) is in Latin America. The top 10 on the LDC list, in descending order from top to 10th, are Afghanistan, Angola, Bangladesh, Benin, Bhutan, Burkina Faso, Burundi, Cambodia, Cape Verde and the Central African Republic. Sixteen of the LDCs are also Landlocked Least Developed Countries (LLDCs) which present them with additional difficulties often due to the high cost of transporting trade goods. Eleven of the LDCs are Small Island Developing States (SIDS), which are often at risk of extreme weather phenomenon (hurricanes, typhoons, Tsunami); have fragile ecosystems; are often dependent on foreign energy sources; can have high disease rates for HIV/AIDS and malaria; and can have poor market access and trade terms.

Legal Description: A description of a property that fulfills legal requirements and properly identifies the property. A legal description is commonly used in conjunction with a boundary survey. Also see "Survey (Real Estate)."

Lifestyle Center: An open-air, highly landscaped configuration of approximately 50 stores. Generally located near upscale neighborhoods, lifestyle centers offer leasable retail area of 150,000 to 500,000 square feet (typically, at least 50,000 square feet are dedicated to upscale national specialty stores).

Limited-Assortment Store: A small, low-priced grocery store that provides limited service, few or no perishables and generally fewer than 2,000 items.

Loan Application Fee: A lender's fee, usually ranging from $75 to $300, which the buyer must pay when applying for a mortgage.

Loan Origination Fee: A fee charged by the lender for processing a mortgage or loan. The mortgage industry standard is 1% of the loan amount, but if the application is taken over the Internet, it is often reduced to 0.5% or even zero, depending on the lender.

Loan-to-Value Ratio (LTV): An underwriting ratio determined by dividing the sales price or appraised value into the loan amount, expressed as a percentage. For example, with a sales price of $100,000 and a mortgage loan of $80,000, the LTV ratio would be 80%. Loans with an LTV over 80% usually require private mortgage insurance. See "Private Mortgage Insurance (PMI)."

Lock (Lock In): A commitment that a borrower obtains from a lender assuring a particular interest rate for a limited time period, such as 30 days. A lock provides protection to the borrower should interest rates rise between the time the borrower applies for a loan, acquires loan approval and, subsequently, closes the purchase.

LOHAS: Lifestyles of Health and Sustainability. A marketing term that refers to consumers who choose to purchase and/or live with items that are natural, organic, less polluting, etc. Such consumers may also prefer products powered by alternative energy, such as hybrid cars.

Low-E: A coating for windows that can prevent warmth from escaping from the inside of a building during the winter, while preventing solar heat from entering the building during the summer. Significant savings in energy usage can result.

Market Segmentation: The division of a consumer market into specific groups of buyers based on demographic factors.

Market Value: An estimation of how much a property or asset would sell for on the open market.

Material Breach: The violation of a contract that causes invalidation or some other penalty.

Megapolitan: Massive corridors comprising several million residents across several cities. Examples include the IH35 Corridor anchored by Dallas and Ft. Worth, Texas, and the Atlantic Coast corridor anchored by Miami, Ft. Lauderdale and Boca Raton, Florida.

Mill: One 10th of a penny. Usually used in real estate taxation.

Millennials: See "Generation Y."

Mineral Interests: The rights of ownership to gas, oil or other minerals as they naturally occur at or below a tract of land. Also known as "mineral rights."

Mineral Rights: See "Mineral Interests."

MIP: See "Mortgage Insurance Premium (MIP)."

Mixed-Use Development: A development that has some combination of residential, office, retail and/or industrial space.

Mortgage Banker: An organization that specializes in underwriting mortgage loans. Mortgage bankers typically sell some or all of their loans to investors but may continue to own and/or service them. Also see "Mortgage Broker."

Mortgage Broker: An organization in the business of arranging funding for a borrower. In contrast to a mortgage banker, a broker does not actually loan the money. Brokers usually charge a fee or receive a commission for their services.

Mortgage Insurance Premium (MIP): Insurance purchased by the borrower to insure the lender against loss should he or she default. MIP is paid on government-insured loans (FHA or VA loans) regardless of the LTV (loan-to-value ratio). Should the borrower pay off a government-insured loan in advance of maturity, he or she may be entitled to a small refund of MIP. Also see "Private Mortgage Insurance (PMI)."

Mortgage REIT: A REIT that specializes in making mortgages rather than owning property outright.

Mortgage-Backed Security (MBS): A security that represents ownership of an undivided interest in a group of mortgages. Mortgage bankers often form MBSs in order to sell pools of their mortgages on the secondary market. Also see "Collateralized Debt Obligation (CDO)."

Municipal Utility District (MUD): A political unit, regulated by state authorities, that has been established to own and operate utilities within its boundaries. These utilities typically include water, sewer and/or drainage. MUDs may be inside or outside of city limits. MUDs generally are empowered to sell bonds in order to raise capital with which to install utility pipes and systems. Property owners within a MUD pay regular fees to the MUD for services.

National Flood Insurance Program (NFIP): A program offered by the Federal Emergency

Management Association (FEMA) that provides flood insurance to individual property owners and tenants.

Negative Amortization: Amortization in which the payments made are insufficient to fund complete repayment of the loan at its termination. This usually occurs when the increase in the monthly payment on an adjustable-rate mortgage (ARM) is limited by a pre-set ceiling. The portion of the payment that should be paid is added to the remaining balance owed. The balance owed may increase rather than decrease at various times in the life of the loan.

Neighborhood Center: A shopping center that includes a supermarket, drugstore, home-improvement center or variety store. Neighborhood centers often include small stores, such as apparel, shoe, camera and other shopping goods stores.

Net Leasable Area (NLA): The square footage in a building or property that is leasable excluding common areas, common hallways, common baths, etc. Also, see "Gross Leasable Area (GLA)."

Net Lease: A lease that requires all maintenance expenses, such as heating, insurance and interior repair, to be paid for by the tenant.

Net Zero: See "Zero Energy Building."

New Urbanism: A relatively new term that refers to neighborhood developments that feature shorter blocks, more sidewalks and pedestrian ways, access to convenient mass transit, bicycle paths and conveniently placed open spaces. The intent is to promote walking and social interaction while decreasing automobile traffic. The concept may also include close proximity to stores and offices that may be reached by walking rather than driving.

Non-Conforming Loan: A loan that is not eligible to be purchased by Fannie Mae or Freddie Mac. These loan agencies have specific upper limits on how large a loan they will buy. These limits are adjusted on a regular basis. Also, several other factors regarding the loan must meet guidelines before they are considered to conform. Also see "Jumbo Loan."

Non-Store Retailing: A form of retailing that is not store-based. Non-store retailing can be conducted through vending machines, direct-selling, direct-marketing, party-based selling, catalogs, television programming, telemarketing and Internet-based selling.

North American Industrial Classification System (NAICS): See "Industry Code."

OECD: See "Organisation for Economic Co-operation and Development (OECD)."

Office of Interstate Land Sales Registration: The HUD agency that is responsible for implementing the Interstate Sales Full Disclosure Act.

Off-Price Retailer: A retailer that offers an inconsistent assortment of brand-name, fashion-oriented soft goods at low prices.

Option ARM: An adjustable rate mortgage that lets the borrower decide how much to pay each month. The borrower may choose between the standard principal and interest amount, or simply pay the interest. Some plans allow the borrower to pay even less, adding the unpaid interest to the total amount due under the note.

Organisation for Economic Co-operation and Development (OECD): A group of more than 30 nations that are strongly committed to the market economy and democracy. Some of the OECD members include Japan, the U.S., Spain, Germany, Australia, Korea, the U.K., Canada and Mexico. Although not members, Estonia, Israel and Russia are invited to member talks; and Brazil, China, India, Indonesia and South Africa have enhanced engagement policies with the OECD. The Organisation provides statistics, as well as social and economic data; and researches social changes, including patterns in evolving fiscal policy, agriculture, technology, trade, the environment and other areas. It publishes over 250 titles annually; publishes a corporate magazine, the OECD Observer; has radio and TV studios; and has centers in Tokyo, Washington, D.C., Berlin and Mexico City that distributed the Organisation's work and organizes events.

Owner Financing: A mortgage transaction whereby the property seller provides some or all of the financing and takes back a mortgage loan at closing.

Pad Site: A location for a free-standing retail building that sits in front of a mall or shopping center.

Participating Mortgage: A mortgage in which the lender is entitled to a share of the income from the property.

Passive Solar: A system in which solar energy (heat from sunlight) alone is used for the transfer of thermal energy. Heat transfer devices that depend on energy other than solar are not used. A good example is a passive solar water heater on the roof of a building.

Pass-Through (Mortgages): See "Collateralized Mortgage Obligation (CMO)."

Patent: An intellectual property right granted by a national government to an inventor to exclude others from making, using, offering for sale, or selling the invention throughout that nation or importing the invention into the nation for a limited time in exchange for public disclosure of the invention when the patent is granted. In addition to national patenting agencies, such as the United States Patent and Trademark Office, and regional organizations such as the European Patent Office, there is a cooperative international patent organization, the World Intellectual Property Organization, or WIPO, established by the United Nations.

Percentage Lease: A lease in which rent payments are based on a store's sales.

Photovoltaic (PV) Cell: An electronic device consisting of layers of semiconductor materials fabricated to form a junction (adjacent layers of materials with different electronic characteristics) and electrical contacts, capable of converting incident light directly into electricity (direct current). Photovoltaic technology works by harnessing the movement of electrons between the layers of a solar cell when the sun strikes the material.

PITI: Principal, interest, taxes and insurance: the elements of a monthly mortgage payment.

Planned Unit Development (PUD): A housing subdivision that contains pre-planned community features, such as parks or adjacent office or retail districts. The common space is owned by a homeowners' association.

Power Shopping Center: A large, open-air shopping center with the majority of space leased to several well-known anchor retail tenants-typically specialty retailers operating large stores that specialize in one type of merchandise. Typical anchor tenants include OfficeMax, Linens 'n Things, Marshall's and Best Buy. Some of these tenants may be on pad sites. Convenient access and parking are emphasized. These centers usually sit on major intersections.

Pre-Boomer: A term occasionally used to describe people who were born between 1935 and 1945. They are somewhat older than Baby Boomers (born between 1946 and 1962). Also see "Baby Boomer."

Prefabricated (Construction): A term for buildings erected onsite from factory-made components, or built entirely offsite and then set down on the property. A mobile home is prefabricated. However, a few manufacturers factory-build certain types of houses and buildings that are not considered in the same class as mobile homes.

Pre-Paid Expenses: At closing of a real estate purchase, monies necessary to create an escrow account. These typically include two month's worth of taxes, hazard insurance, private mortgage insurance and special assessments.

Pre-Paid Interest: The amount of interest paid to cover the period from the closing of a sale until the beginning of the first payment on the mortgage or loan.

Pre-Payment (Prepayment): The unscheduled payment of all or part of the outstanding principal of a loan or mortgage. Pre-payments are typically made, by the borrower, but may also result by foreclosures, condemnations or casualties, such as an insurance settlement resulting from a total loss by fire or flood.

Pre-Payment Penalty (Prepayment Penalty): A penalty found in a promissory note or mortgage, imposed by the lender if the principal of a loan is paid before it is due, thereby reducing the lender's stream of interest earned on the note.

Pre-Payment Privilege: The right to repay the principal of a loan before interest is due.

Pre-Payment Risk: The possibility that the mortgages underlying a mortgage-backed security are repaid faster or more slowly than expected.

Pre-Qualification (Mortgages): The process of determining how much money a prospective property buyer will be eligible to borrow before actually applying for a loan.

Private Mortgage Insurance (PMI): Insurance paid on those loans that are not government-insured when the loan to value ratio is greater than 80%. When you have accumulated 20% of your home's value as equity, your lender may waive PMI at your request. Note that such insurance does not constitute a form of life insurance that pays off the loan in case of death. Also see "Mortgage Insurance Premium (MIP)."

PUD: See "Planned Unit Development (PUD)."

R&D-Flex Building: Industrial-type buildings that are designed to satisfy tenants that require an above-average amount of office space as well as an above-average level of finish that presents a more office-like environment, such as more windows and better landscape. From 30% to 100% of the space in such buildings may be devoted to office or laboratory space, with the balance devoted to light assembly or warehouse space.

Radon: A radioactive gas that may cause health problems. Some state laws require that a property owner disclose any known radon gas conditions. Concerned buyers may include testing for radon in their property inspections.

Real Estate Investment Trust (REIT): Investments that work in a manner very similar to mutual funds in that money from several investors is pooled together to jointly own real estate and/or invest in mortgages. REITs are required to distribute almost all (90%) of their net income annually, directly to shareholders. Many REITs specialize in a specific kind of real estate, such as shopping centers or apartments. Investors generally buy shares in REITs with a two-fold purpose: 1) to earn current income and 2) for long-term capital gains on appreciation in the value of the real estate.

Real Estate Investment Trust Act of 1960: The federal law that established guidelines for establishing REITs.

Real Estate Settlement Procedures Act (RESPA): A federal law that allows consumers to review information on known or estimated settlement costs. A statement of estimated costs must be provided by the lender once after a mortgage application has been completed and once again prior to or at closing.

Refinancing: In mortgages, refinancing is the process of paying off an existing mortgage with a new mortgage, often to attain a certain interest rate or to obtain cash from the transaction by increasing the principal balance.

Regional Center: A shopping center or mall which includes up to three department stores plus shopping or specialty stores rather than convenience stores. Super-regionals are similar but have at least four department stores and at least 1 million square feet of retail space.

Regulation Z: A federal rule that requires a lender to disclose the terms of a loan to the borrower.

REIT Modernization Act of 1999: The federal law that enabled an REIT to own up to 100% of the stock of a taxable subsidiary that provides services to REIT tenants and other customers. The law also changed the minimum distribution requirement from 95% to 90% of a REIT's taxable income.

RESPA: See "Real Estate Settlement Procedures Act (RESPA)."

Retail Chain: A firm that consists of multiple retail units under common ownership and usually has some centralization of decision-making in defining and implementing its strategy.

Return on Investment (ROI): A measure used to determine the efficiency of an investment. It is calculated as (total gain from an investment, minus the cost of that investment), divided by (cost of the investment). ROI may be adjusted to reflect the average yearly return on an investment.

Revenue Per Available Room (REVPAR): A hotel performance measure that divides revenue by the number of available rooms, as opposed to the number of occupied rooms.

Reverse Mortgage: Enables older homeowners (aged 62+) to convert part of the equity in their homes into tax-free income without having to sell the home, give up title, or take on a new monthly mortgage payment. The lender provides either a line of credit, steady monthly payments or one upfront payment to the borrower. In return, the lender receives a right to proceeds upon the eventual sale of the home, typically after the borrower's death. There are no monthly payments to be made by the borrower, and there are no income or health requirements. Sometimes called a reverse annuity mortgage or RAM.

Right of Rescission: The legal right to void or cancel a mortgage contract in such a way as to treat the contract as if it never existed. Right of rescission is not applicable to mortgages made to purchase a home, but may be applicable to other mortgages, such as cash-out refinances.

Robosigning: An illegal practice of signing documents without verifying the information therein. Robosigning was common during the high volume of real estate foreclosures conducted in 2009 and 2010.

R-Value (R Value): A method of measuring the effectiveness of building materials such as insulation. Technically, it is the resistance that a material has to heat flow. The higher the R-Value, the better the insulation provided. It is the inverse of U-Value. See "U-Value (U Value)."

Sales Per Square Foot of Selling Space: Net sales divided by square feet of selling space.

Same-Store Sales: Sales dollars generated only by those stores that have been open more than a year and have historical data to compare this year's sales to the same time-frame last year.

Savings and Loan (S&L, Thrift, Savings Association, Savings Bank): Depository institution historically engaged primarily in accepting consumer savings deposits and in originating and investing in securities and residential mortgage loans; now may offer checking-type deposits and make a wider range of loans. Federally-chartered and federally-insured savings and loans are supervised by the Federal Home Loan Bank Board (FHLBB).

Second Mortgage: A mortgage with collateralization rights that is secondary to a first mortgage. In the event that the borrower defaults, the second mortgage holder may be forced to pay off the first mortgage in order to protect the second mortgage's interests in the property. Second mortgages involve more risks to the mortgage holder and generally are written at higher interest rates. It is also possible to create third or fourth mortgages, etc.

Secondary Market: A market providing for securities to be bought or sold. This market is where the majority of trading occurs. The New York Stock Exchange, all other stock exchanges and the bond markets are secondary markets. In mortgages, large pools of mortgages are bought and sold by major investors in the secondary market.

Securitization: The process of financing a pool of similar assets (such as mortgages, automobile loans, corporate debt instruments or credit card debts) by issuing to investors interests in the funds generated by that pool. Such pools are generally in the range of $100 million or higher, and may represent the debts of dozens of companies or thousands of consumers. Securitization enables banks and other lenders to have ready markets into which they may sell loans that they generate.

Servicing (Mortgages): The steps and operations necessary to manage a mortgage or pool of mortgages, such as the collection of payments and the disbursement of escrowed funds for taxes and insurance. Servicing also includes following up on delinquent borrowers and foreclosure, if necessary. Servicing companies charge a fee to the investor that owns the loan.

Shared-Appreciation Mortgage: A loan that provides a share of the appreciation of the value of the property to the lender. Such loans are typically written at below-market interest rates.

Smart Buildings: Buildings or homes that have been designed with interconnected electronic sensors and electrical systems which can be controlled by computers. Advantages include the ability to turn appliances and systems on or off remotely or on a set schedule, leading to greatly enhanced energy efficiency.

Soft Goods: Apparel and linens.

Software as a Service (SaaS): Refers to the practice of providing users with software applications that are hosted on remote servers and accessed via the Internet. Excellent examples include the CRM (Customer Relationship Management) software provided in SaaS format by Salesforce. An earlier technology that operated in a similar, but less sophisticated, manner was called ASP or Application Service Provider.

Specialty Department Store: A store with a department store format that focuses on apparel and soft home goods (e.g., Neiman Marcus).

Specialty Store: A store specializing in one category of merchandising, frequently fashion-related.

Standard Industrial Classification (SIC): See "Industry Code."

Straight-Lining (Real Estate): Required by GAAP (Generally Accepted Accounting Principles), straight-lining averages a tenant's rent payments over the life of a lease.

Strip Center: A small shopping center that includes several adjacent stores located along a major street or highway.

Subprime: A term used to describe mortgages offered to borrowers with less than perfect credit. Subprime rates are generally higher than typical interest rates.

Subsidiary, Wholly-Owned: A company that is wholly controlled by another company through stock ownership.

Super Warehouse: A large, high-volume warehouse store that offers expanded services similar to a superstore, such as a service deli and bakery. These stores typically focus largely on food and drug items, and offer reduced prices on merchandise.

Super-Regionals: See "Regional Center."

Superstore: A large specialty store, usually over 40,000 square feet. Many superstores focus on a particular field of merchandise. For example, BestBuy is a consumer electronics superstore.

Supply Chain: The complete set of suppliers of goods and services required for a company to operate its business. For example, a manufacturer's supply chain may include providers of raw materials, components, custom-made parts and packaging materials.

Survey (Real Estate): A technical drawing that shows the precise legal boundaries of a property. Also known as a boundary survey. Surveys are generally recognized only when they are drawn by licensed professionals known as registered surveyors. Also see "Legal Description."

Sustainable Development: Development that ensures that the use of resources and the environment today does not impair their availability to be used by future generations.

Sweat Equity: In mortgages, sweat equity created by a purchaser performing work on a property being purchased.

Three/Two (3/2) Option: An alternative mortgage plan that enables households whose earnings are no more than 100% of the median income in their regional area to make a 3% down payment with their own funds, coupled with a 2% gift from a relative or a 2% grant or unsecured loan from a nonprofit agency or state or local government program.

Time-share (Timeshare): A type of joint ownership in which a group of owners share a particular piece of property, agreeing to have use of the property only during set days each year. Typically, time-share properties are vacation properties in such areas as beach resorts or ski resorts. They typically are condominium properties in which each condominium is jointly owned by a large group of people. Each owner typically has access to one week's use per year and pays for a proportionate share of the property's upkeep, taxes and insurance as well as management fees.

Title Company: This is a generic term that generally refers to a company that provides services to buyers and sellers by facilitating the closing of a sale. While a title company charges various fees for services in connection with the closing, it makes most of its revenues from commissions received from the sale of title insurance. The title insurance itself is generally provided and guaranteed by a title insurance underwriter.

Title Examination Fee: A standard fee paid to a title company at closing of a real estate purchase. Also see "Title Search."

Title Insurance (Title Policy): Insurance to protect the lender (referred to as a "lender's policy" or "mortgage policy") or the buyer (an "owner's policy") against losses arising from disputes over ownership or unknown liens on a property.

Title Search: A check of property deed records to ensure that the seller is the legal owner of the property and that there are no unknown liens or other claims outstanding.

TND: Traditional Neighborhood Development. See "New Urbanism."

Traditional Neighborhood Development (TND): See "New Urbanism."

Transportation Oriented Development: The creation of mixed use real estate developments, including homes and offices, adjacent to transportation hubs, such as rapid transit stations. Ideally, the result would be walkable communities with quick access to trains. Also known as Transit Oriented Development. Denver, Colorado's new rapid transit center neighborhood near downtown is a good example.

Truth-in-Lending Act: A federal law that requires lenders to fully disclose, in writing, the terms and conditions of a mortgage or loan, including the annual percentage rate and other charges.

Umbrella Partnership REIT (UPREIT): A real estate trust in which the partners of existing partnerships and a newly formed REIT become partners in a new entity called the operating partnership. The existing partners contribute property interests and the REIT contributes cash (typically proceeds from a public offering of stock).

Underwriter: The issuer of a publicly-held security, a loan or an insurance policy. Also see "Underwriting."

Underwriting: In lending, the reviewing of a loan file to determine the applicant's ability to meet the loan obligation. In insurance, underwriting is the process of reviewing the risk in a given insurance policy. In securities and investments, underwriting is the process of analyzing a bond or stock offering and then completing the necessary paperwork and regulatory filings necessary to offer the bond or stock for sale.

Universal Design: An approach to residential as well as commercial building design that attempts to accommodate as many people as possible, regardless of physical or mental limitations. For example, design elements may include wider doorways and stepless entries that are easy for the physically challenged to navigate.

Urban Development Action Grant (UDAG): A program under the department of Housing and Urban Development that lends money for redevelopment of urban commercial areas.

Urban Renewal: The process of the revitalization of urban areas, usually through redevelopment or older buildings or neighborhoods sponsored by a government or private organization.

Urban Sprawl: The growth of a city in a way that may be viewed by some as unsightly, crowded, unplanned or unproductive.

Urbanism: See "New Urbanism."

U-Value (U Value): A measure of the amount of heat that is transferred into or out of a building. The lower the U-Value, the higher the insulating value of a window or other building material being rated. It is the reciprocal of an R-Value. See "R-Value (R Value)."

VA Mortgage: A mortgage that is guaranteed by the Veterans Administration, a Federal Government agency that provides benefits to U.S. veterans. VA mortgages tend to be made at below-market interest rates and may feature lower down payments or other benefits for the borrower.

Vacancy Rate: The percentage of units available to rent in a given area or building. For example, if 10 units are unrented in a 100-unit apartment building, then the vacancy rate is 10%.

Value Added Tax (VAT): A tax that imposes a levy on businesses at every stage of manufacturing based on the value it adds to a product. Each business in the supply chain pays its own VAT and is subsequently repaid by the next link down the chain; hence, a VAT is ultimately paid by the consumer, being the last link in the supply chain, making it comparable to a sales tax. Generally, VAT only applies to goods bought for consumption within a given country; export goods are exempt from VAT, and purchasers from other countries taking goods back home may apply for a VAT refund.

Variable-Maturity Mortgage: A long-term loan where the due date of the final payment may be changed when the amount of earlier payments has been altered.

Variable-Rate Mortgage: See "Adjustable-Rate Mortgage (ARM)."

Warehouse Store: A discount retailer offering merchandise in a no-frills environment. These stores often cut costs by reducing services and product variety.

Wholesale Club: A retail store that sells a limited assortment of general merchandise to customers who are members of the club. Memberships are generally fee-based, margins are small and there is little customer service provided to the members. Groceries are often sold in bulk sizes. The largest wholesale club chain is Sam's, owned by Wal-Mart. Costco is Sam's biggest competitor in the U.S.

Workout Agreement: An agreement in which a lender and a borrower renegotiate terms on a loan that is in default in order to avoid foreclosure or liquidation. Renegotiated terms generally provide a measure of relief to the borrower in terms of reducing the debt-servicing burden, typically by extending the term of the loan or rescheduling payments. A workout agreement may benefit a lender because it avoids the expense and efforts relating to foreclosure.

World Trade Organization (WTO): One of the only globally active international organizations dealing with the trade rules between nations. Its goal is to assist the free flow of trade goods, ensuring a smooth, predictable supply of goods to help raise the quality of life of member citizens. Members form consensus decisions that are then ratified by their respective parliaments. The WTO's conflict resolution process generally emphasizes interpreting existing commitments and agreements, and discovers how to ensure trade policies to conform to those agreements, with the ultimate aim of avoiding military or political conflict.

Wraparound Mortgage: A transaction whereby an old mortgage is included with a new loan. This is different from a second mortgage in that a second mortgage is legally subordinate to a first mortgage. The borrower makes payments on both loans to the holder of the wraparound.

WTO: See "World Trade Organization (WTO)."

Zero Energy Building: A building that has zero net energy consumption from the traditional electric grid. Sometimes referred to as net zero, it may accomplish this through the generation of its own energy via solar, wind or other means. It may use energy storage devices on-site, or it may draw power from the grid at some times, while selling power back to the grid when it generates more than it requires. Typically, such projects are also designed to create zero net carbon emissions. Some subdivisions have been established with the goal of having all dwellings in the neighborhood be net zero. Also, some commercial buildings are net zero in design

ZigBee: May become the ultimate wireless control system for home and office lighting and entertainment systems. The ZigBee Alliance is an association of companies working together to enable reliable, cost-effective, low-power, wirelessly networked monitoring and control products based on an open global standard, 802.15.4 entertainment systems.

Lightning Source UK Ltd.
Milton Keynes UK
UKOW07f1534070616

275812UK00004B/22/P